Intermolecular Interactions in Crystals

Fundamentals of Crystal Engineering

Intermolecular Interactions in Crystals

Fundamentals of Crystal Engineering

Edited by

Juan J. Novoa

University of Barcelona, Spain
Email: juan.novoa@ub.edu

Print ISBN: 978-1-78262-173-7
EPUB eISBN: 978-1-78801-333-8

A catalogue record for this book is available from the British Library

The Royal Society of Chemistry is a charity, registered in England and Wales, Number 207890, and a company incorporated in England by Royal Charter (Registered No. RC000524), registered office: Burlington House, Piccadilly, London W1J 0BA, UK, Telephone: +44 (0) 207 4378 6556.

Visit our website at www.rsc.org/books

Printed in the United Kingdom by CPI Group (UK) Ltd, Croydon, CR0 4YY, UK

Preface

The aim of this book is to present the current knowledge on the intermolecular interactions found in crystals, including the latest discoveries. But in doing so, we have tried to present the information in a pedagogical form, in order to make its contents understandable for non-specialists. We have tried to make it a textbook for Masters and PhD students who require an in-depth knowledge of inter-molecular interactions, such as those working in Crystal Engineering or in the broader area of Supramolecular Chemistry. We have also tried to make it a reference book for researchers in these two fields who want a reference book containing the state-of-the-art knowledge on intermolecular interactions.

The book contains 20 chapters written by active researchers in the field, in some cases, the originators of important ideas in the field. These chapters cover from fundamental studies on intermolecular interactions, up to special properties and techniques required by their interconnected existence in crystals. A third area is devoted to a review of the state-of-the-art knowledge of the most common types of intermolecular interactions encountered in crystalline molecular solids. Chapters have been grouped into one of these three areas, according to their contents. Within each area, theoretical and experimental chapters have been mixed, ordered in a most-fundamental-first sequence. Although not as complete as originally planned, due to reasons external to this editor, the book covers the major topics on intermolecular interactions, the most common methods and techniques employed in their study, and a description of their most relevant properties, for intermolecular interactions found in molecular crystals with properties of technological interest (such as magnetic, or

Intermolecular Interactions in Crystals: Fundamentals of Crystal Engineering
Edited by Juan J. Novoa
© The Royal Society of Chemistry 2018
Published by the Royal Society of Chemistry, www.rsc.org

conducting, or superconducting properties, to name only the most common).

Intermolecular interactions are the interactions produced between stable molecules oriented in regions of space where spontaneously they do not form new chemical bonds. This definition classifies as intermolecular those produced between two stable radicals at all distances and relative orientations, as in the interaction between two oxygen molecules. It also identifies as intermolecular the interaction between two TCNE$^{\bullet-}$ radical-anions (TCNE = tetracyanoethylene) around 3 Å where TCNE$^{\bullet-}$ neither dissociates nor forms a C–C bond between the radical anions. This is the metastable region of TCNE$^{\bullet-}$ radicals, where the potential energy curve as a function of the shortest separation between the radical anions presents a metastable minimum, a large barrier for the formation of a covalently bonded σ-[TCNE]$_2^{2-}$ molecule, and a smaller barrier towards dissociation (both barriers are overcome in some cases, but not in others). Notice that this definition does not include as intermolecular the interaction between two CH$_3^{\bullet}$ radicals, even if at long distance these radicals exist, because their potential energy curve shows a minimum for the spontaneous formation of ethane; that is, is the curve for the formation of a C–C intramolecular bond.

From a more practical point of view, intermolecular interactions, and particularly intermolecular bonds (loosely considered by this editor as the dominant energetically stabilizing interactions found in supramolecular aggregates), play an essential role in rationalizing in simple terms the structure and relative stability of supramolecular aggregates, not to mention the speed of their transformation into other species. It was said more than 25 years ago that "supramolecular aggregates are to molecules and the intermolecular bond what molecules are to atoms and the covalent bond". In another words, despite their different electronic structures, intermolecular bonds play in supramolecular aggregates the same role as chemical bonds play in molecules. The main difference is in the electronic structure of intra- and intermolecular bonds, which explains the different orders of magnitude of the optimum distance (about 2–3 Å) and interaction energy (in most cases less than 10 kcal mol^{-1}). Improving our control of the outcome of crystallization will only be possible after improving our control of the mechanism by which intermolecular interactions compete.

Intermolecular interactions are also key in determining the structure of the outcome of crystallization and the relative stability against other polymorphic forms. They also determine the speed of the thermodynamically allowed polymorphic transformations. And the structure of the crystal determines other physical properties of the crystal. For instance, it determines the bulk ferromagnetic properties shown by the β-phase crystal of the *p*-nitrophenyl nitronyl nitroxide radical, a property absent in other polymorphs. It can be shown that the observed experimental structure of a crystal corresponds to the geometry of one of the minima on the free energy surface of the crystal (not always the most stable one), and that its relative stability is the energy difference between this minimum and others, while the speed of the polymorphic transformation depends on the height of the transition state connecting the two minima involved. Overall, energy, the free energy of the system, is the language chosen by nature to write the information concerning the main properties of a crystal. For this reason, the use of the term "intermolecular force" as equivalent to "intermolecular interactions" is misleading, as forces are different to energy. For similar reasons, we advocate the use of the term "bonds" (in its energetic interpretation, see above) or others with energetic meaning, rather than alternative concepts where the energetic meaning is lost (as in the term "bridges", based on geometrical concepts). After all, if one has to talk about the energetic stability of a system as the sum of the stability of all its parts (as implied in the use of the Scientific Method), it is going to be more convenient to add components that have an energetic meaning.

Intermolecular interactions are sometimes called non-covalent interactions, as an equivalent name. This is equivalent to saying that covalent bonds are the only type of intramolecular bonds, and the IUPAC Gold Book contemplates a few other intramolecular bonds, among them ionic bonds (one can consider dative bonds, and dipolar bonds, considered to be coordination bonds, therefore to be one type of covalent bond), and also electron-deficient bonds (a two-electron three-center bond). Therefore, we believe that the terms intermolecular bond and intermolecular interaction are better defined. Finally, the term "intermolecular" was that used by the scientist who started the research on the interaction between molecules at the dawn of the 20th Century.

The fast progress in the knowledge on intermolecular interactions that has taken place during the last two decades makes a book collecting this progress very valuable; particularly when each chapter has been written by a specialist caring about delivering all essential information, in a pedagogical form, to the non-specialist. Here are the results. I hope that you like them.

Juan J. Novoa
Professor of Physical Chemistry
University of Barcelona

Contents

Intermolecular Interactions in Crystals: Fundamentals of Crystal Engineering
Edited by Juan J. Novoa
© The Royal Society of Chemistry 2018
Published by the Royal Society of Chemistry, www.rsc.org

Section 2: Spectroscopic and Database Information

Contents xiii

Section 3: Isolated Intermolecular Interactions

10 Intermolecular Interactions in Crystals 377

Peter Politzer, Jane S. Murray and Timothy Clark

11 The Nature of the Hydrogen Bond, from a Theoretical Perspective 410

Steve Scheiner

Section 4: Intermolecular Interactions in Crystals

18   Revealing the Intermolecular Bonds in Molecular Crystals Through Charge Density Methods   617

C. Gatti and A. Forni

19   Noncovalent Interactions in Crystal Structures: Quantifying Cooperativity in Hydrogen and Halogen Bonds   673

Sławomir J. Grabowski

Section 1: Theoretical Foundations of Intermolecular Interactions

1 Bonds and Intermolecular Interactions – The Return of Cohesion to Chemistry

Sason Shaik

The Hebrew University of Jerusalem, Institute of Chemistry and the Lise Meitner-Minerva Center for Computational Quantum Chemistry, Edmond J. Safra Campus, Givat Ram, Jerusalem, 91904 Israel
Email: sason.shaik@gmail.com

1.1 Introduction

One of the fundamental territories of chemistry is the chemical bond, *the glue from which an entire chemical universe is constructed*, resulting in a rich and enticing variety of species held together by stabilizing interactions. Let me first define the interactions and the manner in which they will be classified in this introductory chapter.

There are interactions that pair-up or delocalize electrons over a two or more atoms, and thereby give rise to entities we call *molecules*. This chapter refers to these interactions as *chemical bonds*. The chemical bond defines a *chemical identity*. It also accounts for the 'magic' of chemistry, whereby one molecule disappears and a new one appears. Take for example the soft gray solid, sodium, $(Na)_s$, and mix it with the yellow-greenish gas, chlorine (Cl_2). And within seconds, lo and behold, you get a white solid, $(Na^+Cl^-)_s$. All three species have distinct identities, being defined by their chemical bonds, and as such the magic is a manifestation of breaking bonds and making new

Intermolecular Interactions in Crystals: Fundamentals of Crystal Engineering
Edited by Juan J. Novoa
© The Royal Society of Chemistry 2018
Published by the Royal Society of Chemistry, www.rsc.org

ones. Clearly, then, the term chemical bond defines the molecule, and hence also the chemical identity. Usage of this term is also a tribute to Gilbert Newton Lewis who invented it 100 years ago[1] *as a quantum unit of bonding in molecules.* By so doing, Lewis has ushered "the electronic-structure revolution in chemistry".[2]

There are however, stabilizing interactions also between and among molecules with saturated valences, such as hydrogen bonds, dipole–dipole and dispersion interactions, and so on. These inter-actions are responsible for another type of 'magic'- the magic of ag-gregation, crystallization, gelation, coagulation, micelle formation, self-assembly, and evolution of large mesoscopic bodies. Consider for example discrete water molecules in the gas phase (vapor). As you lower the temperature, gradually you see water appearing as a liquid, and as temperature is further lowered, lo and behold, the liquid be-comes a beautiful translucent solid, ice. Thus, the interactions between the discrete molecules bring about the formation of meso-scopic and macroscopic matter, and as such, we refer to these cohesive forces as *intermolecular interactions.*

Despite the identity of the forces involved in these two kinds of interactions and the fact that their energy representations are entirely identical (Figure 1.1), the distinction of bonds and intermolecular

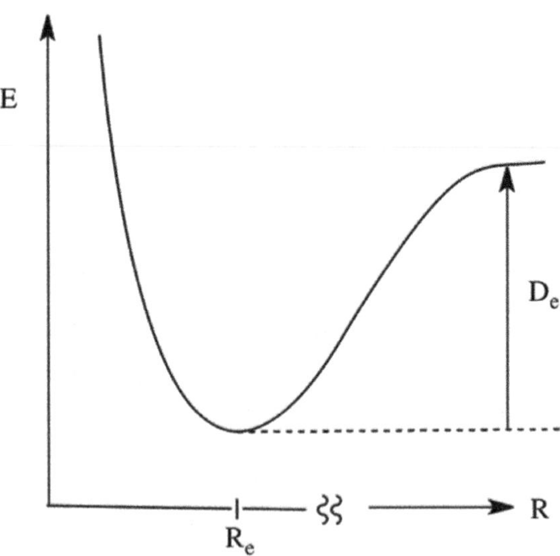

Figure 1.1 One dimensional energy representation of stabilizing interactions, along some general proximity coordinate R, which can be a distance between two bonded atoms or a generalized proximity coordinate between molecules or aggregates. R_e defines the equilibrium structure. D_e is the dissociation energy.

forces is not artificial. Thus, chemical bonding obeys magic numbers (electron pairs, octets, Hückel's $4n + 2$ numbers, *etc.*), and defines chemical identity, while the intermolecular interaction, as a creator of aggregates, is an extensive property that does not obey magic numbers and has much lesser selectivity. In other words, bonds are formed to satisfy the atomic valences, while intermolecular interactions are residual interactions formed between species, which have satisfied their formal valences. Consequently, diatomic bonds are generally stronger than the respective diatomic intermolecular interactions, and as such, 'bonds' make molecules, while 'intermolecular interactions' create aggregates of molecules.

It is deemed necessary therefore to conserve this distinction in view of the current vogues of calling every stabilizing interaction, even as in He_2, a bond; a tendency that has already been criticized.[3] This distinction is also didactic, playing an important role in the comprehension and teaching processes of the structure of chemistry. In fact, it is this distinction that has enabled the development of chemistry as a molecular science.[4]

In many ways, bonding and intermolecular interaction is a field in a midst of a great renaissance.[5,6] There are new interesting theoretical approaches to probe the origins of bonds, many novel bonding motifs,[7,8] and experimental studies that describe imaging of bonds being broken and remade using atomic force microscopy.[9] The bond is becoming again a central intellectual arena, and one can even find allusions to the bond as an elementary particle of chemistry, the so-called "bondon".[10] Likewise, the interest in intermolecular interactions has moved to new frontiers with bonding motifs such as "halogen bonds" and $CH \cdots HC$ interactions, and methods to calculate and analyze them.[11–15] There exist also new chemical bonds that are supported by dispersion interaction, such as very long C–C bonds.[16] There are even some claims of visualizing hydrogen bonds and of bond breakage and remaking. This "return of the bond and intermolecular interactions" to the central arena of chemistry marks the revitalization of chemistry as an intellectual science.

This chapter tries to convey the feeling of the author that the chemical bond and intermolecular interactions are now frontier domains in chemistry. And, by the nature of this goal, the chapter does not present an exhaustive treatise of bonding but rather a bird's eye view with examples, which reflect personal tastes of the author and some reliance of science writers like Ball and Ritter.[5,6] As such, the chapter adopts in places an essay-like style that addresses the target audience, rather than bringing a rigorous equations-laden exposition that would appear in source material.

The chapter starts with a short historical perspective of the evolution of the bond concept from the times of alchemy, through the constitutional revolution by Lavoisier, Cannizzaro, and Dalton, to the electronic structure revolution by Lewis, and the second revolution of quantum mechanics, where valence bond theory (VBT) and molecular orbital theory (MOT) were developed and imported to mainstream chemistry. It continues with the energy perspective of the origins of bonding, and then shows the unity of outlooks on bonding, emerging from VBT, MOT and density functional theory (DFT). The narrative goes on with very brief description of new bonding motifs, and then it switches to intermolecular interactions.

1.2 Historical Perspective of the Evolution of the Bond Concept

With the advent of quantum mechanics there is a growing tendency to annex the chemical bonding territory to physics. Let me start then by establishing a claim on the chemical bond as an original territory of chemistry. As Siegfried musingly defines, the task of the historian is to "play tricks on the dead" (p. 16).[4] To avoid this risk, I shall simply summarize my personal impressions on the matter as shaped by necessarily selective reading.

1.2.1 From Mythical Definitions to Matter with Quantized Weights

Long before Christian era, during the 6th to 4th centuries BCE, the great Greek philosophers (Thales, Anaxagoras, Empedocles, Plato, Aristotle, *etc.*) were postulating the existence of building blocks of matter, calling them *elements* or *atoms*, which were drawn/repelled by cohesive/repulsive (sometimes defining these forces as 'love' and 'hate'). However, the Greek philosophers did not engage in experiments to prove or disprove their ideas.

The first experimental chemists were the alchemists (sometimes called the Neoplatonic philosophers) who succeeded the classical philosophers, at the end of the 4th century BCE during the post-Alexander the Great era in the Greek empire. The alchemists viewed their goal as the attempt to emulate the harmonious changes of Nature and its perfection. In this regard, the idea of "bonding" has alchemical origins, where the *conjuctio* or union of the opposites was considered to be the ultimate synthesis necessary to drive the change of lower matter to gold (including the improvement of the soul).[17] When modern chemistry succeeded alchemy, chemists were trying to grapple with the 'magic of chemistry',[4] the formation of substances and their transmutation. These efforts led eventually to the formulation of theories in which *affinities* were considered to unite substances and to cause selectivity. Affinity and its practical definition are the roots of the modern concept of bonding and selectivity.

In 1675, Lemery published his book *Course de Chymie*[18] and used *elective affinity* to describe the processes in which one metal (or metal ion) is selectively replaced by others in the chemistry of salts (pp. 79, 83, and 94 in ref. 4).[4] In 1718, Etienne Francois Geoffroy systematized this *affinity* phenomenon in his table of *rapports* (pp. 76, 93–96 in ref. 4).[4] This wonderful mnemonic was a precursor of Mendeleyev's periodic table. It allowed Geofrroy to generalize the selectivity relationships between acids, bases (including metals) and salts. The historian Ursula Klein[19] credits Geoffroy as the first to generalize the basic concept of modern chemistry – that of the *compound* held together by *chemical affinity* between the constituents.

In 1744 another French chemist, Guillaume Francois Rouelle, *differentiated the sciences of chemistry and physics* referring to the force that combines chemical elements, and is very different than the *physical forces* of mixing or mechanical forces between bodies as in Newton's theory.[20] The status of chemistry was rather high in those days, judging by the fact that the Swedish chemist Torben Bergman, who assembled thousands of reactions into an elective attraction/ affinity table, presented his table of affinities as a gift to the Duke of Parma. Thus, 18th century chemists were already aware of *chemical identities* of substances and distinguished these identities from physical mixtures where these identities are not lost.

1.2.1.a The Compositional Revolution

At the same period, chemists used an eclectic mixture of terms to describe the formation of substances from simpler building blocks. There were abstract terms, which originated in Aristotle's theory of the four elements, like *protyle*, the hypothetical primitive substance from which all *elements* were made, and subsequently combined to generate matter. There was a term *phlogiston*, which was referred to as a *principle* that causes materials like coal, sulfur, *etc.*, to be inflammable. There was a practical definition of a *simple body*, one that could not be further decomposed to simpler ones by the experimental means of the day. And even *atoms* were invoked to represent material building blocks. All these terms painted a very fuzzy abstract picture of elementary bodies that engaged in elective combinations.[4]

A dramatic change had to occur before the term elective affinity could be defined as the effective concept we now call *the chemical bond*. This was brought about by two consecutive revolutions. The first one was the *compositional revolution*. It started with Lavoisier, who established the material essence and chemical identity of oxygen and other gases (which until then were considered as celestial influences that trickled into earth from heavens). This went on[4] with ideas on compositions of substances and the coining of the term stoichiometry by Richter, and proceeded to the great debate of Proust and Berthollet about whether a chemical substance has definite proportions of its constituent bodies (Proust), or indefinite-variable ones that depend on the relative masses used for the two bodies (Berthollet). Berthollet's idea, apparently influenced by Newton's Law of attraction, would be proved as a wrong one, but at the same time it was a seminal recognition of chemical equilibrium and the mass laws in chemistry. Chemists were groping for significance of their experiments, and eventually all this activity culminated in the atomic hypothesis of Dalton.

The Quantized Weight of Matter: Dalton identified the abstract term *elementary bodies* with the discrete atoms of Democritus. But he went a step further by suggesting that these atoms possess unique weights that could be determined in the laboratory. As such, Dalton showed that matter manifests in *quantized weights* (page 237 in ref. 4),[4] which chemists could determine using the available techniques of the day. He augmented this hypothesis by suggesting compositional rules by which the atoms combine in simple definitive proportions (AB, A_2B, AB_2, *etc.*). *The chemical bond became a LEGO of atoms.*

Still, the LEGO game was somewhat ineffective because the connectivity of the atom was not yet defined, and as a result, the atomic weights seemed to fluctuate, depending on the assumption of the combined ratio of the atoms. In some cases, this indeterminism suggested molecules with fractions of atoms. Avogadro, a physics professor from Turin, tried to resolve these difficulties and postulated in 1814 the existence of molecules, including ones that are made from identical atoms, like H_2 and O_2, which he called *molécules elementaires* (pp. 259–263, in ref. 4).[4] This idea defied all the beliefs of chemists that the atomic combination must involve union of opposites, and the idea was met with strong objections, notably from Dalton himself. The delay in the acceptance of Avogadro's idea continued the misery of chemists, so much so that the great French chemist Dumas exclaimed in 1843, "If I were master I would efface the word atoms from science". Nevertheless, despite being an objectionable 'heresy', some chemists were taken by the concept of the elementary molecules. Thus, Cannizzaro showed in 1860 during the Symposium of Organic Chemistry in Karlsruhe, that Avogadro's molecular hypothesis settled all the problems in the atomic weights and created perfect order (pp. 259–263 in ref. 4).[4] Now, the weights of the building blocks, the atoms, became stable and did not fluctuate wildly as before. Once atomic weights were definite, valences could be defined from the combining weights, and one could understand the existence of multiple molecules made from the same atoms, *e.g.* H_2O and H_2O_2. Thus, Avogadro's notion and its implementation by Cannizzaro formed the basis for an eventual definition of the long sought for *chemical identity*.

1.3   Theories of Chemical Bonding

With matter being quantized into atomic weights, what was still missing was the nature of the forces that caused the union of atoms into molecules, and the ground rules of this game of LEGO. Hence, bonding theories followed and gradually replaced the *affinities* or the elective forces.

1.3.1   The Ionic Bonding Theory and Its Fall Down

The first was the dualistic electrical theory of Davy and Berzelius[21] that sprung from electrochemistry, which postulated attraction between oppositely charged atoms or groups of atoms. A similar theory was proposed later by Thomson, the discoverer of the electron, in which bonding arose from attraction of oppositely charged ions after electron transfer between the atoms. Later, this theory created a vogue of ionic or the so-called *electromer* theory, especially amongst American scientists.[22] However, from their onset, these ionic type theories have been disqualified by *structuralists*, led by Frankland, Kekulé, Couper, Butlerov, and so on, who came up with the concept of valence to describe the 'structure' of organic molecules (the inverted commas are added because this was not at all a physical structure, but something abstract). In a nutshell, the notion of oppositely charge ions could not account for the nonpolar (nonionic) substances, which typify organic molecules. This dichotomy of *two types of compounds* was a call for generalization, which needs a hero who will 'pick up the gauntlet' and meet the challenge. This hero was found in the person of Gilbert Newton Lewis.

1.3.2   The Lewis Hypothesis of Electron-pair-bonding

Heroes are seldom born as such. More commonly they seize the opportunity, when it presents itself. The end of the 19th century and the turn of the 20th century witnessed the rise of physical chemistry, *'the brave new science'* in which chemists like Ostwald, van 't Hoff, Nernst, and Arrhenius performed quantitative measurements of chemical substances in solution.[23,24] These substances were generally strong and weak 'electrolytes', such as NaCl and acetic acid. The conductivity measurements showed the presence of ions in solution, but these were dependent on the strength of the electrolytes. These were exciting days for chemistry where theories of 3-dimensional structure (by van 't Hoff and Le Bel), equilibrium, rates, osmotic pressure, and electrochemistry were being developed. There were many discussions, which urged the community to generalize chemistry and create an *Allgemeine Chemie* (a unified chemistry, p. 4 in ref. 23).[23] This has eventually led to the second chemical revolution, *the electronic structure revolution*, which was ushered by Lewis's hypothesis of the electron-pair bond in 1916 that gave the clue to the LEGO rules of the atomic combinations.

1.3.2.a   *Quantized Bonding Units – The Electronic Structure Revolution*

Lewis seized the opportunity and led the chemical community to the unified land. Initially, Lewis was seeking an understanding of the behavior of strong and weak electrolytes in solution (p. 135 in ref. 23).[23] But like the voyagers to America who sailed seeking the Spice Islands and discovered a new continent, so did Lewis, who was seeking to understand the behavior of electrolytes in solution, find instead the concept of the electron pair bond as an intrinsic property of the molecule that stretches between the covalent and ionic situations.[1]

To appreciate the nature of the discovery, let us recall that Thomson discovered the electron in the early years of the first decade of the 20th century, and he showed that this was an elementary particle of matter. Nevertheless, physicists were relatively latecomers in theorizing an effective chemical bonding mechanism. In fact, as late as 1923, Born writes to Einstein about his perplexing attempts to understand the "homopolar bonding forces" such as those holding H_2, and adds: "Unfortunately, every attempt to clarify the concept fails".[25]

But, Lewis did not try to clarify the forces, nor did he let the awe of the Laws of Classical Physics inhibit his quest. Instead, he let himself be guided by his chemical overview (from the many, many molecules he went through) to hypothesize the electron pairing as *a quantized unit of bonding* which gave the clue to the nature of the atomic combination to molecules.[2] Thus, Lewis was one of the first chemists to realize that the electron, which he referred to as 'the atom of electricity', is an essential building block of the chemical bond. Already in his 1913 paper,[26] Lewis used the electron to re-define the term valence numbers, which were addressed by Abbeg who showed in 1904 that the sum of the maximum positive and negative valences for an atom was eight. Thus, Lewis made the distinction between the number of bonds (or coordination number) and the *oxidation state* of the atom, by using an electron-based language. He then revealed his early insight about the nature of the chemical bond by illustrating the two bonding types in a salt like potassium chloride *vs.* a paraffin hydrocarbon such as methane. As such, Lewis was already then looking at chemistry coming out in two branches separated by a rift, inorganic chemistry and organic chemistry, and trying to unify them in terms of bonding types, which are the ionic and covalent bonds in today's language. In the end of the 1913 article, Lewis added the metallic bond as a third bond-type, in which "the electron is free to move even outside of the molecule".[26]

In his groundbreaking 1916 JACS article, "The Atom and the Molecule",[1] Lewis made a major leap and in a stroke of genius formulated the concept of the electron pair bond. In this paper, he initially discusses the separation of the electrons in an atom to core (kernel) and valence shell, which tends to hold eight electrons, which are normally arranged symmetrically at the eight corners of the cube. This is an iconic representation of the octet rule (which he still calls "the group of eight"). Since in his view the atomic shells are "mutually penetrable" (a term showing he was intuitively cognizant of the modern term overlap), he tries to bind atoms in a manner that satisfies the octet rule, by letting the cubes share corners or edges as shown in Figure 1.2.

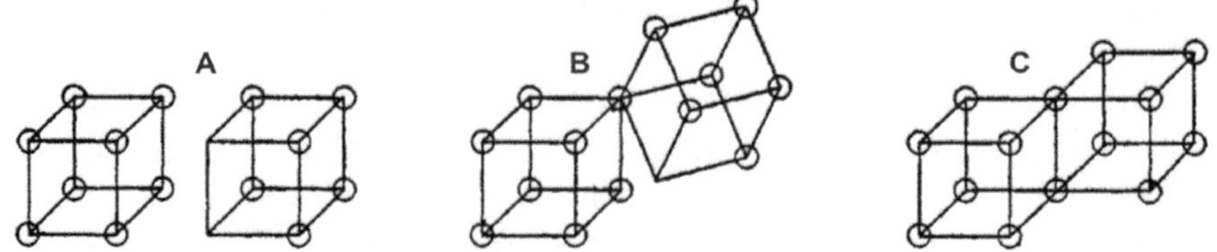

Figure 1.2   Representation of the atoms as cubes, in which electrons (represented by spheres) occupy corners, and which share edges or corners to form bonds in *A–C*. The ionic bond is shown in *A* as two separated cubes. A one-electron bond is shown in *B*, and a covalent bond in *C*.
Reprinted with permission from G. N. Lewis, The Atom and The Molecule, *J. Am. Chem. Soc.* 1916, **38**, 762. Copyright 1916 American Chemical Society.

Lewis calls these shared edges or corners by the name *"shared bonds"* (later to be called "covalent bonds"). Note in Figure 1.2 that Lewis also has an ionic bond in *A*, a shared electron bond in *C*, and a shared one-electron bond *B*. As such, his bonding theory stretches over a gamut of bonding situations.

It is clear that Lewis was aware that a shared bond violates the laws of physics. As he was grappling with the difficulty, he postulated (his 6th postulate)[1] that in the small spaces, as at the size of the atom, the classical laws of physics may not work anymore (he then toys with possible mechanisms of electron pairing like Parson's magnetic model, but he leaves the question open without a commitment). This reflects on Lewis's personality; he is not afraid to abandon a useful law when faced with a situation that demands it. This is the type of courage that was displayed for example by Niels Bohr a few years earlier in suggesting his quantum atom that defied the laws of physics. This is the courage of reformers.

Typically, at some point in the paper, Lewis informs his readers that new measurements by Moseley show "that helium has a total not of eight but of either four or two electrons", and that "from the resemblance of this element to the other inert gases", one may conclude that "here [in Helium] the pair of electrons plays the same role as the group of eight in the heavier elements, and that in the row ... comprising hydrogen and helium we have in place of the rule of eight the rule of two". Later he writes: "and we may question whether in general the pair rather than the group of eight should not be regarded as the fundamental unit". He accordingly draws Cl_2 and Cl:Cl, as represented by the beautiful cartoon of Bill Jensen, in Figure 1.3.

With this single concise pictogram, Lewis uses the Berzelius language of representing molecules (*e.g.*, H_2O) and the structural diagrams of the structuralists, to which he adds the notion of the

Figure 1.3 A caricature of Gilbert Newton Lewis (drawn by W.B. Jensen and reproduced with his permission) holding the cartoon in which Cl_2 is drawn as bound by an electron pair.

electron-pair bond; what an ingeniously portable device! He uses this symbol further to depict the H_2 molecule as H:H and the somewhat polar molecule HCl as H:Cl, where the electron pair is displaced towards Cl. Using the octet rule and the colon symbol, he describes ammonium as a species with four N:H bonds (previously it was represented as ammonia complexed to H^+). The perchlorate ion is drawn in a very modern way (no extension of octet is allowed, as most theoreticians agree...), with four Cl:O bonds with all atoms obeying the octet and having formal charges.

This electron-pair-based pictogram allows Lewis to represent double and triple bonds in compact manners, as $H_2C::CH_2$ and HC:::CH, to arrange electron pairs in space and to anticipate thereby the VSPER rules of 3-dimensional structure. He anticipates resonance theory as well, by shifting electron pairs and forming new structures, which he calls *electronic tautomerism* and uses to explain color due to movement of loose electrons between the two extreme forms, with sufficiently low characteristic frequency to produce color. Here we can see clearly how he linked spectroscopy to electronic structure changes.

Furthermore, Lewis considers in his paper also *reactivity of molecules as a reorganization of electron pairs*. Thus, in a bond with high polarity, the pair will move together with the more electronegative atom, while in a homopolar bond, the pair can move with one atom or the other, depending on the conditions. It is clear that Lewis views the bond as a dynamic entity, and naturally he considers reactivity of molecules as a reorganization of these electron pairs. This is clearly a precursor of the "curved arrow" mnemonics of Robinson and Ingold,[2] and of the subsequent formulation of heterolytic mechanisms like S_N1 by the Ingold–Hughes school. As such, his intellectual impact on chemistry became significant within a few years of the seminal publication. This can be witnessed from the summary of the Faraday Discussion in 1923,[27] where the following introductory statement was made: "It is to Professor G. N. Lewis that we are indebted for a very valuable conception in that he has given us a visual picture of a mode of union between the atoms alternative to . . . the electron type suggested by Sir J. J. Thomson. . . with the aid of two electrons held in common". Lewis has thus mapped chemistry as a science of electronic structure and reorganization.

Lewis's work eventually had its greatest impact in chemistry through the work of Irving Langmuir,[28] who very ably articulated the Lewis concept, coining new and catchy terms, and applying the concept to many branches of chemistry including transition metal complexes. At last, the meaning of the mythical terms *elective affinity* and *elective forces* had an operational and an effective definition in terms of *a quantized unit of bonding* that allowed constructing a chemical universe.[2]

1.4   The Second Electronic Structure Revolution

Despite the great advantage of the Lewis model it still lacked the basic physics of pairing and the quantitative aspect of bonding. This was provided by the quantum mechanical (QM) theory of matter, which sprung up during the early decades of the 20th century, when in 1926–7 both Schrödinger and Heisenberg, published their QM theories of atoms. Even though the two theories looked very different, they were shown later to be different representations of identical physics.

1.4.1   Formulation of the Chemical Bond by Quantum Chemistry

The Schrödinger equation trickled fast into chemistry, soon providing the quantitative and physical basis for the Lewis bond, and eventually it formed the basis for what we call quantum chemistry. Quantum chemistry (QC) is now the second power of chemistry alongside experiment. This is how far the revolution has spread.

1.4.1.a   The Quantum Chemical Trinity

The Schrödinger equation describes the electron as both a particle and a wave, and hence it was called also wave mechanics. Its applications in chemistry have led to three theoretical models, which now encompass virtually all the chemical phenomena. This theoretical trinity is shown in Scheme 1.1.

The first two theories, which were developed in the 1920s, were valence bond theory (VBT) and molecular orbital theory (MOT).[29] The third one is density functional theory (DFT), which was developed in the 1960s, and entered chemistry quite a bit later. I shall briefly describe the historical background of only VBT and MOT, and will restrict this to the development of the notions of the chemical bond. Later on, I shall make comments about the origins of chemical bonding using these three theoretical frames.

1.4.2   The Roots of VB Theory

Initially, the Schrödinger equation was applied to the hydrogen atom but encountered difficulties going even to helium, and it certainly did not show any clear prospects of solving the nature of the chemical bond. At the same time, the overwhelming support of Lewis's idea in chemistry presented an exciting agenda for research directed at understanding the mechanism by which an electron pair could constitute a bond.

1.4.2.a   The Heitler–London VB Theory

The mechanism of electron pairing remained, however, a mystery until 1927 when Walter Heitler and Fritz London went to Zurich to

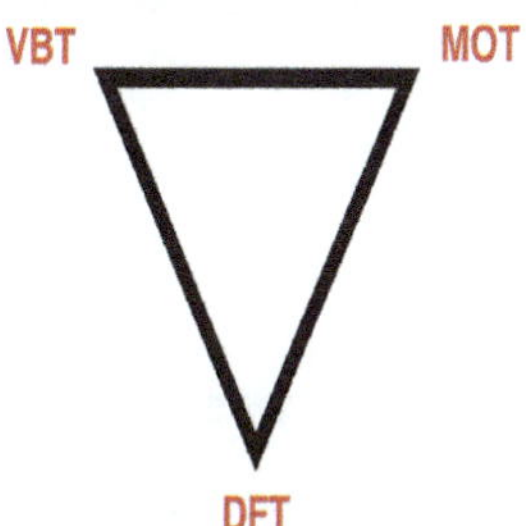

Scheme 1.1   The theoretical trinity, valence-bond theory (VBT), molecular orbital theory (MOT), and density functional theory (DFT).

work with Schrödinger. Schrödinger was not interested in bonding, but they were. In the summer of the same year they published their seminal paper, "Interaction Between Neutral Atoms and Homopolar Binding",[30] in which they showed that the bonding in H_2 originates in the quantum mechanical "resonance" interaction which is contributed as the two electrons are allowed to exchange their positions between the two atoms. This wave function and the notion of resonance were based on the seminal work of Heisenberg,[31] who showed earlier that, since electrons are indistinguishable particles, then for a two electron systems, with two quantum numbers n and m, there exist two wave functions, $[\varphi_n(1)\varphi_m(2)]$ and $[\varphi_n(2)\varphi_m(1)]$, whose positive and negative linear combinations, Ψ_A and Ψ_B, are split in energy by a new energy term which he called *resonance* using a classical analogy of two oscillators that, by virtue of possessing the same frequency, form a resonating situation with characteristic exchange energy.

In modern terms (including the electron's spin), the Heitler–London (HL) bonding in H_2 can be accounted for by the wave function drawn in **1**, in Scheme 1.2. This wave function is a superposition of two covalent situations in which, in the left-hand form one electron has a spin-up (α spin) while the other spin-down (β spin), and *vice versa* in the right-hand form. Thus, the bonding in H_2 arises due to the quantum mechanical resonance interaction between the two patterns of spin arrangement that are required in order to form a

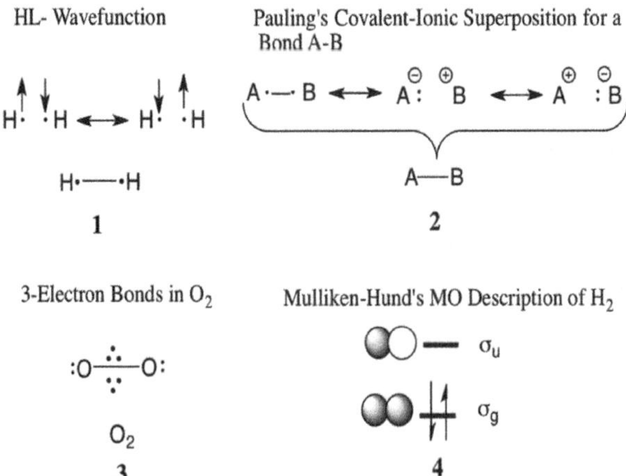

Scheme 1.2 The Heitler–London (HL) wave function (**1**) for H_2. The description of the full bond wave function for A–B as a superposition of a covalent HL-structure and two ionics (**2**). Pauling's description of O_2 as a paramagnetic species with two π-3-electron bonds (**3**). Mulliken-Hund's MO representation of H_2 (**4**).

singlet electron pair. This interaction accounts for about 75% of the total bonding of the molecule, and thereby projects that the wave function in **1**, which is referred to as the HL- or covalent-wave function, can describe the chemical bonding in a satisfactory manner. This origin of the bonding was a remarkable feat of the new quantum theory, since until then it was not obvious how two neutral species could be at all bonded (see above, Born's statement to Einstein from 1923, and recall the objection to Avogadro's hypothesis). The QM phenomenon of resonance or interference provided the new ground rules for interaction of negatively charged particles (electrons) in microscopic systems and in the field of oppositely charged ones (nuclei).

In the winter of 1928, London extended the HL-wave function and drew the general principles of covalent bonding, for which he considered also ionic structures for homopolar bonds, but discarded their mixing as being too small. In essence, the HL theory was a quantum mechanical version of Lewis's electron-pair theory. Thus, in his letter to Lewis,[2] and in his landmark paper,[32] Pauling pointed out that the HL and the London treatments are "entirely equivalent to G. N. Lewis's successful theory of shared electron pair…". Although the final formulation of the chemical bond has a physicists' dress, the origin is clearly the chemical theory of Lewis. *His concept of electron pairing turned out to be the resonance or interference of two electronic structure forms in which the indistinguishable electrons exchange sites and lower the energy.*

The HL-wave function formed the basis for the version of VB theory that became very popular later. In 1929 Slater presented his determinant-based method[33] and in 1931 he generalized the HL model to n-electrons by expressing the total wave function as a product of $n/2$ bond wave functions of the HL type.[34] In 1932 Rumer[35] showed how to write down all the possible bond pairing schemes for n-electrons and avoid linear dependencies between the forms, in order to obtain canonical structures. Further refinements of the new theory of bonding between 1928 and 1933 were mostly quantitative, focusing on improvement of the exponents of the atomic orbitals by Wang,[36] and on the inclusion of polarization function and ionic terms by Rosen[37] and Weinbaum.[38]

1.4.2.b Pauling–Slater's VB Theory

The success of the HL model for H_2 and its relation to Lewis's model posed a wonderful opportunity for the construction of a general quantum chemical theory for polyatomic molecules. Pauling and Slater achieved this import of the HL theory to chemistry. In the same year, 1931, they both published a few seminal papers in which they developed the notion of hybridization, the covalent-ionic superposition, and the resonating benzene picture.[34,39–42] Especially effective were those of Pauling's papers that linked the new theory to the chemical theory of Lewis, and rested on an encyclopedic command of chemical facts, much like the knowledge applied by Lewis to find his ingenious concept 15 years before.[1] In the first paper,[41] Pauling presented the electron pair bond as a superposition of the covalent HL form and the two possible ionic forms of the bond, as shown in **2** in Scheme 1.2, and discussed the transition from a covalent to ionic bonding. He then developed the notion of hybridization and discussed molecular geometries and bond angles in a variety of molecules, ranging from organic to transition metal compounds. In the following paper,[42] Pauling addressed bonding in molecules like diborane, and odd-electron bonds as in the ion molecule H_2^+ and in dioxygen, O_2, which he represented as having two three-electron bonds, **3** in Scheme 1.2. These papers were followed by a stream of five papers, published during 1931–1933 in *Journal of the American Chemical Society*, and entitled "The Nature of the Chemical Bond". This series of papers enabled the description of any bond in any molecule, and culminated in the famous monograph in which all the structural chemistry of the time was treated in terms of the covalent-ionic superposition theory, resonance theory and hybridization theory.[43] The book which was published in 1939 is dedicated to G. N. Lewis, and the 1916 paper of Lewis is the only reference cited in the preface to the first edition. VB theory in Pauling's view is a quantum chemical version of Lewis's theory of valence. As noted by Hager,[44] Pauling's biographer, Pauling discovered Lewis' 1916 paper[1] by reading Langmuir's[28] 1919 paper. Until reading these two papers in 1920, Pauling had been teaching a chemistry course at Oregon Agricultural College in which he used the image of a chemical bond as consisting of hooks and eyes, *e.g.*, with the sodium atom having an eye and chorine having a hook. Thus, the birth of VB theory in chemistry was an ingenious quantum chemical dressing of the Lewis seminal idea by Heitler and London, and then Pauling. At last there was a basis for the unification of chemistry under a single electronic structure theory, as professed by Ostwald, the father of physical chemistry.[23]

1.4.3 The Formulation of MO Theory

At about the same time that Pauling and Slater were developing their VB theory, Mulliken[45–48] and Hund[49,50] were developing an alternative approach called molecular orbital theory (MOT) that has a spectroscopic origin. The term MOT appears only in 1932, but the roots of the method can be traced back to earlier papers from 1928, in which both Hund and Mulliken made spectral and quantum number assignments of electrons in molecules, based on correlation diagrams tracing the energies from separated to united atoms.

In MO theory, the electrons occupy delocalized orbitals made from linear combination of atomic orbitals. Drawing **4**, Scheme 1.2, shows the molecular orbital of the H_2 molecule, and the delocalized σ_g MO can be contrasted with the localized HL description in **1**. While the VB and MO wave functions look very different, in both the origins of the electron-pair-bonding, as in H_2, is the same. In HL theory, it is due to the resonance of the forms, which allows the electrons to exchange positions while in MO theory it is the ability of the two electrons to delocalize over the two centers. Both describe electron sharing à la Lewis.

Eventually, it would be the work of Hückel that would usher MO theory into the mainstream organic chemistry. After a chilly reception,[51] Hückel's theory ultimately gave MO theory an impetus by being a successful and widely applicable tool. In 1930, Hückel used Lennard-Jones's MO ideas[52] on O_2, applied them to C=X (X = C, N, O) double bonds and suggested the σ–π separation.[53] Using this novel treatment Hückel ascribed the restricted rotation in ethylene to the π-type orbital. Equipped with this facility of σ–π separability, Hückel turned to solve the electronic structure of benzene using his new Hückel-MO (HMO) approach.[54] The π-MO picture in the left hand drawing in Scheme 1.3 viewed the molecule as a whole, with a σ-frame dressed by π-electrons that occupy three completely delocalized π-orbitals. The HMO picture revealed a closed-shell π-component with energy being lower relative to that of three isolated π-bonds as in ethylene. In the same paper, Hückel treated the ion molecules of C_5H_5 and C_7H_7 as well as the molecules C_4H_4 (CBD) and C_8H_8 (COT). This allowed him to understand why molecules with six π-electrons had special stability, and why molecules like COT or CBD either did not possess this stability (*i.e.* COT) or had not yet been made (*i.e.* CBD) at his time. Already in this paper and in a subsequent one,[55] Hückel had built the foundations for what would become later known as the *Hückel-Rule* of aromaticity.[29] This rule, its extension to *antiaromaticity*,

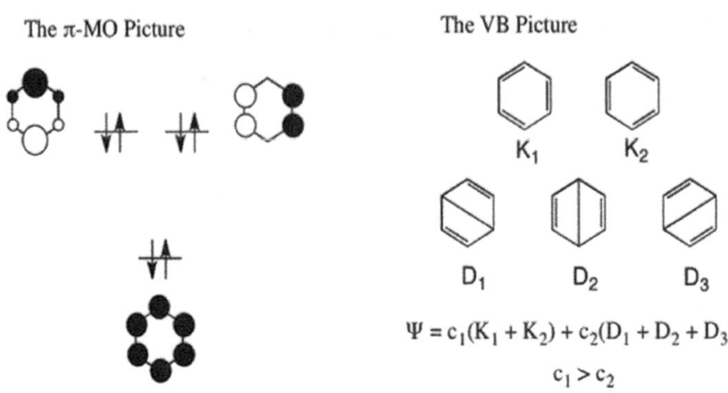

$$\Psi = c_1(K_1 + K_2) + c_2(D_1 + D_2 + D_3)$$

$$c_1 > c_2$$

Scheme 1.3   Hückel's delocalized MO representation of benzene (left) *vs.* Pauling-Wheland's VB representation (right).

and its articulation by organic chemists in the 1950s–1970s constituted major causes for the acceptance of MO theory.[29]

The description of benzene in terms of a superposition (resonance) of two Kekulé structures appeared for the first time in the work of Slater,[39] as a case belonging to a class of species in which each atom possesses more neighbors than electrons it can share with them, much like in metals. Two years later, Pauling and Wheland[56] applied the HL VB theory to benzene, using the five canonical structures in the right hand drawing in Scheme 1.3, and approximating matrix elements between the structures by retaining only close neighbor interactions. Their approach allowed them to extend the treatment to naphthalene and to a great variety of other species. Thus, in the HL VB approach, benzene is described as a *resonance hybrid* of the two Kekulé structures and the three Dewar structures;[56] and the latter had already appeared before in Ingold's idea of mesomerism which itself is rooted in Lewis's concept of electronic tautomerism.[1,2]

1.4.4   From MO–VB Rivalry to the Computational Paradise and the "Return of the Bond"

It is important to stress again that despite the different appearances of the description of *e.g.*, benzene by the two theories, the fact is that in both cases the bonding energy of the six electrons originates from the ability of the electrons to delocalize over six centers, either *via* delocalized multi-center MOs or *via* electron-shifting among the VB structures in Scheme 1.3. Nevertheless, these two seemingly different treatments of benzene faced the chemical community with two alternative descriptions of one of its molecular icons, and this began the VB–MO rivalry.[57] Interestingly, already back in the 1930s, Slater[40] and van Vleck and Sherman[58] stated that since the two methods ultimately converge, it is senseless to quibble on the issue of which one is the better theory. However, this rational attitude does not seem to have made much of an impression on this religious war-like rivalry, which involved most of the prominent chemists in the Pauling–Mulliken generation and beyond.

Although the rivalry makes an interesting story, it has been told before, and will not be repeated herein, as it can be found in a few sources.[29,57,59] What has eventually overshadowed the old conceptual tensions is the development of very effective computational packages. The current arsenal of computational quantum chemistry involves a variety of methods based on MOT, VBT and DFT,[2,7,8,60] and which have brought about discoveries of new bonding motifs (see later).[5–8] Additionally, new techniques have been developed which enable one to 'locate' bonds within the total electron density of the molecule (*e.g.* atoms in molecules (AIM), electron localization function (ELF),[7] *etc.*), and procedures that analyze chemical bonding in terms of energy components (*e.g.* Morokuma analysis, energy decomposition analysis (EDA), block-localized wave function (BLW), natural orbital for chemical valence (NOCV), interacting quantum atoms (IQA), *etc.*). These powerful quantitative possibilities have taken chemistry into the computational paradise, and have led to a renaissance in the chemical bond.[7,8]

1.5 Understanding the Nature of the Chemical Bond

Before discussing the new bonding motifs, it is important first to briefly comment on the origins of the chemical bond, in a manner that does not depend on the bond type or the choice of the representation of the approximate wave function.

1.5.1 The Origins of Bonding

It might be tempting to analyze the chemical bond in terms of forces acting on the atoms and bringing them to an equilibrium distance, where the forces vanish.[61] This pristine idea, correct as it may be, does not provide chemical insight, because it does not differentiate different bond types. A more productive approach is the virial

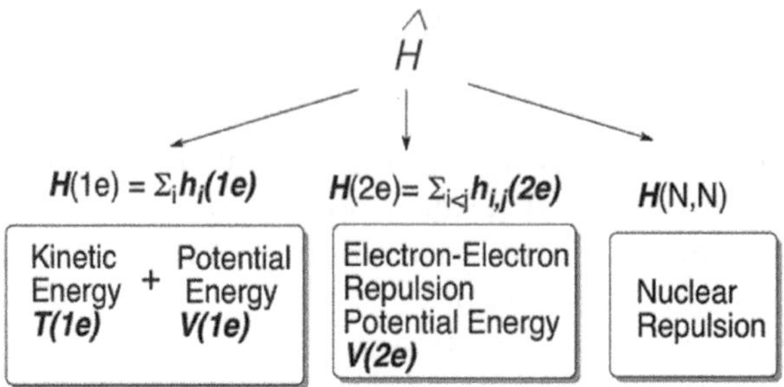

Scheme 1.4   The terms of the Hamiltonian: one- and two-electron parts, and the nuclear repulsion part. The one-electron part is a sum of the kinetic energy (T(1e)) of the electrons, and the potential energy (V(1e)) summed over all the electrons and nuclei. The two electron part is a potential energy term (V(2e)) that accounts for the pairwise electron-electron repulsions.

theorem,[61–67] which deals with the changes of the potential and kinetic energies that bring about the formation of a bond in equilibrium.

Let us digress for a moment, and define some necessary quantities before discussing the theorem. Scheme 1.4 shows the energy terms in the Hamiltonian. There are terms that depend on the coordinates of a single electron, H(1e), terms which depend on the coordinates of two electrons, H(2e), and terms that depend only on the nuclei, H(N,N). In a given geometry, the last term is a constant, and can be excluded from further discussion.

H(1e) is a sum of mono electronic kinetic- and potential-energy terms, T(1e) and V(1e), respectively. T(1e) corresponds to the sum of kinetic energies of the electrons moving in the field of the nuclei, while V(1e) accounts for the sum of attractive terms of the electrons in the field of all the positive nuclei. Finally, H(2e) accounts for the potential energy due to the pairwise electron–electron repulsion in the molecule, V(2e). In an atom there are only one center atomic terms; the expectation values of the kinetic energy term, $\langle T(1e) \rangle$ is positive, the potential energy $\langle V(1e) \rangle$ is negative, and $\langle V(2e) \rangle$ is positive. $\langle V(1e) \rangle$ dominates the energy balance on the atom, and generates a stable atom. However, in a molecule there are interference or resonance terms, which refer to the expectation values of T and V when an electron is allowed to 'travel' between two sites. Consider for example the simplest case of H_2^+ that represents a one electron bond, spanned by the two atomic orbitals (AOs), $\langle a |$ and $\langle b |$, such that in addition to the on-site terms, we have a mono-electronic resonance-interference term which is commonly labeled as β, where β is the reduced resonance integral, which is always negative and stabilizing:

$$\beta = \langle a | \boldsymbol{h(1e)} | b \rangle - \langle a | \boldsymbol{h(1e)} | a \rangle \langle a | b \rangle; \quad \langle a | b \rangle = S_{ab}, \ \beta < 0 \tag{1.1}$$

In turn, β is a sum of kinetic energy and potential energy terms:

$$\beta = \beta_T + \beta_V \tag{1.2}$$

Kutzelnigg showed[67] that β_V is small and slightly repulsive, whereas β_T is dominant and is stabilizing. It may seem counterintuitive that β_V is positive, since it represents the attraction of the overlap population to the two nuclei. However, recalling that the overlap population accumulates by depleting the atomic population, it is clear that the reduced β_V, which is a balance between the very attractive on-site interaction *vs.* the interaction of the overlap population with the nucleus and the smaller inter-site attraction (eqn (1.1)), can be positive. Be that as it may, β_T dominates the resonance term.

Eqn (1.3) shows the expression of the dominant part of the resonance interaction,[67] and its dependence on the interatomic distance R and the orbital exponent ζ:

$$\beta_T = - 1/3[\zeta^4 R^2 e^{-\zeta R}] \tag{1.3}$$

The orbital exponent gauges the compactness of the orbital. The larger the exponent ζ, the more compact is the orbital, the higher is the resonance interaction, and the stronger is the interaction that brings about bonding. Generally speaking, all the interference/resonance terms in MOT, VBT or DFT scale with β, and hence with β_T. In MOT and DFT,[60] these terms are simply the common mono-electron resonance-integral terms of the orbitals. Similarly, in VBT all the mixing terms between VB structures scale with β and hence are dominated by β_T.

Another important effect that scales with β_T is the Pauli repulsion between electrons having identical spins, *e.g.* as we encounter when two closed shell orbitals on two centers (A: :B) overlap, or when three electrons are localized on two centers (A: •B), *etc.* This well-known overlap repulsion is given by:

$$\text{Pauli or Overlap Repulsion} = -n\beta S/(1 - S^2) \tag{1.4}$$

Here n is the number of identical-spin-electrons on the two centers, β the reduced resonance interaction between the two orbitals, and S the corresponding overlap. Since β is dominated by β_T and the latter is negative, Pauli repulsion raises the kinetic energy of the electronic system. This effect is again common to MOT, VBT, and DFT.

1.5.2 The Virial Theorem and Origins of Bonding

The virial theorem[61,62] for two atoms/fragments is expressed in eqn (1.5),

$$-R(dE/dR) = 2T + V \qquad (1.5)$$

where E is the total energy, V and T are respectively the potential and kinetic components, and R is the interatomic distance between the two atoms/fragments.

The term on the left of eqn (1.5) is the force acting on the molecule times the respective length, R (the product is then the 'work' required for stretching or squeezing the distance away from the equilibrium value R_{eq}). At equilibrium, *i.e.* for $R = R_{eq}$ the force $-dE/dR$ is zero, which means that any properly optimized wave function at its own equilibrium distance R_{eq} must obey the following virial ratio of the kinetic to the potential energy:

$$T/(-V) = \tfrac{1}{2} \qquad (1.6)$$

Achieving this ratio will result in energy lowering, namely in bonding, as expressed in eqn (1.7):

$$\Delta E = -\Delta T = \tfrac{1}{2}\Delta V \quad (\Delta V < 0) \qquad (1.7)$$

But, how does bonding actually manifest? When atoms form a covalent bond they shrink by comparison to the free atoms. One can verify this atomic shrinkage by comparing in the periodic table the atomic radius *vs.* the covalent radius; the latter is invariably smaller. To understand this phenomenon, let us imagine two atoms/ fragments, which are infinitely apart, where each is in equilibrium and obeying the virial ratio $T/(-V) = \tfrac{1}{2}$. Let us now shrink their valence orbitals. As a result, the potential energy of the valence electrons in the atom will become more negative (due to larger electron-nuclear attraction). However, the kinetic energy will increase (due to confinement of the electron to a smaller space). The increase of the kinetic energy is steeper because it increases in proportion to the square of the orbital exponent ζ, while the potential energy decreases only in a direct proportion to ζ. Consequently, the ratio $T/(-V)$ of the two shrunk atoms/fragments will become higher than $\tfrac{1}{2}$, due to the excessive T. As such, the two species will be driven to get closer and overlap their valence orbitals, such that the resulting resonance/ interference term β, which as we recall scales as by β_T, will lower their kinetic energy, and restore the virial ratio, giving rise to a bonded species at equilibrium. Thus, the resonance term allows the atoms to shrink, such that in the net balance, the potential energy of the atoms/fragments can decrease and lower the total energy (eqn (1.7)).

Whenever the atoms/fragments possess also lone pairs (or other bond pairs), the shrinkage of the fragments will result in a much higher kinetic energy pressure, which will be augmented by the Pauli repulsions of the spectator lone pairs with the bond pair and among themselves. Therefore, bonds like F–F, O–O, Cl–Cl, S–S, *etc.*, will need a much higher resonance/interference term β to restore the virial ratio and achieve a bond at equilibrium. The manner by which such bonds acquire large resonance energy terms is most apparent by considering the bonding using VBT. But, let us show that VBT, MOT and DFT have the same physical content of bonding.

1.5.3 Bonding in VBT, MOT, and DFT

We limit the following discussion in this section to two-center bonds that occur by sharing of either electron pairs or odd numbers of electrons.

1.5.3.a Electron Pair Bonds

Let us start with the simplest bond in H_2. The energy of the bond is the expectation value of the wave function for H_2 with respect to the Hamiltonian of the molecule, which as we just elaborated has two parts; one is the kinetic energy of the electrons and the potential energy due to their attraction to the nuclei, and the second is the electron–electron repulsion.

Figure 1.4 shows how MOT and VBT describe this bond. Thus, as shown in Figure 1.4a, in MOT the bond is described by the electron pair occupying σ_g MO (g is the orbital symmetry with respect to the inversion point in the mid-bond). To the right of the MO diagram we show shorthand notations of the corresponding Slater determinant Ψ_{MO} with two spin-paired electrons in σ_g, where spin down is indicated by the bar over the orbital while the absence of a bar indicates spin up. The σ_g MO is a bonding orbital, in the sense that its energy is below the AOs of the constituent H atoms, due to the delocalization of the electron over the two centers. This delocalization effect is imparted by the mono-electronic resonance part of the Hamiltonian (the term β in eqn (1.1)). So, MOT shows intuitively and pictorially that having an electron pair in the $\sigma_g{}^2$ configuration binds H_2.

At the same time, we can see that the bonding can be improved if we enable the two electrons in σ_g to get away from each other and

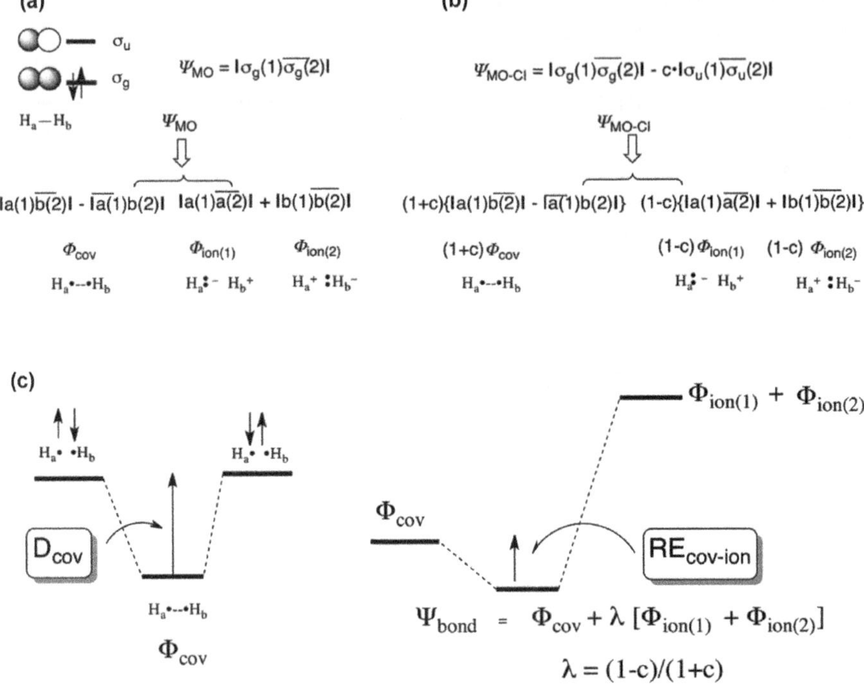

Figure 1.4 The MOT–VBT equivalence: (a) Expansion of the MO wave function to a set of covalent and ionic structures. (b) Correcting the MO wave function by configuration interaction (CI) and expanding the MO-CI wave function to covalent and ionic structures; note that the coefficient of the covalent structure increases $(1+c)$ and the one for the ionic decreases $(1-c)$. (c) Constructing the H_2 bond wave function using VBT, starting from the covalent structure (left) and then mixing the covalent with the ionic structures to gain also covalent-ionic resonance energy. The covalent structure is stabilized by the spin-pairing covalent energy, D_{cov}, while the covalent-ionic mixing leads to resonance energy stabilization, $RE_{cov\text{-}ion}$. The bond energy is the sum of D_{cov} and $RE_{cov\text{-}ion}$. Adapted from ref. 143 with permission from the Royal Society of Chemistry.

lower their Coulomb repulsion. This is achieved by usage of configuration interaction (CI). CI causes the σ_u^2 configuration to mix with σ_g^2, thus generating the MO-CI wave function $(\Psi_{\text{MO-CI}})$ in Figure 1.4b, where c is the mixing coefficient. The mixing is enabled by the bi-electronic part of the Hamiltonian, $H(2e)$, which lowers the electron–electron repulsion in the electron pair, and energetically improves the electron-pair-bonding of H_2.[29,60]

This looks somewhat like a paradox, that adding a high-energy configuration lowers the energy. But it can be immediately elucidated by mapping the Ψ_{MO} and $\Psi_{\text{MO-CI}}$ wave functions to VB ones.[29] This can be done easily, by expressing σ_g as a sum $a+b$ of the corresponding AOs, and σ_u as their difference, $a - b$. Plugging these expressions into the Slater determinants of Ψ_{MO} and $\Psi_{MO\text{-}CI}$ enables one to expand these determinants into AO-based ones.[29] Thus as shown in Figure 1.4a, the Ψ_{MO} gives rise to a mixture of 50% covalent HL VB structure, Φ_{cov}, and 50% ionic structures, $\Phi_{ion(1)}$ and $\Phi_{ion(2)}$, which localize the electron pair once on H_a and the other time on H_b. Repeating the procedure for $\Psi_{MO\text{-}CI}$ (Figure 1.4b) shows that the CI augments the contribution of the covalent structure, which becomes $1 + c$, while diminishing the ionic ones, which becomes $1 - c$. Clearly therefore, the CI brings about correlation between the two electrons, such that they are mostly far apart in the covalent structure and close together in the ionic structures. This is called static correlation or Coulomb correlation (or still, left–right correlation, or Coulomb hole). *It is important to note that $\Psi_{MO\text{-}CI}$ describes the electron pair bond as a covalent-ionic superposition, á la Lewis–Pauling.*

One could do everything in the reverse order: start now from the VB configurations and derive the same wave function as obtained in the MO-CI procedure.[29,60] Thus, as shown in Figure 1.4c, initially we utilize the two determinants needed for constructing Φ_{cov}, which is stabilized by the resonance between the two spin arrangement patterns. This resonance (achieved *via* the mono-electronic part $H(1e)$ of the Hamiltonian) causes the two electrons to become spin-paired to a singlet. The covalent structure, with the electrons being one on each center, exhibits 100% correlation, and this is why the covalent structure is the lowest in energy compared with the ionic structures that keep the electrons together on a single site. We can now bring in the ionic wave function, $\Phi_{ion(1)} + \Phi_{ion(2)}$, and let it mix into Φ_{cov}. This mixture of the covalent and ionic structures will describe the electron-pair-bond in the spirit of Lewis and Pauling. The electron pair is now less correlated compared to the pure Φ_{cov}, but the bond has gained some covalent-ionic resonance energy $(RE_{cov\text{-}ion})$, which over-compensates for the small loss of correlation. This full circle shows that there is no difference between MO-CI and VB when both methods are carried out to their fullest. In fact, one can start from MO-CI and recover both the Φ_{cov} and the $RE_{cov\text{-}ion}$ quantity, which are the elements of VB theory.[68]

When the bond in question is in a more complex molecule *e.g.*, F_2, the bond energy can be further improved by CI that allows the spectator electron pairs (lone pairs), as well as the doubly occupied orbitals in the ionic structures, to adapt themselves to the changes in the static correlation caused in the bonding electrons and the local charges of the ionic structures. This 'grooming' of the electron pairs surrounding the bond in question is called dynamic correlation. Dynamic correlation is rather pictorial in VB, showing how the lone pairs of F–F adapt their shapes and sizes to the local charges on the two atoms, and how they contribute to enhance the bond energy. This differential dynamic correlation can be retrieved in MO-based CI.[68] Thus we can go back and forth and cross the mirrors between the MOT and the VBT worlds, and even retrieve the key concepts of bonding. This realization is helpful, because it allows us to use for any given problem the theory that provides the most lucid insight.

DFT as implemented in computational quantum chemistry, uses Kohn–Sham orbitals,[60] which are pictorially identical to those used in MOT, and differ only in that the KS orbitals include already some dynamic correlation effects due to the exchange–correlation terms. For larger molecules DFT uses the NOCV technique[15] to define the bond orbitals, from fragment orbitals that look very much as in MOT, and in a manner akin to VBT. Thus, the description of the electron-pair-bond in DFT is in no way different than in MOT and VBT; it originates from the one-electronic part of the Hamiltonian augmented by the correlation effect. The terminology is different but the physical essence is the same.[60]

1.5.3.b *Odd-electron Bonds*

Odd-electron bonds are quite common now in the chemistry of main group elements,[69,70] and have been known since the beginning of the 20th century when Lennard-Jones described the O_2 molecule[52] using MOT and subsequently Pauling did the same using VBT (2 in Scheme 1.2). Figure 1.5 shows the description of simple odd-electron bonds by MOT and VBT (DFT is skipped since it suffers from a pathological self-interaction error for these species). Figure 1.5a shows the one-electron bond, H_2^+. Using MOT, the bond is described by the σ_g^1 configuration, which maps directly to the VB

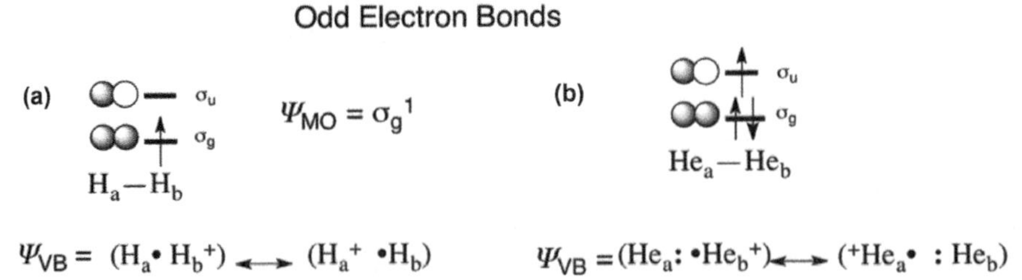

Odd Electron Bonds

(a) σ_u / σ_g, $\Psi_{MO} = \sigma_g^{\ 1}$, H_a—H_b

(b) σ_u / σ_g, He_a—He_b

$$\Psi_{VB} = (H_a\bullet\ H_b^+) \longleftrightarrow (H_a^+\ \bullet H_b)$$

$$\Psi_{VB} = (He_a\!:\bullet He_b^+) \longleftrightarrow (^+He_a\bullet\ :He_b)$$

Figure 1.5 Descriptions of simple odd-electron bonds by MOT and VBT: (a) The one-electron bond in H_2^+. (b) The 3-electron bond in He_2^+.

description as a resonating mixture of two one-electron forms. In both cases, the bond is due to "sharing", *i.e.* delocalization of the single electron over the two centers (the β term). Unlike the electron-pair bond, there is no static/Coulomb correlation to take care of, and the MOT and VBT wave functions are identical.

A prototypical 3e-bond is He_2^+, which is shown in Figure 1.5b. Here too the MOT and VBT wave functions are identical and the bonding is due either to an excess electron in the bonding σ_g orbital, or to the resonance between the two 3e-forms; both are β terms. In more complex species, like F_2^-, dynamic correlation improves the bonding, by allowing the orbitals to respond to the instantaneous charges of the atoms. As such, once again we can see that MOT and VBT describe the same bonding mechanism for archetypical odd-electron bonds: odd electron binding is due to the resonance/interference term β, which scales by β_T, and lowers the kinetic energy at the bonding region, thus restoring the virial ratio.

1.5.4 New Bonding Motifs

In discussing the emergence of new bonding motifs one is limited to a very small selection of topics. As such, I followed with some modifications a recent selection of highlighted topics.[5,71] I duly apologize for any inadvertent omissions in the following discussion.

VBT has given rise to the emergence of new bonding motifs in two-center bonds. Let us therefore start with VBT, while being cognizant that the emerging features are in principle identical to those in MO-CI[68] or DFT, but in the latter approaches these features are not immediately apparent.

1.5.4.a Charge-shift Bonding

Figure 1.6 depicts the essential elements for discussing the emergence of new bonding features using VBT. Figure 1.6a shows the three VB structures needed to describe an electron pair bond between two fragments A and B (where A and B are identical or different). For classical covalent bonds, the covalent structure Φ_{cov} is generally stabilized relative to the dissociation limit $(A^\bullet + {}^\bullet B)$ by the covalent spin pairing energy D_{cov}. We now bring in two ionic structures, $\Phi_{ion(1)}$ and $\Phi_{ion(2)}$, and mix them with the covalent structure. The VB mixing diagram shows that the consequence is the further stabilization of the Lewis bond state $(\Psi_{bond}(A–B))$ by the resonance energy quantity, $RE_{cov\text{-}ion}$, due to the covalent-ionic mixing. Since this mixing is

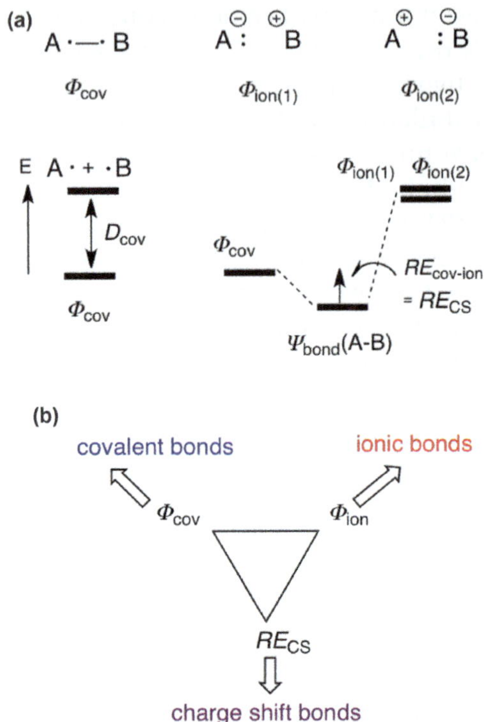

Figure 1.6   (a) Construction of an electron pair bond, A–B, using the VB-structure set of covalent and ionic structure. The general energy relationship between the covalent structure and the dissociation limit is the covalent bond energy contribution D_{cov}. On the right side we show a VB mixing diagram describing the covalent-ionic mixing: RE_{CS} is the charge-shift resonance energy due to the mixing. (b) The emergent bond families from the three independent quantities of the bond; covalent, ionic and charge shift.

common also to homonuclear bonds, we refer to this quantity as the charge-shift resonance energy, RE_{CS}.[72] Of course, when one of the ionic structures is the lowest in energy, the final bond is ionic, which may in principle enjoy some RE_{CS} stabilization too.

The above analysis shows that the electron-pair bond in VBT can be described by three independent quantities, which are indicated in the triangle in Figure 1.6b. Two corners of the triangle are occupied by the covalent and ionic VB structures, while the third corner is the RE_{CS} quantity, due to their mixing. In principle therefore, we might expect three distinct families of electron-pair bonds that emerge from these independent quantities. One family is dominated by the covalent structure and its bond energy is primarily D_{cov}, with a minor contribution of RE_{CS}. The second family is dominated by one of the ionic structures, Φ_{ion}, and its bonding energy is by and large the Coulomb attraction of the opposite charges, with a minor contribution of RE_{CS}. These are the two classical Pauling families of covalent and ionic bonds. Alongside them we expect in principle to find a third family of bonds where most, if not all, of the entire bond energy is provided by the RE_{CS} quantity. This is the charge-shift bonding (CSB) family.[72,73]

Figure 1.7 shows the energy curves of a few bonds plotted against the internuclear distances. For each case one can find the curve for the dominant VB structure in blue, and the one for the 'exact' state of the bond in red. The three bond families manifest nicely in these plots.

Thus, for H_2 (Figure 1.7a) the covalent curve dominates the wave function as well as the bonding energy; the exact curve is very close to it. H–H is a classical covalent bond. By contrast, for F–F (Figure 1.7b) the covalent structure is repulsive throughout the range of internuclear distances. Here, the bond and the bonding energy arise entirely from the RE_{CS} due to the mixing of the ionic structures into the covalent structure. F–F is therefore a CSB. The second panel, in Figure 1.7c and d, shows two heteronuclear bonds B–H and F–H. It is seen that B–H is a classical covalent bond, whereas F–H is a CSB. Finally, the third panel in Figure 1.7 shows the Na–F and Na–Cl bonds. In both of these cases, the dominant VB structure is ionic, and the full exact curve is very close to the ionic curve (RE_{CS} is small). These two bonds belong to the classical family of ionic bonds.

It is important to recognize that CSBs may appear as classical covalent bonds when one examines the weights of the covalent and ionic structures. Thus, for example, if we compare F_2 and H_2 we find that both wave functions are dominated by the covalent structure that has virtually identical weights of 0.76 and 0.73, respectively. This further shows that the bonding in F_2 is not associated with either the covalent or the ionic structures, and is exclusively determined by the RE_{CS} feature. The CSB family is unique.

There are many other bonds which belong to the CSB family: Cl–Cl, Br–Br, O–O, S–S, N–N, N–F, N–O, C–F, Si–Cl, *etc.* All of these bonds are typified by repulsive or weakly bonded covalent structures, whereas the bonding energy is provided exclusively or almost so by RE_{CS}. It is possible to show that the electronegativity scale is proportional to RE_{CS}. *In fact, the origin of the scale is arguably rooted in RE_{CS}.*[73]

Recall that the analysis of the virial theorem led us to conclude that unlike in H_2, the bonding electrons in F_2 suffer severe Pauli repulsion exerted by the lone pairs. As such, the covalent structure in F_2 is destabilized by the Pauli repulsion of the lone pairs with the bond pair.

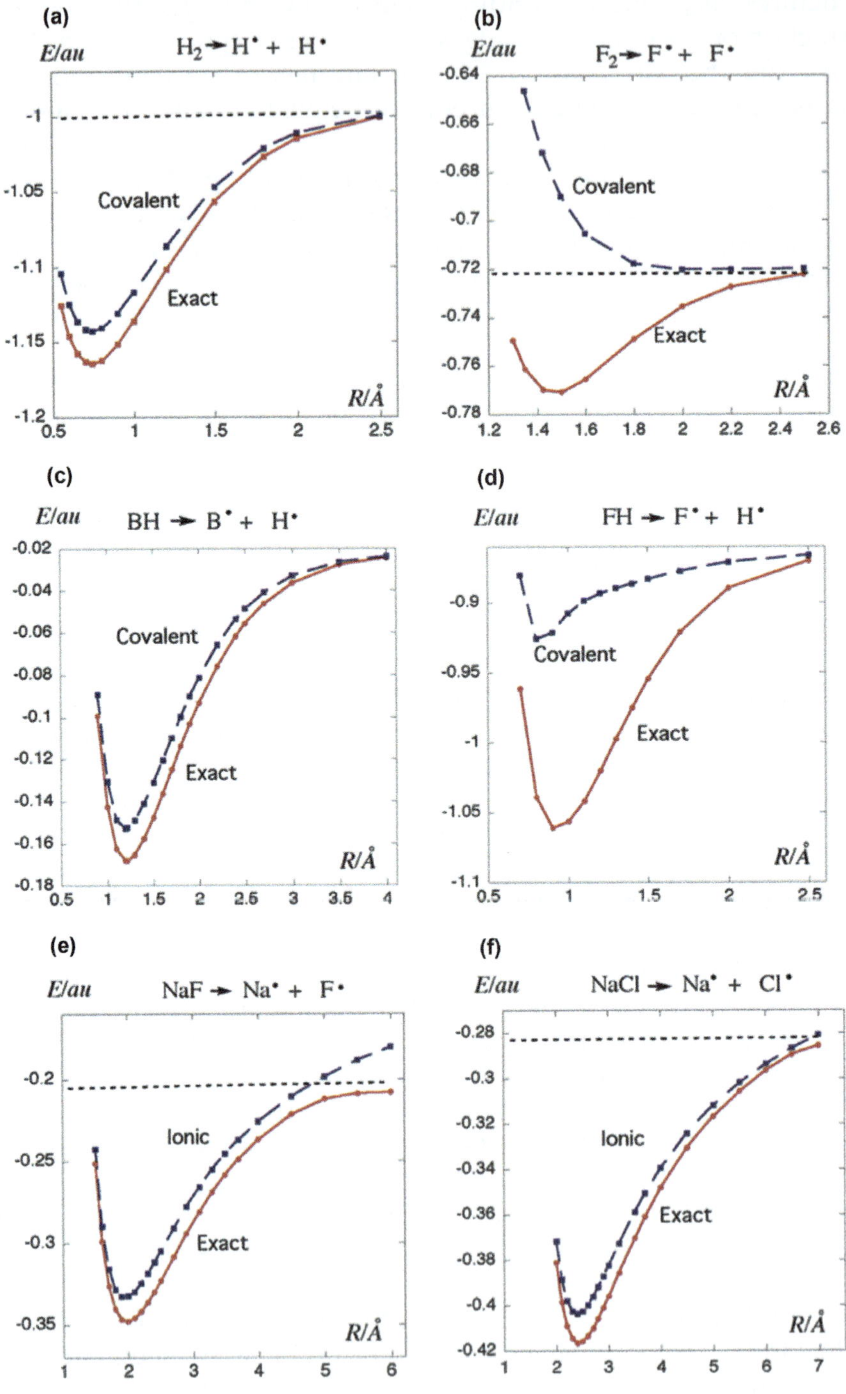
(a)
E/au
H₂ → H• + H•
Covalent
Exact
R/Å

(b)
E/au
F₂ → F• + F•
Covalent
Exact
R/Å

(c)
E/au
BH → B• + H•
Covalent
Exact
R/Å

(d)
E/au
FH → F• + H•
Covalent
Exact
R/Å

(e)
E/au
NaF → Na• + F•
Ionic
Exact
R/Å

(f)
E/au
NaCl → Na• + Cl•
Ionic
Exact
R/Å

This kinetic energy pressure together with the shrinkage of the valence orbitals raise dramatically the kinetic energy of the bond and tip the virial ratio off balance. The requisite kinetic energy lowering and restoration of the virial ratio is achieved by increasing the RE_{CS}, through the covalent-ionic mixing.

As we mentioned before, there are electron density methods that enable one to define and identify bonds. The ELF (electron localization function)[74] method uses a local function related to the Pauli repulsion to find the domains of different spin-paired electron pairs in the molecule. The integration of the electron density of these domains gives the population $\bar{N}$ of these domains.

Another electron density based method is AIM (atoms in molecules).[75] In AIM a bond is characterized by a bond path, which defines a maximum density path connecting the bonded atoms. The point along the path at which the density is at minimum is called the bond critical point (BCP), $\mathbf{r}_c$. One then characterizes the bond type by determining the electron density $\rho(\mathbf{r}_c)$ at the critical point, and the Laplacian of the density at the critical point. The Laplacian is defined in eqn (1.8):

$$\frac{\hbar^2}{4m} \nabla^2 \rho(\mathbf{r}_c) = 2G(\mathbf{r}_c) + V(\mathbf{r}_c) \tag{1.8}$$

It is composed of two terms; $G(\mathbf{r}_c)$, which is the kinetic energy density, and $V(\mathbf{r}_c)$ which is the potential energy density; $G(\mathbf{r}_c) > 0$ and $V(\mathbf{r}_c) < 0$. According to AIM, a significant density and a negative Laplacian at the critical point typify a covalent bond with a shared density. When the Laplacian is negative, this means that the potential energy density dominates it. On the other hand, a positive Laplacian corresponds to a closed shell interaction, which typifies either ionic bonds or cases with closed-shell Pauli repulsion such as He_2. However, as seen from Figure 1.8b, a positive Laplacian occurs also in CSBs.

Figure 1.7   (Figure on previous page). Six frames, (a)–(f), exemplifying the three bond families (indicated in Figure 6b). The frames trace the bond dissociation energies (in au) plots *vs.* the distances R (Å) between the atoms/ fragments, for the bonds indicated by the dissociation processes on the top of the frames. In each case, the red energy curve is the exact bond's curve, while the blue one is the energy of the dominant VB structure. Covalent bonds are shown in (a) and (c), ionic bonds in (e) and (f), while charge-shift bonds (CSBs) in (b) and (d).

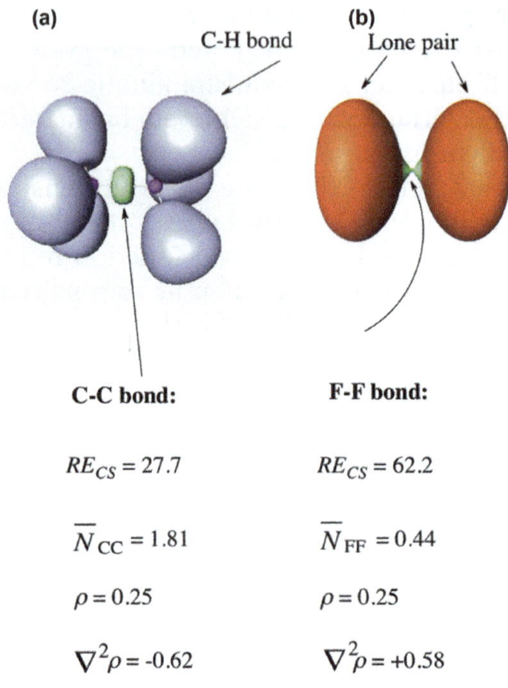

Figure 1.8　Properties of two bonds, (a) the covalent C–C bond in ethane, and (b) the CSB in F_2. RE_{CS} is the charge-shift resonance energy (kcal mol^{-1}), $\overline{N}$ is the integrated electron density in the bond region, ρ is the electron density at the bond critical point, and $\nabla^2\rho$ is the Laplacian of the density (These ELF and AIM representations are limited to homonuclear CBSs). Reproduced with permission from *Nature Chem.* S. Shaik, D. Danovich, W. Wu, P. C. Hibery, Charge-shift bonding and its Manifestations in Chemistry, 2009, **1**, 443. Copyright 2009. Rights Managed by Nature Publishing Group.

Figure 1.8 shows the differences in ELF and AIM qualifiers for the covalent bond C–C, and the CSB F–F. Thus, Figure 1.8a is the C–C bond in ethane, which is characterized as a classical covalent bond in VBT. It is seen that ELF finds a bond region (the green cylinder) between the carbon atoms, and the integrated electron density in this region $\bar{N}$ amounts to 1.81e, close to an electron pair. The AIM analysis shows that C–C has a significant critical point density and a largely negative Laplacian as expected from a covalent bond with a shared density. Thus, similar to VBT, both ELF and AIM characterize C–C as a classical covalent bond.

In a striking contrast, the F–F bond in Figure 1.8b has a severely depleted ELF basin with $\bar{N} = 0.44$e, and the basin looks split with most of the electron density being outside of the bond region. Similarly, in AIM, while F–F has a significant critical point density, its

Laplacian is positive, indicating repulsive Pauli interactions and excess of kinetic energy. Thus, much like VBT, both ELF and AIM characterizes F–F as CSB. As shown recently,[68] the Φ_{cov} wave function, the RE_{CS} quantity, and the contrasting appearance of the H$_2$ *vs.* F$_2$ bonding in Figure 1.8, could all be extracted as well from MO-CI theory.

It is apparent from the above analyses that homonuclear CSBs have very different features than classical covalent ones, even though they may appear covalent by just looking at weights of the covalent structures in the VB wave functions, or by naively considering that a homonuclear bond like F–F must be covalent. In addition to the above computational features, there are quite a few experimental manifestations of CSBs, which set these bonds apart from classical covalent bonds. For example, the RE_{CS} quantity for the dihalogen molecules, X$_2$, can be quantified based on the difference of the barriers for halogen atom transfer *vs.* H atom transfer, $RE_{CS} = 4[\Delta E^{\ddagger}_{HXH} - \Delta E^{\ddagger}_{XHX}]$.[73] Other manifestations concerning the expression of CSB in experimental electron density of bonds in propellanes, and in chemical reactivity have been discussed elsewhere.[73]

CSBs are formed by electronegative and lone-pair rich atoms, or bonds that suffer from Pauli repulsion pressure, such as the internal bond in [1.1.1]propellane.[73] Dative bonds, *e.g.*, the bond between BH$_3$ and an amine, carbocations with amines or water molecules, *etc.*, are CSBs.[76,77] All hypercoordinated species, *e.g.*, XeF$_2$, PCl$_5$, *etc.* are CSBs.[73,78] Odd-electron bonds, such as F$_2^-$, Cl$_2^-$, *etc.* are CSBs.[73] CSB is a new family of bonding alongside the traditional classical covalent and ionic bond families. The ubiquitous CSB family is one of the discoveries of CQC.

1.5.4.b   New Electron-pair Bonding Motifs Discovered by DFT/MOT

The Isolobal Analogy: Perhaps the most productive bonding concept in the past four decades has been the isolobal analogy, developed by Hoffmann.[79] Hoffmann derived this analogy by going back and forth between delocalized and localized MOs of transition metal complexes, in a manner that created a bridge between MOT and VBT, and allowed the predictions of many new organometallic complexes, which were not considered before.

The isolobal analogy implies that the bonding capabilities of transition metal complex fragments can be deduced by comparing their frontier orbital lobes and occupations to those of common organic fragments. Scheme 1.5 exemplifies the essence of the isolobal analogy between transition-metal complex fragments and small organic fragments. Thus for example, $(CO)_5Mn$ is a fragment that possesses an odd electron in a σ-type hybrid on the Mn. As such, the fragment is isolobal to a methyl-radical fragment that has an analogous singly occupied σ-type hybrid orbital. Similarly, $(CO)_4Fe$ has two lobes, one σ- and the other π type, and is therefore isolobal to the CH_2 fragment, and finally $(CO)_3Co$ has one σ- and two π type lobes, all singly occupied, and is hence isolobal to a CH fragment.

Once the bonding capabilities are defined, one can think about new as well as old molecules made from these fragments. For example, $Mn_2(CO)_{10}$ is an analog of ethane having a $\sigma(Mn–Mn)$ bond, $(CO)_5Mn–Mn(CO)_5$, analogous to the $\sigma(C–C)$ bond in ethane. Similarly, by analogy to tetrahedrane which is made of four HC fragments, one can think about the transition-metal based tetrahedrane, $Co_4(CO)_{16}$, constructed from four $(CO)_3Co$ fragments. Mixing of fragments, and devising new ones, gives rise to multitude of new molecules. The isolobal analogy is a bonding concept that emerged from computational quantum chemistry in the late 1970s using extended Hückel

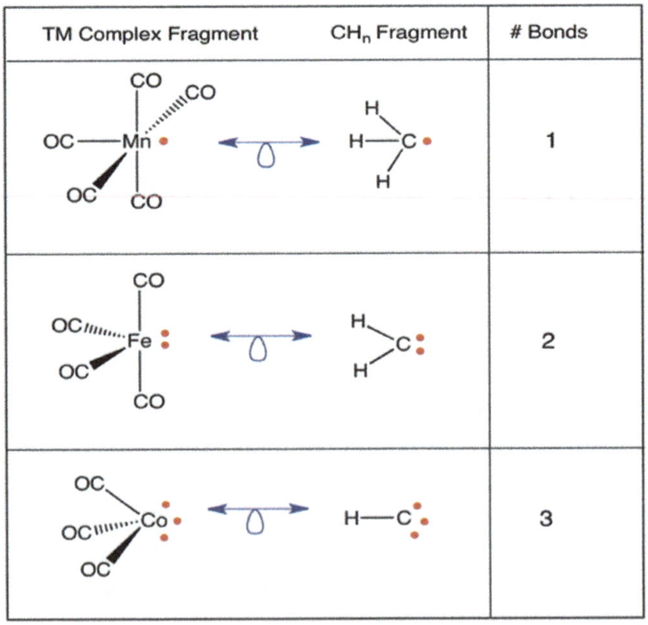

TM Complex Fragment	CH$_n$ Fragment	# Bonds
$(CO)_5Mn\cdot$	$H_3C\cdot$	1
$(CO)_4Fe:$	$H_2C:$	2
$(CO)_3Co\cdot$	$HC:$	3

Scheme 1.5 Fragments of transition metal (TM) complexes, their isolobal CH$_n$ ($n = 1$–3) fragments (isolobality is symbolized by a doubly headed arrow with a lobe), and the bonding capability given by the number of bonds.

theory. Despite the simplicity of the method and the passage of time, the concept is as vibrant as ever, showing how productive theory of chemical bonding can be, when it bridges two branches of chemistry.

Dative Bonds and Bonding in Carbones: Another concept that was imported from coordination chemistry to organic chemistry is the idea of coordinative or dative bonds to carbon atoms. In 1973 Kaska *et al.*[80] described the compound hexaphenylcarbodiphosphorane by analogy to coordinative bonding in transition metal complexes, as being composed of the coordinatively unsaturated C(0) coordinated by two Ph_3P: molecules using dative bonds, which were represented by arrows from the Lewis donor phosphines to the Lewis acid C(0); $Ph_3P: \rightarrow C \leftarrow :PPh_3$. Subsequently, Frenking and coworkers[81] used DFT tools and developed this description into a productive concept that involves the tetreles (Group 14 atoms, $E = C$, Si, Ge, Sn, Pb) and their isoelectronic analogues, *e.g.* N^+. An example of such a compound is shown in Scheme 1.6a, in which the bare C in its ^1D state (*e.g.*, with two lone pairs, $2s^2$ and $2p_x^2$; x is in the molecular plane) acts as a bidentate Lewis acid accepting two dative bonds from appropriate electron-donor ligands, L:. At the same time, its two remaining lone pairs on C, which are the doubly occupied HOMO and HOMO − 1 in Scheme 1.6b, can participate in π-back bonding to the ligands.

Unlike the normal bonding motif that predicts allenic type linear molecules as in Scheme 1.6c, carbones are strongly bent (Scheme 1.6a). Using the idea of dative L: $\rightarrow$ C bonds, Frenking *et al.* showed that in some cases, the central carbon in a carbone has two lone pairs which are largely the two filled 2p orbitals, shown in Scheme 1.6b. As such, the bonding in these compounds appears different than the classical electron-sharing Lewis bonding, which should result in allenic bonding type (as in Scheme 1.6c), with two π

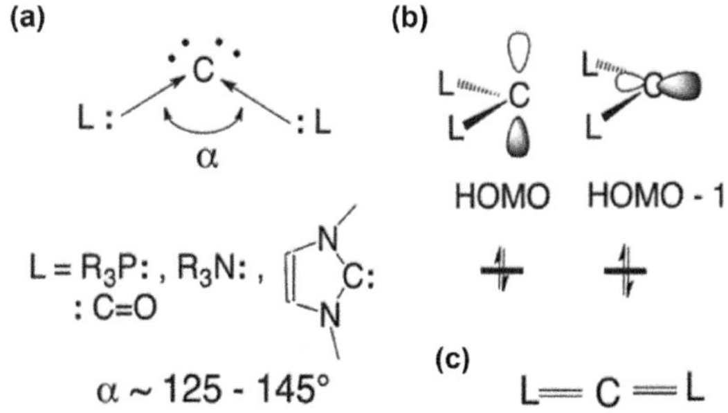

(a)

L :   : L

α

(b)

L   C   L

L   L

HOMO   HOMO - 1

$L = R_3P:$, $R_3N:$,   N␣C: : C=O

α ~ 125 - 145°

(c)   L═C═L

Scheme 1.6 (a) Schematic structure of a carbone with dative bonds from the ligands (L:) to C(0) in its ^1D state. Some ligands are shown below. (b) Simplified representations of the doubly occupied HOMO and HOMO −1 of some carbones. (c) The allenic alternative to carbone.

bonds rather than localized lone pairs. The dative bonding picture led to syntheses of new molecules.[81]

The binding model of carbones was analyzed using DFT computations and EDA analysis, coupled with the NOCV technique,[15] which enables extraction of the energy due to orbital mixing within DFT. It turns out that the orbital mixing term is the dominant one in these bonds, and the σ-orbital mixing is much more significant than the π-back bonding.

The Bonds in Carbones Are CSBs: The usage of arrows, to indicate dative bonds, initiated a dispute regarding the meaning of these arrows *vis-à-vis* normal Lewis bonds. On the one side was Krossing *et al.*, who argued for a traditional Lewis model with covalent and ionic structures,[82] and on the other, Frenking,[83] who insisted that the arrow icon describes the bonding as donor–acceptor bonds that are different to Lewis bonds. In the spirit of building bridges, the author of this chapter agrees with both sides. But how so?

Recall that dative bonds have been predicted to be CSBs,[76] and then shown as such by VB computations.[77] A CSB from an L: donor to an acceptor like C(0) involves mixing of a no-bond structure, Φ_{NB}, with a covalent structure, Φ_{cov}, as shown in the VB mixing diagram in Scheme 1.7, which describes the σ type bonding in a generic carbone. There are other structures wherein the lone pairs of C(0) donate electrons to vacant π type orbitals on L:, but because the σ bonds were shown[81,83] to be dominant, we omit the π back-bonding for the sake of simplicity. Since Φ_{NB} is either nonbonding or repulsive because of Pauli repulsion, *most if not all the bonding in carbones will arise from the VB mixing.* As such the L:$\rightarrow$C(0) bond is a CSB, which does not arise from the stabilization of any one of its constituent VB structures, but rather from their VB mixing. Indeed, the orbital mixing reported

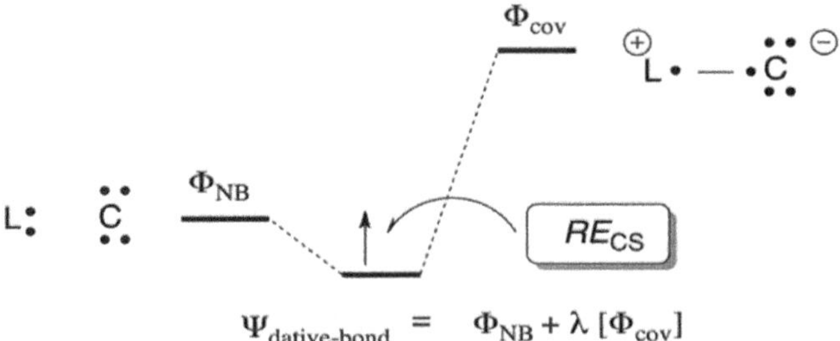

Scheme 1.7   VB mixing diagram for a dative σ-bond in carbones, for a case where the no-bond (Φ_{NB}) structure is more stable than the covalent structure (Φ_{cov}), generated by an electron transfer from L: to C. RE_{CS} is the charge-shift resonance energy responsible for the dative bond. λ is the mixing coefficient.

by Frenking *et al.* is very large and is the origin of the L: → C(0) bond. Furthermore, as the VB mixing is proportional to β we see once again that the driving force for the L: → C(0) CSB is the lowering of the kinetic energy in the bonding region.

When L: is a very good donor (better than L: = R_3N:), the Φ_{cov} structure will be stabilized and intensify the VB mixing and the RE_{CS}. At some critical donor capability, Φ_{cov}, as will be the lowest structure, and the bond will be a CSB similar to other electron-pair bonds like F–F, H–F, *etc.* Thus, as commented above, the present author seems to agree with both sides of the Krossing–Frenking dispute. The L: → C(0) bond is a multi-structure VB bond *á la* Lewis–Pauling, but at the same time, the case drawn in Scheme 1.7 is a CSB, owing its origins to the VB mixing between the structures rather than to either one of them.

Supra-multiple Bonds: By supra-multiple bonds I refer to bond multiplicity that exceeds a triple bond. This glass ceiling has been broken ever since Cotton described the Re–Re quadruple bonding in $[Re_2Cl_8]^{2-}$.[84] Since then many other transition metal and lanthanide/actinide complexes were found to exceed the glass ceiling as well. Thus, based on CASSCF and CASPT2 methods, it has been revealed that Cr_2, W_2, and U_2 complexes and dimers[85,86] reach quintuple and even sextuple bonding. However, usage of effective bond orders (EBO) reduced significantly all these supra-multiple bonds since the formally anti-bonding orbitals, which are the counterparts of the bonding orbitals, are significantly populated. The EBO index has gained much popularity as the means to gauge the number of bonds in the molecule.

Recently, this author and his coworkers proposed,[87] based on both full-CI and VB calculations, that C_2 and its isoelectronic species involve quadruple bonding, with three internal strong bonds (one σ and two π) and a fourth inverted σ bond which is weaker (Scheme 1.8).

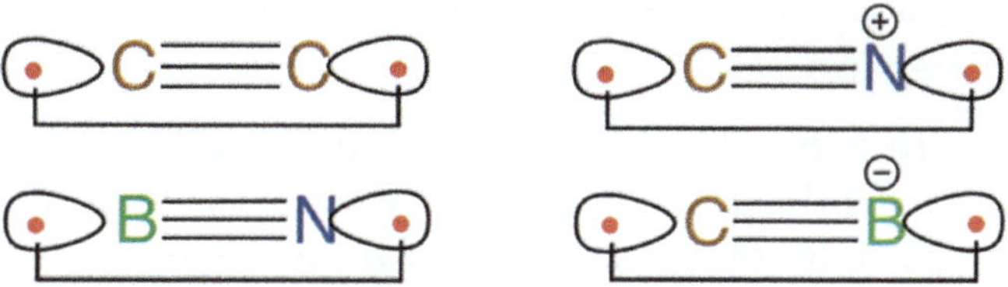

Scheme 1.8   The bonding representation in the ground state of C_2 and some of its isoelectronic molecules. All these species were predicted to possess quadruple bonds, wherein the exo bond is between the inverted hybrids on the atoms.
Reprinted by permission from Macmillan Publishers Ltd: *Nature Chemistry* (ref. 87), copyright 2012.

For C_2, the π bonds and the inverted σ were further shown to have propensity towards CSB.

This idea about C_2 was cordially debated,[88] and recently criticized,[89-91] while the criticism was also responded.[92] One of the lessons from this dispute is that there is a disparity between the number of electron-pair bonds that emerge from direct VB calculations or from localization of a full-valence CASSCF wave function, and the bond indices such as the EBO for CASSCF. Thus, in C_2, VB and the transformed full-valence CASSCF wave functions show the existence of four bonding pairs,[93] whereas the EBO index is only 2.15–2.3 or so. This disparity may well be common to other supra-multiple bonds, many of which have strong multi-reference character.

One might argue that in principle, the EBO has at least two basic blind spots: one has to do with the $2\sigma_u$ orbital of C_2, which is formally antibonding (thus reducing the EBO) but actually this orbital is nonbonding, and hence it should not contribute to the reduction of the EBO. The other blind spot is general and it may trickle into all bonds. Thus, consider the covalent wave function of H_2, which has by itself a strong bonding interaction (95.8 kcal mol^{-1}).[73] The covalent wave function can be expressed in terms of two MO configurations with equal weights (when overlap is neglected; otherwise non-equal weights[68]), one with $\sigma_g{}^2$ configuration, the other with $\sigma_u{}^2$ configuration.[68] As such, the covalent structure of H_2 has by definition EBO $= 0$, despite its very strong bonding interaction. Consequently, the EBO reduction in molecules with strong multi-reference character will be more prominent than in molecules having dominant single-reference character.

The description of supra-multiple bonds is a new frontier area that will continue evolving, as the dispute on the number of bonds eventually settles.

1.5.4.c Discovery of Long and Strong C–C Bonds Supported by Dispersion

A recent exciting bonding motif was found by Schreiner and his group,[16] who synthesized molecules with very long C–C bonds, 1.647–1.704 Å. Two of these are shown in Figure 1.9. Thus the molecule in Figure 1.9a, made of diamantine–diamantine coupling, has a C–C bond 1.647 Å long. Similarly, the molecule in Figure 1.9b, the hexa-(3,5-di-*tert*-butylphenyl)-ethane exists as a stable molecule with a long C–C bond of 1.67 Å. Despite the long bonds, these molecules are stable up to temperatures higher than 200 °C and their

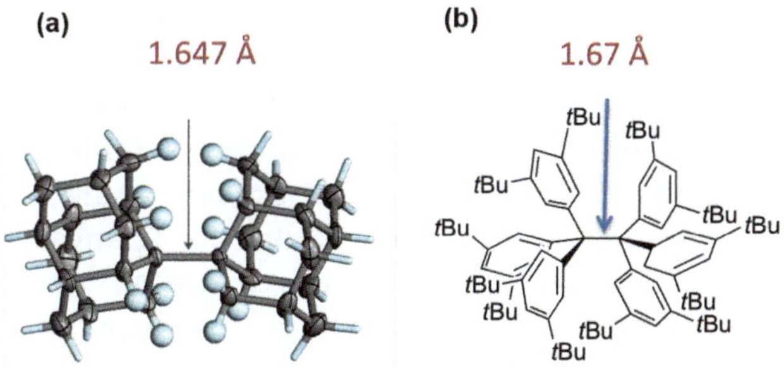

Figure 1.9 Two molecules with long C–C bonds (shown by arrows, in Å): (a) diamantine–diamantine, and (b) hexa-(3,5-di-*tert*-butylphenyl)-ethane. P.R. Schreiner kindly provided the two images.

bond dissociation energies (BDEs) are close to 60 kcal mol^{-1}. Schreiner *et al.*,[16] and subsequently, Grimme and Schreirner[94] showed that approximately 40–50% of the BDEs of these long C–C bonds are contributed by the dispersion interactions due to short and sticky H$\cdots$H contacts (1.94 and 2.28 Å), between the CH$\cdots$HC faces of the diamondoid moieties or of the tertiary-butyl groups of the diarylethane. Grimme[12] and Schriener and Wagner[95] reviewed the many other molecules with bonds supported by dispersion. This new bonding motif was discovered by the interplay of experiment and computational quantum chemistry.

Are the Long C–C Bonds CSBs? These molecules have not yet been computed by higher-level theories like VBT or MO-CI. Looking at these molecules in Figure 1.9, one can see that many lone pairs line up the space of the C–C bond, and it suffers most likely from severe Pauli repulsion. As such, one might speculate that the long C–C bond in the Schreiner compounds is a CSB, and that the dispersion may augment the charge-shift resonance energy to sustain these bonds. This is another frontier.

1.5.4.d Multi-centered Bonds

Multi-centered bonds have been known to occur in organic conductors made from conjugated molecules, such as tetrathiafulvalene (TTF) and tetracyanoquinodimethane (TCNQ), which after undergoing oxidation or reduction form stacks that conduct electricity.[96] For example, reducing TCNQ by alkali metals or by copper, generates solids with stacks of TCNQ$^{\bullet-}$ radical anions, which form dimers at low temperatures. Similarly, oxidation of TTF generates stacks of partially oxidized TTF, which correspond approximately to oxidation of about 50% of the TTF in the stack. Analogous donors form low energy organic superconductors.

The ceramic-based high T superconductors eventually eclipsed the field of organic conductors. However, the interest in the π-overlapping stacks was rekindled,[97] when these species were analyzed as models of multi-centered long bonds, known also as "pancake bonds".[98] What enables the eventual cohesion of the two anions are the counter-ions, which counteract the anion–anion repulsion and hold the "pancake".[97–100] The dimer of tetracyanoethene radical anion $(TCNE^{\bullet-})_2$ was analyzed (Scheme 1.9a) by several groups.

One way to understand this dimer is to regard it as being bound by an electron-pair multi-center bond made from the two overlapping singly occupied π^* orbitals, as shown by the MO in Scheme 1.9b. Using a two configuration CASSCF, which describes this bonding, led to small binding energy, which had to be augmented by multi-reference perturbation theory (MRPT2), which indicated that the bonding must be dominated by dispersion.[99] Tian and Kertesz[98] used MCSCF calculations (including both π and π^* orbitals) augmented by AIM characterization of the Laplacian along the bond paths. They

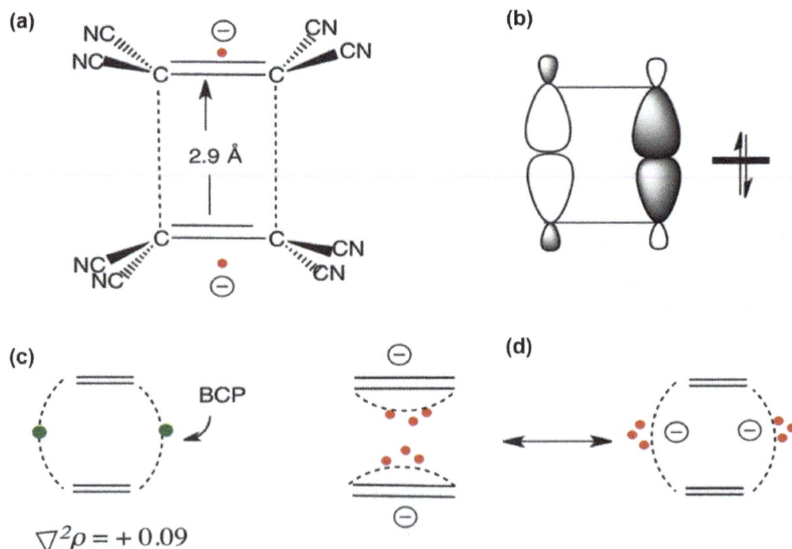

Scheme 1.9 (a) The $(TCNE^{\bullet-})_2$ dimer and its mid-to-mid bond distance of 2.9 Å. (b) The bonding combination of the π^* MOs of the $TCNE^{\bullet-}$ moieties is doubly occupied in the dimer, accounting for a 2-electron/4-centered bond. (c) The positive Laplacian at the bond critical point (BCP) of this multi-centered bond derived from MCSCF calculations, indicates a CSB. (d) The pericyclic 3-electron bonding picture from VBT.

concluded that the electron pair bond between the two $TCNE^{\bullet-}$ radical anion is a CSB, in line with its positive Laplacian, as shown in Scheme 1.9c. More recently, Braida *et al.* used VBSCF calculations[100] augmented by quantum Monte Carlo methods (to retrieve missing correlation). Their conclusion is depicted in Scheme 1.9d, showing that the $(TCNE^{\bullet-})_2$ dimer is bonded by pericyclic 3-electron bonding, which is CSB. These unusual species define also a new frontier for the chemical bond.

1.5.4.e *Discovery of Bonding Based on Bound Triplet Pairs*

Unusual bonding motifs expand the frontiers of the chemical bond. This is the case for bonds based on bound triplets. Bound triplet pairs are uncommon because they violate the Pauli exclusion rule. Nevertheless, VB calculations, which were carried out in 1999,[101] on the weakly bonded $^3Li_2$ $(^3\Sigma_u^+)$ dimer, provided the necessary insight, showing how the Pauli repulsion in the triplet pair is transformed into bonding.[102] The VB structures, which participate in the state-wave-function of $^3Li_2$ (the z-axis is along the Li$\cdots$Li distance), and the respective orbital representations are summarized in Figure 1.10a–d. Figure 1.10a shows that in addition to the fundamental structure, $^3\Phi_{SS}$, there are two ionic structures indicated as $^3\Phi_{SZ}(ion)$, which are generated by transferring an electron from the 2s AO of one Li atom to $2p_z$ of the other, and a covalent configuration, $^3\Phi_{ZZ}(cov)$, in which the electrons are excited from 2s to $2p_z$ on the two atoms ($2p_{x,y}$ are unimportant and can be neglected). The mixing of the latter three configurations into the fundamental and repulsive $^3\Phi_{SS}$ configuration generates some bonding.

The corresponding orbitals of this state are illustrated in Figure 1.10b–d. Thus, by mixing $^3\Phi_{ZZ}(cov)$ into $^3\Phi_{SS}$ (in Figure 1.10b) we get a two VB-configuration wave function. We can transform the orbitals of these two configurations without any change of the energy, and we obtain the hybrid orbitals depicted in Figure 1.10c; one of the hybrid orbitals points in and the other is pointing out. As such, the overlap of these hybrids is reduced compared to the 2s–2s overlap in Figure 1.10b, and hence, populating the triplet electrons, one in each of these two hybrids, reduces the triplet Pauli repulsion. At the same time, the resonance interaction between the two structures, (in/out) $\Leftrightarrow$ (out/in), further stabilizes the triplet pair. Finally, mixing the ionic structure endows each one of these hybrids with a tail on the other atom (Figure 1.10d), and thereby it augments the stabilization of the triplet state, leading to a "bound triplet pair".

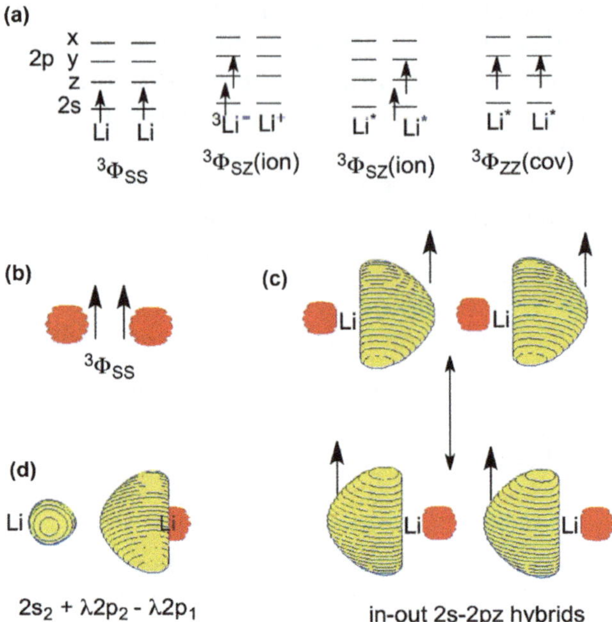

Figure 1.10   The bound triplet pair of Li_2: (a) The VB structures-set for $^3Li_2$ ($^3\Sigma_u^+$) is obtained by distributing the two triplet coupled electrons in the 2s, $2p_z$ AOs of the two Li atoms, where z is the Li$\cdots$Li axis. (b) The pictorial representation of the fundamental structure, $^3\Phi_{ss}$. (c) The hybrid orbitals for the mixed state of $^3\Phi_{ss}$ and $^3\Phi_{zz}$. Note that the hybrids point in/out and out/in in the two resonance structures. (d) After mixing $^3\Phi_{sz}$ into the state in (c), the right hand side lithium hybrid develops a small tail on the other Li; the left hand lithium will have an orbital related by 180° rotation about an axis passing through the mid Li$\cdots$Li distance. Reprinted with permission from D. Danovich and S. Shaik, Bonding with Parallel Spins: High-Spin Clusters of Monovalent Metal Atoms, *Acc. Chem. Res.* 2014, **47**, 417. Copyright 2014 American Chemical Society.

What Happens as the Metal Cluster Grows? As these clusters grow, each Li atom maintains with its close neighbors ionic and covalent configurations, and the total number of VB structures that stabilize the triplet pair increases steeply. Consequently, the bonding energy per single Li atom in the cluster increases also dramatically. The bond energy per one Li atom converges for a cluster of 12 atoms. As shown in Figure 1.11, even without a single electron pair bond, the bond dissociation energy increases from mere 1.7 kcal mol^{-1} for $^3Li_2$ to 145 kcal mol^{-1} for $^{13}Li_{12}$! Thus, the bond energy per atom starts at 0.85 kcal mol^{-1} and converges to about 12 kcal mol^{-1}. This means that the bonding of $^3Li_2$ embedded within $^{13}Li_{12}$ reaches 24 kcal mol^{-1}, which is approximately the bond dissociation energy of a singlet Li–Li

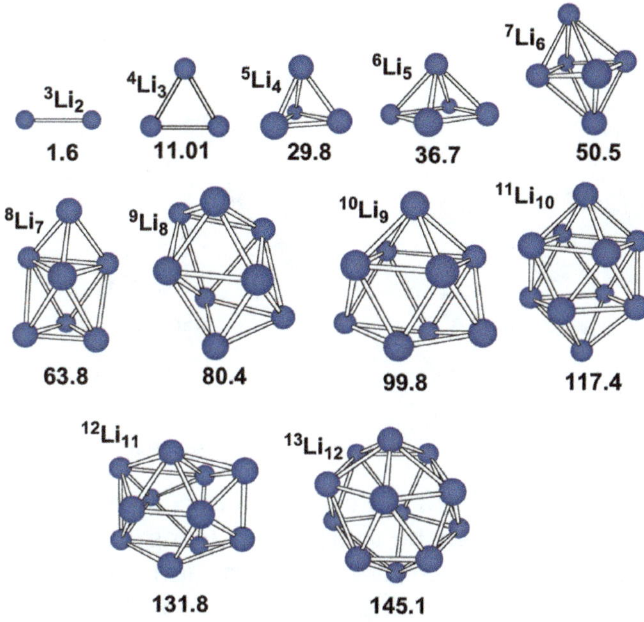

Figure 1.11 Evolution of the $^{n+1}Li_n$ clusters based on bound triplet pairs. The $^3Li_2$ to $^{13}Li_{12}$ clusters along with their bond energies (in $kcal\,mol^{-1}$).
Reprinted with permission from D. Danovich and S. Shaik, Bonding with Parallel Spins: High-Spin Clusters of Monovalent Metal Atoms, *Acc. Chem. Res.* 2014, **47**, 417. Copyright 2014 American Chemical Society.

bond. For the $^{n+1}Au_n$ and $^{n+1}Cu_n$ clusters the pair-bonding energy reaches 33 and 39 $kcal\,mol^{-1}$, respectively.

Since the bonding of the dimers is weak, there has been a tendency to refer to these clusters as van der Waals clusters. However, the bonding energies for all the $^{n+1}M_n$ clusters were calculated with standard DFT methods that lack dispersion interactions. Adding dispersion exaggerates the depth of the minima of the dimers compared with CCSD(T) data and with experiment, but it does not affect the binding energies for larger clusters. Indeed, as was pointed out,[102–104] the bonding in these clusters is highly sensitive to orbital hybridization and orbital mixing effects. VBT provides a clear description of the orbital-related effects and their impact on the bonding energy of these clusters of triplet bound pairs, and the theory further predicts their unique structures. To suit their uniqueness, the bonding in these clusters was termed no-pair ferromagnetic (NPFM) bonding.[102] NPFM bonded clusters can be made to have even stronger bonding and to have chiral structures; chirality and magneticity in the same molecule. Two of these chiral molecules, which were predicted by CQC, are shown in Figure 1.12.

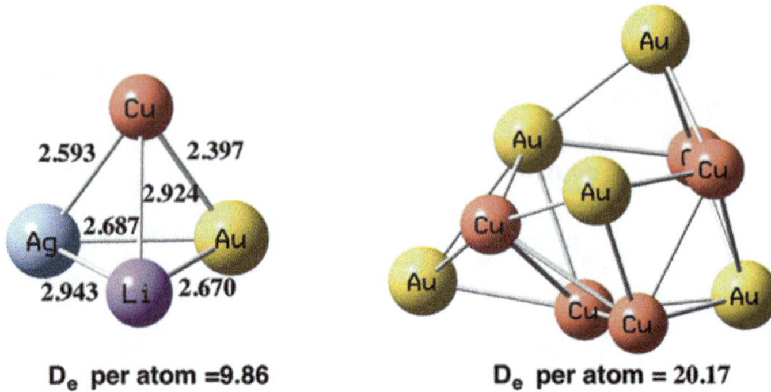

Figure 1.12 Chiral NPFM clusters and their bond energies per atom (in kcal mol^{-1}). Reprinted with permission from D. Danovich and S. Shaik, Bonding with Parallel Spins: High-Spin Clusters of Monovalent Metal Atoms, *Acc. Chem. Res.* 2014, **47**, 417. Copyright 2014 American Chemical Society.

NPFM clusters are related to Bose-Einstein condensates and Fermi-gas condensates, and can be synthesized using helium-droplet isolation.[105] They can be detected by Stern–Gerlach techniques and spectroscopy.[105] NPFM bonding may become a frontier area in the map of chemical bonding.

1.6 Summary of Bonding

The chemical bond is alive and is teeming with novelties. A glimpse into that was allowed above. Many more novelties can be found in the recent perspective, *"Pushing the Limits of the Chemical Bonding"*, written by Ritter,[6] discussing novel aspects of bonding in clusters *e.g.*, TaB_{10}^-, B_{16}^{2-}, Au_{20}, *etc.*, with inorganic aromaticity,[106] and the Jemmis Rules which extend the Wade–Mingos rules to boron-rich clusters.[107] Chemical bonding on surfaces and catalysis,[108] bonding in heavy main group elements[109] and the role of relativistic effects on bond energies and bonding.[110] All these novelties push the frontiers of the chemical bond into the central intellectual arena of chemistry.

1.7 A Brief Historical Background of Intermolecular Interactions

Up to the 19th century there is no distinction between the historical developments of the concepts of bonding and intermolecular interactions. Both started with the question, why does matter stick together? This question can be traced to the Greeks who were the first to ask it in a manner that did not involve the influence of deities.[111,112] The philosophical discussions of the Greeks eventually led to theories of matter in terms of elements or atoms. In the 18th century chemists were seeking what determined chemical identity, and this quest eventually led them at the end of the 19th century to the recognition of molecules made from atoms (the constitutional revolution). This recognition *differentiated the sciences of chemistry and physics*, leading to a revolution of chemistry, which has become a molecular science with clear definitions of *bond* and *structure*.

At the same time, this autonomy of chemistry relegated to physics the topics of cohesion between molecules to form macroscopic matter, and related problems like wetting, capillarity, surface tension, micelles, *etc.* It is from those problems that the notion of inter-molecular interactions evolved and matured into the science we see now. This is a rich science that involves dispersion interaction, dipole–dipole interactions, hydrogen bonds, and many other inter-actions that are crowned by the name Z-bonds; *e.g.*, halogen-bonds, pnictogen-bonds, tetrel-bonds, chalcogen-bonds, *etc.* How did this rich culture evolve?

Here, the historian Rowlinson (chapters 2–4)[111] distinguishes three periods, which he names after Newton, Laplace, and van der Waals and beyond. But eventually, and especially due to the effort to understand water and the discovery of the ubiquitous hydrogen-bonding (H-bonding), the problem returned into the hands of phys-ical chemists, crystallographers, quantum chemists, and it eventually became widespread in mainstream chemistry. To condense the following historic discussion, I shall consider only dispersion and H-bonding since they also represent the major interaction that shape mesoscopic and macroscopic matter in nature.

1.7.1   From Newton Onwards to Laplace

Newton himself was an avid alchemist who carried out many experi-ments including on mercury and its oxide (which according to speculation inflicted him for a period of time with the Mad Hatter Syndrome). Newton believed that both light and matter were par-ticulate. Gravity suggested that particles should attract one another. In contrast, Boyle's gas law suggested that particles of gas repelled each other. Since gases condense to liquids and then to solids, Newton insisted on attraction, but since the solid did not collapse into itself, he concluded that there must also be repulsion at shorter range. He formulated the force between particles as composed of a gravitational attraction given by the inverse-square law $(1/r^2)$ and a repulsion given by $1/r$. However, the inverse-square law diverged when integrated over many particles and long distance, and in any event, Newton did not provide a cause for these forces. Furthermore, the action from a distance required an instantaneous force, which was conceived as being supernatural.

Experiments on the nature of capillarity by contemporaries of Newton (*e.g.*, Newton's demonstrator Hauksbee) led to the recognition that the interparticle forces must be very short ranged, but since liquids rise even in macroscopic tubes, this suggested that the forces are also long-ranged. Despite these conflicting conclusions, the experiments led eventually to the understanding of surface wetting and tension, which later influenced Laplace's theory of capillarity.

The Newtonian legacy continued to the 18th century, but for many reasons, the progress was painfully slow. The experimental infrastructure was too crude to deal with these weak forces. In addition, most natural philosophers of the era had inadequate knowledge of the Newton–Leibnitz calculus and of mechanics to make headways. Importantly, they had difficulties in accepting action at a distance. And finally, the talents of the science preferred the fields of Astronomy, Electricity and so on. Cohesion was not fashionable. Consequently, the Newtonian heirs believed in attraction and repulsion but were unable to define them in a manner that made sense to their contemporaries; to use d'Alembert's words, "an explanation so vague condemns itself".

Nevertheless, at the end of the 18th century, action at a distance started to be tacitly accepted, and physicists like John Leslie and Thomas Young established (1802–1804) good intuitive understanding of capillarity. For example, Leslie explained capillarity as attractive forces between water and glass, and the movement of water particles to be in places close to the glass. Young explained capillarity in terms of surface tension, which manifests as a balance of forces of attraction and repulsion.

At about the same time, the Bernoullis, Euler, d'Alembert, Lagrange and Laplace put the Newtonian mechanics in the form familiar now. And it became evident that objecting to action at a distance was not scientifically productive. These changes caused a transition that was led by French scientists, notably Laplace, who described, in mathematical forms, capillarity, surface tension, and interaction of matter particles.

Laplace was a close friend of Berthollet, who strongly believed that *chemical affinity* was proportional to mass and hence, was gravitational and related to capillarity and surface tension. But, by 1805, Laplace settled on a view that the attractive forces were short-ranged and pairwise, but of unknown nature. He further hypothesized that caloric (the particle of heat that surrounded the particles of matter) was the agent of repulsion that stopped matter from collapsing in by keeping particles apart. These opposite forces served Laplace to develop the mathematical tools for discussing the nature of gases, liquids and solids.

Laplace's model was essentially static. But by the mid-19th century, the new fields of physics, light, electromagnetism, and heat, where field and wave theories reigned, eclipsed this view. Corpuscular/molecular theory and hence also cohesion fell out of vogue! Positivism dominated physics, and all corpuscular phenomena had to be understood in macroscopic visible/measureable terms. The Laplacian physics of interparticle forces lost the battle, but the ideas remained with a few followers like Poisson, and the ideas returned later in the century when van der Waals and others fruitfully united them with a kinetic view of matter.

1.7.2 The Post-Laplacian Intermolecular Interactions

The new way to analyze intermolecular interactions was provided from the concepts of energy and dynamics, which removed the obstacles created by Laplace's static particles and incorrect description of heat (as caloric).[111] Newton wrote, "energy is motion", and although he did not necessarily mean the kinetic energy of gases, what he said inspired others. Already in the 18th century Bernoulli suggested that gases contained particles in rapid movement. Within 1845–1855, two Englishmen, Herapath, a teacher turned journalist, and Waterston, an engineer, published their kinetic theories, disregarding Bernoulli's. This was not the right time, because of the dominant positivism in Physics. Nevertheless, these attempts did not stop, and later Joule and Krönig (from a technical college in Berlin) published in 1856 nonmathematical and intuitive kinetic theories. Krönig's paper induced Clausius and Maxwell into action and they published their own versions of the topic, and within the treatise of energy, they alluded to the interactions between molecules and the behavior of gases and liquids (Chapter 3 in ref. 111).[111]

Thermodynamics was developed in the 19th century also by non-mainstream individuals (Chapter 4 in ref. 111).[111] This started with Carnot's book in 1824, in which he regarded heat as the quantity being conserved. The experiments of Joule in the conversion of heat to mechanical energy, and the calculations of Mayer, both convinced physicists that it was energy not heat that was conserved. This was followed by the derivation of the 1st and 2nd laws of thermodynamics by Clausius and Thomson (William). Helmholtz's publication, *On the Conservation of Force*, was published in 1847 and it sealed the acceptance of the conservation of energy as a law of nature. He postulated that this conservation meant that all forces in nature are attractive or repulsive and acting along lines joining particles of matter. He also introduced the notion of potential energy. Joule, Clausius, Maxwell and Helmholz held that energy originated from motion.

In the meantime, chemists contributed to the corpuscular theory of matter through Dalton's work (itself influenced by the attraction-repulsion mechanism in Laplace's theory of cohesion) and later through the theory of Davy and Berzelius on the nature of the chemical bond as an electrostatic attraction. Dumas presented, in his course in 1836, a theory of attractive forces between particles of matter, in which he made a clear distinction between the '*cohesion of the physicists*' and the '*forces of affinity*' which lead to 'formation of chemical compounds'. There were a few more attempts, but chemists were by and large out of the field of cohesion in the 19th century.

At the end of the 19th century Boyle's Law was extended by Charles, Gay-Lussac, and Dalton, and become the ideal gas law, connecting pressure, volume and temperature:

$$pV = cT \quad (c = \text{constant}) \tag{1.9}$$

Here T is a temperature on a scale whose zero was found to be at approximately $-270\ °C$ (where the pressure p is zero). This equation was in fact a statement that the kinetic energy (cT and later RT) determines the volume and pressure of the gas. For years it was known that the behavior of real gases was non-ideal and the deviation reflected molecular properties. Joule and Thomson explored systematically the deviations and revised eqn (1.9) to (1.10):

$$pV = RT - \alpha p/T^2; \quad \alpha p/T^2 = B(T) \tag{1.10}$$

$B(T)$ is known today as the second virial coefficient and is one of such coefficients associated with intermolecular forces.

For liquids there were known some basic facts about the vapor pressure as a function of temperature, and the change from liquid to vapor involving a large intake of heat (the latent heat of evaporation). The melting of solids involved less intake of heat. The exceptional behavior of ice/water between 0 and 4 $°C$ was also known, but no explanation existed. Heavier vapors such as chlorine, CO_2, and H_2S could be liquefied. Faraday showed (1844) that a variety of gases could be liquefied by combination of pressure and cooling. Andrews (from Queen's College) discussed in 1869 *the critical point where liquid and gas coexist*, and showed the existence of a continuum in the passage vapor→liquid. His interpretation was something close to a hypothesis of the action of 'a molecular force of great attractive power'. Other ideas and developments included Earnshaw's theorem that no static system with inverse-square power law could be in equilibrium, Herapath's derivation of the mean velocity of molecules of a gas from the state equation (eqn (1.9)), and the derivation by Young, Dupré and Waterston, of the range of intermeolcular distance as 1 Å or so; much too short but of the right order of magnitude. In the 1850s and 1870s, Clausius and Maxwell published their new kinetic theories of gases based on molecular motion, rotation and vibration, mean sizes and mean paths, and how those lead to the existence of matter in gaseous and condensed phases. In 1870 Clausius derived the virial theorem, in which positions and speeds were linked. One could then derive for an ideal gas (where all the pairwise intermolecular forces are zero) that the kinetic energy of all the particles was equal to $3pV$, which is equal to $3RT$. In 1868–1871, Boltzmann derived the kinetic energy of molecules as $3RT/2$ using his statistical theory. The corpuscular-molecular view was back in focus.

1.7.3   Van der Waals

Van de Waals, a schoolmaster at The Hague, started his PhD in Leiden in 1871 and defended it in 1873 (Chapter 4 in ref. 111).[111] He was awarded the Nobel Prize for his work in 1910. His conviction of the existence of molecules led him to address the pressure and volume modification in the equation of state (eqn (1.9)) by the intermolecular interaction as well as by the molecular size, by reliance on the virial theorem. He assumed that the temperature gauges the kinetic energy of the molecules, that the effect of intermolecular forces is expressed through the pressure, and that the molecules were hard objects. Thus, he surmised that molecules at the surface of a fluid were pulled inward and thereby affect the pressure in proportion to the number of molecules being pulled per unit volume and to the number doing the pulling in the interior. This correction was a term proportional to the square of the molecular density. The free space available for motion was corrected by taking into account the effective volume (co-volume) occupied by the molecules. His new state equation was given then as follows:

$$(p + a/V^2)(V - b) = RT \tag{1.11}$$

where the constants, *a* and *b*, are related to molecular properties and interactions. He then showed that the new equation applies to gas and liquid, through the critical point where gas and liquid are co-existent, and that this application enabled him to quantify the molecular diameter and the range of intermolecular forces, *e.g.*, 4.0 and 2.9 Å, respectively, for dimethyl ether. Van der Waals still referred to the forces as "Newtonian", not in a gravitational sense, but rather as acting at a distance. In 1896 Boltzmann referred to these as *van der Waals cohesive forces* – a name that has stuck to this day.

1.7.3.a Molecules and Electricity

Molecules became associated with electricity through the works of Faraday, Berzelius and Davy, then by the young school of physical chemistry of Ostwald, van 't Hoff and Arrhenius, and subsequently and mostly so by the work of J. J. Thomson. At the same time, it was clear that the van der Waals equation of state needed many corrections and could not be derived in a closed form, and these corrections defined an agenda for research that tried to derive the intermolecular interaction potential from electrostatic theory.

A first attempt by J. D. van der Waals (the son) to derive the intermolecular interaction in the van der Waals equation of sate by looking at molecules as electrical dipoles and estimating their interaction, failed. In the meantime, Bohr described the atom as a quantal object, with spherical orbits, and the success of this model in predicting the spectrum of the hydrogen atom implied that atoms have no permanent dipoles. This conclusion was further confirmed from experimental evidence of diatomic molecules, such as H_2 and O_2, in electric fields, which showed that these molecules had no dipoles.

The rescue came eventually through considerations of the polarizability of atoms and molecules in electric fields. The behavior of matter in electric fields has been actively studied since the days of Faraday, and the various experiments allowed relating the polarizability of a molecule to the refractive index of the material (or the permittivity of matter *vs.* vacuum) by Clausius and Mossotti and by Maxwell. Then Debye included permanent dipoles, and eventually also quadrupoles and beyond. However, the attempts to relate these electrical properties to the virial coefficients in the van der Waals equation were not entirely successful.

Furthermore, in the first decades of the 20th century, it became apparent that even noble gases had significant interatomic interaction that caused them to liquefy. It was also apparent that classical electrostatics failed to account for these attractive interactions, and the various treatments led to potentials with inverse powers of 1,2,3,4,5,7 and 8, but not 6. In 1927 in the Faraday Discussion meeting on *"cohesion and related problems"* it was clear that there was little progress in the understanding of these intermolecular potentials, and one of the participants (presumably Lennard-Jones) raised a hope that the new quantum mechanical theory (QMT) might solve the problems.

1.7.4 Quantum Mechanics and Archetypical Intermolecular Interactions

The rise of the new QMT in the years 1926–1929, and especially of the Schrödinger equation, revolutionized physics and enabled treatment of molecules and hence also intermolecular interactions, using the Born–Oppenheimer approximation (Chapter 5 in ref. 111).[111]

1.7.4.a van der Waals/London Interaction

An initial treatment of the interaction between hydrogen atoms by Wang in 1927, who was convinced by Debye to try, led to the following expression for the attractive potential at a long distance:

$$U(r) = -C_6 r^{-6} \tag{1.12}$$

The importance of this result was its demonstration that two neutral atoms without permanent dipoles/multipoles could maintain an attractive interaction according to the new QMT.

At the same time, Heitler and London came up with their dramatic discovery of the short-range interaction that leads to the bond between two hydrogen atoms. They also found two new aspects: (i) at very short distances the potential was repulsive due the repulsion between the nuclei, which were not shielded anymore by the electrons, and (ii) when the electrons' spins in the $H\cdots H$ system were parallel there was repulsion throughout. They also showed that the $He\cdots He$ interaction was repulsive, and these electron-electron repulsive terms were soon associated with the Pauli exclusion rule. It became clear therefore that the interaction between molecules must involve an attractive energy varying as r^{-6} and a repulsive energy term that dies faster at long distances. Later, London verified Wang's expression using triplet H$\cdots$H, and this was followed by other verifications. Then Pauling and Beach derived the value $C_6 = 6.499\ 03$ au.[111] The attractive interaction was interpreted to originate from the creation *of instantaneous dipoles* of the two H atoms, and since the induction of a dipole by one atom on the other is proportional to r^{-6}, r being the interatomic distance, then the attractive potential is proportional to r^{-6}.

In 1930 London published a simple derivation of this attractive potential for two oscillating dipoles. Since the frequency of the oscillating dipole is related to the dispersion of light by matter, London dubbed the attractive interaction as 'dispersion energy', which became known also as 'London energy'. As shown by London, the energy associated with the oscillation frequency, $h\nu_0$, reflects the looseness of binding of the valence electrons, and hence it can be replaced by the ionization energy (I) of the molecule and the molecular polarizability, $U(r) = -0.75\ I\alpha^2/r^6$. Slater and Kirkwood further modified the $I\alpha^2$ term to $N^{1/2}I\alpha^{3/2}$, where N is the number of valence electrons. Others modified the potential to include the effect of higher multipoles. Later on, between 1931 and 1938, the potential was modified by Lennard-Jones and later by Buckingham to a 12,6-potential. The efforts after these years were enormous, they included physical chemists, who focused on the Ar$\cdots$Ar problem, using a variety of methods, including simulations, inversion from spectroscopy to potential, and from virial coefficients to pair potentials.

QM calculations were too demanding for this problem, until recently. This fact, and the extreme weakness of the corresponding pairwise interaction energy have held back the acceptance of the van der Waals/Dispersion/London interactions into mainstream chemistry. The weakness of the interaction is not a problem, because the dispersion is cumulative, and if a molecular fragment experiences many such interactions, the total dispersion energy can amount to tens of kcal per mol. Currently there is a good arsenal to calculate van der Waals interactions in 'real' molecules. Quantum chemical methods like CCSD(T) enable quantification of these interactions and have already revealed exciting findings, such as the proposal[113] that the DNA double helix is held mostly by dispersion interactions between the stacked bases. There are also DFT functionals that include dispersion from first principles, as well as empirically corrected functionals of various types.[12,114] The major 'entrance' of dispersion into mainstream chemistry really occurred in the 21st century, much due to the work of Grimme and others, who have shown how ubiquitous dispersion can be, and how significant in fact it is. Grimme's dispersion correction (D3) involves pairwise 6,8 terms and a three body term with $1/r^9$ dependence. The usage of dispersion is now widespread and its impact is found in structure, bonding, and reactivity, in all branches of chemistry. However, understanding of dispersion is still lacking.

1.7.4.b Hydrogen Bonding – A Special Interaction of Permanent Dipoles

Pairwise hydrogen bonding (H-bonding; a term which appeared for the first time in 1923 in Lewis's book, 'Valence') interactions are not as weak as van der Waals interactions. An average H-bond is a few kcal per mol strong. But, in some cases, as in $(FHF)^-$, the interaction becomes a normal covalent bond. The notion of H-bonding emerged in the late 19th century and early 20th century, from the interest in aqueous solutions and the structure of water. The anomalies of water have been known for a long time, and have puzzled the community of physicists and chemists alike (Chapter 5 in ref. 111).[111] In the second decade of the 20th century, it was established that the water molecule has a permanent dipole moment. In 1922 Bragg senior[111] solved the structure of ice and interpreted it *á la* NaCl as a lattice of ions. Chemists on the other hand, described the water molecule with two covalent-polar bonds as Lewis depicted in his 1916 JACS paper.[1] Lewis also mentioned aggregation of water molecules due to polarity of the bonds.

Apparently, the first paper to mention the impact of H-bonding interaction was a 1912 paper by Moore and Winmill,[115] who described the weak basicity of Me_3N as a weak interaction between the base and water, $Me_3N\cdots HOH$. Eight years later, Latimer and Rodebush[116] discussed the importance of this interaction in highly associated liquids, such as water and HF, the formation of dimers by acetic acid, and many other phenomena. In 1933 Bernal and Fowler published a landmark paper[117] on the structure of water, and described ice to be composed of tetrahedral structure involving a central H_2O interacting with four others *via* $O–H\cdots O$ interactions, wherein the three atoms are collinear.

This architectural element of the hydrogen bond (H-bond), its ubiquity in nature, its importance in creating beautiful structures like the α-helix in proteins, its role in the genetic machinery, in crystal engineering and in enzyme catalysis, caused chemists to embrace this intermolecular interaction rather quickly, and very soon it became one of the hottest and most visited topics in chemistry.[118]

1.8　Intermolecular Interactions – A Chemical Field in Eruption

The architecture and character of macroscopic-, mesoscopic- and nano-materials are fashioned by-and-large by intermolecular interactions, made from pairwise weak interactions like dispersion and H-bonding, which accumulate and become hugely forceful (*e.g.*, the Gecko phenomenon). It was immediately recognized that the architectural elements in these interactions and their potential strength make them ideal for design and engineering of new materials. As such, the major interest in these interactions has migrated to chemistry. Chemists like to attach to each interaction a name that serves as a chemical qualifier. To economize the myriad of names, consider the generic symbol, X–Z$\cdots$Y, to represent a generic 'Z-bond', where Z can be hydrogen, halogen, chalcogen, pnicogen, tetrel, *etc.*, and X and Y are molecular fragments that 'donate' and 'accept' Z, respectively.[1] When both Z and Y are hydrogen, the respective interactions are referred to as dihydrogen bonds, as *e.g.*, the CH$\cdots$HC interactions in hydrocarbons. Additionally, Y can be an anion or a cation, and X–Z an aromatic ring, or an olefin, and then we talk about anion–π and cation–π interactions, and so on.

1.8.1　The Chemists' Way of Considering Intermolecular Interactions

The term 'Z-bonds' generally refers to the stabilizing interaction that occurs between atoms or groups, which in the conventional chemical sense have satisfied their formal valence. Chemists, commonly analyze a given interaction using energy decomposition analyses (EDA) of various types (see above). These interactions involve a few common energy terms, which refer to dispersion, electrostatic, promotion and deformation energies, Pauli repulsions, and bonding terms (arising from delocalization and charge transfer). Some of the methods use perturbation theory to calculate the various interactions, and other methods extract these quantities from energy difference calculations of the aggregate *vs.* the separate molecules/species and then extract the separate interaction terms. There are differences and similarities among the various methods, which differ in the manner by which they take the reference state, but there is no general consensus on the "best" method.[14]

Since 'Z-bonds' are formed between molecules/fragments, which have satisfied their valence, one might have thought that the intermolecular interactions should not involve any covalency. This turns out to be a simplistic approach and recent studies have shown that some of these 'Z-bonds' may involve considerable covalency. This issue tends also to be controversial and it is therefore interesting to focus on it. In so doing, I will limit myself to H-bonds (Z = H, X, Y – electron rich centers), Hal-bonds (Z = halogen/halide, X, Y – electron rich), and diH-bonds (CH$\cdots$HC).

1.8.1.a The Nature of H-bonds

How covalent are H-bonds? In his book (p. 452), Pauling[43] estimated that the covalency of the water dimer is 5–6%. In most cases indeed, the consensus is that the weak H-bond is mostly electrostatic.[119] A recent paper by Zhang *et al.*,[120] reported that the H-bond in 8-hydroxyquinoline on a Cu(1.1.1) support could be visualized in real space using non-contact-atomic force microscopy (NC-AFM). In this technique the AFM tip is functionalized with CO that acts as a sensor of the forces exerted by the electron density of the adsorbed molecules. Not long after this publication, Swart *et al.*[121] used a tetramer of bis(para-pyridyl)acetylene, which involves C–H$\cdots$N H-bonds and nonbonded N$\cdots$N moieties at 3 Å apart. They showed that the CO on the AFM tip bends by interacting with intermolecular potential of the molecules, leading to a semblance of electron density also between the nonbonded N$\cdots$N atoms. Thus, generally the Zhang work is being dismissed. Still however, another recent AFM force measurement of the iron-sulfur cluster rubredoxin reveals that the strength of the NH$\cdots$S H-bonding network affects markedly the strength of the Fe–S bond.[122] Moreover, Elgabarty *el al.*,[123] quantified the covalence of the H-bond in water by measuring the anisotropy of the proton magnetic shielding tensor. They reported that the O$\cdots$H H-bond possesses a density of 10me$^-$ and its strength due to this charge transfer is *ca.* 2.4–3.5 kcal mol^{-1}. The prospects of these experimental techniques are exciting despite the cautionary buzz around the Zhang imaging experiment.

How does theory respond to the opening question? In general, the answer remains controversial. Depending on the computational method, the covalence of H-bonds ranges between significant (from NEDA(NBO) analysis) and slight (from other methods, *e.g.*,

Morokuma, BLW, AIM, IQA, *etc.* see above). Nevertheless, all theoretical analyses suggest some covalency, due to charge transfer from the H-acceptor moiety (*e.g.*, $Y:^-$) to the H-donor (X–H).[124,125] Furthermore, the linear structure of the H-bond is considered to arise from this charge transfer/covalency due to the donation from a lone pair orbital on Y: to the σ^*_{XH} orbital of the H-donor.[125]

A recent series of papers[126,127] issued another cautionary alert on the practice of mining the H-bond distances from crystal structures and deducing their relative stability, based on a bond-length-bond-strength (BLBS) principle. For example, in polyprotic acids, such as $K^+[CO_3H]^-$ the H-bonds $O–H^-\cdots O^-$ are actually repulsive and are supported as "short H-bonds" only because of the strong interactions of the anions to the cation. Moreover, using H-bond distances from neutron diffraction data and calculating the bond strengths using a perturbation method[127,128] shows a breakdown of the BLBS relationship due to the fact that the crystallographic data reflect the packing of ions and not necessarily the strength of any specific short distances. A subsequent study[129] showed that when the H-donor (X) and H-acceptor (Y) are ions of opposite charge, the so formed H-bonds are highly covalent with bonding energies between 80 and 210 kcal mol^{-1}.

A resonating VB model[29,43,119,130,131] appears to provide a productive outlook on the problem, by showing how covalency changes from being minimal in weak H-bonds to highly significant in strong H-bonds. Thus, if one views the H-bond as a species along the proton transfer coordinate from X to Y, one requires primarily three structures to model the process. As shown in Scheme 1.10a, these are the two covalent structures of the X–H and H–Y bonds, and a third protonic structure, $^-X: H^+ :Y^-$. Mixing of the two covalent structures in an H-bonding geometry, *e.g.* $X–H\cdots Y^-$, generates some covalency in the H-bond, while a very significant mixing of the protonic structure into both covalent structures contributes to covalency in both X–H and H–Y linkages in a given geometry. Scheme 1.10b shows the example of the formation of the symmetric $(FHF)^-$ species (a "low barrier H-bond") which resides in a deep minimum highly stabilized by the large covalent-ionic resonance energy of the covalent and protonic structures.[130,131] For other systems, where the protonic VB structure is not as low in energy, the H-bond is simply a precursor cluster *en route* of the proton transfer reaction, and as such will possess varying degrees of covalency due to resonance interaction of one covalent structure with the other. This is exemplified in Scheme 1.10c, which describes a proton transfer for X and Y groups

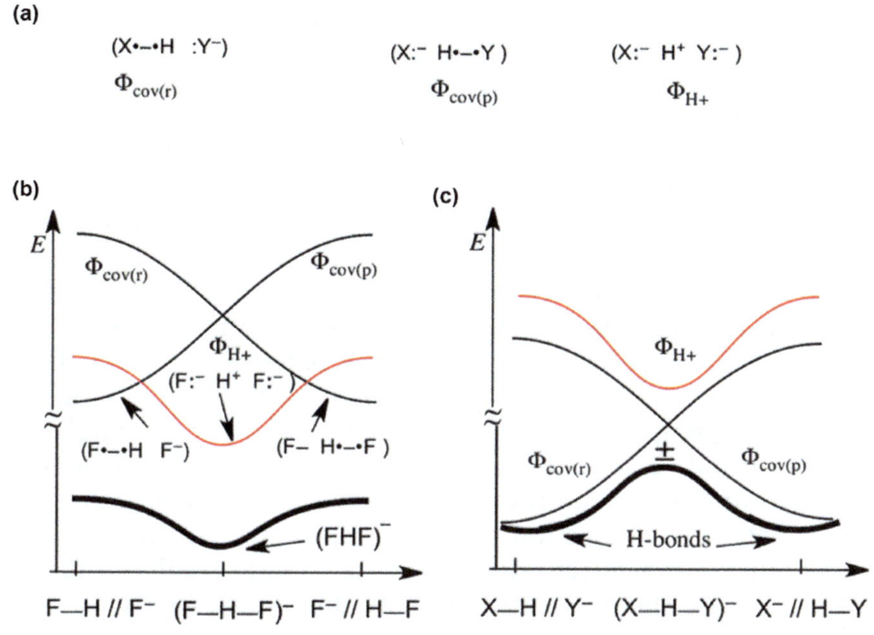

Scheme 1.10 (a) Three VB structures that are required for describing H-bonds as species *en route* to proton transfer from X to Y (the long–bond X• H:⁻ •Y structure is omitted). The bold energy curves in (b) and (c) represent the final states. (b) The case of (FHF)⁻, which represents very strong H-bonds (sometimes called "low barrier hydrogen bonds"). (c) H-bonds for a general case; note that here the final state is close to the VB structures at the H-bonded geometries, indicating a rather small stabilization as opposed to the situation described in part (b).

that are not as electronegative as F. Here the H-bonded clusters are precursors of the transition state for proton transfer.[130,131]

In summary, therefore, despite the fact that the H-bond is now 104 years old, the field is still alive and full with surprises, controversies, and exciting developments.

1.8.1.b The Nature of Hal-bonds

One of the intriguing Z-bonds is the halogen bond (Hal-bond), which involves an interaction of an electron rich center Y: with an electronegative halogen substituent Z in an XZ molecule. In fact, the first Hal-bond was discovered back in 1863 when Guthrie made the ammonia–iodine complex H_3N:· · ·I–I.[132] Subsequently, when Mulliken formulated his charge-transfer (CT) theory, which has provided a framework for the understanding of the unique spectroscopy of these complexes,[133] the Hal-bond was called by the generic name, a charge-transfer complex (CTC). The fascination with the Hal-bond arises from the facts that, (i) the interaction involves two electron rich centers, Y: and the electronegative halogen Z, and (ii) the Hal-bond is highly directional and its XZY angle is close to 180°, and as such it constitutes an architectural element, much like H-bonds. Interest in the subject has hugely surged because of the fact that this 'bond' turns out to be ubiquitous in biological materials[133] such as proteins, nucleic acids, and interactions of drugs with biological objects.

One of the useful theoretical concepts for comprehending the intriguing features of the Hal-bond is electrostatic in origin.[11] Thus, despite the partial negative charge of the halogen (Z), still it has a region of positive electrostatic potential at the head of its lone pair and in the opposite direction to the X–Z axis, called a 'σ-hole'. As such, the σ-hole confers electrostatic and polarization interactions, which account for the ability of the electronegative halogens to accept an interaction from electron rich centers, as well as for the linear X–Z···Y angle. The current discussions of experimental and theoretical studies of Hal-bonds are virtually dominated by the 'σ-hole' notion and its electrostatic effects.

A recent computational study[134,135] showed however, that like the H-bond, here too there is significant covalency. Thus, VB calculations and BLW analysis of over 50 Hal-bonds, led to the results shown in Figure 1.13. Thus, for the great majority of Hal-bonds the major contribution to the bonding interaction (ΔE_b) is due to charge-transfer

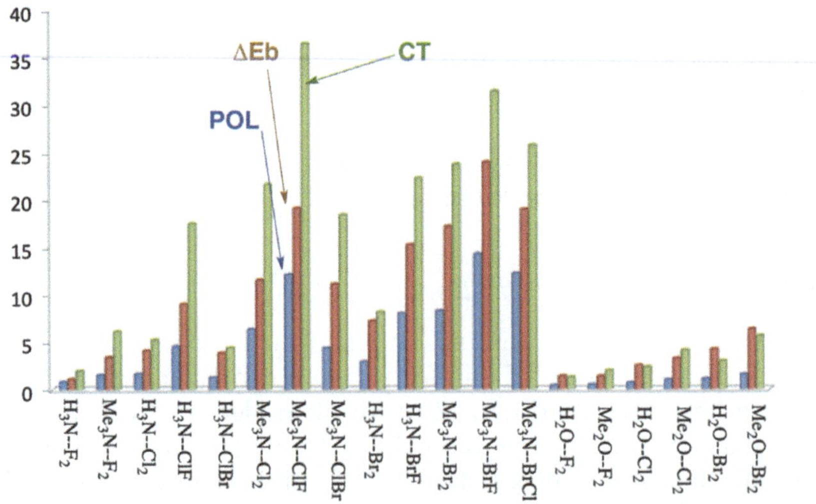

Figure 1.13   A color-coded diagram showing the charge transfer (CT), and polarization (POL) energy components and the bonding energy (ΔE_b) of a variety of Hal-bonds (in kcal mol^{-1}).

(CT)/covalency, as seen by the green columns. On average, the polarization (blue) is much smaller, while electrostatic interaction and long-range dispersion are negligible (not shown), and the Pauli repulsion is destabilizing (not shown). Furthermore, the CT interaction is almost twice as large as the bonding energy of the species (in brown), and one can say therefore that the Hal-bond originates in the CT interaction. Others reached a similar conclusion for other sets of Hal-bonds.[136] *The conclusion is clear-cut, most of the Hal-bonds are held by charge transfer interactions as envisioned more than 60 years ago by Mulliken.*

In a recent study,[135] it was demonstrated that the linear angle of the Hal-bond, $X–Z \cdots Y$, originates also in the CT interaction due to the overlap of the lone-pair orbital on Y with the σ^*_{Z-X} orbital. It is very clear that if at all possible, then real space imaging of Hal-bonds would be an exciting development.

1.8.1.c   The Nature of CH···HC Interactions

As we showed in Section **1.5.4.c**, the seemingly very week $CH \cdots HC$ dispersion interactions can accumulate and contribute greatly to the stabilization of long C–C bonds. Furthermore, this interaction controls the boiling point and melting points of hydrocarbons.[137] $CH \cdots HC$ is by and large a dispersive interaction, but how is it possible to conceptualize it in terms of a clear physical model?

Recalling that the original physical model derived by London involved oscillating dipoles, the appropriate way to conceptualize the effect is to carry our VB calculations of the $CH \cdots HC$ interaction in a series of alkanes and observe the change in the wave function.[137] Each C–H bond is described in VBT as a linear combination of one major covalent structure and two minor ionic structures. The ionic structures of each bond come in two opposing polarities, $C^+ :H^-$ and $C:^- H^+$. At infinite intermolecular separation the weights of the two structures are nearly identical. However, as the two C–H bonds approach one another, one expects the mechanism of oscillating dipoles to be turned on, in order to stabilize the dimer. This should be apparent by looking at the weights of the VB structures with oscillating-opposing dipole combinations $C^+ :H^-//H^+ C:^-$ and $C:^- H^+//:H^- C^+$, wherein the interacting H's have opposite charges, compared with those where the two interacting H's have identical charge. Figure 1.14 shows the interaction of two C–H bonds of two methane molecules. It is seen that at the equilibrium distance ($R_{H \cdots H} = 2.501$ Å) the combined weights of the oscillating-opposing dipole combinations increases to ~1.7 times the weights at infinite distance, while the combined

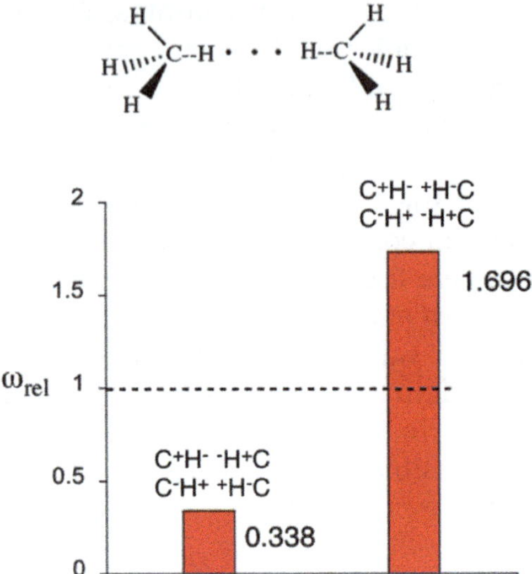

Figure 1.14   Changes in the relative weights (ω_{rel}) of the ionic structures in the C–H bonds of two CH_4 molecules, due to CH$\cdots$HC interactions (at $R_{H\cdots H} = 2.501$ Å), compared to the free molecules. Compared with the free molecules, the relative weight of the VB structures with the favorably oriented ionic charges rises to 1.696 while those of the unfavored charge orientations are reduced to 0.338 ($\omega_{rel} = 1$ represents the free molecules).
Reprinted with permission from D. Danovich, S. Shaik, F. Neese, J. Echeverria, G. Aullon and S. Alvarez, Understanding the Nature of the CH-CH Interactions in Alkanes, *J. Chem. Theor. Comput.* 2013, **9**, 1977. Copyright 2013. American Chemical Society.

weights of the structures with the identical charges on the interacting H's decreases to $\sim$0.34 of the value at infinite separation. This is a vivid demonstration of the oscillating dipoles mechanism.

In $H_3CH\cdots HCH_3$, virtually all the stabilization energy originates from the oscillating dipoles mechanism. Namely, the methane dimer is stabilized by pure dispersion interaction. However, as the alkanes grew, this mechanism continued to flesh out, but it became less and less important. More significantly, the stabilization energy was brought about through some sticky covalent interactions between the two molecules, due to reorganization of the bonding electrons of the two interacting CH bonds *via* recoupling these electrons to H$\cdots$H and C$\cdots$C 'bonds', and charge transfer interactions between the two moieties that create long-range C$\cdots$H bonds. Once again it is seen that the pristine mechanism, derived from the physics of interacting atoms, gets significantly more complex as the molecular species grows.

In fact, the size effect is spectacular.[137,138] Figure 1.15 displays two graphane planes, which like porcupines have axial C–H bonds basically pointing towards each other, and maintaining CH$\cdots$HC interactions (each C–H in one graphane sheet experiences interaction with 3C–H's in the other sheet). The interaction energy is approximately three times the number of carbons in a graphane sheet, reaching 51.6 kcal mol^{-1} for [73]graphane. BLW analysis shows that the major stabilizing factor is the long-range dispersion (presumably between the oscillating dipoles of the C–H bonds in the two sheets). However, there is an important two-way charge transfer energy due to the $\sigma_{CH} \rightarrow \sigma^*_{CH}$ interactions of the CH$\cdots$HC. As seen in Figure 1.15, the ΔE_{CT} term, which accounts for ~15% of the total binding energy, results in the accumulation of electron density in the interface area between two layers. This accumulated electron density thus acts as an electronic "glue" for the graphane layers and constitutes an important driving force in the self-association and stability of graphane at ambient conditions.

The dispersion interactions then are more than the name suggests, and involve in addition orbital interactions of significant size. In

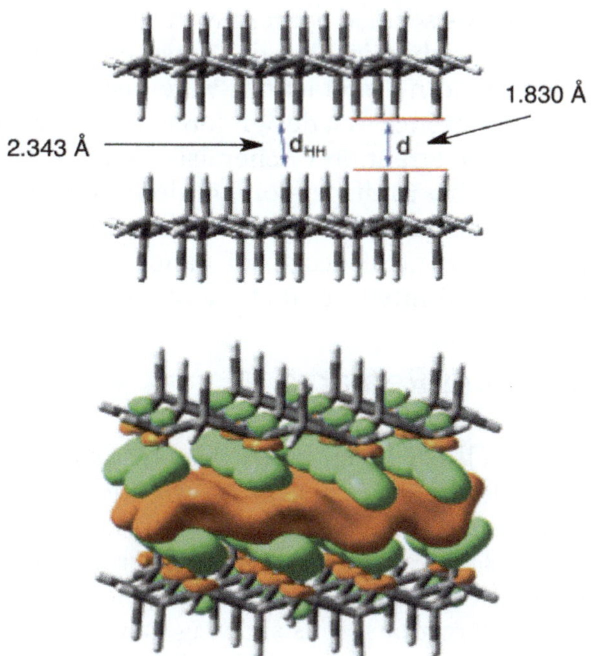

Figure 1.15   The [24]graphane dimer (top), and its electron density difference (EDD) map (bottom) for the charge-transfer effect, with isodensity value 0.0001 a.u. The orange color indicates an increase while the green color a reduction of electron density.
Reprinted with permission from C. Wang, Y. Mo, P. J. Wagner, P. R. Schreiner, E. D. Jemmis, D. Danovich and S. Shaik, The Self-Association of Graphane Is Driven by London Dispersion and Enhanced Orbital Interactions, *J. Chem. Theor. Comput.* 2015, **11**, 1621. Copyright 2015. American Chemical Society.

relating the computed values to experiments, *e.g.*, vaporization energies or heats of melting, one must account also for entropy. Also, recently measured dispersion energies of alkanes in solution appear to be an order of magnitude smaller than the values suggested by calculations and derived from vaporization enthalpies.[139] A possible cause for such a reduction is the dispersion interaction between the alkane and the surrounding solvent molecules, which causes the alkanes to turn on the dipole oscillating interaction with the solvent molecules, at the expense of diminishing the same alkane–alkane interaction due to mismatch of the respective oscillating dipoles. Again, all these uncertainies and complications are a sign of an alive science.

1.9   Summary of Intermolecular Interactions

Intermolecular interactions have been for a long while a territory of physics, while chemists seemed content with their own 'neverland'[140] of molecules and their chemical bonds. However, at turn of the 20th century, hydrogen bonding interactions were discovered and quickly revealed their omnipresence as a design force in matter. Probably this change has turned the attention of chemists to other intermolecular interactions and notably to dispersion/London interactions, halogen bonds, *etc.* The understanding of intermolecular interactions, and their usage as an element of design of nanomaterials and crystals, define a current central intellectual arena of chemistry.

1.10   Concluding Remarks

I tried to describe in this chapter a very brief history of "cohesion" in chemistry by overviewing the conceptual evolution of the bond that glues atoms to molecules, and that of the intermolecular interactions that assemble molecules into larger aggregates of macroscopic matter. Subsequently, I reviewed novel bonding motifs and new aspects of intermolecular interactions, thus showing that the field of cohesion in chemistry is vibrant, exciting, and it still presents problems that await solution. The imaging of bonds, bond breaking and remaking, and perhaps also of hydrogen bonds and maybe in the future of halogen bonds too, mark a degree of hype and excitement in this field. In fact, as I completed this chapter, I noticed two studies which return to the basics, define a generalized electronegativity,[141] and a new parameter that can distinguish bonds and weak interactions.[142] This is high time for bonding and cohesion in chemistry.

Notes Added in Proof

(a) The dative bonds between amines and carbocations were shown to be CSBs in a recent experimental study;[144] (b) the quadruple bonding in C_2 was reproduced in a CASSCF recent study;[145] (c) the inverted bond in C_2 was recently also found in B_2.[146] This calls for a new Aufbau principle for bonding in diatomic molecules.

Acknowledgements

I am thankful to all my friends who participated in the bonding march: W. Wu, C. Wang, Y. Mo, P. C. Hiberty, D. Danovich, B. Braida, P. Su, H. Rzepa, P. R. Schreiner, and S. Alvarez. My special thanks to David Danovich who has read the paper and checked the references and art items. The paper is dedicated to Santiago Alvarez whose gentle coercion brought me to look at dispersion using VB theory, and to Philippe Hiberty my longtime partner in VB theory.

References

1. G. N. Lewis, *J. Am. Chem. Soc.*, 1916, **38**, 762.
2. S. Shaik, *J. Comput. Chem.*, 2007, **28**, 51.
3. C. F. Nejad, S. Shabazian and R. Marek, *Chem. – Eur. J.*, 2014, **20**, 10140.
4. R. Siegfried, *From Elements to Atoms: A History of Chemical Composition*, American Philosophical Society, Philadelphia, 2002.
5. P. Ball, *Chemistry World*, 2014, **January**, 30.
6. S. Ritter, *Chem. Eng. News*, 2014, **September 22**, 10.
7. *The Chemical Bond: Fundamental Aspects of Chemical Bonding*, ed. G. Frenking and S. Shaik, Wiley-VCH, Weinheim, Germany, 2014.
8. *The Chemical Bond: Chemical Bonding Across the Periodic Table*, ed. G. Frenking and S. Shaik, Wiley-VCH, Weinheim, Germany, 2014.
9. D. G. de Oteyza, P. Gorman, Y.-C. Chen, S. Wickenburg, A. Riss, D. J. Mowbray, G. Etkin, Z. Pendramrazi, H.-Z. Tsai, A. Rubio, M. F. Crommie and F. R. Fischer, *Science*, 2013, **340**, 1434.
10. M. V. Putz, *Int. J. Mol. Sci.*, 2010, **11**, 4227.
11. T. Clark, *Directional Electrostatic Bonding*, ch. 18, in Ref. 8, 523.
12. S. Grimme, *Dispersion Interactions and Chemical Bonding*, ch. 16, in Ref. 8, 477.
13. Y. Mo, *The Block Localized (BLW) Perspective of Chemical Bonding*, ch. 6, in Ref. 7, 199.
14. A. M. Pendás, E. Francisco and M. A. Blanco, *J. Phys. Chem. A*, 2006, **110**, 12864.
15. M. Mitoraj, A. Michalak and T. Ziegler, *J. Phys. Chem. A*, 2008, **112**, 1933.
16. R. P. Schreiner, L. V. Chernish, P. A. Gunchenko, E. Y. Tikhonchuk, H. Hausman, M. Serafim, S. Schlecht, J. E. P. Dahl, R. M. K. Carlson and A. A. Fokin, *Nature*, 2011, **477**, 308.
17. A. Roob, *The Hermetic Museum: Alchemy and Mysticism*, Taschen Verlag GmbH, English Translation, S. Whiteside, London, 1997, pp. 132–133.
18. N. A. Lemery, *Course of Chymistry*, 4th English Edition translated from the French edition of 1715, London 1720.

19. U. Klein, *Ambix*, 1995, **42**, 79.
20. B. Bensaude-Vincent and I. Stengers, *A History of Chemistry*, Harvard University Press, English translation, D. van Dam, Cambridge, 1996, pp. 69–70.
21. W. A. Noyes, *J. Am. Chem. Soc.*, 1917, **39**, 879.
22. M. Chayut, *Ann. Sci.*, 1991, **48**, 527.
23. J. W. Servos, *Physical Chemistry from Ostwald to Pauling*, Princeton University Press, Princeton, 1990, pp. 4–5.
24. J. W. Servos, *J. Chem. Educ.*, 1984, **61**, 5.
25. D. A. Davenport, *Bull. Hist. Chem.*, 1996, **19**, 13.
26. G. N. Lewis, *J. Am. Chem. Soc.*, 1913, **35**, 1448.
27. G. N. Lewis, *Trans. Faraday Soc.*, 1923, **19**, 452.
28. I. Langmuir, *J. Am. Chem. Soc.*, 1919, **41**, 868.
29. S. Shaik and P. C. Hiberty, *A Chemist's Guide to Valence Bond Theory*, Wiley-Interscience, 2008, ch. 1, pp. 1–25.
30. W. Heitler and F. London, *Z. Phys.*, 1927, **44**, 455.
31. W. Heisenberg, *Z. Phys.*, 1926, **38**, 411.
32. L. Pauling, *Proc. Natl. Acad. Sci. U. S. A.*, 1928, **14**, 359.
33. J. C. Slater, *Phys. Rev.*, 1929, **34**, 1293.
34. J. C. Slater, *Phys. Rev.*, 1931, **38**, 1109.
35. G. Rumer, *Gottinger Nachr.*, 1932, 337.
36. S. C. Wang, *Phys. Rev.*, 1928, **31**, 579.
37. N. Rosen, *Phys. Rev.*, 1931, **38**, 2099.
38. S. Weinbaum, *J. Chem. Phys.*, 1933, **1**, 593.
39. J. C. Slater, *Phys. Rev.*, 1931, **37**, 481.
40. J. C. Slater, *Phys. Rev.*, 1932, **41**, 255.
41. L. Pauling, *J. Am. Chem. Soc.*, 1931, **53**, 1367.
42. L. Pauling, *J. Am. Chem. Soc.*, 1931, **53**, 3225.
43. L. Pauling, *The Nature of the Chemical Bond*, Cornell University Press, Ithaca New York, 1939 (3rd edn, 1960).
44. T. Hager, *Force of Nature: The Life of Linus Pauling*, Simon and Schuster, New York, pp. 62–63.
45. R. S. Mulliken, *Phys. Rev.*, 1928, **32**, 186.
46. R. S. Mulliken, *Phys. Rev.*, 1928, **32**, 761.
47. R. S. Mulliken, *Phys. Rev.*, 1929, **33**, 730.
48. R. S. Mulliken, *Phys. Rev.*, 1932, **41**, 49.
49. F. Hund, *Z. Phys.*, 1931, **73**, 1.
50. F. Hund, *Z. Phys.*, 1928, **51**, 759.
51. J. A. Berson, *Chemical Creativity. Ideas from the Work of Woodward, Hückel, Meerwein, and Others*, Wiley-VCH, New York, 1999.
52. J. E. Lennard Jones, *Trans. Faraday Soc.*, 1929, **25**, 668.
53. E. Hückel, *Z. Phys.*, 1930, **60**, 423.
54. E. Hückel, *Z. Phys.*, 1931, **70**, 204.
55. E. Hückel, *Z. Phys.*, 1932, **76**, 628.
56. L. Pauling and G. W. Wheland, *J. Chem. Phys.*, 1933, **1**, 362.
57. R. Hoffmann, S. Shaik and P. C. Hiberty, *Acc. Chem. Res.*, 2003, **36**, 750.
58. J. H. van Vleck and A. Sherman, *Rev. Mod. Phys.*, 1935, **7**, 167.
59. S. G. Brush, *Stud. Hist. Philos. Sci.*, 1999, **30**, 263.
60. H. Chen, D. Danovich and S. Shaik, Theoretical Tools for Bioinorganic Complexes: Their Applications and Insights, in *Comprehensive Inorganic Chemistry II*, ed. J. Reedijk and K. Poeppelmeier, 2010, vol. 9 (ed. S. Alvarez), ch. 1, pp. 2–57.
61. T. Bitter, K. Ruedenberg and W. H. E. Schwarz, *J. Comput. Chem.*, 07, **28**, 411.
62. M. W. Schmidt, J. Ivanic and K. Ruedenberg, The Physical Origins of Covalent Bonding, ch. 7, in Ref. 7, pp. 1–65.
63. J. C. Slater, *J. Chem. Phys.*, 1933, **1**, 687.
64. C. Q. Wilson and W. A. Goddard, III, *Theor. Chim. Acta*, 1972, **26**, 195.

65. A. Rozendaal and E. J. Baerends, *Chem. Phys.*, 1985, **95**, 57.
66. R. F. W. Bader, J. Hernandez-Trujillo and F. Cortés-Guzmán, *J. Comput. Chem.*, 2007, **28**, 4.
67. W. Kutzelnigg, in *Theoretical Models of Chemical Bonding*, ed. Z. B. Maksic, Springer-Verlag, New York, 1990, part 2, pp. 1–44.
68. Z. Huaiyu, D. Danovich, W. Wu, B. Briada, P. C. Hiberty and S. Shaik, *J. Chem. Theory Comput.*, 2014, **10**, 2410.
69. P. C. Hiberty, S. Humbel and P. Archirel, *J. Phys. Chem.*, 1994, **98**, 11697.
70. J. F. Berry, *Acc. Chem. Res.*, 2016, **49**, 27.
71. P. Ball, *Chemistry World*, 2013, **10**(4), 41.
72. S. Shaik, P. Mâitre, G. Sini and P. C. Hiberty, *J. Am. Chem. Soc.*, 1992, **114**, 7861.
73. S. Shaik, D. Danovich, W. Wu and P. C. Hibery, *Nat. Chem.*, 2009, **1**, 443.
74. Y. Grin, A. Savin and B. Silvi, *The ELF Perspective of Chemical Bonding*, ch. 20, in Ref. 7, 345.
75. R. F. W. Bader and T. T. Nguyen-Dang, *Adv. Quantum Chem.*, 1981, **14**, 63.
76. S. S. Shaik, Valence Bond Mixing: The LEGO Way. From Resonating Bonds to Resonating Transition States, in *Molecules in Natural Science and Medicine*, ed. Z. B. Maksic and M. E. Maksic, Ellis Horwood, London, 1991, ch. 12, pp. 253–267.
77. A. A. Fiorillo and J. M. Galbraith, *J. Phys. Chem. A*, 2004, **108**, 5126.
78. B. Braida and P. C. Hiberty, *Nat. Chem.*, 2013, **5**, 417.
79. R. Hoffmann, *Angew. Chem., Int. Ed. Engl.*, 1982, **21**, 711.
80. W. C. Kaska, D. K. Mitchell and R. F. Reicheldfeder, *J. Organomet. Chem.*, 1973, **47**, 391.
81. G. Frenking, R. Tonner, S. Klein, N. Takagi, T. Shimizu, A. Krapp, K. K. Pandey and P. Parameswaran, *Chem. Soc. Rev.*, 2014, **43**, 5106.
82. (a) D. Himmel, I. Krossing and A. Schnepf, *Angew. Chem., Int. Ed.*, 2014, **53**, 370; (b) D. Himmel, I. Krossing and A. Schnepf, *Angew. Chem., Int. Ed.*, 2014, **53**, 6047.
83. G. Frenking, *Angew. Chem., Int. Ed.* 2014, **53**, 6040.
84. F. A. Cotton, *Inorg. Chem.*, 1965, **4**, 334.
85. J. E. McGrady, Electronic Structure of Metal-Metal Bonds, in *Computational Inorganic and Bioinorganic Chemistry*, ed. E. I. Solomon, R. A. Scott and R. B. King, Wiley, New York, 2009, pp. 425–431.
86. L. Gagliardi and B. O. Roos, *Nature*, 2005, **433**, 848.
87. S. Shaik, D. Danovich, W. Wu, P. Su, H. S. Rzepa and P. C. Hiberty, *Nat. Chem.*, 2012, **4**, 195.
88. S. Shaik, H. S. Rzepa and R. Hoffmann, *Angew. Chem., Int. Ed.*, 2013, **52**, 3020.
89. W. Zou and D. Cremer, *Chem. – Eur. J.*, 2016, **22**, 4087.
90. M. Piris, X. Loopez and J. M. Uglde, *Chem. – Eur. J.*, 2016, **22**, 4109.
91. M. Hermann and G. Frenking, *Chem. – Eur. J.*, 2016, **22**, 4100.
92. S. Shaik, D. Danovich, B. Braida and P. C. Hiberty, *Chem. – Eur. J.*, 2016, **22**, 4116.
93. R. Zhong, M. Zhang, H. Xu and Z. Su, *Chem. Sci.*, 2016, **7**, 1028.
94. S. Grimme and P. R. Schreiner, *Angew. Chem., Int. Ed.*, 2011, **50**, 12639.
95. J. P. Wagner and P. R. Schreiner, *Angew. Chem., Int. Ed.*, 2015, **54**, 12274.
96. S. S. Shaik and M.-H. Whangbo, *Inorg. Chem.*, 1986, **25**, 1201.
97. J. S. Miller and J. J. Novoa, *Acc. Chem. Res.*, 2007, **40**, 189.
98. Y.-H. Tian and J. M. Kertesz, *J. Phys. Chem. A*, 2011, **115**, 13942.
99. F. Mota, J. S. Miller and J. J. Novoa, *J. Am. Chem. Soc.*, 2009, **131**, 7699.
100. B. Braida, K. Hendrickx, D. Domin, J. Dinnocenzo and P. C. Hiberty, *J. Chem. Theory Comput.*, 2013, **9**, 2276.
101. D. Danovich, W. Wu and S. Shaik, *J. Am. Chem. Soc.*, 1999, **121**, 3165.
102. D. Danovich and S. Shaik, *Acc. Chem. Res.*, 2014, **47**, 417.
103. M. Verdicchio, S. Evangelisti, T. Leininger, J. Sánchez-Marín and A. Monari, *Chem. Phys. Lett.*, 2011, **503**, 215.
104. P. Soldán, M. T. Cvitas and J. M. Huston, *Phys. Rev. A: At., Mol., Opt. Phys.*, 2003, **67**, 054702.

105. N. M. Theisen, F. Lackner and W. E. Ernst, *J. Phys. Chem. A*, 2011, **115**, 7005.
106. I. A. Popov and A. I. Boldyrev, *Chemical Bonding in Inorganic Aromatic Compounds*, 2014, ch. 14, in Ref. 8, pp. 421–444.
107. P. C. Priyakumari and E. D. Jemmis, *Electron Counting Rules in Cluster Bonding – Polyhedral Boranes, Elemental Boron, and Boron-Rich Solids*, 2014, ch. 5, in Ref. 8, pp. 113–148.
108. P. Alemany and E. Canadell, *Chemical Bonding in Solids*, 2014, ch. 15, in Ref. 8, pp. 445–476.
109. M. Kaupp, *Chemical Bonding of Main-Group Elements*, 2014, ch. 1, in Ref. 8, pp. 1–24.
110. H. Schwarz, *Angew. Chem., Int. Ed.*, 2003, **42**, 4442.
111. J. Rowlinson, *Cohesion*, Cambridge University Press, Cambridge, 2002.
112. J. Israelachvili and M. Ruths, *Langmuir*, 2013, **29**, 9605.
113. K. E. Riley and P. Hobza, *Wiley Interdiscip. Rev.: Comput. Mol. Sci.*, 2011, **1**, 3.
114. C. J. Cramer and D. G. Truhlar, *Phys. Chem. Chem. Phys.*, 2009, **11**, 10757.
115. T. S. Moore and T. F. Winmill, *J. Chem. Soc.*, 1912, **101**, 1635.
116. W. M. Latimer and N. H. Rodebush, *J. Am. Chem. Soc.*, 1920, **42**, 1419.
117. J. D. Bernal and R. H. Fowler, *J. Chem. Phys.*, 1933, **1**, 515.
118. G. C. Pimental and A. L. McClellan, *The Hydrogen Bond*, W.H. Freeman & Co., Sans Francisco and London, 1960.
119. G. Gilli and P. Gilli, *J. Mol. Struct.*, 2000, **552**, 1.
120. J. Zhang, P. Chen, W. Ji, Z. Cheng and X. Qiu, *Science*, 2013, **342**, 611.
121. S. K. Hämäläinen, N. van der Heijden, J. van der Lit, S. den Hartog, P. Liljeroth and I. Swart, *Phys. Rev. Lett.*, 2014, **113**, 186102.
122. P. Zheng, S.-i. J. Takayama, A. G. Mauk and H. Li, *J. Am. Chem. Soc.*, 2012, **134**, 4124.
123. H. Elgabarty, R. Z. Khaliullin and T. D. Kühne, *Nat. Commun.*, 2015, **6**, 8318.
124. S. J. Grabowski, *Chem. Rev.*, 2011, **111**, 2597.
125. Y. Mo, C. Wang, L. Guan, B. Braida, P. C. Hiberty and W. Wu, *Chem. - Eur. J.*, 2014, **20**, 8444.
126. J. J. Novoa, I. Nobeli, F. Grepioni and D. Braga, *New J. Chem.*, 2000, **24**, 5.
127. D. Braga, F. Grepioni and J. J. Novoa, *Chem. Commun.*, 1998, 31959.
128. E. D'Oria and J. J. Novoa, *CrystEngComm*, 2004, **6**(64), 367.
129. E. D'Oria and J. J. Novoa, *J. Phys. Chem. A*, 2011, **115**, 13114.
130. S. Shaik and A. Shurki, *Angew. Chem., Int. Ed.*, 1999, **38**, 586.
131. H. Hirao and X. Wang, *Hydrogen Bonding*, 2014, chapter 17, in Ref. 8, pp. 501–522.
132. F. Guthrie, *J. Chem. Soc.*, 1863, **16**, 239.
133. A. C. Legon, *Phys. Chem. Chem. Phys.*, 2010, **12**, 7736.
134. C. Wang, D. Danovich and S. Shaik, *J. Chem. Theory Comput.*, 2014, **10**, 3726.
135. C. Wang, L. Guan, D. Danovich, S. Shaik and Y. Mo, *J. Comput. Chem.*, 2016, **37**, 34.
136. C. Foroutan-Nejad, Z. Badri and R. Marek, *Phys. Chem. Chem. Phys.*, 2015, **17**, 30670.
137. D. Danovich, S. Shaik, F. Neese, J. Echeverria, G. Aullon and S. Alvarez, *J. Chem. Theory Comput.*, 2013, **9**, 1977.
138. C. Wang, Y. Mo, P. J. Wagner, P. R. Schreiner, E. D. Jemmis, D. Danovich and S. Shaik, *J. Chem. Theory Comput.*, 2015, **11**, 1621.
139. L. Yang, C. Adam, G. S. Nichol and S. L. Cockroft, *Nat. Chem.*, 2013, **5**, 1006.
140. The territory of eternal childhood: https://en.wikipedia.org/wiki/Neverland.
141. M. Rahm and R. Hoffmann, *J. Am. Chem. Soc.*, 2015, **137**, 10282.
142. M. Rahm and R. Hoffmann, *J. Am. Chem. Soc.*, 2016, **138**, 3731.
143. S. Shaik, *Phys. Chem. Chem. Phys.*, 2012, **12**, 8706.
144. R. G. Poranne and P. Chen, *Chem. Eur. J.*, 2017, **23**, 4659.
145. A. C. West, M. W. Schmidt, M. S. Gordon and K. Ruedenberg, *J. Phys. Chem. A*, 2017, **121**, 1086.
146. Z. Rashid, J. H. van Lenthe and R. W. A. Havenith, *Comput. Theor. Chem.*, 2017, http://dx.doi.org/10.1016/j.comptc.2017.02.001.

2 Using Computational Quantum Chemistry as a Tool to Understand the Structure of Molecular Crystals and the Nature of their Intermolecular Interactions

Juan J. Novoa[a,b]

[a] Department of Materials Science and Physical Chemistry, Av. Diagonal 645, 08028-Barcelona, Spain; [b] Institute of Theoretical and Computational Chemistry (IQTC), Av. Diagonal 645, 08028-Barcelona, Spain
Email: juan.novoa@ub.edu

2.1 Introduction: Computational Quantum Chemistry as a Tool to Rationalize the Structure of Molecular Crystals

Intermolecular interactions (more specifically, their interaction energy, E_{int}) are the "glue" that hold together the molecules in a molecular crystal. They also determine the relative disposition of the molecules in the molecular crystals and, consequently, the crystal physical properties (conductivity, superconductivity, magnetism, . . .).[1] However, despite the relevant role that some of these molecular crystals play in the research of properties of technological relevance, the rationalization of crystal structures using accurate data obtained from

Intermolecular Interactions in Crystals: Fundamentals of Crystal Engineering
Edited by Juan J. Novoa
© The Royal Society of Chemistry 2018
Published by the Royal Society of Chemistry, www.rsc.org

first-principles calculations is not still a popular procedure with which to rationalize crystals, despite being accurate, general and unbiased (a version of this method was proposed a few years ago,[2] aimed at the analysis of the packing in crystals of small molecules, but the continued increase of computer power makes possible its extended use in crystals of complex molecules). We will refer to such methodology as the Quantum Chemical-Crystal Packing Analysis procedure (in short QC-Crystal Packing Analysis or QC-CPA) because the data needed is computed using the methods of Quantum Chemistry.[3–5] In the following pages, firstly, such methodology is described. Then, it is applied to rationalize the packing of a neutral and an ionic crystal, in order to illustrate the type of problems faced when analyzing the crystal packing in these two families of crystals.

The underlying principle behind the use of QC-Crystal Packing Analysis to rationalize the structure of a crystal is that the only property required for such analysis is the interaction energy of all symmetry-unique intermolecular interactions present in the molecular crystal of interest, values that can be accurately computed using various quantum chemical methods. As will be shown hereafter, the stability of a crystal depends on all symmetry-unique pairs of molecules formed between the molecules of the unit cell. Among these interactions, the most relevant ones are the energetically dominant intermolecular interactions, which are those directly attached to the reference molecule (usually known as the first coordination shell of the reference molecule). Among the energy-dominant interactions, those whose interaction energy (E_{int}) are stabilizing (*i.e.*, $E_{int} < 0$, see Figure 2.1) conform a special subset, called the intermolecular bonds. Obviously, for the crystal to be stable, the sum of E_{int} for all stabilizing interactions should overcome the sum of the destabilizing interactions (*i.e.*, those whose $E_{int} > 0$, see Figure 2.1). It is now clear why intermolecular interactions, and particularly intermolecular bonds are sometimes called the "glue" that keeps together the molecules in a molecular crystal. However, as will be shown below, crystal packing analysis is more informative when one looks at both the stabilizing and destabilizing intermolecular interactions.

As already mentioned, current Computational Quantum Chemistry has become a very powerful tool to get accurate values of E_{int} for all the symmetry-unique intermolecular interactions present in the crystal (before, it was only possible to get estimates of E_{int} using the analytical equations obtained for simple model systems[6]). As will be described more in detail hereafter, the use of quantum chemical

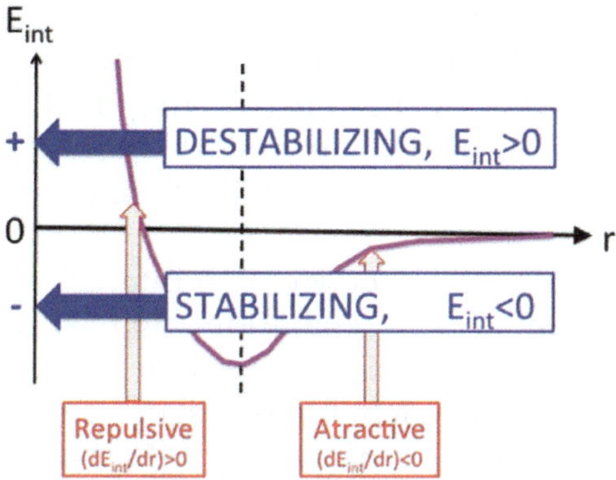

Figure 2.1   Plot of the potential energy curve (that is the E_{int} vs. r curve) for an intermolecular interaction. The stabilizing ($E_{int}<0$) and ($E_{int}>0$) destabilizing regions of the curve are indicated, and so are the attractive ($r>r_{min}$) and repulsive ($r<r_{min}$) regions. (r_{min} is the distance where the potential energy curve has its minimum, marked by the vertical broken line). Note that the stabilizing and destabilizing regions are different than the attractive and repulsive regions.

methods in getting realistic values for the intermolecular interactions found in molecular crystals has been made possible by: (1) the rapid progress in processor performance (it doubles about every two years, while its cost has remained similar, a trend is known as Moore's Law[7]), and (2) the widespread availability of efficient quantum chemical programs allowing the required quantum chemical calculations to be carried out.

2.2   Intermolecular Interactions and Intermolecular Bonds in Molecular Crystals

Intermolecular interactions are a microscopic property. Here we analyze which macroscopic physical properties of the crystal connect with the intermolecular interactions when Statistical Mechanics is applied.

Let's start by saying that the mere existence of molecular solids at room temperature is a consequence of the existence of stabilizing interactions among the molecules that form the crystal, combined with the fact that the sum of the stabilizing ones outweighs the sum of the destabilizing ones with a value that is larger than the thermal energy of the crystal at the working temperature. Otherwise, the molecular solid would become a liquid or a gas at this temperature. Except when dynamical effects are important, the observed experimental structure geometry of the crystal is one of the minima on the free energy surface (G) of the crystal at 0 K. At this temperature, the free energy is equal to the energy computed by any quantum chemical method (E). When dynamical effects are present the description becomes a bit more complex,[13] but the overall message remains.

The variation of the total energy of two interacting molecules with their shortest intermolecular separation (that is, the $E(r)$ potential energy curve for the intermolecular interaction) has a Morse-like shape similar to that for covalent bonds, Figure 2.1, although with two important differences: (a) the minimum of the potential energy curve in intermolecular interactions is located at much larger distances than in covalent bonds; and (b) the energy at that minimum is a lot smaller than that found in covalent bonds. For instance, the minimum for the Ar_2 potential energy curve is located at $r(Ar \cdots Ar) = 7.14033$ Å with an interaction energy of -0.277 kcal mol^{-1}.[8] For comparison, the minimum in the H_2 curve is found at $r(H–H) = 0.7416$ Å[9] with an energy of 104 kcal mol^{-1}.[10] Note that the shape of the $E(r)$ potential curve and the $E_{int}(r)$ curve is the same. The Morse-like shape of the $E(r)$ curve is also found in any stabilizing intermolecular interaction. Also note that the interaction between two molecules is not always stabilizing (for instance, at some orientations of the water molecules in the water $\cdots$ water intermolecular interaction, see Figure 2.2).

The thermodynamic and kinetic properties of chemical substances are all based on energetic considerations. Thus, for instance, the existence of a stable chemical substance is characterized by the presence of a minimum in the free energy (G) surface, and its stability against their transformation into other chemical substances is determined by the difference between its free energy and that of the other substances. Finally, when thermodynamically allowed, the speed at which any of these transformations takes place is determined by the size of the activation energy barrier for the process. It should be clear now that energy is the property used by nature to code the basic properties of chemical substances: existence, relative stability, and the speed of transformation into other substances.

Supramolecular aggregates are considered by the IUPAC as a case of a chemical substance.[11] Furthermore, molecular crystals are supramolecular aggregates having long-distance order. Therefore, it is not surprising that the existence, relative stability, and speed of the energetically allowed transformations satisfy the basic principles described above. For instance, the observed structure of a crystal is a minimum in the free energy surface of the crystal at the working

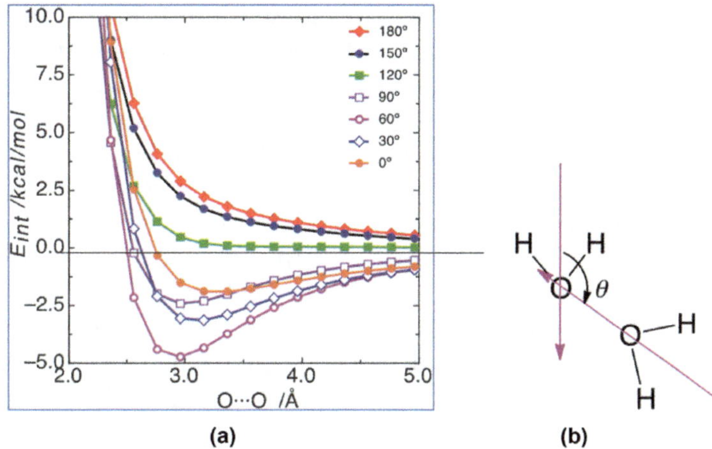

Figure 2.2   (a): Variation of the interaction energy, E_{int}, for a water dimer as a function of the intermolecular $O{\cdots}O$ distance (in Å) shown in (b). (The interaction energy measures the difference between the total energy of the water dimer and that for two dissociated water molecules at rest and whose separation is infinity; the strength is $-E_{int}$). The curves were computed for seven different relative orientations of the two molecules, (measured as a function of the angle θ, see b). All interaction energies were computed at the MP2/aug-cc-pVTZ level, correcting the BSSE error. The minimum for this water dimer is found at $\theta = 60°$, where an $O{-}H{\cdots}O$ hydrogen bond is created. Notice the existence of a destabilizing region for the water$\cdots$water interaction at short distances, whatever the orientation. On the other hand, the curves for $\theta \geq 120°$ are destabilizing at all distances, while those for $\theta < 120°$ are stabilizing close to the minimum of the curve. (b) geometry of the water dimer explored when the angle θ changes from 0 to 180° (the right water is fixed, while the left one is allowed to rotate in the plane of the two molecules.

conditions[12] (in pure substances, completely determined by indicating the working temperature and pressure, T and P). Let's mention in passing that things are not always so simple when dealing with molecular crystals showing dynamical effects.[13]

Statistical Mechanics connects the microscopic and macroscopic worlds rigorously. For instance, it connects minima in the E-surface (a microscopic property) with minima in the G-surface (a macroscopic property). The reason for such behavior originates in the connection that Statistical Mechanics makes between the total energy of a system obtained in quantum chemical calculations (E), and the internal energy (U) employed in thermodynamics, a property connected with the free energy, G ($H = U + PV$; $G = H - TS$, where P, V, H, T and S are the pressure, volume, enthalpy, temperature and entropy, respectively). Only at T and P zero ($P =$ pressure) are the values of G, U and E the same (*i.e*, $G = U = E$). At any other working conditions, $G = H - TS \neq U$. Notice that the evaluation of G requires a previous computation of the value of the entropy,[14] a demanding calculation. It is also important to realize that while the shape of the potential energy surface computed in E (E-surface) is unique, this is not the case when written in terms of G (G-surface), where some unusual properties can appear: (a) the relative stability of the minima, when described in the free energy surface, can change with temperature when the two substances have different entropies (sometimes, even the number of minima is different[15]), and (b) the absolute minimum below a given critical temperature can be different than above such critical temperature.[16]

Finally, we focus our attention in investigating the meaning of E_{int} in polyatomic systems, as in diatomic systems its meaning is clear: the energy gain or loss with respect to the dissociated fragments. In order to give a solid answer to the meaning of E_{int} in polyatomic systems, we will employ the "divide and conquer" method, a well-known procedure to study complex systems in computer programming[17] (therefore, a case of scientific method[18]). According to the "divide and conquer" method, the properties of a complex system can be studied by: (1) decomposing the system into its parts, (2) studying the properties of each part separately, and (3) assuming that the properties of the complex system are equal to the sum of the properties of each isolated part.

Perturbation theory tells us that one cannot apply the "divide and conquer" method to estimate the total energy of a supramolecular aggregate, because the equation $E(\text{total}) = \Sigma_I\, E(i)$ is only satisfied when the parts do not interact. Much better results are obtained when adding interaction energies, as is illustrated in what follows for a supramolecular aggregate of four molecules (with names A, B, C and D). For such a system one has:

$$E_{int}(A,B,C,D) = E_{int}(A,B) + E_{int}(A,C) + E_{int}(A,D)$$
$$+ E_{int}(B,C) + E_{int}(B,D)$$
$$+ E_{int}(C,D) = \Sigma_{I>J}\, E_{int}(I,J), \tag{2.1}$$

whose physical interpretation is: "The total interaction energy of a molecule or supramolecular aggregate can be approximated as the sum, over the non-repeating pairs, of the intermolecular interaction energy". This equation was found to work reasonably well in most cases, although it is well known that tri-body and higher order terms are sometimes needed.

One can simplify eqn (2.1) by only considering the dominant energetic pairs in the right summatory, whatever their sign. It is also possible to simplify eqn (2.1) a bit more, by considering only its *dominant stabilizing* components, as first proposed by Pauling,[19]

$$E_{int} = \Sigma_{I>J}, \text{ and } E_{int} < 0 \; E_{int}(I,J) = \Sigma_m E_{bond}(m). \tag{2.2}$$

Pauling also called each of these dominant stabilizing components a bond, with $E_{bond}(m)$ their bond energy. Thus, the "divide and conquer" method allows connection of the total interaction energy with the bonds of the molecule, or in a supramolecular aggregate with its supramolecular bonds, eqn (2.2). Nowadays, there are methods specially designed to find all intermolecular bonds present in a crystal based on sound theoretical considerations (such as the Atoms-in-Molecules, or AIM, procedure[20]). Alternatively, one can identify good candidates for bonds by applying empirical criteria based on previous studies,[19] or some distance and angularity criteria.[21]

There is an important issue concerning eqn (2.2). It is not exact, as only the dominant stabilizing interactions (in intramolecular or intermolecular bonding) are considered in the first summation. Such error is intrinsic to the approximation and its size depends on: (a) how large the non-dominant stabilizing interactions are compared to the dominant stabilizing interactions (the bonds), and (b) how important the destabilizing interactions are compared to the stabilizing ones. When either (a) or (b) are numerically non-negligible, it will make it difficult to obtain similar values of E_{bond} for the same bond from different molecules.

It is important to keep in mind that, although the observed structure of a molecular crystal is a minimum on its free energy surface, *this does not necessarily imply that all its intermolecular bonds are at their optimum geometry.* Crystal packing is a compromise among all intermolecular bonds, and sometimes some bonds are far from their equilibrium conformation, accommodating the presence of stronger bonds within the crystal. As a direct consequence, empirical rules such as "the closer, the stronger" (the strength–length correlation[22]) are not always obeyed,[23] a breakdown observed in intermolecular[23] and intramolecular bonds.[24] None of this invalidates the objective of rationalizing the structure of molecular crystals by looking at their intermolecular bonds.[25] However, in some cases, a better understanding of the crystal packing is only obtained after considering also the dominant destabilizing interactions[26] (these interactions cannot be called bonds, as such a name can only be applied to stabilizing interactions[19]). The relevant role played sometimes by destabilizing interactions in crystal packing was first pointed out by Dunitz and Gavezzotti,[27] who were the first to state that the most stable relative orientation of the molecules in the crystal must also minimize the destabilizing interactions. Once again, let's keep in mind that for the crystal to be thermodynamically stable, the sum of its stabilizing interactions necessarily has to overcome the sum of the destabilizing ones. Consequently, although a proper rationalization of the crystal structure can be done in terms of intermolecular bonds (that is, by looking only at the dominant stabilizing intermolecular interactions), a more complete rationalization can be obtained by looking at all energetically dominant interactions, whatever their sign.

2.3 The Physical Meaning of Bonds: An In-depth Analysis

Let's start this part by stating that, in our opinion, a proper interpretation of the physical meaning of a bond has to satisfy Pauling's original ideas, Coulson's remarks on the nature of bonds, and use Bader AIM methodology to locate which atoms are involved in the bonds that a chemical substance presents (today's use of AIM methodology to find the atoms involved in the bonds is backed up by the large number of systems where the AIM method has been shown to work well, which shows its basic validity; the limited number of cases where the AIM procedure does not give the expected results suggests the need for some methodological improvements, which would allow AIM to work well in all systems).

Let's first focus our attention on the definition of a bond that Pauling presented in his book,[19] first published in 1939, in our opinion the most fundamental definition of what a bond is. Furthermore, Pauling's definition is that chosen by the IUPAC for the Gold Book:[11] *"There is a chemical bond between two atoms or group of atoms in the case that the forces acting between them are such as to lead to the formation of an aggregate with sufficient stability to make it convenient for the chemist to consider it as an independent molecular species"*. Pauling also indicated a few examples of *intramolecular bonds* known at that time: covalent bond, ionic bond, and coordination bond. He identified the existence of *intermolecular bonds*, stating that: *"In general we do not consider the weak van der Waals forces between molecules as leading to chemical-bond formation; but in exceptional cases, such as the weak bond that holds together the two O_2 molecules in O_4, it may happen that these forces are strong enough to make it convenient to describe the corresponding intermolecular interaction as a* *bond formation*". Notice that such definition places the full weight of the decision on the formation of a new bond on its energetic stability. On top of this, Pauling defined the energetic stability of each bond (the so-called bond energy) in such a way that, when adding all bonds present in a chemical substance, its enthalpy of formation is obtained (measured from its constituent atoms in their normal states, for instance, the O–H bond energy is one half of the enthalpy of formation of $H_2O(g)$ from $2H(g)$ and $O(g)$, equal to 110 kcal mol^{-1}).

There are some aspects concerning Pauling's definition of a bond that should be remarked upon: (1) such a definition does not identify which atoms participate in the bond; (2) it associates the total stability of a chemical substance with the sum of the stability of all its bonds, and (3) of all possible pairs that one could form with the atoms of the chemical substance, it only considers as bonds those pairs expected to present the strongest stabilizing interaction energy (the interaction energy for each bond selected becomes its bond energy). For instance, in an isolated water molecule, where one can form three atomic pairs ($O-H_1$, $O-H_2$, and H_1-H_2) only the two $O-H_i$ interactions are considered as bonds). Notice that Pauling's definition of a bond is thus fully operational only in systems where only one bond can exist (for instance, in diatomic molecules, when looking for their intramolecular bonds). Otherwise, it does not allow identification of which atoms are bonded, or how many bonds are formed. However, such information can be obtained by applying Bader's AIM methodology[28] to the total electron density of the molecule or supramolecular aggregate. This methodology is known to work in most cases, but fails in a few of them,[29] probably suggesting the need of some improvements in order to deal with these exceptional cases. The AIM methodology gives similar results when the electron density is taken from X-ray diffraction experiments,[30,31] or when the electron density is that computed using an accurate method and basis set. Figures 2.3 and 2.4 show two views of the total electron density of the benzene molecule (which has six C–C and six C–H intramolecular bonds) and the $H_2O\cdots H_2O$ dimer (where a lone $O-H\cdots O$ intermolecular bond exists).

Coulson, one of the founders of Quantum Chemistry, added an important piece of information about the nature of bonds that is not always considered properly. Based on the lack of a Hermitian operator whose eigenvalues could be equated to the properties of a bond, he stated in 1955: "*a chemical bond is not a real thing; it does not exist; no one has ever seen it; no one ever can*".[32] As this remark is still valid today, one cannot consider bonds as physically observable objects, as

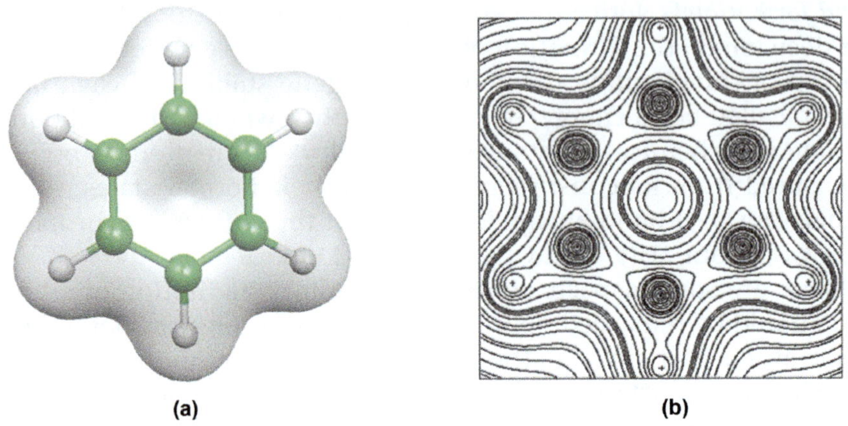

(a)   (b)

Figure 2.3 Two views of the electronic density of benzene. (a): Tridimensional isodensity 3D-surface of 0.002 atomic units (in a transparent representation that allows visualization of the atomic positions); (b): Cut of the same electron density along the plane of the molecule (the bond critical points are placed in the regions of minimum density located in between the C–C and C–H bonds).

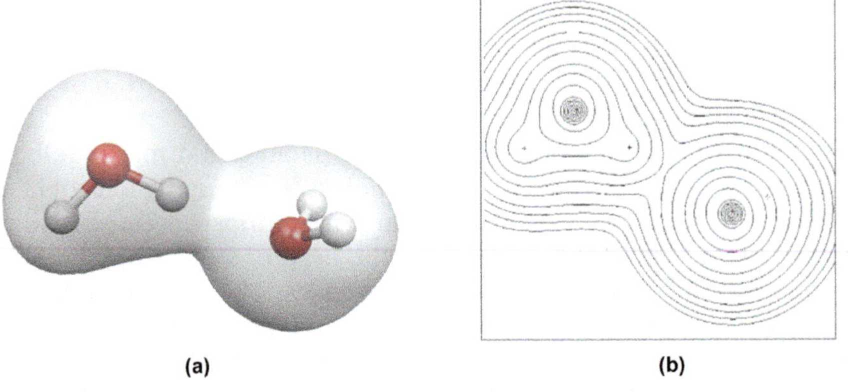

(a)   (b)

Figure 2.4 Two views of the electronic density of the $H_2O\cdots H_2O$ dimer where only a O–H$\cdots$O intermolecular bond can be found. (a): Isodensity 3D-surface of 0.002 a.u. (in a transparent representation that allows a view of the atoms and the dimer graph); (b): Cut of the same electron density along the plane of the left water molecule.

sometimes one can infer after listening to the way chemists talk in terms of bonds in the structures of molecules or supramolecular aggregates. Instead, Coulson's comments force us to consider bonds as *approximate idealized models* of a complex reality, a model capable of reproducing the most relevant features of the structure, thermodynamics and kinetic behavior in chemical systems.

At this point, it is also relevant to the analysis of the physical meaning of bonds to consider the controversial bond-like pictures obtained in some non-contact-AFM (AFM = Atomic Force Microscopy) studies, such as shown in Figure 2.5, where the authors claimed to depict the intramolecular and intermolecular bonds. If such interpretation is proven to be right, Coulson's interpretation about the non-reality of bonds would be proven to be wrong.

These recent non-contact AFM experiments[33] analyzed the aggregates of 8-hydroxyquinoline (8-hq) deposited on a Cu(111) surface, which were imaged with sub-Å precision.[34] Among the variety of aggregates adsorbed on the Cu(111) surface by means of Cu–O and

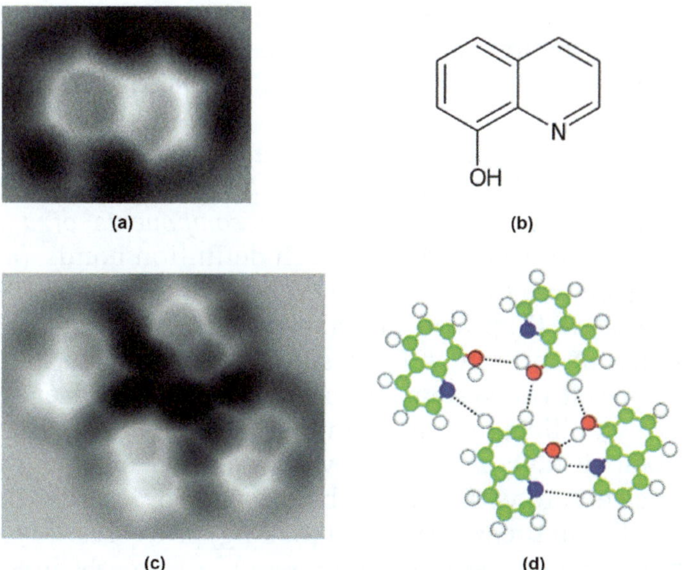

Figure 2.5 Images obtained when doing a non-contact AFM experiment (measured as frequency shifts and using a CO tip) for one of the aggregates formed in an experiment where 8-hydroxyquinoline (8-hq) is deposited on the Cu(111). (a) AFM image of a single 8-hq molecule deposited on the Cu(111) surface; (b) Chemical structure of a single 8-hq molecule (the H atoms bonded to the ring C atoms are not plotted); (c) AFM image of an 8-hq$_4$ aggregate deposited on the Cu(111) surface. The images show the presence of white bands where intramolecular bonds are expected, together with fuzzy weaker bands where intermolecular O–H···O, O–H···N, C–H···O and C–H···N hydrogen bonds could exist; and (d) Chemical structure of the aggregate shown in (c). The color code of the atoms is the following one: Green = carbon; blue = nitrogen; red = oxygen; white = hydrogen.
From J. Zhang, P. Chen, B. Yuan, W. Ji, Z. Cheng, X. Qiu, Science, 342, 611 (2013). Reprinted with permission from AAAS.

Cu–N coordination bonds, they paid special attention to an isolated adsorbed 8-hq molecule (Figure 2.5a and b), and an adsorbed tetrameric aggregate, 8-hq$_4$ (Figure 2.5c and d). For both adsorbed species, the images show the presence of white bands where intramolecular bonds are expected, together with fuzzy weaker bands where intermolecular interactions are also expected (O–H···O, and O–H···N hydrogen bonds, as well as C–H···O and C–H···N hydrogen bonds). However, the origin of these bond-like features is under heavy debate in the AFM community. Thus, there have been reports indicating that the result is an artifact due to the use of a flexible CO tip,[35] but when a rigid Cu tip with an oxygen apex was used the bond-like features were preserved, suggesting a different cause. Therefore, the interpretation of the bond-like features found in non-contact-AFM is still a subject under debate. Thus, until solid evidence about the mechanism that regulates the formation of these images is obtained, one cannot safely conclude that "bonds have been seen", that is, that bonds are real objects.

Finally, we propose a definition of a bond that accommodates all previously described information: *a bond is one of the dominant stabilizing interactions that connect the components of a chemical substance*. Therefore, according to such definition bonds: (a) are not real objects (they are the strongest stabilizing atom–atom pairs in an atom pair model), (b) provide a realistic and easy to catch picture of the structure of molecules (and supramolecular aggregates) that mimics very well the system observed properties (all energetically dominating stabilizing atom–atom pairs are included, and are displayed topologically), and (c) allow association of changes in the molecular structure with specific bonds and to describe them as a sequence of (bond breaking)–(bond making) processes.[36] Notice that such a bond model assumes that bonds are nearly independent entities. As no part of the previous analysis is specifically associated with the intra or intramolecular character of the pair interaction, there is no *"a priori"* reason to believe that it will not work in supramolecular systems and their intermolecular bonds with an accuracy similar to that achieved in molecules.

2.4 Conventions for Naming Intermolecular Interactions and Bonds

There is some confusion in the literature about the names to be used with interactions and bonds. Trying to avoid such confusion, in this chapter we will stick to the following rules in naming intermolecular interactions and bonds:

(a) The term "intermolecular interaction" will be employed when referring to two molecules interacting with each other, with no indication of the sign of E_{int}, or when $E_{\text{int}} > 0$,

(b) The term "intermolecular bond" will be only employed when referring to a dominant stabilizing interaction, which necessarily has $E_{\text{int}} < 0$,

(c) The term "intermolecular bond" will be not substituted by "non-covalent bond", because not all non-covalent bonds are intermolecular bonds (for instance, ionic and metallic bonds are non-covalent but they are not intermolecular),

(d) metal–ligand bonds (also called coordination bonds) will be considered as a case of *intra*molecular bonds, due to their strong similarity with covalent bonds, clearly *intra*molecular bonds.[37]

According to these rules, one can properly speak of a "van der Waals interaction" (when $E_{\text{int}} > 0$ or can have any sign) or speak of a "van der Waals bonds" (when $E_{\text{int}} < 0$) Similarly, for hydrogen bonds (where the bond character always implies that $E_{\text{int}} < 0$) one could use the term "hydrogen bonded interaction" for those interactions that seem hydrogen bonds but do not satisfy their $E_{\text{int}} < 0$ condition (because the natural option, "hydrogen bond interaction", contains a contradiction in its last two terms). Finally, let's remark that interactions and bonds are energetic concepts, thus it is possible to discuss issues related to the aggregate or molecule stability, which cannot be discussed using only purely geometrical-based concepts, such as *contact* or *hydrogen bond bridge*.

2.5 The Accurate Evaluation of the Properties of Intermolecular Interactions and Bonds Found in Molecular Crystals

In order to obtain results for E_{int} that are accurate and reflect the geometry and environment of the intermolecular interaction in the crystal, it is necessary to carefully carry out the following steps:

(a) define an appropriate model system that describes the interaction of interest properly (the complex geometry is that presented by the complex in the crystal; the model system must also reproduce the external field felt by the dimer in the crystal, as well as possible);

(b) select a monoelectronic basis set of good quality (at least, an aug-cc-pVDZ basis set[38,39]);

(c) select a quantum chemical method capable of evaluating E_{int} with enough accuracy–one can do such selection based on previous studies (always search the literature before starting your project)–or compute the components of E_{int} by running IMPT computations, and determine the strongest one by inspection;

(d) either perform IMPT calculations to determine the nature of the interactions of interest, or alternatively analyze the results and check the IMPT results available in the literature for similar intermolecular interactions.

It is important to keep in mind that selecting a proper model system is not always as easy as it seems at first sight. Selecting an inadequate model system can be the source of serious errors in the results. For instance, if one is interested in determining the stability of two $N(C_6H_5)^+$ cations and selects as a model system only two phenyl rings (thus forgetting the net positive charge that these four rings host), the results are going to be seriously wrong (one concludes that they are stabilizing, while they are destabilizing when the right model system is selected[40]).

2.5.1 Choosing the Proper Quantum Chemical Method and Monoelectronic Basis Set

All quantum chemical methods are different approximate ways of solving the time-independent Schrödinger equation:

$$\hat{H}\Psi = E\Psi, \tag{2.3}$$

an eigenvalue–eigenvector equation that allows computation of the energy (E) and the matter-wavefunction (Ψ; usually called the wavefunction) of hydrogen atoms. Once the Hamiltonian operator ($\hat{H}$) is known, the value of the energy and the wavefunction for the chemical substance can be computed.

Table 2.1 collects the most widely used quantum chemical methods (all of them *ab initio* methods[41]) with some suggestions about when they should be preferred over the other methods included in Table 2.1 (note that the selection of the computational method also depends on the size of the model system, the basis set size, and the type of

Table 2.1 Brief description of the characteristics of the quantum chemical methods most commonly used to compute the interaction energy of intermolecular interactions.

Method	Determinants included	Type of intermolecular interactions described properly
Hartree–Fock	– Valid when the Hartree–Fock determinant (the most stable one) dominates the FullCI expansion → monodeterminantal wavefunctions – Valid when dispersion is not important	– Closed shell fragments where the electrostatic term dominates
CASSCF	– Needed when a small number of determinants have similar weight in the FullCI wavefunction → multideterminantal wavefunctions – Valid when dispersion is not important	– Open shell fragments where the electrostatic term is important
MP2	– Monodeterminantal – Perturbationally accounts for the impact on the energy of the diexcitations in the Hartree–Fock wavefunctions – Includes a large part of the dispersion	– Closed-shell fragments where the electrostatic or the dispersion term are important
CASPT2	– Multideterminantal – Perturbationally accounts for the impact on the energy of the diexcitations in the CASSCF wavefunctions – Includes a large part of the dispersion	– Open-shell fragments where the electrostatic or the dispersion term are important
CCSD(T)	– Monodeterminantal – More accurate than MP2 – Variationally treats all single and double excitations and estimates the impact of the triple excitations perturbationally – Includes most of the dispersion	– Closed-shell fragments where the electrostatic or the dispersion term are important – Better results than MP2
DFT	– Depends on the functional – Doesn't describe dispersion properly	– Fast – In many cases gives reliable results – Closed-shell fragments where the electrostatic or the dispersion term are important
FullCI	– All determinants for a given monoelectronic basis set. – Computationally extremely expensive, so usually only used in calibrations – Includes all dispersion	– Gives the exact results for the monoelectronic basis set used

intermolecular interaction being studied). Here we just list them, grouped by their capability to work with monodeterminantal or multideterminantal wavefunctions (the FullCI method has not been included in this list of available methods, as nowadays it is only possible to carry out FullCI computation on very small molecules and aggregates):

(a) *Most commonly used monodeterminantal methods:*
  (1) Hartree–Fock[42] (HF),
  (2) non-parametric DFT functionals,[43]
  (3) MP2 (Second Order Møller–Plesset),[44] and
  (4) CCSD(T) (single and double-excitations Coupled Cluster method with a perturbative estimation of the contribution of the triples).[45]
(b) *Most commonly used multideterminantal methods:*
  (5) CASSCF[46] (almost no dispersion accounted)
  (6) CASPT2[47] (a large part of the dispersion is accounted at the MP2 level)

Previous studies have shown that the quality of the method (measured as the difference between the predicted and the experimental result) depends on the quantum chemical method chosen (with the same model and monoelectronic basis set). For instance, the properties of $S\cdots S$ interactions[48] are not well described when using the Hartree–Fock (HF) method,[49] but they are well characterized by the second-order Møller–Plesset (MP2) method.[50] In contraposition, the HF method describes well the interaction between two ions, such as K^+ and Cl^-, in crystals. Keep in mind that one has to select a method capable of describing the electronic structure of the molecular aggregate and its dissociated fragments for the intermolecular interaction of interest. For such a task, the information collected in Table 2.1 can be of help.

Why is a monoelectronic basis set needed? Because for the numerical resolution in computers of the analytical equations that allow computation of the energy and wavefunction in each approximate method, these equations must be converted into their matrix representation, a fully equivalent representation. Such conversion is done using a monoelectronic basis set. The physics behind the use of a matrix representation can be visualized by looking at the impact of using the LCAO procedure (LCAO = Linear Combination of Atomic Orbitals). This procedure describes the molecular orbitals of the chemical system as a linear combination of the wavefunctions for each atom of the chemical system, that is, in a basis of atomic orbitals, a very similar problem. The LCAO procedure also allows an understanding of why the atomic orbital basis set has to be truncated: the number of atomic orbitals for any atom is infinite, thus the atomic basis sets have an infinite size, so it is necessary to truncate the monoelectronic basis set to the most stable orbitals.

2.5.2 Computing the Interaction Energy of Intermolecular Interactions and Its Angular Dependence

The interaction energy of an intermolecular interaction, E_{int}, is probably the most relevant of the properties of any intermolecular interaction. Physically, E_{int} measures the amount of energy *gained and lost by the aggregate in stabilizing and destabilizing intermolecular interactions.*

The interaction energy of an intermolecular interaction (eqn (2.4)) is computed by subtracting the total energy for the AB complex, that results from the formation of the new A–B bond, from the sum of the total energy of fragments A and B computed with their own basis:

$$E_{\text{int}}(A,B) = E(AB_{AB}) - E(A_A) - E(B_B). \tag{2.4}$$

In molecular crystals the geometry of AB, A and B is that found in the crystal.

When the interaction energy is computed according to eqn (2.4) the result includes an error called the "**B**asis **S**et **S**uperposition **E**rror" (in short, BSSE, a name that results from picking the first letter of each word).[51] The BSSE error is an artifact that originates from the need to truncate the monoelectronic basis sets. Consequently, when the interacting fragments get closer they increasingly borrow the monoelectronic basis set of the other fragment, a fact that results in an increment in the total energy of the AB complex (as required by the variational principle in the systems where such sharing is possible, but not associated to any physically meaningful change). The BSSE can be corrected using the counterpoise method,[52] where the interaction energy is computed by subtracting the energy of both fragments computed using the dimer monoelectronic basis set:

$$E_{\text{int}}(A,B)_{\text{CP}} = E(AB_{AB}) - E(A_{AB}) - E(B_{AB}). \tag{2.5}$$

Therefore, the BSSE can be computed as

$$\text{BSSE}(A,B) = |E_{\text{int}}(A,B) - E_{\text{int}}(A,B)_{\text{CP}}|. \tag{2.6}$$

Empirically, it has been observed that the BSSE computed at the HF level is much smaller than that obtained at the MP2 and CCSD(T) levels, with that obtained at the DFT level the smallest one.[53] The

BSSE decreases as the basis set quality increases, becoming negligible for very extended basis sets.[54]

As shown in Figure 2.2, the molecular complex interaction energy changes with the geometry of the complex (particularly, the relative orientation of the two molecules that form the complex, measured by the angle θ in Figure 2.2, and the intermolecular separation, r). When the E_{int} *vs.* θ curves are plotted, one is mapping the angular dependence of the intermolecular interaction.

2.5.3 Computational Tools to Determine Which Atoms Get Involved in a Bond

Using Bader's AIM methodology,[20] the bond critical point in the total density of the molecule can be found and, depending on its properties, can indicate the presence of a bond. As this technique is discussed in detail elsewhere in this book (Chapters 4, 5, 6 and 18), the interested reader is directed to these chapters and we will not elaborate further on it.

2.5.4 Computational Tools to Analyze the Nature of the Intermolecular Interactions from a Rigorous Quantum Chemical Perspective

A quantum chemical study of the classes of intermolecular interactions allows identification of the similarities among members of the same class, and the differences with members of other classes. It also calls our attention towards the origin of the classes known up to now.

Probably the physically most relevant property of any intermolecular interaction is the sign and size of its E_{int}. Quantum Chemistry has developed the tools that allow the quantitative analysis of the sign and size of the interaction energy for the interaction of two closed-shell molecules, the IMPT (acronym of **Inter Molecular Perturbation Theory**) method[55] and the SAPT[56] (acronym of **Symmetry Adapted Perturbation Theory**) methods. The two are perturbative methods and give similar results. The validity of the IMPT decomposition is demonstrated by comparing the sum of all IMPT energetic components with the MP2 interaction energy. So, except when otherwise indicated from now on, our attention will be focused on the results of the energy decomposition analysis more than on the fact that these are IMPT results.

According to the IMPT method, the interaction energy between two closed-shell molecules is equal to the sum of the following five energetic components:

$$E_{\text{int}} = E_{\text{er}} + E_{\text{el}} + E_{\text{p}} + E_{\text{ct}} + E_{\text{disp}}, \tag{2.7}$$

where each of these components have the following physical meaning:

(1) E_{er} is the *exchange–repulsion component*, a combination of the Pauli repulsion that two electrons experience when forced in the same region of the space, and the attractive exchange component.

$$E_{\text{er}} \approx K \langle \boldsymbol{X}_{\text{A}} | \boldsymbol{X}_{\text{B}} \rangle \approx \exp(-r_{ij}) \tag{2.8}$$

The exchange-repulsion component is always repulsive (Table 2.2), growing exponentially as the distance get smaller (this is the reason for the existence of a component responsible for the existence of the repulsive wall);

(2) E_{el} is the *electrostatic component*, associated with the electrostatic interaction of two non-polarized molecules (having the same electronic distribution as in their isolated molecules);

(3) E_{p} is the *polarization component*, which describes the change in the electrostatic interaction component due to the fragments' polarization when they are close enough. In classical terms, the polarization of each fragment is proportional to its polarizability. The sum of the electrostatic and polarization components results in the *total electrostatic contribution* to the interaction energy.

(4) E_{ct} is the *charge-transfer component*, associated with transfer of electronic charge from one interacting molecule to the other;

(5) E_{disp} is the *dispersion component*, a non-classical term whose existence is due to the correlated motions of the electrons. This component originates from quantum effects that have no exact classical translation. Its closest classical analogy is the interaction between the instantaneous dipole moments of the two molecules, each of these dipoles induced in one fragment by the electron motions in the other fragment.

(6) When the two interacting fragments are radicals, they host unpaired electrons in one or more of the SOMOs (SOMO = Single Occupied Molecular Orbitals). If the SOMOs of one fragment overlap with those of the other fragment and a double occupied bonding component is obtained, with an energy gain due to the pairing of the electrons. This energy is called the bonding component (E_{bond}).

Table 2.2 Values of the interaction energy calculated at the MP2/6-31G(d) level (E_{MP2}, in kcal per mol, BSSE-corrected results) and of the components of the interaction energy (E_{el}, E_{er}, E_p, E_{ct} and E_{disp}, all in kcal per mol) calculated at the IMPT level using the same basis set, for the 15 molecular complexes indicated in the first column (the complexes have been grouped by classes of intermolecular interactions. The quality of the IMPT energy decomposition can be demonstrated by comparing the addition of the IMPT components (E_{tot}, in kcal per mol) with the E_{MP2} interaction energy. The geometry of each complex was previously optimized at the MP2/6-31G(d) level (the value of the optimum intermolecular distance is also collected (r_{opt}, in Å). The dominant IMPT component computed for each complex is also highlighted.

Complexes	r_{opt}	E_{el}	E_{er}	E_p	E_{ct}	E_{disp}	E_{tot}	E_{MP2}
Ionic bonds								
$Na^+\cdots Cl^-$	2.412	−142.3	25.4	−10.5	−1.4	−35.5	−164.3	−128.0
Hydrogen bonds								
(a) Very strong								
$NH_3CH_2COOH^+\cdots SO_4^{2-}$	1.520	−155.1	31.4	−14.4	−6.2	−12.4	−156.6	−160.0
(b) Strong								
BAHLEC[a]	1.473	−36.3	92.2	−10.3	−5.7	−14.8	−86.7	−80.4
$H_3O^+\cdots H_2O$	1.202	−45.5	54.3	−27.4	−9.6	−31.7	−59.9	−31.8
$H_2O\cdots F^-$	1.415	−38.3	30.6	−10.1	−4.3	−8.5	−30.6	−25.3
(c) Moderate								
$FH\cdots H_2O$	1.716	−16.3	12.2	−1.0	−0.6	−3.2	−8.9	−7.9
$CH_4\cdots F^-$	1.873	−8.3	13.7	−6.7	−1.4	−4.0	−6.8	−5.8
$H_2O\cdots H_2O$	1.990	−6.7	5.0	−0.7	−0.5	−2.0	−3.9	−4.2
(d) Weak								
$C_2H_2\cdots H_2O$	2.174	−4.1	2.6	−0.4	−0.2	−1.4	−3.5	−2.7
$CH_4\cdots H_2O$	2.553	−0.9	1.5	−0.2	−0.1	−0.9	−0.6	−0.4
$FH\cdots Ar$	2.634	−0.1	0.5	−0.2	−0.1	−0.4	−0.3	−0.4
$CH_4\cdots Ar$	3.016	−0.5	0.1	0.0	−0.03	−0.6	−0.2	−0.1
van der Waals bonds								
$C_6H_6\cdots C_6H_6$	3.800[b]	1.2	2.7	−0.2	−0.2	−5.2	−1.7	−1.8
$CO_2\cdots CO_2$	3.058	−1.7	1.9	−0.2	−0.1	−1.8	−1.8	−1.0
$Ar\cdots Ar$	3.842	−0.04	0.15	0.00	0.00	−0.28	−0.16	−0.16

[a] $(CH_3)_3N-CH_2-COOH^+\cdots HOOC-(CH_2)_2-COO^-$.
[b] Obtained by doing a partial geometry optimization (freezing the fragments geometry).

An IMPT interaction energy decomposition was carried out for the 15 complexes collected in Table 2.2, which present van der Waals, hydrogen bonded (split by types), and ionic interactions (in fact, they are bonds, as their $E_{int} < 0$). The IMPT decomposition was done on the fully optimized MP2/6-31 + G(d) geometry of these complexes. The analysis of the IMPT energy decomposition results allows the following conclusions to be reached:

(1) The IMPT method reproduces very well the sign and order of magnitude of the MP2 interaction energies for all types of intermolecular interactions,

(2) The energy decomposition gives similar trends for all complexes of the same subclass, a fact that gives a physical meaning to the formation of these groups, and allows the easy classification of any new complex by looking at the largest component in the IMPT energy decomposition analysis,

(3) Except in complexes where both fragments have a high symmetry, the dominant components in closed-shell molecules are E_{el} and E_{disp} (in radical fragments E_{bond} is also important). None of these components becomes close to zero, but its value becomes negligible compared with the other component due to its increment in size,

(4) In general, E_{el} or E_{disp} are both non-negligible, one component dominating over the other,

(5) In all van der Waals complexes $E_{el} \ll E_{disp}$ and E_{disp} is the only dominating energy component,

(6) In weak hydrogen bonds $E_{el} \approx E_{disp}$,

(7) In moderate and strong hydrogen bonds $E_{el} > E_{disp}$ or even $E_{el} \gg E_{disp}$,

(8) In ionic bonds $E_{el} \gg E_{disp}$.

2.6   Types of Intermolecular Interactions

The continued study of the properties of the intermolecular interactions has allowed identification of different families of intermolecular interactions. This variety was clear from the initial studies, and allows a historical classification of the intermolecular interactions, which still nowadays prevails over more physically rooted classifications. Consequently, hereafter the most relevant types of intermolecular interactions will be introduced in the same order in which they were discovered. The most relevant properties of each type of interaction currently known will be briefly described. Finally, an analysis of the components of the interaction energy will allow a quantitative analysis of all types, which will allow the search for trends, both between and within the types of intermolecular interactions.

2.6.1 A Historical Look at Intermolecular Interactions

Although a more physical classification is possible by looking at the dominant components of the interaction energy, we have finally chosen to present them from a historical perspective, as we concluded that this was more informative to the non-specialist reader.

2.6.1.1 *van der Waals Interactions: The Oldest Type of Intermolecular Interactions*

Currently, the term "van der Waals interactions" is reserved for intermolecular interactions produced between two closed-shell molecules that have no net charge or permanent multipoles. In these systems, $E_{el} = 0$ and the only source of stabilizing interactions is E_{disp}.

The first report of the existence of attractive intermolecular interactions between molecules was in 1873 by van der Waals,[57] in his Ph.D. Thesis[58] (for all his contributions on the study of intermolecular interactions, he won the Nobel Prize in Physics in 1910[59]). van der Waals interactions are also known as London interactions, or as dispersive interactions,[60] so called because the dispersion is the strongest component in the interaction energy.

The origin of the stabilizing character of the intermolecular interactions in rare-gas atoms was not understood until 1930 when London, applying second-order perturbation theory, demonstrated that a net attraction existed in dimers of these atoms, where the zero charge, dipole, quadruple and higher multipole moments make negligible any term of the classical electrostatic interaction energy.[61] Such a conclusion was later extended to a pair of stable molecules (Pauling defined[62] a "stable molecule" as a molecule having a closed-shell electronic structure and no net charge or permanent multipoles. Such criteria are satisfied by $H_2 \cdots H_2$ and $Cl_2 \cdots Cl_2$ dimers, but they also apply to any long-living open-shell radical). The energetic stability in $Ar \cdots Ar$, $H_2 \cdots H_2$ and $Cl_2 \cdots Cl_2$ dimers was shown to originate in the instantaneous-dipole generated in each interacting fragment by the electron motions of the other. It can be computed from the difference correlation energy of the complex and its dissociated fragments.

According to the computed E_{int} values (last column in Table 2.2, BSSE-corrected results), van der Waals interactions are weaker than any other type of intermolecular interaction except weak hydrogen bonds, which present a similar interaction energy range of values. However, they increase when the size of the fragments allows the simultaneous overlap of many regions of electron localization. Consequently, values as large as -21 kcal mol^{-1} have been reported for the interaction energy of some van der Waals complexes,[63] and many more where the interaction energy of the complex is about -10 kcal mol^{-1}.[64] There seems to exist a dependence of the complex interaction energy with the number of lone pairs presenting short-distance interactions.

2.6.1.2 Hydrogen Bonds

Pauling defined hydrogen bonds in this form: "under certain conditions an atom of hydrogen is attracted by rather strong forces to two atoms, instead of only one, so that it may be considered as acting as a bond between them. This is called the hydrogen bond". Therefore, he considered an X–H$\cdots$Y interaction where H is bonded to X and Y as a hydrogen bond.

Although hydrogen bonds are nowadays an important field of research, their relevance in Chemistry was recognized very slowly (see Figure 2.6). Just a few years after van der Waals reported his work on dimers of rare-gas atoms, Werner published in 1903 the structure of a

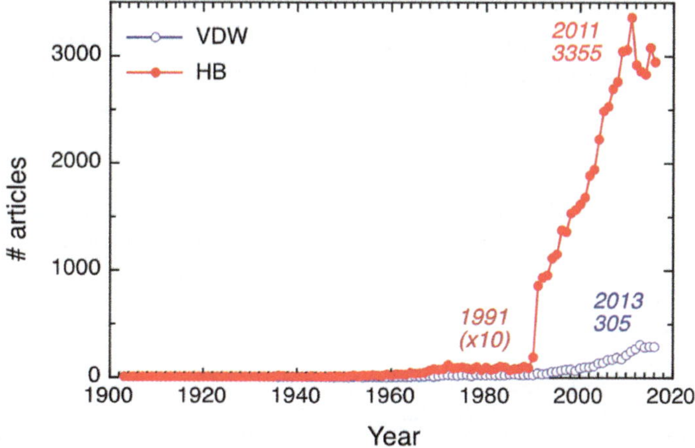

Figure 2.6 Distribution per year, since 1903, of the number of articles indexed in the Web of Science under the keyword "van der Waals" and "hydrogen bonds".

coordination complex,[65] where he proposed the presence of a hydrogen bond. Few years later, in 1912, Moore and Winmill proposed the presence of a hydrogen bond in non-dissociated trimethyl ammoniumhydroxide.[66] But the work that really attracted the chemists' attention to hydrogen bonds was an article published by Latimer and Rodebush in 1920, where they proposed that hydrogen bonds were the reason for the association of water molecules in liquid water, and thus are the origin of the special chemical and physical properties of water.[67]

The study of the nature of the hydrogen bond has been more challenging than that of the ionic, covalent and van der Waals bonds. From their discovery in 1903 up to about 1997, all hydrogen bonds were believed to share the same nature independent of their strength, the general agreement being that their nature was mostly electrostatic but with a non-negligible covalent character. However, probably influenced by the results on the nature of the C–H$\cdots$O hydrogen bonds,[68–71] Jeffrey (in his book "An introduction to hydrogen bonding," published in 1997[72]) proposed to subdivide the hydrogen bond class into three subclasses, according to their range of interaction energies: *weak*, *moderate* and *strong* (see Table 2.3). He based such subdivision on the very wide range of interaction energies that hydrogen bonds presented. This subdivision was quickly accepted by the community, sometimes slightly modified.[83,85] Nowadays, besides the three subclasses proposed by Jeffrey, other hydrogen bond (HB) subclasses have been reported, some based on the electronic structure of its members (charge-assisted HB,[83] resonance-assisted HB,[83] ionic HB[73]), or based on other properties (low-barrier HB,[74] for instance). Note that the members of these classes can be simultaneously classified within any of the three classes proposed by Jeffrey.

Table 2.3 Types of hydrogen bonds according to Jeffrey.

Property	Type of hydrogen bond		
	Weak HB	Moderate HB	Strong HB
Interaction energy (kcal per mol)	<-4	-4 to -15	-15 to -40
Nature, according to Jeffrey	Electrostatic	Mostly electrostatic	Mostly covalent
H$\cdots$Y (Å)	2.2–3.2	1.5–2.2	1.2–1.5
<X–H$\cdots$Y (°)	90–150	130–180	175–180
Examples	C–H$\cdots$O	O–H$\cdots$O	Salts
	Weak acids	Acids	
	Weak bases	Alcohols	

Table 2.2 collects (last column) the interaction energy for a few representative members of the weak, moderate and strong subsets of hydrogen bonds (all hydrogen bonded complexes included in this table present a X–H$\cdots$Y interaction that satisfies the distance and angularity criteria for hydrogen bonds, a covalent X–H bond, and a stabilizing H$\cdots$Y interaction). Their analysis of the classes of HB complexes considered by Jeffrey shows that: (a) the interaction energy of the hydrogen bonded complexes (E_{MP2}) span a much wider range of energies (-0.2 to -80 kcal mol^{-1}) than the van der Waals complexes of Table 2.2 (-0.2 to -2 kcal mol^{-1}, although a value of -21 kcal mol^{-1} has been reported for a large size molecule), (b) the weakest complexes in the moderate class and the strongest one in the weak class have very similar strengths, thus making a classification solely based on the size of E_{int} difficult, (c) the common belief that the strongest possible hydrogen bond is that found in the F–H$\cdots$F$^-$ complex is wrong (its bond dissociation energy in the gas phase is -45.8 ± 1.6 kcal mol^{-1} [75]), and (d) all members of the strong class of hydrogen bonds are ion–molecule complexes.[71] As will be shown below, the classification is less unambiguous when the nature of the hydrogen bond is also considered, that is, after looking at the components of the interaction energy (Table 2.2).

All the previous changes suggested the necessity of re-evaluating the IUPAC definition of what a hydrogen bond was. The result is the IUPAC 2011 definition of a hydrogen bond:[76] "The hydrogen bond is an attractive interaction between a hydrogen atom from a molecule or a molecular fragment X–H in which X is more electronegative than H, and an atom or a group of atoms in the same or a different molecule, in which there is evidence of bond formation". Essentially, it states that a hydrogen bond is a X–H$\cdots$Y interaction where evidence of a bond exists between the H(X–H) and the Y atoms. Then, in the article Supplementary Information, one finds six criteria that were considered as a proof of the existence of a H$\cdots$Y bond, some geometrical, others spectroscopic, and the remaining ones energetic.

A wide variety of effects have been shown to appear due to the formation of hydrogen bonds.[77] Among them, are the following: (a) an increase in the boiling and melting points, (b) an increase in the dielectric constant, (c) a shift in the position of the ν_{AH} vibration observed in IR and Raman spectra, and (d) a shift in the position and relaxation times in the peaks of the NMR spectra. A revision of the early work can be found in the book of Pimentel and McClellan.[77] A modern revision of recent effects can be found in ref. 78

Finally, a measure of the current and past interest in hydrogen bonds can be obtained by comparing the number of articles published in 2016 and during the period 1903–2016 (both years included), indexed in the Web of Science database under the keyword entries "hydrogen bond" and "van der Waals interaction". 3596 articles were indexed under the keyword entry "hydrogen bond", totalling 56848 articles in the period 1903–2016. On the other hand, using the keyword entry "van der Waals interaction": 286 articles in 2016, while 4187 articles were stored during the 1873–2016 period (Figure 2.6 shows the yearly distribution of articles published from 1900 to 2016 with each of these keywords). These numbers speak for themselves about the prominent role played by hydrogen bonds in Crystal Engineering[79–83] and Biochemistry.[84]

2.6.1.3 Ionic Interactions

Although the IUPAC Gold Book only defines the term "ionic bond" ("In strict terms, an ionic bond refers to the electrostatic attraction experienced between the electric charges of a cation and an anion"), from it one can infer that the term ionic interactions includes all electrostatic interactions that electrically charged species can generate of any sign among themselves. Ionic bonds are produced whenever the $\Delta G < 0$ for the reaction:

$$A + C \rightarrow A^{n-} + C^{n+} \tag{2.9}$$

is negative. Such a fact is determined by the difference between the electron affinity of A and the ionization potential of C. When C or A are neutral molecules, the ionic bond is intermolecular (as in potassium oxalate, for instance).

Ionic bonds were already considered by Pauling as a case of possible electrostatic bonds. Its prototypical example is the $Na^+ \cdots Cl^-$ bond produced in the NaCl crystal, which Pauling describes as follows: "We describe the interactions in this crystal by saying that each ion forms ionic bonds with its six neighbors, these bonds combining all of the ions in the crystal into one giant molecule".[85] Pauling also mentions the existence of ion–dipole $(K^+ \cdots H_2O)$, ion–induced dipole $(F–H \cdots Ar)$, and dipole–dipole bonds (water$\cdots$water interactions).

Ionic bonds are the strongest of all intermolecular bonds (see Table 2.2, last column), well above the values of strong hydrogen bonds. Such a fact can be justified by realizing that charge$\cdots$charge interactions are stronger than charge-(neutral molecule) interactions.[4]

2.6.1.4   Ionic $O–H^{(+m)}\cdots O^{(-n)}$ Interactions. New Thoughts on Hydrogen Bonded Interactions

What happens with the destabilizing $X–H\cdots Y$ interactions? Implicit in Pauling's or the 2011 IUPAC definition is the idea that not all short distance $X–H\cdots Y$ interactions are hydrogen bonds. One example is found in some salts of organic acids, where short distance $X–H\cdots Y$ interactions are produced because this is the least repulsive of all orientations of the two like-charged anions.[86–90] These short-distance $X–H\cdots Y$ interactions cannot be a bond, therefore cannot be a hydrogen bond, because the $(-1)\cdots(-1)$ charge$\cdots$charge term generated by the two anions overcomes the dipole$\cdots$dipole term and the anion$\cdots$anion interaction and is destabilizing, thus discarding the possibility that the $O–H^{(-1)}\cdots O^{(-1)}$ interactions produced between these anions could be hydrogen bonds. They are $O–H^{(-1)}\cdots O^{(-1)}$ interactions.

At the opposite end of the energy scale are the $O–H^{(+m)}\cdots O^{(-n)}$ interactions. They have not appeared in any classification of hydrogen bonds despite showing the common properties of all $X–H\cdots Y$ hydrogen bonds. Why? A possible reason could be that the dominant energetic contribution in $O–H^{(+m)}\cdots O^{(-n)}$ interactions is expected to come from the charge–charge electrostatic component rather than from the dipole–dipole term, but there is nothing in the IUPAC 2011 definition of a hydrogen bond that requires such a fact. We will postpone a final conclusion until the methodology to perform a rigorous analysis of the nature of all types of intermolecular interactions is described below. But we can advance that $O–H^{(+m)}\cdots O^{(-n)}$ interactions belong to a new subclass of hydrogen bonds called the very strong hydrogen bonds class.

2.6.1.5   New Types of Intermolecular Interactions

2.6.1.5.a   Halogen Bonds ($D–X\cdots A$, $X = F$, Cl, Br, I)

In 1954, Hassel presented crystallographic evidence of the existence of halogen bonds in the Br_2:1,4-dioxane cocrystal,[91] where $Br\cdots O$ stabilizing interactions appeared at $r_{O\cdots Br} = 2.71$ Å, angle $<Br–Br\cdots O \approx 180°$, the Br atom pointing towards the lone pair of the oxygen atom. The properties of this crystal were explained by Hassel in terms of a donor–acceptor complex.[92] Only after 1978 did the $Br\cdots O$ interaction receive the name of "halogen bond" (represented as $D–X\cdots A$ $X = F$, Cl, Br, and I, see Table 2.4).

Table 2.4 SAPT interaction energy components for the indicated aggregates (at their optimum MP2/aug-cc-pVDZ geometries). The monomer dipole moment is also given, together with E_{tot} (the sum of the energy components) and E_{MP2} (the interaction energy computed at the MP2 level). Distances are given in Å, angles in degrees, energies in kcal per mol and μ in Debyes. The most stable orientation of each $CH_3X\cdots CH_3X$ dimer is marked in italics.

Dimer	r	θ_1	θ_2	μ	E_{el}	E_{ex}	E_{ind}	E_{disp}	E_{tot}	E_{MP2}
$CF_4\cdots CF_4$	3.00	180.0	180.0	0.0	−0.09	0.44	−0.04	−0.48	−0.17	−0.23
$CCl_4\cdots CCl_4$	3.40	180.0	180.0	0.0	0.19	1.32	−0.28	−1.78	−0.54	−0.49
$CBr_4\cdots CBr_4$	3.47	180.0	180.0	0.0						−0.41
$CI_4\cdots CI_4$	3.65	180.0	180.0	0.0	0.96	4.05	−1.97	−4.57	−1.49	−1.42
$CH_3F\cdots CH_3F$	3.00	180.0	180.0	1.96	1.07	0.13	−0.08	−0.27	0.86	0.86
$CH_3F\cdots CH_3F$	3.12	180.0	90.0		0.33	0.48	−0.12	−0.61	0.08	0.10
$CH_3F\cdots CH_3F$	*3.00*	*90.0*	*90.0*		*−1.41*	*1.51*	*−0.26*	*−1.19*	*−1.34*	*−1.30*
$CH_3Cl\cdots CH_3Cl$	3.55	180.0	180.0	1.99	0.17	0.94	−0.13	−1.23	−0.25	−0.22
$CH_3Cl\cdots CH_3Cl$	3.47	180.0	90.0		−0.99	2.40	−0.33	−1.86	−0.77	−0.73
$CH_3Cl\cdots CH_3Cl$	*3.78*	*90.0*	*90.0*		*−0.66*	*1.54*	*−0.21*	*−1.54*	*−0.87*	*−0.82*
$CH_3Br\cdots CH_3Br$	3.60	180.0	180.0	1.92	−0.11	1.97	−0.29	−1.91	−0.34	−0.32
$CH_3Br\cdots CH_3Br$	*3.56*	*180.0*	*90.0*		*−2.29*	*4.57*	*−0.66*	*−2.63*	*−1.01*	*−1.02*
$CH_3Br\cdots CH_3Br$	—	90.0	90.0		−0.75	2.21	−0.25	−1.79	−0.58	−0.55
$CH_3I\cdots CH_3I^a$	3.87	180.0	180.0	1.71	0.22	2.34	−0.66	−2.92	−1.02	−0.95
$CH_3I\cdots CH_3I^a$	*3.82*	*180.0*	*90.0*		*−2.00*	*5.37*	*−1.56*	*−3.97*	*−2.16*	*−2.09*
$CH_3I\cdots CH_3I^a$	4.45	90.0	90.0		−0.25	1.74	−0.29	−2.02	−0.82	−0.78

[a]The SDB-aug-cc-pVTZ basis was used on the iodine atom, which uses pseudopotentials for the inner core electrons.

The properties of halogen bonds have been investigated in detail, both experimentally[93] and theoretically.[94] Remarkably, they present similar properties to hydrogen bonds. This parallelism suggests that hydrogen bonds would be better modeled as donor–acceptor complexes, the H(D–H) atom pointing towards the lone pair on A.

The interaction energy for the C–Cl$\cdots$N halogen bond found in the $CF_3Cl\cdot NH_3$ complex is -3.2 kcal mol^{-1}, in the range of values for the strongest weak hydrogen bonds. For all his work on halogen bonded complexes Hassel received a Chemistry Nobel Prize in 1968.

2.6.1.5.b Dihalogen Bonds (X$\cdots$X, X = F, Cl, Br, and I)

Halogen$\cdots$halogen is another well studied type of intermolecular interaction. An analysis[95] of the crystals deposited in the Cambridge Structural Database (CSD) showed the presence of C–X$\cdots$X–C contacts shorter than the sum of the van der Waals radius for X = F, Cl, Br and

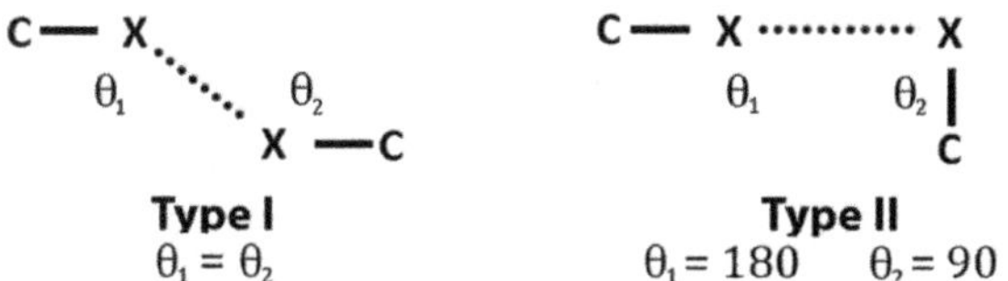

Figure 2.7 Preferred orientations of the short-distance halogen···halogen contacts found for the crystals deposited in the CSD.

I, and their preference for the two orientations shown in Figure 2.7. Dihalogen interactions are a class of van der Waals interactions, since the main driving force behind their attractive nature is expected to originate from the lone-pair···lone-pair interactions, although with some peculiarities, as the electrostatic component is strikingly important.

Previous studies have demonstrated the stabilizing and anisotropic nature of the halogen···halogen (X···X) interactions.[96] Such angular preference can be explained[96,97] by a model in which there are electron-deficient and electron-rich regions of electronic density distribution around each halogen atom that generate bonding attractions between adjacent halogen atoms along some preferred orientations.

MP2 computations of the interaction energy were done on CX_4···CX_4 dimers and on CH_3X···CH_3X dimers, Table 2.4. They confirm the stabilizing energetic character of purely dispersive halogen···halogen interactions (the dipole moment is zero in both interacting fragments due to symmetry, a fact that switches off the electrostatic component). The most stable dimer is CX_4···CX_4 where X = I, whose interaction energy is -1.42 kcal mol^{-1} (see Table 2.4). The study of stability was also done for CH_3X···CH_3X (X = F, Cl, Br and I) dimers. They all present at least one orientation where they are energetically stable (Table 2.4), in good agreement with previous studies.[98,99]

Overall, against previous models obtained from crystallographic analysis, the character of the halogen···halogen interactions should be described as a mixture of electrostatic plus dispersion, where the size of the dispersion component increases as the halogen atomic number increases (Table 2.4). The strongest X···X interaction energy is -2.09 kcal mol^{-1}, computed for the CH_3I···CH_3I dimer (Table 2.4).

2.6.1.5.c π–π Intermolecular Interactions

π–π intermolecular interactions, also called π–stack interactions, are the intermolecular interactions produced when planar molecules that have doubly-occupied π-orbitals are piled up with their molecular planes parallel (or nearly so) to each other, thus allowing the overlap of their doubly-occupied π-orbitals. By definition, π-orbitals change their sign when projected against the plane of the molecule, and have their maximum density above and below the plane, symmetrically related by the plane (they behave as do p_π-orbitals, the simplest example of a π-orbital).

Although the on-top geometrical disposition of the fragment could suggest that the π–π intermolecular interactions are different than any of the previously described intermolecular interactions, in fact they have the same electronic structure as van der Waals interactions, if one just substitutes the overlap of lone pairs by the overlap of doubly-occupied π-orbitals. Their van der Waals nature is consistent with the results of the energy decomposition (see Table 2.2, where the result of the energy decomposition for $Ar\cdots Ar$ and $CO_2\cdots CO_2$, two van der Waals complexes, can be compared with those for the PD conformation of the dimer of benzene (see Figure 2.8a; this conformation is a minimum on the potential energy surface of the dimer of benzene, besides presenting π–π interactions). The results for the PD conformation of the benzene dimer indicate that the dispersion component is the strongest of all components, in fact, the only stabilizing one. Note also that the destabilizing character of the electrostatic component questions the validity of the quadrupole model of π–π interactions,[100] as this model would require a stabilizing character for the electrostatic component.

As shown in Figure 2.8b, the S conformation can transform to the PD conformation without any barrier, while the PD$\leftrightarrow$T transformation

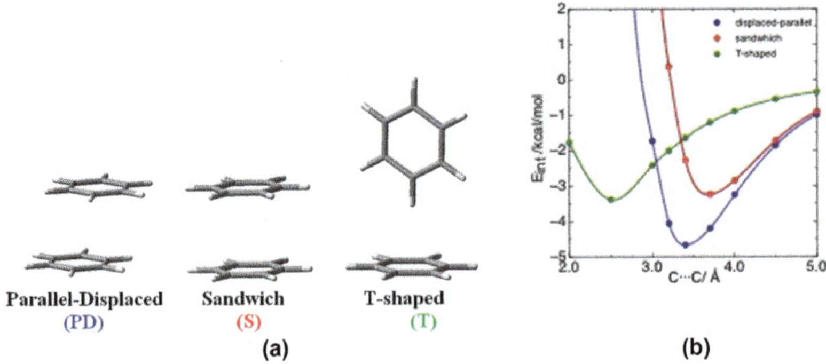

Figure 2.8 (a): Parallel-displaced, Sandwich and T-shaped conformations of the benzene dimer; (b): Potential energy curve computed for the previous three conformations of the benzene$\cdots$benzene dimer (the curves were computed by varying the distance between the two benzene molecules, allowing the rest of the dimer to relax; all calculations of interactions were done at the MP2/aug-cc-pVTZ level, without correcting the BSSE).

requires crossing a barrier, thus suggesting that they should be two separate minima. Extensive SAPT(DFT) calculations[101] (using a triple-zeta quality monoelectronic basis set, to which bond functions were added) have found a minimum for the PD conformation and for a slightly distorted T conformation (with interaction energies of -2.74 and -2.77 kcal mol^{-1}, respectively; there is no BSSE in SAPT computations). The interaction energy of the PD conformation in Table 2.2 is -1.8 kcal mol^{-1}, not far from the CCSD(T)/CBS[102] result of -2.74 kcal mol^{-1}.[103] CCSD(T)/CBS calculations also indicate that T benzene dimers are just slightly more stable than PD benzene dimers (the interaction energy for the T conformer computed by some authors[102] at the CCSD(T)/CBS level is 0.01 kcal mol^{-1} more stable than the PD conformer, while other authors[104] report that T is 0.09 kcal mol^{-1} more stable, using the same computational level). The experimental estimate for the T conformer is -2.1 ± 0.2 kcal mol^{-1}.[105] At this point, one should be aware of the fact that benzene dimers are a particular case of π–π interactions and, as it has been computationally shown, the strength of the π–π interactions increases with the number of overlapping doubly-occupied π electrons in the complex. In fact, interaction energies as large as -21 kcal mol^{-1} have been computed.[63]

Let us also mention that, although the PD and S conformations are commonly used model systems in the study of π–π intermolecular interactions, none of them are detected in the Pbca polymorph of benzene for interfragment distances ($r_{C\cdots C}$) smaller than 5 Å. (see Figure 2.9). The benzene Pbca structure presents two unique planes in its unit cell, one shifted half a unit cell along the *a*- and *b*-axis with respect to the other. Each plane contains only T-shaped benzene dimers (each benzene molecule in this plane acts two times as a hydrogen bond donor and two times as a hydrogen bond acceptor). Adjacent planes are linked by C–H$\cdots$C interactions (indicated by red broken lines in Figure 2.9, while black broken lines have been used for the C$\cdots$C interactions). Benzene dimers with a PD or S orientation are too far away ($r_{C\cdots C} > 5$ Å) to make any sizable overlap between their π orbitals.

2.6.1.5.d Cation–π and Anion–π Intermolecular Interactions

Thanks to theoretical and experimental studies in the gas-phase, in solution, and in the solid state, cation–π interactions, first detected in 1981,[106] play a major role among the relevant intermolecular interactions, particularly those appearing in some biochemical processes and in asymmetric catalysis, where they stabilize some transition states over others that are also possible. The prototypical complex presenting cation–π interactions is the $K_s\cdots C_6H_6$ complex in C_{6v} symmetry. The cation can be any of the alkali metal or quaternary

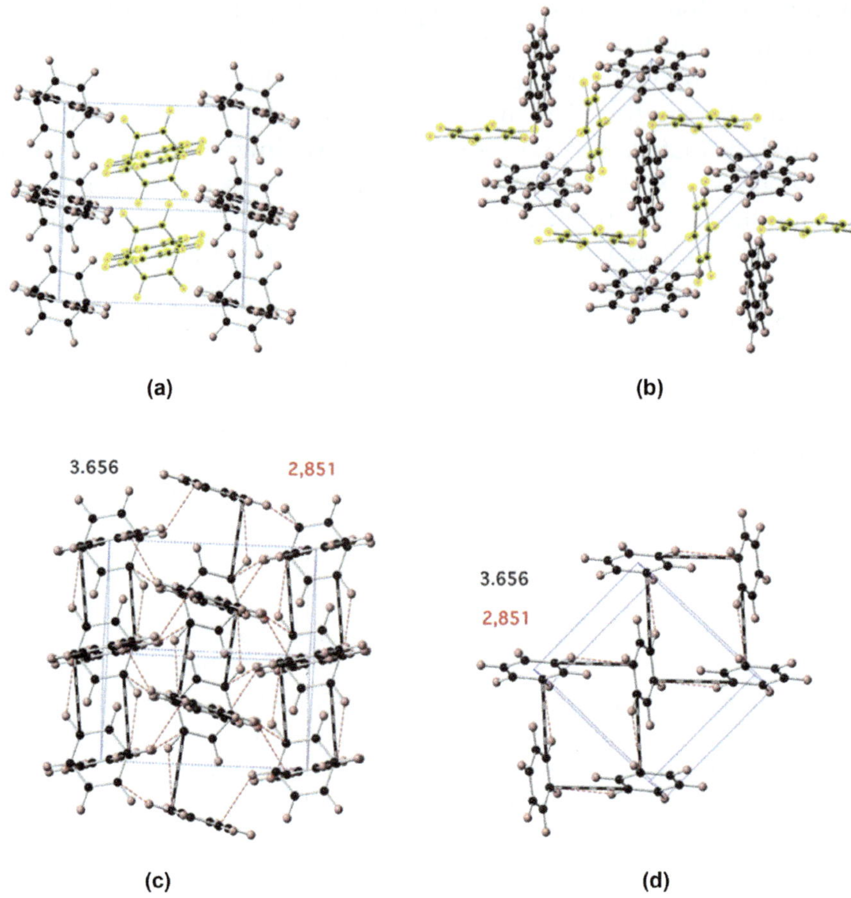

Figure 2.9   (a) View of the Pbca polymorph of benzene along the *c*-axis; (b) View of one plane of two of the same crystal along the *a*-axis (the middle plane is indicated by turning all its atoms yellow, for their easy identification; (c) and (d) are equivalent to (a) and (b) but indicating the shortest C···C and H···C interactions (all of them are *σ* interactions).

ammonium cations, but not cations of transition metals, where the interaction with benzene is of a different nature.

The strength of cation–π interactions has been shown to have values from about -5 to $-50\ \text{kcal}\,\text{mol}^{-1}$,[107] that is, similar to that reported above for strong hydrogen bonds. This suggests the predominant character of cation–π interactions, typical of a charge···(neutral molecule) interaction. A SAPT computation of the components of the interaction energy in K^{+}···benzene confirmed that the electrostatic model is adequate for this complex, as the electrostatic and induction components were both stabilizing and their sum, which is equal to the real electrostatic energy, is stronger than the destabilizing dispersion component ($E_{int} = -17.2$, $E_{er} = 11.8$, $E_{el} = -11.9$, $E_{ind} = -12.8$; $E_{disp} = 11.8$, all values in $kcal\,mol^{-1}$).[108] However, for the PD conformation of the benzene–pyridinium complex (whose interaction energy at the CCSD(T)/CBS level is $-8.34\ kcal\,mol^{-1}$) the dispersion component is slightly stronger than the real electrostatic energy ($E_{el} + E_{ind} = -6.65$; $E_{int} = -8.34$, $E_{er} = 6.02$, $E_{el} = -3.50$, $E_{ind} = -3.15$; $E_{disp} \approx -7.71$, all values in kcal per mol).[109] The optimum distance between K^+ and the centroid of the benzene molecule, along the C_6 axis, was found to be 2.803 Å.

The information on anion$\cdots\pi$-interactions is more limited than on cation$\cdots\pi$ interactions. Anion$\cdots\pi$ interactions[110] were first theoretically proposed in 2002,[111] and then reported in a crystal structure in 2004.[112] There is also evidence that these interactions are produced in solution.[113,114] However, more studies are still required to obtain for anion$\cdots\pi$ interactions the same level of understanding currently available for cation$\cdots\pi$ interactions.

2.6.1.5.e   Recently Proposed Intermolecular Interactions

Nowadays, the list of accepted intermolecular bonds includes hydrogen bonds (A–H$\cdots$D,), halogen bonds (A–X$\cdots$D, X = halogen atom) and dihalogen bonds (X$\cdots$X, X = halogen atom). The addition of the halogen and dihalogen classes is particularly relevant because it is a demonstration that it is possible to make X$\cdots$Y intermolecular bonds between two electronegative atoms (in halogen bonds, X is a halogen atom, which has electronegative properties, while Y is an electronegative atom. As shown in Table 2.5, this has prompted the quest for new types of bonds by changing the halogen atom to a chalcogen, pnicogen, or tetrel atom (C being separately considered in the carbon intermolecular bond). On top of this, lithium and beryllium bonds have recently been added. Readers interested in these new types of intermolecular bonds are directed to Chapter 15.

2.7   Crystal Packing Analysis using Quantum Chemical Information about Intermolecular Interactions

2.7.1   Basic Considerations

It is a procedure aimed at understanding how the structure of a crystal is related to its intermolecular interactions using quantum

Table 2.5　Recently proposed types of intermolecular bond. For their easy comparison, the characteristic values for hydrogen bonds, halogen bonds, and dihalogen bonds are also given.

Type of intermolecular bond	Symbol	Example
Hydrogen bonds[69,90]	A–H$\cdots$D	O–H$\cdots$O
Halogen bonds[93,94]	A–X$\cdots$D X = halogen	C–Br$\cdots$N
Dihalogen bonds[95–99]	A–X$\cdots$X–B X = halogen	C–Br$\cdots$Br–C
Chalcogen bonds[115]	A = Ch$\cdots$Y A_1A_2Ch$\cdots$Y Ch = chalcogen (O, S, Se, Te) Y = electronegative atom	A = O$\cdots$Br–C A_1A_2S$\cdots$Br–C
Dichalcogen bonds[48]	A–Ch$\cdots$Ch–B Ch = chalcogen (O, S, Se, Te)	O$\cdots$O, S$\cdots$O, S$\cdots$S
Pnicogen bonds[116,117]	$A_1A_2A_3$Pn$\cdots$Y Pn = pnicogen (N, P, Se, Te) Y = electronegative atom	N$\cdots$Br–C
Tetrel bond[118]	$A_1A_2A_3A_4$Tr$\cdots$Y Pn = Tetrel (Si, Ge, Sn) Y = electronegative atom	
Carbon bond[119]	$A_1A_2A_3A_4$Tr$\cdots$Y Y = electronegative atom	C$\cdots$O, C$\cdots$Br
Lithium bond[120]	A-Li$\cdots$D	C-Li$\cdots$O
Beryllium bond[121]	Be$\cdots$B (B = NH_3, H_2O, FH, PH_3, SH_2, ClH, BrH)	$(HO)_2$Be$\cdots NH_3$

chemical methods to obtain the needed information about the strength (and any other property) of the intermolecular interactions.

The procedure employs the "divide and conquer" approach to the determination of the interaction energy in a molecular crystal, which gets converted in this way into a sum of two-body terms, the intermolecular interactions, both stabilizing and destabilizing. The latter can play an important role in determining the structure of some crystals,[122] although the energetic stability of the crystal requires that the sum of the stabilizing interactions should prevail over the sum of the destabilizing ones.

The procedure we have followed to rationalize the structure of a molecular crystal consists of sequentially applying the following four steps: (1) identify all energetically dominant intermolecular interactions present in the crystal and, among them, the intermolecular bonds, (2) determine the nature of all dominant intermolecular interactions by doing IMPT or SAPT tests, or based on previous experience, (3) select a proper model system, method and basis set to compute of the strength of the intermolecular interactions *at the geometry they have in the crystal*, (4) compute the strength of all intermolecular interactions identified in the previous steps, and (5) describe how the stability of the crystal relates to that of each individual intermolecular interaction, starting with the most stabilizing interaction, then the second most stable and so on, until all molecules in the crystal are interconnected (if there are unconnected blocks, the crystal should break into these blocks).

As formulated, the procedure does not consider cooperative effects. This is equivalent to assuming that the relative order of stability of the intermolecular interactions does not change when cooperative effects are considered. Let's point here that sometimes crystals present non-negligible cooperative effects. One example of cooperativity is produced in the O–H$\cdots$O hydrogen bonds found in the crystals of ice, where the experimental binding energy per hydrogen bond is -6.7 kcal mol^{-1} in ice,[123] while that for an isolated water dimer is -5.4 ± 0.7 kcal mol^{-1}.[124] Thus, the cooperative effect is 1.3 kcal mol^{-1}, that is, about 20% of the interaction energy of each isolated dimer. However, this difference does not affect the rationalization of the structure of ice, as all four intermolecular bonds created by any water molecule in the I_h crystal of ice are identical by symmetry (two hydrogen bonds of any water molecule act as hydrogen bond donor groups, while in other two act as acceptor groups, but in the diffraction one observes 50% of each possibility[125]).

Let us finally point out here that the *crystal structure analysis* procedure presented is not a cheap alternative to performing *crystal structure prediction* searches. Although both methodologies require an appropriate knowledge of the properties of the intermolecular interactions present in the crystal, their aims are different and so are their computational algorithms. For instance, a quest for the most stable polymorphs of a molecular crystal requires specific subroutines to systematically sample the space of polymorphic structures.[126] Usually a large number of polymorphic structures are obtained within a small window of a few kcal per mol.[122] Once the structure of the lowest energy polymorph has been predicted, one can carry out a crystal packing analysis to understand the origin of their stability.

2.7.2   Selection of the Proper Quantum Chemical Method

This selection will depend on the computational resources available, the size of the model system, and the accuracy required of the results. A good reference method for many cases is the MP2 method and the aug-cc-pVDZ monoelectronic basis set.

2.7.3 Selection of the Proper Model System

For a given model system to be taken as appropriate for the study of a specific intermolecular interaction found in a crystal, such a model must reproduce the environment of the molecular interaction in the crystal. In neutral crystals, where the dominant interactions are hydrogen bonds or van der Waals bonds, the impact produced by the intermolecular interaction on the atoms of the model system only affects the dimer (a part of it or with all its atoms). In ionic crystals, however, the electrostatic interactions are felt at much longer distance than hydrogen bonds or van der Waals interactions and should be included. As shown in Figure 2.10, such long-distance effects are

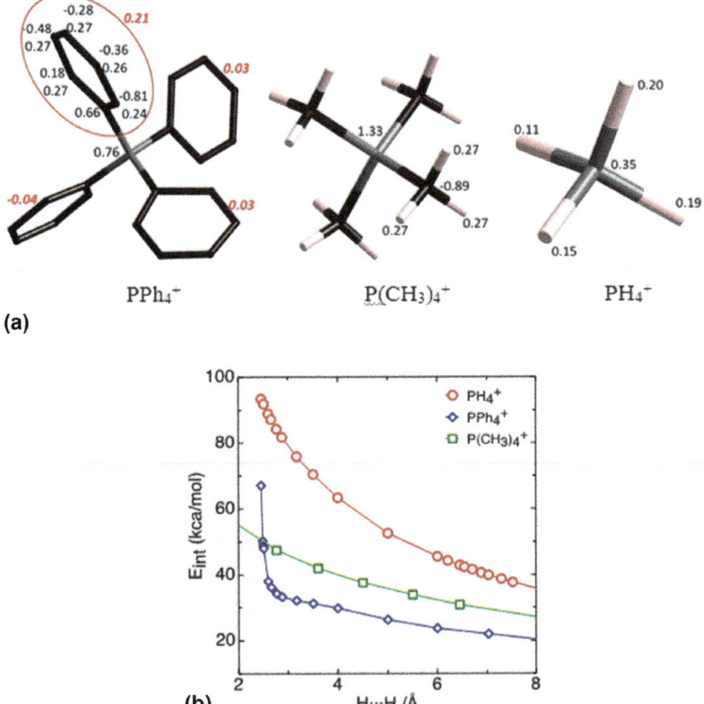

Figure 2.10 **(a):** Mulliken charge distribution (in atomic units) for Phenyl$_4$P$^+$, (left), P(CH$_3$)$_4^+$ (center), and PH$_4^+$ (right) – in Phenyl$_4$P$^+$, all H atoms are omitted in the drawing, for clarity, but their charge is given (on each C(phenyl) the first line is the charge on C and the second line the charge on its bonded H), while the values in red refer to the total charge on each phenyl ring;

(b): Variation of the interaction energy (E_{int}) with the increase in the shortest H···H distance in a PH$_4^+$···PH$_4^+$ dimer, a P(CH$_3$)$_4^+$···P(CH$_3$)$_4^+$ dimer, and a P(C$_6$H$_5$)$_4^+$···P(C$_6$H$_5$)$_4^+$ dimer. All curves and charge distributions were computed at the MP2/6-31 + G(d) level. All dimers were oriented as shown in (a), but changing the phenyl group for a methyl group or an H atom.

created even by large ions, as in the $Phenyl_4P^+$ cation where the net charge is spread over many atoms.

2.7.4 Working with QC-Crystal Packing Analysis: The Acetic Acid Crystal

The analysis of the acetic acid crystal, a neutral molecular crystal whose structure was diffracted at 40 K and ambient pressure (refcode: ACETAC07),[127] is illustrative of the potential of this technique and of the type of difficulties encountered.

In the ACETAC07 crystal the dominant interactions are all hydrogen bonds (they are all plotted in Figure 2.11). As seen in Figure 2.11, the crystal has a complex layered structure, where one only finds up to four H···O short contacts in the range $2.3 \leq r(H···O) \leq 3.1$ Å with the angularity expected for hydrogen bonds, located in two different dimers: dim1 and dim2 (see Figure 2.12). How do we get a handle on

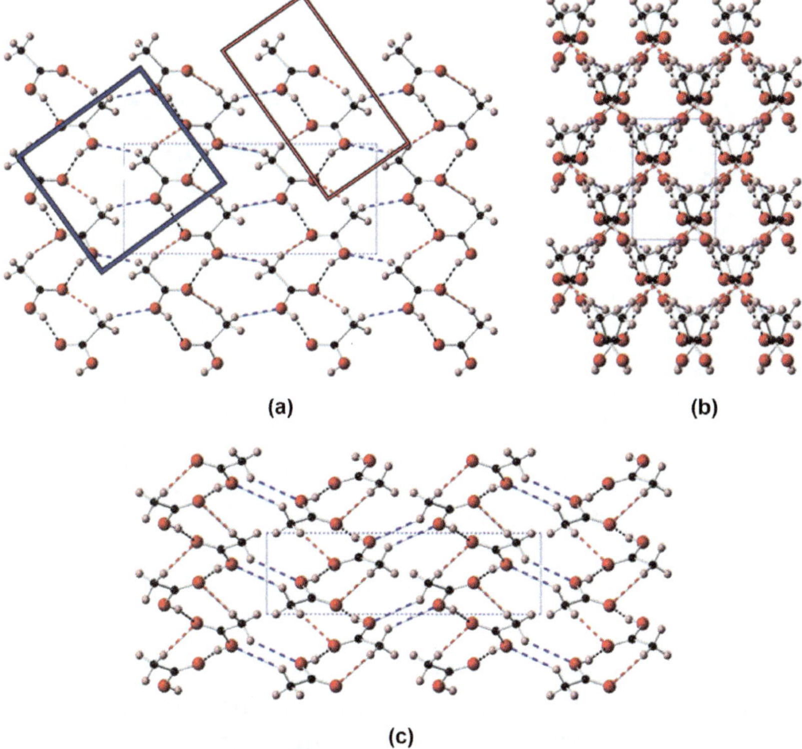

Figure 2.11   (a) View along the *b*-axis of the structure of the acetic acid crystal at $P = 1$ atm and $T = 40$ K; (b) View along the *a*-axis of the same crystal; (c) View along the *c*-axis.

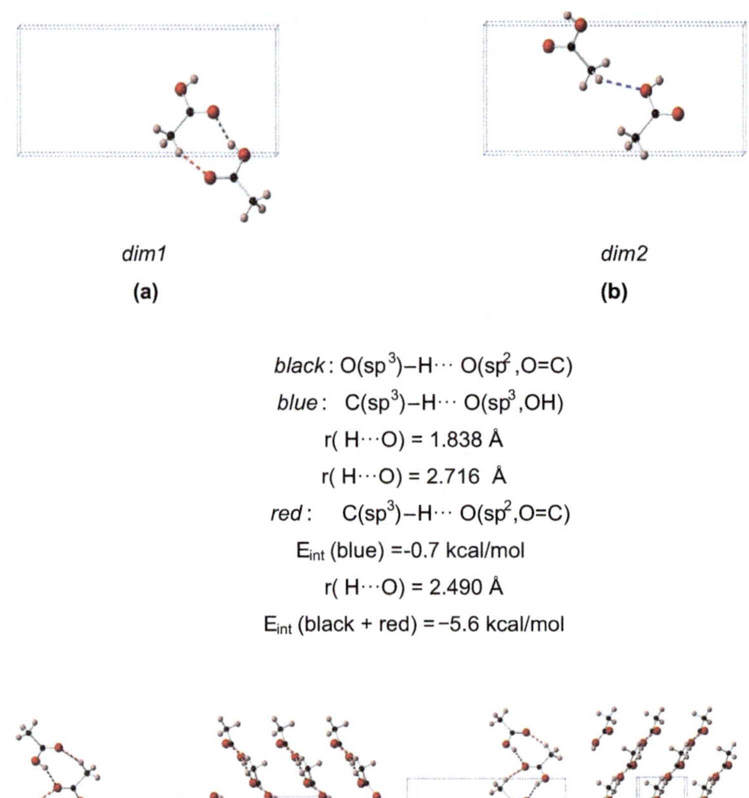

Figure 2.12 (a): View along the *b*-axis of dimer *dim1*; (b): the same for *dim2*; (c): front and side views of the network of hydrogen bonds found in the crystal that are identical to those in *dim1* (only those generated from molecules placed on top of the left molecule of the unit cell are shown); (d): equivalent views for the right molecule of the unit cell. The front and side views of the hydrogen bond of the crystal identical to those in *dim2* are identical to those in parts (a) and (c), and, in consequence, are not plotted again. The value of the interaction energy, E_{int}, for both dimers is given under its name (notice that it is stabilizing in both dimers; the value given is a MP2/aug-cc-pVTZ result and has been corrected by the Basis Set Superposition Error (BSSE).[130]

such complex crystal structure? Our proposal was: look at the energy of the intermolecular interactions, and describe the packing as the result of the successive additions of the most stable attractive interaction (which defines the primary crystal structure), plus the next in stability (which defines the secondary), and so on. In each addition more molecules of the crystal become interconnected forming "islands". The process can be stopped when all molecules of the crystal are interconnected, that is, form a single "island".[128] It is one case of energy-based crystal packing rationalization procedures. Gavezzotti,[129] proposed a similar procedure, but based on looking at the first-neighbor molecule···molecule pairs connected by the symmetry operators of the crystal (these molecular pairs were called by this author crystal-structure determinants), which also helps to rationalize the crystal packing. but doesn't give information about why some pairs are more stable than others. The concept of supramolecular synthons (supramolecular building blocks common to many structures) is closely related, as for the synthon to be found in many structures, it should be energetically more stable than other alternative orientations of the two molecules involved in the synthon.

At $P = 1$ atm and $T = 40$ K, the crystal of acetic acid packs in what it is called a *catemer* motif (indicating that its O–H···O hydrogen bonds[131] form chains). The molecules in this crystal are in their neutral form, where each one hosts two hydrogen bond donor groups $(O(sp^3)$–H and $C(sp^3)$–H) and two hydrogen bond acceptor groups $(O(sp^3, OH)$ and $(O(sp^2, O=C)$. Any combination of these two acceptor and donor groups could be formed in the crystal, but a systematic search of H···O contacts within the $2.2 \leq r(H···O) \leq 3.0$ Å range (Figure 2.11), found three hydrogen bonds located in only two dimers (*dim1*, and *dim2*, see Figure 2.12a and b). The resulting scenario is quite complex: (a) one dimer (*dim1*) has two hydrogen bonds and it is

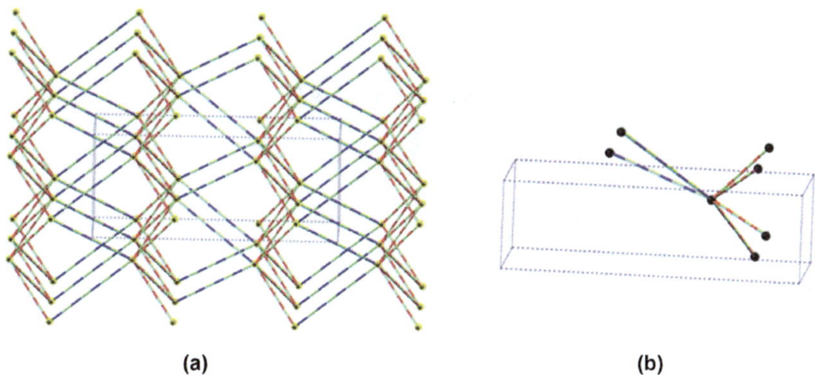

(a)   (b)

Figure 2.13   (a): Network of hydrogen bonds linking the molecules in the acetic acid crystal (at $T = 40$ K, $P = 1$ atm); (b): plot of the coordination of any molecule of the crystal. Red-(light blue) broken lines represent the two hydrogen bonds in dimer *dim1*, while (dark blue)-(light blue) lines represent the hydrogen bond in *dim2*.

very close to the optimum geometry of such dimer when isolated, and (b) *dim2* is far from any minimum for the acetic acid dimer, thus being a demonstration that the observed structure of a crystal (one of minima in the potential energy surface of the crystal) sometimes is a compromise between various intermolecular interactions and some of the first-neighbors dimers can be far from their minima.

The QC-CPA rationalization of the structure of the acetic acid crystal (at $T = 40$ K and $P = 1$), goes as follows: as shown in Figure 2.12, *dim1* is more stable than dimer *dim2*. Thus, *dim1* generates the primary structure of the crystal, shown separately in Figures 2.12(c) and (d), a better visualization of the structure. It is concluded that the primary structure is formed by zig-zag chains, stacked parallel to each other along the *c* and *a*-axis (Figure 2.13). The axis defining how the stacks pile is different for the two molecules in the left unit cell, thus creating a rhombohedral pattern (see Figures 2.11 and 2.12). Each molecule makes four hydrogen bonds of this type on one of its sides. On the other side it makes two hydrogen bonds of the type seen in *dim2,* connecting adjacent stacks of chains. The network of hydrogen bonds that result interconnects all acetic acid molecules, as shown in Figure 2.14, where molecules are represented by only its C atoms bonded to two oxygen atoms; notice the

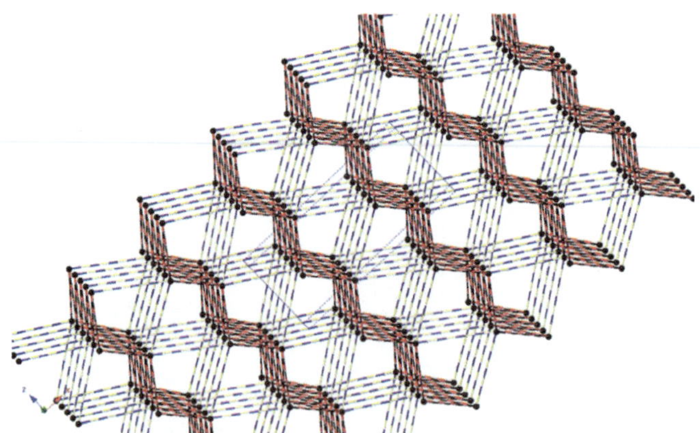

Figure 2.14   The stability of the crystals of acetic acid originates from the two intermolecular interactions shown in Figure 2.12(c) and the only interaction plotted in the dimer of Figure 2.12(d). By application of the symmetry operations in the 3D, the network of intermolecular bonds is plotted in this figure. Molecules are represented by their C atoms linked to the two oxygen atoms, the red-and-black broken lines represent the intermolecular interactions generated in the dimer of Figure 2.12(c), and the blue-white broken lines represent the intermolecular interaction produced by the dimer in Figure 2.12(d). All molecules in this crystal are hexacoordinated.

hexacoordination of the molecules). This completes the structural analysis of the acetic acid crystal at ambient pressure and 40 K.

Acknowledgements

We thank the support of the Spanish Government for support (projects *MAT2014-54025-P*, and a Ph.D. grant to M.C.), the Catalan Government (project 2014 SGR 142) and BSC and CSUC for allocating computer time in their machines.

References

1. *Engineering of Crystalline Materials Properties: State of the Art in Modeling, Design and Applications*, ed. J. J. Novoa, L. Addadi and D. Braga, Elsevier, Dordrecht, 2008.
2. J. J. Novoa, in *Implications of Molecular and Materials Structure for New Technologies*, ed. J. A. K. Howard, F. H. Allen and G. P. Shields, Kluwer Academic, Dordrecht, 1999, p. 235; J. J. Novoa, E. D'Oria and M. A. Carvajal, in *Making Crystals by Design: Methods, Techniques and Applications*, ed. D. G. Braga and Fabrizia, Wiley-VCH, Weinheim, 2007, ch. 2, p. 25.
3. A. J. Stone, *The Theory of Intermolecular Forces*, Clarendon Press, Oxford, 1996.
4. J. Israelachvili, *Intermolecular and Surface Forces*, Academic Press, Boston, 1992.
5. S. Scheiner, *Hydrogen Bonding: A Theoretical Perspective*, Oxford University Press, USA, New York, 1997.
6. G. C. Maitland, M. Rigby, E. B. Smith and W. A. Wakeham, *Intermolecular Forces: Their Origin and Determination*, Oxford University Press, Oxford, 1981.
7. G. E. Moore, *Electronics*, 1965, **38**, 114–117.
8. S. M. Cybulski and R. R. Toczyłowski, *J. Chem. Phys.*, 1999, **111**, 10520.
9. G. Herzberg, *Molecular Spectra and Molecular Structure*, Van Nostrand, Princeton, 1960, vol. I, p. 532.
10. S. J. Blanksby and G. B. Ellison, *Acc. Chem. Res.*, 2003, **36**, 255–263.
11. IUPAC. Compendium of Chemical Terminology, 2nd ed. (the "Gold Book"). Compiled by A. D. McNaught and A. Wilkinson. Blackwell Scientific Publications, Oxford (1997). XML on-line corrected version: http://goldbook.iupac.org (2006-) created by M. Nic, J. Jirat, B. Kosata; updates compiled by A. Jenkins. ISBN 0-9678550-9-8. DOI: 10.1351/goldbook. Last update: 2014-02-24; version: 2.3.3. DOI of this term: DOI: 10.1351/goldbook.S06153.
12. A. I. Kitaigorodsky, *Molecular Crystals and Molecules*, Elsevier, 1973.
13. S. Vela, F. Mota, M. Deumal, R. Suizu, Y. Shuku, A. Mizuno, K. Awaga, M. Shiga, J. J. Novoa and J. Ribas-Arino, *Nat. Commun.*, 2014, **5**, 4411.
14. M. Vasileiadis, *The calculation of the free energy of crystalline solids*, PhD thesis, Imperial College London, 2013.
15. S. L. Price, *Acc. Chem. Res.*, 2009, **42**, 117.
16. For one crystal of this type, see: W. Fujita and K. Awaga, *Science*, 1999, **286**, 261.
17. A. Levitin, *Introduction to the Design & Analysis of Algorithms*, Pearson, Upper Saddle River, 3rd edn, 2011.
18. (a) E. B. Wilson, *An Introduction to Scientific Research*, McGraw-Hill, New York, 1952; (b) R. Feynman, *The Character of Physical Law*, The M.I.T. Press,

Cambridge, Ma, Twelfth printing, 1985; (c) T. S. Kuhn, *The Structure of Scientific Revolutions*, University of Chicago Press, Chicago, IL, 3rd edn, 1996.

19. L. Pauling, *The Nature of the Chemical Bond*, Cornell University Press, Ithaca, 1960, p. 6.
20. R. F. W. Bader, *Atoms in Molecules*, Clarendon Press, Oxford, 1990.
21. H. B. Burgi and J. D. Dunitz, *Acta Crystallogr., Sect. B: Structr. Sci.*, 1988, **44**, 445.
22. G. A. Jeffrey, *An Introduction to Hydrogen Bonding*, Oxford University Press, Oxford, 1997.
23. E. D'Oria and J. J. Novoa, *CrystEngComm*, 2004, **6**, 368.
24. M. Kaupp, B. Metz and H. Stoll, *Angew. Chem., Int. Ed.*, 2000, **39**, 4607.
25. In the Gold Book the term "bond" is defined as in Pauling's book: "There is a chemical bond between two atoms or groups of atoms in the case that the forces acting between them are such as to lead to the formation of an aggregate with sufficient stability to make it convenient for the chemist to consider it as an independent molecular species". See reference 9.
26. See, for instance: J. J. Novoa, I. Nobeli, F. Grepioni and D. Braga, *New J. Chem.*, 2000, **24**, 5; D. Braga, L. Maini, F. Grepioni, F. Mota, C. Rovira and J. J. Novoa, *Chem. - Eur. J.*, 2000, **6**, 4536; D. Braga, E. D'Oria, F. Grepioni, F. Mota, J. J. Novoa and C. Rovira, *Chem. - Eur. J.*, 2002, **8**, 1173–1180
27. J. D. Dunitz and A. Gavezzotti, *Acc. Chem. Res.*, 1999, **32**, 677.
28. R. F. W. Bader, *Atoms in Molecules*, Clarendon Press, Oxford, 1990.
29. For two representative cases, see: (a) S. Pillet, G. Wu, V. Kulsomphob, B. G. Harvey, R. D. Ernst and P. Coppens, *J. Am. Chem. Soc.*, 2003, **125**, 1937; (b) C. Foroutan-Nejad, S. Shahbazian and R. Marek, *Chem. - Eur. J.*, 2014, **20**, 10140.
30. B. Pierre, *Phys. Scr.*, 1977, **15**, 119.
31. P. Coppens, *X-Ray Charge Densities and Chemical Bonding*, International Union of Crystallography, 1997.
32. Coulson stated: "I described a bond, a normal simple chemical bond; and I gave many details of its character (and could have given many more). Sometimes it seems to me that a bond between two atoms has become so real, so tangible, so friendly that I can almost see it. And then I awake with a little shock; for a chemical bond is not a real thing; it does not exist; no one has ever seen it; no one ever can. It is a figment of my own imagination"; See:C. A. Coulson, *J. Chem. Soc.*, 1955, 2069.
33. AFM studies aim at determining the force (F) acting between the tip and the sample. Such force is not measured directly, but calculated by measuring the distance the cantilever is bent (z), knowing the cantilever stiffness (k), as, according to Hooke's law $F = - kz$).
34. J. Zhang, P. Chen, B. Yuan, W. Ji, Z. Cheng and X. Qiu, *Science*, 2013, **342**, 611, DOI: 10.1126/science.1242603.
35. see, for instance: A. J. Lee, Y. Sakai, M. Kim and J. R. Chelikowsky, *Appl. Phys. Lett.*, 2016, **108**, 193102.
36. (a) A. Pross and S. S. Shaik, *Acc. Chem. Res.*, 1983, **16**, 363–370; (b) S. S. Shaik and P. C. Hiberty, *A Chemist's Guide to Valence Bond Theory*, John Wiley & Sons, 2007.
37. Namely, in both bonds their stability originates from the overlap of one orbital from each fragment, which results in the formation of a more stable, double-occupied, bonding combination, and an empty antibonding combination. Coordination bonds only differ from conventional covalent bonds (for instance, the H–H covalent bond formed when two H · radicals approach and form an H_2 molecule) in the occupation of the overlapping orbitals: while in the H–H covalent bond, each overlapping orbital has one electron, in the metal-ligand bond one orbital (the ligand lone-pair) has a pair of electrons, and the other one (the metal orbital) is empty. But the resulting bonding combination of these two orbitals is

also more stable than any of the two overlapping orbitals, and is also double occupied. Thus, there is an energy gain due to the formation of the metal-ligand bond, similar to that obtained in conventional covalent bonds. Therefore, as in the case of conventional covalent bonds, the interaction energy for a metal-ligand bonded dimer is strongly dominated by the bonding component, which becomes larger than the electrostatic, polarization or dispersion components. Consequently, metal-ligand bonds have a similar strength and equilibrium distance to covalent bonds. This explains why it is better considered as an *intra*molecular bond.

38. The aug-cc-pVDZ is a widely used Gaussian monoelectronic basis set of "double zeta quality (DZ)", a member of the "**c**orrelation **c**onsistent-**p**olarized **V**alence" (i.e., cc-pVXZ) monoelectronic basis set (where XZ = DZ, TZ, QZ, 5Z, ...). They were designed to have energies that converge systematically to the complete-basis-set (CBS) limit as the basis set quality is increased from DZ to TZ, to QZ and so on. The "aug-" prefix means that diffuse functions have been added to the basis set, a type of function required to improve the accuracy of the computed interaction energy when working with intermolecular complexes. See: T. H. Dunning, *J. Chem. Phys.*, 1989, **90**, 1007.

39. Web page where it is possible to get this and other basis sets: https://bse.pnl.gov/bse/portal.

40. E. D'Oria, D. Braga and J. J. Novoa, *CrystEngComm*, 2012, **14**, 792.

41. name that indicates that the wavefunction and the energy of the system under study is obtained by solving the equations for the method without: (a) making any assumptions on the properties of the systems studied, and (b) making no use of any parameters obtained from the study of other systems. This avoids that the B3LYP functional could be called an *ab initio* method, but not that the same name could be used with other DFT functionals.

42. (a) D. R. Hartree, *Proc. Cambridge Philos. Soc.*, 1928, **24**, 89–110; (b) V. Fock, *Z. Phys.*, 1930, **61**, 126–148.

43. R. G. Parr and W. Yang, *Density-functional Theory of Atoms and Molecules*, Oxford University Press, New York, Oxford, 1989.

44. C. Møller and M. S. Plesset, *Phys. Rev.*, 1934, **46**, 618.

45. A. Szabo and N. S. Ostlund, *Modern Quantum Chemistry: Introduction to Advanced Electronic Structure Theory*, McGraw-Hill, New York, 1st, Rev. edn., 1989.

46. B. O. Roos, in *Advances in Chemical Physics*, ed. P. K. Lawley, John Wiley & Sons, Inc., 1987, vol. 69, p. 399.

47. (a) K. Andersson, P. Å. Malmqvist and B. O. Roos, *J. Chem. Phys.*, 1992, **96**, 1218; (b) B. O. Roos, K. Andersson, M. P. Fülscher, P.-Â. Malmqvist, L. Serrano-Andrés, K. Pierloot and M. Merchán, *Advances in Chemical Physics*, John Wiley & Sons, Inc., 1996, vol. 93, p. 219.

48. (a) J. J. Novoa, M. H. Whangbo and J. M. Williams, in *Organic Superconductivity*, ed. V. Z. Kresin and W. A. Little, 1990, p. 231; (b) C. Rovira and J. J. Novoa, *Chem. – Eur. J.*, 1999, **5**, 3689.

49. (a) D. R. Hartree, *Proc. Cambridge Philos. Soc.*, 1928, **24**, 89–110; (b) R. Hartree, *Proc. Cambridge Philos. Soc.*, 1928, **24**, 111–132.

50. C. Møller and M. S. Plesset, *Phys. Rev.*, 1934, **46**, 618–622.

51. B. Liu and A. D. McLean, *J. Chem. Phys.*, 1973, **59**, 4557.

52. S. F. Boys and F. Bernardi, *Mol. Phys.*, 1970, **19**, 553.

53. J. J. Novoa and C. Sosa, *J. Phys. Chem.*, 1995, **99**, 1583.

54. (a) F. B. van Duijneveldt, J. G. C. M. van Duijneveldt-van de Rijdt and J. H. van Lenthe, *Chem. Rev.*, 1994, **94**, 1873; (b) J. J. Novoa, M. Planas and M. H. Whangbo, *Chem. Phys. Lett.*, 1994, **225**, 240; (c) M. C. Rovira, J. J. Novoa, M. H. Whangbo and J. M. Williams, *Chem. Phys.*, 1995, **200**, 319; (d) J. J. Novoa, M. Planas and M. C. Rovira, *Chem. Phys. Lett.*, 1996, **251**, 33; (e) J. J. Novoa and M. Planas, *Chem. Phys. Lett.*, 1998, **285**, 186.

55. I. C. Hayes and A. J. Stone, *J. Mol. Phys.*, 1984, **53**, 83.

56. B. Jeziorski, R. Moszynski and K. Szalewicz, *Chem. Rev.*, 1994, **94**, 1887.

57. "The other cause underlying the non-compliance of real gases and liquids with Boyle's law, i.e. the mutual attraction of the molecules", see: J. D. van der Waals - Nobel Lecture: "The Equation of State for Gases and Liquids". Nobelprize.org. Nobel Media AB 2014. Web. 15 Dec 2016. <http://www.nobelprize.org/nobel_prizes/physics/laureates/1910/waals-lecture.html>.

58. J. D. van der Waals, *On the Continuity of the Gas and Liquid State*, Leiden, The Netherlands, 1873.

59. J. D. van der Waals, *1910 Nobel Lecture,* from *Nobel Lectures, Physics 1901-1921*, Elsevier Publishing Company, Amsterdam, 1967.

60. Some authors use the term "van der Waals interaction" as equivalent to all the "intermolecular interactions", thus introducing a lot of confusion in the field. In this work, in good agreement with the most usual convention, the term "van der Waals interaction" is only used for the dispersion-dominated interaction, the case of the Ar · · ·Ar interaction studied in detail by van der Waals.

61. F. London, *Z. Phys.*, 1930, **63**, 245.

62. L. Pauling, *The Nature of the Chemical Bond*, Cornell University Press, Ithaca, 3rd edn, 1960. A first edition of this seminal book was published in 1939.

63. S. Grimme, *Chem. – Eur. J.*, 2012, **18**, 9955.

64. J. Ribas-Arino and J. J. Novoa, *Chem. Commun.*, 2007, 3160.

65. A. Werner, *Ber. Dtsch. Chem. Ges.*, 1903, **36**, 147.

66. T. S. Moore and T. F. Winmill, *J. Chem. Soc.*, 1912, **101**, 1635.

67. W. M. Latimer and W. H. Rodebush, *J. Am. Chem. Soc.*, 1920, **42**, 1419.

68. R. Taylor and O. Kennard, *J. Am. Chem. Soc.*, 1982, **104**, 5063.

69. J. J. Novoa, B. Tarron, M. H. Whangbo and J. M. Williams, *J. Chem. Phys.*, 1991, **95**, 5179.

70. (a) Y. Gu, T. Kar and S. Scheiner, *J. Am. Chem. Soc.*, 1999, **121**, 9411; (b) S. Scheiner, T. Kar and Y. Gu, *J. Biol. Chem.*, 2001, **276**, 9832.

71. (a) J. J. Novoa, P. Constans and M. H. Whangbo, *Angew. Chem., Int. Ed. Engl.*, 1993, **32**, 588; (b) J. J. Novoa and F. Mota, *Chem. Phys. Lett.*, 1997, **266**, 23; (c) J. Novoa, P. Lafuente and F. Mota, *Chem. Phys. Lett.*, 1998, **290**, 519; (d) J. J. Novoa, F. Mota and E. D'Oria, in *Hydrogen Bonding - New Insights*, ed. S. Grabowski, 2006, ch. 5, pp. 193–244.

72. G. A. Jeffrey, *An Introduction to Hydrogen Bonding*, Oxford University Press, New York, 1997.

73. (a) M. Meot-Ner, P. Hamlet, E. P. Hunter and F. H. Field, *J. Am. Chem. Soc.*, 1978, **100**, 5466; (b) M. Meot-Ner, *Chem. Rev.*, 2005, **105**, 213; (c) M. Meot-Ner, *Chem. Rev.*, 2012, **112**, PR22–PR103.

74. G. Gilli and P. Gilli, *J. Mol. Struct.*, 2000, **552**, 1.

75. P. G. Wenthold and R. R. Squires, *J. Phys. Chem.*, 1995, **99**, 2002.

76. E. Arunan, G. R. Desiraju, R. A. Klein, J. Sadlej, S. Scheiner, I. Alkorta, D. C. Clary, R. H. Crabtree, J. J. Dannenberg, P. Hobza, H. G. Kjaergaard, A. C. Legon, B. Mennucci and D. J. Nesbitt, *Pure Appl. Chem.*, 2011, **83**, 1619.

77. G. C. M. Pimentel and L. Aubrey, *The Hydrogen Bond*, W. H. Freeman and Co, San Francisco, 1960.

78. K. Müller-Dethlefs and P. Hobza, *Chem. Rev.*, 2000, **100**, 143.

79. G. R. Desiraju, *Crystal Engineering: The Design of Organic Solids*, Elsevier, Amsterdam, 1989.

80. G. Gilli and P. Gilli, *The Nature of the Hydrogen Bond: Outline of a Comprehensive Hydrogen Bond Theory*, OUP, Oxford, 2009.

81. C. B. Aakeroy and K. R. Seddon, *Chem. Soc. Rev.*, 1993, **22**, 397–407.

82. G. R. Desiraju and T. Steiner, *The Weak Hydrogen Bond: In Structural Chemistry and Biology*, Oxford University Press, 1999.

83. T. Steiner, *Angew. Chem., Int. Ed.*, 2002, **41**, 48–76.
84. G. A. Jeffrey and W. Saenger, *Hydrogen Bonding in Biological Structures*, Springer-Verlag, 1991.
85. Reference 19, page 6.
86. D. Braga, F. Grepioni and J. J. Novoa, *Chem. Commun.*, 1998, 1959.
87. D. Braga, L. Maini, F. Grepioni, F. Mota, C. Rovira and J. J. Novoa, *Chem. Eur. - J.*, 2000, **6**, 4536.
88. J. J. Novoa, I. Nobeli, F. Grepioni and D. Braga, *New J. Chem.*, 2000, **24**, 5.
89. D. Braga, J. J. Novoa and F. Grepioni, *New J. Chem.*, 2001, **25**, 226.
90. D. Braga, E. D'Oria, F. Grepioni, F. Mota, J. J. Novoa and C. Rovira, *Chem. - Eur. J.*, 2002, **8**, 1173.
91. O. Hassel and J. Hvoslef, *Acta Chem. Scand.*, 1954, **8**, 873.
92. (a) R. S. Mulliken, *J. Am. Chem. Soc.*, 1950, **72**, 600; (b) R. S. Mulliken, *J. Am. Chem. Soc.*, 1952, **74**, 811; (c) R. S. Mulliken, *J. Phys. Chem.*, 1952, **56**, 801.
93. A. Priimagi, G. Cavallo, P. Metrangolo and G. Resnati, *Acc. Chem. Res.*, 2013, **46**, 2686.
94. P. Politzer, J. S. Murray and T. Clark, *Phys. Chem. Chem. Phys.*, 2010, **101**, 16789.
95. V. R. Pedireddi, D. S. Reddy, B. S. Goud, D. C. Craig, A. D. Rae and G. R. Desiraju, *J. Chem. Soc., Perkin Trans. 2*, 1994, 2353.
96. T. T. Bui, S. Dahaoui, C. Lecomte, G. R. Desiraju and E. Espinosa, *Angew. Chem., Int. Ed.*, 2009, **48**, 3838.
97. F. F. Awwadi, R. D. Willett, K. A. Peterson and B. Twamley, *Chem. - Eur. J.*, 2006, **12**, 8952.
98. S. J. Grabowski, A. J. Sadlej, W. A. Sokalski and J. Leszczynski, *Chem. Phys.*, 2006, **327**, 151.
99. (a) R. Malave-Osuna, V. Fernandez, J. T. López-Navarrete, E. D'Oria and J. J. Novoa, *Theor. Chem. Acc.*, 2011, **128**, 541; (b) M. Capdevila-Cortada, J. Castello and J. J. Novoa, *CrystEngComm*, 2014, **16**, 8232; (c) M. Capdevila-Cortada and J. J. Novoa, *CrystEngComm*, 2015, **17**, 3354–3365.
100. C. A. Hunter and J. K. M. Sanders, *J. Am. Chem. Soc.*, 1990, **112**, 5525.
101. R. Podeszwa, R. Bukowski and K. Szalewicz, *J. Phys. Chem. A*, 2006, **110**, 10345.
102. CBS is an acronym indicating that the result obtained is an estimate of the Complete Basis Set result, computed by extrapolating the energy computed by doing CCSD(T)/aug-cc-pVXZ calculations using a basis of higher quality (for instance, by making X equal to D, T and Q).
103. P. Hobza, *Acc. Chem. Res.*, 2012, **45**, 663.
104. T. Takatani, E. G. Hohenstein, M. Malagoli, M. S. Marshall and C. D. Sherrill, *J. Chem. Phys.*, 2010, **132**, 144104.
105. H. Krause, B. Ernstberger and H. J. Neusser, *Chem. Phys. Lett.*, 1991, **184**, 411.
106. J. Sunner, K. Nishizawa and P. Kebarle, *J. Phys. Chem.*, 1981, **85**, 1814.
107. J. C. Ma and D. A. Dougherty, *Chem. Rev.*, 1997, **97**, 1303.
108. S. Tsuzuki, M. Yoshida, T. Uchimaru and M. Mikami, *J. Phys. Chem. A*, 2001, **105**, 769.
109. S. Tsuzuki, M. Mikami and S. Yamada, *J. Am. Chem. Soc.*, 2007, **129**, 8656.
110. B. L. Schottel, H. T. Chifotides and K. R. Dunbar, *Chem. Soc. Rev.*, 2008, **37**, 68.
111. (a) I. Alkorta, I. Rozas and J. Elguero, *J. Am. Chem. Soc.*, 2002, **124**, 8593; (b) D. Quiñonero, C. Garau, C. Rotger, A. Frontera, P. Ballester, A. Costa and P. M. Deyà, *Angew. Chem., Int. Ed.*, 2002, **41**, 3389; (c) M. Mascal, A. Armstrong and M. D. Bartberger, *J. Am. Chem. Soc.*, 2002, **124**, 6274.
112. D. Serhiy, D. Sebastian and M. Franc, *J. Am. Chem. Soc.*, 2004, **126**, 4508.
113. M. Hiromitsu, O. Atsuhiro and F. Hiroyuki, *J. Inclusion Phenom.*, 2004, **49**, 33.
114. D.-X. Wang and M.-X. Wang, *J. Am. Chem. Soc.*, 2013, **135**, 892.
115. W. Wang, B. Ji and Y. Zhang, *J. Phys. Chem. A*, 2009, **113**, 8132.

116. J. E. del Bene, I. Alkorta, G. Sanchez-Sanz and J. Elguero, *J. Phys. Chem. A*, 2011, **115**, 137.
117. S. Scheiner, *Acc. Chem. Res.*, 2013, **46**, 280.
118. A. Bauzá, T. J. Mooibroek and A. Frontera, *Angew. Chem., Int. Ed.*, 2013, **52**, 12317.
119. D. Mani and E. Arunan, *Phys. Chem. Chem. Phys.*, 2013, **15**, 14377.
120. A. Shahi and E. Arunan, *Phys. Chem. Chem. Phys.*, 2014, **16**, 22935.
121. M. Yáñez, P. Sanz, O. Mó, I. Alkorta and J. Elguero, *J. Chem. Theory Comput.*, 2009, **5**, 2763.
122. J. D. Dunitz and A. Gavezzotti, *Acc. Chem. Res.*, 1999, **32**, 677.
123. D. Eisenberg and W. Kauzmann, *The Structure and Properties of Water*, Clarendon, Oxford, 1969.
124. S. Suhai, *J. Chem. Phys.*, 1994, **101**, 9766–9782; S. S. Xantheas, *J. Chem. Phys.*, 1994, **100**, 7523–7534.
125. J. D. Bernal and R. H. Fowler, *J. Chem. Phys.*, 1933, **1**, 515.
126. S. L. Price, *Chem. Soc. Rev.*, 2014, **43**, 2098.
127. R. Boese, D. Blaser, R. Latz and A. Baumen, *Acta Crystallogr. Sect. C: Cryst. Struct. Commun.*, 1999, **55**, IUC9900001.
128. J. J. Novoa, E. D'Oria and M. A. Carvajal, in *Making Crystals by Design: Methods, Techniques and Applications*, ed. D. G. Braga and Fabrizia, Wiley-VCH, Weinheim, 2007, ch. 2, pp. 25–57.
129. L. Carlucci and A. Gavezzotti, *Chem. – Eur. J.*, 2005, **11**, 271.
130. By applying the counterpoise procedure of Boys and Bernardi.
131. By convention, in an A–H$\cdots$B hydrogen bond, the A–H group is the donor group, and the B group is the acceptor.

3 Bonding in Organic Molecules and Condensed Phases. The Role of Repulsions

A. Gavezzotti

Dipartimento di Chimica, Università di Milano, via Venezian 21, 20133 Milano, Italy
Email: angelo.gavezzotti@unimi.it

3.1 Introduction

A basic principle of modern chemistry is that the laws that preside over the evolution of molecular systems stem from the analysis and classification of molecular structures. The ultimate terms in which that analysis is conducted and these laws are written are the changes of the relative positions of atomic nuclei and of the distribution of the electron density. The only undisputable principle that applies in such matters is the Feynman condition, stating that equilibrium is established when Coulombic forces at nuclei are null. For the rest, chemists of all sub-disciplines are usually struggling to find simple ways of dealing with the extreme complexity of the rearrangements that occur when chemical systems evolve. Some of the simplest of these evolution modes can be accurately described by the solution of an appropriate Schroedinger equation, but most of the time one has to resort to intuitive shortcuts. This is the origin of the concept of chemical bond: things become enormously easier when recognizable and reproducible attachment modes between atoms and molecules

Intermolecular Interactions in Crystals: Fundamentals of Crystal Engineering
Edited by Juan J. Novoa
© The Royal Society of Chemistry 2018
Published by the Royal Society of Chemistry, www.rsc.org

can be used over classes of chemical transformation. For example, in the reaction of methane with Cl_2 one may say that a C–H bond and a Cl–Cl bond are broken and that a new C–Cl bond is formed, and such a conclusion can be transferred to the class of reactions known as alkane chlorination. By clever use of named reactions, synthetic organic chemistry has been able to synthesize in the last 50 years or so many of the natural compounds it took eons for evolution to provide.

As soon as one adopts a bond-directed attitude there arises the need, at least in theoretical chemistry, to introduce an appropriate terminology. An obvious way of doing so is to define a bond coordinate along which two putative bond partners oscillate under a given potential, with an equilibrium bond distance at which the bond energy is a minimum, and with first and second derivatives that define the forces and the force constant, respectively. In the perspective of this chapter devoted to molecular simulation, potential energy profiles at the same time define and describe the bond, providing in one picture the geometrical framework and the quantitative energetic weighting of the bonding environment. Typically, a bonding relationship may apply between two atoms, the rest of the molecule being kept rigid – the chemical bond "proper". With little changes in terminology one may also define a potential profile for bond bending, or for the relationship between two atomic centers in a 1–4 geometry when the angle between the 1–2–3 and the 2–3–4 planes is varied – bond torsion.

In condensed matter chemistry the main problem is no longer to find how molecules can be synthesized, but to categorize and rationalize the ways in which molecules aggregate from disperse systems, gas or solution, into molecular agglomerates: liquids with molecular rotation and translational freedom, or crystals with librational degrees of freedom only. There is still a wide gap between the brilliant performance of intramolecular synthesis and the opaque achievements of condensed-matter structural prediction. The reasons are clear in terms of elementary physical chemistry, given the wide gap between the corresponding forces, as will be described in the first sections of this chapter. Attempts to transfer the intramolecular bond terminology to the intermolecular arena have been many, assuming that neighboring molecules in condensed systems are kept together by bonds between selected atoms. None of these attempts has been without problems and criticism.

In a different approach, a bonding relationship can be defined between entire molecules in condensed phases: fluxional in liquids, and rigid except for libration in crystals. This might be called the "inter-molecule" bond. A substantial help, as will be demonstrated, comes from a semiempirical subdivision of the calculated cohesive potentials into Coulombic-polarization and dispersion terms, which in spite of some dispute in their definition provide in many cases a better chemical understanding and allow quantitative measures of the strength of the interaction, as well as some means for structural prediction; never more than in crystal engineering is qualitative only bad quantitative.

One of the reasons why intramolecular bonding and reactivity are very different from intermolecular cohesion is that in the intermolecular case one must account for the fact that molecules are in close contact over extended molecular regions, and the structure and properties of the whole system are described and determined also in terms of repulsion and destabilization. There is a tradeoff between local attractions and local repulsions, there being no equilibrium without a balance between the strains of molecular coming together and molecular coming apart. Clearly any interaction scheme based on the localized geometry of single atom–atom intermolecular approaches may never be able to fully account for such a complex situation: one sees two cats bound by their tails and never realizes that they are pushing each other apart by their four paws.

The main core of this chapter is devoted to a description of repulsions in crystals, to an analysis of their effects on bulk crystal properties, and to a discussion of approaches to their simulation by theoretical methods. Some of the results will be unexpected in terms of common thinking.

3.2   Simulation Methods

3.2.1   Methodological Premise

We reserve the term "force" to the dynamic aspect of a bond (attraction or repulsion, with intensity related to the first derivative), while the terms "stabilization" and "destabilization" refer to the system being respectively more stable or less stable than the bonded terms at infinite separation.[1] Figure 3.1 shows the possible combinations of the four terms. When the bonding coordinate is an angle (see below, Figure 3.3) the concept of equilibrium value and force survive, but it is presumably pointless to distinguish between attractive and repulsive. Whenever equilibrium is altered towards either side of the minimum the system is destabilized and the force tends to restore the equilibrium condition.

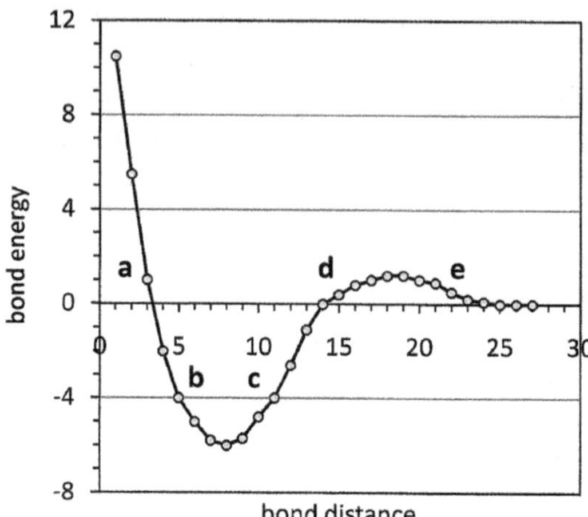

Figure 3.1   Two chemical fragments start at infinite separation (extreme right), climb an activation energy barrier (seldom the case in non-reactive chemistry), and then coalesce at an equilibrium distance R_{eq}, with stabilizing energy ε_0, after which repulsion sets in. $U(R)$ is the potential energy, $F = -dU/dR$ is the force. Region (a) is repulsive ($F > 0$) and destabilizing ($U > 0$), region (b) is repulsive but stabilizing, region (c) is attractive and stabilizing, region (d) is attractive but destabilizing, region (e) is again repulsive and destabilizing.

When two closed-shell organic molecules approach one another in a condensed medium, a convenient partitioning of the interaction energy includes:[2] (i) a *Coulombic electrostatic* term arising from Coulomb potentials among the distributions of nuclei and electrons; this has no minimum and therefore when stabilizing can only be attractive, and when destabilizing can only be repulsive; (ii) a *polarization* term, reflecting the reciprocal influence (distortion of electronic distributions) due to the electric fields of the approaching molecules, and (iii) a *dispersion* term, accounting for the quantum-chemical entanglement of oscillating electrical dipoles (in a London sense), where (ii) and (iii) are always stabilizing and attractive; (iv) a *repulsion* term which arises from the purely quantum mechanical prescription of avoided overlap between closed-shell electrons, as is always the case for the outer electrons in molecular electron distributions. In this respect a distinction must be made between Coulombic repulsion, resulting from charge elements of the same sign, and repulsion stemming from negated overlap; the latter is in fact an aspect of the exclusion principle, and may therefore be called "Pauli" repulsion. Coulombic repulsion is relevant in highly polar molecules, Pauli repulsion is at work wherever there are closed-shell electronic systems – that is, everywhere in crystals. It is not in fact a mechanical force, but rather a prescription that electrons with paired spin avoid one another.

The physical foundations of this partitioning are discussed in detail in a fundamental book by Stone.[2] Although some of its aspects are to some extent parametric, it basically results from second-order Møller–Plesset perturbative expansion, and long experience has shown that it serves and guides chemical intuition pretty well.[3] What has been historically known as "van der Waals" forces and potentials should be more properly renamed the Coulomb–London–Pauli (CLP) scheme.[4] In particular, the dispersion term can be said to account for electron correlation. In a recent formulation of the Density Functional Theory (DFT) computational method,[5] this contribution has been carefully excluded from the quantum chemical treatment and is represented by a semi-empirical correction (DFT-D, dispersion-corrected DFT). The success of this scheme somehow provides dispersion with a better identity and theoretical foundation. Moreover, the semi-empirical correction can be and has been incorporated also in semi-empirical (PIXEL) or even fully empirical, atom–atom computational schemes (see below).

3.2.2 *Ab initio* and PIXEL Methods

In an *ab initio* quantum chemical calculation one obtains a total energy and an electron density distribution for anything from a single atom to a full crystal. When carried out for the stretching or bending or torsional motions, the resulting total energies provide highly reliable energy profiles from which the bond can be defined with its repulsive and attractive branches. In the intermolecular case, the single energy numbers for the interaction between separate molecules do not provide the supplementary insight of the energy partitioning outlined in Section 3.2.1, (i)–(iv).

PIXEL[6] is a partly parametric computational scheme for the evaluation of Coulombic, polarization, dispersion and repulsion components of intermolecular energy between molecules A and B. The exact analytical expression of these components can be found in the documentation of the computer programs for distribution.[7] The evaluation of the intermolecular interaction energy starts with the *ab initio* calculation of the electron densities of two interacting chemical units A and B, $\rho(A)$ and $\rho(B)$, molecules, ions, and even metal atom complexes,[8] provided as a raster of discrete density elements called "pixels"; this term refers to an elementary cell in a two-dimensional picture, so that the more proper name should be "voxels", the three-dimensional equivalent. However, the PIXEL terminology has stuck and can hardly be changed now. This distribution is calculated for a free molecule and the polarization due to the interaction is neglected – a somewhat drastic approximation when molecules are in very close contact or, for example, in crystals at high pressure. For an organic molecule of ordinary size, the number of original pixels is of the order of $N = 10^6$. Since the calculations involve several N^2 terms, to reduce the computational load these are then contracted into $n \times n \times n$ super-cubes with $n = 3$ to 5, so that the final number of pixels, for an accurate performance of what is essentially a numerical integration, becomes of the order of 10^4. Quick but still reliable preliminary estimates can be obtained even with a reduction to 10^3. Then, the Coulombic component is calculated parameter–less for all negative pixels and positive (nuclei) pairs in molecules A and B. The polarization component is a many-body term evaluated by the linear polarization formula, from elementary electrostatics, using the global electric fields acting on the charge distributions. Its parameters here are the atomic polarizabilities and the method for the distribution of polarizability over the molecular sites, an empirical procedure calibrated along with atomic volumes in the molecule.[7] The dispersion component is obtained as pairwise sums of London-type terms between all electron density super-cubes, whose polarizability is again parameterized as above. Finally, the repulsion component is estimated by an empirical formula depending on the overlap integral, evaluated by discrete integration between the original, un-contracted electron density pixels of A and B. Since each charge density element is assigned to an atomic basin, the total overlap is subdivided into contributions from pairs of atomic species m and n, S_{mn}. The expressions are:

$$S_{AB} = \Sigma_{i,A} \, \Sigma_{j,B} \, [\rho_i(A) \, \rho_j(B)] \, V \tag{3.1}$$

$$E_{\text{rep},mn} = (K_1 - K_2 \, \Delta\chi_{mn}) \, S_{mn} \tag{3.2}$$

$$E_{\text{rep,tot}} = \Sigma_{m,n} \, E_{\text{rep},mn} \tag{3.3}$$

where V is the volume of the density pixel, $\Delta\chi_{mn}$ is the difference in Pauling electronegativity between species m and n, and K_1 and K_2 are positive disposable parameters. The overlap integrals are calculated by numerical integration rather than analytically.

By and large the greater advantage of PIXEL is that a molecule is represented by 10^4 point-charge sites rather than by just N sites for a N-atom molecule, as in an atom–atom calculation, with an obvious exponential increase in the accuracy of the description of the electronic structure and with inclusion of penetration energies (see below). The calculation of the interaction energy for an acetic acid dimer takes a few seconds, making possible the calculation of vast potential energy hypersurfaces for dimerization. Lattice energies are obtained as sums of A–B Coulombic, dispersion and repulsion contributions over all molecular pairs in the crystal, while the polarization term involves the overall field from all molecules and is not pairwise-additive. The calculation of the lattice energy of a molecular crystal takes about 30 minutes, allowing statistical sampling of large databases of crystals with their lattice energy components, for example dispersion-dominated in hydrocarbons and moderately polar molecules, or mainly Coulombic-polarization for highly polar compounds or for hydrogen-bonded species. The results have been shown[9,10] to be almost quantitatively comparable with those of refined *ab initio* calculations. The recently proposed q-GRID[11] method uses the same approach but with a subdivision into chemical units of an electron density obtained from the crystal rather than from the isolated molecule, thus better taking into account crystal polarization.

3.2.3 Atom–Atom Empirical Potentials and Monte Carlo Methods

Atom–atom empirical potentials are very coarse-grained tools that with careful use can produce valuable information at a high efficiency/cost ratio. Each chemical unit is represented by one site for each constituent atom, located at the position of the atomic nucleus. The intermolecular potential is a sum over site–site terms, each term being some function of the distance between nuclei. The particular form[4,12] used in this chapter is:

$$E_{i,j} = 1/(4\pi\varepsilon^{\circ})\,(q_i\,q_j)\,R_{i,j}^{-1} - F_P\,P_{i,j}\,R_{i,j}^{-4} - F_D\,D_{i,j}\,R_{i,j}^{-6} + F_R\,T_{i,j}\,R_{i,j}^{-12} \quad (3.4)$$

In this equation $R_{i,j}$ is an internuclear distance, $q_i = F_Q\,q_i^{\circ}$ is the rescaled net charge population on atom i, and F_Q, F_P, F_D, F_R are empirical, disposable scaling parameters for polarization, dispersion and repulsion energies. P, D and T are coefficients calculated on the fly for each pair of atoms in the system, using partial population analysis point-charges, thus adding flexibility to the model.

In comparison with PIXEL, atom–atom point-charge Coulombic energies are largely underestimated due to the lack of the so-called "penetration" energies: these account for the destabilizing interaction between the diffuse outer part of the molecular electron densities, and for the corresponding much larger stabilizing effect arising from interaction between these outer parts and the nuclei of the coupling partners. Point-charge methods, no matter how well parameterized, are intrinsically unable to correctly describe Coulombic interactions due to the charge localization, a very drastic and un-physical approximation. But repulsion energies are also much smaller so that reasonable total energies result from cancellation of two large errors. Conversely, dispersion energies are comparable in the two methods because they both stem from the same London-type parameterization, and, decaying with inverse sixth power of distance, are much less sensitive to delocalization than Coulombic terms. The atom–atom calculation of the lattice energy of large organic molecules typically takes less than one second.

For the present purposes Monte Carlo molecular simulation[12] is adopted for the internal structure and some thermophysical parameters of condensed phases at equilibrium, liquids or crystals. Briefly, the procedure goes through the following steps: (1) prepare a starting configuration (computational box) in the form of atomic coordinates for each chemical item present; (2) compute the starting energy E° by some kind of intermolecular potential, in the present case, eqn (3.4); (3) move one particle at random or change the box volume, and compute E'; (4) accept the move if $E' < E^\circ$ or if $\exp(-\Delta E/kT) > $ a random number between 0 and 1 (the "Metropolis" criterion). For a very large number of moves, after volume and energy have become stationary, the result is a Boltzmann ensemble of equilibrium molecular positions, configurational energies and volume, at the given temperature and pressure. Structural results are usually interpreted in terms of equilibrium distributions of key interatomic distances or of distances between centers of mass, the radial density functions $g(R)$. In the implementation[12,13] for this chapter, molecules are rigid except for torsional degrees of freedom, whose potential energy profiles are evaluated *ab initio* and fitted to cosine or polynomial functions. The enormous advantage of Monte Carlo over quantum chemistry calculations is that one can control temperature and pressure and receives an impression of the amplitude of molecular librations. The obvious disadvantage, at least in the present case, is that only approximate atom–atom potentials can be used.

3.3   Intramolecular Bonds and Bonding

Figure 3.2 shows examples of stretching profiles for what is universally called a "chemical bond" proper. Binding energies are very large and potential wells are very narrow and harmonic, and experimental bond lengths coincide with the predicted value with very narrow esd's. Figure 3.3a shows an example for bond bending, whose energy profile is more open but still almost harmonic. Figure 3.3b shows the corresponding spread of observed bond angle values in crystal structures. The range of experimental values, 105–115°, is strictly parallel to the range allowed by a minimal (of the order of *RT*) energy loss in the potential curve.

Torsional libration produces major conformational changes, with an overall electronic rearrangement along with the rearrangement of nuclei, but one often speaks also of steric clash between distant parts of the molecular system, when they are brought into closer contact by the torsion. Figure 3.4 shows that the HCOC torsional motion occurs at an almost negligible energy price when there is no contact between the extremes (the two esters), but is strongly hindered in methoxybenzene, due to repulsive confrontation between the "closed-shell electrons" of methyl and benzene hydrogen atoms.

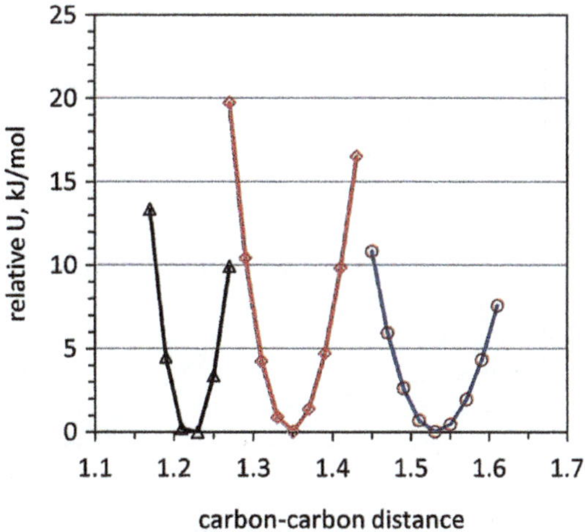

Figure 3.2   Bond stretching profiles for the single, double and triple carbon–carbon bond from MP2/6-31G** *ab initio* calculations on propane, butane, butadiene, butyne. Thermochemical estimates of the absolute binding energies are about 800, 600 and 400 kJ mol^{-1}.

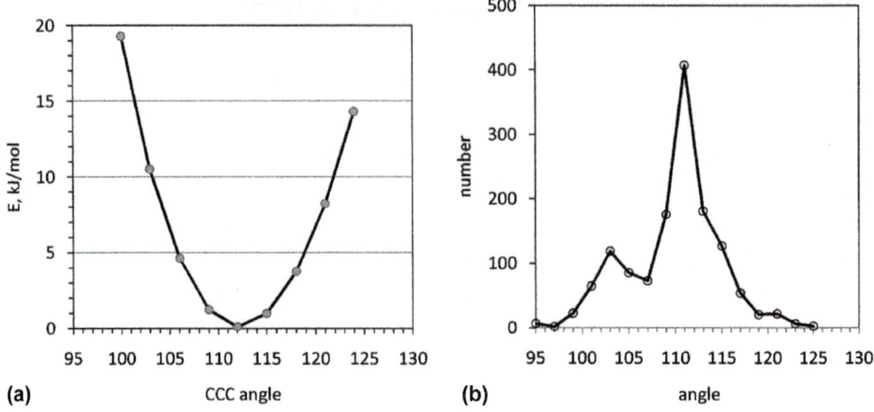

Figure 3.3   (a) *Ab initio* bond bending profile for the C–C–C angle in alkanes. (b) The distribution of alkane CCC angles in a sample of 761 organic crystal structures extracted from the Cambridge Structural Database. The peaks correspond to cyclopentanes (104°) and open C–C–C chains (110°).

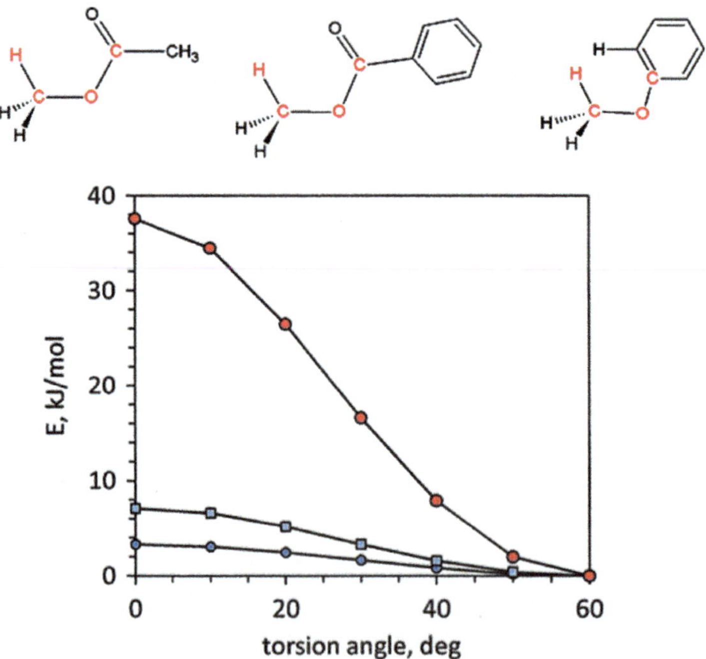

Figure 3.4   Potential energy profiles for the H–C–O–C torsional motions in the upper inset: lower curves, methyl acetate and methyl benzoate; upper curve (red circles), methoxybenzene. The four atoms highlighted in red show the configuration at $\tau(\text{HCOC}) = 0°$. Bond angles are kept fixed. MP2/6-31G** *ab initio* calculations.

With respect to consequences in molecular packing, it is instructive here to compare the energy profiles for the various kinds of intramolecular rearrangements. The intermolecular field has no influence on the very tight bond stretching proper. For bond bending, environmental push or pull may change the bonding geometry in a small but significant way: for example, a change of 5° in the COH angle at a small energy price may shift the target of an O–H···X hydrogen bond by as much as R(O···X) sin 5° ≈ 0.25 Å. Torsional rearrangement poses a different problem. When cohesive forces win over thermal motion, a molecular system condenses and molecules meet their partners in a liquid state. Under the pull of intermolecular adhesive forces molecules remain on average in equilibrium contact, as witnessed by a constant density, but thermal motion also brings them into temporary closer contact. When this occurs, short-range intermolecular repulsion sets in and torsional conformations oscillate in their relatively flat energy wells to respond to forces coming from random directions. Figure 3.5 shows examples of this interplay by means of a Monte Carlo computational experiment: the key torsion angle of the side chain varies within a range of almost 60°. The distribution is even wider under the softer HOCC torsional potential. The HOCC angle may also be influenced by the tendency to optimize

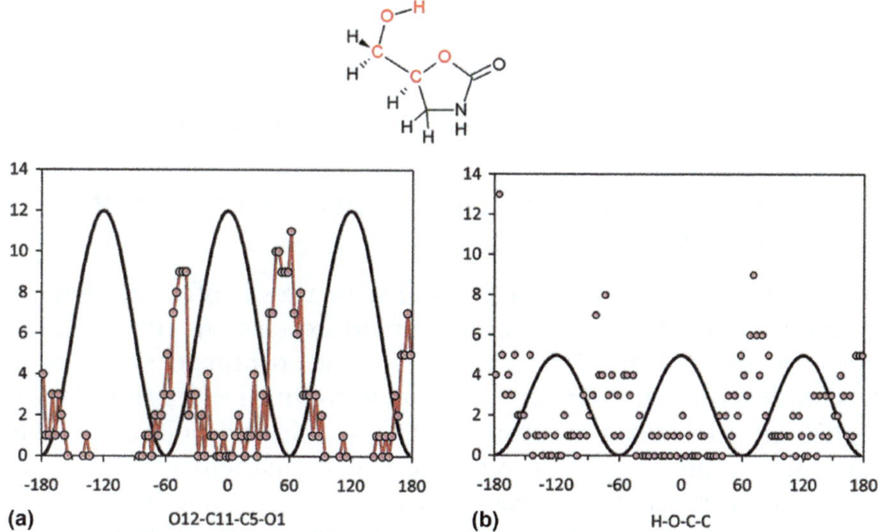

Figure 3.5 Distribution of torsion angles along the chain highlighted in red, in a Monte Carlo simulation of a liquid compound. The black line is the torsional potential, circles denote the number of occurrences of that torsion angle in a computational box of 250 molecules.

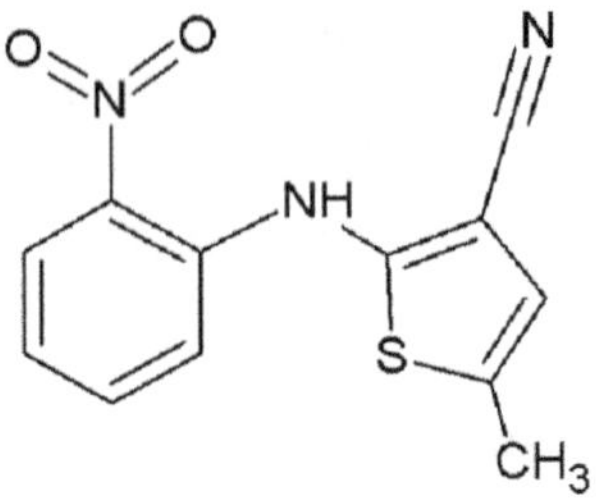

Scheme 3.1

the number of hydrogen bonds, in this case an effect of inter-molecular attraction rather than of repulsion.

As thermal energy decreases further, the system crystallizes and molecules become fixed in space except for residual librational motion. They must find a unique arrangement in which the torsional conformation is frozen so as to optimize the combined effect of environmental repulsion, of environmental attraction (polarization, dispersion and Coulombic) and of intramolecular energy loss. We have here a first hint that molecular arrangements in the solid state may depend on repulsion as well as on attraction, and one also sees the origin of the difficulties in crystal structure prediction. In a typical example the six polymorphs of the thiophene carbonitrile **1** show C–N–C–S torsion angles of −105, −53, +22, +46, +104 and +113° (see Scheme 3.1).[14]

3.4   Repulsion in Intermolecular Bonding

3.4.1   Stabilizing and Destabilizing (Repulsive) Parts of Lattice Energies

The analysis and description of crystal structures is often carried out in generic terms, using for example broad expressions like "van der Waals" *versus* "polar" interactions. Taking advantage of the sub-division described in Section 3.3, some more precise features can be singled out. In a broad assumption, Coulombic and polarization energies are proportional to the amount of permanent charge separation (polarization) in a molecular electron density, and also depend on the extent to which zones of opposite charge are able to get together in the crystal structure, steric influence permitting. While charge polarization in a molecule can be estimated in several ways, including simple population analysis, the best arrangement of opposite charges in different molecules in a crystal is a riddle that still escapes simple intuition. Dispersive energies depend on the number and type of polarizable zones of the electron density, a typical example being the outer electrons of the higher halogens. Dispersive adhesion is anyway present whenever there are electrons, therefore in all kinds of crystals, even ionic crystals – the dispersion term is non-zero even in sodium chloride. The role of Pauli repulsion is very elusive, being related in this empirical model to the amount of formal overlap between neighboring electron clouds. This kind of repulsion increases with the increase of attractive force, granting the equilibrium balance within the structure. The Coulombic contribution to lattice energies is invariably overall attractive, although localized repulsions may appear as described in a further section.

Table 3.1 shows the amounts of the various energy terms in the lattice energies of organic compounds formed by a benzene ring surrounded by all kinds of substituents of different polarity ("electronic demand"). Hydrocarbons are dispersion-dominated with Coulombic terms making up 15–20% of the total cohesive energy. Dispersion increases regularly with the number of methyl substituents, and in *n*-hexane it is even larger than in benzene: contrary to common belief, π-electrons do not seem particularly privileged in the generation of dispersive adhesion. Coulombic energies are small but non-negligible in aromatic hydrocarbons due to positive H-ends interacting with diffuse negative electron clouds, and are in fact much smaller in *n*-hexane. The introduction of aza-substituents decreases dispersion but increases the Coulombic term, due to the C–N charge separation, as expected. In the chlorobenzenes the Coulombic terms show a moderate increase with respect to the hydrocarbons, due to a weak polarity of the C–Cl bond and a minor contribution from $CH\cdots Cl$ contact. The dispersive terms show the expected, massive increase by an almost constant contribution of about 10 kJ mol^{-1} per chlorine atom, irrespective of the substitution pattern. The need to bring chlorine atoms close by to take advantage of the relatively short-range (R^{-6}) dispersion contribution also entails a sharp rise in overlap repulsion in hexachlorobenzene. Interestingly, the Coulombic energy for this crystal structure is destabilizing if calculated by atom–atom point-charge methods, but becomes obviously stabilizing in PIXEL calculations where the contribution of diffuse electron clouds of chlorine is properly taken into account.

In nitro derivatives the Coulombic term rises significantly with respect to hydrocarbons while dispersion stays about constant. The quinone functionality imparts a further increase to the Coulombic

Table 3.1 PIXEL composition (Coulombic, polarization, dispersion, repulsion and total, kJ mol^{-1}) of the lattice energies of substituted benzenes, and experimental sublimation enthalpies.

CSD refcode	Coul.	Pol.	Disp.	Rep.	Total	ΔH_{subl}	
BENZEN07	−16.7	−6.8	−66.1	39.7	−49.9	45	Benzene
TOLUEN	−11.4	−5.5	−66.8	34.4	−49.3	43	Methylbenzene
ZZZITY01	−17.0	−7.6	−86.5	54.2	−56.9	60	1,4-Dimethyl
DURENE05	−17.2	−6.0	−89.2	45.1	−67.3	73	1,2,4,5-Tetramethyl
HEXANE01	−10.7	−4.7	−76.6	44.6	−47.4	51	*n*-Hexane
PYRAZI01	−34.8	−10.6	−61.8	47.1	−60.2	56	Pyrazine (1,4-azobenzene)
TRIZIN01	−25.3	−7.1	−53.9	32.4	−53.9	55	Triazine (1,3,5-triaza)
DCLBEN01	−18.6	−6.4	−86.0	42.3	−68.7	65	1,4-Dichlorobenzene
TCBENZ	−20.9	−7.9	−98.6	51.6	−75.8	75	1,2,3-Trichloro
TCLBZN	−21.5	−9.4	−114.6	64.6	−80.9	80	1,2,3,5-Tetrachloro
PNCLBZ	−24.5	−11.9	−131.8	79.7	−88.4	87	Pentachloro
HCLBNZ11	−23.3	−12.4	−148.3	90.5	−93.5	90	Hexachloro
NITOLU	−24	−7.7	−79.4	37.1	−73.9	79	4-Methylnitrobenzene
DNITBZ11	−38	−10.6	−88.1	45.3	−91.3	96	1,4-Dinitro
ZZZFYW01	−35.5	−9.0	−82.6	38.9	−88.2	87	1,2-Dinitro
BNZQUI03	−40	−10.9	−58.2	48.1	−60.8	68	1,4-Benzoquinone
TCBENQ01	−40.4	−12.8	−102.8	72.2	−83.7	99	Tetrachloro-1,4-benzoquinone
TEPNIT11	−46.7	−11.2	−73.6	47.4	−84.2	89	1,4-Dicyanobenzene
CRESOL01	−53.8	−23.0	−82.5	85.9	−73.4	74	4-Methylphenol
DMEPOL10	−51.9	−24.1	−92.1	85.1	−83.0	76	2,6-Dimethylphenol
NITPOL02	−71.5	−28.8	−90.1	95.8	−94.6	94	4-Nitrophenol
BZAMID01	−74.6	−26.5	−83.8	83.9	−100.9	101	Benzamide
MNIANL02	−50.5	−20.4	−93.2	72.9	−91.3	95	3-Nitroaniline
NANILI02	−64.2	−20.4	−93.7	79.7	−98.6	100	4-Nitroaniline
BENZAC	−85.3	−40.1	−75.0	109.6	−90.8	90	Benzoic acid
PFBZAD10	−91.3	−42.6	−71.1	119.1	−86.0	91	4-Fluorobenzoic
DMBZAC01	−85.1	−41.5	−95.6	124.4	−97.8	103	2,3-Dimethylbenzoic
DMOXBA01	−94.7	−44.7	−102.9	113.8	−128.5	122	2,6-Dimethylbenzoic
BENZDC01	−160.7	−82.6	−99.0	210.2	−132.1	110	1,3-Dicarboxylic
TEPHTH12	−178.6	−86.2	−106.1	223.1	−147.8	135	Terephthalic

energy, and comparison between benzoquinone and its tetrachloro analogue shows no increase in Coulombic energy and the expected increase in dispersion energy by 10 kJ mol^{-1} per chlorine atom. The dicyano derivative sees a further increase in Coulombic contribution (35%) due to the presence of the polar end groups. In a demonstration of complex interplay, tetrachlorobenzoquinone and terephthalonitrile have the same total lattice energy because chlorine atoms bring about larger dispersion but also larger Pauli overlap.

The crystals of hydrogen-bonding derivatives display a quite different picture. In phenols and benzamides the sum of Coulombic and polarization energies is almost on a par with dispersion. In crystals of benzoic acids, Coulombic-polarization terms are predominant. The fluoro derivative is identical to the unsubstituted acid. As could have been anticipated, the Coulombic and polarization energies of crystals of bifunctional acids are roughly twice those of the monofunctional counterparts. At this point however, the effects are mixed and simple speculation in polarity terms runs into trouble. For example, *p*-nitrophenol has a higher Coulombic contribution than *p*-cresol as expected, but the introduction of nitro groups in the benzamide framework brings about a decrease in Coulombic energies. The reader is free to elaborate his or her own explanation in terms of NH···N *versus* NH···O bonding, or to speculate on the reasons why the 2-nitroaniline has a substantially lower Coulombic term than the 4-nitro analogue, or on the reasons for the differences for example between the two dimethyl derivatives of benzoic acid.

These examples illustrate the difficulties one runs into when effects that can be explained singularly in simple terms become competitors – if need be, a further confirmation of the complexity of the crystal prediction problem, where intuition fails and substantial progress has been achieved only at the expense of thousands of hours of brute-force calculation using high-level quantum chemical methods.[15]

3.4.2 Hydrocarbons and Halogenohydrocarbons

As expected and as already noted quantitatively in Section 3.4.1, in crystals of scarcely polar or non-polar molecules dispersion energy plays a predominant role. Therefore, a larger stabilization is expected with molecular systems that include aromatic rings, which are supposed to provide easily polarizable π-electron clouds. This is the basis of the concept of "π–π" interactions, another of the many broad concepts of intermolecular science that have not gone without criticism when carefully examined.[16] In a rather naive assumption one may think that this should lead to aromatic π-stacking in crystals, whereby planar rings condense into parallel arrangements. It is however a fact that none of the simpler flat aromatics like benzene, naphthalene or anthracene, adopt this arrangement in their crystals, preferring what comes to be called the "herring-bone" pattern in which interplanar angles are of the order of 40–60°. Table 3.2 shows an analysis of some π-stacking arrangements with separate contributions to the cohesive energies. These data show that columnar

Table 3.2 PIXEL results for the Coulombic, polarization, dispersion, repulsion and total binding energies (kJ mol^{-1}) of aromatic stacked dimers at an inter-ring distance of 3.6 Å, and of parallel *n*-pentane chains at a distance of 4.8 Å.

Compound	E_{coul}	E_{pol}	E_{disp}	E_{rep}	E_{tot}
C_6H_6–C_6F_6	−7.5	−2.2	−22.2	11.7	−20.2
C_6F_6–C_6F_6	2.7	−2.1	−22.4	10.7	−11.2
C_6H_6–C_6H_6	2.7	−2.1	−22.0	12.9	−8.5
n-Pentane	−1.1	−0.5	−14.8	7.5	−8.9

arrangements of stacked aromatic rings would generate non-negligible repulsive Coulombic interactions as well as substantial Pauli overlap repulsion; what is more, such columns would then have to pack side by side, generating further unfavorable, or at least very scarcely stabilizing, contacts between net positive electron-density regions of neighboring hydrogen atoms. This is just one of the many examples in which reasoning based on bimolecular synthons but neglecting three-dimensional packing requirements often fails. Packing at high interplanar angles is usually explained in terms of stabilizing Coulombic factors between CH groups and π-electron clouds, but repulsion avoidance is an equally important factor. On the other hand, the very common flat arrangement of aromatic rings in benzene-perfluorobenzene co-crystals[17] is nicely explained by the stabilizing Coulombic contribution coming from the reverse polarity of CH and CF bonds.

And yet, most of the time flat aromatic rings are seen to stack in crystal structures. The fact is that in crystals of complex organic molecules, aromatic-substituent rings have to cope with the packing requirements of many other groups – polar groups, hydrogen-bonding functions – and the least disturbing steric arrangement is in stacked pairs. Admittedly there is no definite quantitative proof of this statement, which looks however fairly reasonable. In these provocative terms, much of the literature invoking structure-defining aromatic stacking interactions in organic crystals would be somewhat off the mark. More on the twists of this elusive concept in the next section.

How deceiving common intuition can be in matters of inter-molecular interaction is shown by the last entry in Table 3.2: parallel aliphatic chains are as stabilized as parallel aromatic rings. By a lower Pauli overlap repulsion, owing to the contracted nature of the C–H electron density, plus a minor but stabilizing Coulombic contribution, aliphatic chain stacking gains what it misses in dispersion.

Figure 3.6 and Table 3.3 show an analysis of lateral interactions. The hydrocarbon, as expected, shows a destabilizing Coulombic term, which becomes mildly stabilizing in the fluoro derivative. A decisive stabilization is achieved by the chlorine derivative in terms of dispersion, also as expected. Due to the diffuseness of the bromine electron cloud, a sharp rise in Pauli overlap repulsion destabilizes the bromo derivative dimer faster than it can be stabilized by an increase in dispersion. All considered, these numbers reveal that lateral

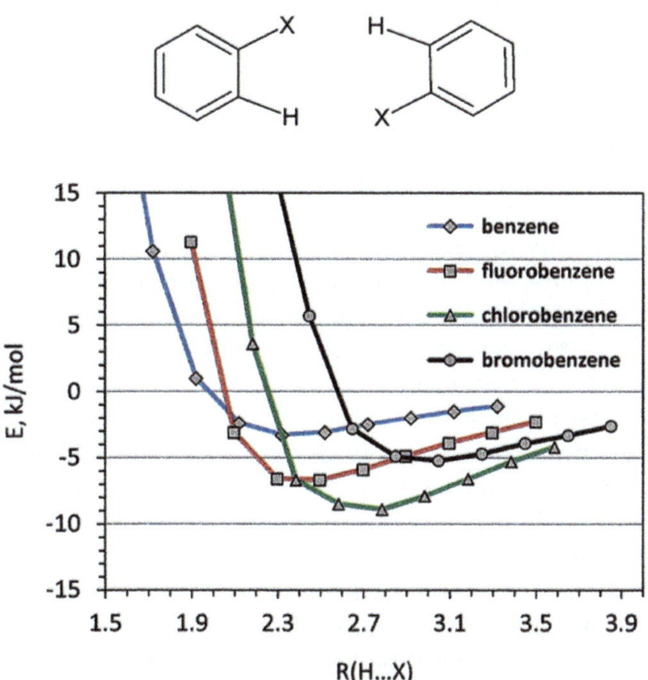

Figure 3.6 The interaction energy profiles for coplanar dimers with X = H, F, Cl and Br. Free stretching oscillation of the bonds at room conditions ($RT = 2.5$ kJ mol^{-1}) is of the order of 1 Å. Compare with the narrow profiles in Figure 3.2 (mean oscillation 0.1–0.2 Å).

Table 3.3 PIXEL results for the binding energies (kJ mol^{-1}) of coplanar dimers of monosubstituted benzenes at the minimum of their interaction energy curves. See also Figure 3.6.

CH$\cdots$X distance		E_{coul}	E_{pol}	E_{disp}	E_{rep}	E_{tot}
X = H	2.32	0.3	−1.1	−8.0	5.5	−3.3
X = F	2.50	−4.3	−1.1	−6.4	5.0	−6.7
X = Cl	2.79	−3.2	−3.2	−16.3	13.8	−8.9
X = Br	2.85	−3.4	−3.2	−16.3	18.0	−4.9

interaction of the C–H···halogen type are very scarcely relevant in terms of cohesive energy, going from about 1.5 kJ mol^{-1} per bond for the hydrocarbon to a maximum of 4.5 kJ mol^{-1} for the chloro derivative. In spite of the conspicuous literature,[18] there seems to be no grounds to claim these interactions as generally relevant or even structure-defining in organic crystals.

3.4.3 Repulsion in Crystals of Neutral but Polar Molecules

Molecules with strongly polar group are to some extent similar to zwitterions (see next section), except that the involved charges are partial and the effects are much smaller because the Coulombic potential goes with the square of the charge. Nevertheless, it is to be expected that the crystal structures of these polar compounds may contain some degree of repulsion–destabilization due to the need to accommodate groups of same charge along with stabilizing interactions involving groups of opposite charge. Dispersion and polarization terms, always stabilizing, can help overcome the consequences of adverse Coulombic terms. Cases with significant repulsion in crystals are not frequent, a reasonable estimate being 10%,[19] but are not negligible either.

For a first example, a methodological problem related to possible repulsions in polar molecules is posed by contacts between oxygen-rich nitro groups. Figure 3.7a shows a comparison of the frequency function for the distance between oxygen nuclei in crystals of nitro compounds and in crystals of general organic molecules.[20] The prominent (DDF $\gg$ 1) peak for nitro compounds clearly indicates that in these crystals oxygen nuclei come into close contact much more frequently than would pertain to a random distribution, and quite often much closer than the standard sum of average atomic radii. In the style of present-day tendencies to nominate atom–atom bonds whenever such a situation occurs, one would be tempted to postulate an oxygen–oxygen intermolecular bond, which seems a sharply counterintuitive assumption. A close look at one typical structure with a very short O···O contact shows (Figure 3.7b) that in all likelihood the contact is forced by the lateral pull of the hydrogen bonds. In fact, the molecular dimer obtained by clipping away the NH(R) groups is barely cohesive at all, while a nitromethane dimer in the configuration of the two nitro groups has a destabilizing Coulombic interaction of 5.0 kJ mol^{-1}.

These findings prompt a further more quantitative analysis involving the detection of destabilizing molecule–molecule energies in

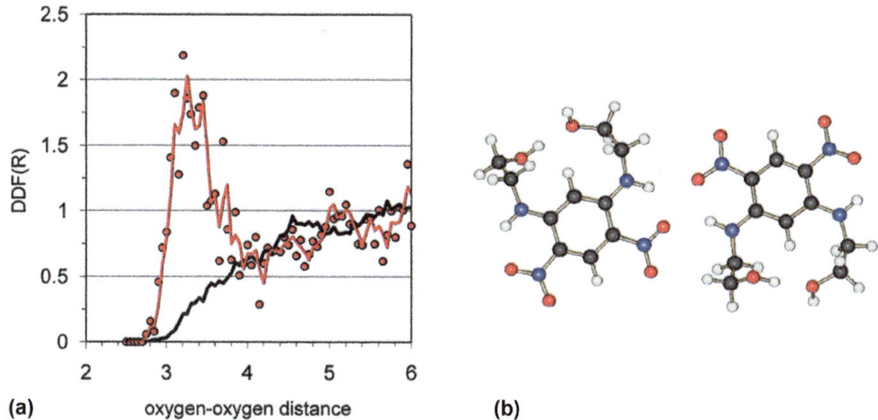

Figure 3.7   (a) The distribution functions for O···O distances in crystals of nitro compounds (red circles) and general organics (black line). The sharp peak indicates frequent O···O contacts. (b) In the central square pattern (oxygen red, nitrogen, blue) a short O···O distance of 2.74 Å (against a sum of standard radii of 3.16 Å) is forced by the concomitant, strongly stabilizing double N–H···O hydrogen bonding. CSD refcode GERPOJ. A nitromethane dimer with the same configuration of NO_2 groups is destabilizing by 5 kJ mol^{-1}.

crystals of polar compounds. Over a sample of 255 crystal structures of variable chemical composition, all molecule–molecule interaction energies destabilizing by more than +10 kJ mol^{-1} have been sorted out by PIXEL analysis. Figure 3.8 shows how the combination of collateral packing potentials can induce destabilizing proximities of nitro groups, more than compensated for by the energetic gain of the more strongly stabilizing partners. Figure 3.9 shows how flat molecules endowed with strongly polar moieties manage a subtle interplay between the need to stack their aromatic planes in parallel, gaining in dispersion contributions, and the 'collateral damage' of pairing strong, parallel local dipoles with repulsive Coulombic interactions, by a modulation of the inter-ring distance. Figure 3.10 shows the extreme case of the interaction between stacked molecules of hydroxypyridine carboxylic acid by calculating the energy profile for displacement of the two molecules along the vector joining their centers of mass. The interaction has a repulsion that increases with decreasing inter-ring distance, then finds a minor help from dispersion generating a minimum in the destabilizing region, after which Pauli repulsion takes over and the repulsive short-distance wall becomes very steep.

These examples highlight the efficiency of theoretical methods that include separate energy contributions in discussing packing effects,

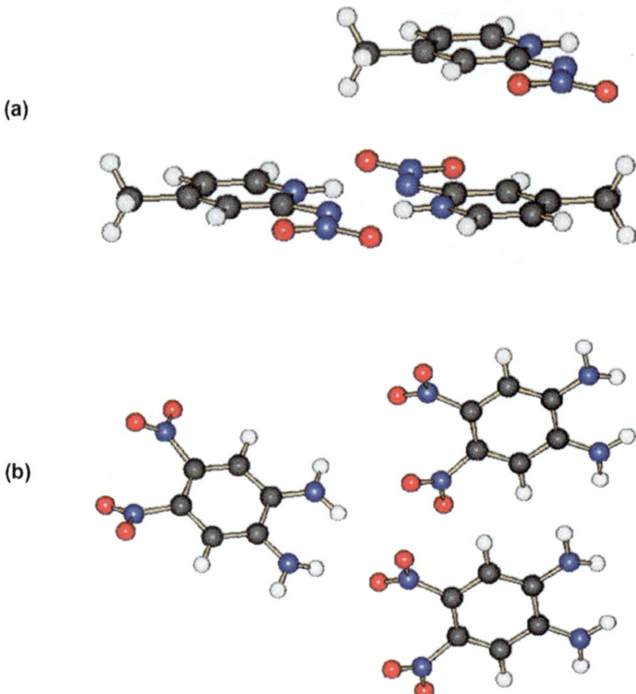

Figure 3.8 Repulsive interactions in nitro compounds: (a) the large stabilization due to close antiparallel vertical stacking and DNA base-type horizontal hydrogen bonding brings about a Coulombic repulsion between the diagonal pair ($+14\ \text{kJ mol}^{-1}$). CSD refcode BENZOK. (b) A two-pronged NH$\cdots$ON hydrogen-bonding pinch and a close repulsive contact ($+15\ \text{kJ mol}^{-1}$) between the two molecules on the right. CSD refcode FELDUW.

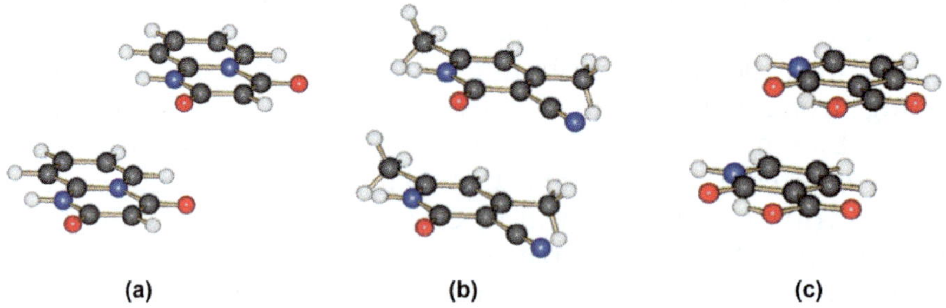

Figure 3.9 Repulsive stacking of parallel polar groups. In (a) the rings are too far for dispersion to help, and the interaction is overall repulsive (Coulombic $+23$, total $+8\ \text{kJ mol}^{-1}$). In (b) the rings are close enough so that a substantial dispersion stabilization (-37) counterbalances the repulsive Coulombic interaction ($+12$, total $-10\ \text{kJ mol}^{-1}$). In (c) rings are close but Coulombic and overlap repulsion keep the overall interaction repulsive by $+3\ \text{kJ mol}^{-1}$ in spite of a significant dispersion stabilization. CSD refcodes CUGBIP, ERISIH and PIMBAP.

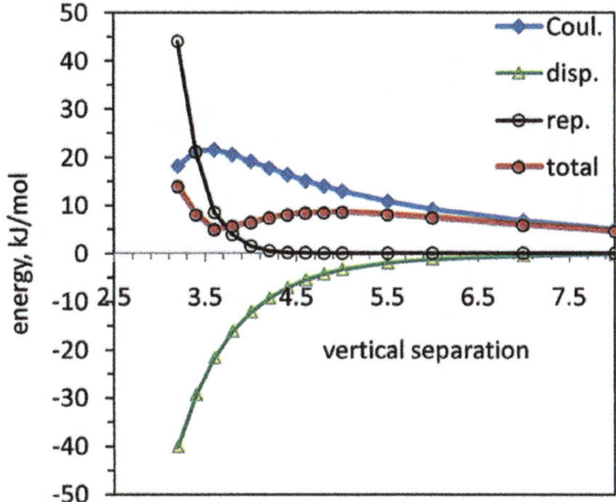

Figure 3.10   The PIXEL energy components for the approach of stacked molecules (as in Figure 3.9c) as a function of inter-ring separation: blue, Coulombic, green, dispersion, black, repulsion, red, total interaction energy. A rare example of the existence of an approach barrier and of a minimum at a destabilizing interaction energy.

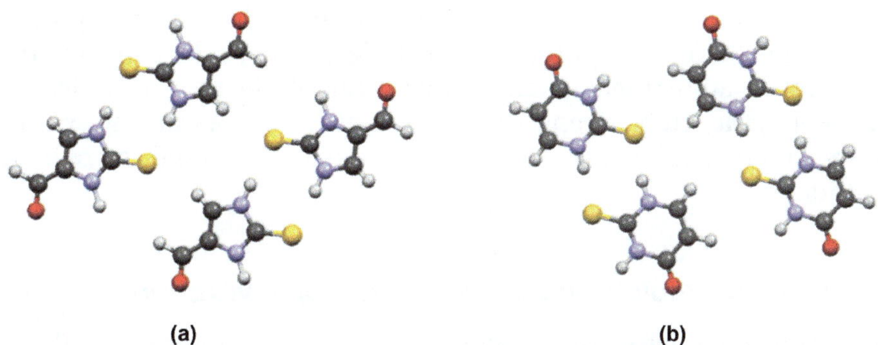

Figure 3.11   Mediating a repulsive S···S interaction: the positive bay formed by NH and CH hydrogens interacts with sulfur atoms and creates a favorable packing environment in spite of repulsive interaction between the two molecules carrying opposing sulfur atoms (+10 to +13 kJ mol^{-1}). CSD refcodes FORIZT10 and TURCIL02.

and issue at the same time a warning against prejudicially judging all stacking interactions as stabilizing.

Figures 3.11 and 3.12 show other examples of repulsive interactions that appear as a result of, or driven by, other, much more favorable, stabilizing contacts. Destabilizing molecule–molecule energies appear with sharply unfavorable S···S and S=O···O=S contacts, as the Coulombic repulsion is overcome by collateral, very favorable

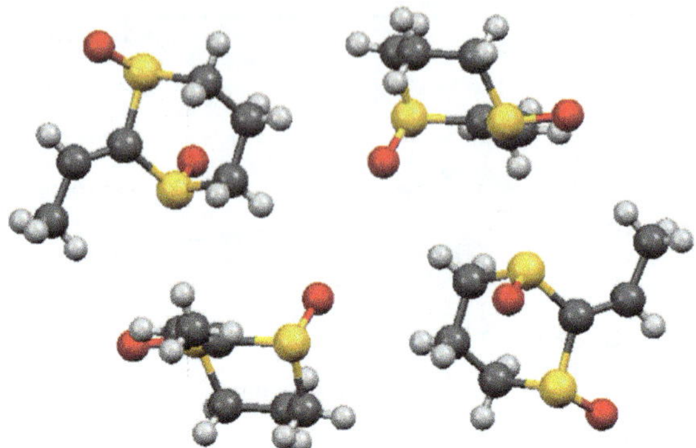

Figure 3.12 Mediating repulsive S=O···O=S interactions: the positive regions of methyl hydrogens interact with oxygens and counteract the destabilizing interaction between molecules carrying opposite oxygen atoms. CSD refcode GAKNIP.

couplings between these negatively charged, repelling groups and an array of positively charged (C)–H or (N)–H hydrogen regions. The pool of all these interactions is overall stabilizing of course, and one may like to sort out single atom–atom contacts and call them hydrogen bonds, sometimes not a simple matter because of the high degree of bifurcation. A more appropriate attitude could be to consider the cooperative adjustment in the whole array of electron densities. In particular, the study of repulsion issues here acts as a warning against hasty interpretations of S···S interactions as crystal stabilizing factors.

3.4.4 Coulombic Repulsions in Ionic or Zwitterionic Solids

In organic chemistry, ionic solids are usually encountered as salts of organic anions with alkali or alkaline-earth metal cations, or as halides of organic cations. In these crystals Coulombic interaction energies are one order of magnitude larger than those in non-ionic crystals, as it must be, recalling that the Coulombic energy between two unit point-charges at a distance R is $E_{Coul} = 1389.34/R$, with R in Å units. Figure 3.13 shows a sample calculation of the intermolecular energies between constituent ions in the salt dipotassium croconate.[21] The pure Coulombic potential has no minimum and therefore stabilizing contacts are always attractive and destabilizing contacts are always repulsive. As a consequence, ionic salt crystals establish an equilibrium between strong repulsion between ions of same sign and strong attraction between ions of opposite sign. It is only through

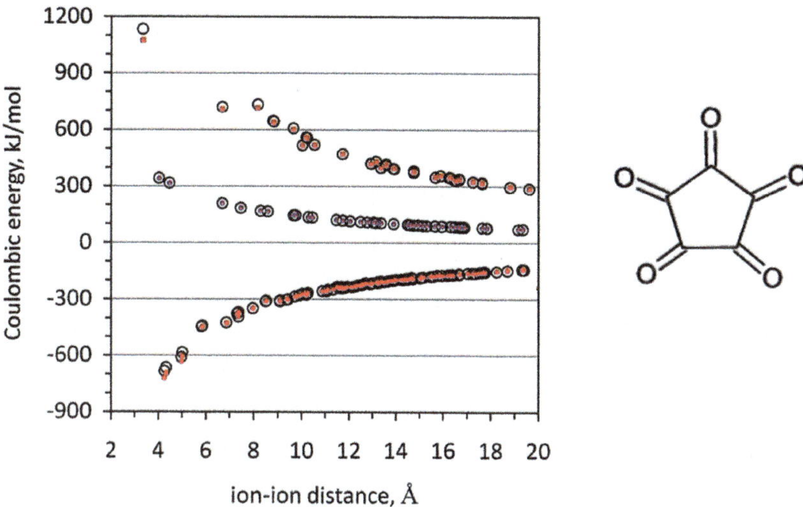

Figure 3.13   Coulombic energies in the ionic crystal of the potassium croconate salt, $K_2(C_5O_5)^{2-}$. Circles: PIXEL energies, red points: point-charge energies. Series are, from top to bottom, anion–anion, cation–cation and anion–cation energies. Croconate anion shown in inset.

Madelung-type sums that such crystals acquire an overall stabilization by the prevalence of terms between opposite charges over terms of same charge.

Zwitterionic organic solids are common because all amino acids crystallize as such.[22] We take the simplest one, glycine, as a test molecule for illustration. Before considering the actual crystal packing in glycine, a few computational experiments using glycine dimers may serve as an illustration of the involved forces. The approach of COO^- to NH_3^+ groups (Figure 3.14a) is strongly stabilizing due to the Coulombic contribution, which is identical to the total cohesive energy because overlap repulsion and dispersion–polarization terms cancel exactly. Interactions between groups of same charge give rise to the expected strongly repulsive profile (Figure 3.14b). When condensing into a crystal the glycine molecules must accommodate somehow a certain amount of repulsive destabilization. In fact, in the actual crystal there are many strongly stabilizing pairs but also a number of moderately repulsive molecular pairs (Table 3.4), the price to be paid for the overall stability of the crystal. Dispersion contributions are small, as well as Pauli overlap terms, which are negligible due to the small size of the molecule and to the contracted electronic shell of the cationic part of the molecule. The repulsion here is always of electrostatic nature.

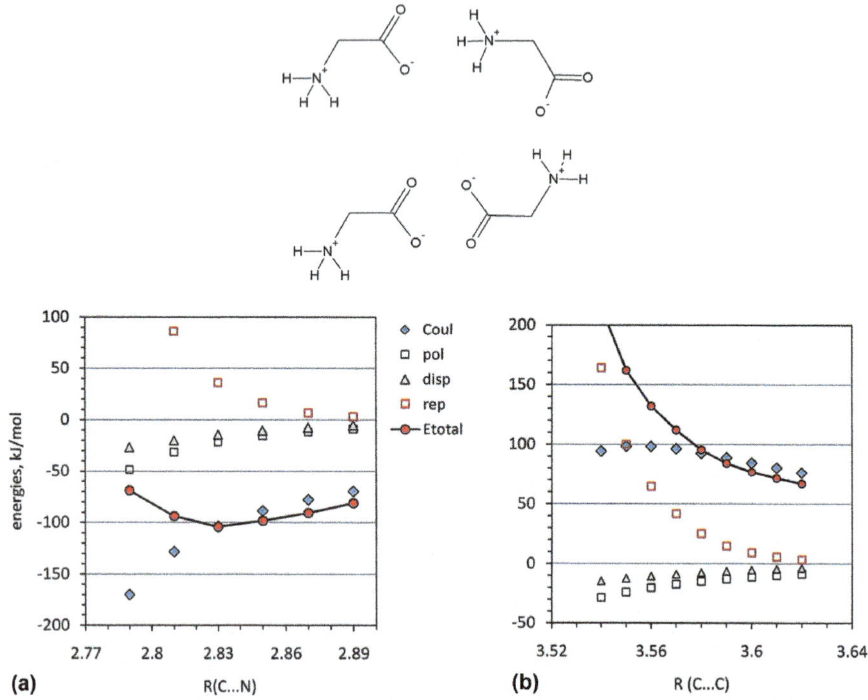

(a) R(C...N)   **(b)** R (C...C)

Figure 3.14 Dimer interaction profiles for glycine: blue, Coulombic, diamonds; dispersion, black squares; polarization; red squares Pauli repulsion. (a): $COO^-\cdots H_3N^+$ confrontation in the zwitterion, strong Coulombic stabilization. (b): confrontation of groups of same charge, strongly destabilizing. Note how the Coulombic term is identical to the total energy until Pauli repulsion sets in.

Table 3.4 Some molecule–molecule PIXEL energies ($kJ\,mol^{-1}$) in the crystal structure of glycine (CSD refcode GLYCIN03): strongly stabilizing but also destabilizing pairs.

Symmetry opn.	Distance	E_{coul}	E_{pol}	E_{disp}	E_{rep}	E_{tot}
$-x,-y,-z$	3.137	−151.8	−40.9	−25.0	41.4	−176.3
$1-x,-y,-z$	4.695	−103.9	−29.4	−18.4	23.5	−128.2
$x,y,1+z$	5.465	−123.2	−41.2	−14.7	74.0	−105.0
$x-1,y,z-1$	5.941	21.1	−2.9	−2.1	0.6	16.8
$-x,-y,1-z$	6.858	26.6	−1.4	−0.4	0.0	24.8
$x-1/2,1/2-y,z-1/2$	4.338	50.3	−10.2	−10.9	9.6	38.8

The case of glycine emphasizes some overlooked dangers in the interpretation of crystal packing in the simple terms of short atom–atom distances, taken to represent stabilizing interactions often on subjective choice. Figure 3.15 shows how misleading this can be, as the order of stabilization inferred from short distances is completely

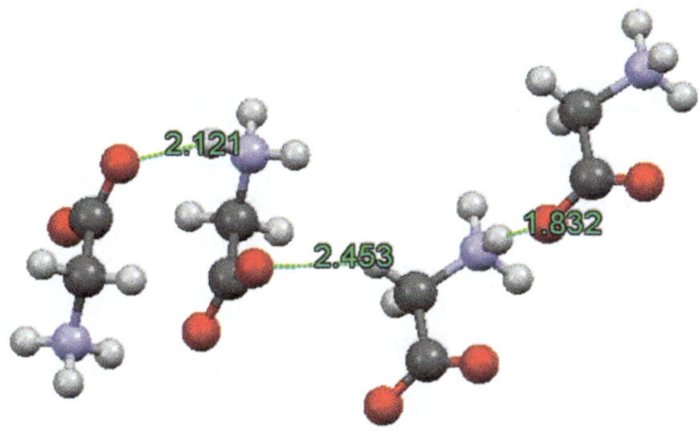

Figure 3.15 Judging glycine crystal stability by atom–atom distances alone would give, left to right, a weaker (longer) N–H···O hydrogen bond, a strong, very short C–H···O bond (sum of atomic radii 2.78 Å), and a stronger (shorter) N–H···O hydrogen bond. In fact, the pair interaction energies (Table 3.4) show that the order of stabilization is reversed and the presumed stabilizing C–H···O bond actually belongs in a destabilizing inter-molecule contact. PIXEL results using a neutron diffraction crystal structure (GLYCIN03).

overturned when full molecule–molecule energies are considered. The key issue here[23] is the lack of appreciation of repulsion.

3.4.5 Effects of External Pressure

In stable, tightly packed organic crystals molecular motion is restricted by repulsion from close neighbors. In proper physical terms one should rather think in terms of phonons, or collective molecular motions. Phonon oscillation that brings lattice molecules into short contact or even collision have higher frequencies and are more energetic. For a simple example consider a row of cylindrical molecules packed perpendicular to the row line. Oscillations along the direction of the molecular axis involve some friction but are less expensive than oscillations along the row line that have molecules bumping into one another.

At standard conditions organic molecules in their crystals never climb too far up the repulsive side of the intermolecular potential. A straightforward way of probing quantitatively the steepness of intermolecular repulsion walls is by compression, that is, by applying an external pressure and studying the variation of cell volume and other thermophysical quantities. Thanks to considerable progress in the quality of experimental apparatus, high-pressure X-ray diffraction is today a fashionable and relatively frequent enterprise.[24] In a comprehensive survey it has been possible to collect from the Cambridge Database[25] a set of 44 room-temperature crystal structure

Table 3.5   List of crystal structures studied as a function of pressure with chemical names or Cambridge Structural Database refcodes. The value of zero-pressure bulk modules, shown in Figure 3.16, increases top down.

Type	
Small molecules, weak or no H-bonds	Dichloroethane, two trichloroethanes, dichloroacetic acid, iodomethane, imidazole (two polymorphs), pyrazole, benzimidazole, pyridine *N*-oxide, cyclohexanedione, nitromethane, aminobenzophenone
Mostly hydrocarbons	Naphthalene, bianthracene, *t*-butylacetophenone (NAPHTA, DANTEN, DECWAJ)
Benzaldehyde oximes	ABULIT, MIGPAU, NIRJAA, NIRJEE, SALOXM
OH···O or NH···O hydrogen bonding	NALCYS, BISMEV, BOQQUT, GOFPUN, HXACAN, SALMID
Zwitterionic amino acids	L- and DL-alanine, L- and DL-serine, L-cysteine, glutamine
Flat selenium heterocycles	AFUDEL, IZOXOL
Quaternary N zwitterions	Betaine, betaine hydrate, sarcosine (KICCOO, WEMWEQ, YIHHON)
Singles	Croconic acid (GUMMUW), a dithiadiazolyl radical (YOMXUT), sucrose, chloropropamide (BEDMIG), oxalic acid hydrate (OXACDH)

determinations for organic molecules, as a function of pressure, with complete determination of space group and molecular coordinates, and such that there be at least three different pressure values up to at least 1 GPa (see list in Table 3.5). For these structures cell volume was plotted against pressure and fitted by a second-order polynomial to determine the zero-pressure experimental compressibility and bulk modulus according to the following equations:

$$V = a\,P^2 + b\,P + c \tag{3.5}$$

$$k = (1/V^\circ)(\mathrm{d}V/\mathrm{d}P)^\circ = (b/c)\ \mathrm{GPa}^{-1} \tag{3.6}$$

$$K^\circ = -1/k\ \mathrm{GPa} \tag{3.7}$$

where the zero superscript indicates zero pressure, k is the compressibility and K° the bulk modulus.

The polynomial fit proved excellent in all cases and the accuracy of the results is only limited by the accuracy of the experimental cell volumes, considering that the high-pressure data are often collected in non-optimal diffraction environments. For a restricted set of 31 of the simplest cases – regular organics without solvates or many molecules in the asymmetric unit – the lattice energies were also calculated at zero pressure.

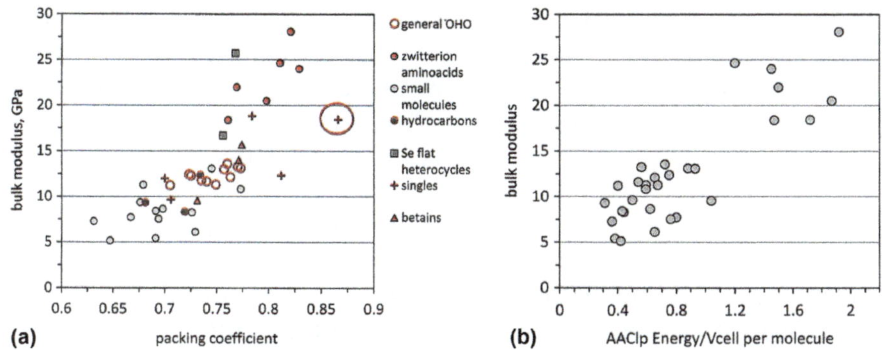

Figure 3.16 (a) A plot of bulk modulus *versus* packing coefficient for 44 crystal structures studied by X-ray diffraction as a function of pressure. (b) For a subset of 31 crystals, a plot of bulk modulus against lattice energy density (total lattice energy divided by the cell volume per molecule).

Finding simple relationships between molecular structure and crystal bulk moduli is not a simple matter. Compressibility seems to depend on a host of electronic and structural factors. Figure 3.16a gives a broad overview of the distribution of bulk moduli in our crystal samples, plotted as a function of the packing coefficient for lack of a better coordinate – one can hardly speak of a correlation in that figure. The distribution is however subdivided into broad zones: a first zone with $K° < 10$ GPa, that is soft crystals, comprises the small molecules like ethanes, the hydrocarbons and some molecules with weak NHN hydrogen bonds (imidazole, pyrazole). A second broad region with $10 < K° < 15$ includes a number of organic molecules of ordinary C–H–O constitution with stronger OHO hydrogen bonds, plus, rather surprisingly, the zwitterionic betaines. The region of higher bulk modulus includes the zwitterionic amino acids, the organoselenium compounds, sucrose and croconic acid: this last crystal structure has a very tight packing of very hard, non-polarizable oxygen atoms strictly joined by hydrogen bonds. For quick reference, the experimental bulk modulus of liquid water (all data in GPa) is about 2, that of steel is 160 and that of diamond is 443. The calculated bulk modulus of the fully ionic crystal disodium oxalate (NAOXAL) is 35 (that of sodium chloride is 24). For a qualitative appreciation of such results, an average bulk modulus of 10, or a compressibility of -0.1 GPa^{-1}, means that an increase in pressure of 1 GPa brings about a 10% reduction in volume, or a modest 3.5% reduction in linear dimension.

Intuitively a high bulk modulus should pertain to strongly bound crystal structures, but this is but a doubtful assumption; for example, intramolecular flexibility could also contribute. Figure 3.16b shows a plot of bulk modulus *versus* the crystal energy density, that is the total calculated lattice energy divided by the cell volume per molecule. Again one can hardly speak of a real trend but the subdivision into general regions is about the same as in Figure 3.16a.

Figure 3.17 shows the PIXEL representation of lattice energy trends as a function of pressure in the X-ray structure determination at high pressures.[26] The resistance of the material to external pressure is mainly exerted by the overlap repulsion term. Coulombic and polarization terms become more stabilizing, because as molecular electron densities become closer the nuclei–electron contribution grows faster than nuclei–nuclei repulsion. The dispersion term also becomes more favorable as distances between electron density elements become smaller. The total energy change is rather minor, with some uncertainty, being the small result of large differences. The PIXEL picture suffers of course of several approximations, but the trend of contributing energies as a function of decreasing distance between intermolecular termini is confirmed by high-level quantum chemical calculations by intermolecular perturbation theory, IMPT and SAPT.[27–29] While Figure 3.17 indicates that the lattice energy is a minimum at room (zero) pressure, in other cases the computed lattice energy shows a minimum at higher pressure. This is understandable in terms of the balance of terms mentioned in the above discussion. Real equilibrium should of course be discussed in terms of free energy, leading to a complex argument which is beyond the scope of the present work.

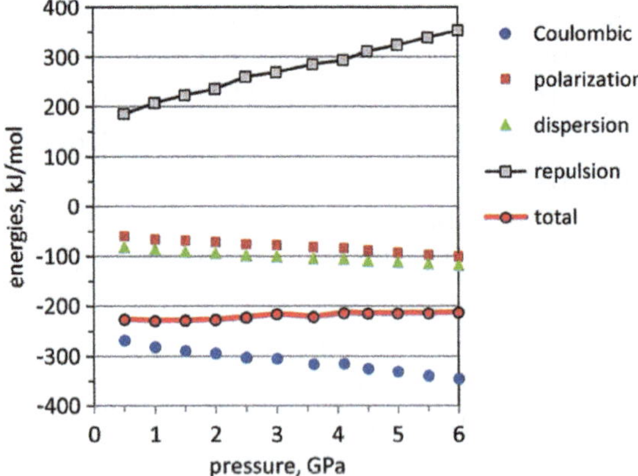

Figure 3.17   The components of the lattice energy of DL-alanine as a function of pressure, as provided by the PIXEL model.

3.4.6 Hydrogen Bonding: Crystal Engineering *versus* Crystal Prediction and Control

The hydrogen bond is generally considered the most reliable "synthon" in crystal structure discussion and prediction. Benchmark *ab initio* calculations at the higher level of theory[30] have established that the stabilization due to single hydrogen bonds in molecular dimers is of the order of 20–60 kJ mol^{-1} depending on the nature of the implied atoms and on the concomitant action of neighboring groups. In the PIXEL approach the analysis of relative contributions shows a very large Coulombic component, contrasted by a large repulsion component. However, a detailed discussion of separate contributions should be taken with care, because the medium-strong hydrogen bond has at least a marginal component of covalency. In the PIXEL approximation, molecular electron densities are calculated over isolated molecules and are kept rigid and undeformed when the coupling pairs approach one another, thus neglecting by definition any effect or contribution due to breakdown of the closed-shell character with incipient formation of a bi-molecular wavefunction. Overall bond energies are on the other hand correct and reproducible due to careful parameterization.

A perspective of hydrogen bond action in crystals can be acquired from systematic studies of existing cases. The present analysis is restricted to the X–H$\cdots$Y case with X and Y = oxygen or nitrogen, and to couplings where the H$\cdots$Y distance is a small fraction (<80%) of the sum of the atomic contact radii, in a way the most classical definition of hydrogen bonding. A database of 11429 crystal structures of hydrogen-bonding, multi-functional organic molecules has been prepared from the Cambridge Structural Database, and the frequencies of appearance of hydrogen bonds between selected donor–acceptor pairs is given in Table 3.6. As a guide for its interpretation, for example, a hydrogen bond between a carbonyl group and an acid hydrogen appears 1048 times in the database, and the bond forms 91% of the times when that pair is present. Overall, carbonyl acceptors are involved in hydrogen bonds 67% of the times when they appear, while acid hydrogen donors are involved in hydrogen bonds for 94% of the cases.

As it appears, 30% of alcohol hydrogens, 28% of amide hydrogens and as many as 52% of amine hydrogens within the database are not involved in H-bonding; from the acceptor side, the situation is even more problematic, as no more than about one third of acceptor nitrogens and alcohol oxygens are involved. Thus, frustrated donors

Table 3.6 Relative frequencies of hydrogen bond appearances in a large database of crystal structures.

| Acceptor | Donor | | | | | | | | Overall % acceptor[c] |
| | O–H Alcohol | | COO–H Acid | | CON–H Amide | | R_2N–H Amine | | |
	N^a	$\%^b$		N%		N%		N%	
=N– sp^2	699	53	123	6	246	24	1257	54	35
(C)≡N nitrile	56	36	—		18	13	141	45	30
O(H) alcohol	2003	45	67	40	194	34	285	39	39
=O carbonyl	1717	70	1048	91	2677	72	1309	63	67
Overall % donor[c]	79		94		77		48		

[a]Number of hydrogen bonds found in the database ($R_{HB} < 0.8$ sum of random contact radii).
[b]Percent of structures with that H-bond over number of structures that contain that pair.
[c]Number of atoms involved in H-bonding over total number of such atoms present in the database. Acceptors not mentioned in the table (*e.g.* ether oxygen) have negligible frequencies.

are rather frequent, and frustrated acceptors are commonplace. Note that this does not mean that the corresponding crystal structure has no hydrogen bonds; it mostly means that some of the hydrogen-bonding termini cannot function because of competition with stronger or better positioned antagonists; competition is in fact the keyword in this matter.

The above facts suggest the following general consideration. Careful design of donor–acceptor pairing singled out at the time of synthesis of the molecular components may sometimes provide reproducible, solid-state "crystal engineering" results over restricted classes of compounds. But in multifunctional systems comprising a mix of several donor and acceptor sites, such as all organic compounds of practical interest generally are, not even hydrogen bonding can provide solid guidelines in "crystal structure prediction". This conclusion delineates the divide between crystal engineering, an area of research restricted to particular cases, and general crystal structure prediction and control: two quite different concepts. The former may succeed by educated guessing and a dash of luck, the latter cannot succeed without extensive, quantitative evaluation of the implied energies.[31]

References

1. J. D. Dunitz and A. Gavezzotti, *Acc. Chem. Res.*, 1999, **32**, 677.
2. A. J. Stone, *The Theory of Intermolecular Forces*, Clarendon Press, Oxford, 2000.
3. A. Gavezzotti, *Molecular Aggregation, Structure Analysis and Molecular Simulation of Crystals and Liquids*, Oxford University Press, 2nd edn, 2013, ch. 12.
4. A. Gavezzotti, *New J. Chem.*, 2011, **35**, 1360.
5. S. Ehrlich, J. Moellemann and S. Grimme, *Acc. Chem. Res.*, 2013, **46**, 916.
6. A. Gavezzotti, *Mol. Phys.*, 2008, **106**, 1473.
7. See at www.angelogavezzotti.it under CLP-PIXEL.
8. A. G. P. Maloney, P. A. Wood and S. Parsons, *CrystEngComm*, 2015, **17**, 9300.
9. L. Maschio, B. Civalleri, P. Ugliengo and A. Gavezzotti, *J. Phys. Chem. A*, 2011, **115**, 11179.
10. J. D. Dunitz and A. Gavezzotti, *Angew. Chem., Int. Ed.*, 2005, **44**, 1766.
11. N. J. J. de Klerk, J. A. van den Ende, R. Bylsma, P. Grancic, G. A. de Wijs, H. M. Cuppen and H. Meekes, *Cryst. Growth Des.*, 2016, **16**, 662.
12. A. Gavezzotti, *New J. Chem.*, 2013, **37**, 2110.
13. A. Gavezzotti, *Cryst. Growth Des.*, 2013, **13**, 3801.
14. S. Chen, I. A. Guzei and L. Yu, *J. Am. Chem. Soc.*, 2005, **127**, 9881.
15. S. L. Price, *Chem. Soc. Rev.*, 2014, **43**, 2098.
16. C. R. Martinez and B. L. Iverson, *Chem. Sci.*, 2012, **3**, 2192.
17. S. Bacchi, M. Benaglia, F. Cozzi, F. Demartin, G. Filippini and A. Gavezzotti, *Chem. - Eur. J.*, 2006, **12**, 3538.
18. D. Chopra and T. N. Guru Row, *CrystEngComm*, 2011, **13**, 2175.
19. A. Gavezzotti, *Acta Crystallogr.*, 2010, **B66**, 396.

20. A. Gavezzotti and C. J. Eckhardt, *J. Phys. Chem.*, 2007, **B111**, 3430.
21. J. D. Dunitz, A. Gavezzotti and S. Rizzato, *Cryst. Growth Des.*, 2014, **14**, 357.
22. A. Gavezzotti and J. D. Dunitz, *J. Phys. Chem. B*, 2012, **116**, 6740.
23. S. A. Moggach, W. G. Marshall, D. M. Rogers and S. Parsons, *CrystEngComm*, 2015, **17**, 5315.
24. M. A. Neumann, J. van de Streek, F. P. A. Fabbiani, P. Hidber and O. Grassmann, *Nat. Commun.*, 2015, **6**, 7793.
25. F. H. Allen, *Acta Crystallogr.*, 2002, **B58**, 380–388.
26. N. A. Tumanov and E. V. Boldyreva, *Acta Crystallogr.*, 2012, **B68**, 412.
27. S. L. Price, A. J. Stone, J. Lucas, R. S. Rowland and A. E. Thornley, *J. Am. Chem. Soc.*, 1994, **116**, 4910.
28. R. Bukowski, K. Szalewicz and C. F. Chabalowski, *J. Phys. Chem. A*, 1999, **103**, 7322.
29. J. C. Flick, D. Kosenkov, E. G. Hohenstein, C. D. Sherrill and L. V. Slipchenko, *J. Chem. Theory Comput.*, 2012, **8**, 2835.
30. P. Jurecka, J. Sponer, J. Cerny and P. Hobza, *Phys. Chem. Chem. Phys.*, 2006, **8**, 1985.
31. J. Kendrick, G. A. Stephenson, M. A. Neumann and F. J. J. Leusen, *Cryst. Growth Des.*, 2013, **13**, 581.

4 On Topological Atoms and Bonds

Paul L. A. Popelier[a,b]

[a] Manchester Institute of Biotechnology (MIB), 131 Princess Street, Manchester M1 7DN, UK; [b] School of Chemistry, University of Manchester, Oxford Road, Manchester M13 9PL, UK
Email: pla@manchester.ac.uk

4.1 Introduction

This chapter will explain some basic concepts behind what is often called the "Quantum Theory of Atoms in Molecules (QTAIM)"[1–4] or just "Atoms in Molecules" (AIM)[†] in short. In 2014, a chapter on QTAIM was published[5] in an edited book on fundamental aspects of chemical bonding. That QTAIM chapter, first of a number of perspectives very recently written by the current author, uses boxes to siphon off the more detailed and mathematical material from the main text. A second chapter[6] presents a philosophical account on QTAIM, or more precisely on Quantum Chemical Topology (QCT).[7,8] The latter research field is a branch of theoretical chemistry that embodies an overarching structure for all methodologies that share the central idea of using the mathematical language of dynamical systems. This second chapter, published in 2016, *advocates much needed*

[†] AIM is used by some authors to refer to *any* method that defines atoms, including non-topological ones. Although this alternative use of the acronym AIM is confusing, one could argue that the name AIM should not have been seized by the originator of QTAIM in the first place. We will always use QTAIM to avoid this confusion.

Intermolecular Interactions in Crystals: Fundamentals of Crystal Engineering
Edited by Juan J. Novoa
© The Royal Society of Chemistry 2018
Published by the Royal Society of Chemistry, www.rsc.org

falsification in the area of chemical interpretation by means of quantum mechanical tools. Falsification has furthered possibly all of Science into the strong bastion that it has become but conceptual theoretical chemistry sadly fails to be exposed to falsification and ultimately benefit from it. This second chapter also attempts to avoid futile discussions by introducing for the first time the concept of the *non-question*. A third chapter,[9] featured in a book edited by Mingos, also published in 2016, this time in honour of Lewis's views on chemical bonding, 100 years after his famous paper on this topic. There, a critical tone is again applied to explain the current status of QCT, emphasise its minimality, and the prospect of its energy partitioning providing urgent clarity to chemical bonding. Finally, a fourth contribution[10] (appearing in 2016) focuses on prediction, again delivered by QCT. Indeed, the systematic work on a QCT-based protein force field finally reached the stage of molecular simulation at finite temperature being possible. This force field, once called QCTFF[11] but now referred to by the more pro-nounceable name FFLUX, does not use any traditional potentials, whether for bonded or non-bonded interactions. Instead, FFLUX works with knowledgeable topological atoms that are endowed with machine learning models, each making a prediction for an atomic property that is crucial in understanding how atoms behave energetically.

The current chapter manages to present the QTAIM (and QCT) material in yet another different way. Other than for new examples and fresh angles of explanation, this chapter progresses by question and answer. This perhaps unusual but effective style serves pedagogy and has two extra advantages: (i) each piece of new information is motivated by a precise question, and (ii) material is sliced in small sections. Both advantages enhance the readability and present an alternative to the previous style of longer text fragments punctuated by "specialised boxes". The material presented here is not exhaustive but should serve as a useful introduction to the chapter by Professor Martín Pendás and that by Professor Gatti, and their respective co-authors. The question-answer style only applies to the middle section of this chapter, which perhaps reads a little like a long argument reminiscent of Nick Lane's book[12] on the Origin of Life.

4.2   Questions and Answers

Can theoretical chemistry provide chemical insight?
The short answer is "yes" but theoretical chemistry best realises this goal *via* computation. Computational chemistry is the embodiment of

the third leg of Science: simulation. This new leg emerged in the second half of the 20th century to take its current strong position next to the two older legs of theory and experiment. In order to handle the complexity of chemical bonding, and be able to predict the overall behaviour of matter at ambient conditions, theoretical chemistry really needs the assistance of powerful computers.

A question that best precedes the title question is if computational chemistry *should* provide chemical insight. The answer to this new question is again "yes", which can be argued as follows. We live in one world only, one in which chemical knowledge must therefore converge. In other words: experiment, theory and simulation must be fully compatible, co-existing in symbiosis. In this important process of reaching perfect co-existence, these three branches must stimulate each other. This co-existence has been achieved to perfection in classical mechanics, for example, which enables us to put a person on the moon or predict the presence of a new planet by studying the motion of detectable bodies that this planet perturbs.

Yet, the state-of-the-art in chemistry has not reached this perfection, unfortunately. There are several examples to support this claim. Firstly, many chemistry teachers admit feeling uncomfortable with the textbook explanations of several chemical observations (*e.g.* the curiously short bond BF bond length[13] in BF_3). Secondly, the concept of chemical hardness has intuitive roots in the work of Pearson,[14] then received some theoretical underpinning[15] in the 1980s but left behind a counterintuitive and contradictory set of results,[16] in which the largest values of the hardness function do not necessarily correspond to the hardest region of the molecule. Thirdly, several population analyses continue to be used simultaneously, even if they give disagreeing results semi-quantitatively, that is, both in magnitude and even sign. Fourthly, traditional force fields are built on unphysical and even "unchemical" electrostatics. Even if these force fields predict within experimental error the self-diffusion coefficient of water in liquid water, for example, one cannot trust the point charges used as correctly expressing the charge transfer within a water molecule. Fifthly, aromaticity is not easy to pinpoint and quantify computationally, to the point that some ask if it should be sustained as a concept in the first place. Sixthly, few experimental chemists have the feeling that their synthetic work has ever been guided by a computational prediction, not in the least in field of drug design. As a final example in this non-exhaustive list we mention the dislocation at the heart of advanced computational chemistry: the more accurate a wavefunction becomes, the less chemical insight it offers.[17]

In summary, we can say that not all is well in chemistry, and that there is a clear and urgent need to improve drastically the symbiosis between theory, experiment and simulation in this important and central branch of Science. In this chapter, we will present a minimal approach that aims to improve the current *status quo*.

What should a promising theory or approach look like?
It speaks for itself that the approach must be free of internal contradictions, that is, it must be internally consistent. Still, two competing approaches (*i.e.* theories) can both be internally consistent but one can be wrong while the other is right. In Science, this outcome is decided by falsification: both theories make contradictory predictions until a pivotal experiment will decide which theory is right. Alas, within conceptual theoretical chemistry this falsification rarely happens. One of the reasons for this otherwise unjustifiable situation is that experiment cannot provide an answer that is independent of a theory or a model. For example, a net atomic charge cannot be measured experimentally. An experimentalist can only isolate some signal but must then invoke a theory or at least an assumption to interpret this signal towards a net atomic charge.

However, within this operational world the next best thing a theoretical chemist can do is *insist that a theory is minimal.* This is the famous principle of Occam's razor (or the Law of Parsimony), which has served physics so well over the centuries. Newton's laws, special and general relativity and Maxwell's equations, for example, all have a compelling compactness about them. They may not be ultimate endpoints governing much of the Universe's behaviour but, at least, they lack the weakness of "if and buts" that plague less powerful theories. If a parameter must pop up in a theory then it should be something beautiful and pleasantly surprising, such as the speed of light in Maxwell's equations. That the cosmological constant was best set to zero, at some point in history of cosmology, was also parsimoniously pleasing (although the full story is more complicated). In summary, good future-proof theories are essentially parameter-free.

Returning to quantum chemistry, it would be nice to have a method that defines atoms, without having to use basically arbitrary atomic radii or reference states, which are also often arbitrary and even artificial. The answer as to what an atom *inside* a system looks like, must be obtained *in a minimal way*. It must be obtained from quantum information that can be calculated, such as the wave function or the electron density denoted ρ. Before we propose how to do that, it is important to understand the difference between *minimal* and *simple.*[18] A *simple* explanation of a natural phenomenon is based on "a superficially attractive but actually inadequate mental picture that is wrongly imposed onto this phenomenon". A very clear example of such an explanation is the Ptolemaic theory, which needed 77 circles to describe the motion of the sun, moon, and the five planets then known. The fact that there are corrections to corrections is a hallmark of simplicity struggling to describe and explain reality. On the other hand, a *minimal* explanation is one that captures the true essence of the phenomenon being studied. Minimality abandons (or better avoids from the start) any misleading mental picture, and instead, boldly introduces a novel and imaginative concept that goes to the core of the phenomenon at hand. In the example above, that concept is of course Kepler's ellipse, which is the correct solution of Newton's law of motion. However, superficially speaking, the ellipse is "uglier" than the circle because it is less symmetric. However, with the beneficial hindsight of a deep understanding, the ellipse is actually *more* beautiful than the circle.

Why is all this relevant to quantum chemistry? Because the tension between simplicity and minimality is universal, it potentially applies to any part of Science. A vivid example of the damage simplicity can cause in quantum chemistry is that of the so-called Stewart atoms. They were constructed[19] in 1996 and briefly analysed in terms of their chemical appearance (atomic charge actually). Stewart atoms are generated *under the assumption that an atom in a molecule is a spherically symmetric function.* Here, it will become clear that simplicity imposes an arbitrary constraint on a situation that does not call for it. An approximate but feasible method[20] to spawn Stewart atoms is to expand the molecular electron density in an auxiliary set of nuclear-centred radial basis functions and directly minimise a general least-squares residual. This procedure forces one to make a choice because these approximate (but practical) Stewart atoms do depend on the particular operator appearing in this residual. It turns out that the charges of Stewart atoms are strongly basis set dependent, ranging[19] from -0.028 to $+0.295$ for the carbon in CH_3F. More worryingly, the charges on the C atoms are surprisingly variable in a variety of hydrocarbons looked at, covering single, double, triple and aromatic CC bond types. The carbon charge in C_2H_6 is $+0.305$, which is alarmingly at odds with any electronegativity scale. This setback is due to the fact that Stewart analysis allocates little electron density to C, which is in turn caused by the fact that C is "buried" within the shells of the H atoms. This example vividly illustrates the price paid for an unwarranted assumption that simplicity introduces: parameter-dependence (*i.e.* choice of operator and basis set) and chemical non-sense results. Therefore, we should abandon simplicity and ask how instead minimality operates in the search for an atom in a molecule.

How can one construct a partitioning of a quantum system in a minimal way?

Let us make a false start first, just to tease out the argument better. Let us work in the Hilbert space of basis functions from which the wave function is constructed. Imagine that a basis function has a centre, which is the case for both a Gaussian (primitive function) and a Slater function but not for a plane wave function. Now suppose that we are interested in finding a numerical value for a (net) atomic (partial) charge. We can all agree that electron density in the immediate vicinity of the nucleus belongs to the atom housing this nucleus. However, questions arise when electron density located in between two nuclei (and quite far away from either nucleus) must be allocated to one atom or the other. The popular but disgraced Mulliken population analysis *uses the centre of a basis function to decide* which piece of molecular electron density belongs to which atom. This population analysis assigns *all* the electron density described by a given Gaussian to the atom that contains the Gaussian centre. This partitioning scheme is troublesome when the Gaussian exponent is small, which is the case for diffuse Gaussians. The latter are then responsible for the build-up of a considerable amount of electron density *far away from the centres* of the diffuse Gaussians. As a result of the centre-based decision, electron density far away from a given nucleus is still allocated to this nucleus. As a result, atomic charge is spuriously allocated and the robustness of the population analysis breaks down. The well-known Distributed Multipole Analysis (DMA)[21] also suffers from this problem.‡ Moreover, this population analysis cannot operate on an electron density expanded in plane wave basis functions, because they have no centres. This means that, Mulliken and DMA are stuck when asked to operate on a solid state electron density, which is typically (but not always) expanded in plane waves. Of course, if a piece of crystalline matter is described by Gaussian basis functions (as is done in several popular *ab initio* programs), then one can proceed with Mulliken or DMA. However, the main point is

---

‡The actual allocation decision is a little more complicated in DMA compared to Mulliken population analysis because the former is built on shifting the origin of a multipole expansion. This decision process again depends on a choice: that of a so-called allocation algorithm.

that any piece of condensed matter should be divisible into atomic electron densities, when starting from the overall electron density itself. In general, the minimal way to make robust progress is to base the sought-after partitioning method on the electron density itself.

Starting from the electron density itself has the added advantage of connecting with experiment. Indeed, high-resolution X-ray crystallography[22] directly accesses the observable that is the electron density. Moreover, an electron-density-based generator of atoms can also be applied when no basis functions at all are used, as is the case for Quantum Monte Carlo calculations. A long time ago it was shown[23] that the minimal method of quantum topological partitioning, to be explained shortly, can operate on a grid of electron density points by means of a numerical algorithm able of interpolation and differentiation.

Although we have made progress in our search for a minimal definition of an atom inside a molecule (or any piece of matter), we still do not know how to partition the electron density, although we recognise it as an advantageous starting point. There is one key concept that is required in order to make final progress, which is the so-called *gradient path*. To understand the essence of this concept, let us start with a metaphor. Imagine a civil war, with two sides only, and imagine that none of the fighters wears a uniform. As a result, it is not straightforward to know which side a given fighter belongs to. However, for the sake of our argument there is a way to figure out this question. We simply ask a fighter who his superior is. Then we ask that new fighter who his own superior is, and so on, until we reach the final warlord who commands the one side of the two warring factions. In other words, we follow a path of hierarchy that always points in the upwards direction of the command chain. Such a path of hierarchy is reminiscent of the gradient path. When applied to each fighter we are able to disentangle the whole system of fighters into two clear groups. Loosely speaking, each fighter stands for a portion of electron density in space, while a warlord corresponds to a nucleus, or more precisely, a maximum in the electron density.

When we apply the idea of the gradient path to the electron density of a diatomic molecule, we obtain two *topological atoms.* There is no need to set a parameter, which serves the cause of minimality. Moreover, we do not invoke a reference electron density in order to reveal the topological atoms, which again serves the cause of minimality. Let us now see how the concept of a gradient path works out in the example of the linear triatomic molecule OCS. We will already give away that the gradient path (in the electron density) is the path of *steepest ascent* in the electron density. At any point in space the path points in the direction of the highest electron density value relative to this point, locally. The gradient, which is an internal difference essentially, is the only concept one needs to answer the title question of this section.

What does a molecular electron density look like?
There are several ways to represent the electron density: by intersections in various dimensions, relief map, four-dimensional colour map and so on. Here, we just focus on a few features of the electron density to pave the way for the topological partitioning.

Let us look at surfaces of constant electron density, or iso-electron-density envelopes. Figure 4.1 shows a convenient number of constant electron density contour lines in two dimensions (2D) only, for the linear molecule OCS. Because of the cylindrical symmetry of this molecule, only two dimensions suffice to portray the full 3D electron density. In other words, any plotting plane containing the molecular axis displays all electron density information we need to know about this linear molecule. By revolving the contour patterns around the molecular axis, one obtains the 3D iso-electron-density envelopes.

The tight concentric contour lines near each nucleus tell us that each atom still retains some atomic identity, even within a molecular system. As expected by the relatively small energies involved in the making of covalent bonds, atoms are not at all destroyed in the process of creating a molecule. Their roughly spherical shape is probably why a simplicity-oriented mind would impose fully spherical

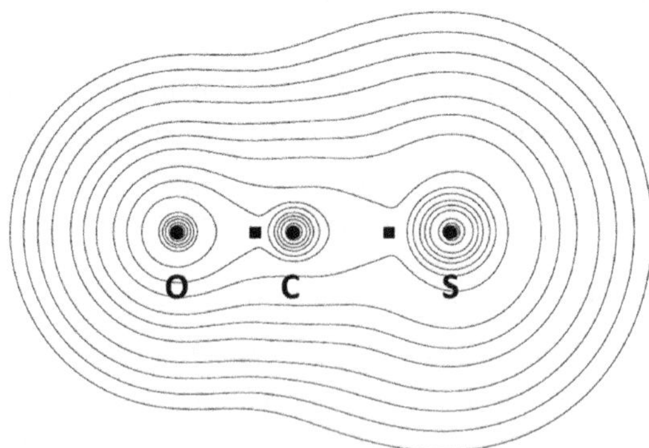

Figure 4.1 Electron density contour plot of OCS. The electron density values are 1×10^{-n}, 2×10^{-n}, 4×10^{-n} and 8×10^{-n} au, where n starts at -3 and increases with unity increments. The small black squares mark the positions where the electron density is a local minimum when seen in one dimension, that is, along the molecular axis.

atoms onto this contour plot. One may get away with this dangerous thought for compact atoms formed by contour lines close to the nuclei. However, a high price will be paid (see Stewart atoms) if this sphericity is maintained far from the nuclei. In summary, we need to let go of this unsafe idea and "look for our ellipse".

We have to decide what the three atoms look like as they reside within the molecule. If we are totally minimality-minded, we strive for the molecule to answer this question itself. In the absence of this self-partitioning as of yet, we progress by noting some remarkable points, which are marked by little black squares in Figure 4.1. Each square marks a local minimum in the electron density, when seen in one dimension only, that is, a minimum along the molecular axis. If the sulfur and the carbon were mountain peaks then we imagine them to be connected by a ridge, which reaches its lowest height at the right-hand square. It is then again tempting to put a plane right through this square and orthogonal to the molecular axis. Such a boundary between C and S would again be a simplicity-minded attempt to start partitioning the molecule. In fact, this is exactly the partitioning that Bader, Beddall and Cade proposed[24] in 1971 for another linear molecule, namely FCN. These authors called it a natural partitioning because *"it is suggested by the nature of the charge distribution itself"*, as stated in the abstract of their JACS paper. Their sister publication[25] focused on the partitioning of CO_2 but again fell in the same trap of simplicity: they partitioned by means of a plane. Who can justify *a priori* the decision that the surface acting as an interatomic boundary is a plane? Imposing such a decision to a situation that does not call for it, smacks of the previously discussed sphericity imposed on atoms within molecules. Both are cases of arbitrarily high symmetry being imposed without any justification other that than it is a simple thing to do. Unfortunately, it turns out that this is not the *minimal* thing to do. So what is?

How does one cut a molecule with interatomic surfaces that draw themselves?
Figure 4.2 offers the solution. Here a curved surface appears between C and S, such that the surface is everywhere orthogonal to the envelopes of constant electron density (as marked by three red arrows in Figure 4.2, each representing a local situation). In the spirit of minimalism, this surface draws itself and is free from any arbitrary decisions. Moreover, the construction of the surface does not require a parameter to be set. In Figure 4.2 we only see the intersection of this surface with the plotting plane but it is easy to envisage this surface in

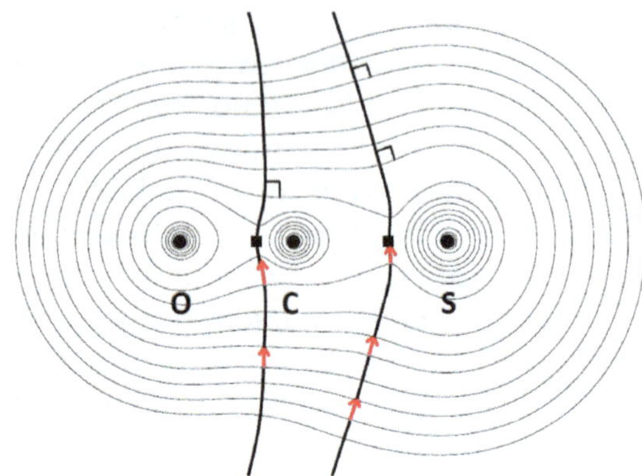

Figure 4.2   Topological partitioning of OCS. Two interatomic surfaces (IAS, in bold, and only the respective intersections with the plotting plane are shown) are superimposed on the same electron density contours as in Figure 4.1. A red arrow marks the local direction of maximum increase in the electron density, which is the gradient vector at that point. A gradient path can be seen as a succession of infinitesimally small gradient vectors.

3D by mentally revolving the bold curved line around the molecular axis. The sharp boundary between two atoms, now revealed for the first time, is called an *interatomic surface (IAS)*. An IAS is an object that naturally follows the shape of the electron density. If the nuclear positions change then the molecular electron density changes, and hence the IASs will change shape.

One can prove that moving in a direction orthogonal to a constant electron density contour is equivalent to moving in the *direction of maximum change* in the electron density. The *gradient vector*, examples of which are marked in red in Figure 4.2, points in the direction of maximum ascent of electron density. In other words, the electron density increases most in this direction. So, a gradient path is a curve that pierces through constant electron density envelopes, being locally orthogonal to each envelope. Alternatively, a gradient path can be defined as a succession of infinitesimally small gradient vectors.

The IAS is a surface of a potentially non-trivial shape in 3D, and consists of a bundle of gradient paths that originate at infinity and terminate in a special point in space, in between two nuclei, which is marked by a black square in Figure 4.2. Figure 4.3 puts the IASs in a wider context, and shows the full *gradient vector field* (of the electron density) of OCS, albeit only as an intersection with a plotting plane. Now we see a multitude (in fact an infinite number) of gradient paths, each originating at infinity and terminating at a nucleus. Together they form the so-called *atomic basin.* While being attracted to the

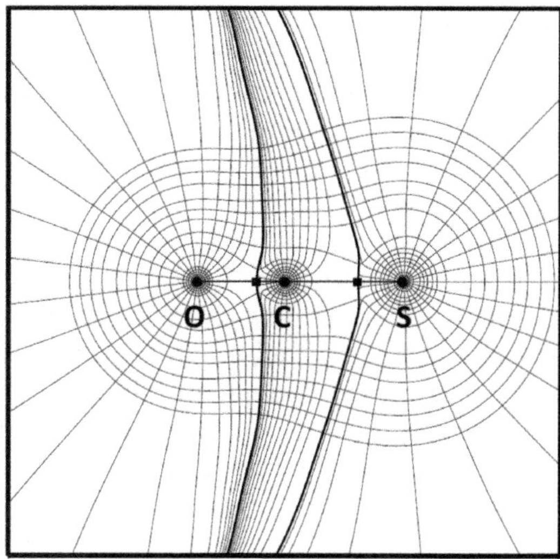

Figure 4.3   Topological partitioning of OCS superimposed on constant electron density contours. An interatomic surface (IAS, curves in bold) acts as the boundary between two topological atoms. The IAS consists of gradient paths that are not attracted to any nucleus but instead to either of the two little black spheres.

same point (*i.e.* the nucleus), the gradient paths carve out a subspace that one associates with the *topological atom*. The reason the gradient paths are attracted to the nucleus is because its position practically coincides with a maximum in the electron density. This maximum is the equivalent of the warlord in our earlier metaphor. A maximum is the endpoint of a "hierarchy of gradient vectors", one vector pointing at the next one, and traversing the electron density landscape in the steepest direction possible (in 3D).

A new observation in Figure 4.3 concerns a gradient path connecting the black square with a nucleus. Such a gradient path originates from the black square[§] and terminates at the nucleus. Taking the black square on the right-hand side of the plot as a reference, there is a gradient path travelling to S and another one to C. This pair of gradient paths is called an *atomic interaction line (AIL)*.[26] Although the AILs trivially lie on the molecular axis, in general they connect the nuclei that are expected to be bonded, as shown in a Lewis diagram. For example, in the water molecule there will be two OH AILs but no HH AIL. The collection of AILs forms a *molecular*

[§]Technically and more rigorously speaking, the gradient path cannot originate exactly from the black square. Instead, the path originates from a point very close to the black square, a point lying on the molecular axis.

graph, which coincides with the Lewis diagram, even in the once controversial case of diborane (B_2H_6). In this molecule there are two bridging hydrogen nuclei, each nucleus connected by two AILs to both boron nuclei. However, there is no AIL connecting these two bridging hydrogen nuclei, nor the boron nuclei in fact.

Figure 4.4 summarises what we have learnt so far, again using the pilot system OCS. This figure shows four views of the same 3D object: the topologically partitioned molecule OCS. The 3D spatial extension of the IAS should now be clear, while a correspondence with the IAS in Figure 4.3 can be made. We emphasise again that the gradient is the only notion required to reveal atoms and bonds. The topological atoms are "malleable boxes" that change shape in response to a change in the nuclear skeleton. If one could travel only along gradient paths, one would be trapped within the atomic subspace. Loosely speaking, this situation is analogous to being trapped by the Earth's gravitational field. If one starts the journey outside the gravitational basin or catchment region, then one will end up at another body in the solar system. However, the analogy is not perfect because electrons do not travel along gradient paths: they are free to cross

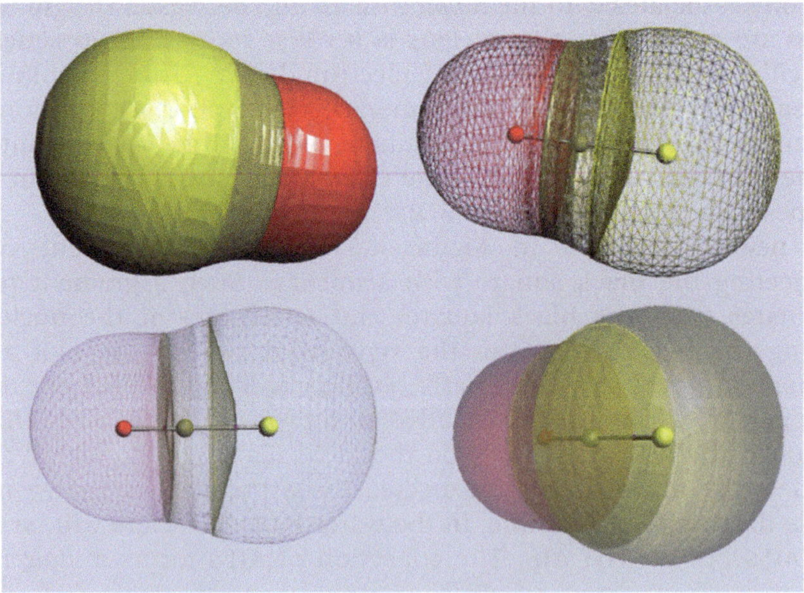

Figure 4.4 Four 3D representations of OCS partitioning into its three topological atoms (S = yellow, C = brown, O = red). The little purple spheres appearing on the molecular axis correspond to the black squares in Figures 4.1–4.3.

the IASs. Indeed, a topological atom can gain or lose electronic charge upon a change in the nuclear geometry.

The nature of the topological partitioning method has been commented on before, at great length. In Section 6 of his chapter in Mingos's book,[9] the current author carefully explained this method's parameter-free, orbital-free and reference-state-free character. Moreover, the fact that topological atoms do not overlap and leave no gaps between them[¶] has been fully discussed in Section 2.8 of ref. 10. The combination of this lack of overlap and lack of gaps is sometimes referred as "space filling". The potential impact of this space filling property on future drug design is currently being researched in our lab but has already been speculated about is some detail in ref. 27. Many metaphors, ranging from colliding galaxies and overlapping clouds in the sky, over standing patterns of running river water, to cell membranes and the "system|surrounding" divide at the core of thermodynamics have been quoted[5,6,8–10] in order to defend the plausibility of sharp atomic boundaries, which still appear alien to so many chemists, regrettably.

Now that topological atoms are defined, how do we calculate their properties?

Atomic properties are obtained by integrating, over the 3D volume of an atom, a property density of interest. For example, if this property density is the charge density (*i.e.* electron density and nuclear charge) then the corresponding atomic property is the net atomic charge. Figure 4.4 clearly shows the volume associated with each of the atomic basins. A 3D integration over this volume, with the electron density as integrand, yields the electronic population of the topological atom in question. Algorithmically, such an integration is challenging but the result is rewarding if one is finally prepared to let go of the *a priori* assumption that net atomic charges always have to be small. QTAIM charges are indeed small when they have to be, in the case of apolar systems. However, in polar systems such as the keto group, the keto carbon is very positive and the keto oxygen very negative according to QTAIM. Does this outcome not help sustain the arrow-pushing game that so many organic chemists play when attacking this keto carbon by a nucleophile?

Other integrands can also be inserted in the 3D volume integral over the topological atom. If one inserts simply unity, then one

---

[¶] The lack of gaps and overlap is sometimes referred to as "space filling" or as "exhaustive partitioning".

obtains the atomic volume. In condensed matter, a topological atom is bounded only by interatomic surfaces and hence there is no need to define a practical but admittedly arbitrary outer boundary. However, introducing the latter boundary is customary for atoms in free gas-phase molecules such as the one in Figure 4.4. Another possible integrand is that of the electron density $\rho(\mathbf{r})$ times a position vector, which is centred at the nucleus. As this little vector sweeps around in the atom's basin, it picks up a varying amount of electron density. In places where there is much electron density, the little vector will be weighted heavily by the large ρ values that it encounters. Essentially being a sum, the integral calculates the mean position of the vector as weighted by the electron density. With a little imagination one can then identify this resulting vector (*i.e.* after completing the integration) with the *intra-atomic dipole moment.* The dipole moment summarises to what extent and in which direction the electron density is shifted away from the nucleus. QTAIM is very clear about the distinction between a contribution to a molecular dipole moment, from intra-atomic dipole moments and inter-atomic dipole moments. The latter are caused by a charge transfer effect, that is, the vectorial separation of two atomic charges. This rigorous and conceptually lucid view on the atomic breakdown of molecular dipole moments perfectly explains[28] why carbon monoxide has a nearly vanishing molecular dipole moment.

Higher multipole moments can also be calculated by volume integration and this is most compactly done[29] by using the spherical tensor formalism rather than the Cartesian formalism.[30] Many other atomic properties can be calculated, including an atom's kinetic energy (which is an important achievement and strong feature of QTAIM, details see Box 8.2 in ref. 5). Atoms can then be characterised by their properties and compared in terms of a Euclidean distance in a property space (typically of dimension higher than 3). When pushed further, this idea can lead to atom types[31] being *calculated* for the first time rather than being determined *a priori* (as done in the parameterisation of the force field AMBER).

Finally, it should be pointed out that a string of so-called atomic theorems have been derived,[32] some of which took pages and pages of hand-written derivation by one of the authors. This work required careful handling of both the real and imaginary part of the atomic expectation value of the commutator $[\hat{H},\hat{G}]$, where $\hat{H}$ is the many-electron Hamiltonian and $\hat{G}$ is a generator, which can be equal to, amongst other possibilities, $\mathbf{r}^2$, $\hat{\mathbf{L}}^2$, $\hat{\mathbf{p}}$, $\hat{\mathbf{r}} \cdot \hat{\mathbf{p}}$, $\hat{\mathbf{r}} \times \hat{\mathbf{p}}$ and $\mathbf{p}^2$. The real parts of the commutator averaged for the last four generators yield the atomic force, virial, torque and power theorems, respectively. The atomic power theorem, for example, enables one to determine the atomic (or molecular source) of the creation and dissipation of power in combustion, or in an explosion.

What is the Topological Bond?

We start this section with the question: *What are these mysterious black squares in Figure 4.2?* We can actually predict the value of the gradient at this remarkable point in space marked by the black square. Take a gradient path that belongs to an IAS. The path is propelled in the direction that the gradient vector indicates, at any point in space. Now, as the path arrives at the black square it stops being propelled. A gradient path can only stop travelling if the gradient vanishes, which is indeed the case at the black square. A point in space at which the gradient vanishes is called a *critical point.* In fact, following the same reasoning, one can deduce that a maximum in the electron density is also a critical point. Both the black square (which is actually called a *bond critical point, BCP*) and the maximum, are two types of a total of four possible types of critical point in 3D.

As expected the bond critical point appears between two bonded nuclei, as a topological signature for all chemical bond types in organic chemistry. All functional groups are recovered, without controversy, from the pattern of bond critical points and AILs that correspond to these groups. In fact, one can characterise these "light atom" bonds by interpreting properties evaluated[33] at the bond critical point, such as the electron density itself. A second popular measure is the Laplacian of the electron density, $\nabla^2\rho$, which gauges the degree of local concentration or depletion of electron density. If positive, then the electron density is locally depleted, which characterises the bond as a so-called *closed-shell interaction.* Ionic bonds, hydrogen bonds and various bonds in van der Waals complexes fall into this category. The opposite situation, in which $\nabla^2\rho < 0$, is a rough and computationally cheap hallmark of bonds with covalent character, which Bader *et al.* called *shared interactions.* More details can be found in Box 8.5 of ref. 5.

In inorganic chemistry, however, chemical bonding becomes more challenging. A sense of the difficulties encountered in transition metal carbonyl clusters, for example, has been extensively discussed in a thorough and critical review.[34] On a positive note, however, this work provides a summary (table 4 in ref. 34) of valuable topological features characterising atomic interactions, covering both light and heavy elements, and demonstrates their novel use. These features are semi-quantitative in nature (*i.e.* small/large, positive or negative sign, ratios smaller or larger than unity, position of bond critical point), and invoke ρ, $\nabla^2\rho$, as well as kinetic energy densities,[35] again evaluated at the bond critical point.

The most involved and intense discussion[36–42] about the meaning of the bond critical point has arisen in the progressively important area of intermolecular interactions, as Chemistry increasingly becomes a "science of the atomistic assembly of matter". The controversy also includes, and was perhaps even triggered by, inclusion compounds (*e.g.* Ne@C_{60}) and so-called repulsive interactions (*e.g.* between *ortho*-hydrogens in biphenyl). Early indications of the complexity of AILs and concomitant bond critical points emerged from a broad study[43] of 36 configurations of 11 van der Waals complexes consisting of two molecules (and one complex with three molecules) including Ar, C_2H_2, CO_2, OCS and SO_2. Atomic interaction lines in the intermolecular region denote the dominant atom–atom interactions underlying the weak interactions. The patterns can be complex, an example of which is shown in Figure 4.5. Here the global energy minimum (of C_{2h} symmetry) of the OCS dimer is held together, as it were, by two AILs. Translating the top OCS molecule to the right and the bottom OCS to the left leads to a local minimum, again of C_{2h} symmetry, but now with only one AIL, which links the oxygen nuclei. Slipping the monomers further over each other in the same direction leads to a C_s transition state where only C and S are linked by an AIL. Previous work on hydrogen-bonded systems established strong correlations between the value of ρ at a BCP (*i.e.* ρ_{BCP}) and the binding energy of the complex (provided comparison is restricted to structures in which the same pairs of atoms are interacting). These correlations between binding energy and ρ_{BCP} are harder to expose with the van der Waals molecules studied in ref. 43.

Essentially a BCP is tasked with summarising, in a very compact way, why and how atoms form a stable system. This task is straightforward for a simple covalent molecule: if atom *A* is bonded to atom *B*, there will be a BCP between them, acting as a signature of their stable relationship. The weaker the interatomic interaction becomes, the more vulnerable this interpretation of a BCP's presence. The increasingly uncertain mapping between a BCP's appearance and its designation of a stable pair of atoms, fuels the heated debate around the meaning of a BCP. Much of the debate involves circular arguments, *a priori* chemical interpretation, or a confusion between *attracting* and *attractive* interactions (the distinction is explained in

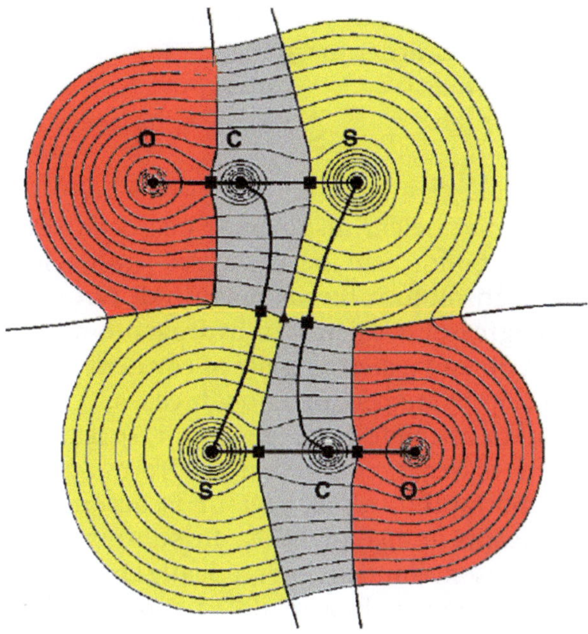

Figure 4.5 The global energy minimum (C_{2h} symmetry) of $(OCS)_2$. The atomic interaction lines (AILs) and interatomic surfaces (IASs) are superimposed onto constant-electron-density contours. The black squares mark bond critical points. The triangle marks a ring critical point, which will be explained in the next question-led section.
(Figure adapted from Bone, R. G. A.; Bader, R. F. W. Identifying and analyzing intermolecular bonding interactions in van der Waals molecules. J. Phys. Chem. 1996, 100, 10892–10911. Copyright 1996 American Chemical Society.)

detail on p. 31 of ref. 8). A way out of the conundrum is by invoking the topological energy analysis (see below). Ultimately, the stability of a molecular system is determined by an intra- and interatomic *energy balance.* In particular, one must find patterns of stability in a potentially complicated balance between various types of energy, and between various atoms. This type of analysis was started[44] in 2001 and perfected[45] in 2005, under the name of *Interacting Quantum Atoms* (IQA).

The full explanation on how IQA can help in settling the controversy around the meaning of the BCP has been published elsewhere (p. 27 of ref. 5) before. Here we repeat the essence of the argument. IQA defines an exchange-delocalisation energy between any two topological atoms A and B, denoted $V_x(A,B)$. This fundamental energy type covers covalency. Using a multipole expansion one can prove that there is a mathematical link[9,46] between $V_x(A,B)$ (which is always negative) and the (topological) delocalisation index ("DI"), which is also known as the bond order (δ^{AB}),

$$V_X^{AB} \cong -\frac{\delta^{AB}}{2R_{AB}} \qquad (4.1)$$

where R_{AB} is the internuclear distance. It has been shown[47] that established covalent bonds can be recovered from clusters of high $V_x(A,B)$ values (in absolute value or magnitude) in a plot of $V_x(A,B)$ *versus* R_{AB}. The degree of covalency in hydrogen bonding can also be measured by $V_x(A,B)$, and a sliding scale of covalent to "non-covalent" bonds can be set up by inspecting enough cases.

The main point to be made here, using $V_x(A,B)$, is that its behaviour with internuclear distance can reveal the forming or breaking of bonds. More importantly, this behaviour is accompanied with the appearance or disappearance of BCPs. In summary, there is a link between a BCP's presence and an energetically established bond. This paramount observation was first made[48] in connection with the formation of water from molecular hydrogen and an oxygen atom. By studying the competition, between atom pairs, for the highest $V_x(A,B)$ values (again, in magnitude), one can rationalise a BCP as a signature of exchange, which is the hallmark of covalency. This principle has been applied and refined[49] in more complicated situations than water.

The author believes that more research is needed before BCPs can be fully confirmed as a flawless compact expression of how a system (molecule or beyond) is held together by exchange correlation. If BCPs are vindicated as doing exactly that then one should open one's mind to the rich and sometimes unexpected patterns they reveal. BCPs will go far beyond the Lewis diagram, which cannot cope with complex bonding patterns anyway. As a result, BCPs should then be welcomed as a valuable asset to modern interpretative quantum chemistry rather than a source of weirdness or a cause of controversy.

Which types of critical point exist other than the bond critical point and the maximum?
The quickest way to explain the full picture is by means of a simple molecule that contains each of the four possible types of critical point in 3D. This molecule is tetrahedrane or C_4H_4, which not surprisingly consists of a tetrahedron of carbon atoms, each of which is bonded to a single hydrogen. Figure 4.6 shows the nuclear positions of tetrahedrane, which (practically) coincide with maxima in the electron density.

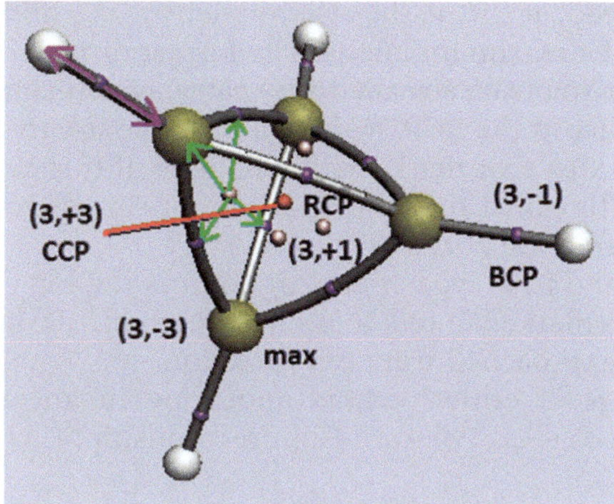

Figure 4.6 The four types of critical point appearing in tetrahedrane (C_4H_4). The 8 maxima or $(3, -3)$ critical points practically coincide with the nuclear positions, the 10 bond critical points (BCP, purple) mark the covalent bonds (CC and CH), the 4 ring critical points (RCP, pink) mark the four (curved) faces while the single cage critical point (CCP, red) marks the (tetrahedral) cage.

As expected, each BCP marks a covalent bond, being the origin for two gradient paths, each of which terminates in a nucleus. The purple arrows show an example of this pattern for a C–H bond. The BCP is also marked with the symbol $(3, -1)$, which can only be understood if the local curvature at the critical point is inspected. How to do this will be outlined without formulae, a few paragraphs later. Here we acknowledge that the BCP is a 3D saddle point, which means that it is a minimum or a maximum depending on the direction in which the electron density is viewed in. In particular, the BCP is a minimum along the AIL, while it is a maximum in any direction orthogonal to the AIL. A rephrasing of the very last statement is that the two directions that span the interatomic surface very near the BCP, are directions in which the electron density is a maximum at the BCP.

The third type of critical point is the ring critical point (RCP), of which there are four in tetrahedrane, one for each of its faces. There is an infinite number of gradient paths originating from a RCP,[||] four of which are marked in green in Figure 4.6. Three green gradient paths terminate in a BCP (purple point) and one path in a carbon nucleus. A RCP is the opposite of (*i.e.* dual of) the BCP in terms of local

---

[||]Again, more rigorously, one should say that no gradient path can escape from the critical point itself because at its location the gradient vanishes. Instead, one should think of a circle of initial points, centred at the critical point and with a very small radius, from which the gradient paths emerge. This circle lies inside the so-called ring plane.

curvature. Indeed, at a RCP, the electron density is a minimum in two directions, and a maximum in one. The former directions are the two directions that span the so-called *ring plane*, which coincides with a tetrahedrane face at the RCP. Note that this face is curved as marked by the curved AILs that mark the boundaries of this face. The latter direction is orthogonal to the ring plane. In this direction two gradient paths terminate at the RCP.

The fourth and final type of critical point is the cage critical point (CCP), which is marked in red. It is a minimum (in 3D) in the electron density when approached from *any* direction.

The numbers of critical points appearing in any molecule are linked by the so-called Poincaré-Hopf relationship,[50] which reads

$$n_{\mathrm{max}} - n_{\mathrm{BCP}} + n_{\mathrm{RCP}} - n_{\mathrm{CCP}} = 1 \qquad (4.2)$$

where n refers to the number of critical points (appearing in the molecule) marked by the subscript. It is readily verified that for tetrahedrane this relationship is valid because $8 - 10 + 4 - 1 = 1$ (use Figure 4.6 to verify the terms of this sum). Note that the first term is written referring to a maximum rather than to a nucleus (n_{max}). While this distinction is irrelevant in the vast majority of molecules, it is technically correct because one may encounter a so-called non-nuclear attractor (NNA). A NNA is a maximum in the electron density that cannot be identified with a nucleus, very nearby. The next question-led section is dedicated to NNAs.

As announced above we now outline, in words only, the meaning of the notation (r,s) used above for critical points. The full mathematical details have been explained before (Box 8.4 in ref. 5) and will therefore not be repeated here. The key is to inspect the local curvature of the electron density ρ at the critical point. This information is given by a 3-by-3 matrix of second derivatives of ρ with respect to the spatial coordinates x, y and z. Unfortunately, this information depends on the choice of axis system, and is therefore arbitrary. Fortunately, there is a mathematical procedure called matrix diagonalisation, which reveals the three *intrinsic* local curvatures contained in this 3-by-3 matrix. Diagonalisation also teases out the three intrinsic directions associated with these three local curvatures. The indicator r in (r,s) is called the *rank* of the critical point and counts the number of non-zero (local intrinsic) curvatures. For the vast majority of critical points the rank's value is 3. When a BCP and a RCP coalesce one obtains a critical point of rank 2. Changes in critical points patterns were already studied[51] in 1981 by the Bader group, using ideas from a mathematical branch called catastrophe theory.[52]

The second indicator s in (r,s) is called the *signature* of the critical point and is a little harder to explain. Essentially, the signature is a single number that summarises the sign pattern of the three local curvatures. For example, for a BCP this pattern is $(-,-,+)$, as explained in detail above. Assigning a (-1) to a negative sign and a $(+1)$ to a positive sign, enables a unique summary of the pattern $(-,-,+)$ by the sum $(-1)+(-1)+(+1)$, which is equal to -1. Hence the signature is -1 and hence the BCP is called a $(3,-1)$ critical point. The other three types of critical point are denoted $(3,-3)$, $(3,+1)$ and $(3,+3)$, which can be easily verified.

Figure 4.7 revisits the topological pattern around one carbon in tetrahedrane, showing the gradient vector field and the critical points in the carbon's vicinity. This type of picture highlights again that an atomic basin is a natural and sophisticated subspace, which traps anyone who could only travel along gradient paths. Box 8.3 of ref. 5 provides more (mathematical) information on this idea.

Note that *a gradient path always connects two critical points.* In fact, one can classify gradient paths by the types of critical points they connect. An exhaustive classification was published[53] in 2001 (see also table B4.1 in Box 8.3 of ref. 5). This classification is universal, beyond that of gradient paths in the electron density, and is valid for other 3D scalar fields, such as the Laplacian of the electron density.

Finally, Figure 4.8 recapitulates what quantum chemical topology reveals in the case of tetrahedrane. The QCT analysis, which coincides

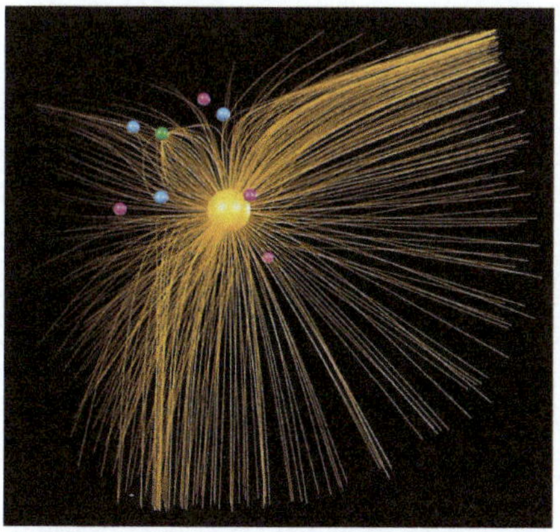

Figure 4.7 Two representations of the same three-dimensional gradient vector field (in gold) of a carbon atom (*i.e.* atomic basin) in tetrahedrane. The blue points are RCPs (one for each of the three faces meeting in the carbon atom), the purple points are BCPs (hence the lone purple point at the bottom right, which corresponds to the CH bond), and the green point is the CCP.

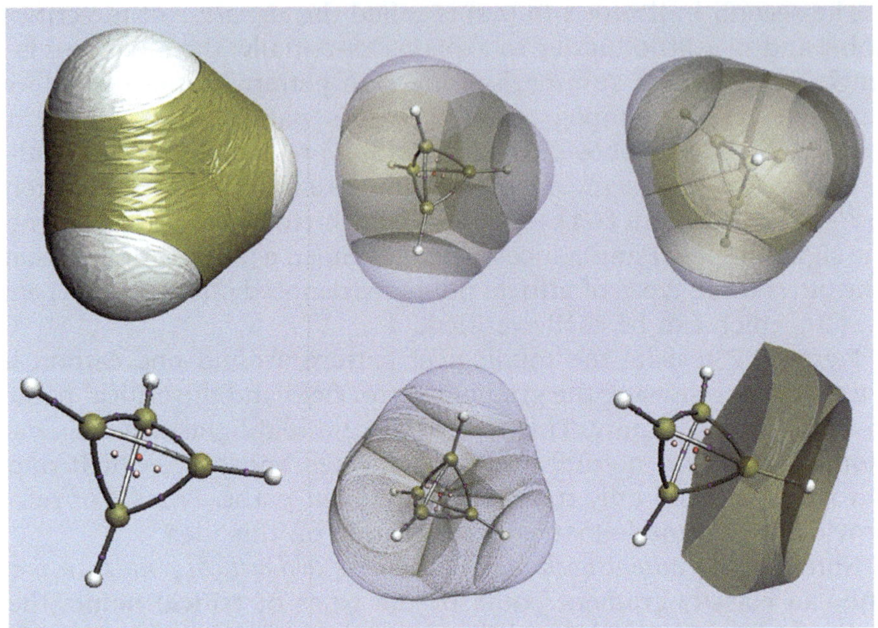

Figure 4.8   Six representations of tetrahedrane, showing all four types of critical point, (topological) atoms atomic interaction lines, interatomic surfaces, molecular graph and constant-electron-density surfaces. Images[54,55] were prepared using the in-house package IRIS.

with that of QTAIM because QCT operates on the electron density, shows atoms and bonds.

In this section on critical points it is appropriate to mention an area of research called Quantum Topological Molecular Similarity (QTMS), which sprung from the idea[56] of the so-called BCP space. The idea is to express each bond in a molecule in an abstract space spanned by properties evaluated at BCPs, such as ρ, its Laplacian and the so-called ellipticity. Each molecule is then represented by a quantum fingerprint essentially inspired by QCT. A set of congeneric series (*i.e.* a common skeleton with systematically varied substituents) can then be ranked according to their similarity as expressed by a Euclidean distance in that space. This method led to a large number of applications[57] including one on ester hydrolysis rate constant prediction.[58]

What is the meaning of a non-nuclear attractor (NNA)?
An NNA is a maximum in the electron density ρ, but in its immediate vicinity there is no nucleus. Such an "atomic" basin cannot be legitimately called a topological atom because there is no nucleus that (practically) coincides with this maximum in ρ. The first question is if

these NNAs are real or an artefact of a poorly calculated electron density. If they indeed turn out to be real, then the next question is what their physical meaning is.

Let us start with the NNA as an artefact. A poorly calculated electron density of a simple molecule such as ethyne shows a spurious NNA between the two carbon atoms. One $(3, -1)$ critical point connects the NNA to one carbon, while another $(3, -1)$ connects the NNA to the other carbon. When calculated with an inadequate basis set, the electron density profile along the molecular axis typically shows a very shallow maximum at the inversion centre of ethyne. However, the fact that, at this inversion centre, ρ is just slightly above the remainder of the electron density, is sufficient to create a maximum critical point. Patterns of critical points have a binary character: they can suddenly appear or disappear, even if the smooth electron density shows an almost invisible ripple, which is a shallow maximum in this case. The volume of the basin associated with such a spurious NNA is typically quite large. As a result, the topological picture of the molecule is visibly perturbed: instead of having four nuclei and four atomic basins, one now has five basins. However, better basis sets make the NNA disappear and restore the topology to its expected pattern, as observed before for Si–Si bonds,[59] for example.

Another well-researched[60] example of an NNA disappearing in electron densities calculated at higher levels of theory is that of dipole-bound water cluster anions $(H_2O)_n^-$. These clusters appear in a variety of forms, where a tube of water trimers or tetramers builds up, layer after layer. At the end of the tube one imagines a diffuse singly occupied molecular orbital that accommodates the excess electron. For the three-layered tube of trimers, there is an NNA at a lower level of theory, which disappears for a higher level of theory. Based on several clusters investigated, one concludes that electron densities obtained with insufficient inclusion of electron correlation effects and tight basis sets show an NNA, which ultimately disappears at a higher level of theory and more diffuse basis sets.

The same study[60] also investigated solvated-electron clusters, where the excess electron is surrounded by solvent molecules. For these systems, the existence of the NNA does not seem to be an artefact of the level of theory used. Here, the NNA is genuine and constitutes an exclusive feature of a confined environment (as opposed to being located at the tip of a dipole-bound cluster where it fails to appear for real).

NNAs have also been attributed[61] a real existence and physical meaning in the context of *F*-centers (called after the German word *Farbe* or colour), which are the simplest and most extensively studied of the colour centers formed in alkali halide crystals. In this study, the removal of a fluorine atom from its central position in a cubic-like $Li_{14}F_{13}$ cluster creates an *F*-center vacancy. If the remaining odd electron possesses a separate identity, then a local maximum in the electron density will be found within the vacancy and the *F*-center will be an NNA. If, however, the density of the odd electron is primarily delocalised onto the neighbouring ions, then a cage critical point, a local minimum in the density, will be found at the center of the vacancy. The topology enables one to make a clear distinction between the two possible forms that the odd electron can assume. Characteristically, the electron density of NNAs is diffuse, exhibits small curvatures, is delocalised, and possesses a low kinetic energy per electron. These are the characteristics of the electron density of the conducting electrons in metals.

In the late 1990s a controversy was resolved[62] concerning the experimental detection of NNAs in solid Be and Si. This study used simple arguments and quantum mechanical calculations to analyse the occurrence of NNAs, which were shown to be a normal step in the evolution of chemical bonding of homonuclear groups as internuclear distances decrease. It was concluded that the shell structure of atoms, contained in promolecular densities, is the basic organizing principle underlying the occurrence of NNAs. Thus, NNAs are not an artefact but a normal step in the chemical bonding of homonuclear groups, if analysed in the appropriate range of internuclear distances. For most elements, however, this range occurs far away from the stable geometry under normal thermodynamic conditions.

Finally, we mention the more recent work[63] of Lain *et al.* who proposed simple procedures to "deal" with NNAs. In particular, the authors used "mappings of some overlap integral sets" *to get rid* of NNAs. The motivation behind this work is curious: either an NNA is real, in which case it should not be removed because it has a physical meaning, or an NNA is an artefact due to an insufficient level of theory and then this level cannot be trusted. Moreover, their proposed "redistribution" procedures are not unique and inspired by Mulliken or Löwdin population analysis, which surely must tarnish their NNA riddance procedure with the well-known problems of Hilbert space partitioning. Note that topological partitioning does not suffer from these problems.[64]

In summary, the occasionally uttered criticism that NNAs are a fatal flaw of QTAIM is thus unfounded. Indeed, if they survive at a sufficiently high level of theory, they reveal a physical entity that should be taken seriously, studied and interpreted as it appears.

What is Quantum Chemical Topology (QCT)?

This is a branch of theoretical chemistry that embodies an over-arching structure for all methods that share the central idea of using the mathematical language of dynamical systems[50] (*e.g.* critical point, saddle point, separatrix, basin, vector field, Poincaré-Hopf relationship). The name QCT was proposed[7] and justified in detail in 2003. Since then, the approach has stood its ground against a non-QCT approach, and a first dedicated symposium has even been held (Mexico City, 2013). Alternative methods of interpretative quantum chemistry[65–71] do not share the central concept of the gradient vector field underpinning QCT. This crucial difference draws together the topological analysis of the various functions under the heading of QCT, which is thus distinct from non-QCT methods. The following non-exhaustive list of QCT subfields shows the breath of activity:

1. electron density $\rho(\mathbf{r})$ (started with ref. 72).
2. the Laplacian of ρ, $\nabla^2\rho(\mathbf{r})$ (started with ref. 73,74 and full topology first explored in ref. 75–77).
3. (bare) nuclear potential $V_{nuc}(r)$ (early start with ref. 78 but later an elaborate and self-contained study[79]).
4. ELF[80] (started with ref. 81 and reviewed in ref. 82).
5. electrostatic potential[83] (started with thorough studies[84,85] and continued with ref. 86, applied in the area of chemical reactions,[87] Lewis acidity[88] or electron diffraction study[89]).
6. virial field (or trace of the Schrödinger stress tensor) (topology explored in ref. 90).
7. magnetically induced molecular current distributions (started with ref. 91).
8. intracule density (started with ref. 92, which reveals correlation cages).
9. Ehrenfest force field (topology first investigated[93] in 2012).
10. Energy partitioning (beyond the kinetic energy and atom virial theorem) (Coulomb potential energy partitioning started with ref. 44 and culminated into the theory of Interacting Quantum Atoms (IQA),[45] leading to energetic underpinning for the topological expression of chemical bonding[48]).

The QTAIM was also extended into the topology of the Laplacian of the electron density, establishing an observational link with the *Valence Shell Electron Pair Repulsion (VSEPR)*[94–96] model of Gillespie and predecessors. It makes sense to view the extensive study of the topology of the Laplacian (as demonstrated by the dozens of papers on it, including a study on its complete[75] topology) as the second branch of Quantum Chemical Topology (QCT).[76] Out of all QCT subfields so far, the Laplacian literature is probably third in size, with ELF probably second in size, and that on the electron density being the largest. The literature using IQA is growing steadily and being contrasted with the older approaches such as Energy Decomposition Analysis (EDA). The latter have been critically reviewed[97] in 2015, showing a raft of issues and problems, again highlighting the vulnerability of non-QCT methods, in spite of their age and entrenchment.

4.3 Conclusion

Quantum Chemical Topology (QCT) was started in 1972, which is the birth year of a *completely novel* way of thinking about how to partition and characterise a quantum mechanical system. This methodology has withstood the attacks over the subsequent decades, and still stands as a robust edifice, one that is often mistaken for just yet another population analysis. However, if QCT was just that, it still avoids the well-known difficulties of other population analyses, even after their recent improvements. Not only does QCT offer a visually appealing and modern picture of atoms and bonds, but it also offers a conceptually lean and powerful energy decomposition analysis called IQA. Finally, QCT underpins a next–next-generation protein force field[10] called FFLUX, which through machine learning teaches topological atoms how to interact with other atoms. This development will enable a new type of geometry optimisation and molecular dynamics simulation, without loss of quantum mechanical insight.

Acknowledgements

I thank the EPSRC for the award of an Established Career Fellowship and I express my thanks to Dr Matthew Mills for the preparation of Figure 4.7.

References

1. R. F. W. Bader, A Quantum-Theory of Molecular-Structure and Its Applications, *Chem. Rev.*, 1991, **91**, 893–928.
2. R. F. W. Bader, *Atoms in Molecules. A Quantum Theory*, Oxford Univ. Press, Oxford, Great Britain, 1990.

3. P. L. A. Popelier, *Atoms in Molecules. An Introduction*, Pearson Education, London, Great Britain, 2000.
4. C. F. Matta and R. J. Boyd, The Quantum Theory of Atoms in Molecules, *From Solid State to DNA and Drug Design*, Wiley-VCH, Weinheim, Germany, 2007.
5. P. L. A. Popelier, The Quantum Theory of Atoms in Molecules, in *The Nature of the Chemical Bond Revisited*, ed. G. Frenking and S. Shaik, Wiley-VCH, 2014, ch. 8, pp. 271–308.
6. P. L. A. Popelier, On Quantum Chemical Topology, in *Challenges and Advances in Computational Chemistry and Physics Dedicated to Applications of Topological Methods in Molecular Chemistry*, ed. R. Chauvin, C. Lepetit, E. Alikhani and B. Silvi, Springer, Switzerland, 2016, pp. 23–52.
7. P. L. A. Popelier and F. M. Aicken, Atomic Properties of Amino Acids: computed Atom Types as a Guide for future Force Field Design, *ChemPhysChem*, 2003, **4**, 824–829.
8. P. L. A. Popelier, Quantum Chemical Topology: On Bonds and Potentials, in *Structure and Bonding. Intermolecular Forces and Clusters*, ed. D. J. Wales, Springer, Heidelberg, Germany, 2005, vol. 115, pp. 1–56.
9. P. L. A. Popelier, Quantum Chemical Topology, in *The Chemical Bond - 100 years old and Getting Stronger*, ed. M. Mingos, Springer: Switzerland, 2016, pp. 71–117.
10. P. L. A. Popelier, Molecular Simulation by Knowledgeable Quantum Atoms, *Phys. Scr.*, 2016, **91**, 033007.
11. P. L. A. Popelier, QCTFF: On the Construction of a Novel Protein Force Field, *Int. J. Quant. Chem.*, 2015, **115**, 1005–1011.
12. N. Lane, *The Vital Question: Why is Life the Way it is?*, Profile Books Ltd, London, Great Britain, 2015.
13. R. J. Gillespie, Covalent and ionic molecules: Why are BeF2 and AlF3 high melting point solids whereas BF3 and SiF4 are gases?, *J. Chem. Educ.*, 1998, **75**, 923–925.
14. R. G. Pearson, *Chemical Hardness*, John Wiley-VCH, Weinheim, Germany, 1997.
15. R. G. Parr and R. G. Pearson, Absolute Hardness - Companion Parameter to Absolute Electronegativity, *J. Am. Chem. Soc.*, 1983, **105**, 7512–7516.
16. M. Torrent-Sucarrat, F. De Proft, P. Geerlings and P. W. Ayers, Do the Local Softness and Hardness Indicate the Softest and Hardest Regions of a Molecule?, *Chem. – Eur. J.*, 2008, **14**, 8652–8660.
17. R. McWeeny, *Coulson's Valence*, Oxford University Press, Oxford, Great Britain, 1979.
18. P. L. A. Popelier, Preface, *Faraday Discuss.*, 2007, **135**, 3–5.
19. P. M. W. Gill, Extraction of Stewart atoms from Electron Densities, *J. Phys. Chem.*, 1996, **100**, 15421–15427.
20. A. T. B. Gilbert, A. M. Lee and P. M. W. Gill, Methods for constructing Stewart Atoms, *J. Molec. Struct.: THEOCHEM*, 2000, **500**, 363–374.
21. A. J. Stone, Distributed Multipole Analysis, or How to Describe A Molecular Charge Distribution, *Chem. Phys. Lett.*, 1981, **83**, 233–239.
22. T. S. Koritsanszky and P. Coppens, Chemical Applications of X-ray Charge-Density Analysis, *Chem. Rev.*, 2001, **101**, 1583–1627.
23. Y. Aray, J. Rodriguez and R. Lopez-Boada, A numerical method for the topological analysis of the Laplacian of the electronic Charge Density in Molecules and Solids, *J. Phys. Chem. A*, 1997, **101**, 2178–2184.
24. R. F. W. Bader, P. M. Beddall and P. E. Cade, Partitioning and Characterization of Molecular Charge disitributions, *J. Am. Chem. Soc.*, 1971, **93**, 3095–3107.
25. R. F. W. Bader and P. M. Beddall, The Spatial Partitioning and Transferability of Molecular Energies, *Chem. Phys. Lett.*, 1971, **8**, 29–36.
26. R. F. W. Bader and H. Essen, The Characterization of Atomic Interactions, *J. Chem. Phys.*, 1984, **80**, 1943–1960.

27. P. L. A. Popelier, New insights in atom-atom interactions for future drug design, *Curr. Topics Med. Chem.*, 2012, **12**, 1924–1934.

28. T. Slee, A. Larouche and R. F. W. Bader, Properties of Atoms in Molecules - Dipole-Moments and Substituent Effects in Ethyl and Carbonyl-Compounds, *J. Phys. Chem.*, 1988, **92**, 6219–6227.

29. P. L. A. Popelier, Integration of Atoms in Molecules - a Critical Examination, *Mol. Phys.*, 1996, **87**, 1169–1187.

30. K. E. Laidig, The Atomic Basis of the Molecular Quadrupole-Moments of Benzene and Hexafluorobenzene, *Chem. Phys. Lett.*, 1991, **185**, 483–489.

31. P. L. A. Popelier and F. M. Aicken, Atomic properties of selected biomolecules: Quantum topological atom types of carbon occuring in natural amino acids and derived molecules, *J. Am. Chem. Soc.*, 2003, **125**, 1284–1292.

32. R. F. W. Bader and P. L. A. Popelier, Atomic Theorems, *Int. J. Quant. Chem.*, 1993, **45**, 189–207.

33. S. E. O'Brien and P. L. A. Popelier, Quantum molecular similarity. Part 2: The relation between properties in BCP space and bond length, *Can. J. Chem.*, 1999, **77**, 28–36.

34. P. Macchi and A. Sironi, Chemical bonding in transition metal carbonyl clusters: complementary analysis of theoretical and experimental electron densities, *Coord. Chem. Rev.*, 2003, **238**, 383–412.

35. R. F. W. Bader and H. J. T. Preston, The Kinetic Energy of Molecular Charge Distributions and Molecular Stability, *Int. J. Quant. Chem.*, 1969, **3**, 327–347.

36. R. F. W. Bader, Bond Paths Are Not Chemical Bonds, *J. Phys. Chem. A*, 2009, **113**, 10391–10396.

37. R. F. W. Bader, Pauli repulsions exist only in the eye of the beholder, *Chem. – Eur. J.*, 2006, **12**, 2896–2901.

38. R. F. W. Bader, A bond path: A universal indicator of bonded interactions, *J. Phys. Chem. A*, 1998, **102**, 7314–7323.

39. E. Cerpa, A. Krapp, R. Flores-Moreno, K. J. Donald and G. Merino, Influence of Endohedral Confinement on the Electronic Interaction between He atoms: A He2@C20H20 Case Study, *Chem.– Eur. J.*, 2009, **15**, 1985–1990.

40. S.-G. Wang, Y.-X. Qiu and W. H. E. Schwarz, Bonding or Nonbonding? Description or Explanation? "Confinement Bonding" of He@adamantane, *Chem. – Eur. J.*, 2009, **15**, 6032–6040.

41. A. Haaland, D. J. Shorokhov and N. V. Tverdova, Topological analysis of electron densities: is the presence of an atomic interaction line in an equilibrium geometry a sufficient condition for the existence of a chemical bond?, *Chem. – Eur. J.*, 2004, **10**, 4416–4421.

42. J. Poater, M. Sola and F. M. Bickelhaupt, Hydrogen–Hydrogen Bonding in Planar Biphenyl, Predicted by Atoms-In-Molecules Theory, Does Not Exist, *Chem. – Eur. J.*, 2006, **12**, 2889–2895.

43. R. G. A. Bone and R. F. W. Bader, Identifying and analyzing intermolecular bonding interactions in van der Waals molecules, *J. Phys. Chem.*, 1996, **100**, 10892–10911.

44. P. L. A. Popelier and D. S. Kosov, Atom-atom partitioning of intramolecular and intermolecular Coulomb energy, *J. Chem. Phys.*, 2001, **114**, 6539–6547.

45. M. A. Blanco, A. Martín Pendás and E. Francisco, Interacting Quantum Atoms: A Correlated Energy Decomposition Scheme Based on the Quantum Theory of Atoms in Molecules, *J. Chem. Theor. Comput.*, 2005, **1**, 1096–1109.

46. M. Rafat and P. L. A. Popelier, Topological atom-atom partitioning of molecular exchange energy and its multipolar convergence, in *Quantum Theory of Atoms in Molecules*, ed. C. F. Matta and R. J. Boyd, Wiley-VCH, Weinheim, Germany, 2007, vol. 5, pp. 121–140.

47. M. Garcia-Revilla, E. Francisco, P. L. A. Popelier and A. Martín Pendás, Domain-averaged exchange correlation energies as a physical underpinning for chemical graphs, *ChemPhysChem*, 2013, **14**, 1211–1218.
48. A. Martín Pendás, E. Francisco, M. A. Blanco and C. Gatti, Bond Paths as Privileged Exchange Channels, *Chem. – Eur. J.*, 2007, **13**, 9362–9371.
49. V. Tognetti and L. Joubert, On the physical role of exchange in the formation of an intramolecular bond path between two electronegative atoms, *J. Chem. Phys.*, 2013, **138**, 024102.
50. K. Collard and G. G. Hall, Orthogonal Trajectories of the Electron Density, *Int. J. Quant. Chem.*, 1977, **12**, 623–637.
51. R. F. W. Bader, T. T. Nguyen-Dang and Y. Tal, A Topological Theory of Molecular-Structure, *Rep. Prog. Phys.*, 1981, **44**, 893–948.
52. R. Thom, *Structural Stability and Morphogenesis*, ed. English, Benjamin, Reading, MA, USA, 1975.
53. N. O. J. Malcolm and P. L. A. Popelier, An Improved Algorithm to Locate Critical Points in a 3D scalar Field as Implemented in the Program MORPHY, *J. Comp. Chem.*, 2003, **24**, 437–442.
54. M. Rafat and P. L. A. Popelier, Visualisation and integration of quantum topological atoms by spatial discretisation into finite elements, *J. Comput. Chem.*, 2007, **28**, 2602–2617.
55. M. Rafat, M. Devereux and P. L. A. Popelier, Rendering of quantum topological atoms and bonds, *J. Mol. Graphics Modell.*, 2005, **24**, 111–120.
56. P. L. A. Popelier, Quantum molecular similarity. 1. BCP space, *J. Phys. Chem. A*, 1999, **103**, 2883–2890.
57. P. L. A. Popelier and P. J. Smith, QSAR models based on Quantum Topological Molecular Similarity, *Eur. J. Med. Chem.*, 2006, **41**, 862–873.
58. U. A. Chaudry and P. L. A. Popelier, Ester hydrolysis rate constant prediction from quantum topological molecular similarity (QTMS) descriptors, *J. Phys. Chem. A*, 2003, **107**, 4578–4582.
59. O. A. Zhikol, A. F. Oshkalo, O. V. Shishkin and O. V. Prezhdo, Non-nuclear attractors on Si–Si bond in quantum-chemical modeling as basis set inadequacy, *Chem. Phys.*, 2003, **288**, 159–169.
60. Q. K. Timerghazin, I. Rizvi and G. H. Peslherbe, Can a dipole-bound electron form a pseudo-atom? An atoms-in-molecule study of the hydrated electron, *J. Phys.Chem. A*, 2011, **115**, 13201–13209.
61. R. F. W. Bader and J. A. Platts, Characterization of an F-center in an alkali halide cluster, *J. Chem. Phys.*, 1997, **107**, 8545–8553.
62. A. Martín Pendás, M. A. Blanco, A. Costales, P. M. Sanchez and V. Luana, Non-nuclear maxima of the electron density, *Phys. Rev. Lett.*, 1999, **83**, 1930–1933.
63. D. R. Alcoba, L. Lain, A. Torre and R. C. Bochicchio, Treatments of non-nuclear attractors within the theory of atoms in molecules, *Chem. Phys. Lett.*, 2005, **407**, 379–383.
64. A. Martín Pendás, M. A. Blanco and E. Francisco, Chemical Fragments in Real Space: Definitions, Properties, and Energetic Decompositions, *J. Comput. Chem.*, 2007, **28**, 161–184.
65. B. Webster, *Chemical Bonding Theory*, Blackwell, Oxford, Great Britain, 1990.
66. A. E. Reed, L. A. Curtiss and F. Weinhold, Intermolecular interactions from a natural bond orbital, donor-acceptor viewpoint, *Chem. Rev.*, 1988, **88**, 899–926.
67. A. J. Stone, Distributed Multipole Analysis, or How to Describe a Molecular Charge-Distribution, *Chem. Phys. Lett.*, 1981, **83**, 233–239.
68. R. S. Mulliken, Electronic Population Analysis on LCAO-MO Molecular Wave Functions. I, *J. Chem. Phys.*, 1955, **23**, 1833–1840.
69. R. G. Pearson, Applying the Concepts of Density Functional Theory to Simple Systems, *Int. J. Quant. Chem.*, 2007, **108**, 821–826.

70. A. Kovacs, C. Esterhuysen and G. Frenking, The Nature of the Chemical Bond Revisited: An Energy-Partitioning Analysis of Nonpolar Bonds, *Chem.- Eur. J.*, 2005, **11**, 1813–1825.
71. R. McWeeny, *Methods of Molecular Quantum Mechanics*, Academic Press, SanDiego, USA, 2nd edn, 1992.
72. R. F. W. Bader and P. M. Beddall, Virial Field Relationship for Molecular Charge Distributions and the Spatial Partitioning of Molecular Properties, *J. Chem. Phys.*, 1972, **56**, 3320–3329.
73. R. F. W. Bader, P. J. MacDougall and C. D. H. Lau, Bonded and Nonbonded Charge Concentrations and Their Relation to Molecular-Geometry and Reactivity, *J. Am. Chem. Soc.*, 1984, **106**, 1594–1605.
74. R. F. W. Bader, R. J. Gillespie and P. J. MacDougall, A Physical Basis For the VSEPR Model of Molecular Geometry, *J. Am. Chem. Soc.*, 1988, **110**, 7329–7336.
75. P. L. A. Popelier, On the Full Topology of the Laplacian of the Electron Density, *Coord. Chem. Rev.*, 2000, **197**, 169–189.
76. N. O. J. Malcolm and P. L. A. Popelier, The full topology of the Laplacian of the electron density: scrutinising a physical basis for the VSEPR model, *Faraday Discuss.*, 2003, **124**, 353–363.
77. N. O. J. Malcolm and P. L. A. Popelier, An improved algorithm to locate critical points in a 3D scalar field as implemented in the program MORPHY, *J. Comp. Chem.*, 2003, **24**, 437–442.
78. Y. Tal, R. F. W. Bader and J. Erkku, Structural homeomorphism between the electron density and the nuclear potential of a molecular system, *Phys. Rev. A*, 1980, **21**, 1–11.
79. P. L. A. Popelier and É. A. G. Brémond, Geometrically faithful homeomorphisms between the electron density and the bare nuclear potential, *Int. J. Quant. Chem.*, 2009, **109**, 2542–2553.
80. A. D. Becke and K. E. Edgecombe, A simple measure of electron localization in atomic and molecular systems, *J. Chem. Phys.*, 1990, **92**, 5397–5403.
81. B. Silvi and A. Savin, Classification of chemical bonds based on topological analysis of electron localization functions, *Nature*, 1994, **371**, 683–686.
82. V. Polo, J. Andres, S. Berski, L. R. Domingo and B. Silvi, Understanding Reaction Mechanisms in Organic Chemistry from Catastrophe Theory Applied to the Electron Localization Function Topology, *J. Phys. Chem. A*, 2008, **112**, 7128–7136.
83. G. Naray-Szabo and G. G. Ferenczy, Molecular Electrostatics, *Chem. Rev.*, 1995, **95**, 829–847.
84. S. R. Gadre, S. A. Kulkarni and I. H. Shrivastava, Molecular Electrostatic Potentials: A topographical study, *J. Chem. Phys.*, 1992, **96**, 5253–5261.
85. S. R. Gadre and I. H. Shrivastava, Shapes and sizes of molecular aniosn via topographical analysis of electrostatic potential, *J. Chem. Phys.*, 1991, **94**, 4384–4391.
86. P. Balanarayan and S. R. Gadre, Topography of molecular scalar fields. I. Algorithm and Poincare-Hopf relation, *J. Chem. Phys.*, 2003, **119**, 5037–5043.
87. P. Balanarayan, R. Kavathekar and S. R. Gadre, Electrostatic Potential Topography for Exploring Electronic Reorganizations in 1,3 Dipolar Cyclo-additions, *J. Phys. Chem. A*, 2007, **111**, 2733–2738.
88. Y. Aray, J. Rodriguez, S. Coll, E. N. Rodrıguez-Arias and D. Vega, Nature of the Lewis Acid Sites on Molybdenum and Ruthenium Sulfides: An Electrostatic Potential Study, *J. Phys. Chem. B*, 2005, **109**, 23564–23570.
89. V. G. Tsirelson, A. S. Avilov, G. G. Lepeshov, A. K. Kulygin, J. Stahn, U. Pietsch and J. C. H. Spence, Quantitative Analysis of the Electrostatic Potential in Rock-Salt Crystals Using Accurate Electron Diffraction Data, *J. Phys. Chem. B*, 2001, **105**, 5068–5074.

90. T. A. Keith, R. F. W. Bader and Y. Aray, Structural homeomorphism between the electron density and the virial field, *Int. J. Quant. Chem.*, 1996, **57**, 183–198.
91. T. A. Keith and R. F. W. Bader, Topological Analysis of Magnetically Induced Molecular Current Distributions, *J. Chem. Phys.*, 1993, **99**, 3669–3682.
92. J. Cioslowski and G. H. Liu, Topology of electron-electron interactions in atoms and molecules. II. The correlation cage, *J. Chem. Phys.*, 1999, **110**, 1882–1887.
93. A. Martín Pendás and J. Hernandez-Trujillo, The Ehrenfest force field: Topology and consequences for the definition of an atom in a molecule, *J. Chem. Phys.*, 2012, **137**, 134101.
94. R. J. Gillespie and E. A. Robinson, Electron Domains and the VSEPR Model of Molecular Geometry, *Angew. Chem., Int. Ed. Engl.*, 1996, **35**, 495–514.
95. R. J. Gillespie and E. A. Robinson, Models of molecular geometry, *Chem. Soc. Rev.*, 2005, **34**, 396–407.
96. R. J. Gillespie, *Molecular Geometry*, Van Nostrand Reinhold, London, 1972.
97. M. J. S. Phipps, T. Fox, C. S. Tautermann and C.-K. Skylaris, Energy decomposition analysis approaches and their evaluation on prototypical protein-drug interaction patterns, *Chem. Soc. Rev.*, 2015, **44**, 3177–3211.

5 Quantitative Determination of the Nature of Intermolecular Bonds by EDA Analysis

A. Martín Pendás, J. L. Casals Sainz and E. Francisco

University of Oviedo, 33006-Oviedo, Spain
*Email: angel@fluor.quimica.uniovi.es; jluiscasalssainz@gmail.com;
evelio@uniovi.es

5.1 Introduction

These are interesting times for physicochemical sciences indeed. On one hand, the basic theory that allows us to rationalize molecular phenomena is close to becoming a hundred years old. Most of our cutting edge technology is described by the same equations we heard about in school, yet only after the advent of high performance computing facilities have we been able to grasp what was actually hidden behind this innocent looking algebra. As funding moves from basic to applied science, researchers face a subtle danger: taking for granted the fully established, basic models or protocols which are used in large scale simulations. In bioscience or crystal engineering, for instance, force fields underlie most simulations, and with them comes a decades old distinction between strong and weak interactions, which leads to completely different parameterization schemes for them. Unfortunately, eagerness for classificatory schemes is a human brain construct which is poorly reflected in nature, so renewed efforts should be directed to build physically, not historically, grounded

Intermolecular Interactions in Crystals: Fundamentals of Crystal Engineering
Edited by Juan J. Novoa
© The Royal Society of Chemistry 2018
Published by the Royal Society of Chemistry, www.rsc.org

models for interatomic or intermolecular interactions. Moreover, as soon as the feeling that strong interactions had been tamed started to spread, scientists turned to the weak interactions realm, where things are considerably more complex and old paradigms may simply not work.

These efforts should start from addressing a not easy to solve conundrum. Quantum mechanical energies are intrinsically non-separable. This means that, strictly speaking, we should solve Schrödinger's equation for every full interacting system we would be interested in. This is not possible either now or in the foreseeable future, and leads to us building approximate methods where such a separability is forced. We call these energy decomposition analyses (EDAs), and a number of them are briefly explained and reviewed in this Chapter. Paraphrasing Stones and Hayes,[12] every EDA should comply with a set of criteria. For instance, it should provide a meaningful partition of interaction energies into physically rooted terms, like electrostatic, or dispersion ones. It should lead to energies agreeing with those obtained from supermolecular calculations, being stable with respect to changes in computational parameters, *i.e.* basis sets. EDAs should also be valid in a wide range of interatomic or intermolecular separations, being free of convergence problems, and should allow for an easy computational implementation. As we will try to show, most methods fail to satisfy one or more of the above criteria.

Interestingly, the requirement of validity at both short and long range distances is one of the most difficult to satisfy. Even electrostatics, which provides the basic language all of us speak when dealing with intermolecular interactions, is subjected to several conflicting interpretations when taken to the short range realm. Our efforts in the present contribution will point towards clarifying the language used in different EDAs, the relation among them, and the problems when interpreting equally named, sometimes vaguely related quantities. In our journey, we will also push towards the generalization of techniques based on the analysis of wavefunctions in real space, which, in our opinion, satisfy the above requirements in a mostly consistent way.

We will first review several well-known approaches, starting from perturbation theory, and then we will proceed by comparing the different energetic terms they propose. We will end with a short overall picture of the different EDAs analyzing a couple of systems. We will use a set of EDA data obtained for a small number of selected examples which have been computed afresh with common basis sets, at different levels of theory. These data are gathered in a few tables that will be used throughout the chapter.

5.2 Intermolecular Interactions and Energy Decompositions

Following the ubiquitous tendency in contemporary science towards specialization and isolation, the traditional approach to inter-molecular, i.e. weak, interactions has given rise to a subdiscipline that has evolved its own theoretical dialect, well separated from that used when dealing with strong chemical bonds. To put it down in black and white, intermolecular interactions have not been considered as full inter-molecular bonds. Historically, (see ref. 41 and references therein) forces among molecules have been understood as arising from two (or more) separated but interacting entities, each con-sisting of a perfectly defined set of nuclei and electrons, and thus Hamiltonians. Within this framework, it is not surprising that perturbation theory, in any of its many flavors, was chosen as the standard theoretical tool to tackle these problems.[22] If we further stick to the common Coulomb approximation, then the interaction Hamiltonian is electrostatic, and for two interacting moieties A and B, with isolated unperturbed Hamiltonians and eigenstates $H^A|i^A\rangle = E_i^A|i^A\rangle$ and $H^B|i^B\rangle = E_i^B|i^B\rangle$ so that $H^0 = H^A + H^B$,

$$H' = -\sum_{i\in A, \beta\in B} \frac{Z_\beta}{r_{i\beta}} - \sum_{j\in B, \alpha\in A} \frac{Z_\alpha}{r_{j\alpha}} + \sum_{i\in A, j\in B} \frac{1}{r_{ij}}. \tag{5.1}$$

This perturbation may be further compacted if we use common labels for nuclei and electrons and appropriate (signed) charges to the following form, $H' = \sum_{i\in A, j\in B} \frac{q_i q_j}{r_{ij}}$. Any text on intermolecular inter-actions will now warn on what comes next. Unfortunately, the wave function of the *AB* system should be fully antisymmetric, thus pre-cluding the use of standard Rayleigh–Schrödinger theory, for in-stance. The simplest product function obeying the particle statistics constraint is built as $\Psi^{AB} = \mathscr{A}\Psi^A\Psi^B$, where $\mathscr{A}$ is a partial anti-symmetrizer that runs over the $N!/(N_A!N_B!)$ intersystem permutations of the electrons.[31]

In the long-range regime, one may invoke that the overlap

$$\int d1\, \Psi^A(1, 2, \ldots, N_A)\Psi^B(1, 2', \ldots, N_B') \tag{5.2}$$

decays exponentially and may be taken safely to equal zero. This is equivalent to Parr's strong orthogonality condition, which effectively annihilates all but the first term in the partial antisymmetrizer above. We should notice, however, that the separation of the wave function associated to the above condition forces the composite system to be an overall insulator, according to Kohn's insights.[21] If we decide to continue with these strong approximations, then Hartree products of the subsystem states are eigenstates of H^0, $H^0 |i^A j^B\rangle = (E_i^A + E_j^B)|i^A j^B\rangle = E_{ij}^0 |i^A j^B\rangle$, and all the perturbation theory machinery is at our disposal. If we take for granted that no degeneracies will occur (which effectively excludes the case $A = B$), and label the ground states of A and B as $|0^A\rangle$, $|0^B\rangle$, respectively, then non-degenerate perturbation theory provides the following results,

$$E_{00}^1 = \langle 0^A 0^B | H' | 0^A 0^B \rangle$$

$$E_{00}^2 = \sum_{ij \neq 00} \frac{\langle 0^A 0^B | H' | i^A j^B \rangle \langle i^A j^B | H' | 0^A 0^B \rangle}{E_{ij}^0 - E_{00}^0}. \tag{5.3}$$

The expressions above are at the root of a large part of the contemporary language on intermolecular interactions, since all the terms coming from eqn (5.3) have simple physical interpretations. They are also the source of much confusion, particularly when the long range approximation is relaxed and antisymmetry enters the game.

For our purposes in this chapter, we should stress that the perturbation expansion provides, directly, an energetic partition or decomposition of the interaction energy $E_{00}^{\text{int}} = E_{00} - E_{00}^0 \approx E_{00}^1 + E_{00}^2$. As in most perturbative applications, the focus is set on the energy, and not on the wave function modifications, for energetic and state vector corrections are not commensurate with each other, $E^n = \langle \Psi^0 | H' | \Psi^{n-1} \rangle$. Since the corrections are built from sums over states, *i.e.* by taking profit of the completeness of the eigensolutions of observables, interpretations will be based on the original static molecules *before interaction takes place*. Although all this is deeply rooted now, it will lead to several problems while comparing with other procedures.

Interpreting the different terms in eqn (5.3) is easy. The first order correction is the expectation value of the intermolecular Coulomb electrostatics,

$$E_{00}^1 = \int dr_1 \, dr_2 \, \rho_t^A(r_1) \rho_t^B(r_2) / r_{12}, \tag{5.4}$$

where ρ_t is the total molecular charge density, including the nuclear contributions. This is also known as the electrostatic energy, E_{elstat}. It is so widely used that it is important to step down at this point and comment on its meaning. E_{elstat} is the electrostatic interaction between two *distinguishable, interpenetrating* charge clouds. Electrons can be unambiguously associated to each molecule: N_A to A, N_B to B. As we leave the long-range regime, this assumption is no longer tenable, and separating the true intermolecular Coulombic contributions from the intramolecular ones becomes increasingly difficult. We will devote some effort below to clarify the situation in different EDAs. At this point it is nevertheless important to notice that E_{elstat}, as defined in eqn (5.4), cannot be unambiguously interpreted as the intermolecular electrostatic energy except in the (very) long range approximation.

The second order corrections embrace the classical induction, when one of the i, j indices in eqn (5.3) is 0, or dispersion terms, when both are non-zero: $E_{00}^2 = E_{\text{ind}}^A + E_{\text{ind}}^B + E_{\text{disp}}^{AB}$. For instance,

$$E_{\text{ind}}^A = -\sum_{i \neq 0} \frac{\langle 0^A 0^B | H' | i^A 0^B \rangle \langle i^A 0^B | H' | 0^A 0^B \rangle}{E_i^A - E_0^A}, \tag{5.5}$$

with a similar expression for E_{ind}^B. Using linear response theory,[23] it can be shown that

$$E_{\text{ind}}^A = \frac{1}{2} \int \mathrm{d}r \Delta \rho^A(r) V^B(r), \tag{5.6}$$

where $\Delta \rho^A$ is the first order change in the density of molecule A due to the potential V^B exerted by molecule B. An equivalent expression is, of course, obtained for E_{ind}^B. As we will show, this provides an easy route to obtaining induction terms in other EDAs.

The plain electrostatic and induction terms have been traditionally obtained from a multipole expansion. This provides us with extra insight: when only a few low ranked molecular multipoles are important, the angular structure and decay rate of the interactions have well known analytic expressions (see below), which have been widely used for building chemical insight about preferred geometrical structures of molecular complexes. At the same time, the multipole expansion is the source of some confusion. Firstly, the series need not be convergent. Actually, the polarization term is, at best, asymptotically convergent. Secondly, when it converges, it does so to a value which is in error with respect to that obtained from exact integration. This is the penetration error, which we will discuss below.

A linear response expression does also exist for the second order dispersion,

$$E_{\text{disp}} = -\sum_{i\neq 0,\, j\neq 0} \frac{\langle 0^A 0^B|H'|i^A j^B\rangle\langle i^A j^B|H'|0^A 0^B\rangle}{E_{ij}^0 - E_{00}^0}, \tag{5.7}$$

in terms of dynamic polarizabilities (or frequency dependent density-density response functions).[31,41] Unfortunately, it is not easy to uncouple electron motion in this case, leaving dispersion as a result of electron correlation that does only admit a per method comparison.

If the long-range regime is abandoned, the antisymmetry (or exchange) requirements may no longer be forgotten, and the perturbation treatment becomes ill-defined. One of the basic problems is easy to catch. If we try to antisymmetrize the set of basis products $|i^A j^B\rangle$, the resulting states are not orthogonal, thus they cannot be eigenstates of the zero-th order Hermitian Hamiltonian H^0. This means that the standard Rayleigh-Schrödinger theory cannot be immediately applied. And when problems arise, multiple solutions do also arise to enforce the correct symmetry. There are several of these symmetric perturbation techniques, that use either non-orthogonal antisymmetrized states, or non-symmetric states.[41] In the former, electrons cease to be assigned to each of the fragments at the cost of cumbersome energetic expressions. In the latter, electrons are still distinguishable, and the correct symmetry is enforced at particular steps in the procedures. Exchange modifies all of the above energy components, including polarization and dispersion. However, they do it differently in different perturbation flavors, giving rise to a zoo of energetic contributions that precludes direct comparison of seemingly equivalent terms.

An example of non-orthogonal treatments is Stone and Hayes intermolecular perturbation theory (IMPT).[12,13] It solves perturbatively the secular configuration interaction equation with non-orthogonal Slater determinants formed with the unperturbed $|i^A j^B\rangle$ states, rearranging the diagonal matrix elements such that Möller–Plesset denominators arise. The overall zeroth plus first order energy in such a treatment is obtained as the energy of the non-orthogonal Slater determinant formed by the juxtaposition of all the occupied molecular orbitals (ϕ) of the fragments in their ground states (SCF zeroth order wavefunctions are assumed):

$$E^0 + E^1 = \sum_{i,j} h_{ij} S_{ji}^{-1} + \frac{1}{2}\sum_{i,k,j,m}\langle\phi_i\phi_k|\phi_j\phi_m\rangle(S_{ji}^{-1}S_{mk}^{-1} - S_{mi}^{-1}S_{jk}^{-1}), \tag{5.8}$$

where $\langle\phi_i\phi_k|\phi_j\phi_m\rangle$ is a generic $\langle\phi_i(1)\phi_k(2)|r_{12}^{-1}|\phi_j(1)\phi_m(2)\rangle$ two-electron repulsion integral. Separating the energy of the isolated fragments as E^0, $E^1 = E_{\text{elstat}} + E_X + E_{\text{rep}}$, where the electrostatic energy is defined as

$$E_{\text{elstat}} = -\sum_{i\in A,\,\beta\in B} \langle\phi_i|Z_\beta/r_{i\beta}|\phi_i\rangle - \sum_{i\in B,\,\alpha\in A} \langle\phi_j|Z_\alpha/r_{j\alpha}|\phi_j\rangle$$
$$+ \sum_{i\in A,\,j\in B} \langle\phi_i\phi_j|\phi_i\phi_j\rangle + \sum_{\alpha,\,\beta\in B} \frac{|Z_\alpha|Z_\beta}{R_{\alpha\beta}}, \tag{5.9}$$

equivalent to eqn (5.4),

$$E_X = -\sum_{i\in A,\,j\in B} \langle\phi_i\phi_j|\phi_j\phi_i\rangle, \tag{5.10}$$

and the repulsion energy, E_{rep}, is a combination of one and two electron corrections to the orthogonal energy as given by eqn (5.8) if $S_{ij} = \delta_{ij}$. With this,

$$E_{\text{rep}} = \sum_{i,j} h_{ij}(S_{ji}^{-1} - \delta_{ji})$$
$$+ \frac{1}{2}\sum_{i,k,j,m} \langle\phi_i\phi_k|\phi_j\phi_m\rangle(S_{ji}^{-1}S_{mk}^{-1} - S_{mi}^{-1}S_{jk}^{-1} - \delta_{ji}\delta_{mk} + \delta_{mi}\delta_{jk}). \tag{5.11}$$

Notice that it is usual practice to join $E_X + E_{\text{rep}} = E_{xr} = E_{\text{exch}}$ and call it exchange-repulsion, or even simply exchange energy as it is customary. The sum is usually positive at standard intermolecular distances, but here we will try to show that it is important that both contributions maintain their individuality. It is also important to stress that the definition of E_X above is consistent with that of E_{elstat} in the sense that both are obtained with unperturbed orbitals and assuming an underlying orthogonality. Looking at eqn (5.11) should convince us that E_{rep} includes antisymmetrization corrections to all the energetic terms in the Hamiltonian: kinetic energy, nuclear attractions, Coulomb, and exchange contributions, both at intra- and inter-monomer levels. We will explain this further below. Most perturbation flavors (and also general EDAs) agree in the first order terms, so the electrostatic energy and exchange, or exchange-repulsion terms are common to a large number of methods. Differences start at second order.

A final point regards the $i\in A$, $j\in B$ labels in eqn (5.10). As far as electrons from different fragments are distinguishable this is not a big issue. However, when this ceases to be so, *i.e.* when orbitals come from a supermolecule calculation, or when orbitals from one fragment clearly invade the other, labelling becomes ill-defined, both in E_X and in E_{elstat}. Moreover, if an antisymmetric state is used, the global energetics is invariant under orbital transformations, but E_X and E_{elstat} are not. This invariance is recovered if the $i=j$ term is retained in the above expressions. The distinction between self-interaction-free and self-interaction-containing energetic contributions will play an important role in our discussion.

In IMPT, second order corrections have cumbersome algebraic expressions that will not be reproduced here.[12,13] They involve Möller–Plesset denominators and single and double excitations in the SCF base determinant, obtained after replacing occupied by virtual orbitals. They are classified according to which virtuals are populated. Single excitations lead to polarization (if *occ* and *virt* belong to the same fragment) or charge transfer (if not). Notice that charge transfer is allowed since electrons lose individuality after antisymmetrization takes place, although distinguishing polarization from charge transfer is rigorously arbitrary, for a complete basis set can be used with primitives centered at only one fragment. Perturbative flavors with explicit distinguishable electrons do not allow for charge transfer terms. Double excitations may involve only one fragment, thus describing intra-fragment electron correlation. If they involve both fragments, they include dispersion, double charge transfers and several other correlation corrections.[12] Despite its cleanness, IMPT suffers from the necessary low quality of the zeroth order wave function.

The use of non-symmetric states has become popular during the last decade thanks to the development of a number of iterative symmetry forcing techniques. One of the most successful is the symmetry adapted perturbation theory (SAPT) developed by Jeziorski, Szalewicz, and co-workers.[17] It is based on performing two Möller–Plesset perturbation expansions for the free monomers that account for intra-monomer correlation, and a third for the intermolecular potential:

$$H = F^A + F^B + \lambda_A W^A + \lambda_B W^B + \varepsilon H', \tag{5.12}$$

where $W^{A,B} = H^{A,B} - F^{A,B}$, F being the standard HF Fockian operator. As a triple perturbative scheme, all energy corrections are characterized by a triple index stating the order of each expansion. It is customary to add the intramolecular indices and use E^{ij} to identify order i in V and j in $W^A + W^B$. All $j=0$ terms have a direct and an exchange term: for instance, $E^{20}_{\text{ind,resp}}$ and $E^{20}_{\text{exch-ind,resp}}$ refer to the direct and exchange induction contributions. It is recommended that coupled perturbation theory is used, so the *resp* label is also added. If $j \neq 0$, the corrections are due to electron correlation. These may be important in many cases. SAPT has become a very complex theory that rivals experimental accuracy in simple complexes. It has also given rise to DFT analogues, known as SAPT-DFT.[32] The first order (eqn (5.10)) direct and exchange terms are identical to E_{elstat} and $E_X + E_{\text{rep}}$ in IMPT. It is not usual to separate the true exchange from repulsion in SAPT. Several production models, ranging from the simple first and second order correction, SAPT0, to the inclusion of coupling terms, SAPT2, or SCF corrections, SCF + SAPT2, have been defined.

The second great category of methods to study intermolecular interactions are based on using standard quantum chemical calculations on the molecular complex and obtaining the interaction energy *a posteriori* by subtracting the isolated energy of the fragments: $E_{\text{int}} = E^{AB} - E^{A} - E^{B}$. This strategy is commonly known as the supermolecule procedure. Since interaction energies are orders of magnitude smaller than total molecular energies, the supermolecular approach faces an important precision problem, usually resting on appropriate cancellation of errors. Potential sources of non-cancelling errors, like the basis set superposition error (BSSE) and the size-consistency problem, are of paramount importance in the accuracy of supermolecular interaction energies, and a huge amount of work has been devoted to minimizing them.[16] The BSSE is very well known. It is due to the enlarged Hilbert space in which a supermolecular calculation is performed with respect to that used for the equivalent computation in each of the fragments. As a result, the monomers-in-the-supermolecule have greater variational flexibility than when isolated, so the E_{int} results are too attractive. After much debate,[34] the counterpoise (CP) correction of Boys and Bernardi,[4] that computes the fragment's energies in the full molecular basis set, is still used as the method of choice to account for BSSE. A solution for size consistency, the ability of a method to dissociate to the sum of the energies computed for the isolated fragments depends, obviously, on the method itself. HF for closed shell fragments, as well as full CI (FCI) or any perturbation scheme based on appropriate size consistent references, like MP_n or coupled cluster expansions (CC), are size consistent. Truncated CIs are not. Standard density functionals are built to ensure size consistency, so this problem is usually not an issue in DFT. A final problem with the supermolecular approach is related to the (in)ability of several methods to adequately capture some parts of the intermolecular E_{int}. In the same way as HF does not include correlation and cannot describe dispersion, most DFT functionals are not asymptotically correct and have the same problem. Several solutions have been proposed,[15] and reasonable functionals suitable for dealing with dispersion exist nowadays.

From the interpretative point of view, the biggest limitation of the bare supermolecular approach stems from its outcome: a plain E_{int} number. If further insight is sought for, an energy decomposition analysis (EDA) is needed. Of course, a plethora of methods have been proposed over the years to perform this task. All of them face, this time upside down, the same basic problem already pointed out for perturbation treatments: given an intrinsically entangled description, *i.e.* wave function, how do we separate contributions coming from each fragment (let us say A or B) and those related to their mutual interaction? Two basic families of approaches have been proposed to answer this problem. Either the A, B partition is performed in Fock (orbital) space, or it is done in real space. To each of these two broad categories we may ascribe a large number of methods, some of them closely related.

Orbital space EDAs share in one way or another the seminal ideas of Morokuma,[33] later modified by Kitaura and Morokuma (KM),[20] always within a HF single determinant approximation.[20] Succinctly, different energetic terms to E_{int} are obtained by performing a series of calculations. First the free monomers are distorted to the in-the-molecule geometries. This has an energetic cost, the preparation energy, E_{prep}. The classical electrostatic energy E_{elstat} is then obtained as shown in eqn (5.4). In the original implementation, the Hartree-product is then optimized avoiding all mixing of A and B primitives. This includes the electrostatic and induction components. An antisymmetrized undistorted state is also evaluated, now including E_{elstat} and E_X. From the final supermolecular energy one obtains the charge transfer energy, E_{CT}. In order to avoid the use of non-antisymmetric states to evaluate the induction energy, KM proposed to obtain all the energetic components by deleting appropriate blocks of the full supermolecular Fock operator.

A series of steps that exclude particular blocks of F leads to the KM partition, in which E_{int} is decomposed as a sum of electrostatic, induction (or polarization), exchange-repulsion, charge-transfer, exchange-polarization, and residual (mix) energetic terms. In a KM decomposition, besides the standard interpretation of the electrostatic and exchange terms, polarization is attributed to the interaction of occupied orbitals of a fragment with its own virtuals, and charge transfer appears when the virtuals belong the other fragment. The KM procedure is clearly affected by BSSE, and several methods have been proposed to correct it.[4] The same remarks relative to the rigorous meaning of E_{CT} made when presenting the perturbative approaches apply here unchanged. Notice that besides the possibly intuitive meaning of these steps, no formal justification exists for any of the energetic quantities used. The KM EDA uses states that are either non-stationary (the antisymmetric product), or completely fictitious (obtained by deleting matrix blocks). Be it as it may, the KM general strategy is at the root of many orbital based EDAs. In practice, KM is limited by its difficult generalization to general correlated, *i.e.* multi-determinant, descriptions, but retains its academic and historical importance.

A KM inspired method that has been used extensively in the last years was proposed by Bickelhaupt and Baerends,[2] with roots in the somewhat older Ziegler–Rauk decompositions.[45] Its power emanates from its easy adaptation to DFT frameworks. It is usually known as EDA, but to distinguish it from our general EDA definition, we will name it here as BEDA. In BEDA, the fragments are first prepared (E_{prep}), and their standard electrostatic interaction is obtained, E_{elstat}. Then an antisymmetric HF or Kohn–Sham (KS) determinant is formed from those of the fragments: $\Phi^0 = \mathscr{A}\Phi^A\Phi^B$, and its energy, measured with respect to the prepared fragments, is ascribed to the joint effect of Pauli repulsion and the previously computed E_{elstat}:

$$E^0 - (E_{\text{prep}}^A + E_{\text{prep}}^B) = \Delta E^0 = E_{\text{elstat}} + \Delta E_{\text{Pauli}}. \tag{5.13}$$

The Pauli repulsion term is a catch-all that may be assimilated to the E_{exch} contribution in perturbative approaches. It contains kinetic, pure exchange and other potential terms, and it is sometimes partitioned in this way. Notice, however, that if BEDA is used (as usual) in a DFT context, the exchange and kinetic terms are those of the non-interacting KS system, usually far from the real interacting electrons. Notwithstanding this fact, ΔE_{Pauli} is usually equated to other exchange-repulsions. Finally, Φ^0 is relaxed to the final stationary supermolecular state, and the drop of energy is known as the orbital interaction term, ΔE_{orb}, which may be further decomposed into symmetry (*e.g.* σ and π) contributions or into kinetic, exchange-correlation and Coulombic terms. Overall,

$$E_{\text{int}} = E_{\text{prep}} + E_{\text{elstat}} + \Delta E_{\text{Pauli}} + \Delta E_{\text{orb}}. \tag{5.14}$$

Another popular approach (LMOEDA)[42] uses a mixture of a KM-like decomposition and a Hayes Stone non-orthogonal treatment. It starts from an HF treatment of the fragments, constructing a supermolecular determinant with all the occupied orbitals. Orthogonal energy formulae are used to obtain E_{elstat} and the pure exchange E_X. The non-orthogonal energy is then obtained and subtracted from the orthogonal one to define the repulsion energy E_{rep}. For restricted HF cases, all these components are equivalent to KM. The polarization contribution (E_{pol}) is obtained as the difference between the relaxed HF function and the antisymmetrized state, and gathers the polarization, charge transfer and mixing terms of the KM approach. It is thus equivalent to the ΔE_{orb} term in BEDA. Finally, the dispersion contribution is obtained from the difference between the relaxed HF interaction energy and that of a CCSD(T), or MP2, for instance, calculation. Minor adaptations, that include the use of non-orthogonal expressions for the densities of the fragments in the non-orthogonal determinant allow use of the method in DFT. Efficient implementations of the method use localized orbitals, and write the total interaction energy as $E_{\text{int}} = E_{\text{elstat}} + E_X + E_{\text{rep}} + E_{\text{pol}} + E_{\text{disp}}$. CP corrections are easily introduced to account for BSSE.

It is relevant to notice that all KM-inspired methods may be applied from the weak intermolecular regime to the strong covalent one. In the short range case, particularly but not exclusively when open-shell fragments interact, they suffer from an important arbitrariness related to the reference states used to describe the fragments, which may possibly be chosen in many ways.

A different, although related, orbital EDA is the Natural Energy Decomposition Analysis (NEDA) of Glendening and Streitwieser.[11] This is based on constructing supermolecular antisymmetric states from the natural bond orbitals (NBO) of Weinhold and co-workers.[19,36–38,44] The NBO procedure is an efficient, involved localization scheme that yields close to minimal atomic basis sets (Natural Atomic Orbitals, NAOs) from HF or DFT descriptions. NAOs are later used to build core, lone pair, and two-center (and further if necessary) bonding functions called natural bond orbitals. NBO descriptions provide palatable Lewis-like images, and are widely used in the literature. In NEDA, all off-diagonal matrix elements of the Fock matrix written in the NBO basis that couple any two orbitals residing on different fragments are deleted in a first step. These elements are associated to charge transfer. Diagonalization of this CT-free F leads to a new set of (orthogonal) orbitals. With them, deformed wave functions are constructed for the fragments, $\Phi_{\text{def}}^{A,B}$, from which a single determinant supermolecule wave function, called the localized state $\Phi_{\text{loc}} = \mathcal{A}\,\Phi_{\text{def}}^A\,\Phi_{\text{def}}^B$ is also built. The charge transfer energy is defined as the energy distance from the final relaxed function to the energy of this localized state, $E_{CT} = E(\Phi) - E(\Phi_{loc})$. The joint electrostatic, polarization, and exchange is obtained as $E_{elstat} + E_{pol} + E_X = E(\Phi_{loc}) - E(\Phi_{def}^A) - E(\Phi_{def}^B)$. Notice that, since the orbitals of the localized determinant and those of the deformed fragments are the same (and are orthogonal), the above difference may be partitioned à la IMPT. E_{elstat} is obtained from eqn (5.9) with the orbitals of the infinitely separated fragments, and is equivalent to CP corrected KM. E_{pol} is the difference between E_{elstat} computed with the localized orbitals of Φ_{loc} and E_{elstat} obtained with the unperturbed orbitals, and E_X is obtained from eqn (5.10) with the localized orbitals, thus being different from the standard KM-like exchange energy. At variance with IMPT, SAPT, or BEDA, the polarization or exchange part of NEDA is not obtained from non-interacting densities, but from antisymmetric ones. Moreover, since the fragments are modified, the energy required to deform them, $E(\Phi_{def}^{A,B}) - E(\Phi^{A,B})$ is called the deformation energy, $E_{def}^{A,B}$. Deformation energies are destabilizing contributions, since the $\Phi_{def}^{A,B}$ are not variational states for the fragments. Sometimes they are self-interaction corrected.

BSSE is taken into account *via* CP corrections automatically. Overall, the NEDA interaction energy is $E_{int} = E_{elstat} + E_{pol} + E_X + E_{CT} + E_{def}$.

A pertinent remark concerns the isolation of fragment terms in an antisymmetrized state. Once a determinant has been built from a set of orbitals, orthogonal or not, the individuality of electrons is lost and, strictly speaking, the ascription of these orbitals to fragments is no longer possible. This may be understood by noticing that the state is invariant under orbital transformations, which may become localized differently.

A completely different approach to energy partitions follows from real space analyses. The basic tenet in these techniques is that all energetic quantities should be invariant under orbital transformations, being thus built from reduced density matrices (RDMs). Since molecules are recognizable in real, not momentum space, only RDMs in real space are useful for EDAs. Once this is assumed, the only problem left is partitioning the physical space into regions associated to the molecules that interact. Quite a large number of possibilities have been proposed, but we will only comment on the partition provided by the Quantum Theory of Atoms in Molecules (QTAIM) of Bader and co-workers.[1] Although the QTAIM is deeply rooted in quantum mechanics, it suffices our purpose here to indicate that it provides a way to exhaustively divide the space into non-overlapping regions (basins or domains) associated to atoms. The union of all these domains fills completely the space: $\bigcup_A \Omega_A = R^3$, and any collection of atomic basins, *e.g.* those forming a molecule, is also a proper QTAIM region. All the terms in the standard Coulomb Hamiltonian have well defined expectation values over QTAIM basins (or pair of basins). A suitable EDA within the QTAIM useful for intermolecular interactions is the Interacting Quantum Atoms (IQA) approach.[3,8,27] It starts from recognizing that only the 1- and 2-particle RDMs are needed to recover the energy of a molecular system:

$$E = h + V_{ee} + V_{nn} = \int_\infty h\rho_1(r_1; r_1')|_{r_1' \to r_1} \, dr_1$$

$$+ \frac{1}{2}\int_\infty\int_\infty \rho_2(r_1, r_2)r_{12}^{-1}dr_1dr_2 \tag{5.15}$$

$$+ \frac{1}{2}\sum_A\sum_{B \neq A}\frac{Z_A Z_B}{R_{AB}}.$$

Here h is the one-electron Hamiltonian, $h = t - \sum_A Z_A/r_A$, with kinetic and electron- nucleus terms, and ρ_1 and ρ_2 are the spinless 1- and 2RDMs. Their diagonal parts, or reduced densities, are the electron and pair densities, respectively.

After partitioning the space into QTAIM atomic basins, IQA proceeds by partitioning all the integrals above into one- (for one-electron operators) and two-basin contributions (in the case of interelectron repulsion):

$$E = \sum_A T_A + \sum_{A,B} V_{en}^{AB} + \sum_{A>B} V_{nn}^{AB} + \sum_{A>B} V_{ee}^{AB}, \tag{5.16}$$

in a self-explanatory notation. All intra-atomic components are gathered together to form the atomic self-energy, E_{self}^A,

$$E_{self}^A = T^A + V_{en}^{AA} + V_{ee}^{AA}, \tag{5.17}$$

while all the inter-basin ones form the inter-basin interaction energy, E_{int}^{AB},

$$E_{int}^{AB} = V_{en}^{AB} + V_{en}^{BA} + V_{ee}^{AB} + V_{ee}^{AB}. \tag{5.18}$$

For a set of isolated, non-interacting atoms, atomic self-energies tend to the free atomic energies, and interaction energies vanish. The total energy acquires a familiar expression in atomistic simulations,

$$E = \sum_A E_{self}^A + \sum_{A>B} E_{int}^{AB}. \tag{5.19}$$

The above expressions are immediately generalized when A is a group of basins, the group self-energy now containing all the energetic terms in which only the atomic labels of the atoms in the group are found, and the interaction energy between two groups gathering all terms with mixed atomic labels. Atomic or group self-energies measure the proper energy of a component of the system in-the-molecule. As in NEDA, these groups are deformed with respect to their *in vacuo* states, so a deformation energy may immediately be defined given such reference states,

$$E_{\text{def}}^A = E_{\text{self}}^A - E_{\text{vac}}^A. \tag{5.20}$$

With this,

$$E_{\text{int}} = \sum_A E_{\text{def}}^A + \sum_{A>B} E_{\text{int}}^{AB}. \tag{5.21}$$

Notice that the introduction of a reference is not needed for the EDA. IQA provides a reference-free energy partition, being one element of a very limited category of methods with this property. References are only used to decompose interaction energies. As a by-product, BSSE errors are immediately CP corrected. They only affect the deformation terms, and never the interaction ones.

A further point is needed. The electronic repulsion terms contain all types of contributions: Coulombic, exchange, and correlation. It is thus profitable to separate them. This can be done by separating in ρ^2 the Coulombic part, $\rho_2(r_1, r_2) = \rho(r_1)\rho(r_2) + \rho_{xc}(r_1, r_2)$, where ρ_{xc} is the exchange-correlation density. Exchange itself is not well-defined in general correlated cases, but can be obtained in an invariant manner from the Fock–Dirac expression, $\rho_X(r_1, r_2) = \rho_1(r_1; r_2)\rho_1(r_2; r_1)$, so that $\rho_{xc}(r_1, r_2) = \rho_X(r_1, r_2) + \rho_{\text{corr}}(r_1, r_2)$. This partition allows us to write

$$V_{ee}^{AB} = V_C^{AB} + V_X^{AB} + V_{\text{corr}}^{AB}. \tag{5.22}$$

If we gather all the classical interaction terms and define a classical interaction, $V_{cl}^{AB} = V_{nn}^{AB} + V_{ne}^{AB} + V_{en}^{AB} + V_C^{AB}$, then

$$E_{\text{int}}^{AB} = V_{cl}^{AB} + V_{xc}^{AB}. \tag{5.23}$$

In terms of these quantities, $E_{\text{int}} = E_{\text{def}}^A + E_{\text{def}}^B + V_{xc}^{AB} + V_{cl}^{AB}$ for a two-fragment complex. Notice that, in the absence of correlation, $V_{xc}^{AB} = V_X^{AB}$.

Several important comments are due. First, all IQA energy terms have a clear, unique interpretation, being (domain) expectation values of physical interactions. There is no ambiguity in their meaning whatsoever. No fictitious states at any steps. All of the energy components are obtained from the analysis of a single supermolecular wavefunction. Only if binding energies are necessary are the energies of the isolated fragments also requested. Electrons in IQA are fully indistinguishable, so no fragment labels may be attached to them. They "belong" to a fragment when they are located within that fragment, in real space. As a consequence, IQA may go all through dissociation curves, from the long-range to the covalent regimes. IQA fragments do not have an integer number of electrons. Charge transfer is intrinsically built-in, and affects all the energetics. In a CT complex with full electron transfer, the self-energy of the fragments will tend to those of a cation and an anion. No perturbative method is able to describe this type of situation.

Electron repulsion, in any method that uses invariant RDMs, faces the problem of electrostatic self-interaction. Take the intermolecular Coulomb repulsion, $\int dr_1\, dr_2\, \rho^A(r_1)\rho^B(r_2)/r_{12}$. As already stated, this expression assumes that electrons in different fragments are distinguishable. In orbital-free techniques, only the full $\rho(r)$ is available and the A, B labels above are dropped. This means that a given electron at r_1 will interact with itself when at r_2. In orbital parlance, $i=j$ terms are included in eqn (5.9). Coulombic self-interaction is corrected by the invariant definition of exchange, which is thus also affected by self-interaction (again, $i=j$ terms in eqn (5.10)). In fact, a large part of the Fermi hole that builds around an electron is due to the self-interaction correction, or proper electron counting. There is no way out: using invariant densities implies self-interaction. Comparison with results obtained from Pauli violating densities needs care. As an example, V_X^{AB} will be non-zero even in the H_2 molecule, where an equal spin pair does not exist but a self-interaction one does. In strong bonding interactions, V_{xc}^{AB}, being the non-classical interaction energy, measures covalency, while V_{cl} describes ionicity.

5.3   The Energetic Terms: A Comparison

After the general presentation of the previous section, we will compare now the different energy terms as obtained in different methods. To fully understand their meaning, besides actual calculations in very simple systems, we will use the triplet state of the H_2 molecule as an algebraic laboratory. As possibly the simplest interacting system with Pauli repulsion, it will serve well our didactic aim. Moreover, this system has been repeatedly used as a test. Good accounts may be found in the books of Stone or Magnasco.[24,41]

Let us then consider the dissociative Heitler–London H2 triplet as two interacting fragments, a and b. Their isolated H wave functions are described by pure 1s orbitals in the ground state, $\phi(r) = e^{-r}/\sqrt{\pi}$. The Heitler–London two-electron function is

$$\Phi^{ab} = \frac{1}{\sqrt{2(1-S^2)}} (\phi_a(r_1)\phi_b(r_2) - \phi_b(r_1)\phi_a(r_2))\alpha(1)\alpha(2). \tag{5.24}$$

S is the $\langle \phi_a | \phi_b \rangle$ orbital overlap. Let us write the Hamiltonian as $H = h_1 + h_2 + r_{12}^{-1} + 1/R_{ab}$, with $h = t - 1/r_a - 1/r_b$. Notice that, for instance, ϕ_a is an eigenfunction of a $h_a = t - 1/r_a$ with eigenvalue equal to $E_a^0 = -1/2$ a.u. The energy of this state, $\langle \Phi^{ab} | H | \Phi^{ab} \rangle$ equals

$$E = \frac{1}{1-S^2} (h_{aa} + h_{bb} - 2h_{ab}S + J_{ab} - K_{ab}) + \frac{1}{R_{ab}}, \tag{5.25}$$

where J_{ab} and K_{ab} are the standard Coulomb and Exchange two-electron integrals,

$$J_{ab} = \langle \phi_a(r_1)\phi_b(r_2) | r_{12}^{-1} | \phi_a(r_1)\phi_b(r_2) \rangle,$$
$$K_{ab} = \langle \phi_a(r_1)\phi_b(r_2) | r_{12}^{-1} | \phi_a(r_1)\phi_b(r_2) \rangle. \tag{5.26}$$

All the integrals may be expressed in terms of known functions, see for instance the books by Slater[40] or Magnasco.[24] The energy of the isolated fragments may be directly extracted from eqn (5.25) by noticing the aforementioned eigenvector character of ϕ, recovering the textbook energy expression

$$E = E_a^0 + E_b^0 + \frac{J-K}{1-S^2} + \frac{1}{R_{ab}}, \tag{5.27}$$

where J and K are the Heitler–London direct (or Coulomb) and exchange energies,

$$J = J_{ab} - \langle \phi_a | 1/r_b | \phi_a \rangle - \langle \phi_b | 1/r_a | \phi_b \rangle,$$
$$K = K_{ab} - S(\langle \phi_a | 1/r_b | \phi_b \rangle + \langle \phi_b | 1/r_a | \phi_a \rangle). \tag{5.28}$$

J and K should not be confused with J_{ab} and K_{ab}. The former contain all classical and antisymmetry related terms, respectively, and the latter are pure two-electron integrals. It is very instructive to recall that the correctly normalized Slater determinant formed with the non-orthogonal ϕ_a, ϕ_b orbitals is exactly the Heitler–London state:

$$\Phi^{ab} = \frac{1}{\det(\mathscr{S})} \frac{1}{\sqrt{2}} |\phi_a \phi_b|, \tag{5.29}$$

where $\mathscr{S}$ is the orbital overlap matrix, so that the energy expression in eqn (5.8) coincides with eqn (5.27).

It is also important in what follows to recognize that Φ^{ab} may be also interpreted as the antisymmetrized Hartree product of the monomers' functions, $\Phi^{ab} \equiv \Phi^0 = \mathscr{A}\phi_a\phi_b$, so that

$$E = E^0 + \frac{\langle \phi_a \phi_b | H' | \mathscr{A} \phi_a \phi_b \rangle}{\langle \phi_a \phi_b | \mathscr{A} \phi_a \phi_b \rangle}, \tag{5.30}$$

and, as in SAPT, $E = E^0 + E^1$. Since $J + 1/R_{ab}$ is immediately identified as E_{elstat}, then $E_X = -K_{ab}$, and

$$E_{\text{rep}} = \frac{S^2 J - K}{1 - S^2} + K_{ab}, \tag{5.31}$$

$$E_{\text{exch}}^{10} = E_X + E_{\text{rep}}.$$

It is quite clear that in this simple approximation the monomers' functions are not relaxed, so that the energy is equal to the zeroth and first order perturbative terms. No second order contributions appear, so no polarization or dispersion (*i.e.* electron correlation) terms are included. Expressions for the density and the pair densities may immediately be obtained using eqn (5.8) with the appropriate one- and two-electron operators:

$$\rho(r_1) = \sum_{i,j} \phi_i(r_1)\phi_j(r_1)_{ij} S_{ji}^{-1},$$

$$\rho_2(r_1 r_2) = \sum_{i,k,j,m} \phi_i(r_1)\phi_k(r_2)\phi_j(r_1)\phi_m(r_2)(S_{ji}^{-1} S_{mk}^{-1} - S_{mi}^{-1} S_{jk}^{-1}). \tag{5.32}$$

The power of such a simple model is that we can also read it using supermolecular eyes. This offers a different perspective. If we look at eqn (5.25) instead of at eqn (5.27), we see that upon building Φ^{ab}, *i.e.* upon antisymmetrizing, all the molecular energy components change, and that the kinetic energy, electron-nucleus attraction, Coulomb, and exchange terms in-the-molecule are all modified. Separating them into fragment contributions is what a supermolecular EDA achieves. Direct substitution in eqn (5.32) shows that, to first order in S, the change in the density is $\Delta\rho(r1) = -2S\phi_a(r_1)\phi_b(r_1)$, so density is pumped from the internuclear region towards the rear parts of the internuclear axis, as all textbooks explain and relate to Pauli exclusion. This gives interesting clues about the nature of the energy changes. For instance, the contraction of the density (together with the introduction of the spatial node in Φ_{ab}) will increase the kinetic energy. Density contraction caused by Pauli exclusion also forces a considerably more negative electron-own-nucleus interaction, and a much less attractive electron-other-nucleus contribution. However, the increased kinetic energy exactly compensates the extra electron-own-nucleus stabilization, leading to $E_a^0 + E_b^0$ in eqn (5.27), where $\phi_{a,b}$ are eigenfunctions of $h_{a,b}$. This cancellation annihilates large energy variations, and lies behind much of the magic of the small energy differences of chemistry. Similarly, electron repulsion in-the-molecule is greatly affected by antisymmetrization. From eqn (5.27), its value is scaled from $J_{ab} - K_{ab}$ to $(J_{ab} - K_{ab})/(1 - S^2)$. This is a large effect.

Under the BEDA perspective, for instance, E_{elstat} is as just written above, and $\Delta E_{Pauli} = E_{exch}^{10}$ exactly. No ΔE_{orb} appears, since orbitals have not been relaxed. The HF SCF solution would end up with a set of modified orbitals that we could localize to make them resemble the atomic ones, but eqn (5.25) would survive unchanged. The orbital interaction term is thus a complex mixture of kinetic and potential terms, much as ΔE_{Pauli} is. KM, LMOEDA, and NEDA decompositions lead to identical results, since no orbital mixing beyond that forced by symmetry is included.

In real space, Φ^{ab} is analyzed *per se*. Due to symmetry, any exhaustive partition of space into two regions $R^3 = \Omega_a + \Omega_b$ will result in atomic domains which are semi-spaces separated by the plane that mid-intersects the internuclear axis orthogonally. Electrons to the left of that plane (the QTAIM interatomic surface) will be considered as belonging to atom H_a and *vice versa*. Now consider eqn (5.25) and separate each integral into atomic domains. One-electron integrals will become sums of domain contributions, *e.g.* $h_{aa} = h_{aa}^a + h_{aa}^b$, where the $h_{aa} = \int_\infty \phi_a \hat{h} \phi_a dr_1$ has been split with $h_{aa}^{a,b} = \int_{\Omega_{a,b}} \phi_a \hat{h} \phi_a dr_1$. Two electron terms like J_{ab} will be split into *aa, ab, ba,* and *bb* terms. By symmetry, the atomic kinetic energy is half the total kinetic energy, $T^a = T/2$. At large internuclear distances it is clear that E_{self}^A will tend to E_a^0 in IQA, and that E_{int}^a will tend to zero. As S increases, T^a will increase, as we have seen, and V_{ne}^{aa} will become more negative. They do not compensate exactly, however. Notice also that the electrons of H_a, for instance, will never sense the nucleus of H_b at small distances as it would occur if described by an orbital with a tail invading that nucleus. They are always at a distance greater than half the internuclear separation. Besides this, now there is a non-zero probability that the two electrons lie in Ω_a. Electrons are identified by their positions, not by their orbitals. This means that there will be an intra-atomic electron repulsion. In other words, the J_{ab}, and K_{ab} terms do not only account for the inter-fragment *interaction*. They include now intra-atomic contributions. The Coulomb repulsion (and exchange attraction) are part intra- and part inter-atomic. As a result, the truly interatomic terms will be smaller in magnitude than the (scaled) J_{ab}, K_{ab} integrals. Finally, when separating electron repulsion into Coulomb and exchange-correlation contributions, we should recall that we can only use ρ and ρ_{xc}. In consequence, the density at a point in region a contains contributions of both electrons. Then $\int dr_1\, dr_2\, \rho(r_1)\rho(r_2)/r_{12}$ will include self-interaction terms, and the same applies to the exchange-correlation term. If we add together all classical interaction terms to build the IQA V_{cl}, the non-interpenetrating character of the atomic densities makes this term positive, *i.e.* destabilizing. This is at variance with the standard E_{elstat}, and may be shown to be always the case in symmetric interactions. Since the magnitudes of both self-interaction and intra-atomic electron repulsion components depend on the probability of finding the two electrons in a region, *i.e.* on the delocalization of the electrons which is controlled by S, they decay exponentially with the internuclear distance. In the long-range limit, IQA interactions converge on their perturbative or orbital-EDA counterparts. There is a notorious difference, however. Real space techniques like IQA do not need any external reference to obtain interactions. They are obtained directly from the wave function.

After this general discussion, we will center on comparing different energy components for different methods, in a set of very simple dimers, going from fully covalent links to weak intermolecular interactions. We have performed SAPT calculations in the dimer centered basis using the SAPT12 code,[5] KM, and LMOEDA analyses using the GAMESS package,[39] BEDA decompositions with Gaussian basis sets with our IMOLINT program,[7] NEDA partitions with the NBO5 code,[10] and IQA/QTAIM real space analyses with our PRO-MOLDEN[28] program. Suitable restricted Hartree–Fock (RHF) or open-shell RHF, plus full CI, CAS multireference calculations, and CCSD or MP2 expansions have also been performed. In all the cases we have fully optimized geometries with common aug-cc-pVnZ ($n =$ D, T) basis sets. CP corrections have also been applied selectively. All the energetic quantities and charges evaluated with the different EDAs and levels of calculation are collected in Tables 5.1 (LMOEDA), 5.2 (IQA RHF), 5.3 (IQA RHF H-H EDA like), 5.4 (NEDA), 5.5 (BEDA), 5.6 (SAPT0), 5.7 (IQA FCI), and 5.8 (IQA CAS).

Table 5.1 LMOEDA results.[a]

	E_{elstat}	E_X	E_{rep}	E_{pol}	$E_{\text{disp–MP2}}$	$E_{\text{disp–CCSD}}$	E_{HF}	E_{MP2}	E_{CCSD}
H_2 (S)	−2.63	0.00	0.00	−80.96		−24.93	−83.58		−108.52
H_2 (T)	−2.63	−199.43	427.82	−89.99	−2.07		135.78	133.71	
HF	−115.39	−229.74	588.73	−343.53		−35.91	−99.92		−135.83
HF (ion)	−233.57	0.00	0.00	−149.55	11.73	6.63	−383.12	−371.39	−376.49
HCl	−66.56	−105.33	238.63	−146.74		−22.41	−80.00		−102.42
HCl (ion)	−151.31	0.00	0.00	−183.85	1.95	−1.19	−335.16	−333.21	−336.35
LiH	−7.22	−5.27	15.62	−37.21		−22.91	−34.08		−56.99
LiH (ion)	−175.04	−17.85	56.26	−28.23	−0.12	1.37	−164.86	−164.98	−163.49
LiF	−38.93	−58.76	135.68	−130.79		−39.21	−92.81		−132.02
LiF (ion)	−207.49	−22.70	−66.31	−21.67	4.75	2.78	−185.55	−180.80	−182.77
F_2	−153.54	−310.05	786.88	−295.83		−52.98	27.46		−25.53
Cl_2	−53.00	−129.95	281.24	−123.08		−23.37	−24.79		−48.17
ClF	−52.34	−147.82	337.00	−145.45		−35.03	−8.60		−43.63
HF–HF	−5.41	−3.64	6.83	−1.42	−0.31	−0.32	−3.64	−3.96	−3.96
HF–ClF	−2.67	−2.11	3.69	−0.53	−0.40	−0.25	−1.61	−2.01	−1.86
BH_3–NH_3	−77.90	−115.90	222.86	−62.44	−7.40	−5.36	−33.38	−40.78	−38.74
H_2O–H_2O	−6.43	−5.29	9.40	−1.40	−0.57	−0.45	−3.72	−4.30	−4.17
LiH–LiH (D_{2h})	−83.70	−63.22	119.98	−21.59	−0.55	0.48	−48.53	−49.08	−48.05
CO^2–CO_2 (C_{2h})	−1.56	−1.58	2.78	−0.11	−0.11	−0.12	−0.48	−0.59	−0.60

[a]All units in kcal per mol. All data are CP corrected and ion means an ionic reference. Dispersion energies are obtained (see the main text) with respect to correlated supermolecular calculations, in this case MP2 and CCSD.

Table 5.2 IQA RHF results.[a]

	QA	QB	E_{def}^A	E_{def}^B	E_{int}^{AB}	E_{int}	V_{xc}^{AB}	V_{cl}^{AB}
H_2 (S)	0.0000	0.0000	27.42	27.42	−138.41	−83.56	−163.63	25.22
H_2 (T)	0.0006	0.0006	73.36	73.36	−10.96	135.76	−70.97	60.01
H–F	0.7714	−0.7711	194.71	21.86	−316.51	−99.94	−81.50	−235.01
H–F (ion)	0.7714	−0.7711	−118.95	52.32	−316.51	−383.14	−81.50	−235.01
H–Cl	0.2984	−0.2984	77.52	−13.48	−143.98	−79.93	−163.68	19.70
H–Cl (ion)	0.2984	−0.2984	−236.13	45.02	−143.98	−335.09	−163.68	19.70
Li–H	0.8569	−0.8571	120.02	11.53	−165.64	−34.09	−30.87	−134.77
Li–H (ion)	0.8569	−0.8571	−3.17	3.93	−165.64	−164.87	−30.87	−134.77
Li–F	0.9174	−0.9173	141.15	−38.10	−195.54	−92.49	−29.64	−165.91
Li–F (ion)	0.9174	−0.9173	17.97	−7.66	−195.54	−185.23	−29.64	−165.91
F_2	0.0000	0.0000	110.32	110.32	−193.14	27.50	−238.34	45.20
Cl_2	0.0000	0.0000	69.62	69.62	−145.76	−26.21	−195.38	29.93
Cl–F	0.5051	−0.5051	216.32	70.90	−295.77	−8.55	−179.08	−116.69
HF–HF	−0.0061	0.0066	3.59	6.21	−13.40	−3.60	−8.44	−4.96
HF–ClF	−0.0024	0.0024	5.34	2.07	−9.24	−1.84	−6.44	−2.81
BH_3–NH_3	−0.0858	0.0858	46.65	89.80	−170.70	−34.25	−98.46	−72.24
H^2O–H_2O	−0.0077	0.0077	4.71	6.25	−15.22	−4.26	−9.49	−5.72
LiH–LiH (D_{2h})	0.0000	0.0000	20.06	29.06	−106.71	−48.60	−40.53	−66.18

[a]All units in kcal per mol. All data are CP corrected and ion means an ionic reference.

Table 5.3 IQA Hartree–Fock results for the singlet and triplet states of H_2.[a]

	Q_A	Q_B	ΔE	E_{def}^A	E_{def}^B	E_{kin}^A	E_{kin}^B	E_{int}^{AB}	E_{int}	V_X^{AB}	V_{cl}^{AB}
$\mathscr{A}\Phi_A^0\mathscr{A}\Phi_B^0$ (S)	0.0000	0.0000	81.0000	22.82	22.82	0.50	0.50	−48.22	−2.57	−96.00	47.79
$\mathscr{A}\Phi_A^0\mathscr{A}\Phi_B^0$ (S)	0.0000	0.0000	81.0000	22.82	22.82	0.50	0.50	−48.22	−2.57	−96.00	47.79
Φ (S)	0.0000	0.0000	0.0000	27.42	27.42	0.56	0.56	−138.41	−83.56	−163.63	25.22
$\mathscr{A}\Phi_A^0\mathscr{A}\Phi_B^0$ (T)	0.0000	0.0000	−337.6952	−35.79	−35.79	0.50	0.50	−130.36	−201.94	−178.15	47.79
$\mathscr{A}\Phi_A^0\mathscr{A}\Phi_B^0$ (T)	0.0000	0.0000	90.2104	101.22	101.22	0.77	0.77	23.51	225.95	−32.54	56.05
Φ (T)	0.0005	0.0005	0.0000	73.37	73.37	0.58	0.58	−11.00	135.74	−71.00	60.00

[a]All units in kcal per mol, except kinetic energies in Hartree. All data is CP corrected. S means singlet and T triplet. Results obtained at the equilibrium geometry of the singlet.

Table 5.4 NEDA.[a]

	Q_A	Q_B	E_{def}^{A} (SE)	E_{def}^{B} (SE)	E_{CT}	E_{elstat}	E_{pol}	E_X	E_{int}
H_2 (S)	0.0000	0.0000	78.51(−20.70)	78.51(−20.70)	−314.90	−2.63	76.92	0.00	−83.58
H_2 (T)	0.0000	0.0000	84.77(−10.66)	84.77(−10.66)	−68.54	−2.63	47.37	−9.96	135.78
HF	0.5494	−0.5494	308.77(−10.58)	139.54(−17.48)	−524.72	−62.68	54.80	−12.67	−96.96
HF (ion)	0.5494	−0.5494	0.00(0.00)	254.21(−8.03)	−419.82	−233.57	16.06	0.00	−383.12
HCl	0.2472	−0.2472	248.73(20.51)	125.07(−17.54)	−359.62	−66.94	−6.10	−17.52	−76.37
HCl (ion)	0.2472	−0.2472	0.00(0.00)	187.14(−35.55)	−442.09	−151.31	71.10	0.00	−335.16
LiH	0.8446	−0.8446	63.58(−5.25)	19.81(−4.70)	−146.93	−7.22	15.55	−1.32	−34.07
LiH (ion)	0.8446	−0.8446	13.64(0.78)	40.01(4.34)	−30.77	−175.04	−10.06	−2.63	−164.86
LiF	0.9778	−0.9778	116.12(−10.30)	68.57(2.58)	−933.44	−39.03	13.43	−4.71	−89.78
LiF (ion)	0.9778	−0.9778	28.91(6.70)	38.66(5.44)	−16.73	−207.49	−24.08	−4.82	−185.55
F_2	0.0000	0.0000	253.69(29.59)	253.69(29.59)	−277.09	−48.67	−118.33	−29.88	33.41
Cl_2	0.0000	0.0000	206.12(26.60)	206.12(26.60)	−241.01	−52.66	−106.83	−29.25	−17.53
ClF	0.3973	−0.3973	268.88(30.85)	225.59(37.62)	−275.98	−52.19	−136.67	−32.76	−3.13
HF–HF	−0.0041	0.0041	3.52(0.77)	4.22(0.21)	−3.36	−5.41	−1.89	−0.72	−3.64
HF–ClF	0.0028	−0.0028	5.84(1.90)	4.14(1.46)	−0.98	−2.67	−6.70	−1.25	−1.61
BH_3–NH_3	−0.3108	0.3108	91.87(14.34)	188.38(−10.17)	−209.11	−77.90	−9.97	−16.65	−33.38
H_2O–H_2O	−0.0064	0.0064	4.94(1.48)	7.31(0.25)	−4.83	−6.43	−3.45	−1.26	−3.72
LiH–LiH (D_{2h})	0.0000	0.0000	36.86(−0.74)	36.86(−0.74)	−36.85	−83.70	3.95	−5.64	−48.53
CO_2–CO_2 (C_{2h})	0.0000	0.0000	2.33(0.51)	2.33(0.51)	−0.91	−1.56	−2.05	−0.62	−0.48

[a]All units in kcal per mol. SE stands for self-interaction corrected. All data are also CP corrected. Ion means an ionic reference.

Table 5.5 BEDA results.[a]

	ΔV_{Ele}	ΔV_{Pauli}	ΔT_{Pauli}	$\Delta E_{\text{orb}}^{\text{kin}}$	$\Delta E_{\text{orb}}^{xc}$	$\Delta E_{\text{orb}}^{\text{Coul}}$	$\Delta E_{\text{orb}}^{\text{Total}}$	E_{int}
H_2 (S)	−2.63	0.00	0.00	71.42	−18.89	115.58	−80.96	−83.58
H_2 (T)	−2.63	−116.29	344.68	−241.56	55.62	−88.82	−89.99	135.78
HF	−115.39	−543.61	902.60	−808.91	22.65	593.89	−343.52	−99.92
HF (ion)	−233.57	0.00	0.00	376.09	−102.51	625.99	−149.55	−383.12
HCl	−95.37	−464.83	747.14	−661.07	40.74	282.41	−266.95	−80.01
HCl (ion)	−151.31	0.00	0.00	333.44	−92.18	815.10	−183.85	−335.16
LiH	−7.21	−17.56	27.90	4.12	−28.15	218.49	−37.21	−34.08
LiH (ion)	−175.04	−47.40	85.80	77.20	−51.75	210.80	−28.23	−164.86
LiF	−36.81	−137.73	205.88	−143.60	−85.63	1909.72	−123.81	−92.47
LiF (ion)	−202.23	−111.96	148.82	4.46	−20.54	205.53	−19.83	−185.20
F_2	−153.54	−1514.13	2001.33	−2039.32	247.94	−1182.59	−306.21	27.46
Cl_2	−53.00	−585.39	736.77	−726.14	68.01	27.44	−123.08	−24.80
ClF	−52.34	−671.55	860.73	−876.96	93.37	−131.13	−145.45	−8.60
HF–HF	−5.41	−29.55	32.74	−20.32	−0.28	25.35	−1.42	−3.64
HF–ClF	−2.67	−20.16	21.75	−14.17	−0.30	6.41	−0.53	−1.61
BH_3–NH_3	−77.90	−244.36	351.32	−260.13	29.76	−137.16	−62.44	−33.38
H_2O–H_2O	−6.43	−33.59	37.70	−27.21	1.53	2.22	−1.40	−3.72
LiH–LiH (D_{2h})	−83.70	−78.71	135.48	10.99	−22.60	102.40	−21.59	−48.53
CO_2–CO_2 (C_{2h})	−1.56	−11.61	12.80	−9.27	0.00	−5.66	−0.11	−0.48

[a]All units in kcal per mol. All data are CP corrected. Ion means an ionic reference.

Table 5.6 SAPT0 results.[a]

	$E_{\text{ele}}^{(10)}$	$E_{x}^{(10)}$	$E_{\text{ind,resp}}^{(20)}$	$E_{\text{ind}-x,\text{resp}}^{(20)}$	$E_{\text{disp}}^{(20)}$	$E_{\text{disp}-x}^{(20)}$	E_{SAPT0}
LiH (ion)	−175.04	38.22	−91.40	51.33	−0.93	0.28	−176.89
LiF (ion)	−202.23	36.63	−42.31	27.89	−1.09	0.14	−180.02
HF–HF	−5.41	3.18	−1.74	0.75	−1.29	0.17	−3.23
HF–ClF	−2.67	1.58	−0.85	0.54	−1.08	0.09	−1.40
BH_3–NH_3	−77.90	101.11	−92.72	48.29	−19.25	4.11	−21.22
H_2O–H_2O	−6.43	4.08	−1.69	0.82	−1.57	0.24	−3.22
LiH–LiH (D_{2h})	−83.70	53.98	−48.08	26.49	−5.29	0.57	−51.31
CO_2–CO_2 (C_{2h})	−1.56	1.19	−0.46	0.36	−0.77	0.07	−0.69

[a]All units in kcal per mol. Data using $SAPT_{\text{resp}}$. All data are CP corrected. Ion means an ionic reference.

Table 5.7 IQA FCI results.[a]

	QA	QB	E_{def}^{A}	E_{def}^{B}	E_{int}^{AB}	E_{int}	V_{xc}^{AB}	V_{cl}^{AB}
H–H (S)	0.0000	0.0000	7.44	7.44	−123.58	−108.70	−148.72	25.14
H–H (T)	0.0006	0.0006	71.54	71.54	−10.85	132.23	−70.92	60.06
Li–H	0.8979	−0.8968	109.63	−2.55	−176.69	−69.61	−26.84	−149.85
Li–H (ion)	0.8979	−0.8968	−11.38	14.11	−176.6	−173.96	−26.84	−149.85
LiH–LiH (D_{2h})	0.0001	0.0001	30.53	30.53	−111.46	−50.40	−38.14	−73.32

[a]All units in kcal per mol. Data are not CP corrected. Ion means an ionic reference.

5.3.1 The Coulombic, or Electrostatic Interaction

As it has become clear in our previous analysis, there is general consensus about obtaining the electrostatic interaction between fragments from eqn (5.4). This has its roots in the long-range perturbation expansion, and has survived unchanged in almost every EDA. E_{elstat} may be obtained directly, *i.e.* from one- and two-electron integrals between orbitals, or may be Laplace expanded to give rise to the multipolar series. Locating two parallel reference frames centered on fragments A and B and separated by the vector $\boldsymbol{R} = R\hat{R}$,

$$E_{\text{elstat}}^{m} = \sum_{l_1 m_1} \sum_{l_2 m_2} C_{l_1 m_1, l_2 m_2}(\hat{R}) \frac{Q_{l_1 m_1}^{A} Q_{l_2 m_2}^{B}}{R^{l_1 + l_2 + 1}}, \tag{5.33}$$

where the C coefficient is a tensor coupling term only dependent on the orientation of both frames and the Q's are the fragments' multipole moments,

$$Q_{lm}^{A} = \sqrt{\frac{4\pi}{2l+1}} \int \mathrm{d}r \rho_t^{A}(r) r^l s_{lm}(\hat{r}) \tag{5.34}$$

Table 5.8   IQA CAS results.[a]

	Q_A	Q_B	E_{def}^A	E_{def}^B	E_{int}^{AB}	E_{int}	V_{xc}^{AB}	V_{cl}^{AB}
H_2 (S)	0.0000	0.0000	9.55	9.55	−114.42	−95.33	−139.55	25.13
H_2 (T)	0.0005	0.0005	73.37	73.37	−10.99	135.76	−70.99	60.00
H–F	0.7627	−0.7627	189.53	−22.56	−306.47	−139.49	−81.81	−224.66
H–F (ion)	0.7627	−0.7627	−124.11	56.10	−306.47	−374.48	−81.81	−224.66
H–Cl	0.2864	−0.2864	73.20	−22.46	−146.21	−95.46	−167.97	21.77
H–Cl (ion)	0.2864	−0.2864	−240.44	37.04	−146.21	−349.61	−167.97	21.77
Li–H	0.8679	−0.8680	117.00	6.60	−168.38	−44.77	−27.14	−141.23
Li–H (ion)	0.8679	−0.8680	−6.05	14.46	−168.38	−159.97	−27.14	−141.23
Li–F	0.9152	−0.9152	145.71	−100.42	−194.56	−149.27	−29.80	−164.77
Li–F (ion)	0.9152	−0.9152	22.66	−29.49	−194.56	−201.39	−29.80	−164.77
F_2	0.0000	0.0000	52.96	52.96	−143.43	−37.51	−191.61	48.18
Cl_2	0.0000	0.0000	50.32	50.32	−145.76	−45.12	−174.63	28.87
Cl–F	0.4492	−0.4492	183.48	3.19	−249.71	−63.04	−169.42	−80.30
HF–HF	−0.0075	0.0079	3.96	6.36	−14.00	−3.68	−9.57	−4.44
H_2O–H_2O	−0.0107	0.0110	4.99	7.21	−15.98	−3.78	−10.79	−5.19

[a]All units in kcal per mol. Data are not CP corrected. Ion means an ionic reference.

with S_{lm} denoting a real spherical harmonic.[14] E_{elstat} may thus be obtained very quickly at different geometries, if the expansion converges, and the angular and distance dependence written easily and understood in terms of the moments. Unfortunately, the expansion has been shown to be only asymptotically convergent[41] in most of the interesting situations. Even when strictly convergent, E_{elstat}^m is in error with respect to the exact value, the difference being called penetration error: $E_{pen} = E_{elstat} - E_{elstat}^m$. In our Heitler–London H_2 triplet, for instance, the isolated fragment density has spherical symmetry, so the only non-zero $Q_{00} = N - Z = 0$. $E_{elstat}^m = 0$ at any distance, and the full E_{elstat} value is due to penetration. Much of the power of the multipole expansion depends on the magnitude of E_{pen}. At the singlet aug-cc-pVTZ//FCI geometry, this is equal to -2.58 kcal mol^{-1} (dropping to -2.63 if CP corrected) a relatively small value. Here we find the first interpretation problem with E_{elstat}. It does not depend on the specific molecular state we are examining. It is seen in Table 5.1 that E_{elstat} is exactly the same for the (dissociative) triplet and the (bonding) singlet of H_2. Although this is perfectly admissible if we think of non-interacting fragments in the long-range, it is slightly disturbing outside it.

In real space, V_{cl}^{AB} admits exactly the same expression as E_{elstat} in eqn (5.4) if we notice that we can define in-the-molecule fragment densities as $\rho_t^A(r) = \rho_t(r)\,\theta(r, \Omega_A)$, where θ is a step function equal to 1 within the fragment domain and vanishing outside. With this definition,

$$V_{cl}^{AB} = \int_{\Omega_A} dr1 \int_{\Omega_B} dr_2 \frac{\rho_t(r_1)\rho_t(r_2)}{r_{12}}, \tag{5.35}$$

may be written as $V_{cl}^{AB} = \int dr_1\, dr_2\, \rho_t^A(r_1)\rho_t^B(r_2)/r_{12} \equiv E_{elstat}$. Let us not forget that these real space ρ_t^A's contain self-interaction terms. The above expression shows that an equivalent multipole expansion for E_{elstat} exists in real space. We just need to use eqn (5.34) with real space ρ_t^A's. The Q_{lm}^A multipoles are usually obtained for QTAIM atomic domains, and are known as atomic multipoles. This approach has been pursued successfully by Paul Popelier, among others.[35] Needless to say, QTAIM atomic multipoles contain self-interaction and reflect the changes suffered by the fragments' electronic distribution upon interaction.

It is worthwhile mentioning that if instead of the Laplace formula we use the bipolar series[18,25] of r_{12}^{-1} we arrive at a convergent expansion that provides the exact E_{elstat}. It uses a triple lm expansion,

$$E_{\text{elstat}} = \sum_{l_1 m_1} \sum_{l_2 m_2} \sum_{l_3 = |l_1 - l_2|}^{l_1 + l_2'} \sum_{m_3} S_{l_3 m_3}(\hat{R}) D_{l_1 m_1 l_2 m_2}^{l_3 m_3} I,$$

$$I = \int_0^\infty r_1^2 \mathrm{d}r_1 \int_0^\infty r_2^2 \mathrm{d}r_2 \rho_{l_1 m_1}^A(r_1) V_{l_1 l_2 l_3}(r_1, r_2, R) \rho_{l_2 m_2}^B(r_2), \qquad (5.36)$$

$$\rho_{lm}^A(r) = \int \sin(\theta) \mathrm{d}\theta \mathrm{d}\phi \, \rho_t^A(r, \theta, \phi) S_{lm}(\theta, \phi).$$

Notice that the ρ_{lm}^A functions are fine-grained fragment multipoles, since $Q_{lm}^A = \int_0^\infty \mathrm{d}r \, r^2 r^l \rho_{lm}^A(r)$. The D coefficients are Gaunt coupling terms, and V are functions which depend on the relative values of r_1, r_2, and R. They divide the r_1, r_2, R space into four different regions. If $r_1 + r_2 < R$ always, *i.e.* when the densities do not overlap spherically, then it is easy to show that V contains a $\delta_{l_3, l_1 + l_2}$ term, and that the above expression collapses into the standard multipole expansion. The non-overlapping condition may be fulfilled with QTAIM densities whenever R is large enough, but will never be attained with overlapping fragments. Another point is that the bipolar expansion contains an explicit expression for E_{pen}.

A HF IQA analysis for the H_2 triplet gives $V_{\text{cl}}^{AB} = 60.01$ kcal mol^{-1}, that only changes to 60.06 at the FCI level. The multipole series fails to converge at this short range, and the QTAIM atomic multipoles in-the-molecule are large and l-divergent. For the left hand side H, $Q_{10} = -0.65$, $Q_{20} = 0.41$, $Q_{30} = 0.13$, $Q_{40} = -0.99$, all in a.u. The large V_{cl} compared to the non-interacting fragment E_{elstat} is basically due to the self-interaction. The probability of finding the two electrons in one of the atomic domains is about 0.13, so although we tend to think of the triplet as a system that excludes heavily the two electrons, there is still a considerable delocalization in real space. In other words, *if the electrons and nuclei would be completely localized in their atomic regions* and described by the molecular density ρ, then the classical interaction between them would be about $+60$ kcal mol^{-1}. This is an invariant result, that does not depend on orbital transformations. We will delay more comments to the section on exchange.

Let us now consider a more interesting example, like the LiH molecule, that gives rise to an important question. *How to choose the states of the fragments?* This may seem in principle unimportant when dealing with true intermolecular interactions, but care is to be taken: long-range charge transfer complexes in crystals may defy our intuition. In LiH, any reference-dependent method will provide very different answers if choosing, for instance, an atomic or an ionic reference. In the former case, E_{elstat} at the CCSD geometry is equal to -7.22 kcal mol^{-1} from an LMOEDA calculation (with an HF reference), changing to -175.04 if the ionic reference is chosen. This is the same value obtained in SAPT, for instance. Obviously, the rest of the energy components will change accordingly, and the interpretation of the analyses will be close to opposite. In IQA, the HF V_{cl} value is -134.77 kcal mol^{-1}, changing to -141.23 (-145.75) at the CAS (FCI) level. There is no reference here, so not much doubt should remain about which reference is better. In this method we have also the atomic multipoles in-the-molecule at our disposal. The net charge (Q_{00}) of the Li atom is 0.857 $|e|$ in the HF and 0.899 in the CI case, close to the ionic limit. A simple scaling of Vcl to complete ionization leads to about -183, -187 kcal mol^{-1}, similar to the SAPT or LMOEDA values. Moreover, a real space technique like IQA can be applied at any internuclear distance, V_{cl} changing automatically from the atomic-like long-range limit to the ionic situation at equilibrium. Let us notice that the multipole expansion with IQA multipoles does not converge in LiH, despite the large ionic character. This is due to the large deformability of the hydride anion. In fact, the Q_{l0} of the H moiety diverge with l, being equal to 0.44, -0.90, -5.15 a.u. for $l=1$, 2, 3. A HF IQA calculation in LiF leads to $V_{cl}^{AB} = -165.90$ kcal mol^{-1}, with a Li net charge equal to 0.917 a.u. The multipole expansion is now conditionally convergent, changing along the series $V_{cl}^{AB} = -173.93$, -184.29, -176.65, -166.60, -159.12 when we include $l=0$, ... 4-poles. Notice the importance of the octupolar terms in such a spherical environment.

A similar scenario can only be found in NEDA, which also uses deformed fragments. NEDA net charges, 0.8446 and 0.9778 a.u. in LiH and LiF, respectively, are very close to the IQA results, and do not depend on the reference. However, once the choice of obtaining NEDA's E_{elstat} from the isolated fragments' orbitals was taken, all the other energy components become affected by the reference.

Summarizing, tradition rooted in perturbation theory and inherited in orbital EDAs obtains E_{elstat} from the Coulomb interaction between the interpenetrating, distinguishable densities of isolated fragments. This is not a tenable assumption in the short range regime, where considerable charge transfer may have taken place, as shown in our limiting ionic examples above. In these cases, choosing the appropriate reference gets crucial. Electrostatic energies obtained with this prescription are not affected by self-interaction corrections, and are negatively defined due to inter-penetration. Real space approaches like IQA compute these energies from densities in-the-molecule, without any recourse to references. They include self-interaction corrections. IQA V_{cl}^{AB} have a naïve ionic interpretation, depending on the multipolar interactions between the fragments: if they are charged, they will be dominated by their monopolar energy (as shown), if not by their dipoles, *etc.* Since they have no definite sign, they signal electrostatically favorable interactions when negative. A quick look at, for instance, Tables 5.1 and 5.2 is useful to gain intuition. Strong covalent interactions have destabilizing V_{cl}^{AB} values around tens of kcal per mol. Strong ionic bonds stabilizing ones as large as -200 kcal mol^{-1} for nominally singly charged ions, and hydrogen bonded dimers, like the water dimer, small stabilizing terms of about a few kcal per mol, in agreement with standard wisdom. Analysis of non-obvious cases is illuminating. We will just consider the F2 and HCl molecules. The E_{elstat} contribution in F_2 is -153.54 kcal mol^{-1} if CP corrected. This stabilizing contribution is too big if we notice that the molecule is unbound at the HF level. It is just revealing that the isolated densities overlap too much due to the short distance and the large number of electrons. The IQA V_{cl} value is destabilizing, as in a broadly speaking covalent interaction: 48.2 kcal mol^{-1} at the CAS level, a bit larger, but not so different from the value in singlet H_2, around 25 kcal mol^{-1}. This shows how artificial standard E_{elstat} values may be. Similarly, E_{elstat} in HCl is -66.56 and -151.31 kcal mol^{-1} if an atomic or ionic reference is chosen, respectively. Contrarily, $V_{cl} = 21.77$ kcal mol^{-1} for an IQA/CAS calculation. This value is compatible with the highly covalent character of this molecule, and contrasts with $V_{cl} = -224.66$ kcal mol^{-1} in the HF molecule. As we shift from the short to the long range, and as long as charge transfer is not dominant, orbital and real space Coulomb energies tend to resemble each other. For instance, in the BH3-NH$_3$ and $(H_2O)_2$ complexes, E_{elstat} at the HF level are -77.90, -6.43 kcal mol^{-1}, respectively, while $V_{cl} = -72.24$, -5.72 kcal mol^{-1}.

5.3.2 Exchange and Antisymmetry

The difficulties posed by the appropriate consideration of anti-symmetry have led to much confusion. As briefly reviewed, most perturbation methods agree in the first order energy correction, which defines the exchange-repulsion energy when the long-range E_{elstat} is subtracted. The use of orthogonal orbital algebra leads a separation of this into exchange E_X and repulsion E_{rep} components, and much of this machinery has been inherited by a whole class of orbital based EDAs. Many of them use, directly or indirectly, a fictitious antisymmetric $\Phi^0_{AB} = \mathscr{A}\,\Phi^0_A\Phi^0_B$ state, that coincides with the full wavefunction in our triplet H2 model. As already recognized, this induces a considerable energy rise, loosely ascribed to Pauli repulsion. However, all the energy components and not, as sometimes stated, the kinetic energy rise, are responsible for this effect and need be properly understood.

Figure 5.1 shows how the different energetic terms behave with distance in the Heitler–London triplet. If we interpret the interaction energy in terms of H', as shown in eqn (5.27), then it is clear that direct terms are stabilizing, the contrary being true for those coming from antisymmetry (the J *versus* the K in the equation). Notice that J_{ab} and K_{ab} have contrary signs to J and K, so it is the one-electron electron-other-nucleus attraction terms that determine the sign of the latter. This is a very unfortunate notation, in our opinion. When E_{elstat}

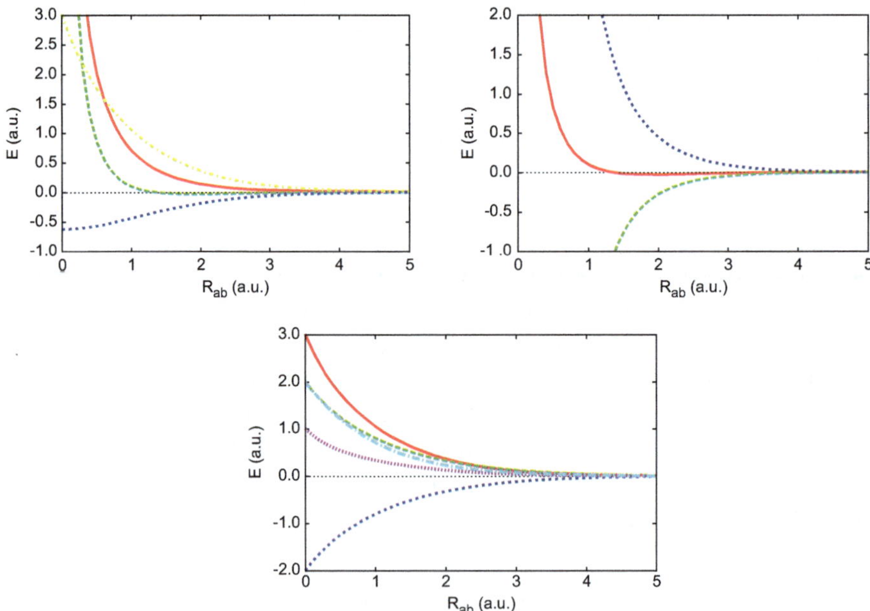

Figure 5.1 Contributions to the repulsive wall in the Heitler–London H$_2$ triplet. From left to right, top to bottom: (a) Total binding energy (red) separated into its E_{elstat} (green), E_X (blue), and E_{rep} (yellow) components. (b) Partition into E_{elstat} (red), direct (J, green), and exchange (K, blue) terms. (c) Decomposition of E_{rep} (red) into one- and two-electron contributions: kinetic (green), electron-own-nucleus (blue), electron-other-nucleus (cyan), two-electron terms (magenta).

and E_X are isolated, and E_{rep} is analyzed, then again a confusing contrary image is obtained: the two-electron corrections are mainly stabilizing, while the polarization of the density induced by the antisymmetrization causes the decrease in the electron-other-nucleus attraction to be the main ingredient of the repulsion, a change that can be understood by the retraction of the electron density induced by antisymmetrization. More noise may be added. Recall that eqn (5.27), in which only potential energy terms are present, is possible due to the fact that the unperturbed orbitals are eigenvectors of the non-interacting hydrogen atom Hamiltonian. Upon antisymmetrizing, the opposing changes in kinetic energy and electron-own-nucleus attraction cancel. However, if eqn (5.25) is used to interpret results, the change in kinetic energy is an important ingredient in the game. However, as shown in Figure 5.1, it is by no means dominant, as usually stated.

If we stick to tradition, and try to analyze the sources of the total E_{rep}, the first interesting point is that E_{rep} tends to 3 a.u. at $R = 0$, *i.e.* the *repulsion* brought about by antisymmetry does not diverge. As shown, both one- and two-electron terms are destabilizing. The first dominate at all distances, and since kinetic and electron-own-nucleus corrections cancel, the repulsion ends up being dominated by the destabilizing corrections coming from electron-other-nucleus attraction. Actually, as Figure 5.1 shows, the latter are about twice as big as the former. All the terms behave in the same qualitative way, the kinetic, electron-own-nucleus, electron-other-nucleus, and two-electron repulsions tending to 2, -2, 2, 1 a.u. at $R = 0$, respectively. Notice that we can also almost cancel out the electron-nucleus corrections, leaving in this other view the kinetic and two-electron contributions as creators of the repulsion, this time with the kinetic correction dominating the change. We would like to stress that all these important effects end up hidden in a single number E_{exch} or ΔE_{Pauli}.

The effect of antisymmetry in real space EDAs is much the same. In IQA we have used both the Hartree product $\Phi_A^0 \Phi_B^0$ and the antisymmetrized Φ_{AB}^0, at the singlet equilibrium geometry. The results are in Table 5.3. Notice how V_X^{AB} is quenched upon symmetrizing, an effect that is also clear in the NEDA decomposition, where $E_X = -9.96$ kcal mol^{-1}. Antisymmetrization also induces a large increase in the atomic kinetic energy with does not cancel out the naïve extra electron-own-nucleus attraction of the Heitler–London model. As explained, real space exchange includes self-interaction, and part of it is intra- ($V_X^{AA} = -208$ kcal mol^{-1}, not in the table) and part interatomic (-32.5 kcal mol^{-1}). $E_{\text{int}}^{AB} = 23.5$ kcal mol^{-1}, which together with the large deformation energy, $E_{\text{def}}^{A} = 101.2$ kcal mol^{-1}, agrees with the traditional Pauli image. Orbital relaxation reduces the deformation energy to 73.4 kcal mol^{-1}, increases E_X^{AB} to -71.0 kcal mol^{-1} and makes $E_{\text{int}}^{AB} = -11.00$ kcal mol^{-1}, a stabilizing value that is not able to overcome the large deformation. Again, we can see how all the energy components become heavily modified in Φ_{AB}^0. From this simple analysis, it is quite obvious that the use of orthogonal algebra to obtain V_X in standard orbital EDAs or perturbation expansions greatly exaggerates exchange in the antisymmetric state, thus equally enhancing in an artificial manner the E_{rep} values. Exchange itself, both in a self-interaction free method like NEDA or in IQA turns out to be a much more important component of EDAs than usually stated.

A look at Tables 5.1, 5.2, and 5.4 corroborates the conclusions from the H2 model. Notice how E_X is a large quantity that is overcompensated by E_{rep}. For instance, in the CP F$_2$ molecule, the orthogonal $E_X = -310.05$ and $E_{\text{rep}} = 786.88$ kcal mol^{-1}. In a simple BEDA context, $\Delta T_{\text{Pauli}} = 2001.33$, and $\Delta V_{\text{Pauli}} = -1514.13$ kcal mol^{-1}, so the kinetic energy term is clearly not the only factor shaping E_{rep}. An HF IQA calculation gives $V_X^{AB} = -238.34$ kcal mol^{-1} with large atomic deformations, $E_{\text{def}}^{A} = 110.32$ kcal mol^{-1}. The NEDA deformation is even larger, 253.69 kcal mol^{-1}, with a self-interaction free $E_X = -29.88$ kcal mol^{-1}. Again, artificial antisymmetric states constructed with unperturbed orbitals exacerbate the magnitude of the energy components, leading to a distorted view of strong interactions. As we move towards the long-range regime, the different methods tend to converge, as with E_{elstat}. In the HF–ClF dimer, for instance, $EX = -2.11$, and E_{rep}, as usual, almost doubles this value, $E_{\text{rep}} = 3.69$ kcal mol^{-1}. In BEDA, $\Delta T_{\text{Pauli}} = 21.75$ and $\Delta V_{\text{Pauli}} = -20.16$. The quenched NEDA exchange is -1.25 kcal mol^{-1}, with deformationns of 5.84 and 4.14 kcal mol^{-1} for the fragments. In the same order, the Φ_{AB}^0 antisymmetrized IQA values are -6.22, 5.52, 2.19 kcal mol^{-1}, that change, when relaxed to the final HF wave function to -6.44, 5.34, 2.07 kcal mol^{-1}. Notice how the IQA exchange corresponds to the sum of CT, polarization, and exchange NEDA contributions, as already noticed.[26] Similar insights may be obtained from examining other systems.

5.3.3 Polarization, Charge Transfer

Polarization and charge transfer are two quantities so diversely defined in different EDAs that appropriate comparison becomes a difficult task. Apart from the long range regime (ignoring the possible convergence problems), in which the perturbation series is well defined, the proper management of antisymmetry mixes polarization with other terms in SAPT, and makes it extremely dependent on basis sets in ordinary orbital EDAs. In these methods it is also very difficult to separate from CT, since populating a virtual orbital of the other fragment may be seen as a lack of variational flexibility, as already explained. Moreover, pure charge transfer is not possible in standard perturbation expansions, that start from neutral fragments, or in BEDA or EDA treatments, since the antisymmetric state is not partitioned into fragments. In the latter, the polarization is due to orbital relaxation, and hides all kind of energetic variations, including those related to exchange in second order. LMOEDA E_{pol} and BEDA ΔE_{orb} are exactly the same quantities. A comparison of Tables 5.1, 5.4–5.6 shows very clearly that in strong interactions all BEDA, SAPT, and NEDA give very different results, and that in the long-range case, only BEDA and SAPT resemble each other. This is, in part, due to cancellation of residual terms. Inclusion of the $E_{\mathrm{ind}-x}^{20}$ term in SAPT into the induction energy worsens the agreement between BEDA or LMOEDA polarizations and SAPT induction.

NEDA's E_{pol} is close to the definition of induction contained in eqn (5.6), measuring the difference in electrostatic energy brought about by the interaction of the fragments. This includes the density distortion due to antisymmetry, which is generally included in $\Delta E_{\mathrm{Pauli}}$ in BEDA, in E_{rep} in LMOEDA, or in E_{exch}^{10} in SAPT. As a result, in weak complexes NEDA's polarization energies are larger. They may also be stabilizing or destabilizing, at variance with perturbative expansions. This is clearly shown in strong bonding cases. For instance, Table 5.4 shows that $E_{\mathrm{pol}} > 0$ in the H_2, HCl, HF, and LiH molecules. This is related to the large retraction suffered by the hydrogen density in the localized description, which decreases considerably their mutual interpenetration.

Although NEDA is one of the few orbital EDAs in which charge transfer has a physically sound definition, its E_{CT} keeps on measuring electron delocalization, and not the electrostatic component due to pure electron transfer between the fragments. This is particularly clear in Table 5.4, where this term accounts for most of the bonding in all the covalent homodiatomics, where strict charge transfer is null. The use of a contracted localized state Ψ_{AB}^{loc} makes this component considerably larger than the orbital interaction term in BEDA, exacerbating delocalization. All this may be observed when comparing Tables 5.4 and 5.5.

Eqn (5.6) is a viable way to estimate polarization corrections in real space EDAs. In IQA, for instance, V_{cl}^{AB} may be obtained from the non-interacting densities, so that its difference with respect to the in-the-molecule V_{cl}^{AB} actually measures the energy of distortion. Some values are found in Table 5.3. In the triplet state of H_2 (at the singlet geometry), the non-interacting V_{cl}^{AB} is equal to 47.79 kcal mol^{-1}, so that $E_{pol} = 12.20$ kcal mol^{-1}, positive as rationalized in the previous paragraph. This value is considerably smaller than the one in NEDA (47.37 kcal mol^{-1}), showing, once again, that using unphysical states instead of in-the-molecule ones exacerbates many interaction components. We can also decompose E_{pol} into two steps, one due to antisymmetrization and the other to relaxation. As expected, the first dominates, contributing with 8.26 kcal mol^{-1} to E_{pol}. It is interesting to notice that both steps have destabilizing contributions. Antisymmetrization retracts, as already commented, the density, and relaxation increases this effect. Contrarily, in ionic interactions with ionic references, the antisymmetrization component of E_{pol} plays a minor role, since the exclusion effect is small. In LiF, for instance, it equals -2.29 kcal mol^{-1}, the relaxation component being much larger, -21.13 kcal mol^{-1}. Interestingly, now both E_{pol} components are stabilizing. For weak interactions, like in the HF dimer or the HF–ClF complex, the two components are small. In the first, $+0.16$, and -0.99 kcal mol^{-1}, respectively. In the second, -0.04, and -0.34 kcal mol^{-1}. It is instructive to compare the total IQA E_{pol} with the sum of the SAPT $E_{ind}^{20} + E_{ind-x}^{20}$. In the LiF, HF–HF and HF–ClF cases they equal -14.42, -0.99, and -0.31 kcal mol^{-1}, respectively.

Taking into account that IQA automatically selects the charge state in-the-molecule, which is about 0.05 electrons away from the full ions in LiF, this agreement shows that real space EDAs may capture all the effects of perturbative and orbital EDAs at the same time.

Finally, we should add that CT in real space EDAs is pure and reflected in the net charges of the fragments. It is however not easy to associate an energetic quantity to its role. CT permeates all the energy components in IQA, for instance, through the altered number of electrons in the atomic or fragment domains. Estimating CT energies is still possible through recourse to the theory of resonance structures in real space,[6,9,29,30] in which we can extract the energy of having a definite number of electrons in each region of space, but we will not pursue this here any further. We will simply note in passing that IQA charges resemble those of NEDA, and that all the IQA energy components are consistent with them. Net charges immediately classify systems into ionic and covalent, as we can see in Table 5.2. NEDA, on the contrary, provides fragment charges, but computes its energy decomposition through states which are not consistent with them.

5.3.4　Dispersion

As commented, dispersion is well defined as a second order correction in long-range perturbation theory. Antisymmetry introduces several complications through various couplings but, overall, SAPT E_{disp}^{20} and $E_{\text{disp}-x}^{20}$ can be routinely computed. Some numbers are found in Table 5.6. Orbital EDAs rely on the construction, explicitly or implicitly, of an antisymmetric Φ_{AB}^{0} from the fragments. This is easy in single determinant expansions (or in DFT methods), but not in the general case, and has led to inclusion of dispersion indirectly, as DFT based BEDAs, or through corrections, like in LMOEDA. In the first case, ΔE_{orb} includes dispersion, but the energy components, *i.e.* ΔT_{Pauli}, are not exact, being computed through the pseudo Kohn–Sham determinant of the non-interacting system. In the second case, dispersion is obtained as the difference in correlation energies $E_{\text{disp}} = E_{AB}^{\text{corr}} - E_{A}^{\text{corr}} - E_{B}^{\text{corr}}$. Now it is all the other energy components that do not include the effect of correlation. For example, E_{elstat} has to be computed from the HF densities. This means that LMOEDA's E_{disp} will not coincide with SAPT or perturbative dispersion values. If we compare Tables 5.1 and 5.6 we clearly see that SAPT dispersion energies are larger than LMOEDA ones in all the cases. This effect is possibly related to the different quality of CCSD and MP2-like correlation energies. Let us also notice that E_{disp} has no definite sign in LMOEDA, and that ionic systems with ionic references show small positive dispersion corrections. This is not possible in a perturbative method. Dispersion in IQA is difficult to estimate, since wavefunction methods require large determinant expansions which become easily intractable from the numerical point of view. In principle, interfragment dispersion may be traced to the correlation part of V_{xc}^{AB}, that may be obtained from the Fock–Dirac partition of the exchange-correlation density into an exchange and a correlation part. We cannot forget, anyway, that interfragment interaction will also change the intrafragment correlation energies. Table 5.8 shows some IQA results obtained with complete active space (CAS) descriptions. We clearly see that correlation (even if limited) tends to decrease deformation energies and makes E_{int} to be more stabilizing in ionic interactions and less stabilizing in covalent ones. This may be traced successfully to the increased (decreased) electron delocalization caused by correlation in both cases. In weak interactions, like the water or HF dimers, the difference between V_{xc}^{AB} in the correlated and non-correlated descriptions compares reasonably well with the SAPT $E_{\text{disp}}^{20} + E_{\text{disp}-x}^{20}$ sum. In an FCI calculation in LiH (Table 5.7), the IQA E_{disp} calculated with the above prescription is equal to $+4.02$ kcal mol^{-1}, in qualitative agreement with our LMOEDA discussion. Another route, that we shall not explore here is to correct non-correlated IQA data à *la* LMOEDA, including the energetic change upon inclusion of correlation as dispersion.

5.4   The Full Picture

It is now time to examine the global image provided by the different EDAs that we have tried to examine. We have chosen the HF and water dimers, the BH_3NH_3 adduct, and the HF–ClF complex. CCSD interaction energies are found in Table 5.1. SAPT0 shows that, except in the stronger BH_3NH_3 system, dispersion contributes about -1 kcal mol^{-1} in all the systems. Induction is small in the HF–ClF complex, with a small ClF dipole, and about -1 kcal mol^{-1} in the H-bonded dimers, due to dipole-dipole terms. In the HF and water complexes exchange-repulsion is smaller in magnitude (about 60%) than the electrostatic attraction. The latter is itself a gross estimate of the total interaction energy, so broadly speaking, exchange-repulsion is mostly cancelled by induction and dispersion. In BH_3NH_3 this repulsion is considerably larger than E_{elstat}, so we cannot understand its bonding without strong polarization and dispersion energies (-44.4, -15.4 kcal mol^{-1}, respectively).

Orbital EDAs allow us to decompose the SAPT0 exchange-repulsion into exchange and repulsion separately. In LMOEDA, for instance, we can easily verify that (the self-interaction-less) E_X is a very important ingredient that cannot be overlooked. In the adduct it is considerably larger than E_{elstat} itself, and in the other dimers comparable to it. In weak complexes, E_X turns out also to be a good estimate of the total interaction energy. We have to notice that LMOEDA's E_{pol} gathers induction (with exchange) and all other bonding effects. In the weak interactions the first effect dominates, while in stronger ones, like in BH_3NH_3, the second cannot be overlooked, and its comparison (-63.27 kcal mol^{-1}) with SAPT0 inductions is considerably more difficult. It is also interesting to consider the standard BEDA. ΔE_{Pauli} is slightly dominated by the change in kinetic energy over the potential one. In the HF dimer, 32.74 *vs.* -29.55 kcal mol^{-1}, respectively. The latter has a very negative electron-nucleus contribution equal to -54.39 kcal mol^{-1}. As the ΔE_{orb} or polarization term is considered, relaxation decreases the kinetic energy by 20.32 kcal mol^{-1}, at the expense of increasing the Coulomb repulsion by 18.90 kcal mol^{-1} and making more stabilizing the electron-nucleus attraction. As previously commented, by no means is ΔT_{Pauli} the largest contribution to ΔE_{Pauli}. Exchange can also be obtained after antisymmetrizing. In the HF dimer it is equal to -6.20 kcal mol^{-1}, 2.57 kcal mol^{-1} below the standard E_X obtained from the non-interacting orbitals. This is in agreement with our detailed discussion of the triplet. Polarization or orbital interaction has a minor role, decreasing it by only 0.28 kcal mol^{-1}. Some new elements appear in BH_3NH_3, and again, the more interesting are related to exchange. Now anti-symmetrization decreases its magnitude from -115.90 to -70.85 kcal mol^{-1}. Orbital relaxation decreases exchange even more to -33.38 kcal mol^{-1}, with a large decrease in electron repulsion coupled to a more stabilizing electron-nucleus attraction.

NEDA and IQA provide a relatively common framework, although the use of non-interacting orbitals to obtain E_{elstat} and of localized states to compute E_X in NEDA makes a direct comparison not straightforward. NEDA's E_{CT} hides all relaxation effects from the localized state. As seen above with the BEDA discussion, this will include density distortions with their Coulombic effects, plus exchange modifications and kinetic energy modifications. In this sense, $E_{\text{pol}} + E_{\text{elstat}}$ in NEDA should contain a large part of the total electrostatic interaction V_{cl}^{AB} in IQA. The use of interpenetrating orbitals in NEDA will make this sum larger in magnitude than in IQA, as we can see. Similarly, NEDA's exchange is much smaller than in IQA due to self-interaction. The kinetic energy relaxation (which is large and negative, as seen in BEDA) in E_{CT} makes NEDA's deformation energies usually larger than those in IQA. Notice that, formally, the CT term in NEDA is equivalent to the orbital relaxation in BEDA, being the relaxation with respect to an antisymmetric state. However, the use of the NBO localized orbitals to construct it makes the comparison rather useless, and E_{CT} turns out to be considerably larger in NEDA. In other words, NEDA's antisymmetric state usually lies considerably above the standard Φ_{AB}^0 constructed from non-interacting orbitals.

This last story summarizes the problem of using artificial states in EDA-like research. Even with the same formal definitions, different prescriptions lead to quite different final energetic contributions. The absence of both references and spurious states in real space methods endows them with a very desirable simplicity, at the cost of computational complexity. IQA, as an example, does only use terms with a clear physical meaning. Since these methods analyze wavefunctions, we can still use spurious states, if we like. This allows us to compare results with other prescriptions, as we have tried to show. But we need not use them.

A final point should be added. Real space methods can be applied equally well to strong covalent interactions, like the H_2 singlet, where perturbation theory fails and most orbital EDAs have interpretative problems. A NEDA calculation of this singlet gives $E_{\mathrm{def}}^A = 78.51$ kcal mol^{-1}, with a very positive $E_{\mathrm{pol}} = 76.92$ kcal mol^{-1}. $E_{\mathrm{elstat}} = -2.63$, and $E_{\mathrm{CT}} = -314.90$ kcal mol^{-1}. The large polarization energy shows again that calculating E_{elstat} from isolated densities is not reasonable in the covalent regime. A LMOEDA treatment gives $E_{\mathrm{pol}} = -80.96$ and $E_{\mathrm{disp}} = -24.93$ kcal mol^{-1}, respectively, with an obvious $E_X = 0$ self-interaction-less exchange. Bonding is orbital interaction, and although $\Delta T_{\mathrm{Pauli}} = 0$ in this system, there is a considerable rise in kinetic energy, $\Delta E_{\mathrm{kin}}^{\mathrm{orb}} = 71.47$ kcal mol^{-1}, to be compared with the total kinetic energy rise in the triplet, equal to 103.12 kcal mol^{-1}. The NEDA triplet provides 84.77, 47.37, -2.63, -68.54 kcal mol^{-1} for E_{def}^A E_{pol}, E_{elstat}, and E_{CT}, respectively, with $E_X = -9.96$ kcal mol^{-1}. IQA, on the contrary, provides a deceptively simple picture in the singlet. At the FCI level, $E_{\mathrm{def}}^A = 7.44$, $V_{\mathrm{xc}}^{AB} = -148.72$, and $V_{\mathrm{cl}}^{AB} = 25.14$ kcal mol^{-1}, respectively. Deformations are very small, *i.e.* atoms-in-the-molecule show resilient energetics, and exchange-correlation (with self-interaction) is responsible for the bonding. In other words, electron delocalization, as measured by V_{xc}, is to blame. The same components for the FCI triplet provide 71.54, -70.91, 60.06 kcal mol^{-1}. Notice that the deformation of the triplet in IQA is much, much larger than in the singlet, at variance with NEDA. NEDA's localized states are too perturbed, offering a reasonable image in the triplet, but not in the singlet.

5.5 In Support of Real Space Energetic Decompositions

After presenting a relatively self-contained journey through several well-known EDAs, taking special care in showing the similarities and differences among them, we would like in this short section to advocate real space partitionings. To that end we will briefly discuss two very different systems, the acetamide–SO_2 dimer, and the square water tetramer. Both have been optimized at the CCSD//cc-pVDZ level, and their geometries are shown in Figure 5.2. Both pose non-trivial cases for EDAs. In the acetamide-SO_2 case we may ask whether real charge transfer is or not an issue, for instance. Similarly, in the water

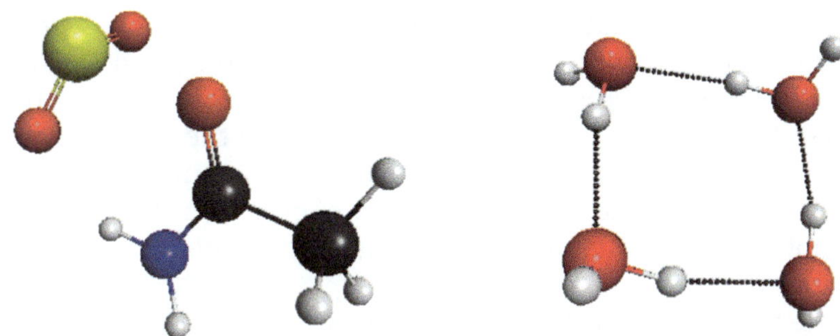

Figure 5.2 CCSD//cc-pVDZ geometries of the acetamide–SO$_2$ dimer (left) and the square-like C_2 symmetry conformer of the water tetramer (right).

tetramer we would like to differentiate the 6 interactions between pairs of H$_2$O moieties.

In the acetamide(A)–SO$_2$(B) dimer, the CP-corrected LMOEDA E_{elstat}, E_X, E_{rep}, E_{pol}, $E_{\text{Disp–CCSD}}$, and E_{CCSD} energies are -24.06, -31.67, 58.70, -9.83, 2.57, and -4.29 kcal mol^{-1}, respectively. BEDA shows that $\Delta E_{\text{Pauli}} = 27.03$ kcal mol^{-1}, while ΔE_{orb} accounts for -9.83 kcal mol^{-1}. Similarly, SAPT0 provides, $E_x^{(10)} = 26.54$, $E_{\text{ind}}^{(20)} = -8.93$, and $E_{\text{disp}}^{(20)} = -4.92$ kcal mol^{-1}. Given the possibility of formation of a hydrogen bond between the amine group and the SO$_2$ molecule, we may ask whether the above numbers support some transfer of charge between the two molecules. NEDA says that $Q_A = 0.0025$ e, relatively small, with deformation energies of the A and B fragments of 39.87 and 37.05 kcal mol^{-1}, respectively. The NEDA exchange is small, -6.84 kcal mol^{-1}, almost equal to the HF binding energy, -6.86 kcal mol^{-1}. Interestingly, the second order dispersion in SAPT also coincides with the SAPT0 binding energy, -4.92 kcal mol^{-1}. Does a simple, invariant picture emerge from this bunch of data? Probably not, for all dispersion, polarization, and electrostatics seem to be necessary ingredients in this complex interaction. As different methods provide different answers to each of these physical components of the interaction energy, differently biased minds would ascribe binding equally well to orbital interactions, dispersion, or polarization. A real space perspective will probably create another bias, but at least all energy components in a method such as IQA may be obtained without external references and from a panoply of underlying theoretical procedures. IQA in this case provides a considerably larger charge transfer than NEDA, $Q_A = 0.031$ e, so as expected, the SO$_2$ molecule extracts some electrons from acetamide. IQA deformation energies of the A and B fragments are 25.07 and 23.03 kcal mol^{-1}, and $V_{\text{xc}}^{AB}, V_{\text{cl}}^{AB}$ are -36.20, -18.99 kcal mol^{-1}, respectively. Notice that in IQA we can proceed further, dissecting all those numbers into atom-atom specific interactions, and atomic deformation energies. Thus the contribution to binding of a purported hydrogen bond can be unmasked, if present, and the role of each pair interaction examined. In this sense, real space technologies allow for an extremely thin decomposition of the energy of a system, that reveals molecular interactions at different zoom levels, all at the same computational price.

The last ideas are easy to grasp when we examine the water tetramer. In the standard implementations of EDAs, SAPT is difficult to examine beyond the two-fragment paradigm. Similar objections apply in LMOEDA or BEDA. NEDA defines, in general, the deformation energies of each of the constituents of a multi- fragment system, but usually gathers all charge transfer, electrostatic, polarization, and exchange components together. In some of the above methods, available implementations are restricted to HF or KS levels. In others, correlation can only be included *via* MP2 or CCSD corrections. Real space EDAs are, in principle, free of those restrictions, and can be obtained at very different levels of theory. Table 5.9 contains IQA data at the HF, CCSD, and CCSD(T) levels. The CC IQA calculations have been obtained through a PROMOLDEN interface to the pyscf code.[43] Notice that we can now explore each of the intermolecular interactions separately. Charge transfer is negligible, and has not been included. As expected, adjacent water molecules interact most. We can see that the each of the molecules of this latter pair are not exactly equally deformed, but that the difference is very small. Saturation of correlation decreases the electrostatic interaction, as expected due to an enhanced electron localization, and increases the exchange–correlation one. Although we will not pursue it any further, the difference between the HF and CC V_{xc}'s is a direct measure of

Table 5.9 IQA results in the square-like C_2 water tetramer.[a]

Interaction	E_{def}^{A}	E_{def}^{B}	E_{int}^{AB}	V_{xc}^{AB}	V_{cl}^{AB}
1,2		19.06	−23.62	−14.70	−8.92
		21.11	−25.82	−17.78	−8.04
		20.98	−25.87	−17.94	−7.94
1,3	19.07		−1.87	−0.09	−1.79
	21.11		−1.89	−0.27	−1.61
	20.98		−1.90	−0.30	−1.60

[a]1,2 refers to adjacent and 1,3 to non-adjacent H_2O molecules; A, B labels the first and the second water molecules for each (1,2 or 1,3) interaction. First, second, and third lines correspond to HF, CCSD, and CCSD(T) data, respectively. All units in kcal per mol, and data are CP corrected.

inter-molecular electron correlation (*i.e.* dispersion). This accounts for about -3.2 and -0.21 kcal mol^{-1} for the 1,2 and 1,3 interactions. Such spatially resolved data are difficult to obtain from other schemes. Let us also point out that pure 1,3 exchange energies, although small, are not negligible.

5.6 Final Remarks

We expect to have shown in the pages above that the energetic decomposition of intermolecular interactions is a far from settled problem. Except in the (very) long-range limit, where most methods converge and the different contributions become well defined, we face a zoo of terms, many times with the same name, that unfortunately may have quite different physical interpretations. Practical advice for the non-specialist is nevertheless possible: use whatever method you feel most comfortable with, but clearly state what you are doing, always bearing in mind the meaning of any contribution you will be using. Be very careful when criticizing the results of others: they may be referring to quantities with very different meaning! In passing, partitionings that use real space quantities, with well-defined energetic contributions at the strong and weak interacting limits may turn out to be useful alternatives to standard EDAs in the long run.

Acknowledgements

The authors thank the Spanish MINECO for funding (grant CTQ2012-31174).

References

1. R. F. W. Bader, *Atoms in Molecules*, Oxford University Press, Oxford, 1990.
2. F. M. Bickelhaupt and E. J. Baerends, Kohn-Sham density functional theory: Predicting and understanding chemistry, *Rev. Comput. Chem.*, 2007, **15**, 1–86.
3. M. A. Blanco, A. Martín Pendás and E. Francisco, Interacting quantum atoms: a correlated energy decomposition scheme based on the quantum theory of atoms in molecules, *J. Chem. Theory Comput.*, 2005, **1**, 1096–1109.
4. S. F. Boys and F. Bernardi, The calculation of small molecular interactions by the differences Def separate total energies. some procedures with reduced errors, *Mol. Phys.*, 1970, **19**, 553.
5. R. Bukowski, W. Cencek, P. Jankowski, M. Jeziorska, B. Jeziorski, S. A. Kucharski, V. F. Lotrich, A. J. Misquitta, R. Moszyński, K. Patkowski, R. Podeszwa, F. Rob, S. Rybak, K. Szalewicz, H. L. Williams, R. J. Wheatley, P. E. S. Wormer and P. S. Żuchowski, SAPT2012: An *ab initio* program for many-body

symmetry-adapted perturbation theory calculations of intermolecular inter-action energies. University of Delaware and University of Warsaw (2012). URL http://www.physics.udel.edu/šzalewic/SAPT/SAPT.html.

6. E. Francisco and Martín Pendás, Towards an energy partition into real space resonance structures: 1- and 2-particle density matrix decomposition, *Mol. Phys.*, 2015, DOI: 10.1080/002689762015.1121296.

7. E. Francisco, A. Martín Pendás: IMOLINT (Unpublished).

8. E. Francisco, A. Martín Pendás and M. A. Blanco, A molecular energy decomposition scheme for atoms in molecules, *J. Chem. Theory Comput.*, 2006, **2**, 90–102.

9. E. Francisco, A. Martín Pendás and M. A. Blanco, Electron number probability distributions for correlated wave functions, *J. Chem. Phys.*, 2007, **126**(094), 102.

10. E. D. Glendening, J. K. Badenhoop, A. E. Reed, J. E. Carpenter, J. A. Bohmann, C. M. Morales, F. Weinhold, Nbo 5.0. (Theoretical Chemistry Institute, University of Wisconsin, Madison (2001).

11. E. D. Glendening and J. Streitwieser, Natural energy decomposition analysis: An energy par- titioning procedure for molecular interactions with application to weak hydrogen bonding, strong ionic, and moderate donoracceptor interactions, *J. Chem. Phys.*, 1994, **100**, 2900.

12. I. C. Hayes and A. J. Stone, An intermolecular perturbation theory for the region of moderate overlap, *Mol. Phys.*, 1984, **53**, 83–105.

13. I. C. Hayes and A. J. Stone, Matrix elements between determinantal wavefunc-tions of non- orthogonal orbitals, *Mol. Phys.*, 1984, **53**, 69–82.

14. T. Helgaker, J. Olsen and P. Jorgensen, *Molecular Electronic-Structure Theory*, John Wiley & Sons, Ltd, New York, 2000, ch. 6.

15. A. Hesselmann, G. Jansen and M. Schütz, Density-functional theory-symmetry-adapted inter- molecular perturbation theory with density fitting: A new efficient method to study intermolecular interaction energies, *J. Chem. Phys.*, 2006, **122**(014), 103.

16. P. Hobza and K. Müller-Dethlefs, *Non-covalent Interactions: Theory and Experi-ment*, Cambridge University Press, 2010.

17. B. Jeziorski, R. Moszynski and K. Szalewicz, Perturbation theory approach to intermolecular potential energy surfaces of van der Waals complexes, *Chem. Rev.*, 1994, **94**, 1887–1930.

18. K. G. Kay, H. D. Todd and H. J. Silverstone, Bipolar expansion of $r_{12}{}^n y_l{}^m (\theta_{12}, \phi_{12})$, *J. Chem. Phys.*, 1969, **51**, 2363–2367.

19. B. King and F. Weinhold, Structure and spectroscopy of (hcn)n clusters: Cooperative and electronic delocalization effects in chn hydrogen bonding, *J. Chem. Phys.*, 1995, **103**, 333; K. Kitaura and K. Morokuma, A new energy decomposition scheme for molecular interactions within the Hartree-Fock ap-proximation, *Int. J. Quantum Chem.*, 1976, **10**, 325.

20. K. Kitaura and K. Morokuma, A new energy decomposition scheme for molecular interactions within the Hartree-Fock approximation, *Int. J. Quantum Chem.*, 1976, **10**, 325.

21. W. Kohn, Theory of the insulating state, *Phys. Rev. Lett.*, 1964, **133**, A171.

22. F. London, The general theory of molecular forces, *Trans. Faraday Soc.*, 1937, **33**, 8–26.

23. H. C. Longuet-Higgins, Intermolecular forces, *Discuss. Faraday Soc.*, 1965, **40**, 7–27.

24. V. Magnasco, *Elementary Molecular Quantum Mechanics: Mathematical Methods and Applications*, Elsevier, Amsterdam, 2013.

25. A. Martín Pendás, M. A. Blanco and E. Francisco, Two-electron integrations in the quantum theory of atoms in molecules, *J. Chem. Phys.*, 2004, **120**, 4581.

26. A. Martín Pendás, M. A. Blanco and E. Francisco, The nature of the hydrogen bond: A synthesis from the interacting quantum atoms picture, *J. Chem. Phys.*, 2006, **125**(184), 112.

27. A. Martín Pendás, M. A. Blanco and E. Francisco, Chemical fragments in real space: Definitions, properties, and energetic decompositions, *J. Comput. Chem.*, 2007, **28**, 161–184.
28. A. Martín Pendás and E. Francisco: PROMOLDEN: A QTAIM/IQA code (Unpublished).
29. A. Martín Pendás, E. Francisco and M. A. Blanco, Charge transfer, chemical potentials, and the nature of functional groups: answers from quantum chemical topology, *Faraday Discuss.*, 2007, **135**, 423.
30. A. Martín Pendás, E. Francisco and M. A. Blanco, Pauling resonant structures in real space through electron number probability distributions, *J. Phys. Chem. A*, 2007, **111**, 1084.
31. R. McWeeny, *Methods of Molecular Quantum Mechanics*, Academic Press, London, 2nd edn, 1992.
32. A. J. Misquitta, R. Podeszwa, B. Jeziorski and K. Szalewicz, Intermolecular potentials based on symmetry-adapted perturbation theory with dispersion energies from time-dependent density-functional calculations, *J. Chem. Phys.*, 2005, **123**(214), 103.
33. K. Morokuma, Molecular orbital studies of hydrogen bonds. iii, *J. Chem. Phys.*, 1971, **55**, 1236–1244.
34. J. J. Novoa, M. Planas and M. H. Whangbo, A numerical evaluation of the counterpoise method on hydrogen bond complexes using near complete basis sets, *Chem. Phys. Lett.*, 1994, **225**, 240–246.
35. P. L. A. Popelier and D. S. Kosov, Atom-atom partitioning of intramolecular and intermolecular coulomb energy, *J. Chem. Phys.*, 2001, **114**, 6539.
36. A. R. Reed, L. A. Curtiss and F. Weinhold, Intermolecular interactions from a natural bond orbital, donor-acceptor viewpoint, *Chem. Rev.*, 1988, **88**, 899.
37. A. R. Reed and F. Weinhold, Natural bond orbital analysis of nearhartreefock water dimer, *J. Chem. Phys.*, 1983, **78**, 4066.
38. A. R. Reed, F. Weinhold, L. A. Curtiss and D. J. Pochatko, Natural bond orbital analysis of molecular interactions: Theoretical studies of binary complexes of HF, H2O, NH3, N2, O2, F2, CO, and CO2 with HF, H2O, and NH3, *J. Chem. Phys.*, 1986, **84**, 5687.
39. M. W. Schmidt, K. K. Baldridge, J. A. Boatz, S. T. Elbert, M. S. Gordon, J. H. Jensen, S. Koseki, N. Matsunaga, K. A. Nguyen, S. J. Su, T. L. Windus, M. Dupuis and J. A. Montgomery, General atomic and molecular electronic structure system, *J. Comput. Chem.*, 1993, **14**, 1347.
40. J. C. Slater, *Quantum Theory of Molecules and Solids*, McGraw-Hill, New York, 1965, vol. 1.
41. A. J. Stone, *The Theory of Intermolecular Forces*, Oxford University Press, 2013.
42. P. Su and H. Li, Energy decomposition analysis of covalent bonds and intermolecular interactions, *J. Chem. Phys.*, 2009, **131**(014), 102.
43. Q. Sun, pyscf: GNU Python module for quantum chemistry. https://github.com/sunqm/pyscf/ (2015).
44. F. Weinhold, *J. Mol. Struct.*, 1997, **399**, 191.
45. T. Ziegler and A. Rauk, On the calculation of bonding energies by the Hartree Fock Slater method, *Theor. Chem. Acc.*, 1977, **46**, 1.

6 Beyond QTAIM: NCI Indexes as a Tool to Reveal Intermolecular Bonds in Molecular Aggregates

Roberto A. Boto[a,b] and Julia Contreras-García*[a,b]

[a] UPMC Univ Paris 06, CNRS, UMR 7616, Laboratoire de Chimie Théorique, case courrier 137, 4 place Jussieu, F-75005, Paris, France; [b] Sorbonne Universités, UPMC Univ Paris 06, ICS, F-75005, Paris, France
*Email: contrera@lct.jussieu.fr

6.1 Introduction

Weak interactions span a wide range of binding energies, and encompass hydrogen bonding (HB), dipole–dipole interactions[1] and London dispersion as well as more up to date interactions such as halogen bonds,[2] $CH\cdots\pi$[3] and $\pi\cdots\pi$[4] interactions. Repulsive interactions, also known as steric clashes, should not be disregarded either.

It is commonly understood that bonds refer to strong covalent interactions, and their rearrangement determines the way along a chemical reaction. However, in spite, or rather because of their weak strength, weak interactions are of utmost importance to our world as we conceive it. Chemical interactions between a protein and a drug, or a catalyst and its substrate, self-assembly of nanomaterials,[5,6] and even some chemical reactions,[7,8] are dominated by non-covalent interactions.

Intermolecular Interactions in Crystals: Fundamentals of Crystal Engineering
Edited by Juan J. Novoa

More specifically, non-covalent interactions are of paramount importance in chemistry and especially in bio-disciplines,[9,10] since they set up the force field scenario through which chemical species interact with each other without a significant electron sharing between them. They represent, in fact, the machinery through which molecules recognize themselves and establish how molecules will approach and eventually pack together.

During the last decade, non-covalent interactions have also raised a great deal of interest in the context of materials through self-assembly[11] and crystallization,[12] whose underlying general rules are at the moment too faraway to be fully rationalized and understood.[13] Knowledge of such rules would in principle allow to build from scratch (even complex) materials exhibiting the desired properties.[5,14,15] Non-covalent interactions are also extremely relevant at the intra-molecular level. Although it cannot be ignored that a given observed structure is generally the outcome of a "drawing" among a plethora of energetically similar, but structurally dissimilar options,[16] understanding intermolecular non-covalent interactions and their mutual interplay in the supramolecular assemblies is nonetheless a fundamental step in making progress in structural prediction and evolution.

In spite of the relevance of these interactions in chemistry and related disciplines, identifying and quantifying non-covalent interactions remains a difficult issue. Given how useful and central this concept is to chemistry, several approaches have been developed to provide a quantum mechanical equivalent of such "fuzzy" concepts as the chemical bond in Hilbert and in real space. Whereas Hilbert space approaches are based on some unitary transformation of the occupied orbitals of the system, they are somehow biased since they depend on the initial set of orbitals. Real space approaches exploit the topological properties of some scalar field. Whereas the most common topological approach to interactions in general is the Quantum Theory of Atoms in Molecules (QTAIM), it does not always agree with the original concepts and experimental results. The Electron Localisation Function (ELF) in turn, enables a clear visualization of electron localisation, yielding bonds, lone pairs and atomic shells. However, non-covalent features require a deeper analysis of the topology and cannot be directly visualized with ELF.

A recently introduced tool, the reduced density gradient enables us to overcome these drawbacks. From the theoretical point of view, we will show how it works and its theoretical basis, emphasizing how it actually provides a generalized approach to QTAIM and ELF. Indeed, the reduced density gradient recovers the whole QTAIM topology, but in addition it also reveals a structure similar to ELF: lone pairs and atomic shells can be visualized. And most importantly, critical points appear for those weak interactions where QTAIM fails, yielding a better agreement with text-book examples and with subtle experiments.

6.2   The Need for a New Approach

Covalent bonds are intuitively represented using conventional Lewis structures.[17] Purely electrostatic interactions can be analyzed using electrostatic potential maps.[18] Non-covalent interactions are frequently visualized using distance-dependent contacts, generally without consideration of hydrogen atoms.[19–21] Hydrogen-bonds can be identified from the molecular geometry,[22] while grid-based calculations based on classical force fields are used to model other van der Waals interactions.[23]

From the topological point of view, interactions can be visualized from properties of the electron density with modern quantum-mechanical models of bonding, such as ELF[24,25] and QTAIM theory.[26,27] Let's look at an example to simultaneously visualize the information coveyed by QTAIM and ELF considering both strong and weak interactions. The image provided by QTAIM and ELF of benzene dimer is provided in Table 6.1. Let's first focus on a given benzene molecule: the electron density shows maxima (cusps) for the C and H atoms, whereas C–H and C–C bonds are represented by bond critical points (BCPs) in red. ELF, instead, provides a picture based on

Table 6.1   Comparison between QTAIM and ELF topologies in benzene dimer. Carbon atoms in dark grey, hydrogen atoms in light grey, bcps in red.

Method	QTAIM	ELF
Function	Electron density	Pauli kinetic energy density
Example	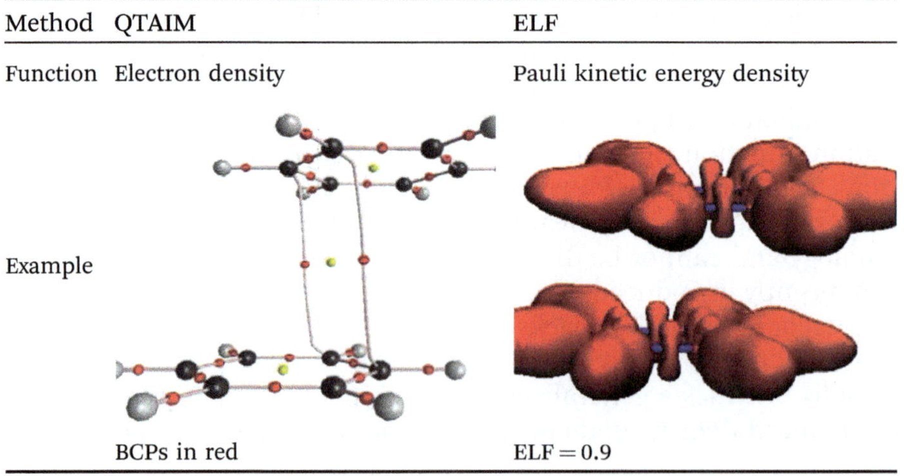 BCPs in red	ELF = 0.9

electron localization, so that isosurfaces appear around atoms and C–H bonds. In both cases the chemical structure (intramolecular covalent bonds) is revealed as expected from chemical intuition. However, the ELF picture is more visual thanks to the isosurfaces.

Now, if we take a look at the complete system, the stacking dimer, we can see that intermolecular BCPs appear along with a ring critical point (second order saddle point, in yellow). The fact that interactions are related to saddle points, locates them on the interatomic surfaces, so that they highlight interatomic contact, but they do not have an associated region within this approach. Moreover, the critical points unite pairs of C atoms, which is not the chemical picture we have of a stacking interaction: it should appear as a benzene to benzene interaction. The VSEPR regions of benzene dimer are clearly identified by ELF, but nothing is seen for the inter-benzene stacking interaction.

Looking at the benzene dimer examples for QTAIM and ELF, we can see that they provide a very good description of localized covalent bonds. However, they do not provide a clear picture of delocalized interactions. In other words, a visual quantum chemical approach to non-covalent interactions is conspicuously missing in the quantum topology scenario.

It is worth noting that the electron density has a fundamental advantage over MO-based descriptors because it is an experimentally accessible scalar field and a local function defined within the exact many body theory and supported by the Hohenberg-Kohn theorems.[28] Furthermore because chemical reactions proceed by $\rho(\mathbf{r})$ redistributions, methods that analyse $\rho(\mathbf{r})$ distributions should help us to understand the electronic structure of molecules and thus chemical reactivity. Hence it would be advisable to include a complementary tool for describing non-covalent interactions which is dependent on the electron density and thus invariant under unitary transformations.

The NCI approach provides such an option. It is a function that depends on the density and its derivatives, allowing simultaneous analysis and visualisation of a wide range of non-covalent interaction types as real space surfaces, adding an important tool to the chemist's arsenal.[29–31] We will review its physical meaning, its characteristics and it use in molecular solids in the coming sections.

6.3 From Kinetic Energy Densities to Non Covalent Interactions

In order to develop a topological tool better adapted to reveal non covalent interaction, let's analyze the ingredients of the ones that are already known. Scalar fields based on any of the forms of the kinetic energy density are specially useful for the visual analysis of chemical interactions. The success of this family of scalar fields rests on the decrease in the interatomic kinetic energy density during the formation of a covalent interaction;[32–34] therefore any function able to map this variation should be a good candidate as a real space bonding indicator.

However, one should start by recalling that there is an infinity of kinetic energy densities; the only requirement is that they integrate to the total kinetic energy of the system, T:[35,36]

$$T = \int \tau_g(\mathbf{r})d\mathbf{r}, \tag{6.1}$$

where $\tau_g(\mathbf{r})$ is any form of the kinetic energy density. Two particular forms of τ_g have been widely used in the literature $\tau(\mathbf{r})$ and $K(\mathbf{r})$

$$K(\mathbf{r}) = -\frac{1}{2}\sum_i^{occ} n_i \psi_i(\mathbf{r})\nabla^2\psi_i(\mathbf{r}) \tag{6.2}$$

$$\tau(\mathbf{r}) = \frac{1}{2}\sum_i^{occ} n_i \nabla\psi_i(\mathbf{r})\nabla\psi_i(\mathbf{r}), \tag{6.3}$$

where ψ_i are real natural orbitals and n_i their occupation numbers such that $\rho(\mathbf{r}) = \sum_i n_i \psi_i(\mathbf{r})\psi_i(\mathbf{r})$. Both of these functions integrate to the same total kinetic energy, but they differ locally. Whereas $\tau(\mathbf{r})$ is positive everywhere, $K(\mathbf{r})$ (the definition used by Schrödinger) exhibits positive and negative values. Both definitions are related *via* the Laplacian of the electron density.

$$K(\mathbf{r}) = -\frac{1}{4}\nabla^2\rho(\mathbf{r}) + \tau(\mathbf{r}). \tag{6.4}$$

Since $\nabla^2\rho(\mathbf{r})$ is a divergence term, its net contribution to T is cancelled out when integrated over the whole space (and also within AIM basins)

$$\int \nabla^2\rho(\mathbf{r})d\mathbf{r} = 0. \tag{6.5}$$

Thus, the total kinetic energy of a system, T, is well defined and equivalent from the various kinetic energy density definitions

$$T = \int K(\mathbf{r})d\mathbf{r} = \int \tau(\mathbf{r})d\mathbf{r} \tag{6.6}$$

The positive definite condition of $\tau(\mathbf{r})$ enables the interpretation of its local behaviour in terms of classical effects, and thus most chemical bonding descriptors have chosen this definition. It is worth noticing that for a Hartree–Fock wave function, eqn (6.2) becomes

$$\tau(\mathbf{r}) = \frac{1}{2} \sum_{i=1}^{\text{occ}} |\nabla \psi_i(\mathbf{r})|^2, \tag{6.7}$$

where $\psi_i(\mathbf{r})$ are Hartree–Fock orbitals.

The limiting behaviour of $\tau(\mathbf{r})$ at $\mathbf{r} = 0$ and $\mathbf{r} \to \infty$ is determined by the von Weizäcker kinetic energy density,[37] $\tau_w(\mathbf{r})$

$$\tau_w(\mathbf{r}) = \frac{1}{8} \frac{\nabla^2 \rho(\mathbf{r})}{\rho(\mathbf{r})}. \tag{6.8}$$

This is the kinetic energy density associated with a system of the same electron density as the original one, but where the Pauli repulsion has been turned off. That is, a bosonic system where all electrons occupy the same orbital. Thus, it constitutes a rigorous lower boundary for $\tau(\mathbf{r})$. The difference between $\tau(\mathbf{r})$ and $\tau_w(\mathbf{r})$ can be understood as a measure of the excess of kinetic energy density due to the fermionic nature of electrons, *i.e.* due to the Pauli principle. This difference is known as the Pauli kinetic energy density, $t_p(\mathbf{r})$[38]

$$t_p = \tau(\mathbf{r}) - \tau_w(\mathbf{r}) \tag{6.9}$$

It is important to note that the right hand term of eqn (6.9) matches the Laplacian of the conditional probability density $D(\mathbf{r})$ introduced to define ELF.[24] So that the ELF kernel, χ, can be viewed in kinetic energy terms as t_p divided by the Thomas–Fermi kinetic energy density, $\tau_{TF}(\mathbf{r}) = \frac{3}{5}(6\pi^2)^{2/3} \rho_\sigma^{5/3}$.[38]

$$\chi = \frac{\tau(\mathbf{r}) - \tau_w(\mathbf{r})}{\tau_{TF}(\mathbf{r})} \tag{6.10}$$

$$\text{ELF} = \frac{1}{1 + \chi^2} \tag{6.11}$$

In other words, from the kinetic energy density point of view, ELF may be understood as the excess of kinetic energy compared with a system of bosons of the same density due to the Pauli principle, all of it scaled by the Thomas–Fermi term.

Reordering the terms in eqn (6.10), we may also understand the ELF kernel as the difference:

$$\chi(\mathbf{r}) = \frac{\tau(\mathbf{r})}{\tau_{TF}(\mathbf{r})} - \frac{\tau_w(\mathbf{r})}{\tau_{TF}(\mathbf{r})} = t_{\text{LOL}} - t_{\text{bose}} \tag{6.12}$$

The first term is known as localised-orbital locator (LOL), $t_{\mathrm{LOL}} = \dfrac{\tau(\mathbf{r})}{\tau_{\mathrm{TF}}(\mathbf{r})}$.
It is bounded by zero from below, but has no an upper boundary:

$$0 \leq t_{\mathrm{LOL}} < \infty, \tag{6.13}$$

At the positions of the stationary points of localized orbitals, $t_{\mathrm{LOL}}(\mathbf{r})$ is driven to small values.[39] In regions dominated by the overlap of localized orbitals, $t_{\mathrm{LOL}}(\mathbf{r})$ attains large values. This basically means that the chemical content and topological features of LOL are similar to those of the ELF kernel, $\chi(\mathbf{r})$. Indeed, the topological features of LOL and the ELF kernel are very close.

What is the topology and the chemical content included in t_{bose}? Let's start with the Weizäcker term. Although $\tau_w(\mathbf{r})$ is not a kinetic energy density in the sense that it integrates to the von Weizsäcker correction to the Thomas–Fermi model, rather than to the total kinetic energy, recent studies have shown it is a useful tool for revealing atomic and molecular structure. This is because electrons forming pairs (low $t_p(\mathbf{r})$) or alone (low $\rho(\mathbf{r})$) both approach the bosonic behavior. Moreover $\tau_w(\mathbf{r})$ is exact for any system described by a single spatial orbital.[40] Additionally, $\tau_w(\mathbf{r})$ is the kinetic energy density of the marginal probability amplitude introduced by Hunter some time ago.[41] $\tau_w(\mathbf{r})$ may be thereby understood as a measure of the single particle character of the system. Because a localized electron pair behaves as a single particle, namely as a boson, its kinetic energy density is given by $\tau_w(\mathbf{r})$.[42]

The problem with $\tau_w(\mathbf{r})$ is that it depends on the electron density, and thus, its behavior is mainly marked by the core. It is advisable to get rid of this dependence to reveal chemical structure by dividing by the Thomas–Fermi kinetic energy density, $\tau_{\mathrm{TF}}(\mathbf{r})$. This gives rise to a dimensionless variable $t_{\mathrm{bose}}(\mathbf{r})$:[43]

$$t_{\mathrm{bose}}(\mathbf{r}) = \frac{\tau_w(\mathbf{r})}{\tau_{TF}(\mathbf{r})}, \tag{6.14}$$

t_{bose} may be understood as a locator of the stationary points of the electron density. As explained by Savin,[44] at regions where localized orbitals attain their maxima, $\nabla\rho(\mathbf{r})$ is expected to be close to 0, and t_{bose} is driven to small values. Thus, one requirement (though not sufficient) for electron localization is small values of $\nabla\rho(\mathbf{r})$, and therefore low values of $\tau_w(\mathbf{r})$ as well as t_{bose}.

In order to yield pictures similar to ELF, we follow Schmider and Becke,[45] who proposed to map LOL onto the range [0,1], ν:

$$\nu(\mathbf{r}) = \frac{t(\mathbf{r})}{1 + t(\mathbf{r})} = \frac{1}{1 + \dfrac{\tau(\mathbf{r})}{\tau_{\mathrm{TF}}(\mathbf{r})}}. \tag{6.15}$$

We also define $\beta(\mathbf{r}) = 1/1 + t_{\mathrm{bose}}(\mathbf{r})$. With these definitions, β becomes an upper limit to $\nu(\mathbf{r})$ since the total kinetic energy density is always greater than the bosonic one (Figure 6.1). ν shows maxima at regions where the Pauli repulsion, accounted for by the Pauli kinetic energy density, is relatively low: at infinity and in electron pairs (N lone pairs, N cores, N–N bond). The chemical picture obtained by t_{bose} is the same; maxima account for regions of electron-pair localisation, *i.e.* cores, lone and bonding electron pairs.

The important point here is that the opposite is not true: not all regions of low values of $\nabla\rho(\mathbf{r})$ involve maxima of localized orbitals. In other words, $t_{\mathrm{bose}}(\mathbf{r})$ contains more chemical information than $\mathrm{LOL} = t(\mathbf{r})/t_{\mathrm{TF}}(\mathbf{r})$. Whereas covalent bonding and Lewis structure in general are related to electron pairing as induced by the Pauli principle, non-covalent interactions appear in regions of interaction at low density and in the absence of electron pairing. Contrary to the bonding descriptors based on $\tau(r)$, $\tau_w(r)$ bonding descriptors have the unique feature of revealing not only atomic shells, bonding and lone electron pairs, but also non-covalent interactions.

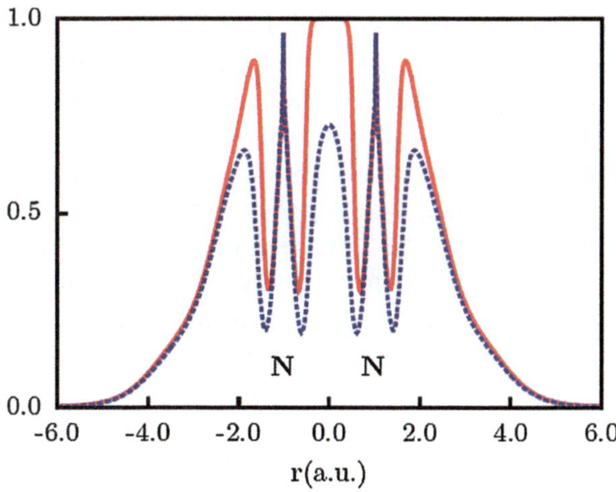

Figure 6.1 $\beta(r)$ (solid red line) and ν (dashed blue line) along internuclear axis for N_2. The zero was set at the BCP position.

Historically, t_{bose} has not been used in the non-covalent region. Instead, a function related to it, the reduced density gradient, $s(\mathbf{r})$

$$s(\mathbf{r}) = \frac{1}{C_s} \frac{|\nabla\rho(\mathbf{r})|}{\rho(\mathbf{r})^{4/3}}, \tag{6.16}$$

has been used for visualizing weak interactions, with $C_s = 2(3/\pi^2)^{1/3}$. It is easy to see that these two functions are deeply related:

$$t_{\text{bose}}(\mathbf{r}) = \frac{\tau_w(\mathbf{r})}{\tau_{\text{TF}}(\mathbf{r})} = \left(\frac{5}{3}\right) s(\mathbf{r})^2. \tag{6.17}$$

$s(\mathbf{r})$ is a fundamental dimensionless quantity in DFT used to describe the deviation from a homogeneous electron distribution. Properties of $s(\mathbf{r})$ have been investigated in depth in the process of developing increasingly accurate functionals.[46–50]

The origin of $s(\mathbf{r})$ can be traced back to the generalised gradient contribution to the Generalized Gradient Approximation (GGA) exchange energy E_x^{GGA}:[48]

$$E_x^{\text{GGA}} - E_x^{\text{LDA}} = -\sum \int F(s)\rho^{4/3}(\mathbf{r})\mathrm{d}\mathbf{r}, \tag{6.18}$$

where $F(s)$ is a function of $s(\mathbf{r})$ for a given spin, and $\mathrm{E}_x^{\text{LDA}}$ stands for the Local Density Approximation (LDA) exchange energy.

The way $s(\mathbf{r})$ was initially introduced to analyze non covalent interactions is as follows. The reduced density gradient accounts for local density inhomogeneities due to its differential behaviour depending on the chemical region of the molecule. The reduced density gradient assumes large values in the exponentially-decaying density tails far from the nuclei. However, in the presence of a non-covalent interaction, s approaches zero. This effect is so dramatic that is is extremely easy to visualize, when $s(\mathbf{r})$ is plotted as a function of the density. Graphs of $s(\mathbf{r})$ *versus* $\rho(\mathbf{r})$ assume the form $s(\mathbf{r}) = a\rho(\mathbf{r})^{-1/3}$, where a is a constant. This can be easily proven from a Slater orbital model density (STO). For a single atomic orbital $\psi(\mathbf{r}) = \mathrm{e}^{-\alpha r}$, the density is $\rho(\mathbf{r}) = \mathrm{e}^{-2\alpha r}$ and the gradient is $\nabla\rho(\mathbf{r}) = -2\alpha\rho(\mathbf{r})$, such that

$$s^{\text{STO}}(\mathbf{r}) = \frac{1}{C_s} \frac{2\alpha\rho(\mathbf{r})}{\rho(\mathbf{r})^{4/3}} = \frac{2\alpha}{C_s}\rho(\mathbf{r})^{-1/3}. \tag{6.19}$$

When interference between atomic clouds occurs, peaks appear in the $s(\rho)$ plot that reveal the presence of an interaction. Figure 6.2a shows such behavior for an atomic hydrogen density.[29] When two

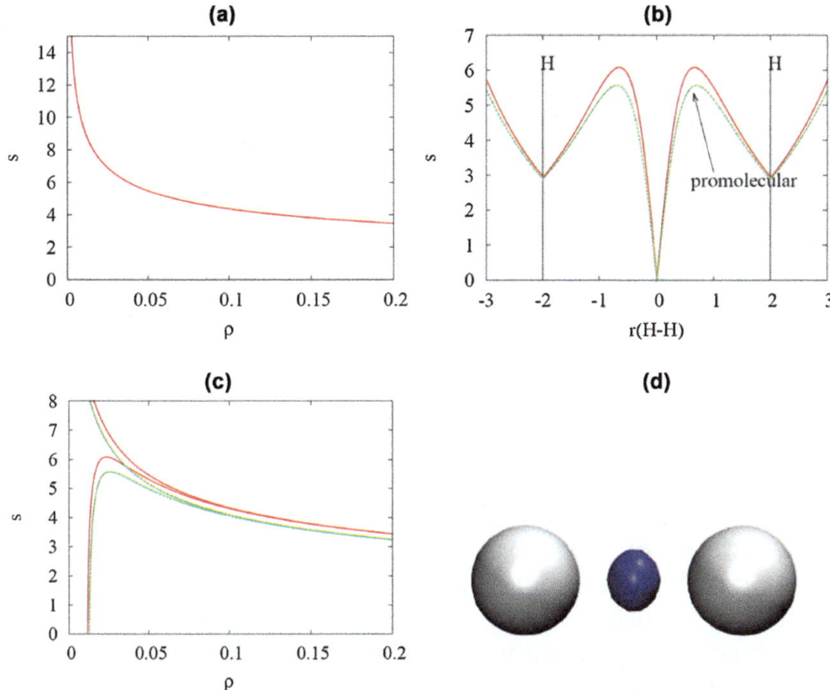

Figure 6.2 Behavior of s for the hydrogen atom and the hydrogen molecule. Green correponds to the relaxed density, whereas red corresponds to a model density $\rho = e^{-\alpha_r}$ with $\alpha = 0.5288$ (hydrogen[29]). (a) $s(\rho)$ atomic (b–d) diatomic considering $R = 1$ Å, (b) s as a funtion of the H–H distance; (c) $s(r)$; and (d) isosurface for $s = 0.2$. $s = 0$ corresponds to the position of the BCP within the AIM approach.

atoms come together, a critical point of the density appears ($\nabla \rho = 0$) which gives rise to $s = 0$, which in the $s(\rho)$ diagram is translated into a peak. Figure 6.2b shows the result from two promolecular hydrogen densities along the H–H bonding line, which gives rise to the $s(\rho)$ diagram in Figure 6.2c.

Plotting s isosurfaces around these minima, non-covalent interactions appear as very intuitive isosurfaces (Figure 6.2d). This procedure is able to reveal non-covalent interactions. As we shall explain later, interaction types are differentiated by a color code: blue for attractive such as hydrogen bonds (HBs), green for very weak interactions such as van der Waals (vdW) and red for steric repulsion. The first two are illustrated in Figure 6.3.

Since AIM has probably been the most extended topolopogical analysis, AIM critical points have been included to show the relationship between AIM and NCI fields. There are three $(3, -1)$ critical

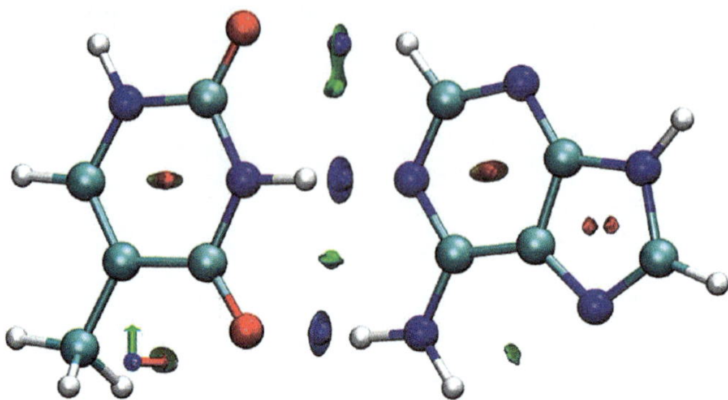

Figure 6.3   Adenine-thymine complex with NCI surfaces ($s = 0.3$, color range: -0.03 to 0.03 a.u.). AIM critical points have been highlighted for clarity: $(3, -1)$ in blue, intermolecular $(3, +1)$ in green and intramolecular $(3, +1)$ in red.

points corresponding to hydrogen bonds and two $(3, +1)$ points. NCI isosurfaces recover AIM results. They highlight the two strong and localized hydrogen bonds (N–H$\cdots$H and NH$\cdots$O) with blue and disk-shaped isosurfaces, and weaker dispersive interactions around the $(3, +1)$ critical points and the non coventional C–H$\cdots$O hydrogen bond. This hydrogen bond is weaker and embedded into the same NCI isosurface as the dispersive interaction. More quantitative results will be shown in the coming sections: NCI critical points recover AIM results, but also ELF localization and even very weak interactions not revealed by AIM.

6.4   Critical Points

The critical points of $s(\mathbf{r})$ and $t_{\mathrm{bose}}(\mathbf{r})$ are identical both in location and in nature. Thus, we will limit to the analysis of the critical points of t_{bose}.

In one direction, critical points appear at those points in space where the gradient cancels out, hence, we want to analyze those points where $\boldsymbol{\nabla} t_{\mathrm{bose}}(\mathbf{r}) = \mathbf{0}$ with

$$\boldsymbol{\nabla} t_{\mathrm{bose}}(\mathbf{r}) = \frac{1}{4C_F \rho(\mathbf{r})^{5/3}} \boldsymbol{\nabla}\rho(\mathbf{r}) \left[\frac{\nabla^2 \rho(\mathbf{r})}{\rho(\mathbf{r})} - \frac{4}{3}\frac{(\boldsymbol{\nabla}\rho(\mathbf{r}))^2}{\rho(\mathbf{r})^2} \right], \qquad (6.20)$$

From this expression we may differentiate two different situations where $\boldsymbol{\nabla} t_{\mathrm{bose}}(\mathbf{r})$ is cancelled:

1. AIM-CPs: CPs of $\rho(\mathbf{r})$, for which $\boldsymbol{\nabla}\rho(\mathbf{r}) = \mathbf{0}$.

2. Non-AIM-CPs: CPs for which

$$\frac{\nabla^2\rho(\mathbf{r})}{\rho(\mathbf{r})} - \frac{4}{3}\frac{(\boldsymbol{\nabla}\rho(\mathbf{r}))^2}{\rho(\mathbf{r})^2} = \mathbf{0} \tag{6.21}$$

These CPs are related to the Lewis picture (lone pairs, shells) and very weak interactions that cannot be revealed by the gradient of the electron density.

6.4.1 AIM-CPs

AIM-CPs owe their name to QTAIM, which focuses on critical points of $\rho(\mathbf{r})$. At these CPs $\nabla^2 t_{\mathrm{bose}}(\mathbf{r})$ takes the following form

$$\nabla^2 t_{\mathrm{bose}}(\mathbf{r}) = \frac{1}{4C_F}\frac{(\nabla^2\rho(\mathbf{r}))^2}{\rho(\mathbf{r})^{8/3}}, \tag{6.22}$$

As pointed out above, these positions correspond to minima of $t_{bose}(\mathbf{r})$ and $s(\mathbf{r})$. Since $s \geq 0$, s attains zero in AIM-CPs. This can happen in many different QTAIM casuistics which we detail below.

6.4.1.1 *Attractive localized BCP*

The easiest cases are those where we have a localized attractive interaction identified by the presence of a BCP ($\boldsymbol{\nabla}\rho(\mathbf{r}_{\mathrm{bcp}}) = \mathbf{0}$) clearly associated with a two-body interaction. A typical example is in conventional hydrogen bonds, like the one shown in Figure 6.4 for the hydrogen bond in the water dimer.

These interactions show a disk-shaped surface (Figure 6.4 bottom). This shape is characteristic of localized interactions, highlighting that it mainly takes place between a pair of atoms, O and H in this case. According to QTAIM, this is a BCP (one positive eigenvalue, λ_3, and two negative ones, λ_1 and λ_2), yielding a $(3, -1)$ CP with negative Laplacian. The electron density and the sign of λ_2 are used to identify this interaction type as attractive through the magnitude $\mathrm{sign}(\lambda_2)\rho \ll 0$. This is related to the peak appearing in the negative part of the plot in Figure 6.4 top (note that in many cases a smaller peak on the opposite site of the $s(\rho)$ plot accompanies the main one).

6.4.1.2 *Delocalized Attractive Interactions*

Dispersion interactions are due to time-dependent perturbations of the electron density and imply correlations among distant electrons.

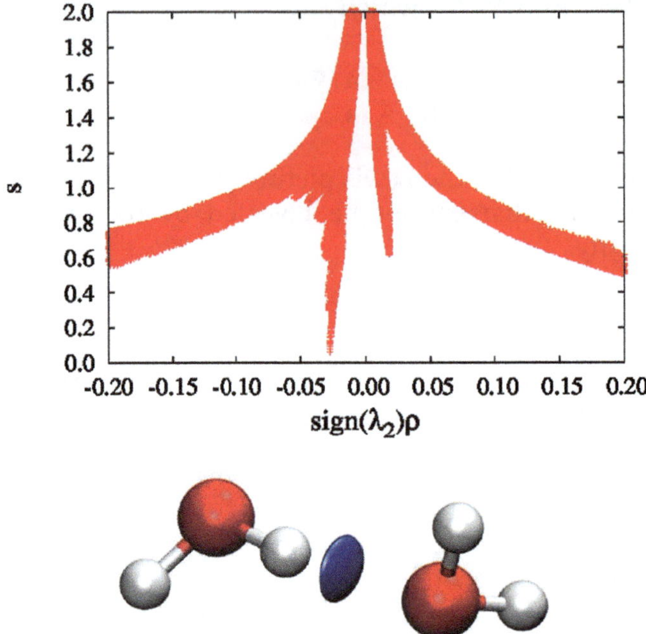

Figure 6.4 2D (top) and 3D (bottom) NCI plots of water dimer. The gradient isosurfaces ($s = 0.5$ a.u.) are colored on a BGR scale according to the sign(λ_2)ρ over the range -0.03 to 0.03 a.u.

Therefore, they cannot be correctly taken into account by ground state adiabatic methods, such as standard DFT theory, where the exchange-correlation potential is estimated on the basis of a finite number of static electron density derivatives.[51,52] Moreover, dispersion interactions are inherently non-local in nature, as they correlate the overall charge density distributions of individual molecular moieties to each other.

However, they leave an easy-to-identify trace on the electron density map. Van der Waals dominated compounds show very flat densities, with very low values of ρ and positive Laplacians. It should be noted that these interactions are not clearly pictured by QTAIM. BCPs are by definition pair interactions. Remember the benzene dimer example in Table 6.1, where the stacking was represented by two C-to-C interactions. BCPs do not give an intuitive picture of the delocalized and additive nature of dispersive interactions.

The flatness of the electron density profile is probably the easiest way to identify the presence of these interactions. This is easily cast by NCI, which yields characteristic large and flat NCI isosurfaces (see

Figure 6.5 bottom). According to the above color coding, we are now in the middle of the scale ($\rho \simeq 0$), so surfaces appear in green. From the electron density point of view, the different critical points (bonds, rings are cages) are extremely interrelated and embedded into the same NCI isosurface. This highlights that they all have a chemical meaning as a whole and that the van der Waals interaction takes place between monomers and not single atoms. From a topological point of view, this is related with the fact that all the density critical points in the NCI surface have a very low persistance, *i.e.* they have very similar densities.

A warning should be added here. NCI highlights atomic contacts, and not the energetic nature of dispersive interactions. The presence of different critical points yields an oscillating sign(λ_2)ρ, sometimes positive, sometimes negative (see Figure 6.5 top), which should not be taken as an indication of repulsion due to the fact that it oscillates very close to zero.

6.4.1.3 Repulsive Interactions

As reminded by Bader,[53–55] there are no net attractive or repulsive forces acting on a field-free quantum system at equilibrium. In this case, the overall balance of the quantum-mechanical Ehrenfest and Feynman forces (acting respectively on electrons and nuclei) is exactly zero.

However, there are known situations in chemistry where there are local hindrances that render the molecule very unstable and reactive. For an accurate characterization of non-covalent interactions, these steric clashes cannot be disregarded.

Let's look at a typical example in organic chemistry: bicyclo[2,2,2]octene (Figure 6.6). NCI features typically appear in regions of steric clash like the one in the center of bicyclooctene. They are typically related to the clash of several atoms, which yield ring or cage critical points in the density. These points, classified as $(3, +1)$ and $(3, +3)$, respectively, can be differentiated from BCPs, which are attractive, by the sign of λ_2. The reader now understands the interest in using the quantity sign(λ_2)ρ: attractive interactions appear at negative values whereas repulsive ones appear at the positive end of the scale. This explains the red color within the BGR scheme in Figure 6.6 bottom.

It should be noted that ring and cage critical points have been rarely used in the literature to conceal clashes, where most of the attention has been paid to bonded contacts as revealed by BCPs.

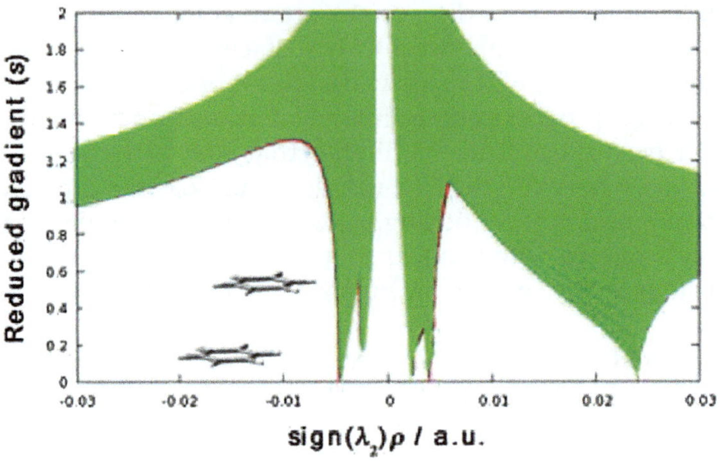

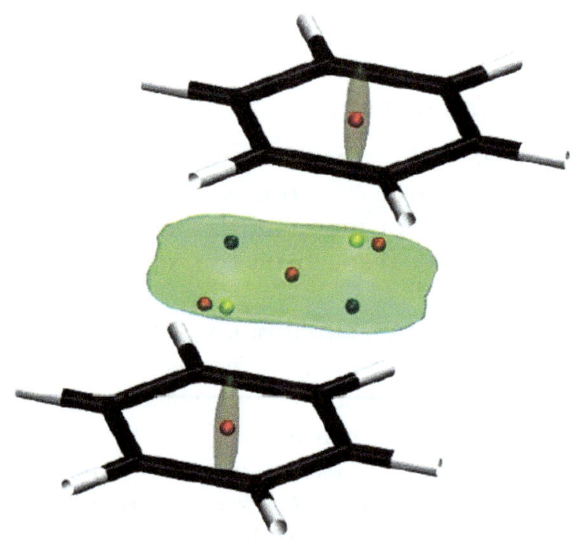

Figure 6.5 2D (top) and 3D (bottom) NCI plots of benzene dimer. The gradient isosurfaces ($s = 0.6$ a.u.) are colored on a BGR scale according to the sign(λ_2)ρ over the range -0.04 to 0.04 a.u. Critical points of the density have been included for comparison.
Reprinted with permission from ref. 31. Copyright (2011) American Chemical Society.

However, among non-covalent interactions, steric clashes can be even more relevant than attractive ones. They are crucial when determining conformational stability, enantiomeric selectivity and other subtle reactivities. Looking then at all critical points becomes of utmost importance.

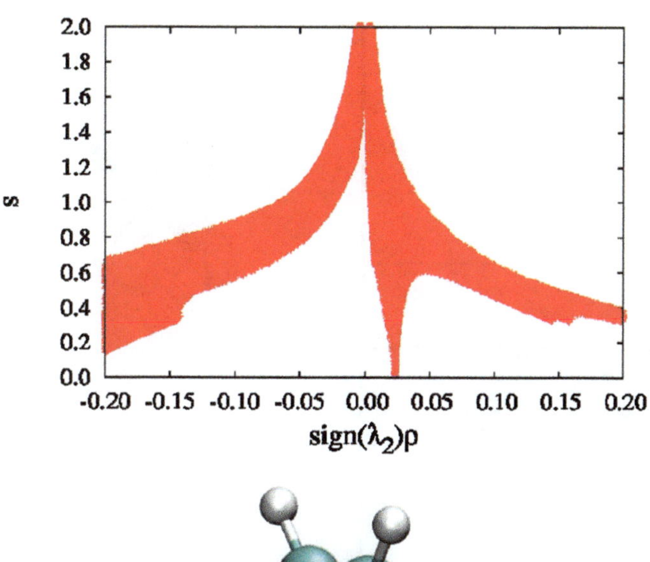

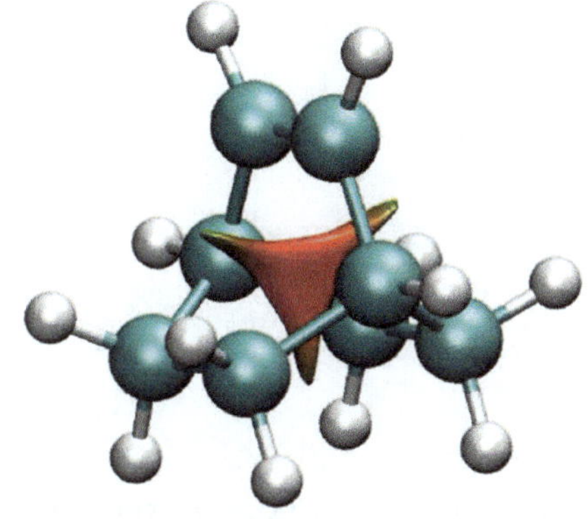

Figure 6.6 2D (top) and 3D (bottom) NCI plots of bicyclooctene. The gradient isosurfaces ($s = 0.5$ a.u.) are colored on a BGR scale according to the sign(λ_2)ρ over the range -0.03 to 0.03 a.u.

6.4.1.4 Mixed Example

Let's now put all the above considerations together. Figure 6.7 shows the results for the phenol dimer. This dimer is a hydrogen-bonded complex that also exhibits non-bonding interactions within each benzene ring and a stacking interaction between the benzene rings. We thus have the three main types of interactions.

Characteristic densities of van der Waals interactions are much smaller than densities at which hydrogen bonds and steric clashes appear. These two interaction types, steric clashes and hydrogen bonds, span similar density ranges, as can be seen in Figure 6.7 top left. This is illustrated in Figure 6.7 bottom left which shows a modification of the $s(\mathbf{r})$ plot, such that the ordinate is now sign(λ_2)ρ.

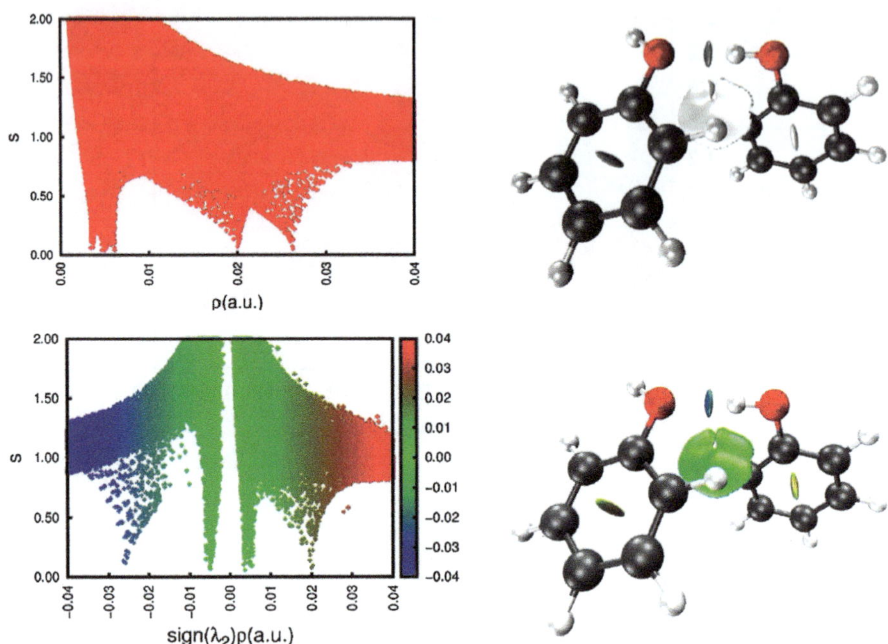

Figure 6.7 NCI analysis for the phenol dimer. (Top) $s(\mathbf{r})$ *versus* ρ diagram (left) and $s = 0.5$ isosurface (right). (Bottom) $s(\mathbf{r})$ *versus* $\mathrm{sign}(\lambda_2)\rho$ diagram (left) and $s(\mathbf{r}) = 0.5$ isosurface coloured over the range -0.04(blue) $< \mathrm{sign}(\lambda_2)\rho < 0.04$(red).

When the Hessian eigenvalue is considered, the different nature of these interactions is made clear: the benzene-ring interactions remain at positive values, whereas the hydrogen bond now lies at negative values, *i.e.* within the attractive region. The NCI spikes nearest zero density correspond to weakly-attractive dispersion interactions between the phenyl rings.

The 3D spatial visualisation of the non-covalent interactions as defined above is done using the data from the 2D plots as input to construct 3D plots composed of reduced density gradient isosurfaces. In a nutshell, a cut-off value of s close to zero, typically $s < 0.5$ is chosen in order to recover all the non-covalent interactions in the systems, *i.e.* all the spikes in the 2D plots. The corresponding reduced density gradient isosurfaces give rise to closed domains in the molecular space which highlight the spatial localisation of the interactions within the system. Since 3D isosurfaces are, by definition, regions of low reduced gradient, the density is nearly constant within them.

The density oriented by the sign of λ_2 is further used (as in the 2D plot). A RGB (red-green-blue) colouring scheme is chosen to rank interactions, where red is used for destabilising interactions, blue for stabilising interactions and green for delocalised weak interactions.

Figure 6.7 bottom right highlights 3D NCI surfaces along with the BGR coloring in the $s(\mathbf{r})$ to ease the interpretation of the 3D graph (compare with non-colored isosurfaces in Figure 6.7 top right). The hydrogen bond appearing in between alcohols appears in blue, the steric repulsion in red and the van der Waals interaction in green. The latter appears at both positive and negative λ_2, as can be seen in the 2D graph in Figure 6.7 bottom left. This color code has been applied throughout the whole chapter.

6.4.1.5 Controversies

Several controversies have shaken the QTAIM approach during the last decades. Let's review probably two of the most important ones with the NCI approach.

1. Repulsive BCPs?

   One of the most important questions that has struck the QTAIM community is the possibility that BCPs also appear in repulsive, and not only attractive, cases. The most typical example for this is biphenyl. The most stable structure of biphenyl is twisted ($\phi = 44°$). It lies 1.4 kcal/mol lower in energy than the planar geometry ($\phi = 0°$) and 1.6 kcal mol^{-1} lower in energy than the perpendicular geometry ($\phi = 90°$).[56,57] From a chemical point of view, this can be understood as a balance between aromaticity and steric repulsion. Whereas the perpendicular geometry breaks the delocalization of the π density over the two rings, the planar structure is subject to steric non-bonded repulsion between adjacent hydrogen atoms of the bridge. These two factors roughly balance, giving rise to stabilization of the twisted structure.

   This has caused much controversy with respect to the classical QTAIM approach.[56,57] In spite of showing a bond point between the ortho hydrogens, the parallel structure is not favored. On the one hand, Bickelhaupt *et al.* believe that the bond point in between hydrogens should be associated with steric repulsion instead of a bonding interaction.[56] Bader and coworkers, instead, think that the H–H contribution to energy is stabilizing, whereas the destabilization comes from elongation of the bridging C–C bond.[57]

The picture provided by NCI is able to reconcile both points of view. Figure 6.8 shows the NCI surfaces for orthogonal, twisted, and parallel biphenyl geometries. It can be seen that the absence of non-covalent interactions in the orthogonal arrangement (Figure 6.8a) is in agreement with it being the highest-energy conformer. The parallel conformer, instead, shows both stabilizing H–H interactions as well as strong steric clashes (Figure 6.8c). These do not arise from the H–H direction, which is indeed a direction of density accumulation, but rather from the whole non-bonding region, where H–C–C–H densities overlap. This destabilizing interaction can be softened by slightly twisting the benzene rings. At 40° both interactions diminish, giving rise to the balance that characterizes the stable conformation (Figure 6.8b). Thus, the H–H interaction alone is indeed slightly stabilizing, but a greater steric repulsion arises in the area due to the non-bonded overlap of densities. This is actually reflected in the increase of the C–C bonding distance from 2.808 bohr in the stable structure to 2.824 bohr in the parallel structure.

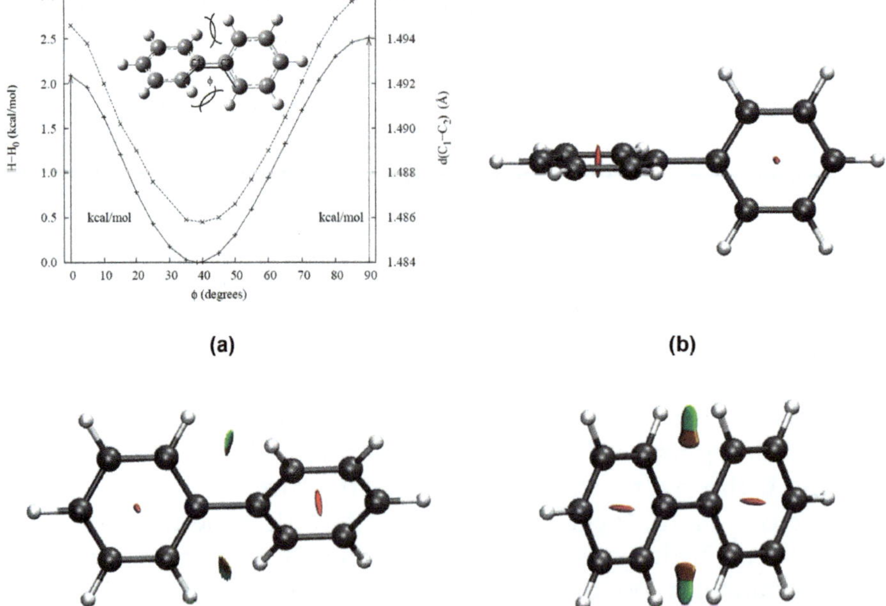

Figure 6.8   (a) Energetic profile along biphenyl rotation, along with changes in C_1–C_2 distance. (b–d) Biphenyl NCI surfaces at 90° (b), 40° (c) and 0° (d). $\phi = 0°$ and $\phi = 90°$ are TS, whereas $\phi = 40°$ is the minimum.

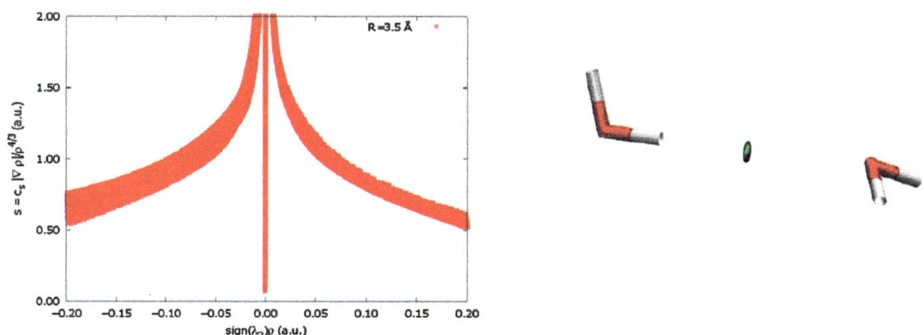

Figure 6.9 $s(\rho)$ diagrams (left) and NCI isosurface (right) in the non interacting region ($d = 3.5$ Å). The isosurface was generated for $s = 0.6$ au and $-0.04 < \text{sign}(\lambda_2)\rho < 0.04$ au.
Reprinted with permission from J. Contreras-Garcia, E. R. Johnson, W. Yang, "Analysis of Hydrogen-Bond Interaction Potentials from the Electron Density: Integration of Noncovalent Interaction Regions", J. Phys. Chem. A 115, 12983 (2011). Copyright 2016 American Chemical Society.

2. Infinite distance

One of the problems of associating interactions with critical points is that some critical points are imposed by symmetry constraints, beyond their physico-chemical meaning. The most common example of such a case is the stretching of a molecule to infinity. Even though at very large distances molecules will not interact with one another, symmetry imposes the appearance of at least one critical point between them. The use of surfaces and volumes instead of critical points to identify interactions naturally overcomes this limitation.

Figure 6.9 shows $s(\rho)$ diagrams and NCI isosurfaces for the water dimer at very long separations ($d = 3.5$ Å). At this distance, the NCI region reduces to the BCP: the $s(\rho)$ peak is extremely sharp, such that the NCI surface and critical point basically coincide. The stabilization of the complex does not extend to other regions of space beyond the critical point itself. Since the volume of the NCI region can be associated with the degree of stabilization, NCI is able to supersede symmetry constraints that compel the appearance of a critical point and identify that the monomers are not interacting at infinite separation.

6.4.2 Non-AIM-CPs

In some cases, experimental evidence has been found for interactions that are not revealed by AIM BCPs.[58] This is most typically the case in intramolecular interactions, where the geometry is constrained and the BCP has typically collapsed with a ring critical point. NCI is able to reveal these points. As a reminder (eqn 6.21), they appear in one dimension when:

$$\frac{\nabla^2 \rho(\mathbf{r})}{\rho(\mathbf{r})} - \frac{4}{3} \frac{(\nabla \rho(\mathbf{r}))^2}{\rho(\mathbf{r})^2} = 0 \tag{6.23}$$

Thus, approximating along the radial direction, they involve a balance between $\nabla^2 \rho(\mathbf{r})/\rho(\mathbf{r})$ and a term proportional to the local von Weizsäcker kinetic energy density $8\tau_w(\mathbf{r})/\rho(\mathbf{r}) = (\nabla \rho(\mathbf{r}))^2/\rho^2(\mathbf{r})$. It is also interesting to note that in these points s does not attain zero and that $\nabla^2 \rho(\mathbf{r})$ is positive

$$\nabla^2 \rho(\mathbf{r}) = \frac{4}{3} \frac{(\nabla \rho(\mathbf{r}))^2}{\rho(\mathbf{r})} > 0 \tag{6.24}$$

which means that these critical points appear in regions of electron density depletion.

Two well differentiated cases can be identified: intraatomic and interatomic. When appearing in intraatomic regions, they reveal the shell structure, whereas they identify weak chemical bonding to which QTAIM is blind in the interatomic case.

6.4.2.1 Intraatomic

The variables in eqn (6.23) for the non-AIM-CPs are also involved in the one-electron potential $(\text{OEP}(\mathbf{r}))$[41,59]

$$\text{OEP}(\mathbf{r}) = \frac{1}{4} \left[\frac{\nabla^2 \rho(\mathbf{r})}{\rho(\mathbf{r})} - \frac{1}{2} \left(\frac{\nabla \rho(\mathbf{r})}{\rho(\mathbf{r})} \right)^2 \right], \tag{6.25}$$

$$= \frac{\nabla^2 \sqrt{\rho(\mathbf{r})}}{2\sqrt{\rho(\mathbf{r})}}, \tag{6.26}$$

Regions of negative $\text{OEP}(\mathbf{r})$ have been identified as classically allowed regions, in the sense that the kinetic energy takes positive values. Conversely, the regions where $\text{OEP}(\mathbf{r})$ attains positive values, have negative kinetic energy, and therefore have been identified as classically forbidden regions. This separation has been used to identify atomic shells and bonding regions as the classically allowed ones. Transition between both regions occurs when $\text{OEP}(\mathbf{r})$ cancels. At these points, the ratio between $\nabla^2 \rho(\mathbf{r})/\rho(\mathbf{r})$ and $(\nabla \rho(\mathbf{r}))^2/\rho^2(\mathbf{r})$ is equal to 1/2,

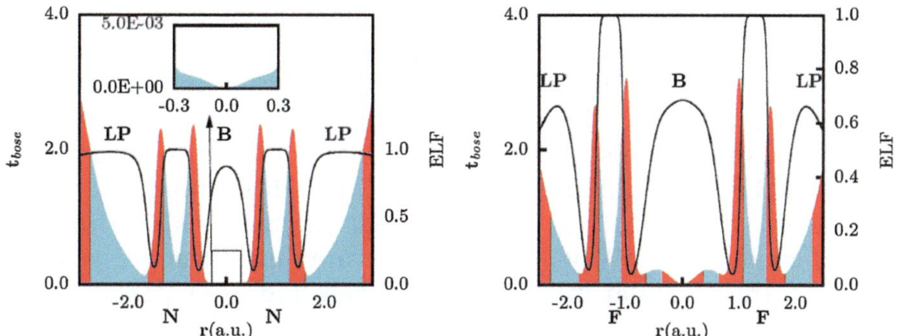

Figure 6.10 $t_{bose}(\mathbf{r})$ along with ELF values (solid black line) for N_2(a) and F_2(b). Negative (classically allowed) and positive (classically forbidden) regions of OEP are displayed as cyan and red-colored areas respectively. Labels B and LP stand for bond and lone pair ELF basins respectively.

whereas it is 4/3 when there is a non-AIM-CP (*i.e.* when eqn (6.23) is satisfied). Thus, any non-AIM-CP of $t_{bose}(\mathbf{r})$ is anticipated by a zero of OEP($\mathbf{r}$), and therefore by a transition from a classically allowed to a forbidden region.

Figure 6.10 displays $t_{bose}(\mathbf{r})$, $t_{LOL}(\mathbf{r})$ and $s(\mathbf{r})$ along the internuclear axis for F_2 and N_2. The origin was set at the BCP. $t_{bose}(\mathbf{r})$ and $t_{LOL}(\mathbf{r})$ differentiate the core, lone-pairs and interatomic bonding regions as minima separated by maxima. As revealed by ELF maxima, these minima correspond to regions of large electron pair localisation. Nuclear and bond critical points of $\rho(\mathbf{r})$ are identified as zeros of $t_{bose}(\mathbf{r})$. Conversely lone pairs are not revealed by critical points of $\rho(\mathbf{r})$, but by critical points of the Laplacian of the electron density. $t_{bose}(\mathbf{r})$ shows minima at such positions driven by the non-AIM-CP condition (eqn (6.23)), following thereby a transition from a classically forbidden region to a classically allowed region, as may be seen in Figure 6.10.

It is interesting to note that the BCP for F_2 is localised at a region of positive OEP. It is well known that F_2 exhibits a positive value of the Laplacian of the electron density at the BCP, being thereby identified as a region of electron depletion. Because the signs of OEP and $\nabla^2\rho(\mathbf{r})$ are the same at CPs of $\rho(\mathbf{r})$, the BCP for F_2 is localized at a classically forbidden region. This characterizes the charge-shift bond present in F_2.

6.4.2.2   Interatomic

Non-AIM-CPs can also appear in interatomic regions, that is, in regions of chemical interactions. This means that we are looking at interactions without a BCP (or any other sort of CP) from the QTAIM point of view. In other words, the absence of a BCP should not necessarily be considered evidence of the absence of interaction.

These cases have also shaken the QTAIM community for a while, and can be easily identified within the NCI approach. Most commonly, they occur in intramolecular interactions due to geometric constraints. We will show two examples: attractive and repulsive intramolecular interactions not detected by QTAIM.

6.4.2.3 Attractive

To illustrate the occurance of NCI attractive features in the absence of BCP, we consider a series of 1,n-alkanediols: 1,2-ethanediol (**ED**), 1,3-propanediol (**PD**) and 1,4-butanediol (**BD**). Experimental data suggest that the strength of the intramolecular hydrogen bonding interactions in these three molecules increases as the alkane chain length increases and the OH$\cdots$O angle becomes more linear. For example, the OH-stretching vibrational mode of the hydroxyl group involved in hydrogen bonding becomes progressively more red shifted in the fundamental and overtone regions, with a corresponding increase in intensity in the fundamental region and decrease in intensity in the first overtone region from **ED** to **BD**. However, when QTAIM theory is applied to this 1,n-alkanediol series, only **PD** and **BD** are found to exhibit BCPs but not **ED**.

Figure 6.11 illustrates the results for the diol series. The image for **BD** (Figure 6.11 bottom) corresponds to a typical strong hydrogen bond within the NCI framework. Blue in color and very disk-shaped, that is, a very localized interaction, which expands from the BCP (red little sphere). This interaction becomes weaker in **PD**, and even more so in **ED**. In the latter case the BCP is not present anymore, but there are critical points of the reduced density gradient, *i.e.* non-AIM-CPs, that highlight the presence of the interaction in agreement with the experiment.

Figure 6.12 highlights the fact that these non-AIM-CPs are easily identified because they appear for $s \neq 0$. In all three molecules, two spikes are obtained for the interaction, with one spike at negative λ_2 corresponding to an attractive hydrogen bonding interaction and a second spike at positive λ_2 corresponding to a repulsive steric interaction due to ring formation. Differences arise in the minimum values of s attained in the graphs: whereas BD and PD attain zero (AIM-CPs), ED does not (non-AIM-CP).

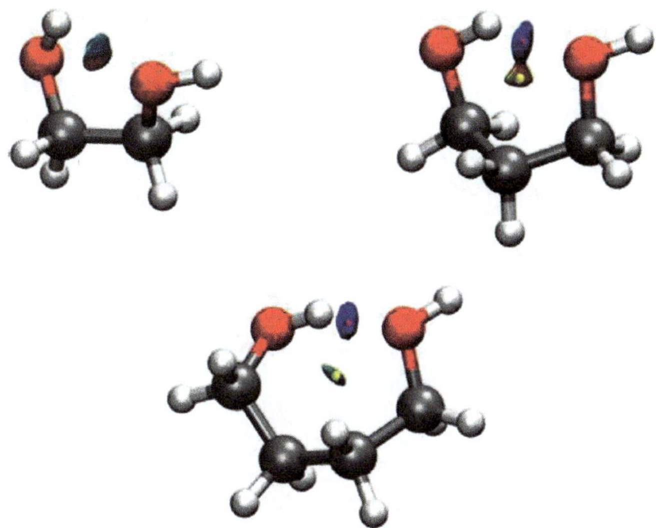

Figure 6.11   NCI isosurfaces for **ED** (top left), **PD** (top right), and **BD** (bottom): $s = 0.5$ and a color scale from $-0.02 < \text{sign}(\lambda_2)\rho(\mathbf{r}) < +0.02$ au. Small spheres represent critical points: BCPs in purple and RCPs in yellow. A continuous change is observed in the isosurfaces from **ED** to **BD**. Reprinted with permission from J. Chem. Theory Comp. 9, 3263 (2013). Copyright 2016 American Chemical Society.

6.4.2.4   *Repulsive*

Figure 6.13a shows the $s(\rho)$ diagram for the neopentane molecule, $C(CH_3)_4$. A deviation from the $\rho^{-1/3}$ decay appears at low density and low s, on the positive side of the diagram (*i.e.* the non-bonding region). However, the peak does not reach $s = 0$, which means that no density critical point (ring or cage, since we are facing the repulsive region) is associated with this interaction.

In real space, the points giving rise to this peak appear between the methyl groups and correspond to the expected regions of steric hindrance. In Figure 6.13c van der Waals spheres have been added to highlight the localization of NCI surfaces in the interacting region. Since there is no density critical point, this text-book intramolecular steric hindrance can be cast with NCI, but not with QTAIM.

6.4.3   Summary

Putting together the results from the previous sections, we have shown that the reduced density gradient is able to reveal

- atomic structure (shells, lone pairs) like in ELF
- covalent bonding, like both ELF and QTAIM

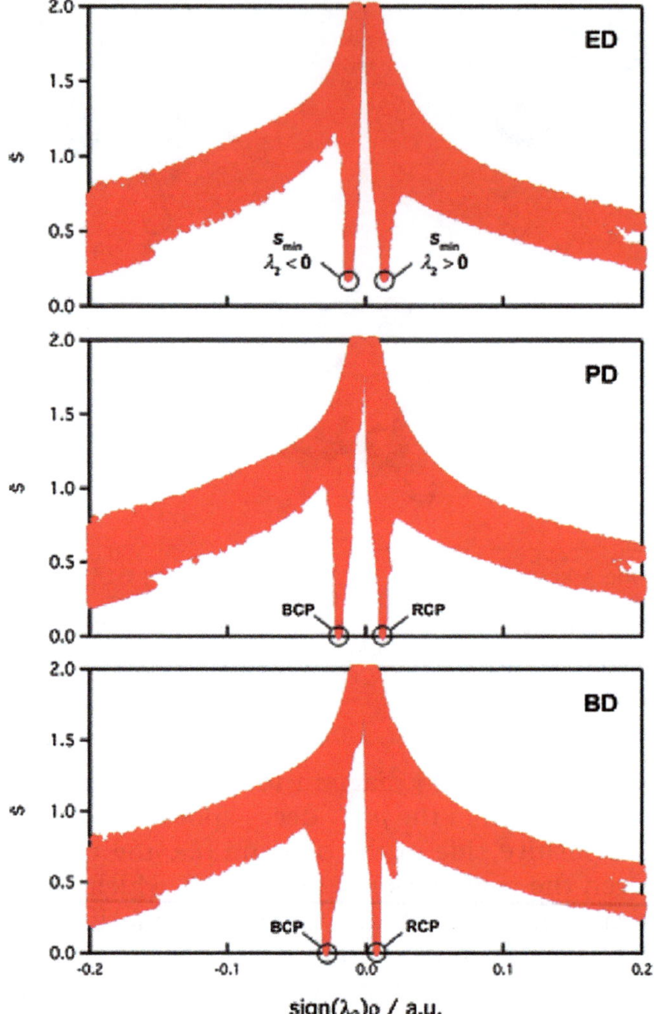

Figure 6.12 Plot of the reduced density gradient s and sign(λ_2)$\times\rho$ for **ED** (top), **PD** (middle), and **BD** (bottom). The density features are qualitatively the same with a continuous change observed from **ED** to **BD**.
Reprinted with permission from J. Chem. Theory Comp. 9, 3263 (2013). Copyright 2016 American Chemical Society.

- non covalent bonding, like QTAIM, with a visual differentiation of delocalized and localized interaction; as well as a clear cast to steric clashes, not usually analyzed within QTAIM
- and finally weak intramolecular bonding that cannot be detected with QTAIM.

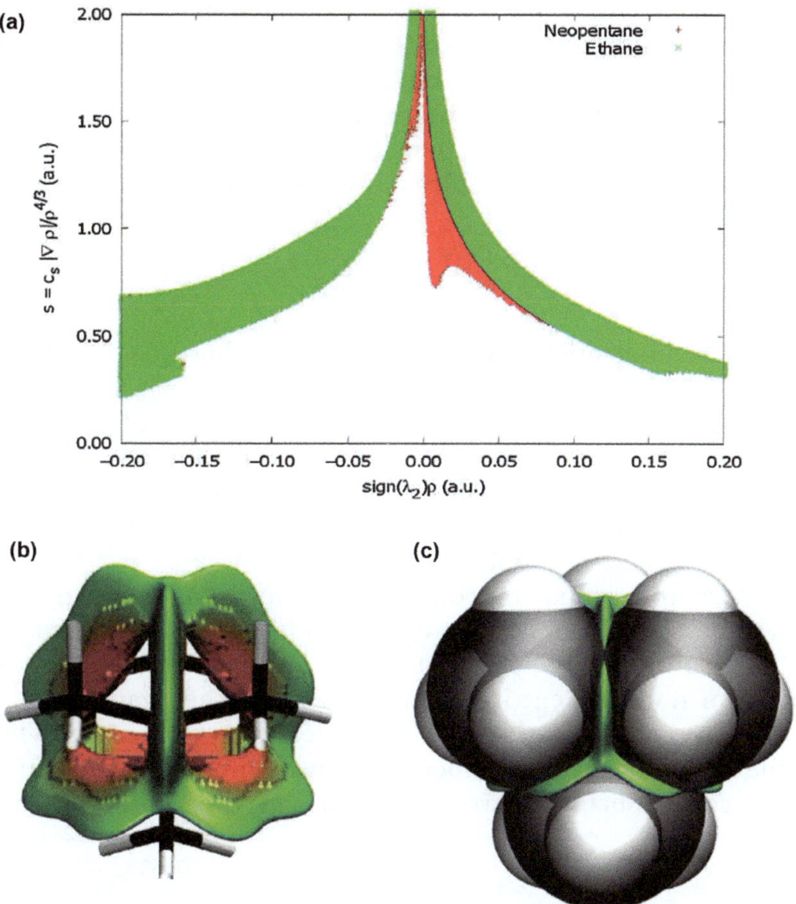

Figure 6.13 Plot of $s(\rho)$ (a) and NCI isosurfaces (b,c) for neopentane, $C(CH_3)_4$. Isosurfaces with $s = 1.2$ au were generated for the region below the lower edge of the ethane curve. A scale $-0.03 < \text{sign}\,(\lambda_2)\rho < 0.03$ au was used to color the isosurfaces. In (c), van der Waals spheres have been added to highlight the steric nature of the interaction.
Reprinted with permission from ref. 31. Copyright (2011) American Chemical Society.

This makes the reduced density gradient a nice tool to analyze molecular aggregates, which will be the aim of the next section.

6.5 Applications to Solid State

6.5.1 From the Resolved Structure

Understanding of non-covalent interactions is crucial for the comprehension of the 3D structure and, thus, of the activity of biosystems.[60,61] However, the calculation of the electron density in these systems is totally impossible from quantum mechanical calculations. Approximations need to be sought.

Densities are stable to such an extent that they are already contained in the sum of atomic densities, ρ^{at}. The resulting molecular density, also known as promolecular density, ρ^{pro}, is then given by:

$$\rho^{pro}(\mathbf{r}) = \sum_i \rho_i^{at}(\mathbf{r}), \qquad (6.27)$$

Promolecular densities obviously lack relaxation; however, the promolecular densities are extremely useful in biomolecular systems, such as proteins or DNA. Because the calculation of the electron density in these systems becomes extremely computationally expensive, the promolecular density becomes an attractive option: non-covalent interactions can be analyzed with only the molecular geometry required as input.

As an example, let us now consider the interaction between a ligand and a protein active site. The low-gradient isosurface for a tetracycline inhibitor bound to the tetR protein in Figure 6.14 shows a complex network of non-covalent interactions between the ligand and active site. When analyzing non-covalent interactions in protein–ligand complexes, it is usually assumed that these interactions are due to a specific contact between two atoms.[21] However, it is clearly seen in

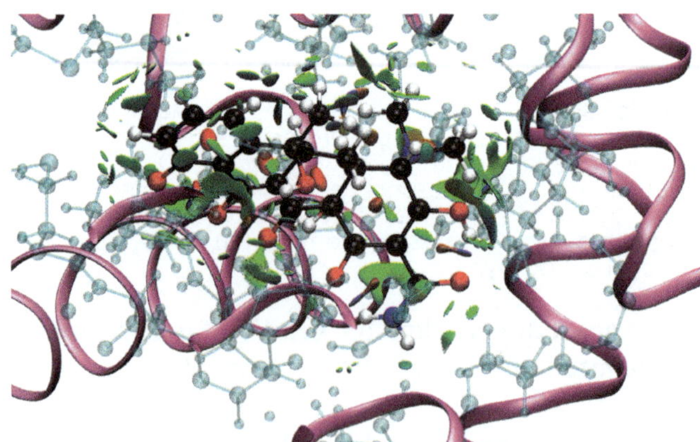

Figure 6.14   Gradient isosurfaces ($s = 0.35$) for interaction between the tetR protein and tetracycline inhibitor. Surfaces colored in the sign(λ_2)ρ range from -0.06 to $+0.05$ au.
Adapted with permission from E. R. Johnson, S. Keinan, P. Mori-Sanchez, J. Contreras-Garcia, A. J. Cohen, and W. Yang, "Revealing Noncovalent Interactions", J. Am. Chem. Soc. 132, 6498 (2010). Copyright 2016 American Chemical Society.

Figure 6.14 that this assumption is only partially correct. Hydrogen bonds, such as those between the tetracycline amine groups and two water molecules, are directional and specific. Conversely, van der Waals, dipole–dipole, and hydrophobic interactions, such as those between the tetracycline and the Leu61, Val91, Ile136, and Val166 residues, are not atom-specific and occupy broader regions in space. The figure reveals some steric clashes (orange and red regions of the isosurface) that must be offset by stronger, attractive interactions to give binding in this crystal structure. A ligand "fits" the geometry of the active site, and the interaction energy between the ligand and protein is comprised of many small contributions. When trying to design a new ligand to fit a specific active site, one should consider all such interactions, which can be easily calculated directly from the geometric position of atoms (*e.g.* xyz, pdb files).

6.5.2   From Structure Factors

From the experimental point of view, one usually counts with structure factor data rather than calculated densities. We have implemented NCI for experimental electron densities reconstructed from structure factors by multipole refinements using the XD2006 program package.[62]

In order to highlight the advantages of using NCI in molecular aggregates we will analyze the interactions in the benzene crystal, where conventional and non conventional hydrogen bonds appear. Diffraction data were all collected at low temperature (100 K) (see ref. 63 for refinement details). The crystal packing of aromatic and polyaromatic compounds was largely investigated by Desiraju and Gavezzotti.[64] In particular, the typical packing motif in crystalline benzene was classified as a herringbone structure. This packing was found to maximize the number of C–H$\cdots\pi$ and CH$\cdots$C interactions, and this feature is clearly reflected by NCI. For comparison, Table 6.2 reports the topological properties of the electron density BCPs. In this case, we have used a color scale between -0.03 (red) to 0.035 (violet) au.

Table 6.2   Geometrical and topological (BCP) data for selected intermolecular CH$\cdots$C and CH$\cdots$H interactions in the benzene crystal.

Bond	type	ρ_{BCP} (e/Å^{-3})	$\nabla^2\rho_{\mathrm{BCP}}$ (e/Å^{-5})	dH$\cdots$C (Å)
C3$\cdots$H2	CH$\cdots\pi$	0.030(4)	0.256(1)	3.065
C1$\cdots$H3	CH$\cdots$C	0.042(6)	0.371(2)	2.862

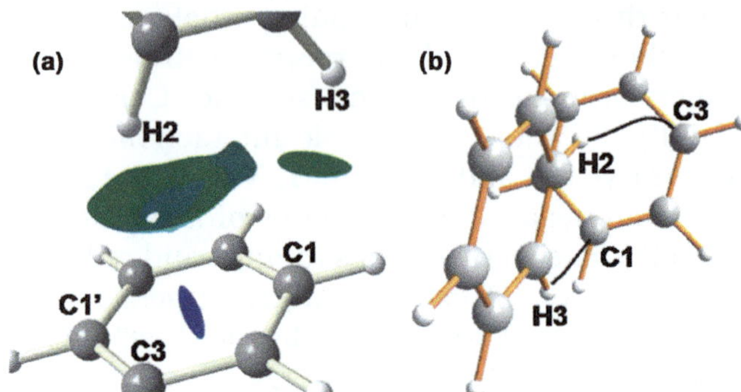

Figure 6.15   (a) NCI isosurfaces for a molecular pair extracted from the benzene crystal. (b) Intermolecular bond paths for the same molecular pair. Reproduced from G. Saleh, C. Gatti, L. Lo Presti and J. Contreras-García, Revealing Non-covalent Interactions in Molecular Crystals through Their Experimental Electron Densities, *Chemistry: A European Journal*, John Wiley and Sons, © 2012 WILEY-VCH Verlag GmbH & Co. KGaA, Weinheim.

Note that this color code convention does not correspond to the common BGR one.

Figure 6.15b shows the QTAIM results for a pair of benzenes extracted from the crystal. Two BCPs link **C3** and **C1** with **H2** and **H3** atoms, respectively. From the QTAIM perspective, this should indicate two well-defined CH···C interactions. However, a closer look shows that this is not the case for the **H2**···**C3** interaction, where the bond path is significantly bent. Moreover, the **H2** atom is roughly equidistant from all the carbon atoms belonging to the other molecule of the pair: the H···C distances range from 3.065 Å (**H2**···**C3**) up to 3.109 Å (**H3**···**C1**). These features are typical of a system showing a CH···π interaction,[65] and suggest that **H2** is involved in a CH···π attractive contact.

This is extremely easy to identify from NCI. A large isosurface (Figure 6.15a) almost entirely covering the hydrocarbon ring of the other molecule is obtained. This picture clearly clashes with the one provided by the BCP analysis, which is found to favour the interaction of the H with just one single atom of the ring. It is worth noting that the **H2**···**C3** NCI isosurface looks to some extent similar to the broad surfaces associated with dispersive interactions. In fact, CH···C interactions are essentially dispersive in nature.[65]

The NCI isosurface of the **H2**···**C1** interaction looks much smaller, disk-shaped, and only slightly negative. All these features comply with a conventional, very weak HB, as anticipated by the largely asymmetric location of **H3** with respect to the ring atoms of the other molecule in the pair and by the almost straight bond path and density properties at the BCP (Table 6.2). The NCI and bond critical point pictures nicely match in this case.

All in all, NCI provides a fast an intuitive picture of conventional and non conventional hydrogen bonds, which can be obtained not only from calculations or the geometry alone, but also from X-ray densities.

6.5.3 Using NCI to understand the properties of molecular aggregates

Pressure is a thermodynamic variable that allows tuning of inter-atomic distances and consequently is a powerful tool in the study of atomic interactions and the connectivity of different molecular units. It also offers the possibility of trapping small atoms or molecules in structural voids, an option that has received a lot of interest from the scientific community in the last decade. Molecular trapping has important industrial applications that cover hydrogen storage[66] and CO_2 segregation from other organic compounds,[67] among others.

Arsenolite, As_4O_6, is composed of pseudotetrahedral units consisting of an arsenic atom surrounded by three oxygen ligands and a lone pair. Pseudotetrahedral units in arsenolite are configured in closed-compact adamantane-type As_4O_6 molecular cages (Figure 6.16 left) bonded together due to weak van der Waals

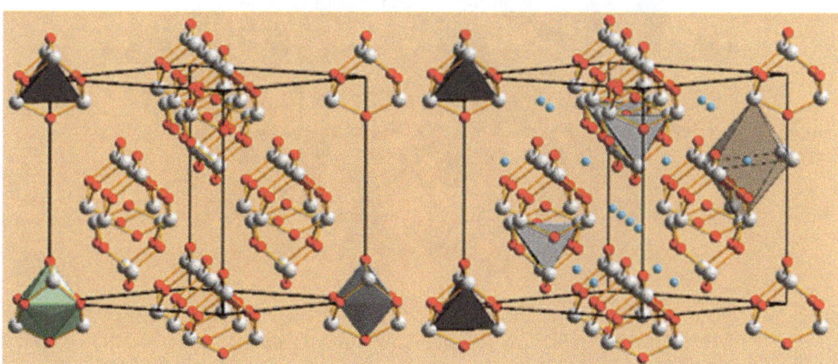

Figure 6.16 Detail of the cubic unit cell of arsenolite (As_4O_6) (left) and $As_4O_6 \cdot 2He$ (right). $As_4O_6 \cdot 2He$ is a new compound formed above 3 GPa by arsenolite with He atoms inserted into the voids. Big gray balls, medium-size red balls and small blue balls represent As, O, and He atoms, respectively.
Reprinted with permission from Juan A. Sans, Francisco J. Manjón, Catalín Popescu, Vanesa P. Cuenca-Gotor, Oscar Gomis, Alfonso Muñoz, Plácida Rodriguez-Hernandez, Julia Contreras-García, Julio Pellicer-Porres, Andre L. J. Pereira, David Santamaría-Pérez and Alfredo Segura, Physical Review B, "Ordered helium trapping and bonding in compressed arsenolite: Synthesis of $As_4O_6 \cdot 2He$, 93, 054102, 2016. Copyright 2016 by the American Physical Society."

forces with lone pairs pointing towards the exterior of the molecular unit (Figure 6.17 top).

When compressed, arsenolite suffers amorphization at around 15 GPa. This is not the case when compression takes place with He as a pressure transmitting medium. Recent results have shown that this

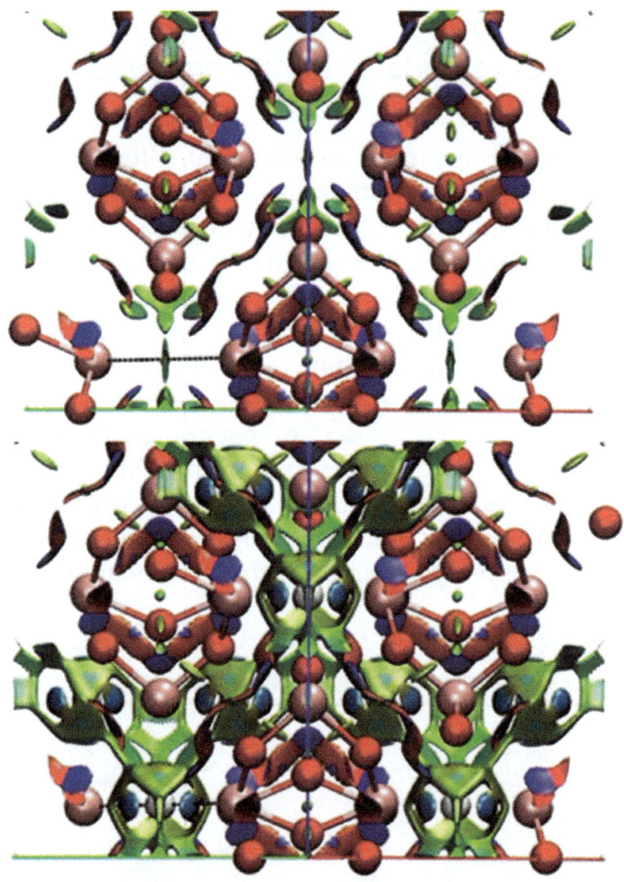

Figure 6.17 Isosurfaces of the reduced density gradient evidencing non-covalent interactions in As_4O_6 (top) and $As_4O_6 \cdot 2He$ (bottom) near 16 GPa. As expected for a noble gas, van der Waals (green) interactions around the He atoms are revealed. However, unexpectedly localized interactions (in turquoise blue) also make their appearance along the He-As interaction lines.
Reprinted with permission from Juan A. Sans, Francisco J. Manjón, Catalín Popescu, Vanesa P. Cuenca-Gotor, Oscar Gomis, Alfonso Muñoz, Plácida Rodriguez-Hernandez, Julia Contreras-García, Julio Pellicer-Porres, Andre L. J. Pereira, David Santamaría-Pérez and Alfredo Segura, Physical Review B, "Ordered helium trapping and bonding in compressed arsenolite: Synthesis of $As_4O_6 \cdot 2He$, 93, 054102, 2016. Copyright 2016 by the American Physical Society."

is due to an ordered helium being trapped in the voids of the structure (Figure 6.16 right).

Figure 6.17 (bottom) shows NCI when He is incorporated. Rather localized structural bonds between He and arsenic appear, forming a compound with stoichiometry $As_4O_6 \cdot 2He$. These bonds stabilize the behavior of the voids upon compression and explain the absence of amorphization.

This way, a simple simulation helps us understand the reason behind the new properties of our material, and even to uncover the possibility of strong He "non-inertivity", which can become very relevant in pressure induced experiments for void-containing samples.[68]

6.6 Conclusions

We have introduced a new tool based on the bosonic kinetic energy density which is able to reveal all features of chemical structure as intuitive isosurfaces. This tool recovers ELF and QTAIM results, but also provides a global approach to the electron density topology which enables us to overcome old problems and controversies, providing a semi-quantitative picture of delocalized interactions, steric clashes, and very weak intramolecular interactions.

In other words, this function merges previous topological approaches and even overcomes deficiencies identified within them. Thus, the user does not need to know in advance which bonding types/interactions are present in the solid when choosing the tool of analysis. The reduced density gradient is a tool that can reveal localization and chemical change at all levels (from covalent to non-covalent), which should help in the understanding of strong and weak interactions in solids, providing valuable insight in important processes in material science, from packing and self-assembly to adsorption, the formation of cocrystals, *etc.*

We have implemented NCI for electron densities derived from single-crystal X-ray diffraction data. NCI is shown to be a valuable descriptor also for experimental electron densities, able to highlight their different nature and strength. This approach could help us to more deeply understand the role of weak interactions in proteins using structural and experimental electron density. All in all, NCI is an ideal tool for the analysis of molecular aggregates and the understanding of their properties.

References

1. P. A. Kollman, Noncovalent interactions, *Chem. Rev.*, 1977, **10**, 365.
2. P. Metrangolo, F. Meyer, T. Pilati, G. Resnati and G. Terraneo, Halogen bonding in supramolecular chemistry, *Angew. Chem., Int. Ed.*, 2008, **47**, 6114–6127.
3. M. Brandl, M. S. Weiss, A. Jabs, J. Sühnel and R. Hilgenfeld, C-H...pi-interactions in proteins, *J. Mol. Biol.*, 2001, **307**, 357–377.
4. B. G. McGaughey, M. Gagné and A. K. Rappé, π-stacking interactions, *J. Biol. Chem.*, 1998, **273**, 15458–15463.
5. H. Fenniri, M. Packiarajan, K. L. Vidale, D. M. Sherman, K. Hallenga, K. V. Wood and J. G. Stowell, Helical rosette nanotubes: design, self-assembly, and characterization, *J. Am. Chem. Soc.*, 2001, **123**, 3854.
6. P. Kruse, E. R. Johnson, G. A. DiLabio and R. A. Wolkow, Patterning of vinyl-ferrocene on H-Si (100) via self-directed growth of molecular lines and stm-induced decomposition, *Nano Lett.*, 2002, **2**, 807.
7. S. S. Sheiko, F. C. Sun, A. Randall, D. Shirvanyants, M. Rubinstein, H. Lee and K. Matyjaszewski, Adsorption-induced scission of carboncarbon bonds, *Nature*, 2006, **440**, 191.
8. G. A. DiLabio, P. G. Piva, P. Kruse and R. A. Wolkow, Dispersion interactions enable the self-directed growth of linear alkane nanostructures covalently bound to silicon, *J. Am. Chem. Soc.*, 2004, **126**, 16048.
9. J. Cerniý and P. Hobza, Non-covalent interactions in biomacromolecules, *Phys. Chem. Chem. Phys.*, 2007, **9**, 5291–5303.
10. A. L. Lehninger, D. L. Nelson and M. M. Cox, *Principles of Biochemistry*, Worth Publishers, Inc, 2nd edn, 1993.
11. N. Krishnamoorthy, M. H. Yacoub and S. N. Yaliraki, A computational modeling approach for enhancing self-assembly and biofunctionalisation of collagen biomimetic peptides, *Biomaterials*, 2011, **32**, 7275–7285.
12. A. Dutta, A. D. Jana, S. Gangopadhyay, K. Kumar Das, J. Marek, R. Marek, J. Brus and M. Ali, Unprecedented π-π interaction between an aromatic ring and a pseudo-aromatic ring formed through intramolecular h-bonding in a bidentate Schiff base ligand: crystal structure and DFT calculations, *Phys. Chem. Chem. Phys.*, 2011, **13**, 15845.
13. A. Gavezzotti, Molecular Aggregation. Structure Analysis and Molecular Simulation of Crystals and Liquids, *IUCr Monographs on Crystallography n. 19*, Oxford University Press, 2007.
14. S. Keinan, M. A. Ratner and T. J. Marks, Molecular zippers–designing a supramolecular system, *J. Chem. Phys. Lett.*, 2004, **392**, 291–296.
15. G. R. Desiraju, Crystal Engeneering, *The Design of Organic Solids*, Elsevier, 1989.
16. G. M. Day, T. G. Cooper, A. J. Cruz-Cabeza, K. E. Hejczyk, H. L. Ammon, S. X. M. Boerrigter, J. S. Tan, E. Venuti, R. G. Della Valle, J. Jose, S. R. Gadre, G. R. Desiraju, T. S. Thakur, B. P. van Eijck, J. C. Facelli, V. E. Bazterra, M. B. Ferraro, D. W. M. Hofmann, M. A. Neumann, F. J. J. Leusen, J. Kendrick, S. L. Price, A. J. Misquitta, P. G. Karamertzanis, G. W. A. Welch, H. A. Scheraga, Y. A. Arnautova, M. U. Schmidt, J. van de Streek, A. K. Wolf and B. Schweizer, Significant progress in predicting the crystal structures of small organic molecules a report on the fourth blind test, *Acta Cryst. B*, 2009, **65**, 107–125.
17. G. N. Lewis, The atom and the molecule, *J. Am. Chem. Soc.*, 1916, **38**, 762.
18. B. Honig and A. Nicholls, Classical electrostatics in biology and chemistry, *Science*, 1995, **268**, 1144.
19. J. M. Word, S. C. Lovell, T. H. LaBean, H. C. Taylorand, M. E. Zalis, B. K. Presley, J. S. Richardson and D. C. Richardson, Asparagine and glutamine: using

hydrogen atom contacts in the choice of side-chain amide orientation, *J. Mol. Biol.*, 1999, **285**, 1711.

20. I. W. Davis, A. Leaver-Fay, V. B. Chen, J. N. Block, G. J. Kapral, X. Wang, L. W. Murray, W. B. Arendall III, J. Snoeyink, J. S. Richardson and D. C. Richardson, Molprobity: all-atom contacts and structure validation for proteins and nucleic acids, *Nucleic Acids Res.*, 2007, **35**, W375.
21. V. Sobolev, A. Sorokine, J. Prilusky, E. E. Abola and M. Edelman, Automated analysis of interatomic contacts in proteins, *Bioinformatics*, 1999, **15**, 327.
22. I. K. McDonald and J. M. Thornton, Satisfying hydrogen bonding potential in proteins, *J. Mol. Biol.*, 1994, **238**, 777.
23. R. D. Cramer III, D. E. Patterson and J. D. Bunce, Comparative molecular field analysis (comfa). 1. effect of shape on binding of steroids to carrier proteins, *J. Am. Chem. Soc.*, 1988, **110**, 5959.
24. A. D. Becke and K. E. Edgecombe, A simple measure of electron localization in atomic and molecular systems, *J. Chem. Phys.*, 1990, **92**, 5397.
25. S. Silvi and A. Savin, Classification of chemical bonds based on topological analysis of electron localization functions, *Nature*, 1994, **45**, 683.
26. R. F. W. Bader, A quantum theory of molecular structure and its applications, *Chem. Rev.*, 1991, **91**, 893.
27. C. F. Matta and R. J. Boyd, *The Quantum Theory of Atoms in Molecules*, Wiley-VCH, New York, 2007, pp. 1–34.
28. P. Hohenberg and W. Kohn, Inhomogeneous electron gas, *Phys. Rev.*, 1964, **136**(3B), B864.
29. E. R. Johnson, S. Keinan, P. Mori-Sanchez, J. Contreras-Garcia, A. Cohen and W. Yang, Revealing noncovalent interactions, *J. Am. Chem. Soc.*, 2010, **132**(18), 6498.
30. J. Contreras-García, E. Johnson, S. Keinan, R. Chaudret, J.-P. Piquemal, D. N. Beratan and W. Yang, Nciplot: a program for plotting noncovalent interaction regions, *J. Chem. Theory Comput.*, 2011, **7**(3), 625–632.
31. J. Contreras-García, W. Yang and E. R. Johnson, Analysis of hydrogen-bond interaction potentials from the electron density: integration of noncovalent interaction regions, *J. Phys. Chem. A*, 2011, **115**(45), 12983.
32. W. Kutzelnigg, Chemical bonding in higher main group elements, *Angew. Chem., Int. Ed. Engl.*, 1984, **23**(4), 272–295.
33. K. Ruedenberg and M. W. Schmidt, Why does electron sharing lead to covalent bonding? a variational analysis, *J. Comput. Chem.*, 2007, **28**(1), 391–410.
34. T. Bitter, K. Ruedenberg and W. H. E. Schwarz, Toward a physical understanding of electron-sharing two-center bonds. i. general aspects, *J. Comput. Chem.*, 2007, **28**(1).
35. L. Cohen, Local kinetic energy in quantum mechanics, *J. Chem. Phys.*, 1979, **70**(2), 788–789.
36. L. Cohen, Representable local kinetic energy, *J. Chem. Phys.*, 1984, **80**(9), 4277–4279.
37. C. F. von Weizsäcker, On the theory on nuclear masses, *J. Phys.*, 1935, **96**.
38. A. Savin, O. Jepsen, J. Flad, O. K. Andersen, H. Preuss and H. G. von Schnering, Electron localization in solid-state structures of the elements: the diamond structure, *Angew. Chem., Int. Ed. Engl.*, 1992, **31**(2), 187.
39. K. Finzel, ELF and its relatives a detailed study about the robustness of the atomic shell structure in real space, *Int. J. Quant. Chem.*, 2014, **114**(22), 1546.
40. Y. Tal and R. F. W. Bader, Studies of the energy density functional approach. i. kinetic energy, *Int. J. Quant. Chem.*, 1978, **14**(S12), 153–168.
41. G. Hunter, The exact one-electron model of molecular structure, *Int. J. Quant. Chem.*, 1986, **29**(2), 197.
42. H. J. Bohórquez and R. J. Boyd, A localized electrons detector for atomic and molecular systems, *Theor. Chem. Acc.*, 2010, **127**(4), 393.

43. R. A. Boto, J. Contreras-García, J. Tierny and J.-P. Piquemal, Interpretation of the reduced density gradient, *Mol. Phys.*, 2015, 1–9.
44. A. Savin, The electron localization function (elf) and its relatives: interpretations and difficulties, *J. Mol. Struct.: THEOCHEM*, 2005, **727**(1), 127.
45. H. L. Schmider and A. D. Becke, Two functions of the density matrix and their relation to the chemical bond, *J. Chem. Phys.*, 2002, **116**(8), 3184.
46. V. Sahni, J. Gruenebaum and J. P. Perdew, Study of the density-gradient expansion for the exchange energy, *Phys. Rev. B*, 1982, **26**(8), 4371.
47. E. W. Pearson and R. G. Gordon, Local asymptotic gradient corrections to the energy functional of an electron gas, *J. Chem. Phys.*, 1985, **82**(2), 881.
48. J. P. Perdew, K. Burke and M. Ernzerhof, Generalized gradient approximation made simple, *Phys. Rev. Lett.*, 1996, **77**(18), 3865.
49. A. Zupan, K. Burke, M. Ernzerhof and J. P. Perdew, Distributions and averages of electron density parameters: Explaining the effects of gradient corrections, *J. Chem. Phys.*, 1997, **106**(24), 10184.
50. V. Tognetti, P. Cortona and C. Adamo, A new parameter-free correlation functional based on an average atomic reduced density gradient analysis, *J. Chem. Phys.*, 2008, **128**(3), 034101.
51. J. F. Dobson, K. McLennan, A. Rubio, J. Wang, T. Gould, H. M. Lee and B. P. Dinte, Prediction of dispersion forces: Is there a problem? *Aust. J. Chem.*, 2001, **54**, 513.
52. Density Functional Theory: A Practical Introduction D. S. Sholl and J. A. Steckel, *Density Functional Theory: A Practical Introduction*, Wiley, New York, 2009.
53. R. F. W. Bader, *Atoms in Molecules. A Quantum Theory*, Clarendon Press, Oxford, UK, 1990.
54. R. F. W. Bader, Bond paths are not chemical bonds, *J. Phys. Chem. A*, 2009, **113**, 10391.
55. R. F. W. Bader, Definition of molecular structure: By choice or by appeal to observation? *J. Phys. Chem. A*, 2010, **114**, 7431.
56. J. Cioslowski and S. T. Mixon, Universality among topological properties of electron density associated with the hydrogen-hydrogen nonbonding interactions, *Can. J. Chem.*, 1992, **70**, 443.
57. C. F. Matta, J. Hernández-Trujillo, T. Tang and R. F. W. Bader, Hydrogen-hydrogen bonding: A stabilizing interaction in molecules and crystals, *Chem. – Eur. J.*, 2003, **9**, 1940.
58. J. R. Lane, J. Contreras-García, J.-P. Piquemal, B. J. Miller and H. G. Kjaergaard, Are bond critical points really critical for hydrogen bonding? *J. Chem. Theory Comput.*, 2013, **9**(8), 3263.
59. R. P. Sagar, A. C. T. Ku and H. Vedene Jr., An examination of the shell structure of atoms and ions as revealed by the one-electron potential, *Can. J. Chem.*, 1988, **66**(4), 1005.
60. S. Fiedler, J. Broecker and S. Keller, Protein folding in membranes, *Cell. Mol. Life Sci.*, 2010, **67**, 1779.
61. K. A. Dill, Dominant forces in protein folding, *Biochemistry*, 1990, **29**, 7133.
62. A. Volkov, P. Macchi, L. J. Farrugia, C. Gatti, P. Mallinson, T. Richter and T. Koritsanszky, XD2006-A Computer Program Package for Multipole Refinement, Topological Analysis of Charge Densities and Evaluation of Intermolecular Energies from Experimental and Theoretical Structure Factors, 2006.
63. H.-S. Bürgi, S. C. Capelli, A. E. Goeta, J. A. K. Howard, M. A. Spackman and D. S. Yufit, Electron distribution and molecular motion in crystalline benzene: An accurate experimental study combining ccd X-ray data on C_6H_6 with multitemperature neutron-diffraction results on C_6D_6, *Chem. – Eur. J.*, 2002, **8**, 3512.

64. G. R. Desiraju and A. Gavezzotti, Crystal structures of polynuclear aromatic hydrocarbons. classification, rationalization and prediction from molecular structure, *Acta Crystallogr. Sect. B*, 1989, **45**, 473.
65. M. Nishio, The CH/π hydrogen bond in chemistry. Conformation, supramolecules, optical resolution and interactions involving carbohydrates, *Phys. Chem. Chem. Phys.*, 2011, **13**, 13873.
66. Y. Song, New perspectives on potential hydrogen storage materials using high pressure, *Phys. Chem. Chem. Phys.*, 2013, **15**, 14524.
67. S. Xiang, Y. He, Z. Zhang, H. Wu, W. Zhou and R. K. Chen, Microporous metal-organic framework with potential for carbon dioxide capture at ambient conditions, *Nat. Commun.*, 2012, **3**, 954.
68. J. A. Sans, F. J. Manjon, C. Popescu, V. P. Cuenca-Gotor, O. Gomis, A. Munoz, P. Rodriguez-Hernandez, J. Contreras-Garcia, J. Pellicer-Porres, A. L. J. Pereira, D. Santamaria-Perez and A. Segura, Ordered helium trapping and bonding in compressed arsenolite: Synthesis of $As_4O_6 \cdot 2He$, *Phys. Rev. B*, 2016, **93**, 054102.

Section 2: Spectroscopic and Database Information

7 Molecular Beam and Spectroscopic Techniques: Towards Fundamental Understanding of Intermolecular Interactions/Bonds

Sharon Priya Gnanasekar and Elangannan Arunan*

Department of Inorganic and Physical Chemistry, Indian Institute of Science, Bangalore 560012, India
*Email: arunan@ipc.iisc.ernet.in

7.1 Introduction

7.1.1 Intermolecular Interactions/Intermolecular Bonds

An invitation to write this Chapter on fundamentals of intermolecular interactions from a gas phase spectroscopist's point of view was surprising, given the title of this book. During the 19th International Conference on Chemistry of Organic Solid State held in Bangalore (ICCOSS 2012), a similar invitation to give a talk was accepted. The title chosen then was 'Structure of molecular complexes in the gas phase: Are they relevant to solid state?'. It was well received by the participants. We hope this Chapter proves useful to the intended audience and gives a flavor of what 'gas phase spectroscopists' do in unravelling the intricacies of intermolecular interactions. These

Intermolecular Interactions in Crystals: Fundamentals of Crystal Engineering
Edited by Juan J. Novoa
© The Royal Society of Chemistry 2018
Published by the Royal Society of Chemistry, www.rsc.org

interactions control the properties of the condensed phase, for example water and DNA, and hence our life. This chapter covers some important results from our laboratory and several others mostly from the last decade or so. Some examples from the earlier literature which, in our view, are important for anyone learning about inter-molecular interactions have been included as well. However, it is not intended to be an exhaustive review of results in this field and we apologize to those authors whose work has not been included.

Deviations from the ideal gas behavior led van der Waals to rectify the state equation by accounting for the possible intermolecular interactions between gaseous atoms/molecules, without which the condensed phase would not exist.[1] Due to this, all intermolecular forces were called van der Waals forces. van der Waals did not specify, nor perhaps know, the details of the intermolecular forces, which were to be discovered in the early part of 20th Century by Keesom, Debye and London. They derived expressions for dipole–dipole, dipole-induced dipole and dispersive (instantaneous dipole-induced dipole) forces, respectively and one could say van der Waals forces are Keesom-Debye-London forces. The book titled 'Molecular Forces' based on the Baker lectures by Debye, authored by Chu,[2] provides a rigorous summary of the physical forces responsible for intermolecular interactions. However, this does not seem to be popular among chemists or even physical chemists. Another book[3] based on the Baker Lectures by Pauling, also authored by him, became quite popular and every potential reader of this chapter will be aware of the book: The Nature of the Chemical Bond and Structure of Molecules and Crystals!

Pauling's book had a chapter titled 'The Hydrogen Bond' and it is widely accepted as the first to recognize the importance of hydrogen bonding. Chu's book based on Debye's lectures was published in 1967, nearly three decades after the first edition of Pauling's book, but does not mention hydrogen bonding anywhere. It could be considered blasphemous now to write a book on intermolecular interactions without mentioning the hydrogen bond.

Prominent among the material presented in Pauling's book discussing the 'hydrogen bond', in our view, are the structures of H_2O and H_2S, when they are frozen. Figure 7.1, taken from ref. 4, shows a pictorial representation of the structures. In ice *i.e.* H_2O, each molecule is surrounded by four other molecules to which the central H_2O molecule donates two hydrogen bonds and accepts two hydrogen bonds, which may be represented as O–H$\cdots$O. In solid hydrogen sulfide, each central molecule is surrounded by 12 other molecules. During Pauling's time only the heavy atom distances were known.

Figure 7.1 Schematic structure of ice (solid H_2O) and solid H_2S. In ice, each water molecule is surrounded by four others donating and accepting two hydrogen bonds. In solid H_2S, each molecule is surrounded by 12 others appearing spherical, as shown by a pack of oranges.
Reproduced from ref. 4 with permission from the Royal Society of Chemistry.

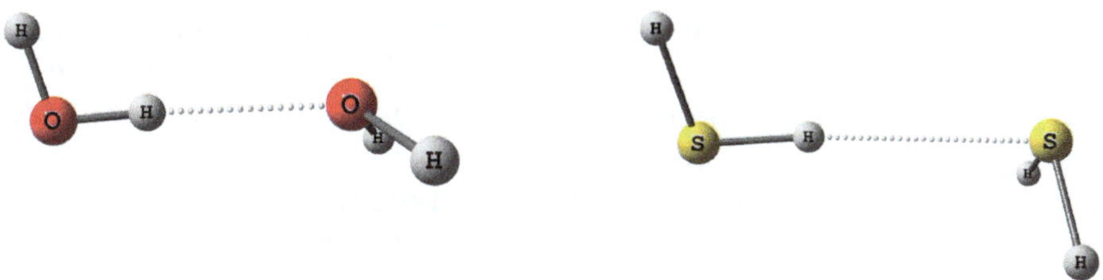

Figure 7.2 Structure of $(H_2O)_2$ and $(H_2S)_2$ as observed in a molecular beam, both showing one hydrogen bond.
Reproduced from ref. 4 with permission from the Royal Society of Chemistry.

Hence, the information available was that each O was surrounded by four other O atoms in H_2O and each S was surrounded by 12 other S atoms in H_2S. It was Pauling's genius to conclude that the O–H bond in H_2O was unlikely to be very different between gaseous, liquid and solid H_2O and hence each water had two O–H covalent bonds and two O$\cdots$H hydrogen bonds.

What is the nature of force acting between the H_2S molecules in solid H_2S? The structures of solid H_2O and H_2S were so distinctly different, 'intermolecular bonds' in H_2O were considered to be hydrogen bonds and the intermolecular 'interactions' in H_2S were left as 'van der Waals interactions'. Gas phase studies have proved that the structures of $(H_2O)_2$ and $(H_2S)_2$, shown in Figure 7.2, are very similar and they are both hydrogen bonded![4]

Now we know that fundamentally, there is no difference between intermolecular forces in H_2O and H_2S and the recent IUPAC recommendation[5] and technical report[6] on hydrogen bonds explicitly mentions this. A perspective article[7] from our laboratory pointed out that barriers for torsional/internal rotational motions which destroy the orientational preference, as found in H_2O, are the key to distinguishing 'hydrogen bonding' from 'van der Waals interactions'. Intermolecular interactions permeate every sphere of science and their study cannot be undermined. From something as innocuous as the gecko walking on the ceiling wall to something as intricate as DNA replication, all require a deeper understanding of the weak intermolecular interactions.

Clearly, the intermolecular interactions are manifestation of the various intermolecular forces such as electrostatics, induction, exchange repulsion, and dispersion, and also some charge transfer/covalency, which was largely considered for intramolecular bonds initially. In 1999, NMR spectroscopic[8] and Compton scattering[9] experiments showed evidence for partial, about 5–10%, covalency in hydrogen bonds. Accurate treatment of electrostatics needs to consider a multipole expansion beyond the dipole moment, as proved by the structure of HF dimer in the gas phase.[10] This was the first observation of hydrogen bonding in the gas phase and it showed that hydrogen bonding goes beyond dipole–dipole interactions. However, even today, it is not uncommon to find hydrogen bonding described as simple dipole–dipole interaction.

Over the last decade, it has been confirmed that hydrogen bonding is not unique. Naturally, there have been significant efforts to study halogen/lithium bonding as hydrogen has been grouped with alkali and halogen groups in the periodic table. Our perspective article on hydrogen/halogen/lithium bonding appeared recently and it highlights that hydrogen bonding and halogen bonding are similar but lithium bonding is distinctly different.[11] Similar interactions involving other group elements in the periodic table have now been found. Some of these interactions have been named as halogen bonding,[12–14] chalcogen bonding,[15,16] pnicogen bonding[17] or pnictogen bonding,[18] carbon bonding[19–21]/tetrel bonding,[22,23] triel bonding[24] and beryllium bonding[25]! It appears that both pnicogen and pnictogen are in use to describe elements of Group 15. Clearly some questions arise! When is it reasonable to give a common name for the intermolecular bonding involving the whole group of elements? When is it necessary to have unique name for every element? Clearly no one would question the 'hydrogen bond'. Do we need 'carbon bonding' specifically or is tetrel bonding enough to describe the interactions involving all group IV elements? In all these discoveries, gas phase spectroscopy, solid state crystallography and theoretical calculations have all played synergetic roles. This chapter will highlight some contributions from gas phase spectroscopic studies towards the fundamental understanding of intermolecular interactions/bonds.

7.1.2 What Can Spectroscopy Give Us?

This Chapter discusses recent results from the following experimental methods: (1) Molecular beam microwave spectroscopy;[26] (2) IR/UV double resonance[27] and Fluorescence/ion detected infrared spectroscopy;[28] (3) Zero kinetic energy photoelectron spectroscopy;[29] (4) Matrix isolation infrared spectroscopy;[30] and (5) Velocity map imaging.[31] Detailed descriptions about these techniques are given in the references cited. All these spectroscopic techniques are coupled with molecular beam machines which has been described in detail in a two volume book on this topic edited by Scoles.[32] Considering the length of this Chapter, it has been decided to focus on specific examples and highlight how the structure of molecular complexes is determined using experimental data and how it helps in understanding intermolecular interactions.

Section 7.2 discusses structural determination using spectroscopy with some specific examples and some general results. That microwave spectroscopy can give an accurate structure of molecules/complexes in the gas phase is common knowledge. Infrared spectroscopy can give indirect structural information and in particular, the red-shift in X–H stretching frequency has been used as a measure of hydrogen bond strength. Molecular beam techniques aid in the observation of the global minimum structure of the complex formed, though the choice of the carrier gas influences the probability of observing other minima structures. The matrix isolation technique inherently traps local minima structures along with the global minimum. Section 7.2.1 will discuss the structure of the phenylacetylene (PHA)–water complex in detail, where all these techniques have been used to identify the global minimum and also observe some local minimum structures.[33–35] Microwave spectroscopy gives precise positions of the atoms present in a molecule or a complex. However, it does not tell us how the molecules or complexes are bound! A molecule such as ethane can be drawn with one C–C and six C–H bonds. In the complex between PHA and water, can we draw one or more bonds between specific atoms from the two molecules? This can be done by Atoms in Molecules (AIM) theoretical analysis[36,37] which computes the electron density topology to obtain critical points. The advantage of the AIM theory is that it computes an experimentally measurable quantity, *i.e.* electron density, and it is particularly useful for comparison in solids as electron densities can be determined following X-ray diffraction experiments.[38] The AIM theoretical analysis has been used to shed light on the intermolecular bonds in this complex.[34]

That the red-shift observed in X–H stretching vibration following formation of an X–H···Y hydrogen bond is correlated with the hydrogen bond energy is well known.[5,6,39,40] The IUPAC recommendation, which insists on 'evidence for bond formation', explicitly mentions IR and NMR spectroscopic evidence. However, for a molecule like ethane-1,2-diol not only microwave spectroscopy[41] but also NMR spectroscopy[42] appear incapable of deciding on whether an intramolecular hydrogen bond exists. A small red-shift of about 40 cm^{-1} is expected for one of the two OH groups that could form this intramolecular bond. However, as this molecule has 10 conformers, with several of them contributing to the spectrum at ambient conditions, interpreting the infrared spectrum is not trivial.[43] However, advances in overtone spectroscopy, which is inherently weak, comes to our aid, as the shift scales with the vibrational quantum number and can be readily observed.[44] More interestingly, the AIM analysis[45] on this molecule suggests that there is no intramolecular hydrogen bond, contradicting the marginal evidence from the IR overtone spectroscopy. Just in time to settle this question, theoreticians improved the sensitivity of detecting such weak bonds and a non-covalent index[46] based on reduced electron density offered more insight.[47] These will be discussed in detail in Section 7.2.2

Section 7.3 discusses red- and blue-shifting hydrogen bonds. Interestingly, in recent times, blue-shifting of the order of 50 cm^{-1} was predicted based on theoretical calculations on the T-shaped structure of the benzene dimer and benzene-HCCl$_3$ complexes. While a small blue-shift was indeed observed by experiment on the benzene–HCCl$_3$ complex, experiments on the benzene dimer showed a small red-shift. However, the computational paper generated a lot of interest and it eventually led to the observation of several examples, almost all having C–H···Y hydrogen bonds.[48–51]

Discussion on intermolecular bonds beyond hydrogen is given in Section 7.4. In this section, two examples are discussed which show how microwave spectroscopic studies on molecular complexes of academic interest led to discovering the 'halogen bond'[52] and the 'carbon bond'.[53] This section also discusses the recent advances in 'intermolecular bonding', highlighting results from various spectroscopic and theoretical techniques. This section ends with a discussion on the very recent Resonant Ion Dip Infrared Spectroscopy (RIDIR) experiment which has given the first spectroscopic evidence for n–π^* interaction, proposed as a new 'non-covalent' interaction.[54]

Microwave spectroscopy can also shed some light on the strength and nature of bonds and this is not widely appreciated. The bond strength can be estimated through the experimentally measured centrifugal distortion constants and the nature of the bonds can be ascertained through quadrupole coupling constants. Legon and co-workers have done systematic work on a series of molecular complexes exploiting both.[55–57] Spectroscopy can not only be used to get an indirect estimate of bond strength; it can be used to measure the accurate bond energy. Recent experiments based on the velocity map imaging technique have provided the first direct measurement of hydrogen bond energy[58,59] in $(H_2O)_2$. Moreover, observation of vibrational progressions can provide another way to determine the bond energy using the Birge-Sponer plot.[60] This method has been used to determine the hydrogen bond energy between a carboxylic acid and water[61] and also for S center hydrogen bonds.[62] More interestingly, traditional intensity measurements on two monomers and the complex by infrared spectroscopy have been used to determine the Gibbs energy change on complex formation.[43,63] Thus, spectroscopy can not only be useful in determining the structure resulting from intermolecular interaction, it can also be used to determine the nature and strength of intermolecular bonding and also allow direct determination of the intermolecular bond energy. However, this chapter focuses on structural determination and inferences on intermolecular interactions and the reader is referred to the articles cited for other applications of spectroscopy. Finally, Section 7.5 gives our summary and outlook

7.2 Structural Determination by Spectroscopy

In this section, we discuss two representative examples in detail, both of which have multiple minima in the potential energy hypersurface. Both of these needed the data from several spectroscopic techniques which had to be supplemented with results from a variety of theoretical approaches. One is the weakly bound complex of phenylacetylene (PHA) and H_2O. Another is the deceptively simple ethylene glycol also known as 1,2-ethanediol. The former has two monomers, both of which could be potential hydrogen bond donors and acceptors. The latter has three single bonds (two C–O and one C–C) about which low barrier internal rotation can happen, leading to 10 unique conformers. Some of these conformers have the potential for an intramolecular hydrogen bond, which has attracted enormous interest. The debate about whether an intramolecular hydrogen bond is present in this molecule is still continuing. In our view, this question has not yet been unambiguously settled and we discuss all the available results.

7.2.1   Phenylacetylene Complexes

Phenylacetylene (PHA) is a multifunctional molecule, it comprises of two π-systems with an aromatic phenyl ring and an acetylenic group, both of which could accept a hydrogen bond. It also has an acetylenic hydrogen which is moderately acidic and which could be a hydrogen bond donor. The hydrogens in the phenyl ring could in principle be hydrogen bond donors as well, though this would be less stable. Although H_2O appears to be a much simpler molecule, it has the same level of complexity as it can form four hydrogen bonds, two as donors and two as acceptors. To guide us or perhaps confuse us, structures of benzene–H_2O and acetylene–H_2O were solved by molecular beam spectroscopy decades ago. The former has a structure in which O–H from water forms a hydrogen bond with the benzene acting as a π acceptor.[64,65] It is likely that a local minimum having C–H$\cdots$O hydrogen bonding exists and to the best of our knowledge, there is no experimental data showing this structure yet. The latter does have a C–H$\cdots$O interaction, with the acetylenic C–H donating a hydrogen bond to the lone pair acceptor in H_2O.[66,67] Thus in these two complexes, one has the water as donor and the π system as acceptor and the other has the H_2O as acceptor and the acetylene as the donor. Would the PHA–H_2O complex have a structure similar to the former or the latter?

Figure 7.3 shows the three different structures predicted by *ab initio* calculations. The first structure has the water interacting with the π cloud of the phenyl ring *via* an O–H$\cdots\pi$ interaction, similar to that of benzene–H_2O. The second structure is one in which the PHA molecule acts as the hydrogen bond donor, with the acidic hydrogen present on the acetylenic moiety forming a C–H$\cdots$O bond with water, resembling acetylene–H_2O. The third structure is quasi-planar with water forming two hydrogen bonds with PHA in a cyclic manner.

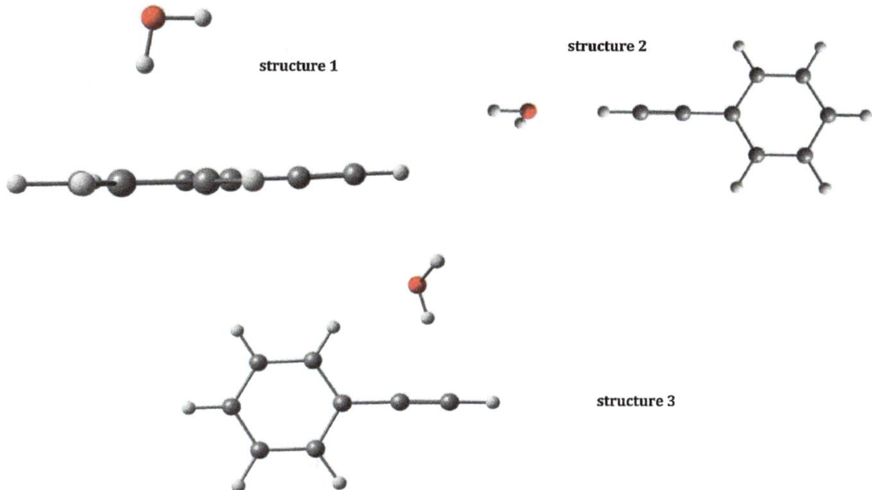

Figure 7.3 Structures of the PHA–water complex.
Reproduced from ref. 34 with permission from the Royal Society of Chemistry.

Water as an acceptor forms a C–H$\cdots$O hydrogen bond with the ortho hydrogen of the phenyl moiety and as a donor forms an O–H$\cdots\pi$ hydrogen bond with the acetylenic π system. It turns out the third structure is the global minimum. Can spectroscopy distinguish the three possibilities and give unambiguous evidence for one or more structures?

7.2.1.1 IR-UV Double Resonance Spectroscopy

Of course, microwave spectroscopy can give unambiguous structural evidence. However, this problem was initially tackled using the IR-UV double resonance technique by Patwari and co-workers.[33] Their experimental data convinced them that structure 3 was formed in the molecular beam used to produce this complex. How did they come to this conclusion? Typically, IR-UV double resonance experiments rely on comparison of the experimental spectrum to the predictions for the various possible structures from *ab initio* calculations using an harmonic oscillator model, with some scaling factor. As hydrogen bonding typically leads to significant red-shift in the X–H stretching frequency, often this can give reasonable conclusions. Some molecules have unique spectroscopic characteristics which offer more clues to structural determination by infrared spectroscopy and PHA is one. It has Fermi resonance which enables some overtone and combination bands to gain intensity from a fundamental transition which happens to be close in energy *i.e.* near resonance. In general, such overtone and combination bands are significantly weaker than fundamental transitions.

Let us discuss the infrared spectrum of PHA first. This molecule has only one acetylenic C–H group and one would expect a single peak corresponding to the C–H stretching vibrational mode around 3300 cm^{-1} (closer to that of acetylene) and it is well separated from the aromatic C–H stretching modes expected around 3000 cm^{-1}. However, Stearns and Zwier[68] noted three strong peaks centered around 3330 cm^{-1}; see Figure 7.4. They realized that the C–H stretching mode has strong Fermi resonance with a combination mode having one quantum of C≡C stretch and two quanta of C≡C–H bending vibrational modes.

In a complex formed by PHA, if either of the C≡C stretching or C–H stretching modes are affected due to interaction with the other molecule, it would affect the Fermi resonance. One can see that in the structure 1 for the PHA–H$_2$O complex, Fermi resonance is unlikely to be affected and in both structures 2 and 3, Fermi resonance will be affected and the C–H stretching region in the spectrum would look

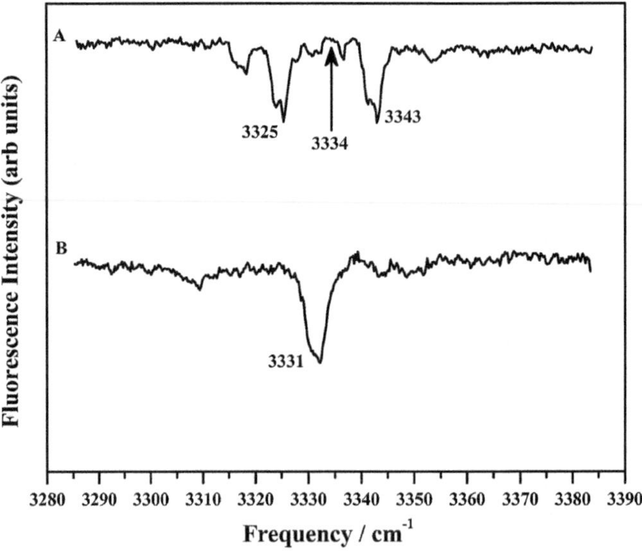

Figure 7.4 The acetylenic C–H stretching region in the fluorescence detected infrared spectrum of PHA (A) and PHA–water complex (B).
Reprinted with permission from P. C. Singh, B. Bandyopadhyay, and G. N. Patwari, Structure of the Phenylacetylene–Water Complex as Revealed by Infrared-Ultraviolet Double Resonance Spectroscopy, J. Phys. Chem. A, 2008, **112**, 3360–3363. Copyright (2008) American Chemical Society.

distinctly different. This region of the infrared spectrum for the complex and the PHA monomer is shown in Figure 7.4. Clearly, structure 1 can be ruled out immediately as the Fermi resonance has disappeared and only one peak appears in this region. To differentiate between, structures 2 and 3, the acetylenic C–H stretching peak can be used. Structure 2 is expected to have a significant red-shift compared to the monomer. For the acetylene–water complex, the observed red-shift[67] is about 60 cm^{-1}. For PHA-water complex, scaled harmonic calculations at MP2/aug-cc-pVDZ level, predict the C–H stretching mode at 3333, 3274 and 3328 cm^{-1} for structures 1, 2 and 3, respectively.[33] It should be pointed out that harmonic calculations do not cover overtone and combination bands and would predict only one C–H stretching peak at these wave numbers. The same level calculations predict this mode at 3335 cm^{-1} for the PHA monomer and hence, structures 1 and 3 have a small red-shift of 2 and 7 cm^{-1}, respectively. For structure 2, the predicted red-shift is 61 cm^{-1}, close to what has been experimentally measured for the H$_2$O–HCCH complex. The fact that Fermi resonance of the C–H stretching mode has been affected and it appears very close to the expected monomer frequency suggests that structure 3 has been observed in the molecular beam experiment. Microwave experiments[34] published in 2011, unambiguously confirm this and a matrix isolation experiment,[35] reported in 2016, observed structure 2 and confirmed the red-shift to be 61 cm^{-1}. These experiments are discussed next.

7.2.1.2 Molecular Beam Microwave Spectroscopy

The IR-UV double resonance experiments had suggested that only the global minimum was produced in the molecular beam and it was most likely to have structure 3. A key piece of evidence from this experiment is the loss of Fermi resonance, which was somewhat fortuitous. If this molecule didn't have Fermi resonance, experimentally observed C–H stretching frequencies could not have differentiated between structures 1 and 3. Moreover, the O–H stretching modes in the complex were determined to be at 3772 and 3644 cm^{-1}. These may be compared to the scaled harmonic frequencies for structures 1 and 3: 3744 and 3621 for 1 and 3735 and 3587 for 3! While it is difficult to choose between these two, the experimentalist is more likely to have assumed that structure 1 has been observed. A careful look at Figure 7.5 will convince the reader. One can see that the observed peaks are closer to the two modes corresponding to 'free O–H' and 'bonded O–H' stretching in structure 1.

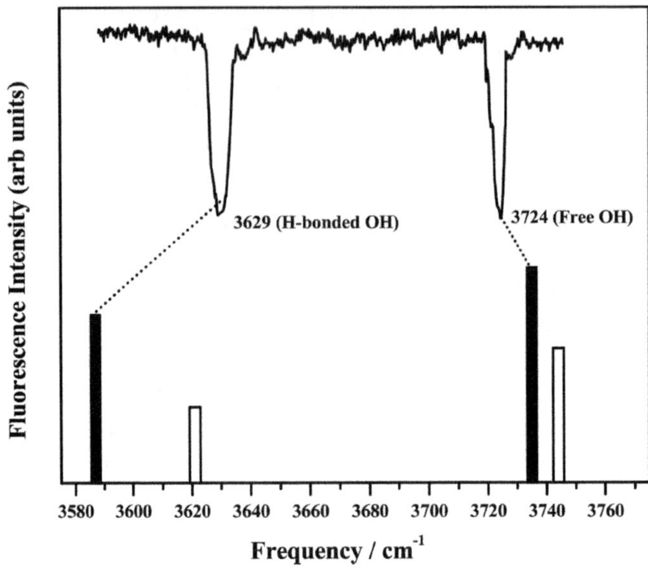

Figure 7.5   The O–H stretching region in the fluorescence detected infrared spectrum of PHA–water complex with predictions for structure 3 (solid bar) and structure 1 (open bar). Though the agreement for structure 1 appears better, structure 3 was confirmed based on Figure 7.4. In structure 1, the Fermi resonance would not be affected.
Reprinted with permission from P. C. Singh, B. Bandyopadhyay, and G. N. Patwari, Structure of the Phenylacetylene-Water Complex as Revealed by Infrared-Ultraviolet Double Resonance Spectroscopy, J. Phys. Chem. A, 2008, **112**, 3360–3363. Copyright (2008) American Chemical Society.

When several structures are possible, infrared spectroscopy alone may not be able to provide an unambiguous answer in many instances. Even in this case, if the experiment had given evidence for structure 2, that would be unambiguous as there is a large red-shift in C–H stretching mode, as found later in a matrix isolation experiment. Microwave spectroscopy gives unambiguous evidence for structural determination in most cases.

Goswami and Arunan reported the microwave spectrum of this complex in 2011.[34] Table 7.1 gives the rotational constants of the three possible structures along with the experimentally determined rotational constants. Clearly, there is no ambiguity about which structure has been experimentally observed.

While the rotational constants given in Table 7.1, already yield unambiguous evidence for the structure, microwave spectroscopy offers more. The inertial defect, $\Delta = I_C - I_A - I_B$, is also given in Table 7.1. If a molecule were to be planar, Δ would be exactly zero. Structure 1 has a large inertial defect and both structures 2 and 3 have

Table 7.1 *Ab initio* derived rotational constants for the three geometries of $C_6H_5CCH\cdots H_2O$ at MP2/aug-cc-pVDZ level from ref. 33 compared with experimental values from ref. 34. See text for details.

Rotational constants	Structure 1	Structure 2	Structure 3	Experiment
A (MHz)	2083	5518	2678	2672.092(3)
B (MHz)	1132	506	998	996.3581(8)
C (MHz)	995	464	729	731.7055(4)
Δ(amu Å^2)	-181	-1.2	-1.9	-5.672
ΔE (kJ mol^{-1})	-6.5	-6.2	-7.1	—

small inertial defect. The experimental inertial defect is slightly larger as the zero point vibrational motions along some modes, such as out of plane bending modes, can contribute to the inertial defect. The small value, still, indicates that the equilibrium structure is nearly planar.

In this work, microwave spectra of several isotopologues, such as $C_6H_5-C\equiv C-H\cdots H_2O$, $C_6H_5-C\equiv C-H\cdots DOH$, $C_6H_5-C\equiv C-H\cdots D_2O$ $C_6H_5-C\equiv C-D\cdots H_2O$, $C_6H_5-C\equiv C-H\cdots H_2{}^{18}O$ were measured. Rotational constants of these isotopologues can be used to determine several structural parameters directly. The location of the substituted atoms can be determined by Kraitchman analysis.[69] In general, the results from H/D substitution are less accurate compared to the heavier atom substitutions, such as ${}^{16}O/{}^{18}O$ due to the significant differences in the zero point vibrations. Using the rotational constants for the two isotopologues having ${}^{16}O/{}^{18}O$, the location of O atom in the complex can be calculated.

Moreover, a simultaneous fit of the rotational constants for these five isotopologues, five structural parameters could be determined using STRFIT programme.[70] Figure 7.6 shows the reference structure of the structure 3 and Table 7.2 lists the structural parameters derived.

In Table 7.2, the fitted parameters were varied to match the five sets of experimental rotational constants and the uncertainties follow from the experimental uncertainties in rotational constants. These parameters can be used in turn to derive other structural parameters and the authors chose the five derived parameters listed in this table. These include the distance between the π centre and the H from OH, $\pi(C\equiv C)\cdots H_{16}$, as 2.477 Å and the $\angle C_3H_{11}O_{15}$, the angle between the ortho C–H in phenyl group and the O in H_2O as 145.6°. These are of direct relevance in identifying the O–H$\cdots\pi$ and C–H$\cdots$O as hydrogen bonds using the most popular distance and angle criteria.[5] However, microwave spectroscopy cannot give any direct evidence about whether these two interactions are 'hydrogen bonds'. Infrared

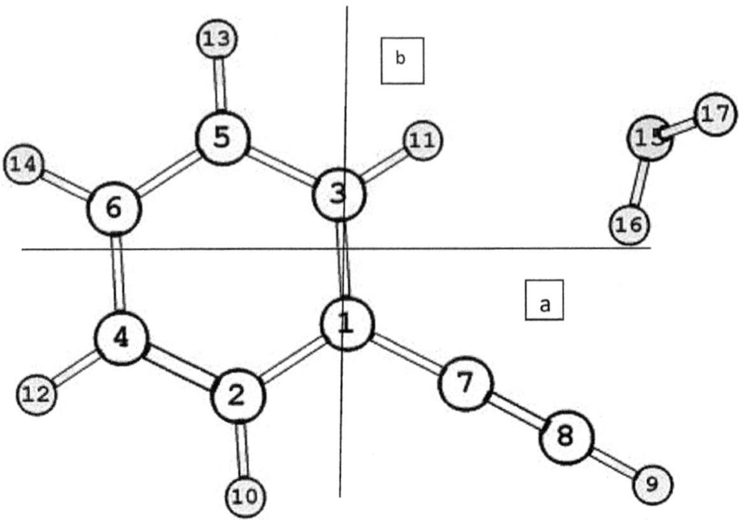

Figure 7.6 The structure observed in molecular beam experiments for the PHA–H_2O complex with all atoms numbered for defining some structural parameters, given in Table 7.2.
Reproduced from ref. 34 with permission from the Royal Society of Chemistry.

Table 7.2 Fitted and derived structural parameters from the five sets of experimental rotational constants for the five isotopologues.

Fitted parameter	Fitted value	Derived parameter	Value
$r(O_{15}C_7)$ (Å)	3.423(4)	$r(\pi(C\equiv C)\cdots H_{16})$	2.477
$\angle O_{15}C_7C_8$ (degree)	80.4(1)	$\angle O_{15}C_7C_8$	2.544
$D(O_{15}C_7C_8C_3)$ (degree)	11.1(6)	$D(O_{15}C_7C_8C_3)$	154.9
$\angle(H_{16}O_{15}C_7)$ (degree)	25(3)	$\angle C_3H_{11}O_{15}$	145.6
$D(H_{16}O_{15}C_7C_3)$ (degree)	226(12)	$D(H_{16}O_{15}H_{11}C_3)$	45.8

spectroscopy can indeed give direct evidence, but in this particular case, as evident from Figure 7.5, it is not unambiguous.

This is when Atoms in Molecules theory comes in handy as it analyzes the total electron density in the complex.[36,37] The molecular graph obtained for this complex is shown in Figure 7.7.

We pointed out the various spectroscopic methods for determining the intermolecular hydrogen bond energy in Section 7.1.2. Determining the intramolecular hydrogen bond energy is not trivial. The electron density at the bond critical point (BCP) has been shown to have a strong correlation with hydrogen bond energy.[11,71–73] This fact has been used to estimate the intramolecular hydrogen bond energy from experimental/theoretical electron density found at the bond

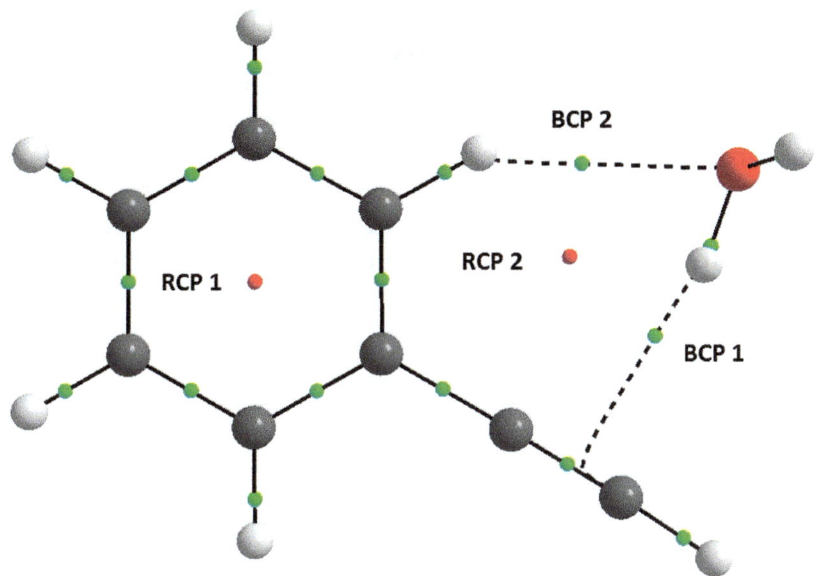

Figure 7.7 The molecular graph obtained using AIM analysis of the wave function obtained at MP2/aug-cc-pVTZ level calculations. Red dots denote bond critical points and one can see both O–H$\cdots\pi$ and C–H$\cdots$O hydrogen bonds in the global minimum of the PHA–H_2O complex.
Reproduced from ref. 34 with permission from the Royal Society of Chemistry.

critical points. Somewhat curiously, the electron densities at the BCPs corresponding to O–H$\cdots\pi$ and C–H$\cdots$O hydrogen bonds turned out to be very close, 0.013 and 0.011 au, respectively. Clearly, both these hydrogen bonds are of the same strength. A strong hydrogen bond donor (OH) with a weaker acceptor (π) and a weak hydrogen bond donor (CH) with a strong acceptor (O), both lead to similar strength hydrogen bonds.

7.2.1.3 Matrix Isolation Infrared Spectroscopy

With the IR-UV double resonance and the microwave spectroscopy experiments positively identifying the planar structure as the global minimum and not giving any evidence for any other structures, could we conclude that it would be the only structure for the complex? Matrix isolation and Helium nano-droplet experiments have succeeded in identifying local minimum structures and the complex formed between HF and HCN offers a good test case. An initial microwave spectroscopy experiment[74] identified only the global minimum with the linear structure HCN$\cdots$HF. A few years later, a matrix isolation experiment[75] positively identified two structures for the dimer, HCN··· HF and HF··· HCN based on infrared spectroscopy. More recently, Helium nano-droplet experiments have also identified the HF··· HCN structure by rotationally resolved vibrational spectroscopy.[76]

A matrix isolation experiment on the PHA–H_2O complex was carried out only recently by Viswanathan and co-workers[35] and they could identify structure 2 positively from their data. The infrared spectrum of the complex recorded in the matrix is shown in Figure 7.8. It

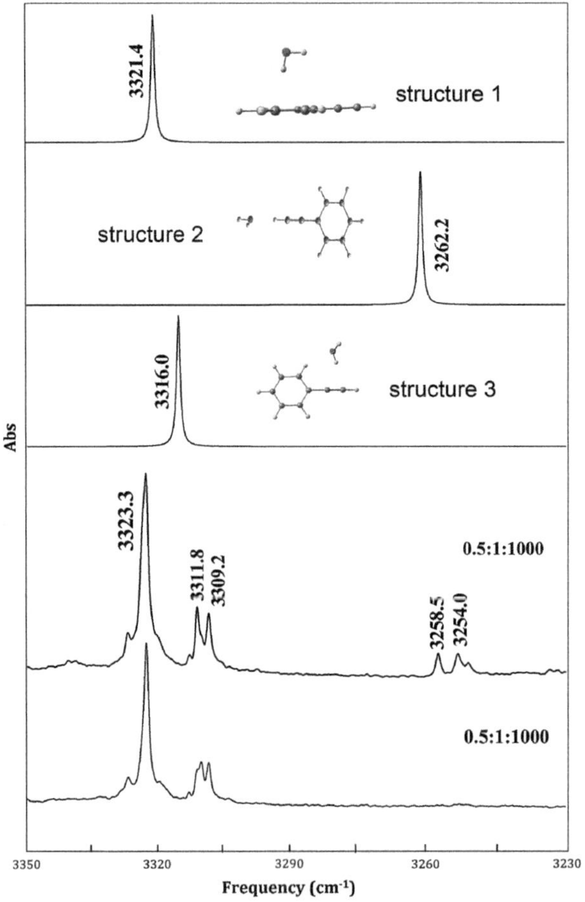

Figure 7.8 Matrix isolation infrared spectrum of phenylacetylene and water containing an N_2 matrix at 12 K. From bottom to top: (1) Experimental spectrum of PHA–H_2O (0.5 : 1 : 1000) recorded at 12 K in an N_2 matrix; (2) Spectrum after annealing at 27 K; (3) Computed spectra of complex 1; (4) Computed spectra of complex 2; (5) Computed spectra of complex 3.
Reprinted from Journal of Molecular Structure, 1107, G. Karir and K. S. Viswanathan, Phenylacetylene–water complex: Is it n···σ or H···π in the matrix? 145–156. Copyright 2016 with permission from Elsevier.

positively identifies the structure in which the acetylenic C–H acts as the hydrogen bond donor, and the C–H$\cdots$O hydrogen bond leads to a 62 cm^{-1} red-shift in the C–H stretching frequency. The IR-UV double resonance experiments have mass resolution and so it is possible to get the IR spectrum of the monomer and complexes separately. In the matrix isolation experiments, the monomer and various complexes that are formed will all appear in the infrared spectrum. Hence, it is not straightforward to conclude anything about the presence/absence of other structures from Figure 7.8.

In Figure 7.8, it may be noted that the Fermi resonance in PHA is preserved in the matrix. It should be pointed out that the computed spectra used a harmonic approximation and so the Fermi resonance would not be there. As discussed earlier, structure 1 would have this region of the spectrum unaffected and could overlap with the monomer peaks observed. Hence, the matrix isolation infrared spectroscopy has given positive evidence for the presence of structure 2 but does not rule out the presence/absence of the other structures. So far, in this section, we have discussed how a variety of spectroscopic techniques, aided by computational tools, have been used to characterize a complex formed between PHA and water. We discuss other related studies on PHA complexes before moving on to the next system.

7.2.1.4 Other PHA Complexes

Patwari and co-workers have exploited the Fermi resonance in PHA as a tool to deduce the structures of several of its complexes, for example with Ar, NH_3, CH_3OH and CH_3NH_2.[77] Microwave spectroscopy results are available for PHA–H_2S complex. These complexes offer an interesting summary and show how the global minimum can change with the partnering molecules. For H_2S and CH_3OH, structure 1 becomes the global minimum and for NH_3 and CH_3NH_2, structure 2 is the global minimum! We give a brief summary of these results in this section. The C–H stretching region of the IR spectrum is shown in Figure 7.9 for all these complexes and some details are discussed next.

PHA-Ar

The vibrational spectrum for the PHA-Ar dimer shows no change compared to the monomer PHA spectrum. The shifts observed are within experimental uncertainties suggesting there is no interaction between Ar and the acetylenic moiety. Hence, this supports the structure in which Ar interacts with the phenyl π electrons as observed in Ar–benzene complex.[78]

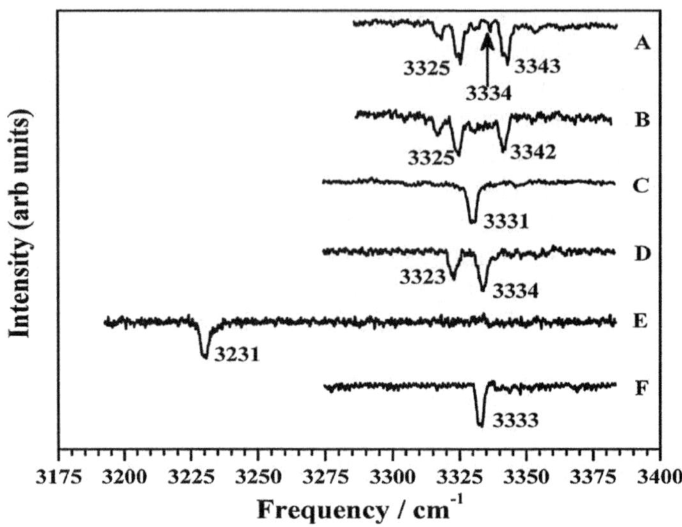

Figure 7.9   The acetylenic C — H stretching region of the IR spectrum of (A) PHA, (B) PHA — Ar, (C) PHA — H_2O, (D) PHA — MeOH, (E) PHA — NH_3, and (F) PHA — $MeNH_2$. In A, the arrow indicates the position of the unperturbed C–H oscillator of PHA evaluated using the two-state deperturbation model.
Reprinted with permission from R. Sedlak, P. Hobza and G. N. Patwari, Hydrogen-Bonded Complexes of Phenylacetylene with Water, Methanol, Ammonia, and Methylamine. The Origin of Methyl Group-Induced Hydrogen Bond Switching, J. Phys. Chem. A, 2009, **113**, 6620. Copyright (2009) American Chemical Society.

PHA-CH₃OH and PHA-H₂S

The methanol complex with **PHA** can provide information of the effect of methyl substitution. The spectrum shows two peaks corresponding to Fermi resonance, though the intensity and position are different from that of the **PHA** molecule.[79] The interaction is between the methanolic O–H and the phenyl ring (confirmed by the O–H stretch), this causes some changes in the zero point frequencies and coupling constants leading to the observed vibrational spectrum. The computational *ab initio* calculation shows that in CH_3OH, the methyl group has an inductive effect making the O more polarizable. The phenyl moiety which has a large interaction cross section can easily interact with the CH_3 group. Thus we can see a change in H-bond interaction due to the addition of a methyl group to the hydrogen bond donor.

Interestingly, we note that the PHA–H_2S has the global minimum structure very similar to that of PHA–CH_3OH. Though, IR-UV double resonance measurements on this complex have not been done, microwave spectroscopic[80] results unambiguously prove the structure

Table 7.3 *Ab initio* derived rotational constants for the three geometries of $C_6H_5CCH\cdots H_2S$ at MP2/aug-cc-pVDZ level compared with experimental values from ref. 79. The three structures are similar to $C_6H_5CCH\cdots H_2O$ shown in Figure 7.1. Binding energies are given in the last row.

Rotational constants	Structure 1	Structure 2	Structure 3	Experiment
A (MHz)	1279	5374	2209	1206.551(6)
B (MHz)	1176	313	652	1134.152(6)
C (MHz)	782	296	506	732.192(6)
ΔE (kJ mol^{-1})	-10.2	-1.9	-6.5	—

of the complex in which the S–H interacts with the phenyl ring and has a structure very similar to benzene–H_2S.[81] Table 7.3 gives the rotational constants of three equivalent structures of PHA–H_2S and the experimental values and the conclusion is obvious. It is interesting to note that the structures of PHA–Ar/PHA–H_2S and PHA–CH_3OH are very similar and they resemble those of the respective complexes with benzene. It was not the case for PHA–H_2O, where the two hydrogen bonds tilt the favor towards the planar structure. Patwari and co-workers have carried out an energy decomposition analysis and point out that electrostatic contributions dominate the PHA–H_2O interaction and dispersion dominates in the PHA–CH_3OH interaction, in determining the global minimum.[77–79] For PHA–CH_3OH, the planar structure is not a minimum as the CH_3 group would have steric repulsion and the structure having O–H$\cdots\pi$ (acetylenic) interaction also has the CH_3OH above the plane with additional C–H$\cdots\pi$(phenyl) interaction. The energy differences between these three minima are indeed too small for the PHA–H_2O (see Table 7.1) and PHA–CH_3OH complexes.[79] As the matrix isolation experiments on PHA–H_2O show, it is only a matter of time and technique which allows observation of other structural isomers. Moreover, as discussed next, in the case of PHA–CH_3NH_2 complex, IR-UV double resonance experiments did identify two conformers.

PHA-NH$_3$ and PHA-CH$_3$NH$_2$

The IR-UV double resonance experiments identified only one structure for the PHA–NH_3 complex and it shows a large red shift in acetylenic C–H stretch and the disappearance of Fermi resonance. This suggests that the structure has an acetylenic C–H$\cdots$N hydrogen bond. *Ab initio* calculations predict this structure to be the global minimum as well and the three structures (similar to those given for PHA–H_2O in Figure 7.1) have binding energies of 8.2, 3.2 and 5.9 kJ mol^{-1}, respectively. What is remarkable about PHA is that it has three equivalent minima in complexes with $H_2O/H_2S/NH_3$ and these three complexes have unique global minimum structures. These have all been unambiguously characterized by microwave and infrared spectroscopy experiments and in every case, electronic structure theory predicts the global minimum accurately. The energy differences between these structural isomers are rather small and the fact that experimental and theoretical techniques are able to characterize them unambiguously augers well for the field.

The global minimum structures of PHA–H_2O and PHA–CH_3OH were so different and only one structure was observed in IR-UV double resonance experiments, Patwari and co-workers called it an 'alkyl group induced hydrogen bond switching'.[79] That these are too subtle features to be generalized became obvious when they characterized the structure of PHA–CH_3NH_2. They could optimize five different structural isomers for this complex and it turned out the global minimum was indeed similar to that of PHA–NH_3 complex. The red-shift in the acetylenic C–H stretching for the PHA–CH_3NH_2 complex (139 cm^{-1}) was found to be larger than that for the PHA–NH_3 complex (103 cm^{-1}). These can be compared to the red-shift of about 60 cm^{-1} found for the equivalent PHA–H_2O complex. More interestingly, they were able to identify two different structural isomers using the laser induced fluorescence and 1 colour – resonant two photon ionization techniques. The second structure they identified turned out to be similar to structure 1 for PHA–H_2O complex, having N–H$\cdots\pi$ (phenyl) and in addition C–H$\cdots\pi$ (acetylenic) and it was about 3 kJ mol^{-1} higher in energy.

7.2.2 Ethane-1,2-diol or Ethylene Glycol Structure and Interactions

In the previous section, we discussed intermolecular interactions. In the gas phase, that two molecules are interacting is never in doubt, if one can form a stable complex as discussed above. From Figure 7.1, it is straightforward to conclude that PHA and H_2O have attractive interactions in all three structures. In a crystal, it may not be straightforward to conclude that two atoms that happen to be close in proximity have any attractive interaction. Are the two atoms closer because of any direct interaction or are they close because of packing effects? This situation can be easily understood even for an isolated molecule such as H_2O. We all know it has two O–H covalent bonds. Does it also have an H$\cdots$H bond? Most chemists would dismiss this question with contempt. The distance between the two H atoms in an H_2O molecule is about 1.5 Å and it is significantly less than the sum of their van der Waals radii, 2.4 Å. Often, crystallographers end up using such a distance criterion to conclude that two atoms have an attractive interaction. Over the last decade, ethane-1,2-diol, $HOCH_2$-CH_2OH, has attracted significant interest. It will be denoted as EG following its common name, ethylene glycol, from now onwards.

There are three single bonds in EG, two C–O and one C–C, and the molecule could be rotated about these three axes resulting in several conformational isomers. Each of these rotations can result in 3 conformations and so there are 27 potential conformers (3^3). However, several of them are degenerate and there are 10 unique conformers as shown in Figure 7.10. The conformers are labelled based on the torsional angle and it is t/T for 180° (trans), g/G or g′/G′ for +60° or −60° (gauche) with an allowance of about 30°. Calculations show that the global minimum is tGg′. This has been attributed to intramolecular O–H···O hydrogen bonding and it is a text book example of the same, finding a place in the popular book by Eliel on the 'Stereochemistry of Carbon Compounds'.[82] Klein has questioned this interpretation claiming that the intramolecular hydrogen bond in EG has always been ascertained but never proved.[83] One of the major concerns is the angle ∠O–H···O being 108°, lower than the lower limit for hydrogen bond angle (110°) suggested by various authors including the recent IUPAC recommendation.[5,6] In this section, we discuss all the spectroscopic and computational results published recently.

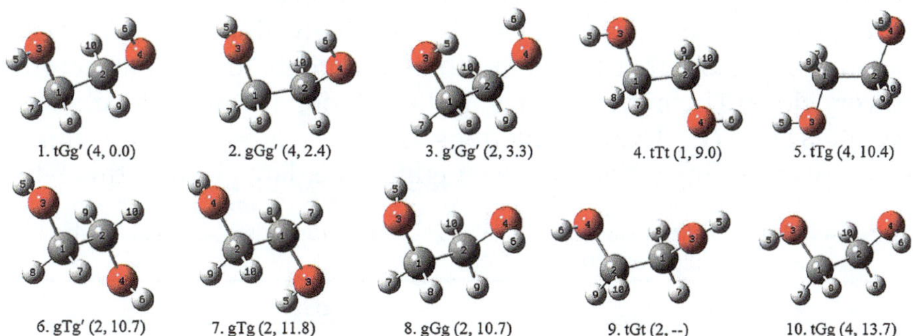

Figure 7.10 Structures of the 10 unique conformers of EG optimized at B3LYP/ 6-311++G** level of calculations. Numbers in parenthesis indicate the number of equivalent structures adding up to 27 and the relative energy in kJ mol⁻¹. Structure 9 is a saddle point at this level and its energy is in between those of structures 8 and 10.
Reprinted with permission from P. Das, P. K. Das and E. Arunan, Conformational Stability and Intramolecular Hydrogen Bonding in 1,2-Ethanediol and 1,4-Butanediol, J. Phys. Chem. A, 2015, 119, 3710–3720. Copyright (2015) American Chemical Society.

7.2.2.1 Microwave Spectroscopy

The microwave spectrum of EG was first recorded in 1974[84] but the transitions could not be assigned as the large amplitude motions led to tunnelling splitting, of the order of 7 GHz. This motion involves the concerted large amplitude torsions of the two hydroxyl groups in a double minimum potential. Most readers may be familiar with the umbrella inversion of NH_3 which is comparable to such large amplitude motions. These motions can couple with overall rotational motion leading to splitting of energy levels. Assigning the microwave spectrum in such molecules is a major challenge faced by spectroscopists and the details may be out of place in this article. Almost 2 decades later, molecular beam microwave spectroscopy combined with MW-MW double resonance[85] was used to assign the transitions of the global minimum, tGg' (referred to in the paper as g'Ga). Another decade later, microwave and millimeter wave spectroscopies were combined to assign the transitions from the second lowest conformer, gGg'.[86] As the intensity of microwave transitions directly depends on the electric dipole moment, the Stark effect can be used to determine the dipole moments of the molecule. Comparison of the experimental dipole moment about the three principal axes, with theoretical estimates, also helps in confirming the structure. Table 7.4 gives the experimental and theoretical values of the dipole moments and rotational constants determined for the two conformers.

The dipole moment data given in Table 7.4, taken from ref. 86, shows the importance of measuring the dipole moment along the three principle axes, which can be routinely done in microwave spectroscopy. Note that the total dipole moments for both structures are very close. Theoretical values of the rotational constants are from B3LYP/6311++G** level calculations and these are for the optimized structure excluding vibrational averaging. The data given in this table

Table 7.4 Rotational constants (in MHz) and dipole moment (in Debye) for the two most stable conformers of ethylene glycol.

Constants	tGg'		gGg'	
	Expt.	Theory	Expt.	Theory
A	15361.1856(3)	15373.3994	15210.0578(4)	15151.3401
B	5588.24272(7)	5548.0085	5542.4340(2)	5417.3679
C	4614.48957(8)	4587.9469	4595.4449(1)	4469.4096
μ_A	2.080	2.446	1.30	1.372
μ_B	0.936	1.093	1.37	1.503
μ_C	0.470	0.672	1.42	1.42
μ_T	2.329	2.762	2.36	2.608

are enough to confirm the structures of the two most stable conformers. Are they enough to conclude that there is intramolecular hydrogen bonding in these two conformers of EG?

7.2.2.2 *Experimental and Theoretical Electron Density*

As with the example for the PHA–H_2O complex, it would be interesting to look at the electron density topology for the various conformers. Such analysis has been done by several groups starting with Klein.[83,87–89] All these investigations have not been able to identify a BCP between the H and O atoms, expected for an intramolecular hydrogen bond! Klein argued that the absence of BCP is enough to conclude that there is no intramolecular hydrogen bond in the EG structures. The molecular graph obtained for an isolated EG, tGg′ structure is shown in Figure 7.11 and it does not show a BCP. For comparison, similar results for 1,4-butanediol (BD) are shown as well, clearly identifying the intramolecular hydrogen bond. Could we rely on this theoretical model to conclude the absence of intramolecular hydrogen bond in EG?

Experimental confirmation would be essential. X-ray diffraction experiments can indeed give the electron density.[38] However, the total electron density obtained experimentally has not been found to be useful for interpretation and so a core electron density for every atom, determined from a model, is subtracted. The deformation density still yields information similar to what AIM theoretical analysis does. We are not aware of such results on gas phase molecules. Chopra *et al.*

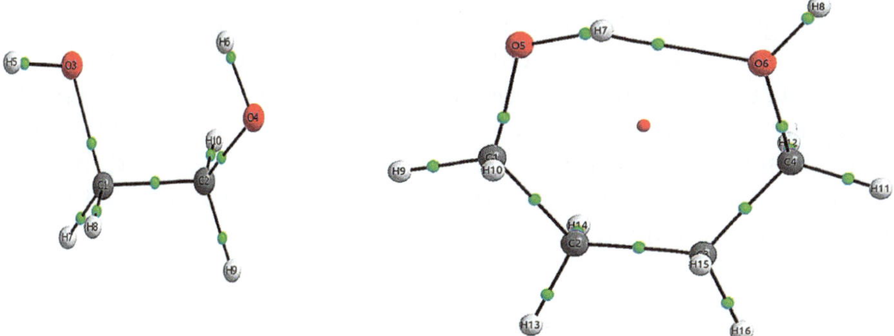

Figure 7.11 Molecular graph of the tGg′ conformer of EG and gGGGt conformer of 1,4-butanediol (BD) showing all the BCPs corresponding to the covalent bonds. In addition, there is a BCP corresponding to intramolecular hydrogen bond in BD and a resultant ring critical point but none corresponding to the intramolecular hydrogen bond in EG.

determined the crystal structure and also deformation density for the EG using cryo-crystallography.[90] In the crystal, they found that the EG existed only in the tGg$'$ geometry. However, the O$\cdots$H distance had increased by 0.45 Å, compared to the isolated molecule in the gas phase. Experimental deformation density also did not show any evidence for intramolecular hydrogen bond. However, their work led to the observation of C–H$\cdots$O hydrogen bonds, which were not recognized in previous crystallographic studies,[91] in addition to the well-recognized intermolecular O–H$\cdots$O hydrogen bonds, both contributing to the overall crystal stability.

7.2.2.3 NMR Experiments in the Liquid Phase

Numerous NMR experiments on EG solutions in various solvents have yielded contradicting conclusions.[42] Several investigations on this molecule used the vicinal H,H couplings, $^3J_{1,3}$ and $^3J_{1,4}$ to address the question. Pearce and Sanders had earlier analyzed the proton coupling of EG in chloroform solvent and concluded that intramolecular hydrogen bonding in EG is significant.[92] In this solvent, the proton coupling turned out to be very slow enabling the observation of the splitting of the hydroxyl proton by methylene protons. Roberts and co-workers[42] have determined the dilution shift of EG and related molecules in chloroform. For EG, the dilution shift for the hydroxyl proton is larger, going from 4.14 ppm (in saturated 0.02M solution) to 1.78 ppm (in 1.79×10^{-4} M), compared to 2-methoxyethanol, for which the OH proton shift is 3.5 ppm (3 M) and 1.93 ppm (6×10^{-4} M). Based on this, they argue that the intramolecular hydrogen bond is absent in both these molecules. All the NMR investigations on EG conclude that the geometry observed in solution is predominantly gauche. Roberts and co-workers conclude that this preference is not due to intramolecular hydrogen bonding but due to a 'gauche effect'. Though, this conclusion does not reveal much about the conformational preference, their interpretation of all the NMR data led them to conclude that intramolecular hydrogen bonding does not exist in EG in solutions, perhaps until infinite dilution!

7.2.2.4 Infrared Spectroscopy

The most significant experimental evidence for 'intramolecular hydrogen bonding' is from infrared spectroscopy. However, it is not unambiguous. Typically, at room temperature, EG could exist as one of the 10 conformers and the infrared spectrum will have contribution from all of them. Infrared spectrum of EG at room temperature was reported recently[43] and compared to 1,4-butanediol (BD). While EG has 10 unique conformers, BD has 70 unique conformers and interpreting the spectra of these two molecules involves simulation of the spectra of all the conformers and using their thermal population to average. The IR spectra of both EG and BD are shown in Figure 7.12.

Interpreting the spectra given in Figure 7.12 and concluding about whether or not an intramolecular hydrogen bond is present in EG is far from trivial. The simulated spectrum (at 303 K) of EG has contributions from the various conformers as follows: tGg′, gGg′, g′Gg′, tTt, tTg, gTg′, gTg, gGg, tGt, and tGg: 55.5, 22.6, 13.4, 0.5, 2.8, 1.3, 0.4, 1.0, 1.0, and 1.5%, respectively. These thermal populations were derived using Boltzmann distribution based on the Gibbs energies calculated at B3LYP/aug-cc-pVdZ level. A careful reader might note that the conformers have been ordered based on the energies calculated at B3LYP/6-311++G**, as shown in Figure 7.10. Not surprisingly, the relative energies of the higher energy conformers change with the level of calculations, but the three most stable conformers remain in the same order. Despite this, the simulated spectra with population from B3LYP/6-311++G** level energies did not match with the experimental spectrum (not shown here) and if one were to take these results at face value, some wrong conclusions could have been made.

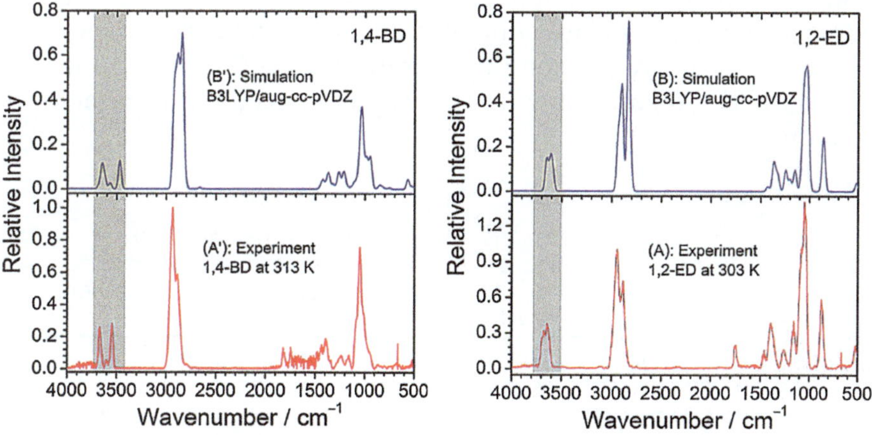

Figure 7.12 Experimental and simulated infrared spectra of 1,2-ethanediol and 1,4-butanediol. The grey region shows the OH stretching region. Reprinted with permission from P. Das, P. K. Das and E. Arunan, Conformational Stability and Intramolecular Hydrogen Bonding in 1,2-Ethanediol and 1,4-Butanediol, J. Phys. Chem. A, 2015, 119, 3710–3720. Copyright (2015) American Chemical Society.

The agreement between experimental and the simulated spectra at B3LYP/aug-cc-pVDZ level, especially in the O–H peaks, look much better, see Figure 7.12. The simulated spectrum for BD included contributions from the 10 most stable conformers.

Though, the low resolution spectra and the numerous conformers make it difficult to make rigorous conclusions, they do reveal some important information. For EG, the difference in wave numbers between the two O–H groups is significantly smaller compared to that found in BD. Though from the spectra it is not obvious, the intensity of the red-shifted peak in EG is comparable or lower than that for the other peak for EG and for BD, the red-shift is significant and there is an intensity enhancement of the red-shifted peak. Both are signatures for hydrogen bonding. The peak positions and relative intensities for some of the conformers of EG and BD, in which intramolecular hydrogen bond is expected, are given in Table 7.5. These are from calculations and the agreement between experimental and simulated spectra indicate that these are reasonable.

From Table 7.5, it can be seen that the red-shift for the $\nu(OH)_b$ in EG is of the order of 23–48 cm^{-1} and the two levels of calculations lead to slightly different results for the tGg$'$ and gGg$'$ conformers which could have intramolecular hydrogen bonding. The g$'$Gg$'$ conformer cannot have this and the two peaks are close and the two levels of calculations have their order reversed. For the most stable conformer, both levels of calculations predict the intensity of $\nu(OH)_b$ to be slightly smaller than that for the $\nu(OH)_f$. Especially for the O–H$\cdots$O hydrogen bonds, intensity enhancement was considered a key piece of evidence for hydrogen bonding. For BD, the gG$'$G$'$Gt conformer has a red-shift of 170 cm^{-1} and an intensity enhancement by a factor of more than 8!

Table 7.5 Calculated wave numbers and intensities (in parentheses) for the two O–H stretching vibrations in EG (top 3 rows) and BD (bottom three rows). The $\nu(OH)_b$ is for the OH group which could be the hydrogen bond donor and the $\nu(OH)_f$ is for the free OH.

Conformer	$\nu(OH)_b$		Higher $\nu(OH)_f$	
EG/BD	6-311++G**	aug-cc-pVDZ	6-311++G**	aug-cc-pVDZ
tGg$'$	3636.1 (34.3)	3605.8 (31.5)	3658.9 (35.3)	3653.2 (33.1)
gGg$'$	3600.4 (36.6)	3586.1 (33.5)	3645.2 (29.0)	3618.0 (21.2)
g$'$Gg$'$	3682.2 (15.7)	3638.0 (6.2)	3684.2 (56.8)	3635.8 (39.0)
g$'$GG$'$Gt	3503.6(256.1)	3470.7 (258.3)	3674.3(31.3)	3648.0 (30.8)
gG$'$G$'$Gt	3582.6(96.4)	3568.7(86.3)	3663.5(28.6)	3640.2 (27.2)
tG$'$TGt	3665.5(51.1)	3684.5 (48.5)	3665.6(0.0)	3684.5 (0.0)

Another interesting observation can be made based on the work of Ioganssen.[40] Based on the data from a large database, he derived an empirical relation between frequency shift and hydrogen bond enthalpy:

Interestingly, a 40 cm^{-1} shift corresponds to zero enthalpy for hydrogen bond formation according to this empirical relationship. Das *et al.*[43] have calculated the enthalpy of hydrogen bond formation based on this relationship for BD to be 16.2 kJ mol^{-1}. For the tGg′ conformer of EG, at the B3LYP/aug-cc-pVDZ level, it is 3.8 kJ mol^{-1}. For the same conformer, with the shift predicted at B3LYP/6-311+G**, this relation would not work as the shift is below 40 cm^{-1}. Previous experimental measurements[93–95] in the gas phase, solution and in an argon matrix have given values ranging from 33–40 cm^{-1}.

7.2.2.5 Vibrational Overtone Spectroscopy

Kjaergaard and co-workers have measured the overtone spectra for EG and BD.[44,96] The red-shift increases linearly from fundamental to overtones ($\Delta v = 1$, 2, 3, 4 for fundamental, first, second and third overtone) and so the $v(OH)_f$ and $v(OH)_b$ should be well separated in the overtones. However, in general overtones are weaker than fundamentals. More importantly, it has been observed that hydrogen bonding leads to further weakening of overtones and often the absence of overtones has been used as evidence for hydrogen bonding.[3,6] This did not deter Kjaergaard and co-workers from looking for higher overtones in these interesting molecules. The first overtone spectrum in the $v(OH)$ region is shown in Figure 7.13. This spectrum was collected at room temperature and they had to simulate the spectra including contributions from various conformers.[44] They used B3LYP/aug-cc-pVTZ level calculations to determine the populations of various conformers and their results are similar to those from the more recent work by Das *et al.* discussed in the previous section. The three most stable conformers are the same and they contribute close to 94% as compared to 90% in Das *et al.*'s calculation. Their simulation included only these three conformers. As we can see from Figure 7.13, the OH stretching peaks are well resolved compared to the fundamental transition shown in Figure 7.12 for EG.

However, whether there is an intramolecular hydrogen bond in EG is not directly determined in this work, as the calculated red-shift in the fundamental is 48 cm^{-1} in this work as well. The intensity ratio of the two peaks in the fundamental region is 1.2, again very similar to that reported by Das *et al.* They investigated the overtone spectra of

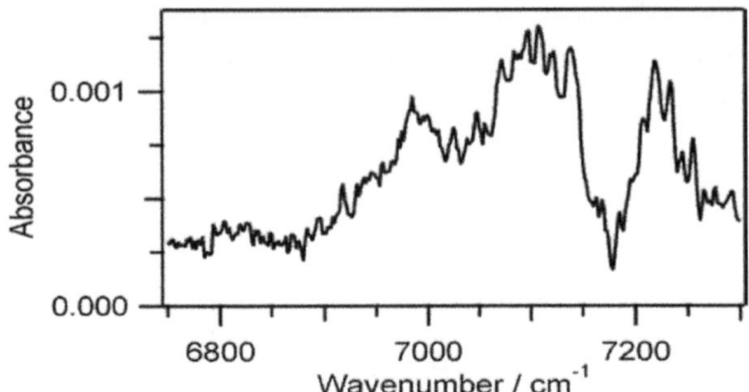

Figure 7.13   First OH overtone spectra of EG at room temperature measured using photoacoustic spectroscopy.
Reprinted with permission from D. L. Howard, P. Jorgensen, and H. G.Kjaergaard, Weak Intramolecular Interactions in Ethylene Glycol Identified by Vapor Phase OH-Stretching Overtone Spectroscopy, J. Am. Chem. Soc., 2005, 127, 17096–17103. Copyright (2005) American Chemical Society.

EG, 1,3-propanediol (PD) and BD later and found that the overtone intensities of PD and BD were significantly less compared to that observed in EG (see Figure 7.14). They again carried out extensive anharmonic local mode calculations and showed that the loss of intensity could be due to line broadening resulting from intramolecular vibrational relaxation.[96] While their experimental and theoretical efforts in recording and simulating these spectra are commendable, it is not clear if these results prove that there is intramolecular hydrogen bonding in EG. As mentioned below, empirical observations had shown earlier that hydrogen bonds lead to significant reduction in overtone intensities. This is what has been observed for both PD and BD and in both these cases, the intramolecular hydrogen bonding is well established based on geometrical, spectroscopic and electron density analysis. The fact that overtones have appreciable intensity for EG can be used as an argument against hydrogen bonding!

7.2.2.6   *Reduced Density Gradient of EG and BD*

Experimental and theoretical methods keep advancing. Atoms in Molecules theory has given results that were at odds with the 'gut feeling' of chemists leading to papers titled 'Chemical bonds without chemical bonding'[97] and 'Hydrogen bonding with a hydrogen bond'[98] depending on whether or not a BCP was found between two atoms that were expected or not expected to have a 'bond'. In the former example, the authors did the AIM analysis on iron trimethylene-methane complex and found that there was no BCP connecting iron

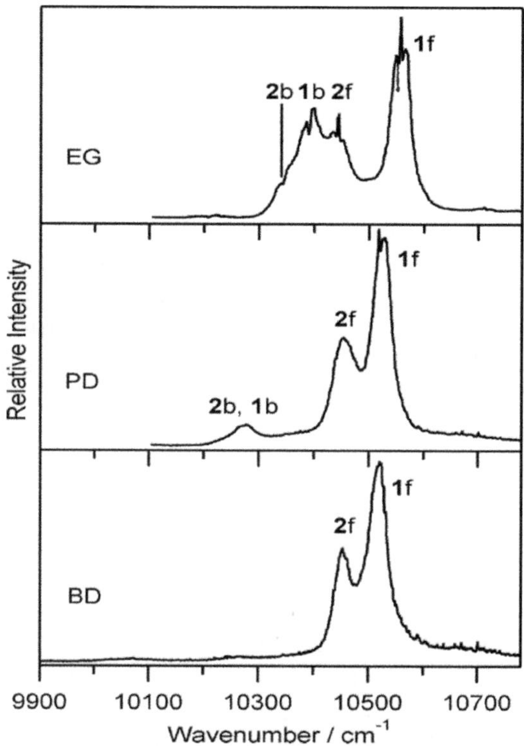

Figure 7.14 Third overtone spectra of EG, PD and BD. 1 and 2 represent the two OH groups and b and f represent 'hydrogen bonded' and 'free'.
Reprinted with permission from D. L. Howard and H. G. Kjaergaard, Influence of Intramolecular Hydrogen Bond Strength on OH-Stretching Overtones, J. Phys. Chem. A, 2006, 110, 10245–10250. Copyright (2006) American Chemical Society.

to the three C atoms, as had been concluded by chemists based on the barrier to rotation of the trimethylenemethane group in this complex. The analysis, however, showed one BCP connecting iron to the central carbon. In the latter example, a BCP was found between an H of H_2O and the C of CH_4 in a complex formed between H_2O and CH_4. Most of the theoretical studies on the CH_4–H_2O complex focused on C–H$\cdots$O hydrogen bond geometry, though earlier microwave spectroscopic investigations pointed to the structure in which O–H from H_2O was pointing towards the centre of a tetrahedral plane in CH_4. This geometry was considered to result primarily from dispersive forces and not hydrogen bonding. Raghavendra and Arunan used AIM analysis to show that there was indeed a BCP connecting H and C forming an O–H$\cdots$C hydrogen bond and the geometry is in fact evidence for a pentacoordinate C *i.e. en route* to the elusive methanium,

$(CH_5)^+$ ion.[98] This was later observed experimentally in OH interactions with CH_3 groups adsorbed on a surface.[99] In a way, this is an ideal example to show that an interaction was identified in the gas phase microwave study, clarified by theory and eventually observed in a solid surface. Later studies showed that O–H$\cdots$C interactions are stronger than C–H$\cdots$O interactions for larger hydrocarbons as well.[100]

From the beginning, the AIM theory has had its critics and there have been attempts to look for more variables and parameters but that is not our main concern here. A non-covalent index (NCI) has recently been introduced to identify interacting atoms and centers even in the case where there is no BCP.[46] In this analysis, a normalized and reduced (dimensionless) electron density gradient is defined as the NCI index, s, as given below:

$$s = \frac{1}{\left[2((3\pi)^2)^{\frac{1}{3}}\right]} \frac{|\nabla\rho|}{\rho^{\frac{4}{3}}}$$

Electron density topology has indeed been accepted as a quantity directly relatable to the state of the system. The authors of the NCI index argue that a more general quantity describing the changes in electron density would be a better choice than simply its local values. It is important to point out that the Laplacian, $\nabla\rho$, is the second derivative of the electron density. A bond critical point is a saddle point in electron density plotted in three dimensions, being minimum along the bond path connecting the two atoms and maximum along the perpendicular direction. Thus a BCP represents a critical point and hence the local value of the electron density should be an extremum in all directions. On the other hand, examination of the NCI index, s, and looking at its isosurfaces can reveal how the electron density varies over all the regions. The plot of NCI isosurfaces, now known as an NCI plot, has now become popular among scientists investigating weak interactions. Johnson and co-workers also realized that the plot of s *vs.* sign(λ_2)ρ is more instructive. It is to be noted that λ_1, λ_2 and λ_3 are the three eigenvalues of the Laplacian and for a BCP, λ_2 is negative and for a ring critical point (RCP), it is positive. Such plots for EG and BD are shown in Figure 7.15. For BD, electron density is reduced drastically and reaches zero (resulting in a BCP) and for EG, electron density is reduced drastically, but does not reach zero.

The data presented in Figure 7.15 have convinced Kjaergaard and co-workers about the presence of intramolecular hydrogen bonding in both EG and BD.

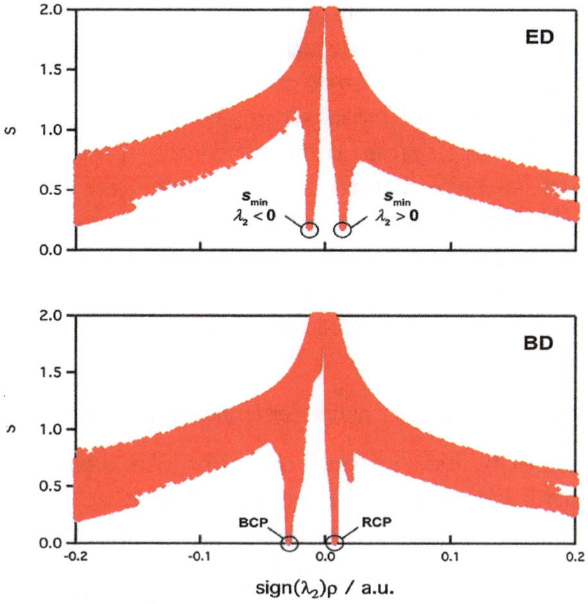

Figure 7.15 The reduced density gradient plotted against sign(λ_2)ρ. On the left, $s = 0$ points denote BCPs and on the right they denote RCPs. Note the similarity between the two plots, though for EG, no BCP or RCP are found.
Reprinted with permission from J. R. Laane, J. Contreras-Garcia, J. P. Piquemal, B. J.Miller, H. G. Kjaergaard, Are Bond Critical Points Really Critical for Hydrogen Bonding?, J. Chem. Theory Comput., 2013, 9, 3263–3266. Copyright (2013) American Chemical Society.

7.2.2.7 *Valence Internal Coordinate Analysis of EG and BD*

For a simple diatomic molecule having one vibration, it is well known that the stretching frequency is related to the force constant, which can also be calculated as the second derivative of the potential energy *vs.* bond distance plot. The force constant is a measure of the bond strength. For a polyatomic molecule with N atoms, having $3N - 6$ vibrational degrees of freedom, harmonic approximation leads to $3N - 6$ normal coordinates of vibrations. Hence, for a molecule like EG or BD, it is not straightforward to relate the force constants of the normal modes to individual bond strengths. However, the normal coordinates can be transformed to local or internal coordinates and a valence internal coordinate can be formed in which the transformed 'relaxed force constants' could become a meaningful measure of individual bond strengths. Brandhorst and Grunenberg showed this for a series of covalent molecules and also extended it for hydrogen bonded complexes.[101] This approach can work for hydrogen bonds only if they are included as part of the valence internal coordinates, as this is not usually the case. Manogaran and co-workers have shown that hydrogen bonds can be included in 'equivalent valence internal coordinates' and the relaxed force constants from this analysis would be a direct measure of the bond strength.[102]

Pandey *et al.*[103] transformed the normal coordinates for EG and BD into equivalent valence internal coordinates including an 'intra-molecular hydrogen bond' coordinate. From their analysis, they determined the relaxed force constant for EG and BD as 0.23 mdyn/Å and 0.08 mdyn/Å. Clearly, the 'intramolecular hydrogen bond' in EG is significantly weaker making it difficult to characterize.

7.2.2.8   *Dynamics of an Intramolecular Hydrogen Bond*

In the Introduction, we pointed out that the barrier for torsional motion is more important for characterizing a 'hydrogen bond'. Thermal energy along intermolecular torsional coordinates is enough to make solid H_2S appear like a sphere misleading everyone, including Pauling, to conclude that H_2S has van der Waals interactions while H_2O has hydrogen bonds (see Figure 7.1). This is a case where intermolecular hydrogen bonding is broken due to thermal energy along torsional coordinates, but H_2S still remained a solid as energy along the stretching coordinates was not enough to liberate molecules into the gas phase. Among the normal coordinates for EG, there is indeed one that may have significant contribution from $O\cdots O$ stretching. In fact, as discussed above, Manogaran and co-workers transformed the normal coordinates into valence internal coordinates that included the intramolecular hydrogen bond stretching. The question we address in this section is how close the two OH groups would come, during the thermal vibrational motion. Looking at the subtle difference between EG and BD, one might ask, if a BCP, if not a bond, would appear corresponding to an intramolecular hydrogen bond during the course of this vibration. See Figure 7.16.

Arunan and Mani[4] carried out normal mode analysis on the tGg' conformer of EG and identified a mode at 173 cm^{-1}, nominally corresponding to $O\cdots O$ stretch. During this vibration, the two O–H groups do indeed come closer. They calculated the potential energy along this coordinate and it is shown in Figure 7.16. Moreover, at every point along this coordinate, they carried out AIM analysis. Interestingly, the structure observed just near the $v = 1$ vibrational level in this mode, exhibited the BCP corresponding to an intra-molecular hydrogen bond. Of course, on the other side of the vibra-tions, when the $O\cdots O$ distance increases, no BCP appears. The

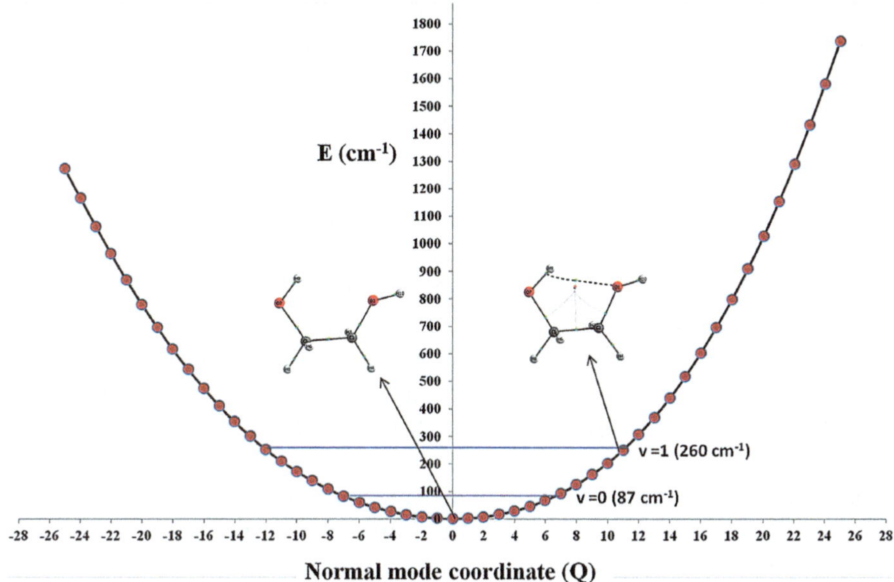

Figure 7.16 Potential energy along the nominal O···O normal coordinate of tGg′ conformer of EG. AIM analysis was carried out along this coordinate and a BCP corresponding to an intramolecular hydrogen bond appears at about the $v = 1$ level along this coordinate.
Reproduced from ref. 4 with permission from the Royal Society of Chemistry.

thermal population at $v = 1$ level in this low frequency vibration is more than 40% and so EG will be sampling this region of the geometry. Within the **AIM** framework, an intramolecular hydrogen bond is formed and broken during this vibration. Can this be verified experimentally? Zewail and co-workers have developed 4D electron diffraction and microscopy techniques to get time resolved images of atoms in motion.[104] These techniques have been used to capture reactants, transition states and products during the course of a chemical reaction.[105] However, they may still not be able to provide time-resolved electron density maps.

7.2.2.9 Does EG Have an Intramolecular Hydrogen Bond? Does it Really Matter?

In this section, so far we have discussed results from microwave, infrared, and vibrational overtone spectroscopy of gaseous EG, NMR spectroscopy of EG in solution and X-ray diffraction from crystalline EG. We have also covered *ab initio* electronic structure theory, harmonic and anharmonic frequency calculations at advanced levels of theory, and AIM and NCI analysis of the total electron density. These methods have all yielded varying answers to this question. That the tGg' conformer is the global minimum for EG has been proved beyond doubt by all methods. Roberts and co-workers attribute it to the 'gauche effect' but this is almost like not offering an explanation. It is also at odds with the fact that the three highest energy conformers happen to be all 'gauche', gGg, tGt, and tGg, with the two hydroxyl groups in gauche conformation, but there is no possibility of an intramolecular hydrogen bond. The recent IUPAC technical report has given a nearly all-inclusive definition of hydrogen bonding in terms of donors and acceptors but insisted on some 'evidence for bond formation'. The two experimental techniques suggested are IR and NMR and these two techniques have yielded contradictory conclusions for EG.

Does it matter whether EG has an intramolecular hydrogen bond? Is this only a question of semantics? It turns out EG is one of the training set for parametrizing molecular mechanics programs[106] and also a ubiquitous synthon in sugar.[107] Molecular mechanics calculations have identified the symmetric g'Gg' or gG'g conformer as the global minimum perhaps due to poor parametrization.[108] However, recent accurate *ab initio* calculations show it to be third in energy with only about 13% population at room temperature.[43,44] Based on all the experimental and computational results presented in this section, one could conclude that intramolecular hydrogen bond in EG is very weak and could play a role in the stabilization of the global minimum structure. Chopra *et al.* had expressed a concern that without a bond critical point, there could not be an 'electron wire' or the phenomenon of a cooperative network as in liquid water.[90] Interestingly, as discussed in the previous section, now we know that thermal vibration is enough to bring in this continuity.

7.3   Red, Blue and No-shifting Hydrogen Bonds: Spectroscopy and Theory

The discussion in Section 7.2.2 focused on the marginal red-shift in O–H stretching frequency as evidence for a hydrogen bond. The recent IUPAC report also suggests this as key experimental evidence for hydrogen bonding. However, blue-shifting hydrogen bonds have also been identified causing a lot of confusion and debate in the first decade of this millennium. Initially, they were called an 'anti-hydrogen bond'[109] and an 'improper hydrogen bond'[48] before they could be well understood. When the dust settled, it was realized that these classes of hydrogen bonds are just like any other hydrogen bonds. Moreover, they have been known for much longer. The initial reports on blue-shifting hydrogen bonds came many decades ago, but just as a matter of fact. There has been unambiguous experimental proof of blue-shifting hydrogen bonds in recent years and some examples are discussed in this section, following a brief history.

Blue-shifting hydrogen bonds came in to focus in 1998 following some moderate level theoretical calculations by Hobza and co-workers on the benzene dimer and it was termed anti-hydrogen bonding.[109] Somewhat coincidentally, this term was previously used by Klemperer and co-workers[110] to describe the $HF \cdots ClF$ complex, as they expected it to form a hydrogen bonded $ClF \cdots HF$ complex, and it will be discussed in the next section in the context of halogen bonds. Hence, in the following publications, Hobza and co-workers termed it an 'improper hydrogen bond'.[48] Coming back to the blue-shift in the benzene dimer, MP2/6-31G** level calculations predicted a 42 cm^{-1} blue-shift in one of the C–H stretching frequencies. As none of the C–H stretching vibrations would be localized in a specific C–H bond in benzene, this should have raised some doubts initially about this characterization of the T-shaped benzene dimer, having $C-H \cdots \pi$ hydrogen bonding. However, they also reported a blue-shift of about 50 cm^{-1} in the benzene–$HCCl_3$ complex and as there is only a single C–H bond in the $HCCl_3$, identifying the blue-shift in hydrogen bond donor is not ambiguous. Not surprisingly, the unambiguous experimental verification of a blue-shift of a moderate 20 cm^{-1} was reported for a complex between HCF_3 and C_6H_5F in early 2001.[49] This was done using infrared ion-depletion spectroscopy.

It took a few more years to record the infrared spectrum of the benzene dimer using ion-dip spectroscopy[111] and the result disproved the predictions of a large blue-shift and showed instead a much smaller red-shift of about 3 cm^{-1}. This small red-shift was confirmed later by direct infrared absorption spectroscopy of the benzene dimer by Chandrasekaran *et al.*[112] Although the initial prediction of the blue-shift for the benzene dimer turned out to be incorrect, the 1998 paper by Hobza and co-workers created interest and eventually led to the observation of several blue-shifting hydrogen bonds. In most cases, the blue-shift is observed when the hydrogen bond donor is C–H. Van der Veken and co-workers[113] recorded the infrared spectra of cryo-solutions in liquid Ar and reported a blue-shift of about 18 cm^{-1} for the $(CH_3)_2O \cdots HCF_3$ complex. Wategaonkar and co-workers have reported a blue-shift of 16 cm^{-1} in the 3-methylindole-HCF_3complex.[51]

Interestingly, the 3-methylindole-HCCl$_3$ complex showed a moderate 2 cm^{-1} shift. These very small blue-shifts suggest that there may be hydrogen bonded compelxes having zero-shift in X–H stretching frequency. In fact, Joseph and Jemmis used a simple model of X–H$\cdots$Y hydrogen bonds and predicted red-, blue- and zero-shift hydrogen bonds.[50] For a series of hydrogen bonded complexes, they varied the H$\cdots$Y distance and optimized all the coordinates to minimize the interaction energy. They found that the energy minima in different complexes happened when the X–H distances were longer (typical red-shifted), shorter (blue-shifted) or remain unchanged (zero-shift). There have been numerous papers explaining the blue-shift in hydrogen bonds[114–116] and from a spectroscopist's point of view, the work of Hermansson deserves mention. In typical hydrogen bonds, the X–H bond length increases and it leads to an increase in the dipole moment of X–H and to stronger dipole–dipole interaction. In the blue-shifting hydrogen bonds, Hermansson showed that the dipole moment of X–H increased when the X–H bond distance decreased.[117]

7.4 Spectroscopic Investigations on Non-covalent Bonds from Other Elements

7.4.1 Halogen Bonds by Microwave Spectroscopy

In the last three sections, we have discussed the spectroscopic characterization of inter- and intramolecular hydrogen bonds. While similar interactions by other elements have been thought about for a long time, the last decade has seen a sudden spurt of work in this area. In our view, Klemperer and co-workers' characterization of the HF$\cdots$ClF complex[110] in 1976 is the first direct spectroscopic observation of a 'halogen bond', as we understand it today. Interestingly, their motivation to study this complex arose as they expected the 'hydrogen bond' (ClF$\cdots$HF) to win over the Lewis acid–base structure, HF$\cdots$ClF, and found the opposite to be the case. They used molecular beam electric resonance to investigate the complex and the experimental rotational constants led them to identify the structure as HF$\cdots$ClF. As they had already solved the HF$\cdots$HF structure they noted that the 'bonding in both complexes is quite similar' and chose to call it an 'anti-hydrogen bond' rather than a 'chlorine or halogen bond'. Within a few years, they went on to look at the HF$\cdots$Cl$_2$ complex.[118] This time they expected the structure to be of the Lewis acid–base type, similar to that of HF$\cdots$ClF and microwave spectroscopy indeed confirmed this. They compared these results with Hassel's extensive crystallographic studies[119] on complexes in which halogens interacted with Lewis bases! They also pointed out the extensive spectroscopic studies on halogen complexes in solution, which provided evidence for Mulliken to classify them as 'charge transfer complexes'.[120] These were not called halogen bonds in those days though the nature of interaction was well understood as arising from the overlap of a Lewis base (typically a lone pair or π electrons) with the empty σ^* orbital. We will denote them as halogen bonded from now on.

Legon and co-workers deserve credit for not only extending these studies but also for comprehensive investigations on a series of B$\cdots$ClF and B$\cdots$Cl$_2$ complexes, in which B is a Lewis base. They also investigated numerous complexes having Br and I as the 'halogen' atom interacting with the base. Legon summarized these studies in exhaustive reviews and identified them as 'halogen bonded'[12,121] This helped them in showing that the halogen bond angles in B$\cdots$ClF complexes are invariably more linear than the corresponding hydrogen bond angle in B$\cdots$HF complexes. The B$\cdots$H distances are usually much smaller, enabling secondary interaction between F and some atoms in B and the angle $\angle$B–H–F is reduced from 180°. However, the B$\cdots$Cl distances are significantly larger and the secondary interactions nearly vanish.

7.4.2 Halogen Bonds by Infrared Spectroscopy

For a hydrogen bond, the red-shift in X–H stretching frequency has become a hallmark for identification and there has been a plethora of gas phase infrared studies on hydrogen bonded complexes. To the best of our knowledge, there appears to be no gas-phase infrared spectroscopic investigations on halogen bonded complexes! A comprehensive review on the halogen bond published this year in Chemical Reviews refers to infrared studies mostly in solid phases and to a limited extent in the liquid phase.[122] We focus in this section on infrared experiments in a rare gas matrix and on cryo-solutions.

Hunt and Andrews investigated halogen–HF interactions in a solid Ar/Ne matrix[123] a few years after Klemperer's investigations in the molecular beam. They were aware that in a matrix, higher energy minima could be trapped, and they could indeed identify both the 'halogen bonded' (HF$\cdots$Cl$_2$) and 'hydrogen bonded' (Cl$_2\cdots$HF) structures. The matrix experiments revealed two strong additional peaks in the ν_{HF} region, one red-shifted by about 50 cm^{-1} and the other red-shifted by about 15 cm^{-1}. The former was assigned to the hydrogen bonded structure and the latter to the 'halogen bonded' structure. As mentioned earlier, formation of hydrogen bonds leads to an intensity enhancement and this is not expected for the ν_{HF} peak in halogen bonded structures. The fact that both are strong implies that the halogen bonded structure is more abundant. They also investigated Br$_2$/HF and F$_2$/HF complexes. For the former, both hydrogen bonded and halogen bonded structures were identified and for the latter only the hydrogen bonded structure was present. To this day 'halogen bonds' with F are rare and there have been only a few reports in which F has been shown to be a halogen bond donor in crystals.[124,125] The red-shift in ν_{HF} for F–F$\cdots$HF, ClF$\cdots$HF, Cl–Cl$\cdots$HF, BrF$\cdots$HF, and Br–Br$\cdots$HF were determined to be 4 cm^{-1}, 50 cm^{-1}, 59 cm^{-1}, 62 cm^{-1} and 68 cm^{-1}, respectively. These shifts are correlated with the bond strength and one can see that commonly held views about hydrogen bonding would not hold in this series, as the least electronegative and non-polar Br$_2$ forms a stronger hydrogen bond than the polar ClF molecule which has the most electronegative F as hydrogen bond acceptor.

One important thing to note here is that Hunt and Andrews did not report the assignment or analysis of the vibrational modes in the halogen molecule. The H–X vibration remains as a pure H–X vibration in the hydrogen bonded complex and it can be used as a measure of bond strength. However, the X–Y vibrations in the halogen molecule, being much lower in frequency can get mixed with other low frequency vibrations in the Lewis base. For example, in a FCl$\cdots$CH$_3$ halogen bonded complex, Raghavendra and Arunan[73] noted a large blue-shift in the nominal F–Cl stretching following the complex formation, in their computational study. However, a careful analysis of the vibrational mode indicated that normal modes having significant contributions from Cl–F stretch also had contributions from CH$_3$ deformation modes. They called it an anomalous one-electron blue-shifting halogen bond. They also showed that frequency shifts cannot be directly correlated with the bond strength for halogen bonded complexes.

Van der Veken, Herrebout and co-workers have carried out comprehensive studies on halogen bonded complexes in cryo-solutions *i.e.* in liquid krypton. Herrebout has recently written a comprehensive review of their work.[126] In this section, we discuss their work on halogen bonded complexes with S as an acceptor, as a representative example.[127] Infrared and Raman spectra of dimethyl sulfide complexes with CF_3I, CF_3Br or CF_3Cl were recorded in liquid krypton. They also recorded the spectra of the two molecules in liquid krypton independently and identified the new peaks that are observed in the mixed solution. One thing that was striking about their results and discussion is that all the vibrational modes of both molecules could be listed in tables (not included here) for the monomer and compared with those assigned for the halogen bonded complex. This may be contrasted with the infrared spectroscopic investigations on hydrogen bonded complexes, in which the X–H stretching frequency alone is used as evidence. One of the main reasons for this is that, in halogen bonded complexes, the vibrational modes are not localized in one bond, such as X–H. These vibrational frequencies can, however, be computed for both monomer and complex and they offer further evidence for the assignment of complex frequencies. The shifts between the monomer and complex for fundamental transitions are small and vary from -13 cm^{-1} to $+5$ cm^{-1}, with several of the bands showing no shift.

They were able to identify a $2:1$ complex of $(CH_3I)_2 : (CH_3)_2S$ as well, the structure of which is shown in Figure 7.17. This assignment depended on the fact that the complex has nearly a local center of inversion and the ν_3 (C–I stretch) could have symmetric and asymmetric combinations in the complex, the former appearing strong in Raman and the latter appearing strong in infrared spectra. Most readers may be familiar with the mutual exclusion principle and in particular the example of CO_2, which has a center of inversion and the symmetric stretch is Raman active while the asymmetric stretch is IR active.

Another notable feature of this work is that they were able to determine the intermolecular stretching frequency for the $1:1$ complex to be 100 cm^{-1} and 121 cm^{-1} for the $CF_3I : S(CH_3)_2$ and $CF_3Br : S(CH_3)_2$, respectively.

Figure 7.17 Structure of the doubly halogen bonded $CF_3I : (CH_3)_2S : CF_3I$ complex determined from infrared and Raman spectra observed in a mixed solution in liquid krypton and ab initio calculations.
Reproduced from ref. 127 with permission from the Royal Society of Chemistry.

7.4.3 Carbon Bonds by Microwave Spectroscopy

Mani and Arunan[19] proposed a 'carbon bond' analogous to the 'hydrogen bond' in 2013, when halogen, chalcogen and pnicogen bonds were already known. However, their proposal did not result from a careful analysis of intermolecular bonds across the periodic table but it was an offshoot of microwave spectroscopic investigations on the argon-propargyl alcohol complex,[53] which may not be of much general interest. Like PHA, propargyl alcohol (PA from now on) is a multifunctional molecule having an H–C≡C group and an OH group, both of which could act as hydrogen bond donors or acceptors, or in other words as electron acceptors or donors. The gauche form has the OH group pointing towards the π system and is more stable than the trans. It appears that no one has proposed an intramolecular hydrogen bond to account for this stability. Both AIM and NCI analysis do not show any specific attractive interaction in the gauche form, though it is still more stable than the trans form and we may call this the 'gauche effect' until a better understanding emerges. Maybe there is still some room for advances in experimental and theoretical techniques to look for hydrogen bonding.

During this investigation, the structure of the Ar–PA complex with PA in gauche and trans forms were both optimized by theory. Experiment gave evidence only for the complex in which PA was in gauche form. In it, as shown in Figure 7.18, the Ar atom interacts with both the OH group and the π electrons, as might be expected from the structures of Ar–CH$_3$OH and Ar–HCCH. As the microwave spectrum does not reveal the nature of intermolecular interactions directly, the authors carried out AIM theoretical analysis on both forms. As they

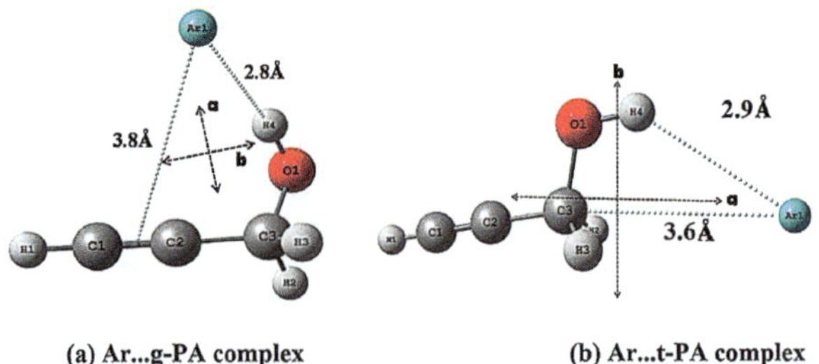

(a) Ar...g-PA complex   (b) Ar...t-PA complex

Figure 7.18   Optimized structures of the Ar–PA complex having PA in gauche (a) and trans (b) forms. The AIM analysis (not shown here) confirms the interactions indicated by the dotted lines.
Reprinted from ChemPhysChem, D. Mani and E. Arunan, Microwave Spectroscopic and Atoms in Molecules Theoretical Investigations on the Ar-Propargyl Alcohol Complex: Ar-H2O, Ar–π and Ar-C Interactions, 14, 754–763. Copyright (2013), with permission from Elsevier.

expected, the global minimum structure had O–H$\cdots$Ar and Ar$\cdots\pi$ interactions and two corresponding BCPs were observed. The AIM analysis on the Ar-*trans*-PA complex revealed something unexpected. It showed an expected BCP connecting O–H to argon and also an unexpected BCP between Ar and the methylene C! Although it is not unusual to find such BCPs in unexpected locations in a molecule, instead of ignoring this, they did an AIM analysis on the CH_3OH–Ar complex and found two similar BCPs.

Raghavendra and Arunan[98] had earlier used AIM theory to confirm the O–H$\cdots$C hydrogen bond in the H_2O–CH_4 complex and rationalized it based on the electrostatic potential (ESP) in CH_4 which showed negative potential in the four face centers of the tetrahedron. The ESP of methanol was calculated and it was found that the tetrahedron face opposite to the OH group in CH_3OH was positive. It is obvious that if H_2O were to interact with this face of CH_3OH, then it could be a Lewis base bringing its lone pair of electrons towards this electron deficient region. They optimized the H_2O–CH_3OH complex and could identify two structures in which the two molecules switched their role as Lewis acid or base, electron acceptor or donor, or hydrogen bonded or carbon bonded! One can see the similarity between this and Klemperer's investigation on the HF–ClF complex, nearly four decades ago.[110] Mani and Arunan carried out AIM analysis on H_2O–HF, H_2O–ClF and H_2O–CH_3F and the results are shown in Figure 7.19. One can see the similarity between hydrogen bond, halogen bond and carbon bond.

Preliminary microwave spectroscopic experiments on H_2O–CH_3F have so far not yielded any evidence of a carbon bonded structure.[128] However, it turns out that microwave spectroscopy had already yielded evidence for a 'carbon bond' decades ago, in work by Klemperer's group again, except they did not call it so. They reported[129] the structure of the complex between HCN and CO_2 in 1984. It turned out to be T-shaped with the N end of HCN pointing towards C in CO_2, but again they were looking for the linear hydrogen bonded OCO–HCN structure. A few years later, during the search for $HCN(CO_2)_2$ trimer, Gutowsky's group[130] identified the spectrum of linear OCO:HCN. In those days, molecular beam experiments invariably gave only the global minimum structure of complexes. The main difference between the experiments carried out in Gutowsky's and Klemperer's laboratories were the carrier gas used to produce the molecular beam. Klemperer's group used argon. Gutowsky's group had used first run neon (a 70:30 mixture of He and Ne). Later on, Ruoff *et al.* used the carrier gas dependence to explore the kinetics of complex formation.[131]

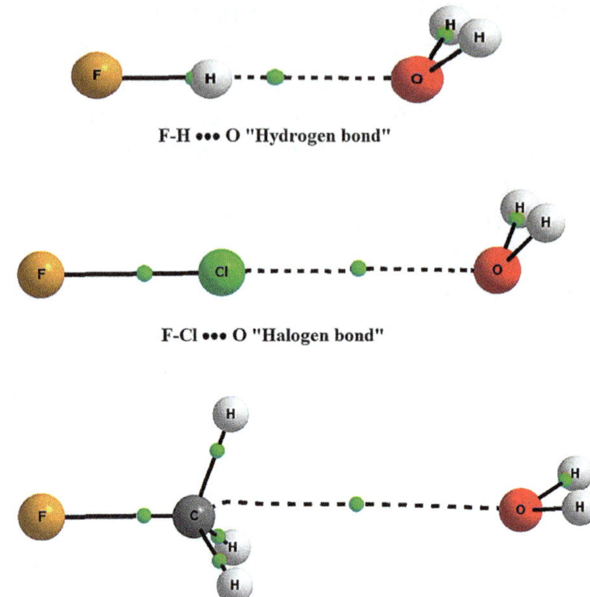

Figure 7.19   AIM theoretical analysis on H_2O complexes with $HF/ClF/CH_3F$ reveal the similarities between hydrogen, halogen and carbon bonds. The structure of the H_2O–CH_3F complex would normally be shown as representative of the ill-defined hydrophobic interactions.
Reproduced from ref. 34 with permission from the Royal Society of Chemistry.

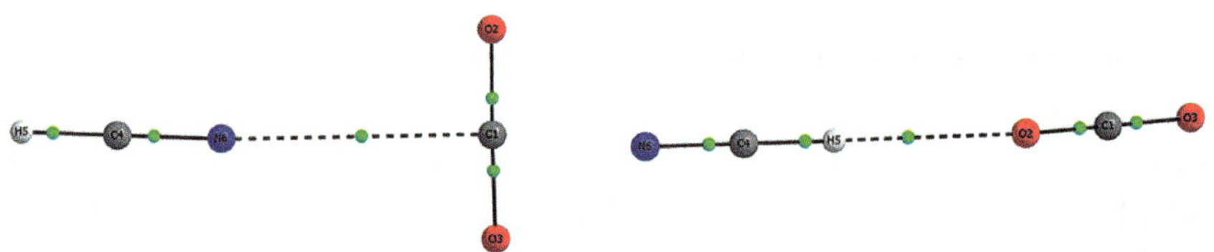

Figure 7.20   AIM analysis on the two structures of the $HCN:CO_2$ complex determined by microwave spectroscopy. The T-shaped structure is carbon bonded and the linear structure is hydrogen bonded.

We carried out AIM analysis on the two structures of $HCN:CO_2$ complexes and the result is shown in Figure 7.20. Now we can call the linear structure hydrogen bonded and the T-shaped structure carbon bonded.

An attempt to observe the red-shift in C–F stretching frequency following the carbon bond formation in H_2O–CH_3F has not succeeded yet.[128] However, Guru Row and co-workers identified carbon bonding in crystal structures, a few months after the suggestion by Mani and

Arunan.[20] Southern and Bryce provided NMR evidence for carbon bonding in 2015.[21] More recently, Frontera and co-workers have analyzed the protein data bank and investigated XCH_3 interactions with O atoms.[132] They specifically addressed the question of whether they are C–H$\cdots$O hydrogen bonds or X–C$\cdots$O carbon bonds. Their extensive analysis led them to conclude the importance of carbon bonding in two protein substrate complexes.

7.4.4 n–π* Interaction by Infrared Spectroscopy: Is it a Carbon Bond too?

A new type of non-covalent interaction, called an n–π* interaction, has been discussed in this millennium. It perhaps started with the NMR investigation on collagen stability published in 2002 by Raines, Markley and co-workers.[133] They pointed out that there is an attractive interaction between two carbonyl groups in collagen such that the π* orbital of one carbonyl interacts with the lone pair from the other carbonyl. A closer look at a pictorial description shows that the C from one C=O interacts with the O from the other and they are almost at right angles, *i.e.* 90°. As discussed earlier, hydrogen bonding and halogen bonding are n–σ* interactions, as the H/X bond donors almost always have a single sigma bond i.e. D–X$\cdots$Y. Unlike hydrogen or halogen atoms, the C atom can appear in diverse environments and have an X–C single bond, an X=C double bond or an X≡C triple bond. When it appears in a C=O group, C happens to be the positive end and it interacts with the electron donor and this is precisely what we see in an X–C$\cdots$Y carbon bond. The lone pair on Y interacts with the σ* orbital of X–C, providing the molecular orbital view for a carbon bond. In the proposed n–π* interaction, a lone pair from Y is interacting with the π* orbital leading to an X=C$\cdots$Y interaction. This n–π* interaction has attracted significant interest over the last decade and most of the publications in this area have discussed results from the condensed phase.[133–135]

Earlier this year, Das and co-workers[54] studied phenyl formate using ion detected infrared spectroscopy and observed a red-shift of about 30 cm^{-1} in the *cis* form, which could have an n–π* interaction, unlike the *trans* form, where such an interaction is not possible. They claimed this to be the first spectroscopic confirmation of n–π* interaction. However, in this case the π* orbital is localized in the phenyl ring unlike in the case of collagen. We carried out AIM theoretical analysis on the two structures and the results are shown in Figure 7.21. As with the EG conformers, the AIM analysis does not

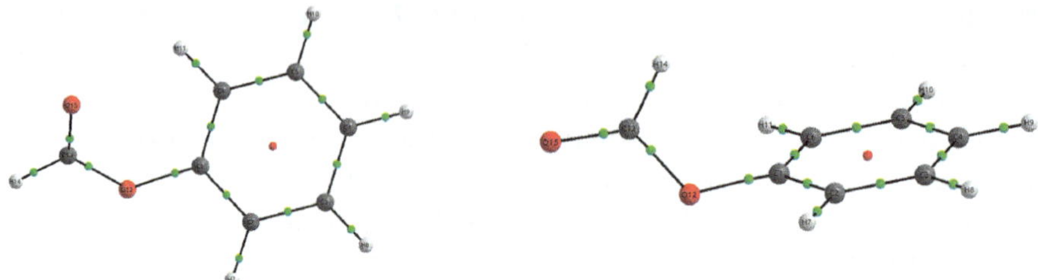

Figure 7.21 Structure of the cis (left) and trans (right) phenyl formate shown as an AIM molecular graph. The cis form is more stable by 5 kJ mol^{-1} which is attributed to an n–π* interaction between the lone pair in the carbonyl oxygen and an aromatic π* orbital. Clearly the AIM theory does not show any intramolecular interactions in either conformer.

reveal any evidence for such an intramolecular interaction in phenyl formate. However, out of curiosity, the NCI analysis was done as well and indeed it shows some evidence for an n–π* interaction.

7.4.5 Spectroscopic Investigations on other Intermolecular Bonds

It appears that most of the studies on other intermolecular bonds are either in the condensed phase or computational studies on isolated molecular complexes. There are very few spectroscopic investigations on these systems. As this chapter was nearing conclusion, we came across a paper by Sundararajan and co-workers.[136] They have investigated the PCl_3–benzene complex using matrix isolation infrared spectroscopy and concluded there is a 'pnicogen bond' between the P and a π centre. However, their conclusions relied on computational predictions, as the shift in vibrational frequencies between monomer and complex was small. Considering the recent advances in intermolecular bonding, we are convinced that gas phase spectroscopy can still contribute significantly to this field.

7.5 Concluding Remarks

In this chapter, we have discussed how spectroscopy of molecular complexes in a cold environment, such as in a molecular beam, rare gas matrix or rare gas liquid, has contributed to our fundamental understanding of intermolecular interactions. When Pauling concluded that water was bound by 'hydrogen bonding' and H_2S had van der Waals interactions, our knowledge about intermolecular interactions was in its infancy. Now we know that both H_2O and H_2S can form hydrogen bonds. We also know that O and S can form similar bonds that have been termed chalcogen bonds. Particularly over the last decade, both experimental and theoretical advances have helped us investigate intermolecular interactions involving many other atoms beyond hydrogen. There appears to be a frenzy now in coming up with names for these intermolecular bonds! While some purists find these new names unnecessary, we do feel that it is important to recognize these interactions. It is too early to decide if any or all of these names will survive in Science. With specific reference to the interaction shown between the CH_3 tetrahedral face in CH_3OH and the O of H_2O, without electron density analysis, many would have concluded it to be a trifurcated hydrogen bond. Following such an analysis, one could see the $O-C\cdots O$ carbon bond. Why is it important? Chemists learn about S_N2 reactions from early days. When an X approaches CH_3Y, it does not abstract an H atom but bonds with C and displaces Y. One can see that 'carbon bonded complexes' are stable intermediates along the reaction coordinate for S_N2 reactions. Hydrogen bonding has been known to form intermediates in proton transfer reactions. Along the proton transfer coordinate, one end has an $X-H\cdots Y$ hydrogen bond and the other end has $X\cdots H-Y$ hydrogen bond. Similarly, along the S_N2 reaction coordinate, there is a stable intermediate having an $X\cdots C-Y$ carbon bond on one end and a stable intermediate having an $X-C\cdots Y$ carbon bond on the other end.

In closing, we quote Richard Feynman on naming things. He said:[137] *You can know the name of a bird in all the languages of the world, but when you're finished, you'll know absolutely nothing whatever about the bird... So let's look at the bird and see what it's doing – that's what counts. I learned very early the difference between knowing the name of something and knowing something.* Spectroscopy has indeed helped in knowing intermolecular bonds!

Acknowledgements

The authors thank the Department of Science and Technology, Council of Scientific and Industrial Research, Indian Institute of Science for financial support. They acknowledge the works of past doctoral students from their group including Dr Mausumi Goswami, Dr Prasanta Das and Dr Devendra Mani, which have all been published and referred to here. They thank Prof. K. S. Viswanathan for providing a modified version of Figure 7.8.

References

1. J. D. van der Waals, PhD Thesis, University of Leiden, 1873.
2. B. Chu, *Molecular Forces, Based on the Baker Lectures of Peter J W Debye*, John Wiley & Sons, New York, 1967.
3. L. Pauling, *The Nature of the Chemical Bond and the Structure of Molecules and Crystals; An Introduction to Modern Structural Chemistry*, Cornell University Press, Ithaca, New York, 1960.
4. E. Arunan and D. Mani, *Faraday Discuss.*, 2015, **177**, 51–64.
5. E. Arunan, G. R. Desiraju, R. A. Klein, J. Sadlej, S. Scheiner, I. Alkorta, D. C. Clary, R. H. Crabtree, J. J. Dannenberg, P. Hobza, H. G. Kjaergaard, A. C. Legon, B. Mennucci and D. J. Nesbitt, *Pure Appl. Chem.*, 2011, **83**, 1619–1636.
6. E. Arunan, G. R. Desiraju, R. A. Klein, J. Sadlej, S. Scheiner, I. Alkorta, D. C. Clary, R. H. Crabtree, J. J. Dannenberg, P. Hobza, H. G. Kjaergaard, A. C. Legon, B. Mennucci and D. J. Nesbitt, *Pure Appl. Chem.*, 2011, **83**, 1637–1641.
7. M. Goswami and E. Arunan, *Phys. Chem. Chem. Phys.*, 2009, **11**, 8974–8983.
8. A. J. Dingley and S. Grzesiek, *J. Am. Chem. Soc.*, 1998, **120**, 8293–8297.
9. E. D. Isaacs, A. Shukla, P. M. Platzman, D. R. Hamann, B. Barbiellini and C. A. Tulk, *Phys. Rev. Lett.*, 1999, **82**, 600–603.
10. T. R. Dyke, B. J. Howard and W. Klemperer, *J. Chem. Phys.*, 1972, **56**, 2442–2454.
11. A. Shahi and E. Arunan, *Phys. Chem. Chem. Phys.*, 2014, **16**, 22935–22952.
12. A. C. Legon, *Angew. Chem., Int. Ed.*, 1999, **38**, 2686–2714.
13. P. Politzer, J. S. Murray and T. Clark, *Phys. Chem. Chem. Phys.*, 2013, **15**, 11178–11189.
14. *Halogen Bonding*, ed. P. Metrangolo and G. Resnati, SpringerVerlag, Berlin, 2008.
15. W. Wang, B. Ji and Y. Zhang, *J. Phys. Chem. A*, 2009, **113**, 8132–8135.
16. D. Manna and G. Mugesh, *J. Am. Chem. Soc.*, 2012, **134**, 4269–4279.
17. S. Scheiner, *J. Chem. Phys.*, 2011, **134**, 094315–094319.
18. S. J. Grabowski, I. Alkorta and J. Elguero, *J. Phys. Chem. A*, 2013, **117**, 3243–3251.
19. D. Mani and E. Arunan, *Phys. Chem. Chem. Phys.*, 2013, **15**, 14377–14382.
20. S. P. Thomas, M. S. Pavan and T. N. Guru Row, *Chem. Commun.*, 2014, **50**, 49–51.
21. S. A. Southern and D. L. Bryce, *J. Phys. Chem. A*, 2015, **119**, 11891–11899.
22. S. J. Grabowski, *Phys. Chem. Chem. Phys.*, 2014, **16**, 1824–1834.
23. A. Bauza, T. J. Mooibroek and A. Frontera, *Angew. Chem., Int. Ed.*, 2013, **52**, 12317–12321.
24. S. J. Grabowski, *ChemPhysChem*, 2015, **16**, 1470–1479.
25. L. Albrecht, R. J. Boyd, O. Mo and M. Yáñez, *Phys. Chem. Chem. Phys.*, 2012, **14**, 14540–14547.
26. E. Arunan, S. Dev and P. K. Mandal, *Appl. Spectrosc. Rev.*, 2004, **39**, 131–181.
27. R. H. Page, Y. R. Shen and Y. T. Lee, *J. Chem. Phys.*, 1988, **88**, 4621–4636.
28. A. Fujii, G. N. Patwari, T. Ebata and N. Mikami, *Int. J. Mass Spectrom.*, 2002, **220**, 289–312.
29. K. Mueller-Dethlefs, O. Dopfer and T. G. Wright, *Chem. Rev.*, 1994, **94**, 1845–1871.
30. H. J. Jodl, in Chemistry and Physics of Matrix-isolated Species, ed. L. Andrews and M. Moskovitz, North Holland, Amsterdam, 1989, p. 343.
31. H. Reisler, *Annu. Rev. Phys. Chem.*, 2009, **60**, 39–59.
32. *Atomic and Molecular Beam Methods*, ed. G. Scoles, Oxford University Press, New York, 1992, vol. 1, 1988 and vol. 2.
33. P. C. Singh, B. Bandyopadhyay and G. N. Patwari, *J. Phys. Chem. A*, 2008, **112**, 3360–3363.

34. M. Goswami and E. Arunan, *Phys. Chem. Chem. Phys.*, 2011, **13**, 14153–14162.
35. G. Karir and K. S. Viswanathan, *J. Mol. Struct.*, 2016, **1107**, 145–156.
36. R. F. W. Bader, *Atoms in Molecules: A Quantum Theory*, Oxford University Press, Oxford, 1990.
37. P. L. A. Popelier, *Atoms in Molecules. An Introduction*, Pearson Education, 2000.
38. P. Coppens, *X-Ray Charge Densities and Chemical Bonding*, Oxford University Press, Oxford, 1997.
39. G. C. Pimentel and McClellan, *The Hydrogen Bond*, Freeman, San Francisco, 1960.
40. A. V. Ioganssen, *Spectrochim. Acta A*, 1999, **55**, 1585–1612.
41. D. Christen, L. H. Coudert, J. A. Larsson and D. Cremer, *J. Mol. Spectrosc.*, 2001, **205**, 185–196.
42. K. A. Petterson, R. S. Stein, M. D. Drake and J. D. Roberts, *Magn. Reson. Chem.*, 2005, **43**, 225–230.
43. P. Das, P. K. Das and E. Arunan, *J. Phys. Chem. A*, 2015, **119**, 3710–3720.
44. D. L. Howard, P. Jorgensen and H. G. Kjaergaard, *J. Am. Chem. Soc.*, 2005, **127**, 17096–17103.
45. R. A. Klein, *Chem. Phys. Lett.*, 2006, **429**, 633–637.
46. E. R. Johnson, S. Keinan, P. Mori-Sánchez, J. Contreras-García, A. J. Cohen and W. Yang, *J. Am. Chem. Soc.*, 2010, **132**, 6498–6506.
47. J. R. Laane, J. Contreras-Garcia, J. P. Piquemal, B. J. Miller and H. G. Kjaergaard, *J. Chem. Theory Comput.*, 2013, **9**, 3263–3266.
48. P. Hobza and Z. Havlas, *Chem. Rev.*, 2000, **100**, 4253–4264.
49. B. J. van der Veken, W. A. Herrebout, R. Szostak, D. N. Shchepkin, Z. Havlas and P. Hobza, *J. Am. Chem. Soc.*, 2001, **123**, 12290–12293.
50. J. Joseph and E. D. Jemmis, *J. Am. Chem. Soc.*, 2007, **129**, 4620–4632.
51. P. R. Shirhatti and S. J. Wategaonkar, *Phys. Chem. Chem. Phys.*, 2010, **12**, 6650–6659.
52. S. J. Harris, S. E. Novick and W. Klemperer, *J. Chem. Phys.*, 1974, **60**, 3208–3209.
53. D. Mani and E. Arunan, *ChemPhysChem*, 2013, **14**, 754–763.
54. S. K. Singh, K. K. Mishra, N. Sharma and A. Das, *Angew. Chem., Int. Ed.*, 2016, **55**, 7801–7805.
55. E. J. Goodwin and A. C. Legon, *J. Chem. Phys.*, 1986, **84**, 1988–1995.
56. A. C. Legon and D. J. Millen, *Chem. Rev.*, 1986, **86**, 635–657.
57. A. C. Legon and D. G. Lister, *Chem. Phys. Lett.*, 1995, **238**, 156–162.
58. B. E. Rocher-Casterline, L. C. Ch'ng, A. K. Mollner and H. Reisler, *J. Chem. Phys.*, 2011, **134**, 211101.
59. L. C. Ch'ng, A. K. Samanta, Y. Wang, J. M. Bowman and H. Reisler, *J. Phys. Chem. A*, 2013, **117**, 7207–7216.
60. R. T. Birge and H. Sponer, *Phys. Rev.*, 1926, **28**, 0259–0260.
61. Q. Gu and J. L. Knee, *J. Chem. Phys.*, 2013, **136**, 171101.
62. S. Ghosh, S. Bhattacharya and S. Wategaonkar, *J. Phys. Chem. A*, 2015, **119**, 10863–10870.
63. L. Du, K. Mackeprang and H. G. Kjaergaard, *Phys. Chem. Chem. Phys.*, 2013, **15**, 10194–10206.
64. S. Suzuki, P. G. Green, R. E. Bumgarner, S. Dasgupta, W. A. Goddard III and G. A. Blake, *Science*, 1992, **257**, 942–945.
65. H. S. Gutowsky, T. Emilsson and E. Arunan, *J. Chem. Phys.*, 1993, **99**, 4883–4893.
66. K. I. Petersen and W. Klemperer, *J. Chem. Phys.*, 1984, **81**, 3842–3845.
67. A. Engdahl and B. Nelander, *Chem. Phys. Lett.*, 1983, **100**, 129–132.
68. J. A. Stearns and T. S. Zwier, *J. Phys. Chem. A*, 2003, **107**, 10717–10724.
69. W. Gordy and R. L. Cook, *Microwave Molecular Spectra*, John Wiley & Sons, New York, 1984.
70. Z. Kisiel, *J. Mol. Spectrosc.*, 2003, **218**, 58–67.

71. U. Koch and P. L. A. Popelier, *J. Phys. Chem.*, 1995, **99**, 9747–9754.
72. R. Parthasarathi, V. Subramanian and N. Sathyamurthy, *J. Phys. Chem. A*, 2006, **110**, 3349–3351.
73. B. Raghavendra and E. Arunan, *J. Phys. Chem. A*, 2007, **111**, 9699–9706.
74. A. C. Legon, D. J. Millen and S. C. Rogers, *Proc. R. Soc. London, Ser, A*, 1980, **370**, 213–237.
75. G. L. Johnson and L. Andrews, *J. Am. Chem. Soc.*, 1983, **105**, 163–168.
76. G. E. Douberly and R. E. Miller, *J. Chem. Phys.*, 2005, **122**, 024306.
77. R. Sedlak, P. Hobza and G. N. Patwari, *J. Phys. Chem. A*, 2009, **113**, 6620.
78. T. Brupbacher, J. Makarewicz and A. Bauder, *J. Chem. Phys.*, 1994, **92**, 90.
79. P. C. Singh and G. N. Patwari, *J. Phys. Chem. A*, 2008, **112**, 5121–5125.
80. M. Goswami and E. Arunan, *J. Mol. Spectrosc.*, 2011, **268**, 147–156.
81. E. Arunan, T. Emilsson, H. S. Gutowsky, G. T. Fraser, G. De Oliveira and C. E. Dykstra, *J. Chem. Phys.*, 2002, **117**, 9766–9776.
82. E. L. Eliel, *Stereochemistry of Carbon Compounds* McGraw-Hill, New York, 1962, p. 131: "…ethylene glycol may exist in two distinct stable conformations…, the anti conformation and gauche conformation (of which there are two enantiomers). Only in the latter are the hydroxyl groups close enough together to give rise to an intra-molecular bond…".
83. R. A. Klein, *J. Comput. Chem.*, 2003, **24**, 1120.
84. K. M. Marstokk and H. Møllendal, *J. Mol. Struct.: THEOCHEM*, 1974, **22**, 301–303.
85. D. Christen, L. H. Coudert, R. D. Suenram and F. Lovas, *J. Mol. Spectrosc.*, 1995, **172**, 57–77.
86. H. S. P. Müller and D. Christen, *J. Mol. Spectrosc.*, 2004, **228**, 298–307.
87. C. Trindle, P. Crum and K. Douglass, *J. Phys. Chem. A*, 2003, **107**, 6236–6242.
88. M. M. Deshmukh, N. V. Sastry and S. R. Gadre, *J. Chem. Phys.*, 2004, **121**, 12402–12410.
89. M. Mandado, A. M. Grana and R. A. Mosquera, *Phys. Chem. Chem. Phys.*, 2004, **6**, 4391–4396.
90. D. Chopra, T. N. Guru Row, E. Arunan and R. A. Klein, *J. Mol. Struct.*, 2010, **964**, 126–133.
91. I. Bako, T. Grosz, G. Palinkas and M. C. Bellissent-Funel, *J. Chem. Phys.*, 2003, **118**, 3215.
92. C. M. Pearce and J. K. M. Sanders, *J. Chem. Soc., Perkin Trans.*, 1994, **1**, 1119.
93. P. Buckley and P. A. Giguère, *Can. J. Chem.*, 1967, **45**, 397–407.
94. L. J. Bellamy, *The Infrared Spectra of Complex Molecules*, Chapman and Hall, London. 3rd ed., 1975, vol. 1.
95. B. Renault, E. Cloutet, H. Cramail, T. Tassaing and M. Besnard, *J. Phys. Chem. A*, 2007, **111**, 4181.
96. D. L. Howard and H. G. Kjaergaard, *J. Phys. Chem. A*, 2006, **110**, 10245–10250.
97. L. J. Farrugia, C. Evans and M. Tegel, *J. Phys. Chem. A*, 2006, **110**, 7952–7961.
98. B. Raghavendra and E. Arunan, *Chem. Phys. Lett.*, 2008, **467**, 37–40.
99. A. Ambrosetti, F. Costanzo and P. L. Silvestrelli, *J. Phys. Chem. C*, 2011, **115**, 12121.
100. R. Parajuli and E. Arunan, *J. Chem. Sci.*, 2015, **127**, 1035.
101. K. Brandhorst and J. Grunenberg, *Chem. Soc. Rev.*, 2008, **37**, 1558.
102. M. Majumder and S. Manogaran, *J. Chem. Sci.*, 2013, **125**, 9.
103. S. K. Pandey, P. Das, P. K. Das, E. Arunan and S. Manogaran, *J. Chem. Sci.*, 2015, **127**, 1127.
104. A. H. Zewail, *Annu. Rev. Phys. Chem.*, 2006, **57**, 65.
105. M. P. Minitti, J. M. Budarz, A. Kirrander, J. S. Robinson, D. Ratner, T. J. Lane, D. Zhu, J. M. Glownia, M. Kozina, H. T. Lemke, M. Sikorski, Y. Feng, S. Nelson, K. Saita, B. Stankus, T. Northey, J. B. Hastings and P. M. Weber, *Phys. Rev. Lett.*, 2015, **114**, 255501.

106. S. O. Jonsdottir and R. A. Klein, *Fluid Phase Equilib.*, 1997, **132**, 117–127.
107. Y. L. Li, X. L. Sun and Y. L. Wu, *Tetrahedron*, 1994, **50**, 10727–10738.
108. S. O. Jonsdottir, W. J. Welsh, K. Rasmussen and R. A. Klein, *New J. Chem.*, 1999, 153.
109. P. Hobza, V. Špirko, H. L. Selzle and E. W. Schlag, *J. Phys. Chem. A*, 1998, **102**, 2501.
110. S. E. Novick, K. C. Janda and W. Klemperer, *J. Chem. Phys.*, 1976, **65**, 5115.
111. U. Erlekam, Frankowski, G. Meijer and G. von Helden, *J. Chem. Phys.*, 2006, **124**, 171101.
112. V. Chandrasekaran, L. Biennier, E. Arunan, D. Talbi and R. Georges, *J. Phys. Chem. A*, 2011, **115**, 11263.
113. D. Hauchecorne, R. Szostak, W. A. Herrebout and B. J. van der Veken, *ChemPhysChem*, 2009, **10**, 2105.
114. I. V. Alabugin, M. Manoharan, S. Peabody and F. Weinhold, *J. Am. Chem. Soc.*, 2003, **125**, 5973.
115. S. Scheiner and T. Kar, *J. Phys. Chem. A*, 2002, **106**, 1784.
116. A. Karpfen and E. S. Kryachko, *J. Phys. Chem. A*, 2007, **111**, 8177.
117. K. Hermanssonn, *J. Phys. Chem. A*, 2002, **106**, 4695.
118. F. A. Baiocchi, T. Dixon and W. Klemperer, *J. Chem. Phys.*, 1982, **77**, 1632.
119. O. Hassel, *Q. Rev., Chem. Soc.*, 1962, **16**, 1.
120. R. S. Mulliken, *J. Am. Chem. Soc.*, 1950, **72**, 600.
121. A. C. Legon, *Phys. Chem. Chem. Phys.*, 2010, **12**, 7736.
122. G. Cavello, P. Metrangolo, R. Milani, T. Pilati, A. Priimagi, G. Resnati and G. Terraneo, *Chem. Rev.*, 2016, **116**, 2478.
123. R. D. Hunt and L. Andrews, *J. Phys. Chem.*, 1988, **92**, 3769.
124. M. S. Pavan, K. D. Prasad and T. N. Guru Row, *Chem. Comm.*, 2013, **49**, 7558.
125. P. Metrangolo, J. S. Murray, G. Pulati, P. Politzer, G. Resnati and G. Terraneo, *CrysEngComm*, 2011, **13**, 6593.
126. W. Herrebout, in *Cryogenic Solutions. In Halogen Bonding I: Impact on Materials Chemistry and Life Science*, ed. P. Metrangolo and G. Resnati, Springer International Publishing, Cham, Switzerland, 2015, vol. 79.
127. D. Hauchecorne, A. Moiana, B. J. van der Veken and W. A. Herrebout, *Phys. Chem. Chem. Phys.*, 2011, **13**, 10204–10213.
128. S. P. Gnanasekar, M. Goubet, R. Georges and E. Arunan, unpublished work.
129. K. R. Leopold, G. T. Fraser and W. Klemperer, *J. Chem. Phys.*, 1984, **80**, 1039.
130. T. D. Klots, R. S. Ruoff and H. S. Gutowsky, *J. Chem. Phys.*, 1989, **90**, 4216.
131. R. S. Ruoff, T. D. Klots, T. Emilsson and H. S. Gutowsky, *J. Chem. Phys.*, 1990, **93**, 3142.
132. A. Bazua and A. Frontera, *Crystals*, 2016, **6**, UNSP26.
133. M. L. DeRider, S. J. Wilkens, M. J. Waddell, L. E. Bretscher, F. Weinhold, R. T. Raines and J. L. Markley, *J. Am. Chem. Soc.*, 2002, **124**, 2497–2505.
134. G. J. Bartlett, A. Choudhary, R. T. Raines and D. N. Woolfson, *Nat. Chem. Biol.*, 2010, **6**, 615.
135. B. C. Gorske, B. L. Bastian, G. D. Geske and H. E. Blackwell, *J. Am. Chem. Soc.*, 2007, **129**, 8928.
136. N. Ramanathan, K. Sankaran and K. Sundararajan, *Phys. Chem. Chem. Phys.*, 2016, **18**, 19350–19358.
137. R. Feynman's quote found in http://www.quotationspage.com/quote/26933.html Accessed on 18 July 2016.

8 Solid-state NMR Techniques for the Study of Intermolecular Interactions

P. Cerreia Vioglio, M. R. Chierotti and R. Gobetto*

University of Torino, Department of Chemistry and NIS Centre, Via P. Giuria n° 7, Torino I-10125, Italy
*Email: roberto.gobetto@unito.it

8.1 Introduction

In the last decades solid-state NMR (SSNMR) has made a large and increasing contribution to the structural knowledge of crystalline and non-crystalline solids.[1] The information provided by 1D and 2D spectra of NMR active nuclides such as ^1H, ^{13}C, ^{15}N, ^{31}P, *etc.*, represent an important tool for the investigation of the microscopic environment at each individual site in a solid material. One of its advantages is that the solid material can be in powder form which in principle requires neither isotopic labeling nor long range order.

The local environment, *i.e.* chemical structure, weak interactions and crystal packing, does affect the chemical shift of the nuclei, but also a wealth of other local parameters such as, for example, chemical shift anisotropy (CSA), dipolar and quadrupolar interactions, *etc.*, directly reports on the geometric and electronic structure for the complete characterization of the material. Rapid developments in SSNMR methodology propelled by new pulse sequences applied on more sensitive probes, have boosted this technique into a highly

Intermolecular Interactions in Crystals: Fundamentals of Crystal Engineering
Edited by Juan J. Novoa
© The Royal Society of Chemistry 2018
Published by the Royal Society of Chemistry, www.rsc.org

versatile tool for investigating structure, packing and weak interactions in organic, inorganic and biological systems.

The possibility of determining by SSNMR techniques homo- and heteronuclear proximities has been used to provide restraints in the analysis of powder diffraction patterns to give full crystal structures. It is clear that such information is complementary not only to that obtained from diffraction studies but also from other spectroscopic techniques. The final result of such combination of techniques is a deeper knowledge of intermolecular interactions, hydrogen bonds (HBs) and hydrogen atom locations. Quantum mechanical computations of shielding have also been used to assist in the choice of the correct structure in polymorphs, stoichiometric and non-stoichiometric solvates, cocrystals, as well as amorphous forms and formulations.

The aim of this chapter is the description of the potential of SSNMR techniques to access information at molecular level on intermolecular interactions in molecular crystals. In distinct sections, intriguing examples will be presented in order to provide a selected overview of the information obtainable from SSNMR parameters by focusing our attention, in particular, on detection and characterization (strength, network...) of hydrogen and halogen bonds (hereafter HB and XB). Our main goal is to offer the state-of-the-art description of these subjects in order to stimulate further future work in the supramolecular field.

8.2   NMR Parameters Involved in the Study of Weak Interactions

Orientation-dependent interactions, which are usually averaged out in NMR spectra of solutions, are present in solid-state spectra causing substantial broadening of the lines and a loss of spectral resolution. High resolution can be obtained by rapid sample rotation about the magic angle ($54.7°$), Magic Angle Spinning (MAS), which is effective in averaging orientation-dependent interactions. Often only moderate spinning speeds are required for dilute nuclei such as ^{13}C whereas very fast MAS exceeding 30 kHz is needed in order to narrow dipolar-broadened proton spectra. In order to achieve a better signal to noise ratio for nuclei having long relaxation times, Cross Polarization (CP) experiments are usually performed. In this case magnetic polarization is transferred from abundant nuclei like ^1H or ^{19}F to dilute, rare nuclei like ^{13}C, ^{15}N, ^{29}Si, *etc.* This is achieved by applying strong radio-frequency (RF) fields along the rotating frame x axis to both types of spins following a ^{1}H 90° excitation pulse along the y axis. Transverse magnetizations of both spins are "spin-locked" along the rotating frame x axis. If the Hartmann–Hahn ($\gamma_I B_{1I} = \gamma_S B_{1S}$) conditions are achieved, the precession frequency in the rotating frame of both types of spin will be equal, allowing for transfer of the abundant spin polarization to the dilute nuclei. ^{13}C polarization increases approximately by a factor of 3–4 whereas a gain of about 9–10 times is achieved in ^{15}N polarization. Additionally, since the ^{1}H nuclei are abundant, their T_1 relaxation time will often be much faster than ^{15}N or ^{13}C and the experimental repetition rate may be increased significantly. Combining cross polarization with magic-angle spinning (CPMAS) has become a standard and routine experiment in most laboratories.

8.2.1 Chemical Shift and Chemical Shift Tensors

Chemical shift values obtained from Cross-Polarization Magic Angle Spinning (CPMAS) or Magic Angle Spinning (MAS) SSNMR spectra can provide valuable information on solid compounds. In general, the chemical shift of a given nucleus in a molecule depends on the cumulative effects of the electrons of the neighboring atoms and the nature of the interaction between these electrons and the nucleus of interest. Thus, it clearly provides information on the local environment at each site such as atom hybridization, coordination geometry, weak interactions, crystal packing, surrounding substituents, *etc.* It also allows crystallographic asymmetric units, space group and molecular symmetry to be determined. Chemical shift is obviously affected by local environment, but, even more informative is the evaluation of the chemical shift principal values since they can provide information on the electron distributions around a local space in the vicinity of the resonant atom. Chemical shift is a second-rank tensor specified by the three principal components (δ_{XX}, δ_{YY}, δ_{ZZ}) in a frame of reference defined by axes X, Y, and Z in the principal axis system (PAS). δ_{iso}, the isotropic average of the tensor, is given by

$$\delta_{iso} = 1/3(\delta_{XX} + \delta_{YY} + \delta_{ZZ})$$

In the literature two main notations have been reported, the "Haeberlen notation" and the "Mehring notation". In the Haeberlen notation, the three components are defined by:

$$|\delta_{ZZ} - \delta_{iso}| \geq |\delta_{XX} - \delta_{iso}| \geq |\delta_{YY} - \delta_{iso}|$$

δ_{ZZ} is then the principal component farthest from δ_{iso}, whereas δ_{YY} is the component closest to the isotropic value. The reduced anisotropy is defined by: $\delta = \delta_{ZZ} - \delta_{iso}$ and the anisotropy by:

$$\Delta\delta = \delta_{ZZ} - 1/2(\delta_{XX} + \delta_{YY}) = 3\delta/2$$

whereas the shielding asymmetry is:

$$\eta = (\delta_{YY} - \delta_{XX})/\delta(0 \leq \eta \leq +1)$$

In the "Mehring notation" the three principal components are indicated as:

$$\delta_{33} \leq \delta_{22} \leq \delta_{11}$$

In this notation $\delta_{iso} = 1/3(\delta_{11} + \delta_{22} + \delta_{33})$, whereas the convention for anisotropy and asymmetry are substituted with the span (Ω) and skew (κ) parameters defined respectively as:

$$\Omega = \delta_{11} - \delta_{33}(\Omega \geq 0)$$

$$\kappa = 3(\delta_{22} - \delta_{iso})/\Omega(-1 \leq \kappa \leq +1)$$

In the literature many authors present their data utilizing shielding constants [*i.e.* the difference in shielding between the frequency of the bare nucleus, ν_{nucl}, and the frequency of the same nucleus in the molecule under investigation, ν_m: $\sigma/\text{ppm} = 10^{6*}(\nu_{nucl} - \nu_m)/\nu_{nucl}$] instead of chemical shifts [*i.e.* the difference in shielding between the nucleus in the molecule under investigation, σ_m, and the shielding of the same nucleus in a reference compound, σ_{ref}: $\delta/\text{ppm} = 10^{6*}(\sigma_{ref} - \sigma_m)/(1 - \sigma_{ref})$]. Since σ_{ref} is often a small number compared to 1, one can approximate using $\Delta\delta = -\Delta\sigma$ and $\sigma_{33} \geq \sigma_{22} \geq \sigma_{11}$.

Chemical shift tensors change significantly with the nucleus investigated, the nature of the functional group and its mobility, the local environment and the weak interactions such as, for example, the protonation state of a carboxylic group.[2] The chemical shift tensors can be determined on a single crystal by static NMR measurements[3] or by analysis of spinning sideband manifolds in low spinning samples.[4] Several procedures have been developed for recovering CSA information on fast-spinning MAS NMR experiments without sacrificing sensitivity and resolution. Such CSA recoupling procedures are based on multidimensional experiments able to separate the isotropic part in one dimension and to generate spinning sidebands (the anisotropic part) in the indirect dimension by the use of π pulse timing within the rotor period.[5]

The increasing ability to relate chemical shifts (including the tensor components) to the crystallographic location of relevant atoms in the unit cell *via* computational methods has added significantly to the practice of NMR crystallography. Such computations are commonly carried out within the well-established density functional theory (DFT), using the plane waves approach as implemented in the gauge-including projected augmented wave (GIPAW) method.[6,7] Although the earlier developed gauge-including atomic orbitals (GIAO) method is still popular among SSNMR users,[8] the GIPAW approach makes it possible to include the whole crystalline unit cell into calculation, thus accounting for the periodicity of a crystal, which provides accurate results when strong intermolecular interactions (*i.e.* HBs) are present.[9] In the case of calculations involving nuclei for which relativistic effects cannot be neglected (*e.g.* halogen nuclei, transition metals), it is necessary to employ a quantum-chemical approach incorporating relativistic corrections.[10] Of particular importance is the zero-th order regular approximation (ZORA) theory,[11] which has been extensively used to calculate the chemical shifts of heavy nuclei and the effects of the latter in their surroundings in a more accurate way, at affordable computational costs.[12] The synergy between experimental data and quantum chemical computations enables excellent structure prediction: for instance, NMR crystallography allows for an accurate localization of hydrogen atoms in HB motifs.[13] For this reason, the proton CSA has been successfully used as a sensitive probe to determine the packing interactions present in the supramolecular arrangements of both organic and inorganic solids.[14] Likewise, chemical shift parameters of other nuclei such as ^{13}C, ^{15}N, ^{11}B are also important in gaining structural information.[15–17]

Several NMR techniques have been developed to obtain other parameters, such as intermolecular distances,[18] ^{1}H–^{1}H distances,[19] or torsion angle restraints[20] in order to obtain a more direct approach to crystal structure determination. Some of them will be discussed in the next paragraph.

8.2.2 Dipolar Interaction

Internuclear dipolar coupling, D, depends on the gyromagnetic ratios (γ) of the two spins I and S and the inverse third power of the distance r_{IS}, according to:

$$D = \frac{\mu_D}{4\pi} \frac{h\gamma_I\gamma_S}{r_{IS}^3}$$

allowing internuclear distances to be obtained or connectivities for structure elucidation to be established. At 5 Å distance the dipolar couplings are, respectively, 61 Hz for ^{13}C–^{13}C, 24 Hz for ^{13}C–^{15}N, and 99 Hz for ^{13}C–^{31}P spin pairs. Under MAS conditions, dipolar couplings are normally averaged out owing to its dependence on the geometric factor $(3\cos^2\theta - 1)$, but they can be reintroduced in the spectra by applying a series of radio-frequency pulses during the evolution period.[21]

The rotational-echo double resonance (REDOR) technique consists of two π pulses applied on spin S per rotor period, which prevents the refocalization of the I–S dipolar coupling at the end of each full rotation.[22] The difference between the intensity of I recorded with this pulse sequence and that obtained in a reference experiment where no pulses are applied to spin S is reported as a function of rotation period. Internuclear distances between atoms such as ^{13}C and ^{15}N (up to about 5 Å), ^{13}C and ^{31}P (up to about 6 Å) and ^{13}C and ^{19}F (up to about 8 Å) can be obtained. Nevertheless, the ability to measure weak dipolar couplings is intrinsically limited by the transverse (T_2) relaxation of the nuclear spins. Several strategies to increase the distance reach by MAS SSNMR spectra have been developed to measure distances up to 15–20 Å by taking advantage of the high gyromagnetic ratios of ^{1}H and ^{19}F spins, multispin effects that speed up dipolar dephasing, and ^{1}H and ^{19}F spin diffusion that probes distances in the nanometer range.[23]

Other popular recoupling sequences are Dipolar Recovery at the Magic Angle (DRAMA),[24] Transferred Echo Double Resonance (TEDOR),[25] Radio-Frequency Driven Recoupling (RFDR)[26] and symmetry-based sequences introduced by Lewitt and co-workers.[27]

Evaluation of dipolar couplings between spin-1/2 and quadrupolar nuclei can be obtained by Transfer of Population in Double Resonance (TRAPDOR)[28] and Rotational-Echo, Adiabatic Passage, Double Resonance (REAPDOR).[29]

Rotor-synchronized two-dimensional ^{1}H double quantum (DQ) MAS experiment acquired at fast MAS (>30 kHz) is able to generate DQ coherence (DQC) between two dipolar-coupled ^{1}H nuclei affording specific information about intra- and intermolecular proton–proton spatial proximities.[30] In this kind of experiment, a DQC between two unlike protons A and B gives rise to two cross-peaks arranged symmetrically on either side of the diagonal at frequencies $(\omega_A, \omega_A + \omega_B)$ and $(\omega_B, \omega_A + \omega_B)$, whereas a DQC between two like protons gives rise to a diagonal peak at the positions $(\omega_A, 2\omega_A)$. Finally, Combined Rotation And Multiple Pulse Sequences (CRAMPS) obtained by the combination of physical rotation of the sample under MAS with the application of carefully synchronized RF pulses (PMLG27 or DUMBO28) afford considerable line-narrowing in the ^{1}H solid-state spectra compared to the use of fast MAS alone.[31]

8.2.3 Quadrupolar Interaction

For nuclei with spin $> 1/2$, the most important interaction is the quadrupole interaction which dominates the spectrum, determining signal line shapes. Its magnitude can be of the order of tens of MHz. The nuclear quadrupolar interaction arises from the coupling between the nuclear quadrupole moment, Q, and the electric field gradient (EFG) at the nuclear position. The EFG is described by a traceless second-rank tensor, whose principal components are defined as $|V_{zz}| > |V_{yy}| > |V_{xx}|$. Other important NMR parameters used to describe a quadrupolar coupling tensor are the quadrupole coupling constant (QCC), $\chi = e^2 Q V_{zz}/h$, and the asymmetry parameter, $\eta = (V_{xx} - V_{yy})/V_{zz}$. As for the chemical shift tensors, the QCC of a nucleus can provide several pieces of information concerning the surrounding environment such as coordination geometry, weak interactions, *etc.*

Several are the developed experimental techniques for the study of quadrupolar nuclei with half-integer spins. Double rotation (DOR) or dynamic angle spinning (DAS) rely on the possibility of spinning the sample simultaneously or sequentially about two angles (54.7 and 30.56 or 70.12°) which allow for an efficient averaging of the Hamiltonian since this depends on two geometry factors: $(3\cos^2\theta - 1)$ and $(35\cos^4\theta - 30\cos^2\theta + 3)$.

Recently a new experimental method has been introduced, Quadrupolar Carr–Purcell–Meiboom–Gill (QCPMG), for determination of magnitudes and relative orientation of chemical shielding and quadrupolar coupling tensors of half-integer quadrupolar nuclei exhibiting relatively large quadrupole coupling and anisotropic chemical shielding interactions. It relies on acquisition of the quadrupolar-echo spectrum during a CPMG (Carr–Purcell–Meiboom–Gill) train of selective π pulses such that the second-order quadrupolar line shape for the central transition is split into a comb of sidebands, leading to a considerable increase in the sensitivity compared to a conventional quadrupole echo sequence. On the other hand, Multiple-Quantum Magic Angle Spinning (MQMAS) methodology[32,33] is an echo experiment involving multiple quantum (MQ) transitions during the excitation of the spin system by a first pulse. The MQ coherences generated are converted back to single quantum (SQ) coherence as an echo and an antiecho by a second pulse.

The result is a 2D spectrum able to separate the isotropic part from the anisotropic part which contains QCC and CSA information.

8.2.4   Correlation of SSNMR Parameters and XRD Data

The high sensitivity of SSNMR to local structural details can be combined with the accuracy in detecting long-range ordering and crystal symmetries shown by XRD and the potential of first-principles calculations to correlate several computed and experimental parameters. An interesting example is represented by the structure of the 1 : 1 cocrystal formed by 3,5-dimethyl-1*H*-pyrazole (dmpz) and 4,5-dimethyl-1*H*-imidazole (dmim), determined by means of powder X-ray diffraction data combined with data obtained by 1D/2D ^{1}H, ^{13}C and ^{15}N high-resolution SSNMR techniques (Figures 8.1 and 8.2).[34]

SSNMR data provided structural insight on local length scales revealing internuclear proximities which helped to build the structure model and also to validate the proposed final structure. Molecular

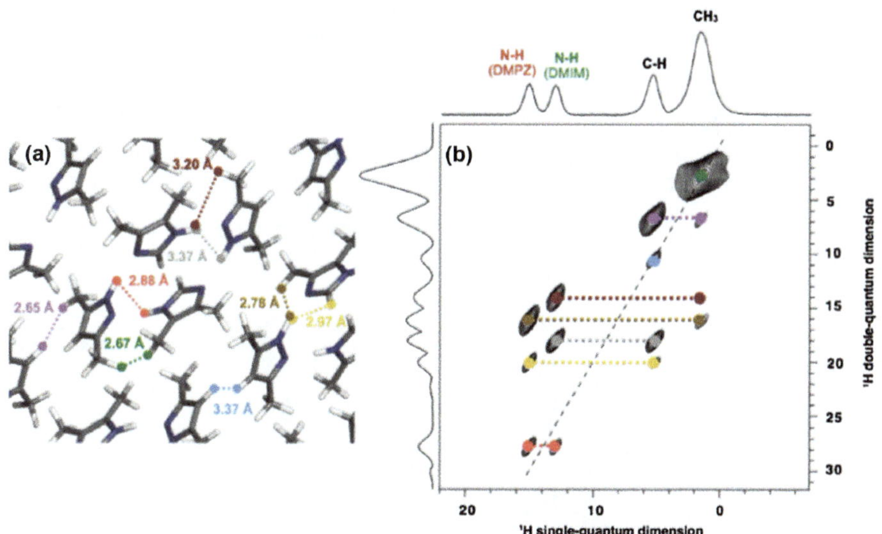

Figure 8.1   (Left) Crystal structure scheme illustrating the intermolecular contacts, with corresponding distances, of correlations highlighted in the 2D ^{1}H DQ CRAMPS spectrum (right) of 1 : 1 cocrystal formed by 3,5-dimethyl-1*H*-pyrazole (dmpz) and 4,5-dimethyl-1*H*-imidazole (dmim) ($B_0 = 16.4$ T, MAS rate of 26 kHz).
(Reprinted from Solid State Nuclear Magnetic Resonance, 65, M. Sardo, S.M. Santos, A.A. Babaryk, C. López, I. Alkorta, J. Elguero, R.M. Claramunt and L. Mafra., Diazole-based powdered cocrystal featuring a helical hydrogen-bonded network: Structure determination from PXRD, Solid-state NMR and computer modeling, 49–63, Copyright 2015 with permission from Elsevier.)

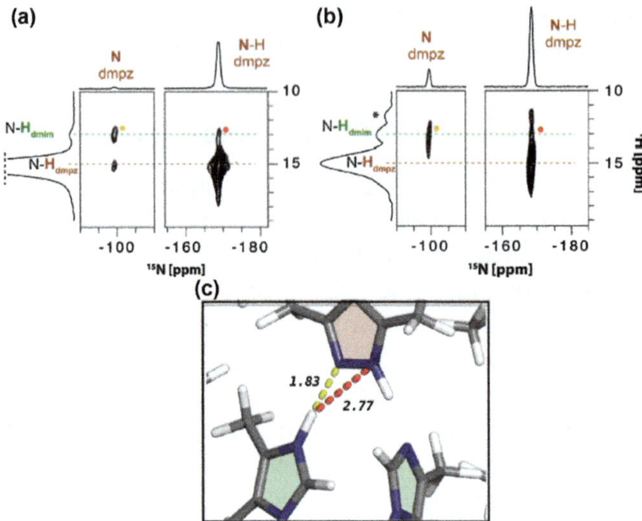

Figure 8.2 (a) 2D ^{1}H–^{15}N CP HETCOR spectrum of 1 : 1 cocrystal formed by 3,5-dimethyl-1*H*-pyrazole (dmpz) and 4,5-dimethyl-1*H*-imidazole (dmim) ($B_0 = 16.4$ T, MAS rate = 25 kHz and 1.0 ms mixing time). (b) 2D ^{1}H–^{15}N LG-CP HETCOR spectrum of 1 : 1 cocrystal formed by 3,5-dimethyl-1*H*-pyrazole (dmpz) and 4,5-dimethyl-1*H*-imidazole (dmim) ($B_0 = 9.4$ T, MAS rate = 10 kHz and 0.5 ms mixing time). (c) Part of the crystal structure highlighting the intermolecular contacts observed in both spectra: Ndmpz–H–Ndmim in yellow, 1.83 Å and H–Ndmpz–H–Ndmim in red, 2.77 Å.
(Reprinted from Solid State Nuclear Magnetic Resonance, 65, M. Sardo, S.M. Santos, A.A. Babaryk, C. López, I. Alkorta, J. Elguero, R.M. Claramunt and L. Mafra., Diazole-based powdered cocrystal featuring a helical hydrogen-bonded network: Structure determination from PXRD, Solid-state NMR and computer modeling, 49–63, Copyright 2015 with permission from Elsevier.[34])

Scheme 8.1   β-ʟ-aspartyl-ʟ-alanine.

modeling and DFT calculations were also employed to generate meaningful structures.

L. Emsley and coworkers[35] were able to develop a new method of structure determination of an organic compound, β-ʟ-aspartyl-ʟ-alanine (1) (Scheme 8.1), in powder form and at natural isotopic abundance, that combines molecular modeling with experimental spin diffusion data obtained from high-resolution ^{1}H SSNMR, and which allows the determination of the three-dimensional structure. Proton spin diffusion is referred to the magnetization exchange process that occurs in solids between protons according to a phenomenon driven by the internuclear distance dependent dipolar coupling. A phenomenological multispin kinetic rate matrix approach has been proposed that circumvents complications due to the orientation of the internuclear vector in the sample, the details of the anisotropic chemical shifts of the two coupled nuclei, the coupling to other protons, and experimental factors such as the MAS rate. This simplified model for spin diffusion, within its validity, provides a way of back calculating spectra from trial structures and, therefore, a way of determining structures by evaluating the best fit between the data and trial structures (Figures 8.3 and 8.4).

8.3   SSNMR and Hydrogen Bonds

Among all weak interactions, the HB still plays the most important role in the supramolecular field.[36] Its directionality, specificity, strength and selectivity make it the most used cement in Crystal Engineering syntheses.[37] In addition, characteristic qualities of transferability (from crystal to crystal), reproducibility (in terms of cohesive contribution to crystal packing) and ease of chemical manipulation (molecular functionalization with HB donor/acceptor systems) allow self-organization of molecules into one-, two- or three-dimensional H-bonded architectures to be easily designed and achieved.[38] However, the intrinsic limitations of X-ray diffraction techniques in detecting the hydrogen atom positions prompted scientists to explore and develop spectroscopic tools to study this interaction. Among all of these, SSNMR represents one of the most powerful techniques. One of the important aspects of SSNMR is the multinuclear approach. Owing to its ability to probe either carbon, nitrogen or even directly the hydrogen atoms, it gives information of several types of contacts according to their strength and geometry. Strong (single well and low barrier), moderate and weak HBs have been easily detected and characterized (Scheme 8.2).[39]

The main SSNMR parameters that can provide insights on the HB interactions in the solid state are (a) the chemical shift, (b) the dipolar interaction, (c) the CSA, (d) the relaxation parameters and (e) the isotopic effect. All these parameters can be obtained and evaluated for each NMR active nucleus involved in the weak contact whether ^{1}H,

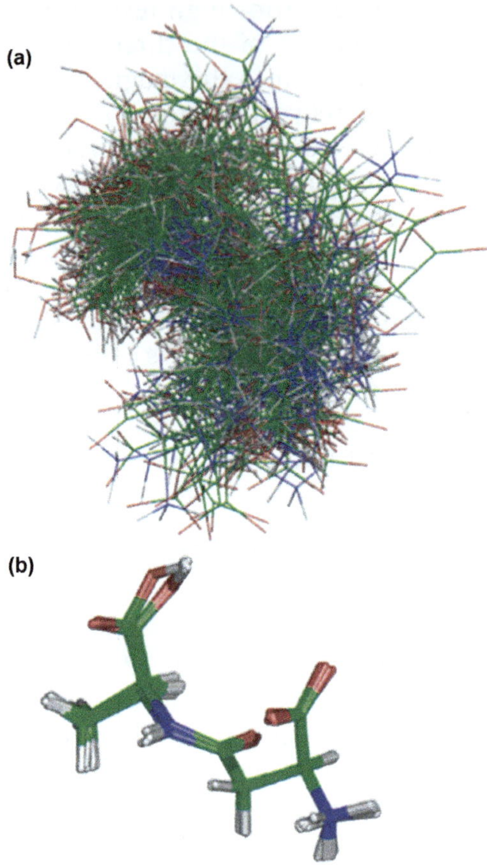

Figure 8.3   (a) A set of 150 structures from the ensemble of 3000 random structures used as the starting point for the structure refinement. (b) The 16 structures determined with the lowest EPSD values after the optimization procedure described in the text.
(Reprinted with permission from B. Elena, G. Pintacuda, N. Mifsud and L. Emsley., Molecular Structure Determination in Powders by NMR Crystallography from Proton Spin Diffusion, *J. Am. Chem. Soc.*, 2006, 128, 9555–9560. Copyright (2006) American Chemical Society).

^{19}F, ^{13}C or ^{15}N. The influence of HBs on these nuclei have been recently reviewed.[40]

8.3.1   Chemical Shift, Chemical Shift Anisotropy and Hydrogen Bonds

^{1}H SSNMR spectra of rigid solids suffer from poor resolution and severe peak overlap caused by the strong ^{1}H–^{1}H homonuclear dipolar couplings and narrow ^{1}H chemical shift ranges, which render it

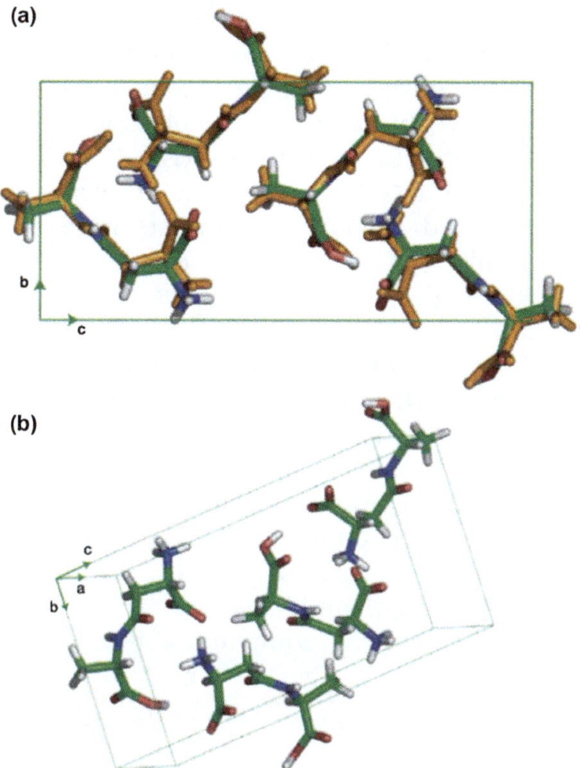

Figure 8.4   (a) Comparison between the known crystal structure of 1 in orange and the structure obtained from the average coordinates of the 16 structures determined here with the lowest EPSD values as described in the text. The view is taken along the a axis of the unit cell. (b) A second view of the average structure determined here, illustrating the packing arrangement from another angle.
(Reprinted with permission from B. Elena, G. Pintacuda, N. Mifsud and L. Emsley., Molecular Structure Determination in Powders by NMR Crystallography from Proton Spin Diffusion, *J. Am. Chem. Soc.*, 2006, 128, 9555–9560. Copyright (2006) American Chemical Society).

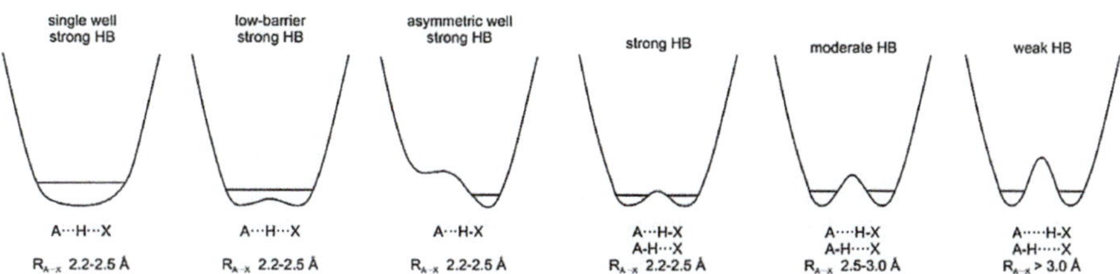

Scheme 8.2   Different types of HB according to their strength. A··· X distances are according to Jeffrey's classification from ref. 39.

difficult to individuate specific proton sites in standard experiments. However, ^{1}H SSNMR spectroscopy has recently attracted much attention owing to the significant enhancement in spectral resolution afforded by the remarkable advances in ultrafast MAS capabilities that can rotate the rotor up to 110–120 kHz. Combined with high magnetic fields (up to 1 GHz) to enhance spectral resolution and sensitivity, proton-based SSNMR experiments have become an attractive choice. Applications of the latest developed high-resolution ^{1}H SSNMR technologies and techniques have been recently reviewed.[41]

Concerning ^{1}H spectra of H-bonded supramolecular systems, since the chemical shift is directly related to the polarization of the covalent X–H bond, signals related to the H-bonded protons are increasingly shifted toward higher frequencies (around 10–20 ppm), far from the aliphatic and aromatic regions, according to the strength of the interaction (Scheme 8.3). The magnitude of the shift is directly correlated with the strength/length of the HB.[41] Direct relationships between δ ^{1}H and HB strength and between δ ^{1}H and X–H distance for different classes of H-bonded compounds have been reported.[42] Similar trends have been observed when O$\cdots$O, O$\cdots$N, or N$\cdots$N distances have been considered, with shifts that span 3–10 ppm.[43]

Similarly, the ^{13}C and ^{15}N chemical shifts are also correlated to the presence and strength of HBs accordingly to the network of HBs and

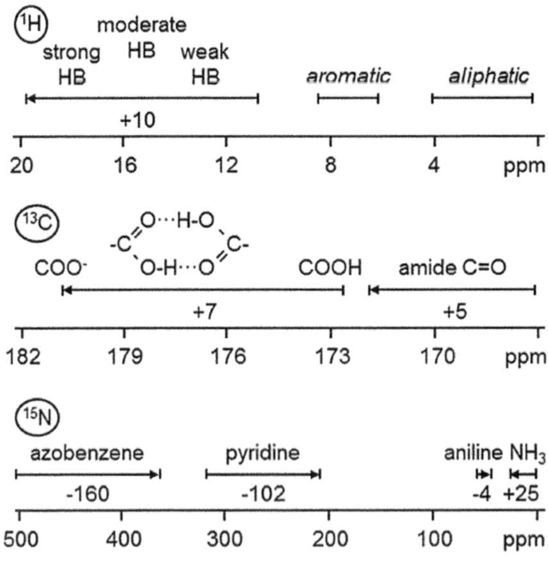

Scheme 8.3 ^{1}H, [13]C (for carboxylic and amide carbonyl only) and ^{15}N SSNMR shifts due to HB formation.

functional groups, whether carboxylic or carbonyl for ^{13}C and aliphatic or aromatic for ^{15}N (Scheme 8.3).

In the case of ^{13}C signals, the shift upon HB formation is of about 3–7 ppm always towards higher frequencies. In cases of both carboxylic and carbonyl groups there are approximately linear relationships between δ and X···O distances (X = N or O) (Å) in the HB length range considered. For carboxylic groups, isotropic ^{13}C chemical shifts linearly increase with decreasing N···O distances. If a proton transfer occurs across the HB, with formation of a carboxylate group and a charge-assisted N^+–H···O^- HB, the high-frequency shift is maximised (around 6–7 ppm).[44] Very strong HBs, such as those formed in the cyclic dimerization of the COOH functions, usually present chemical shifts similar to that of carboxylate groups ($\sim$2 ppm lower).[45] For carbonyls in peptides and polypeptides, it is found that a decrease in N···O distances leads to a high-frequency shift ($\sim$5 ppm) of the amide C=O signal.[46]

Concerning ^{15}N spectra, HB formation produces high- or low-frequency shifts in the chemical shift values according to type of nitrogen atom and interaction (Scheme 8.3). The protonation induced shifts are of the order of 100 ppm towards lower frequencies for aromatic amines, and of about 25 ppm towards higher frequencies for aliphatic amines, in agreement with the minor contribution of the lone pair to σ^P_{loc} due to its removal by quaternization. A similar effect is also observed upon coordination of the nitrogen atom to metals.[47]

Thus, ^1H, ^{13}C and ^{15}N SSNMR shift data can be used to detect and to estimate the presence and the strength of HBs as well as the occurrence of proton transfer from the acid to the base with formation of charge-assisted HBs.

Owing to tensorial nature of the chemical shift, the isotropic chemical shift (δ_{iso}) represents an average of three components, the chemical shift tensors δ_{11}, δ_{22} and δ_{33}. This suggests that variations in individual principal components might be more sensitive indicators of changes in the HB environment than the isotropic chemical shift alone.

Knowledge of the chemical shift tensor values and their orientation with respect to a molecular axis system has at least two important consequences: (a) they can be correlated with geometry of chemical bonds, *i.e.* with the structure of a molecular fragment, since strictly related to the distribution of electron density around nuclei; (b) more important for application of SSNMR, their values and orientation for some functional groups are rather conservative parameters and can be fixed relative to a molecular reference frame for a small fragment.

Useful examples of this are the ^{13}C chemical shift tensors of carbonyl groups in peptides and proteins, which do not vary much (*ca.* ±5°) in proteins adopting different secondary structures.

It is worth noting that the chemical shift tensors can be obtained for every nucleus involved in an HB interaction: ^{1}H, ^{13}C, ^{15}N or even ^{17}O. However, measurements of the ^{1}H shift tensor in solids are challenging because of strong ^{1}H homonuclear dipolar couplings, which broaden the lines and reduce resolution, and the relatively small size of the ^{1}H chemical shift interaction. Some progress has now been made towards general and robust methodologies, based on two-dimensional NMR experiments, which correlate the anisotropic and isotropic parts of the chemical shift interaction. Most of these rely on fast MAS or MAS coupled with homonuclear decoupling schemes to resolve different ^{1}H sites *via* their isotropic shifts in the detection dimension and a method of reintroducing or "recoupling" the MAS averaged ^{1}H CSA during the evolution time.[48]

With these technical improvements, proton CSA has become an important tool for obtaining specific insights into inter/intra-molecular HBs. Indeed, as confirmed by *ab initio* calculations, the main origin of the proton high-frequency shift upon HB formation is the deshielding of the principal component of the ^{1}H chemical shift tensor perpendicular to the HB that occurs as the bond length decreases.[49] For instance, the calculated values of the ^{1}H CSA were found to be linearly correlated to the O–H distances in galactose (Figure 8.5).[50] A similar correlation was identified between the span and the skew of the ^{1}H chemical shift tensor and both HB length and geometry (O–H· · ·O angle).[51]

These approaches are generally well suited to recording proton CSA parameters even in more complex biological and organic systems, including protein assemblies and nucleic acids. For instance, accurate ^{1}H chemical shift tensor parameters have been extracted for an 89-residue protein, U-^{13}C,^{15}N-CAP-Gly domain of dynactin.[52] The CSA parameters correlate with the HB distances (Figure 8.6), and the trends are in excellent agreement with the prior solution from NMR results.

These results suggest that variations in individual principal components are more sensitive indicators of changes in the HB environment than the ^{1}H isotropic shift.

In a similar way, ^{13}C chemical shift tensor values and orientations also provide further insights on the strength and geometry of the HB interaction. Differently to ^{1}H, they can be easily obtained from the sideband analysis of low spinning speed MAS NMR spectra. When the

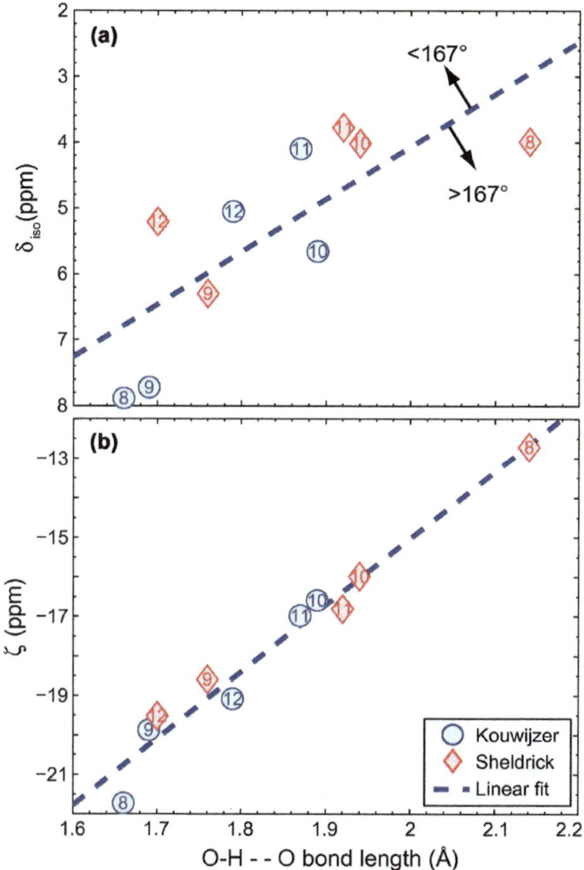

Figure 8.5   Plots of ^{1}H (a) isotropic shift δ_{iso} and (b) reduced anisotropy ζ, calculated for partially optimized Kouwijzer (blue circles) and Sheldrick (red diamonds) α-D-galactose structures using the PBE functional against the corresponding O–H$\cdots$O bond length. The result of a linear regression analysis is also shown in each case (blue dashed line).
(Reprinted from Chemical Physics Letters, 498, M. Kibalchenko, D. Lee, L. Shao, M.C. Payne, J.J. Titman and J.R. Yates., Distinguishing hydrogen bonding networks in α-d-Galactose using NMR experiments and first principles calculations, 270–276, Copyright 2010 with permission from Elsevier.)

number of peaks leads to a strong spinning sideband overlap, the 2D PASS (phase adjustment spinning sideband) technique is the preferred choice since it can separate the isotropic chemical shift in the direct dimension from the anisotropic pattern associated at each distinct site in the indirect dimension.[53] Their orientation for generic amide carbonyl and carboxylic groups is depicted in Scheme 8.4. In the case of amide carbonyl groups, studies of model peptides indicate that δ_{33} is perpendicular to the carbonyl sp^2 plane, δ_{22} is *ca.* 10° off the

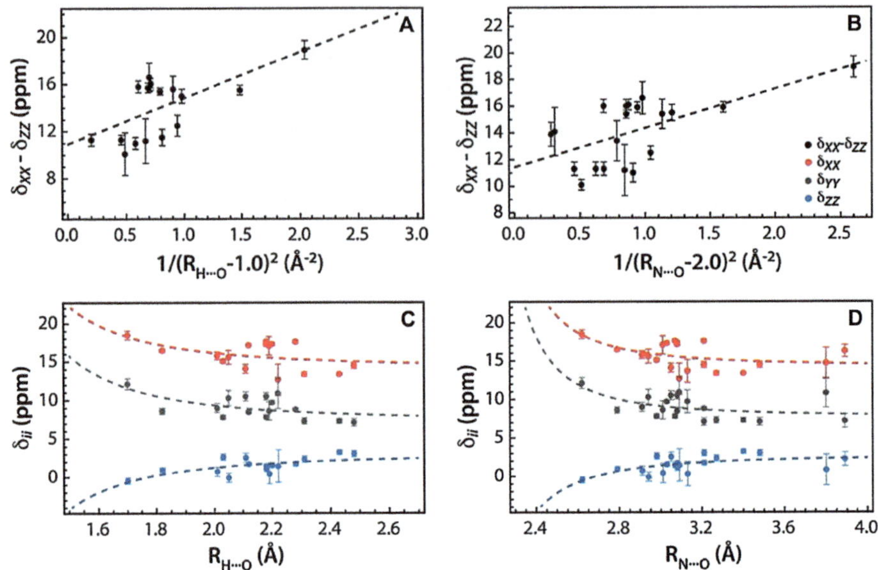

Figure 8.6 Correlation between principal components of ^{1}H CSA tensors and HB length in U-$^{13}C,^{15}N$-CAP-Gly domain of dynactin. (A) and (B) show the correlation between the span, $\delta_{XX}-\delta_{ZZ}$ (black), and the H···O ($R_{H···O}$) and N···O ($R_{N···O}$) distances, respectively. In (C) and (D), correlations of the principal components δ_{XX} (red), δ_{YY} (gray), and δ_{ZZ} (blue) are shown with the H···O and N···O distances, respectively. The H···O and N···O distances are extracted from the MAS NMR structure of CAP-Gly (PDB code 2m02).
(Reprinted with permission from G. Hou, S. Paramasivam, S. Yan, T. Polenova and A.J. Vega., Multidimensional Magic Angle Spinning NMR Spectroscopy for Site-Resolved Measurement of Proton Chemical Shift Anisotropy in Biological Solids, *J. Am. Chem. Soc.* 2013, 135, 1358–1368. Copyright (2013) American Chemical Society).

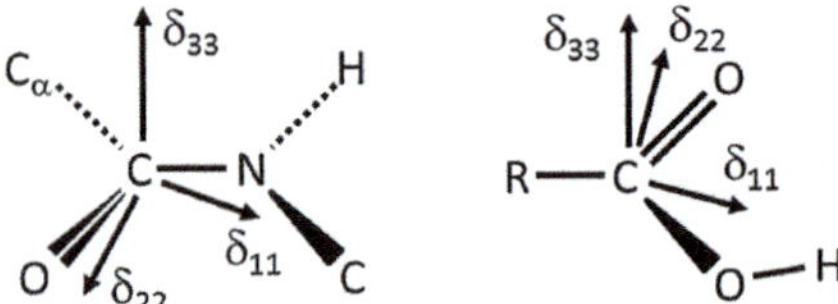

Scheme 8.4 Orientation of ^{13}C chemical shift tensors (δ_{11}, δ_{22} and δ_{33}) in amide and carboxylic groups.

C=O bond and the angle between δ_{11} and the C–N bond is approximately $40°$ (*ca.* $\pm 5°$) in proteins adopting different secondary structures. The ^{13}C chemical shift tensors have been determined for the glycine C=O in series of peptides of the general form *N*-acetyl[1-^{13}C]glycyl-X-amide, where X is [^{15}N]glycine, DL-[^{15}N]-tyrosine, L-[^{15}N]phenylalanine and DL-[^{15}N]alanine. On the basis of SSNMR studies on single crystals[54] and powder samples[55] of peptides, it has been shown that a consistent high-frequency shift for δ_{22}, a low-frequency shift for δ_{11} and no change for δ_{33} are expected to result from a decrease in the N$\cdots$O distance in N–H$\cdots$O HBs. It has been shown that the range of isotropic chemical shift δ_{iso} from the carbonyl carbons in proteins predominantly arises from the dependence of δ_{22} on the secondary structure.[56]

Concerning the carboxylic groups, as shown in Scheme 8.4, δ_{11} and δ_{22} lie on the carboxylic plane, with the former oriented in the middle of the O–C=O angle, while δ_{33} is perpendicular to δ_{11} and δ_{22}.

Studies of protonated and deprotonated carboxylic groups in amino acids have shown that the values of the principal elements of the nuclear shielding tensor change significantly upon HB formation and with the protonation state of the carboxylic groups.[57] A typical example is represented by the H-bonded supramolecular adducts between dicarboxylic acids of variable chain length and the diamine DABCO. The series has been fully investigated by ^{1}H, ^{15}N and ^{13}C SSNMR[44,58] and the ^{13}C CSA tensor components have been obtained by Herzfeld–Berger analysis[4a] of the spinning sidebands. For instance, the ^{13}C CPMAS spectrum of the DABCO–glutaric acid adduct obtained at low spinning speed (Figure 8.7) is indicative of the difference in the sideband pattern between the COOH ($\delta_{iso} = 176.5$ ppm) and the COO groups ($\delta_{iso} = 181.8$ ppm).

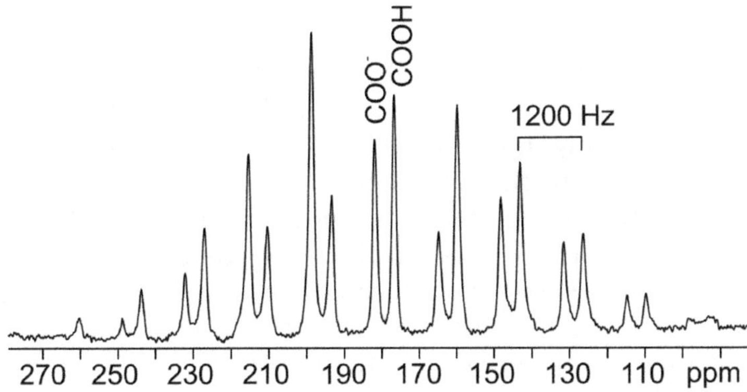

Figure 8.7 Carboxylic region of the ^{13}C CPMAS spectrum of DABCO–glutaric acid adduct recorded at a spinning speed of 1.2 kHz. It shows a carboxylic and a carboxylate signal with the respective spinning sideband patterns. (Reproduced from D. Braga, L. Maini, G. de Sanctis, K. Rubini, F. Grepioni, M.R. Chierotti, R. Gobetto., Mechanochemical Preparation of Hydrogen-Bonded Adducts Between the Diamine 1,4-Diazabicyclo[2.2.2]octane and Dicarboxylic Acids of Variable Chain Length: An X-ray Diffraction and Solid-State NMR Study, *Chem. Eur. J.* 2003, 9, 5538–5548, with permission from John Wiley and Sons. Copyright © 2001, John Wiley and Sons).

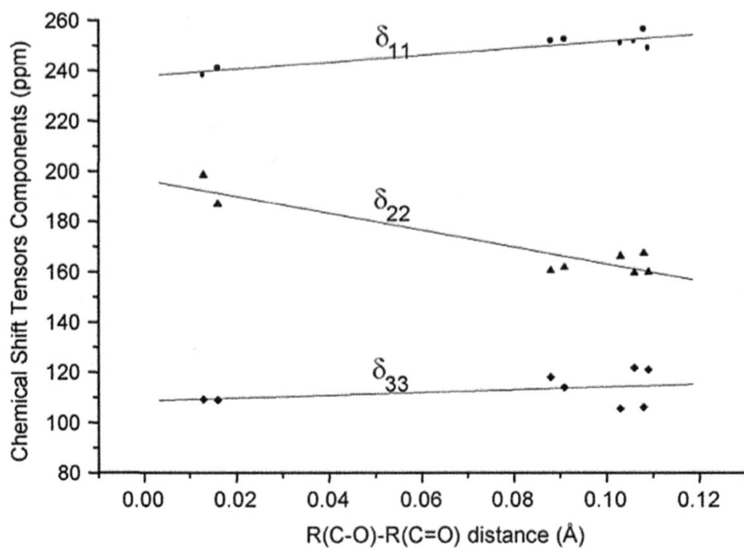

Figure 8.8 Chemical shift tensor components (δ_{11}, δ_{22} and δ_{33}) of DABCO-dicarboxylic acid compounds plotted as function of the difference between the C–O and C=O bond lengths from crystallographic data. (Reproduced from D. Braga, L. Maini, G. de Sanctis, K. Rubini, F. Grepioni, M.R. Chierotti, R. Gobetto., Mechanochemical Preparation of Hydrogen-Bonded Adducts Between the Diamine 1,4-Diazabicyclo[2.2.2]octane and Dicarboxylic Acids of Variable Chain Length: An X-ray Diffraction and Solid-State NMR Study, *Chem. Eur. J.* 2003, 9, 5538–5548, with permission from John Wiley and Sons. Copyright © 2001, John Wiley and Sons).

The comparison between the chemical shift tensors and the X-ray data for the series (Figure 8.8) shows that δ_{22}, perpendicular to δ_{11} in the plane containing the carbon and oxygen atoms, is the most sensitive parameter to the localization of the HB with values in the range of 177 ± 10 ppm for the deprotonated form and 155 ± 20 ppm for the protonated form. It is worth noting that often the isotropic chemical shift δ_{iso} shows only a small increase in shielding upon protonation since this information is intrinsically limited by the fact that the changes in δ_{11} and δ_{22} are in opposite directions, while δ_{33} is not particularly influenced.

Compared to ^{13}C CSA, dealing with ^{15}N CSA tensors is more challenging because of the very poor sensitivity of the ^{15}N resonance due to the small γ value and the low natural abundance ($\sim$0.4%). Indeed, their determination usually requires very expensive selectively ^{15}N-enriched samples. However, the principal values and the orientation of the ^{15}N chemical shift tensors (Scheme 8.5) could be derived for different classes of nitrogen-containing compounds such as

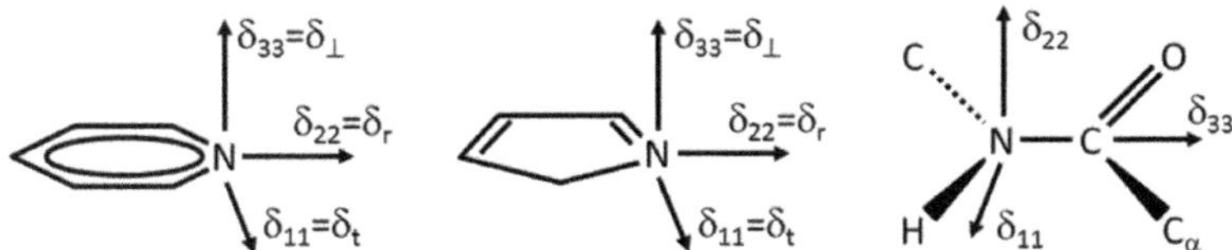

Scheme 8.5   Orientation of ^{15}N chemical shift tensors (δ_{11}, δ_{22} and δ_{33}) in pyridine and pyrrole/pyrazole/imidazole derivatives and amide groups.

amine, imine, pyridine-like, and pyrazole/imidazole nitrogen atoms. The tensor values were extracted by ^{15}N lineshape simulation of static powder spectra recorded under conditions of ^{1}H–^{15}N CP, as reported for the acid-base adducts between 2,4,6-trimethylpyridine (collidine) and several benzoic acid derivatives (Figure 8.9).[59]

For heterocyclic nitrogen atoms, the principal chemical shift tensors are also labeled according to the literature as a tangential component $\delta_{11} = \delta_{t}$, a radial component $\delta_{22} = \delta_{r}$ and a perpendicular component $\delta_{33} = \delta_{\perp}$. Solum *et al.* showed that for pyridine and pyridinium the values of δ_{t} and δ_{r} are inversed.[60]

In general, for heterocyclic systems, it has been reported that δ_{22} and δ_{33} are not very sensitive to the HB geometry and move towards one another, whereas δ_{11} is strongly shifted to lower frequencies when the proton approaches nitrogen.[61]

Concerning amide groups in crystalline histidine and histidine-containing peptides, systematic trends in values of imidazole CSA with changes in HBs have been documented. A correlation was found between the ^{15}N δ_{11} tensor value and the HB length for cationic species. As the HB distance decreases, the δ_{22} tensor value shifts toward higher frequencies.[62]

8.3.2   Dipolar Interactions and Hydrogen Bonds

A major problem concerning HBs is the localization of the hydrogen atom along the interaction.[63] This is due to the high proton mobility and to the intrinsic limitations in the determination of bond lengths involving hydrogen by X-ray diffraction.[64]

From a theoretical point of view, all methods which provide information on the localization and transfer of protons along HBs are of great interest. In the case of weak HBs, such as double-well HBs, variable temperature CPMAS NMR experiments have proven to be useful for the study of rate constants of proton transfer. This method relies on modulations of the isotropic chemical shifts of nuclei such as ^{13}C or ^{15}N, *etc.*, during the transfer processes studied. Otherwise,

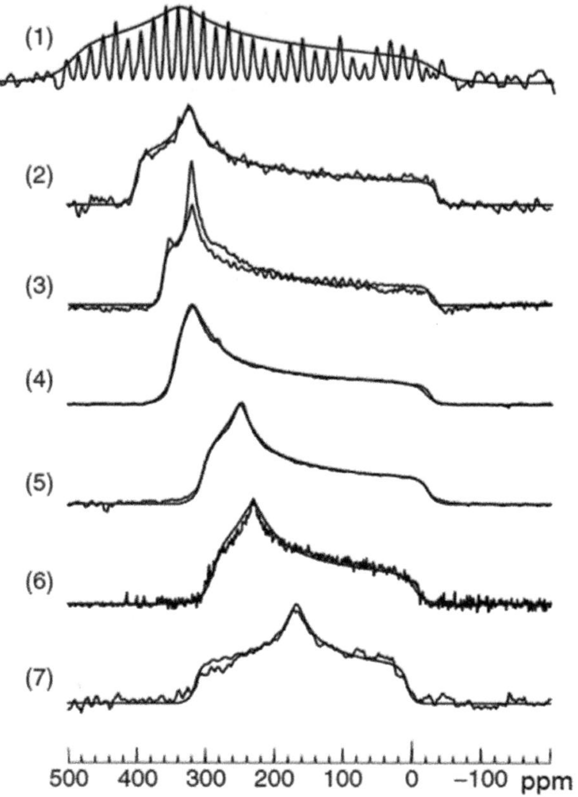

Figure 8.9 Superposed experimental and calculated ^{15}N CP NMR static spectra of collidine (**1**) and H-bonded adducts between collidine and benzoic acid (**2**), 4-nitrobenzoic acid (**3**), 4-chloro-3-nitrobenzoic acid (**4**), 3,5-dinitrobenzoic acid (**5**), 2-nitrobenzoic acid (**6**) and 2,4,6-trimethyl-piridiniumtetraphenilborate (**7**), respectively. The spectrum of **1** was obtained by slow MAS at 140 K with 540 Hz spinning speed. The spectra are referenced to external solid ^{15}NH$_4$Cl.
(Reproduced from P. Lorente, I. G. Shenderovich, N. S. Golubev, G. S. Denisov, G. Buntkowsky and H.-H. Limbach, 1H/15N NMR chemical shielding, dipolar 15N,2H coupling and hydrogen bond geometry correlations in a novel series of hydrogen-bonded acid-base complexes of collidine with carboxylic acids, Magn. Reson. Chem. 2001; 39, S18–S29 with permission from John Wiley and Sons. Copyright © 2001 John Wiley and Sons).

the analysis of the spinning sideband pattern of a particular group can provide information on its protonation state, as in the case of carboxylic/carboxylate groups (see above). However, NMR offers additional possibilities which may be potentially useful in the study of H-bonded systems, even in cases where the barrier of proton transfer is small or absent: mainly the possibility of obtaining homo- or heteronuclear distances by studying specific nuclear dipolar interactions D between the spins S and I in the solid state. Indeed, as stated above, D is proportional to r_{SI}^{-3}, where r_{SI} represents the distance between the two spins, so the dipolar coupling constant is the parameter with the highest crystallographic content.[65]

For instance, ^{1}H DQ MAS spectroscopy represents a powerful method that relies on the generation of DQCs between proximal ^{1}H spins, covering a [1]H–[1]H distance range of up to approximately 3.5 Å, thereby probing the spatial arrangement of protons based on the strength of the through-space dipolar coupling.[66] By employing advanced techniques in homonuclear ^{1}H decoupling that deliver high-resolution ^{1}H spectra, such as ^{1}H DQ CRAMPS,[67] relevant information on HB networks can be extracted that is not usually available from other experiments.[68] The applications include distinguishing a pairwise and chainlike structure of benzoxazine oligomers,[69] study of structure and dynamics in H-bonded polymers,[70] proton-conducting materials,[71] the identification of the keto or enol tautomers in a series of barbituric and thiobarbituric acid polymorphs,[72] biological molecules,[73] π–π stacked polycyclic aromatic systems,[74] polyoxoniobate materials,[75] and surface organometallic species.[76]

For instance, the influence of N–H···O, N–H···N, O–H···N, O–H···O, and CH–π interactions in stabilizing the ribbon-like supramolecular structures of three different guanosine derivatives (guanosine dihydrate, 3′,5′-*O*-dipropanolyl deoxyguanosine, and 3′,5′-*O*-isopropylideneguanosine hemihydrate) have been demonstrated by combining ^{1}H DQ MAS, [13]C–[1]H refocused INEPT, [14]N–[1]H HMQC spectra and DFT calculations.[77]

In the polymer field, the presence, relative strengths and network of poly(methacrylic acid)'s HBs, that form dimers or complexes with complementary polymers, was assessed by combining ^{13}C CPMAS NMR, [1]H–[13]C HETCOR, 1D and 2D ^{1}H DQ MAS NMR experiments.[78] Three possible HB arrangements within poly(methacrylic acid) have been predicted and deduced from ^{1}H DQ MAS spectra (Figure 8.10). Furthermore, the analyses of ^{1}H DQ spinning sideband patterns provided estimates of the proton–proton distances (Figure 8.11).

In a similar way, in the organometallic field, ^{1}H DQ CRAMPS spectra were fundamental in determining the HB network and, thus, the self-assembly of several organometallic Ru(ɪɪ) complexes: three polymorphs (α, β and γ) and one hydrate of [(*p*-cymene)Ru(κN-INA)Cl₂] (where INA is isonicotinic acid), [(*p*-cymene)Ru(κN-A4AB)Cl₂] (where A4AB is 4-aminobenzoic acid) (Scheme 8.6) and the free ligands INA and A4AB.[45] The presence or not of peculiar DQ auto-peaks in the ^{1}H DQ CRAMPS spectra (Figure 8.12) allowed the conclusion

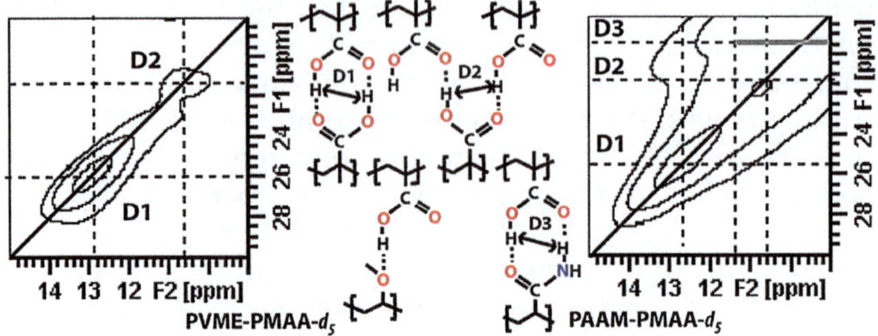

Figure 8.10   HB regions, SQ/DQ, 9–15 ppm/18–30 ppm, of the 2D ^{1}H DQ MAS of the (left) poly(vinyl methylether)–poly(methacrylic acid) dimer and the (right) poly(acrylamide)–poly(methacrylic acid) complex. The labels D1 and D2 identify the auto-peaks of cyclic and open dimers of poly(methacrylic acid), respectively, while the D3 cross-peak indicates the poly(acrylamide)–poly(methacrylic acid) hetero dimer.
(Reproduced with permission from B. Fortier-McGill, V. Toader and L. Reven., 1H Solid State NMR Study of Poly(methacrylic acid) Hydrogen-Bonded Complexes, *Macromol.* 2012, **45**, 6015. Copyright (2012) American Chemical Society).

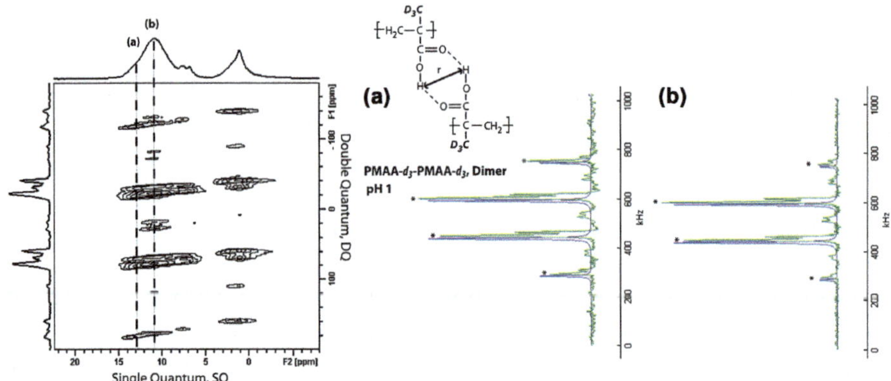

Figure 8.11   (Left) 2D ^{1}H DQ MAS NMR spectrum of poly(methacrylic acid)-d_3, using $N = 4$ cycle excitation/reconversion BABA sequence at $vr = 30$ kHz. (Right) Experimental (green) and simulated (blue) 1D ^{1}H DQ MAS NMR sideband spectra, for the slices taken at 12.5 ppm (a) and 10.5 ppm (b). The asterisk denotes the DQ spinning sidebands of interest as the unlabeled peaks are the result of the overlapping DQC signal between the H-bonded COOH and aliphatic protons.
(Reproduced with permission from B. Fortier-McGill, V. Toader and L. Reven., 1H Solid State NMR Study of Poly(methacrylic acid) Hydrogen-Bonded Complexes, *Macromol.* 2012, **45**, 6015. Copyright (2012) American Chemical Society).

that only [(p-cymene)Ru(κN-A4AB)Cl$_2$] shows the expected supramolecular cyclic dimerization of the carboxylic functions of the ligand, because, in [(p-cymene)Ru(κN-INA)Cl$_2$], the Cl atoms act as stronger HB acceptors.

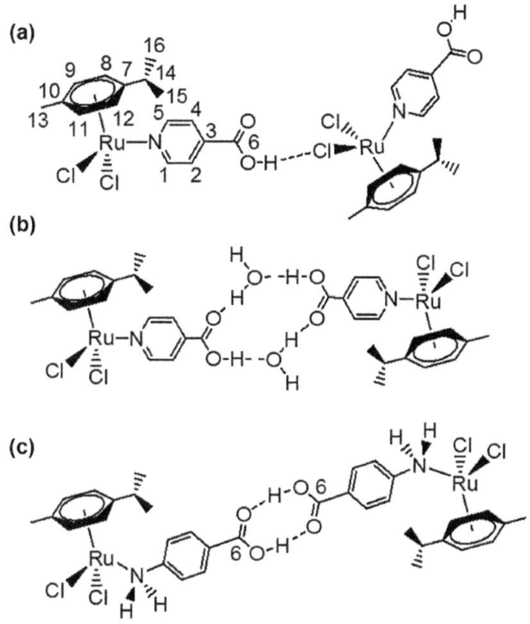

Scheme 8.6 HB arrangements of [(*p*-cymene)Ru(κN-INA)Cl$_2$] (a) α, β, γ, (b) the hydrate form and (c) of [(*p*-cymene)Ru(κN-A4AB)Cl$_2$].

Besides the possibility of probing proton–proton proximities through homonuclear dipolar coupling, SSNMR has provided tools for measuring the heteronuclear dipolar coupling and thus determining the average lengths of the X–H bonds (X usually $=$ N), as well as the motional width and time-scale(s). Applications of SSNMR experiments have expanded significantly in scope and utility with the development of methods for reintroducing and measuring specific dipolar coupling frequencies in both static and under fast MAS samples.

In the case of a series of benzoxazine oligomers, fast MAS ^{1}H NMR and ^{1}H DQ MAS NMR spectra allowed the resonances of the protons forming the HBs to be assigned and used for validating and refining the structure, by means of DFT-based geometry optimizations and ^{1}H chemical-shift calculations.[69] Additionally, quantitative ^{1}H–^{15}N distance measurements obtained by analysis of dipolar spinning sideband patterns of ^{15}N–^{1}H inverse CP-REPT-HDOR spectra (Figure 8.13) confirm the optimized geometry of the tetramer. This particular HB arrangement, in which the N$\cdots$H$\cdots$O and O$\cdots$H$\cdots$O interactions are protected on the inside of a helix, accounts for many of the exemplary chemical properties of polybenzoxazine materials.

In a series of lidocaine-dicarboxylic acid cocrystals, the analysis of ^{1}H–^{13}C proximities achieved from ^{1}H–^{13}C on- and off-resonance

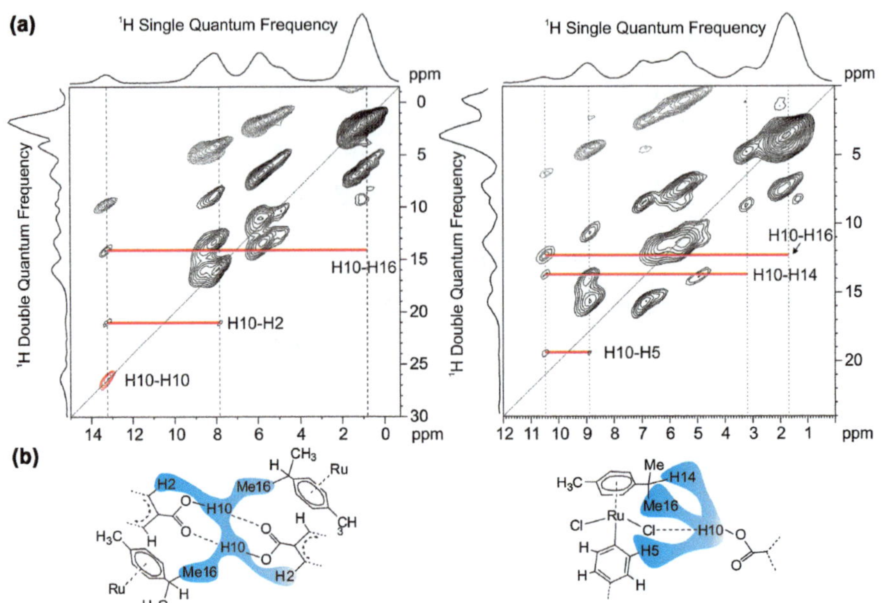

Figure 8.12 ^{1}H DQ CRAMPS (PMLG5–POSTC7–wPMLG5) spectra of (left) [(*p*-cymene)Ru(κN-A4AB)Cl$_2$]) and (right) [(*p*-cymene)Ru(κN-INA)Cl$_2$], together with skyline projections recorded at 12.5 kHz MAS. Solid red horizontal bars indicate specific DQCs between OH (H10) and nearby protons. The OH diagonal peak indicating the cyclic dimerization of the COOH groups is highlighted in red. (b) Representation of the crystal structures showing the main inter- and intramolecular proton–proton proximities.
(Reproduced with permission from M.R. Chierotti, R. Gobetto, C. Nervi, A. Bacchi, P. Pelagatti, V. Colombo and A. Sironi., Probing Hydrogen Bond Networks in Half-Sandwich Ru(II) Building Blocks by a Combined 1H DQ CRAMPS Solid-State NMR, XRPD, and DFT Approach, Inorg. Chem. 2014, 53, 139–146. Copyright (2014) American Chemical Society).

FSLG HETCOR combined with data from ^{1}H DQ MAS and CRAMPS allowed the salt or cocrystal nature and HB network to be elucidated.[79] In particular, the structure of lidocainium hydrogen succinate was determined from X-ray powder diffraction assisted by SSNMR data, which provided information on the number of independent molecules, HB network, and crystal packing through the analysis of ^{1}H–^{1}H or ^{1}H–^{13}C proximities.

With a similar approach, topological details of HB connectivities and weak interactions, such as C–H···π contacts of a 1:1 cocrystal formed by 3,5-dimethyl-1*H*-pyrazole and 4,5-dimethyl-1*H*-imidazole, have been determined. The use of various 1D/2D ^{13}C, ^{15}N and ^{1}H high-resolution SSNMR techniques provided structural insight on local length scales revealing internuclear proximities and relative orientations of the two

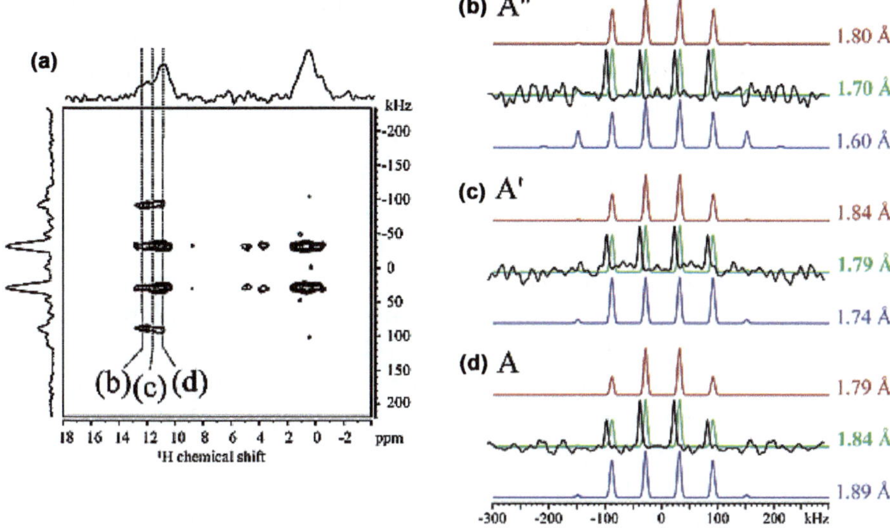

Figure 8.13   (a) ^{15}N–^{1}H inverse CP-REPT-HDOR spectra for the ^{15}N-enriched benzoxazine tetramer, collected using $\tau_{exc} = 20\,\tau_r$ under MAS at 30 kHz. The experimental 1D patterns shown in (b, c, d) were taken from the 2D spectrum shown in (a) at the resonance positions of the three different NH protons. The colored patterns in (b, c, d) are calculated using the N–H distances given on the right. The red and blue patterns represent the upper and lower limits of the error margin, while the green pattern fits best to the experimental data.
(Reproduced with permission from G.R. Goward, D.l. Sebastiani, I. Schnell, H.W. Spiess, H.-D. Kim and H. Ishida., Benzoxazine Oligomers: Evidence for a Helical Structure from Solid-State NMR Spectroscopy and DFT-Based Dynamics and Chemical Shift Calculations, *J. Am. Chem. Soc.* 2003, **125**, 5792–5800. Copyright (2003) American Chemical Society).

molecular building blocks of the studied cocrystal. In particular, the 2D ^{1}H–^{15}N HETCOR highlighted the intermolecular N–H$\cdots$N contact through different ^{1}H–^{15}N correlation peaks.[34]

Opella and co-worker demonstrated that two- and three-dimensional proton-detected experiments can provide high spectral resolution and accurate measurements of ^{1}H–^{15}N heteronuclear dipolar coupling frequencies under fast MAS on a non-crystalline sample of the coat protein in intact Pf1 bacteriophage particles.[80]

8.3.3   Quadrupolar Interactions and Hydrogen Bonds

Solid-state ^{17}O data have been obtained for a series of carboxylic compounds, namely maleic acid, chloromaleic acid, KH maleate, KH chloromaleate, K_2 chloromaleate, and LiH phthalate $\cdot$ MeOH.[81]

The experimental and theoretical results indicate that both ^{17}O chemical shifts and EFG tensors of the carboxylic functional group are sensitive to the $C=O\cdots H-N$ hydrogen-bonding environment, suggesting the possibility of estimating HB information from ^{17}O NMR data. The calculations also reveal intermolecular HB effects on the ^{17}O NMR shielding tensors. It is found that the δ_{11} and δ_{22} components of the chemical shift tensor at O–H and C=O, respectively, are aligned nearly parallel with the strong HB and shift away from this direction as the HB interaction weakens. Other very interesting results have been obtained in the case of four polymorphs of the model peptide system Gly-(Gly-^{17}O)-Gly.[82]

In this case the effect of ion binding can be compared with HB interactions since two polymorphs form antiparallel β-sheets and their central carbonyl groups form C=O–H–N intermolecular HBs with different O–H distances, whereas in the other two polymorphs interactions with Li^+ or Ca^{2+} ions are present. Combined experimental and theoretical results demonstrate an excellent sensitivity of the ^{17}O QCC and chemical shift interactions to ion binding, whereas hydrogen bonding has a very small impact on chemical shift properties. Thus, ^{17}O NMR can be used as a very sensitive tool for the characterization of ion binding.

8.4   SSNMR and Halogen Bonds

Besides HBs, in recent years there has been a growing focus on XBs thanks to their possible applications spanning a wide number of different fields, *e.g.* crystal engineering, materials chemistry and biochemistry. The success of this specific non-covalent interaction comes from its unique physico-chemical properties namely its strength, selectivity, tunability and directionality. These, together with bonding energies comparable or even stronger than the commonly used HB, enable the predictable alignment of molecular building blocks in crystalline architectures and functional materials such as polymers, photoactive compounds, pharmaceuticals, porous materials and so on.

The evidence for the occurrence of the XB may be obtained by a number of analytical techniques. Since the first studies in the 1960s, the most commonly employed ones are single crystal X-ray diffraction (SCXRD), NMR, vibrational and UV-visible spectroscopies and thermal analysis.[83] XBs have long been studied in the solid state, mainly by X-ray diffraction,[84] while NMR spectroscopy has been traditionally adopted to shed light on the solution behavior.[85] Among all of the different techniques able to characterize it, NMR plays a key role for the information it can provide. NMR spectroscopy has already proven to be a powerful and versatile tool with which to investigate non-covalent interactions. In the case of HBs, it may probe either the hydrogen atom directly involved in the interaction, or the atoms within, or in close proximity of, the HB donor and acceptor groups (see above).[59] The same method applies in the case of XB, in which NMR succeeds in collecting detailed information about molecular structure, dynamics of interaction formation and quantitative relationships between NMR parameters and the local structure. It is against this background that IUPAC states that "the X···Y XB usually affects the nuclear magnetic resonance observables (*e.g.*, chemical shift values) of nuclei in both R–X and Y, both in solution and in the solid state".[86] Indeed, each of the three atoms involved in the interaction is in principle amenable to be studied by NMR (see Scheme 8.7). In practice, unless one has access to very high magnetic fields (≥20 T), halogen atoms are not easily probed. The NMR active isotopes of halogens possessing quantum spin number greater than 1/2: $^{35/37}$Cl, $^{79/81}$Br and ^{127}I, are all quadrupolar nuclei. The NMR spectra of such nuclei display very broad lines due to the rapid quadrupolar relaxation; in some cases, the broadening is so large that no signal is observable. For these reasons the acquisition and interpretation of such spectra is extremely challenging, with only a few studies published in the literature.[87] This helps to explain why NMR studies on XBs are focused mainly on probing the atoms in the R and Y moieties. Table 8.1 summarizes the nuclei of interest for NMR spectroscopy that may be present in the structural unit of the XB.

In spite of the difficulties stemming from quadrupolar nuclei, NMR offers unique insights into XB. Solution NMR spectroscopy has proven to be a convenient and sensitive tool with which to establish the occurrence of XBs, to rank the ability of XB donors and acceptors to be involved in the interaction and even to describe thermodynamics and geometric characteristics of adducts.[88] Solution NMR has resorted to

R – X···Y

R can typically be: C, N, H, P, F, Cl, Br, I

X can typically be: Cl, Br, I

Y can typically be: N, O, S, P, As, Se, I^-, Br^-, Cl^-, F^-

Scheme 8.7 Non-exhaustive list of atoms commonly present in XB donors and acceptors.

Table 8.1 Summary of nuclei amenable to study by NMR in the three fragments of the XB interaction.

R, (spin)	X, (spin)	Y, (spin)
^{1}H (1/2)	^{35}Cl (3/2)	^{14}N (1)
^{13}C (1/2)	^{37}Cl (3/2)	^{15}N (1/2)
^{19}F (1/2)	^{79}Br (3/2)	^{31}P (1/2)
^{31}P (1/2)	^{81}Br (3/2)	^{77}Se (1/2)
^{14}N (1)	^{127}I (5/2)	^{35}Cl (3/2)
^{15}N (1/2)		^{37}Cl (3/2)
		^{79}Br (3/2)
		^{81}Br (3/2)
		^{127}I (5/2)

the analysis of ^{19}F, ^{1}H, ^{13}C and $^{14/15}$N. In particular, in the early 2000's the systematic recourse to haloperfluorocarbons for the synthesis of X-bonded adducts led to a heavy utilization of ^{19}F NMR, which in turn provided evidence for the occurrence of the XB through the analysis of chemical shift changes of the interacting compounds relative to the isolated compounds.[88h] Furthermore, trends were obtained (the stronger the bond, the larger the shift) and a strength scale of donors (I > Br > Cl).[88a] Unlike solution NMR, detailed studies of XBs by SSNMR have been carried out only in recent years, although some ^{13}C CPMAS data of a complex of 1,4-diazabicyclo[2.2.2]octane and CBr$_4$ were reported by Kochi in 1987.[89] Recent efforts are directed to in-depth, specialized SSNMR analysis, with multinuclear approaches and the systematic use of quantum chemical calculations with DFT-based methods. The following sections outline the main results obtainable by SSNMR techniques, examining selected NMR interactions and parameters such as chemical shifts, CSA, dipolar and quadrupolar interactions.

8.4.1 Chemical Shift, Chemical Shift Anisotropy and Halogen Bonds

SSNMR chemical shift is by far the most used parameter to evaluate the occurrence and the strength of the XB. Among all nuclei listed in Table 8.1, the ^{13}C nucleus is perhaps the most studied one in the solid state, because of its extensive presence in molecules that are commonly used to synthesize X-bonded adducts. Furthermore, CP techniques allow probing the carbon nucleus in a quicker way by enhancing its sensitivity; if CP is not accessible, because of lack of adjacent and suitable nuclei from which polarization transfer occurs (*e.g.* ^{1}H, ^{19}F), one can always rely upon direct excitation experiments.

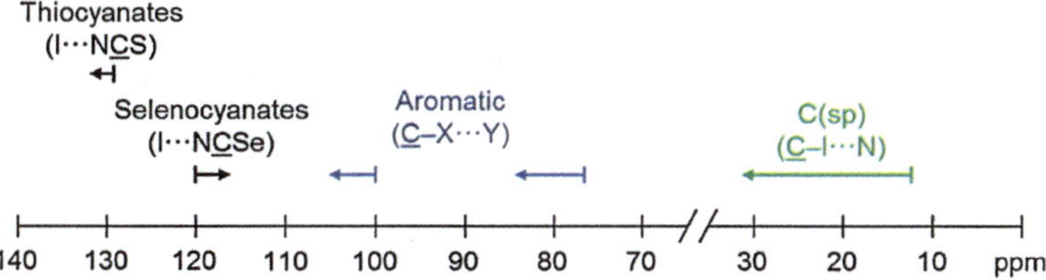

Scheme 8.8   Schematic representation of XB-induced changes in ^{13}C SSNMR chemical shifts. The changes are referred to carbon atoms of either the acceptor (black), or the donor (blue and green). Data are from ref. 88*b*, 90, 91 and 93.

SSNMR experiments can concern the carbon atoms of either the XB acceptor (Y) or the donor (R–X). However, while the changes in the $\delta_{iso}(^{13}C)$ of several acceptors have been reported to be in the range of 1–3 ppm,[4b,90] the extent of the shift is much wider from the donor point of view (see Scheme 8.8). Bryce and co-workers have put a lot of effort into studying the changes in the $\delta_{iso}(^{13}C)$ of several donors such as 1,4-diiodotetrafluorobenzene (*p*-DITFB), 1,2-diiodotetrafluorobenzene (*o*-DITFB) and 1,4-diiodobenzene (DIB), by characterizing their co-crystals with ammonium or phosphonium halide salts.[87b,91] They have observed a systematic increase (2–10 ppm) in $\delta_{iso}(^{13}C)$ for the C–I resonance in the presence of XB relative to the pure compounds. It must be noted that such spectra are challenging because MAS does not entirely average the dipolar coupling between carbon, which is a spin 1/2 nucleus, and iodine, which is quadrupolar. Thus, residual dipolar broadening of the ^{13}C signals is observed. Nevertheless, the C–I resonance can be unambiguously assigned owing to the unusual shielding of the chemical shift of such carbon atom: this is due to a well-known spin–orbit-induced heavy atom effect of iodine.[92] Our group also has recently published[88b,93] several $\delta_{iso}(^{13}C)$ values for the C(sp)–I resonance in two pharmaceutical compounds, X-bonded with different acceptors such as 4,4′-bipyridine, 1,2-bis-4(pyridyl)ethane and alkyl ammonium halide salts. Consistently with the literature results discussed above, we have observed an increase (7–15 ppm) of the $\delta_{iso}(^{13}C)$ values upon XB formation. Figure 8.14 shows the ^{13}C CPMAS and Non Quaternary Suppression (NQS) NMR spectra of one of the two pharmaceutical compounds (namely 3-Iodo-2-propynyl-*N*-butylcarbamate, IPBC) and its cocrystals. NQS experiments were crucial to unambiguously assign the chemical shifts of the overlapped C(sp)–I resonances.

Nitrogen chemical shifts are also important in the study of XB, due to the fact that this nucleus is often directly involved in the interaction. However, as already seen (above), the non-quadrupolar isotope, ^{15}N, has a low natural abundance and low sensitivity: therefore,

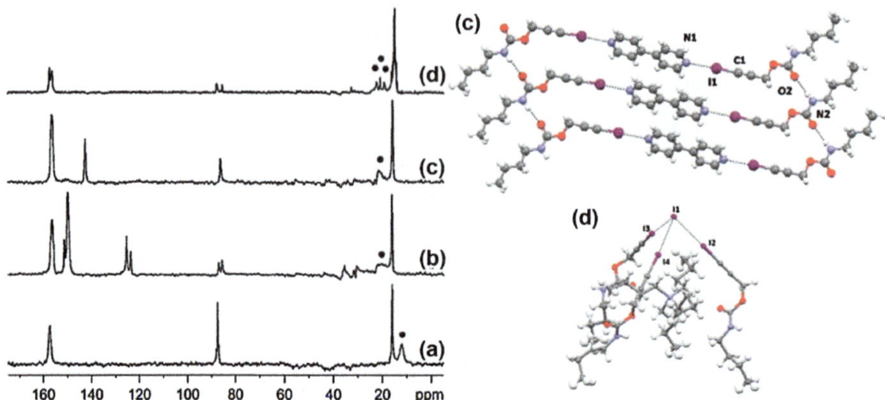

Figure 8.14   (Left) ^{13}C (100 MHz) NQS spectra of IPBC (a), and its X-bonded adducts with 1,2-bis-4(pyridyl)ethane, (b) 4,4′-bipyridine (c), and tetrabutylammonium iodide (d). C(sp)–I signals are marked with a black circle. (Right) ball and stick representation of the crystal packing of (c) and (d). Color code: gray, C; red, O; blue, N; purple, I; white; H.
(Reproduced with permission from M. Baldrighi, G. Cavallo, M.R. Chierotti, R. Gobetto, P. Metrangolo, T. Pilati, G. Resnati and G. Terraneo., Halogen Bonding and Pharmaceutical Cocrystals: The Case of a Widely Used Preservative, *Mol. Pharm.*, 2013, 10, 1760. Copyright (2013) American Chemical Society).

a ^{15}N CPMAS spectrum may take a very long acquisition time, in the order of 15–20 hours (or days at worst), to achieve an acceptable signal-to-noise ratio. Such long collection can be significantly reduced using ^{15}N-enriched compounds, from which one can also obtain information on the CSA, as reported by Bryce and co-workers for a tetramethylammonium thiocyanate sample.[4h] They have observed a small, but measurable shift (2.5 ppm) to low frequencies of the thiocyanate $\delta_{iso}(^{15}N)$ upon XB formation; subsequently they have examined the nitrogen tensors, observing an increase of 60 ppm in the span value relative to the pure compound. Unfortunately, the disorder of the thiocyanate anion and the small changes of the tensor components prevented any further interpretation of the results. Dipyridylic nitrogens are another class of Lewis bases commonly used in XB structures. Their $\delta_{iso}(^{15}N)$ values move toward lower frequencies when a XB occurs; for example, $\delta_{iso}(^{15}N)$ of 4,4′-bipyridine and 1,2-bis(4-pyridyl)ethane undergoes a remarkable change of 10 and 16 ppm, respectively, when forming N···I motifs.[88b,93] Figure 8.15 shows the ^{15}N CPMAS NMR spectra for the bipyridine series, where the low frequency shift is clearly measurable.

Results from ^{15}N investigations are outlined in Scheme 8.9; the shift direction depends upon the type of nitrogen, but the shielding is

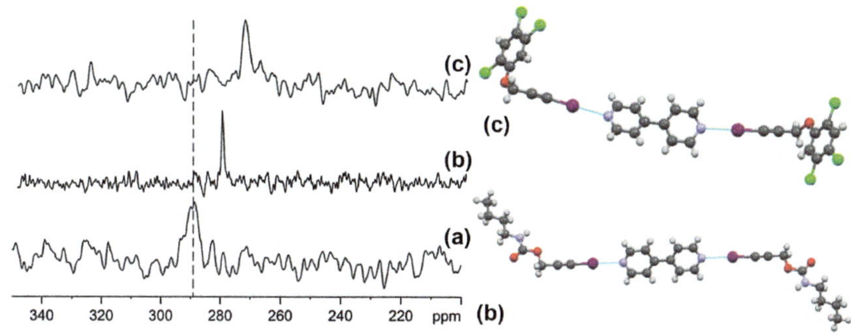

Figure 8.15   Left: ^{15}N (40 MHz) CPMAS spectra of pure 4,4′-bipyridine (a), and its cocrystals with IPBC (b) and haloprogin (c) recorded at 9 kHz. The vertical dotted line highlights the change of the chemical shifts upon XB formation. Right: ball and stick representation of the halogen-bond pattern in (b) and (c); the color code is the same as in Figure 8.14.

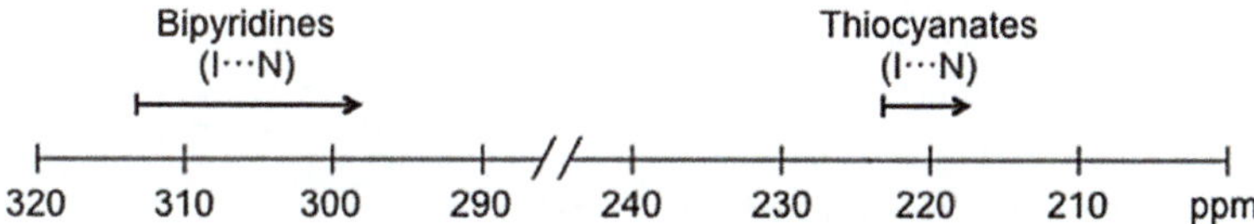

Scheme 8.9   Schematic representation of XB-induced changes in ^{15}N SSNMR chemical shifts. The scale is referenced to liquid NH$_3$ (i.e., $\delta(^{15}$N, NH$_3) \equiv 0$ ppm).
Data are from ref. 4b and 88b.

consistent with the lone pair removal in pyridine-like nitrogen atoms as already observed for HB formation.[94]

A few other studies have addressed different kinds of spin 1/2 nuclei occurring in XB. For example, a complete CSA study of ^{77}Se has been carried out by Viger-Gravel *et al.*, exploiting the Se···I interaction between tetramethylammonium selenocyanate and *o*-DITFB or *p*-DITFB.[4b] The authors related the change of the smallest component of the chemical shift tensor, δ_{33}, which in turn represents the chemical shift along the axis of the selenocyanate anion, to the perturbations of molecular orbitals in the plane perpendicular to the anion's axis. This plane is that in which iodine engages in XB with the selenium center; for this reason, the δ_{33} value is a good measure of the extent of the XB in $^{-}$NCSe···I adducts.

Quadrupolar nuclei chemical shift and CSA have been reported for ^{35}Cl and ^{81}Br in several works by Bryce and co-workers.[87,90] The analyses have been carried out for halide anions only, because the acquisition of covalently-bonded halogen spectra is still highly impractical.[95] The results of these pioneering studies show the

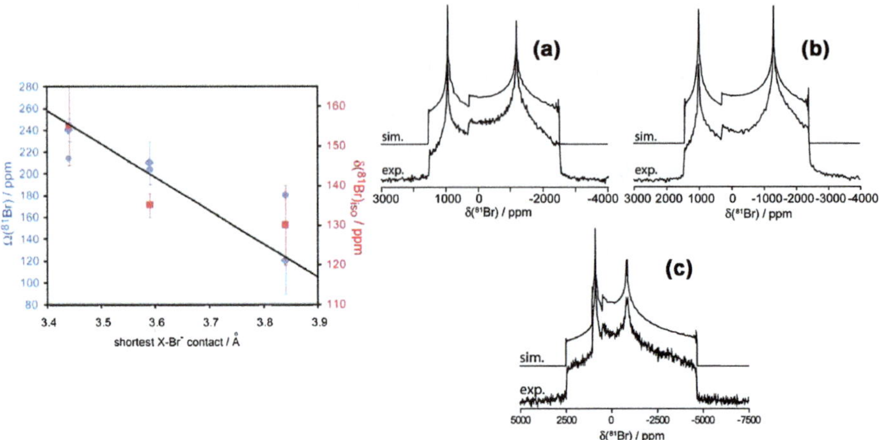

Figure 8.16 (Left) plot of experimental chemical shifts (red squares), bromine chemical shift tensor spans (blue diamonds), and GIPAW-DFT calculated spans (blue circles) as a function of the shortest X–Br distance in a series of haloanilinium bromides. The line indicates a linear fit to the experimental span data (Ω/ppm = $-310d + 1313$; $R^2 = 0.99$). (Right) ^{81}Br SSNMR spectra, recorded at 21.1 T, of 2-chloroanilinium bromide (a), 2-bromoanilinium bromide (b) and 3-chloroanilinium bromide (c), from which $\delta_{iso}(^{81}$Br) and $\Omega(^{81}$Br) are taken.
(Reproduced with permission from R.J. Attrell, C.M. Widdifield, I. Korobkov and D.L. Bryce., Weak Halogen Bonding in Solid Haloanilinium Halides Probed Directly via Chlorine-35, Bromine-81, and Iodine-127 NMR Spectroscopy, *Cryst. Growth Des.*, 2012, 12, 1641. Copyright (2012) American Chemical Society).

importance of the EFG and chemical shift tensor parameters in describing the XB environment: in particular, the C_Q of a chloride anion may be used to differentiate between two distinct ion coordination environments; moreover, smaller $\delta_{iso}(^{81}$Br) and ^{81}Br span are observed when XB becomes weaker (see Figure 8.16). SSNMR data for ^{127}I have also been reported;[87a,90] however, due do the extremely high difficulty in collecting such spectra even at very high magnetic fields, along with an unsatisfactory data quality, authors were careful in not attempting hasty conclusions.

Finally, it is worth noting that the chemical shift tensor analysis heavily relies upon quantum chemical calculations, which are fundamental in validating experimental spectra line shape, thus affording a reliable measure of the NMR parameters. However, having to deal with heavy atoms, EFG tensors and magnetic shielding tensor calculations require the inclusion of relativistic effects. The calculation of relativistic effects in the periodic boundary conditions (GIPAW) is still an ongoing field of research; hence calculations are actually carried out using cluster models, for which scalar and spin–orbit effects can be accounted for.

8.4.2 Dipolar Interactions and Halogen Bonds

Other NMR evidence of XBs in the solid state can be obtained by internuclear dipolar coupling. This through–space interaction has been exploited by Tekely and co-workers[96] to determine the N$\cdots$I length in a ^{15}N-enriched benzyl-di(4-iodobenzyl)-amine. They used rotary resonance recoupling to measure the dipolar interaction between ^{15}N and ^{127}I. The numerical value of D was obtained by fitting the rotary resonance lineshapes with simulated ones, which in turn were obtained from experimental CSA parameters. This method afforded an internuclear distance significantly shorter than the sum of the van der Waals radii, and it is particularly useful when the quality of single crystals is not suitable for XRD analysis.

8.4.3 Correlation of SSNMR Parameters and XRD Data in Halogen Bond Study

SSNMR studies on XBs may be coupled to XRD data in order to explore how NMR parameters are related to the structural and/or geometric features of the XB. Viger-Gravel *et al.* have established a correlation between $\delta_{\mathrm{iso}}(^{13}\mathrm{C})$ values and the C–I distance in a series of *p*-DITFB complexes:[91] as the C–I distance increases, the chemical shift also increases. The general trend was corroborated by GIPAW-DFT calculations, using a cluster model in order to take account of the relativistic effects. Surprisingly, correlations between the chemical shifts and R_{XB} were not observed experimentally; however, considering that the C–I distance scales directly with the XB strength, the authors concluded that ^{13}C chemical shifts are diagnostic of the strength of the XB. The same trend was confirmed in a later work by the same authors.[12]

Relationships between NMR parameters of quadrupolar nuclei and the XB environment have also been reported, affording further chemical insight into the nature of the interaction. For example, there is a clear correlation between the ^{35}Cl and ^{81}Br quadrupolar asymmetry parameter η_Q and the I$\cdots$X$^-\cdots$I angle, for a broad range of geometries.[87b] This result is in qualitative agreement with previous corresponding data available for the HB. An additional relationship has been found for ^{35}Cl and ^{81}Br C_Q values, which are well correlated with the $I \cdots X^- \cdots I$ angles by the largest principal component of the EFG, that is, V_{zz}: when the angle of the interaction increases, V_{zz} also increases.

8.5 Conclusions

In this chapter we demonstrated through selected examples how SSNMR can contribute to understanding the structure and properties of supramolecular architectures, focusing mainly on HB and XB detection and characterization.

The use of a multinuclear and multiparametric approach in the SSNMR investigation allows information at the local level to be obtained, opening new perspectives in the prediction and design of supramolecular systems. In particular, HBs and XBs and intermolecular packing arrangements can be investigated in great detail, in most cases without the need for highly crystalline samples or special sample preparations.

Correlations between the chemical structure and the SSNMR parameters have been extensively established. The isotropic chemical shift, the CSA, the dipolar interaction, the quadrupolar interaction and the spin diffusion are the main parameters that allow correlations with specific structure features to be established for an increasing number of complex supramolecular architectures. Moreover, quantitative correlations between the SSNMR parameters and specific bond angles and bond lengths have been obtained.

Instrumental and technical improvements play an essential role in the development of SSNMR spectroscopy. Both are far from being complete and several further improvements are expected in the coming years. If SSNMR is already fundamental in the supramolecular and crystal engineering fields now, it is envisaged to become an indispensable tool in the near future.

References

1. K. J. D. MacKenzie and M. E. Smith, in *Multinuclear Solid-State NMR of Inorganic Materials*, ed. R. W. Cahn, Pergamon, Oxford, 2002.
2. (a) Z. Gu and A. McDermott, *J. Am. Chem. Soc.*, 1995, **115**, 4262; (b) A. Naito, S. Ganapathy, K. Aakasaka and C. J. McDowell, *J. Chem. Phys.*, 1981, **74**, 3198; (c) R. Haberkom, R. Stark, H. van Willigen and R. Griffin, *J. Am. Chem. Soc.*, 1981, **103**, 2534; (d) N. James, S. Ganapathy and E. Oldfield, *J. Magn. Reson.*, 1983, **54**, 111; (e) R. Griffin, A. Pines, S. Pausak and J. Waugh, *J. Phys. Chem.*, 1975, **65**, 1267; (f) W. Veeman, *Prog. NMR Spectrosc.*, 1984, **16**, 193.

3. R. K. Harris, *Solid State Sci.*, 2004, **6**, 1025–1037.
4. (a) J. Herzfeld and A. E. Berger, *Chem. Phys.*, 1980, **73**, 6021–6030; (b) J. Viger-Gravel, I. Korobkov and D. L. Bryce, *Cryst. Growth Des.*, 2011, **11**, 4984–4995.
5. (a) S. M. de Paul, K. Saalwachter, R. Graf and H. W. Spiess, *J. Magn. Reson.*, 2000, **146**, 140–196; (b) Y. Wei, D.-K. Lee, A. E. McDermott and A. Ramamoorthya, *J. Magn. Reson.*, 2002, **158**, 23–35.
6. C. J. Pickard and F. Mauri, *Phys. Rev.*, 2001, **B63**, 245101.
7. J. R. Yates, C. J. Pickard and F. Mauri, *Phys. Rev.*, 2007, **B76**, 024401.
8. (a) D. Lüdeker and G. Brunklaus, *Solid State Nucl. Magn. Reson.*, 2015, **65**, 29; (b) M. Khan, V. Enkelmann and G. Brunklaus, *CrystEngComm*, 2011, **13**, 3213; (c) F. Pourpoint, J. Yehl, M. Li, R. Gupta, J. Trébosc, O. Lafon, J. P. Amoureux and T. Polenova, *ChemPhysChem*, 2015, **16**, 1619.
9. (a) L. Szeleszczuk, D. M. Pisklak, M. Zielińska-Pisklak and I. Wawer, *Chem. Phys. Lett.*, 2016, **653**, 35; (b) T. Charpentier, *Solid State Nucl. Magn. Reson.*, 2011, **40**, 1.
10. (a) J. R. Yates, C. J. Pickard, M. C. Payne and F. Mauri, *J. Chem. Phys.*, 2003, **118**(13), 5746; (b) J. Autschbach and T. Ziegler, *J. Chem. Phys.*, 2000, **113**(3), 936.
11. R. Bouten, E. J. Baerends, E. van Lenthe, L. Visscher, G. Schreckenbach and T. Ziegler, *J. Phys. Chem. A*, 2000, **104**, 5600.
12. (a) C. Bonhomme, C. Gervais, F. Babonneau, C. Coelho, F. Pourpoint, T. Azaïs, S. E. Ashbrook, J. M. Griffin, J. R. Yates, F. Mauri and C. J. Pickard, *Chem. Rev.*, 2012, **112**, 5733; (b) T. F. G. Green and J. R. Yates, *J. Chem. Phys.*, 2014, **140**, 234106; (c) A. C. de Dios and C. J. Jameson, *Annu. Rep. NMR Spectrosc.*, 2012, **77**, 1.
13. S. P. Brown, in *NMR Crystallography*, ed. R. K. Harris, R. E. Wasylishen and M. J. Duer, John Wiley & Sons, Chichester, 2009, p. 321.
14. (a) S. P. C. Gervais, C. Coelho, T. Azaïs, J. Maquet, G. Laurent, F. Pourpoint, C. Bonhomme, P. Florian, B. Alonso, G. Guerrero, P. H. Mutin and F. Mauri, *J. Magn. Reson.*, 2007, **187**, 131; (b) R. Dervişoğlu, D. S. Middlemiss, F. Blanc, Y. L. Lee, D. Morgan and C. P. Grey, *Chem. Mater.*, 2015, **27**, 3861; (c) J. A. Fernandes, M. Sardo, L. Mafra, D. Choquesillo-Lazarte and N. Masciocchi, *Cryst. Growth Des.*, 2015, **15**, 3674.
15. D. V. Dudenko, P. A. Williams, C. E. Hughes, O. N. Antzutkin, S. P. Velaga, S. P. Brown and K. D. M. Harris, *J. Phys. Chem.*, 2013, **C117**, 12258–12265.
16. E. D. L. Smith, R. B. Hammond, M. J. Jones, K. J. Roberts, J. B. O. Mitchell, S. L. Price, R. K. Harris and D. Apperley, *J. Phys. Chem.*, 2001, **B105**, 5818–5826.
17. F. Franco, M. Baricco, M. R. Chierotti, R. Gobetto and C. Nervi, *J. Phys. Chem. C*, 2013, **117**, 9991–9998.
18. M. Aluas, C. Tripon, J. M. Griffin, X. Filip, V. Ladizhansky, R. G. Griffin, S. P. Brown and C. Filip, *J. Magn. Reson.*, 2009, **199**, 173–187.
19. E. Salager, R. Stein, C. J. Pickard, B. N. D. Elena and L. Emsley, *Phys. Chem. Chem. Phys.*, 2009, **11**, 2610–2621.
20. D. A. Middleton, X. Peng, D. Saunders, K. Shankland, W. I. F. David and A. J. Markvardsen, *Chem. Commun.*, 2002, 1976–1977.
21. (a) G. De Paepe, J. R. Lewandowski and R. G. Griffin, *J. Chem. Phys.*, 2008, **128**, 124503; (b) B. Hu, J. Trebosc and J. P. Amoureux, *J. Magn. Reson.*, 2008, **192**, 112–122.
22. T. Gullion and J. Schaefer, *J. Magn. Reson.*, 1989, **81**, 196–200.
23. M. Hong and K. Schmidt-Rohr, *Acc. Chem. Res.*, 2013, **46**, 2154–2163.
24. R. Tycko and G. Dabbagh, *Chem. Phys. Lett.*, 1990, **173**, 461–465.
25. A. W. Hing, S. Vega and J. Schaefer, *J. Magn. Reson., Ser. A*, 1993, **103**, 151–162.
26. (a) A. E. Bennett, J. H. Ok, R. G. Griffin and S. Vega, *J. Chem. Phys.*, 1992, **96**, 8624–8627; (b) D. K. Sodickson, M. H. Levitt, S. Vega and R. G. Griffin, *J. Chem. Phys.*, 1993, **98**, 6742–6748.

27. M. H. Levitt, in *Encyclopedia of Nuclear Magnetic Resonance*, ed. D. M. Grant and R. K. Harris, John Wiley & Sons, Ltd, Chichester, 2002, vol. 9, pp. 165–196.
28. C. P. Grey, W. S. Veeman and A. J. Vega, *J. Chem. Phys.*, 1993, **98**, 7711–7724.
29. E. Hughes, T. Gullion, A. Goldbourt, S. Vega and A. J. Vega, *J. Magn. Reson.*, 2002, **156**, 230–241.
30. D. Braga, G. Palladino, M. Polito, K. Rubini, F. Grepioni, M. R. Chierotti and R. Gobetto, *Chem. – Eur. J.*, 2008, **14**, 10149–10159.
31. S. P. Brown, *Prog. Nucl. Magn. Reson. Spectrosc.*, 2007, **50**, 199–251.
32. L. Frydman and J. S. Harwood, *J. Am. Chem. Soc.*, 1995, **117**, 5367.
33. A. Medek, J. S. Harwood and L. Frydman, *J. Am. Chem. Soc.*, 1995, **117**, 12779.
34. M. Sardo, S. M. Santos, A. A. Babaryk, C. López, I. Alkorta, J. Elguero, R. M. Claramunt and L. Mafra, *Solid State Nucl. Magn. Reson.*, 2015, **65**, 49–63.
35. (a) B. Elena and L. Emsley, *J. Am. Chem. Soc.*, 2005, **127**, 9140–9146; (b) B. Elena, G. Pintacuda, N. Mifsud and L. Emsley, *J. Am. Chem. Soc.*, 2006, **128**, 9555–9560.
36. G. R. Desiraju, *Angew. Chem.*, 1995, **107**, 2541–2558; *Angew. Chem., Int. Ed. Engl.*, 1995, **34**, 2311–2327.
37. G. R. Desiraju, *Cryst. Growth Des.*, 2011, **11**, 896–898.
38. (a) G. R. Desiraju, *Crystal Engineering. The Design of Organic Solids*, Elsevier, Amsterdam, 1989; (b) G. R. Desiraju, *Chem. Commun.*, 1997, 1475–1482; (c) D. Braga, F. Grepioni and L. Maini, *Chem. Commun.*, 2010, **46**, 6232–6242.
39. T. Steiner, *Angew. Chem., Int. Ed.*, 2002, **41**, 48–76.
40. M. R. Chierotti and R. Gobetto, *Chem. Commun.*, 2008, 1621–1634.
41. S. P. Brown, *Solid State Nucl. Magn. Reson.*, 2012, **41**, 1–27.
42. U. Sternberg and E. Brunner, *J. Magn. Reson., Ser. A*, 1994, **108**, 142.
43. K. Yamauchi, S. Kuroki and I. Ando, *J. Mol. Struct.*, 2002, **602–603**, 9–16.
44. D. Braga, L. Maini, G. de Sanctis, K. Rubini, F. Grepioni, M. R. Chierotti and R. Gobetto, *Chem. – Eur. J.*, 2003, **9**, 5538–5548.
45. (a) A. Bacchi, G. Cantoni, M. R. Chierotti, A. Girlando, R. Gobetto, G. Lapadula, P. Pelagatti, A. Sironi and M. Zecchini, *CrystEngComm.*, 2011, **13**, 4365; (b) M. R. Chierotti, R. Gobetto, C. Nervi, A. Bacchi, P. Pelagatti, V. Colombo and A. Sironi, *Inorg. Chem.*, 2014, **53**, 139–146.
46. K. Tsuchiya, A. Takahashi, N. Takeda, N. Asakawa, S. Kuroki, I. Ando, A. Shoji and T. Ozaki, *J. Mol. Struct.*, 1995, **350**, 233–240.
47. (a) F. Marchetti, J. Palmucci, C. Pettinari, R. Pettinari, F. Condello, S. Ferraro, M. Marangoni, A. Crispini, S. Scuri, I. Grappasonni, M. Cocchioni, M. Nabissi, M. R. Chierotti and R. Gobetto, *Chem. – Eur. J.*, 2015, **21**, 836–850; (b) R. Pettinari, C. Pettinari, F. Marchetti, R. Gobetto, C. Nervi, M. R. Chierotti, E. J. Chan, B. W. Skelton and A. H. White, *Inorg. Chem.*, 2010, **49**, 11205–11215.
48. (a) D. H. Brouwer and J. A. Ripmeester, *J. Magn. Reson.*, 2007, **185**, 173–178; (b) M. Carravetta, M. Eden, X. Zhao, A. Brinkmann and M. H. Levitt, *Chem. Phys. Lett.*, 2000, **321**, 205–215; (c) M. H. Levitt, *J. Chem. Phys.*, 2008, **128**, 052205; (d) L. Duma, D. Abergel, P. Tekely and G. Bodenhausen, *Chem. Commun.*, 2008, 2361–2363; (e) E. R. Andrew, A. Bradbury, R. G. Eades and V. T. Wynn, *Phys. Lett.*, 1963, **4**, 99–100; (f) T. G. Oas, R. G. Griffin and M. H. Levitt, *J. Chem. Phys.*, 1988, **89**, 692–695; (g) Z. Gan, D. M. Grant and R. R. Ernst, *Chem. Phys. Lett.*, 1996, **254**, 349–357.
49. C. M. Rohlfing, L. C. Allen and R. Ditchfield, *J. Chem. Phys.*, 1983, **79**, 4958–4966.
50. M. Kibalchenko, D. Lee, L. Shao, M. C. Payne, J. J. Titman and J. R. Yates, *Chem. Phys. Lett.*, 2010, **498**, 270–276.
51. G. Wu, C. J. Freure and E. Verdurand, *J. Am. Chem. Soc.*, 1998, **120**, 13187–13193.
52. G. Hou, S. Paramasivam, S. Yan, T. Polenova and A. J. Vega, *J. Am. Chem. Soc.*, 2013, **135**, 1358–1368.
53. (a) O. N. Anzutkin, S. C. Shekar and M. H. Levitt, *J. Magn. Reson. A*, 1995, **115**, 7; (b) O. N. Anzutkin, Y. K. Lee and M. H. Levitt, *J. Magn. Reson.*, 1998, **135**, 144.

54. (a) N. Takeda, S. Kuroki, H. Kurosu and I. Ando, *Biopolymers*, 1999, **50**, 61; (b) G. S. Harbison, L. W. Jelinski, R. E. Stark, D. A. Torchia, J. Herzfeld and R. G. Griffin, *J. Magn. Reson.*, 1984, **60**, 79.
55. (a) C. J. Hatzell, M. Whitfield, T. G. Oas and G. P. Drobny, *J. Am. Chem. Soc.*, 1987, **109**, 5966; (b) N. Asakawa, S. Kuroki, I. Ando, A. Shoji and T. Ozaki, *J. Am. Chem. Soc.*, 1992, **114**, 3261; (c) Q. Teng, M. Iqbal and T. A. Cross, *J. Am. Chem. Soc.*, 1992, **114**, 5312; (d) M. Hong, *J. Am. Chem. Soc.*, 2000, **122**, 3762.
56. T. Kameda, N. Takeda, S. Kuroki, H. Kurosu, S. Ando, I. Ando, A. Shoji and T. Ozaki, *J. Mol. Struct.*, 1996, **384**, 17.
57. (a) R. Haberkom, R. Stark, H. van Willigen and R. Griffin, *J. Am. Chem. Soc.*, 1981, **103**, 2534; (b) S. N. Smirnov, N. S. Golubev, G. S. Denisov, H. Benedict, P. Schah-Mohammedi and H.-H. Limbach, *J. Am. Chem. Soc.*, 1996, **118**, 4094–4101.
58. (a) R. Gobetto, C. Nervi, E. Valfrè, M. R. Chierotti, D. Braga, L. Maini, F. Grepioni, R. K. Harris and P. Y. Ghi, *Chem. Mater.*, 2005, **17**, 1457–1466; (b) R. Gobetto, C. Nervi, M. R. Chierotti, D. Braga, L. Maini, F. Grepioni, R. K. Harris and P. Hodgkinson, *Chem. – Eur. J.*, 2005, **11**, 7461–7471.
59. A. E. Aliev and K. D. M. Harris, in *Supramolecular Assembly Via Hydrogen Bonds I*, ed. D. M. P. Mingos and M. Alajarin, Springer, Heidelberg, 2004.
60. (a) M. S. Solum, K. L. Altmann, M. Strohmeier, D. A. Berges, Y. Zhang, J. C. Facelli, R. J. Pugmire and D. M. Grant, *J. Am. Chem. Soc.*, 1997, **119**, 9804; (b) D. Schweitzer and H. W. Spiess, *J. Magn. Reson.*, 1974, **15**, 529; (c) R. W. Schurko and R. E. Wasylishen, *J. Phys. Chem. A*, 2000, **104**, 3410.
61. P. Lorente, I. G. Shenderovich, N. S. Golubev, G. S. Denisov, G. Buntkowsky and H.-H. Limbach, *Magn. Reson. Chem.*, 2001, **39**, S18–S29.
62. Y. Wei, A. C. de Dios and A. E. McDermott, *J. Am. Chem. Soc.*, 1999, **121**, 10389–10394.
63. *The Hydrogen Bond*, ed. P. Schuster, G. Zundel and C. Sandorfy, North Holland Publ. Co., Amsterdam, 1976.
64. (a) I. Olovsson and P. G. Jonsson, *The Hydrogen Bond*, ed. P. Schuster, G. Zundel and C. Sandorfy, North Holland Publ. Co., Amsterdam, 1976, ch. 8, pp. 393–456; (b) D. J. Jones, I. Brach and J. Roziere, *J. Chem. Soc., Dalton Trans.*, 1984, 1795.
65. M. R. Chierotti and R. Gobetto, *CrystEngComm*, 2013, **15**, 8599.
66. (a) I. Schnell and H. W. Spiess, *J. Magn. Reson.*, 2001, **151**, 153–227; (b) Y. K. Lee, N. D. Kurur, M. Helmle, O. G. Johannessen, N. C. Nielsen and M. H. Levitt, *Chem. Phys. Lett.*, 1995, **242**, 304–309; (c) A. Brinkmann, M. Eden and M. H. Levitt, *J. Chem. Phys.*, 2000, **112**, 8539–8554; (d) M. Hohwy, H. J. Jakobsen, M. Eden, M. H. Levitt and N. C. Nielsen, *J. Chem. Phys.*, 1998, **108**, 2686–2694.
67. (a) I. Schnell, A. Lupulescu, S. Hafner, D. E. Demco and H. W. Spiess, *J. Magn. Reson.*, 1998, **133**, 61–69; (b) L. Mafra, R. Siegel, C. Fernandez, D. Schneider, F. Aussenac and J. Rocha, *J. Magn. Reson.*, 2009, **199**, 111–114.
68. R. K. Harris, P. Hodgkinson, V. Zorin, J.-N. Dumez, B. E. Hermmann, L. Emsley, E. Salager and R. S. Stein, *Magn. Reson. Chem.*, 2010, **48**, S103–S112.
69. G. R. Goward, D. L. Sebastiani, I. Schnell, H. W. Spiess, H.-D. Kim and H. Ishida, *J. Am. Chem. Soc.*, 2003, **125**, 5792–5800.
70. I. Schnell, S. P. Brown, H. Y. Low, H. Ishida and H. W. Spiess, *J. Am. Chem. Soc.*, 1998, **120**, 11784.
71. J. W. Traer, E. Montoneri, A. Samoson, J. Past, T. Tuherm and G. R. Goward, *Chem. Mater.*, 2006, **18**, 4747.
72. (a) M. U. Schmidt, J. Brüning, J. Glinnemann, M. W. Hützler, P. Mörschel, S. N. Ivashevskaya, J. Van de Streek, D. Braga, L. Maini, M. R. Chierotti and R. Gobetto, *Angew. Chem., Int. Ed*, 2011, **50**, 7924–7926; (b) D. M. Toebbens, J. Glinnemann, M. R. Chierotti, J. Van de Streek and D. Sheptyakov, *CrystEngComm*, 2012, **14**, 3046–3055; (c) M. R. Chierotti, L. Ferrero, N. Garino,

R. Gobetto, L. Pellegrino, D. Braga, F. Grepioni and L. Maini, *Chem. - Eur. J.*, 2010, **16**, 4347–4358.

73. S. P. Brown, X. X. Zhu, K. Saalwachter and H. W. Spiess, *J. Am. Chem. Soc.*, 2001, **123**, 4275.

74. (a) S. P. Brown, I. Schnell, J. D. Brand, K. Mullen and H. W. Spiess, *J. Am. Chem. Soc.*, 1999, **121**, 6712; (b) S. P. Brown, I. Schnell, J. D. Brand, K. Mullen and H. W. Spiess, *J. Mol. Struct.*, 2000, **521**, 179; (c) S. P. Brown, I. Schnell, J. D. Brand, K. Mullen and H. W. Spiess, *Phys. Chem. Chem. Phys.*, 2000, **2**, 1735.

75. T. M. Alam, M. Nyman, B. R. Cherry, J. M. Segall and L. E. Lybarger, *J. Am. Chem. Soc.*, 2004, **126**, 5610.

76. (a) F. Rataboul, A. Baudouin, C. Thieuleux, L. Veyre, C. Coperet, J. Thivolle-Cazat, J. M. Basset, A. Lesage and L. Emsley, *J. Am. Chem. Soc.*, 2004, **126**, 12541; (b) P. Avenier, A. Lesage, M. Taoufik, A. Baudouin, A. De Mallmann, S. Fiddy, M. Vautier, L. Veyre, J. M. Basset, L. Emsley and E. A. Quadrelli, *J. Am. Chem. Soc.*, 2007, **129**, 176.

77. G. N. M. Reddy, A. Marsh, J. T. Davis, S. Masiero and S. P. Brown, *Cryst. Growth Des.*, 2015, **15**, 5945–5954.

78. B. Fortier-McGill, V. Toader and L. Reven, *Macromolecules*, 2012, **45**, 6015.

79. D. Braga, L. Chelazzi, F. Grepioni, E. Dichiarante, M. R. Chierotti and R. Gobetto, *Cryst. Growth Des.*, 2013, **13**, 2564–2572.

80. S. H. Park, C. Yang, S. J. Opella and L. J. Mueller, *J. Magn. Reson.*, 2013, **237**, 164–168.

81. A. Wong, K. J. Pike, R. Jenkins, G. J. Clarkson, T. Anupold, A. P. Howes, D. H. G. Crout, A. Samoson, R. Dupree and M. E. Smith, *J. Phys. Chem. A*, 2006, **110**, 1824–1835.

82. E. Y. Chekmenev, K. W. Waddell, J. Hu, Z. Gan, R. J. Wittebort and T. A. Cross, *J. Am. Chem. Soc.*, 2006, **128**, 9849–9855.

83. (a) D. W. Larsen and A. L. Allred, *J. Phys. Chem.*, 1965, **69**, 2400; (b) H. A. Bent, *Chem. Rev.*, 1968, **68**, 587; (c) J.-M. Dumas and M. Gomel, *J. Chem. Phys.*, 1975, **72**, 953; (d) R. D. Bailey, G. W. Drake, M. Grabarczyk, T. W. Hanks, L. L. Hook and W. T. Pennington, *J. Chem. Soc., Perkin Trans. 2*, 1997, 2773.

84. (a) P. Metrangolo and G. Resnati, T. Pilati and S. Biella, in *Halogen Bonding: Fundamentals and Applications*, ed. P. Metrangolo and G. Resnati, Springer-Verlag Berlin, Berlin, 2008, vol. 126, p. 105; (b) P. Metrangolo, F. Meyer, T. Pilati, G. Resnati and G. Terraneo, *Angew. Chem., Int. Ed.*, 2008, **47**, 6114; (c) A. Mukherjee, S. Tothadi and G. R. Desiraju, *Acc. Chem. Res.*, 2014, **47**, 2514.

85. R. A. Thorson, G. R. Woller, Z. L. Driscoll, B. E. Geiger, C. A. Moss, A. L. Schlapper, E. D. Speetzen, E. Bosch, M. Erdelyi and N. P. Bowling, *Eur. J. Org. Chem.*, 2015, 1685.

86. G. R. Desiraju, P. S. Ho, L. Kloo, A. C. Legon, R. Marquardt, P. Metrangolo, P. Politzer, G. Resnati and K. Rissanen, *Pure Appl. Chem.*, 2013, **85**(8), 1711.

87. (a) R. J. Attrell, C. M. Widdifield, I. Korobkov and D. L. Bryce, *Cryst. Growth Des.*, 2012, **12**, 1641; (b) J. Viger-Gravel, S. Leclerc, I. Korobkov and D. L. Bryce, *J. Am. Chem. Soc.*, 2014, **136**(19), 6929.

88. (a) P. Metrangolo and G. Resnati, *Chem. - Eur. J.*, 2001, **7**, 2511; (b) M. Baldrighi, G. Cavallo, M. R. Chierotti, R. Gobetto, P. Metrangolo, T. Pilati, G. Resnati and G. Terraneo, *Mol. Pharm.*, 2013, **10**, 1760; (c) J. F. Bertrán and M. Rodríguez, *Org. Magn. Reson.*, 1979, **12**, 92; (d) M. Rodríguez and J. F. Bertrán, *Org. Magn. Reson.*, 1980, **14**, 244; (e) P. Cardillo, E. Corradi, A. Lunghi, S. Valdo Meille, M. Teresa Messina, P. Metrangolo and G. Resnati, *Tetrahedron*, 2000, **56**, 5535; (f) P. Metrangolo, W. Panzeri, F. Recupero and G. Resnati, *J. Fluor. Chem.*, 2002, **114**, 27; (g) A. De Santis, A. Forni, R. Liantonio, P. Metrangolo, T. Pilati and G. Resnati, *Chem. - Eur. J.*, 2003, **9**, 3974; (h) P. Metrangolo, T. Pilati, G. Resnati and A. Stevenazzi, *Curr. Opin. Colloid Interface Sci.*, 2003, **8**, 215; (i) E. Dimitrijević, O. Kvak and M. S. Taylor, *Chem. Commun.*, 2010, **46**, 9025; (j) M. G. Sarwar,

B. Dragisic, L. J. Salsberg, C. Gouliaras and M. S. Taylor, *J. Am. Chem. Soc.*, 2010, **132**, 1646; (k) S. Libri, N. A. Jasim, R. N. Perutz and L. Brammer, *J. Am. Chem. Soc.*, 2008, **130**, 7842.

89. S. C. Blackstock, J. P. Lorand and J. K. Kochi, *J. Org. Chem.*, 1987, **52**, 1451.

90. C. M. Widdifield, G. Cavallo, G. A. Facey, T. Pilati, J. Lin, P. Metrangolo, G. Resnati and D. L. Bryce, *Chem. – Eur. J.*, 2013, **19**, 11949.

91. J. Viger-Gravel, S. Leclerc, I. Korobkov and D. L. Bryce, *CrystEngComm*, 2013, **15**, 3168.

92. R. Glaser, N. Chen, H. Wu, N. Knotts and M. Kaupp, *J. Am. Chem. Soc.*, 2004, **126**, 4412.

93. M. Baldrighi, D. Bartesaghi, G. Cavallo, M. R. Chierotti, R. Gobetto, P. Metrangolo, T. Pilati, G. Resnati and G. Terraneo, *CrystEngComm*, 2014, **16**, 5897.

94. G. C. Levy and R. L. Lichter, *Nitrogen-15 Nuclear Magnetic Resonance Spectroscopy*, John Wiley & Sons, New York, 1979, p. 7.

95. D. L. Bryce and J. Viger-Gravel, in *Halogen Bonding I*, ed. P. Metrangolo and G. Resnati, Springer International Publishing, 2014, p. 183.

96. M. Weingarth, N. Raouafi, B. Jouvelet, L. Duma, G. Bodenhausen, K. Boujlel, B. Schöllhorn and P. Tekely, *Chem. Commun.*, 2008, 5981.

9 The Use of Databases in the Study of Intermolecular Interactions

Alessia Bacchi

Dipartimento di Scienze Chimiche, della Vita e della Sostenibilità Ambientale, University of Parma, Italy
Email: alessia.bacchi@unipr.it

9.1 Definition of Database

The term database defines a set of related data and its organization. Access to these data is usually possible through a database management system, which is software that interacts with one or more databases allowing access to the data contained in the database. Databases may be built according to different schemes: the *flat* model consists simply of a two-dimensional array of data, like for instance a table where covalent radius, first ionization potential and electronegativity could be columns and each chemical element would occupy one row; this is the organization of a spreadsheet. In the *network* model multiple tables are linked together by the use of pointers, and some columns contain pointers to different tables of data. *Relational databases* also comprise multiple tables, where any column may be used to set a relationship between two or more tables by writing new queries.

Databases must ensure data integrity, meaning consistency, accuracy, and correctness of stored data.

Intermolecular Interactions in Crystals: Fundamentals of Crystal Engineering
Edited by Juan J. Novoa
© The Royal Society of Chemistry 2018
Published by the Royal Society of Chemistry, www.rsc.org

9.1.1 Crystallographic Databases

Primary crystallographic data are extracted from published scientific articles and supplementary material. Modern crystallographic databases are built on the relational database model, which allows efficient cross-referencing of tables. Cross-referencing is useful to obtain derived data and improve the search efficiency of the database.

The role of databases in crystallographic research has had a tremendous impact in the last 40 years, following the parallel development of computing power and information technology: electronic exchange of information has revolutionized the communication of data between researchers and journals, and the adoption of the CIF format has standardized data deposition.[1] Moreover, the amount of crystallographic data produced per year by crystallographers worldwide has increased enormously due to increased speed and accessibility of data collection, and to increased computing power for crystallographic analysis. Additionally, software applications for data retrieval and processing have been constantly upgraded and refined, allowing more and more users to access sophisticated data elaborations by using standard packages, that can be operated directly online. Nowadays the accessibility to freshly published structural data is almost immediate, and virtually every researcher can make their data publicly available using the most popular information channels. However, it must be stressed that scientists need well-established databases to guarantee data integrity and validation. Crystallographic data submitted directly by authors or retrieved from scientific literature are in fact generally checked and validated before being stored and made accessible.[2]

There are a number of major crystallographic and solid state properties databases continuously maintained and updated. An exhaustive overview of crystallographic databases can be found in the special issues B58 and D58 of Acta Crystallographica Sections B and D (2002), and in other recent reviews.[3-8]

A brief outline of the most used databases for crystallographers follows, and an extended list of secondary databases can be retrieved on the IUCr web site at http://www.iucr.org/resources/data.

Bilbao Crystallographic Server of crystallographic symmetry information (http://www.cryst.ehu.es/) is an open access website offering online crystallographic database and programs aimed at analyzing, calculating and visualizing problems of structural and mathematical crystallography, solid state physics and structural chemistry.

Biological Macromolecule Crystallization Database (http://xpdb.nist. gov:8060/BMCD4/index.faces/) contains molecule, crystal and crystallization data for macromolecules for which diffraction quality crystals have been obtained.

Cambridge Structural Database (http://www.ccdc.cam.ac.uk/) contains information on over 800 000 experimentally determined (X-ray or neutron diffraction) crystal structures of organic and organometallic compounds, compiled since 1923. Each crystallographic entry in the CSD provides bibliographic information, chemical connectivity, and numeric data.

CRYSTMET is a relational database of critically evaluated crystallographic data for metals, alloys, intermetallics, and minerals, accompanied by pertinent physical, chemical, and bibliographic information since 1913.

Inorganic Crystal Structure Database (http://icsd.ill.fr/icsd/) is a comprehensive compilation of crystal structure data of inorganic compounds. All information has been obtained from the original sources and checked to assure high quality. The literature is covered back to 1915. The database is now updated by direct scanning of major journals.

Nucleic Acids Database (http://ndbserver.rutgers.edu/NDB/) contains three-dimensional structural information on RNA and DNA oligonucleotides that have been obtained from X-ray crystallographic experiments. The NDB allows retrieval of coordinates, of information about the conditions used to derive the coordinates, and of the structural information derived from the coordinates.

NIST Crystallographic Databases (http://www.nist.gov/mml/mmsd/ materials-structure/crystallographic-databases.cfm) provide critically evaluated, comprehensive crystal-structure databases that enable the phase identification required for the development of advanced inorganic materials and devices depending on them. NIST designs, populates, evaluates and disseminates NIST Standard Reference Databases SRD 3 (NIST Crystal Data), SRD 83 (NIST Metals Structural Database, NSD) and SRD 84 (FIZ/NIST Inorganic Crystal Structure Database, ICSD).

Powder Diffraction File (http://www.icdd.com/) is the most comprehensive database for single phase X-ray powder diffraction patterns, updated every year. The primary use of the PDF is to identify "fingerprints" of unknown materials by matching the spacings of the unknown material's diffraction patterns with the spacings of known substances.

Protein Data Bank (http://www.rcsb.org/pdb/) is an electronic archive of experimentally determined 3-dimensional structures of proteins, nucleic acids, and other biological macromolecules. It contains atomic coordinates, bibliographic citations, primary and secondary structure information, as well as crystallographic structure factors and nuclear magnetic resonance (NMR) experimental data of proteins, nucleic-acids, and viruses. PDB is the single global archive of such data.

Structural Classification of Proteins (http://scop.mrc-lmb.cam.ac.uk/scop/) provides a detailed and comprehensive description of the structural and evolutionary relationships between all proteins whose structure is known. It provides a broad survey of all known protein folds, detailed information about the close relatives of any particular protein, and a framework for future research and classification.

NIST Chemistry WebBook (http://webbook.nist.gov/chemistry/) provides access to chemical and physical property data for chemical species through the internet. Particularly of interest in crystal engineering (CE) are the condensed phase thermochemistry data and the phase change data.

Database of Zeolites (http://www.iza-structure.org/databases/) provides structural information on all of the zeolite framework types that have been approved by the Structure Commission of the International Zeolite Association (IZA-SC).

The Crystallography Open Database (COD) is a recent tool offered to the scientific community on the Web at http://www.crystallography.net/. The COD is a project that aims to gather all available inorganic, metal–organic and small organic molecule structural data in one database. The database adopts an open-access model. The COD currently contains more than 300 000 entries in crystallographic information file format, with nearly full coverage of the International Union of Crystallography publications.[9]

Databases contain important information hidden in the relationships between the data, such as similarities and dissimilarities, which may reveal important new chemical knowledge. Finding these hidden relationships is sometimes called data mining or knowledge discovery. In the last decades, database developers have also conceived several applications which pre-digest database contents and present the information at a higher level of organization, the so-called knowledge-based applications.

9.2 Early Applications of Crystallographic Databases

As soon as the first embryos of crystallographic databases were established, back in 1970's, the scarce but precious amount of structural data encrypted in the deposited files were used to estimate the average molecular bond lengths and angles observed in the solid state. The dawn of quantitative structural chemistry had been marked.[11] Following these pioneering works, lists of expected values for commonly found bond distances and angles were compiled in the following years, both for organic and inorganic systems. These represented a valuable tool for building reliable starting geometries in molecular modelling, and for validation of structural results, especially in macromolecular crystallography.[12–14] As soon as more data began to be available, correlations between intramolecular bonding parameters were investigated, in order to develop models of conformational dynamics and chemical reactivity from the comparison of similar crystal structures: the *structure correlation* principles had been invented.[15,16] Once the subject of intramolecular bonding had been explored, and the tools for the analysis of molecular bond geometry were well assessed, the attention of the cutting edge structural chemistry focused on molecular recognition: statistical studies were employed to categorize the role of hydrogen bonding in this.[17–19]

Structural descriptors were treated with statistical methods optimized especially to deal with crystallographic parameters like circular statistics to cope with torsion angles, univariate tests to assess the relative importance of experimental error and crystal packing effects, the use of parametric or non-parametric tests to evaluate significant differences of mean values, chi-square tests to probe non-normal spatial distributions in intermolecular contacts, regression analysis to estimate linear correlations between parameters, principal component analysis and cluster analysis to recognize patterns in structure–properties relationship studies.[16]

With the advent of crystal engineering as a new scientific paradigm for chemical crystallographers and material scientists, in the last thirty years the focus has moved to the use of crystallographic information as a design tool for rationalizing polymorphism, co-crystallization, inclusion compounds, and reticular chemistry. A number of key fundamental papers tackled the analysis of the correlations between molecular volume, density, packing coefficient, calculated lattice energy,[20–23] the analysis of crystal symmetry[24–31] and the individuation and characterization of particularly robust interaction patterns between functional groups.[32–34] For an exhaustive compilation of scientific papers concerning the use of the Cambridge Structural Database, one can refer to https://www.ccdc.cam.ac.uk/researchandconsultancy/ccdcresearch/

Crystallographic databases cover different fields of solid state chemistry; the Cambridge Structural Database has been up to now the most relevant for researches dealing with the study of intermolecular interactions and crystal engineering[35–37] because it collects exhaustively all the known experimental structural data both in organic and in organometallic systems, among them: molecular crystals and co-crystals, polymorphs, solvates, nanostructures, metallo-organic-frameworks (MOFs) and coordination polymers.

9.3 Intermolecular Interactions

'We had a passionate belief that the collective use of data would lead to the discovery of new knowledge which transcends the results of individual experiments.' (Kennard, 1997).[38]

Databases are a valuable goldmine of experimental knowledge on intermolecular interactions of all types. Primary crystallographic databases contain atomic coordinates and cell and space group information; from these, all the possible derived structural information (bonding geometry, intermolecular distances, *etc.*) can be computed by appropriate software associated with the database.

Generally intermolecular interactions are studied by defining their paradigmatic structure associated with some useful geometric descriptors, and then retrieving the experimental values of the descriptors by searching the database. The procedure is (a) to define the fragments involved in the interaction; (b) to define one or more descriptors that give information on the geometry of the interaction; (c) select appropriate thresholds that identify the experimental occurrence of the interaction; (d) run a query to retrieve all the instances of the interaction; (e) choose appropriate statistics and graphic representations to analyse the distribution of the geometric parameters in the dataset (Figure 9.1). Such analyses have been used to study supramolecular recognition and aggregation for the most common and known interactions.

9.3.1 Hydrogen Bond Geometry

The knowledge of the geometry and preferential spatial environment of hydrogen bonds is of paramount importance in supramolecular chemistry and molecular biology. A survey of all the instances of spatial organization of hydrogen bonds in the crystalline state provides information on the nature of this interaction that is fundamental in

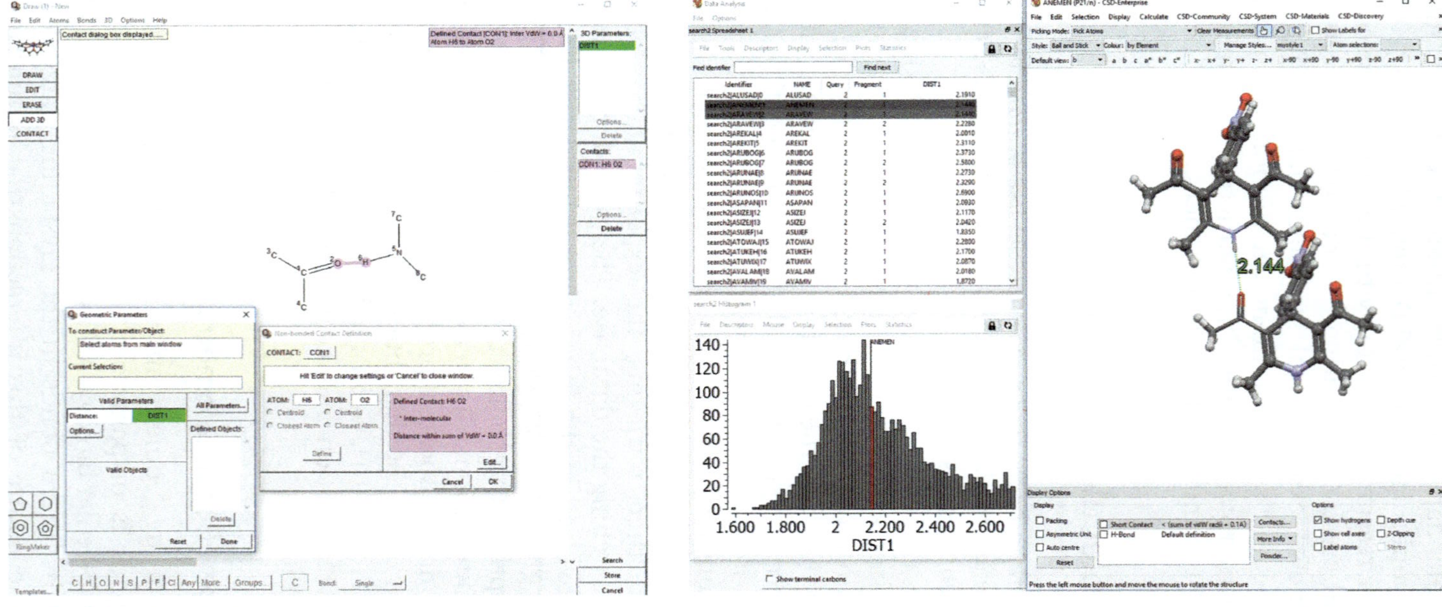

Figure 9.1 Typical CSD Query for intermolecular contacts. Left: a fragment is defined (here the hydrogen bond between a secondary amine and a ketone), and a geometrical descriptor is selected, with acceptable value range (here H···O distance shorter than van der Waals contact); right: all the occurrences of the contact are retrieved, visualized and a pertinent statistic (here a histogram) is obtained.

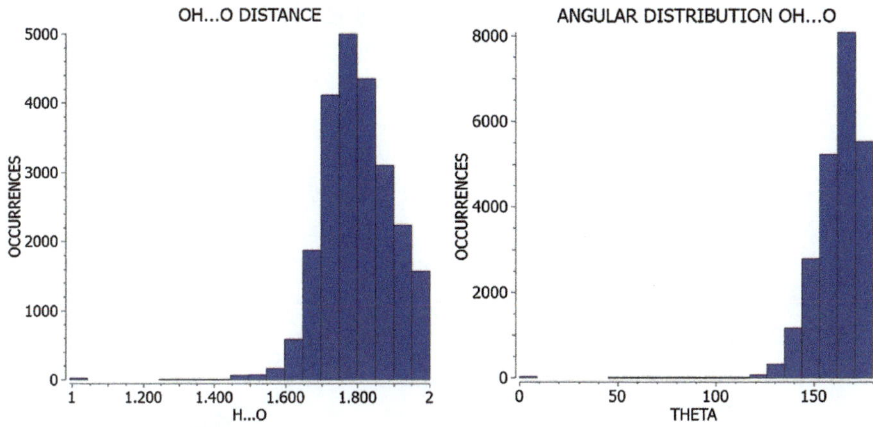

Figure 9.2 Histograms of distances (Å) and angles theta (°) (no cone correction) for carbohydrates retrieved from the CSD.

chemistry. Databases contain a vast amount of experimental data that allows such analysis to be performed with appropriate statistical methods.[39] The Cambridge Structural Database has been used extensively to define the most frequent occurrence of strong hydrogen bonds having N or O as donors and acceptors. In particular, the examination of the distribution of H-bond distances and angles is useful in assessing the average modes of geometric arrangement for this interaction,[17,39] (Figure 9.2) but also to study H-bond lone-pair directionality at the acceptor,[33] and resonance assisted and resonance-induced H-bonds.[19] Moreover, the investigations on a wide range of interactions involving weak donors and strong acceptors (*e.g.* C–H$\cdots$O, C–H$\cdots$N), strong donors and weak acceptors (*e.g.* O,N–H$\cdots$Cl, O,N–H$\cdots\pi$), and weak donors and weak acceptors (*e.g.* C–H$\cdots$Cl, C–H$\cdots\pi$) have been fundamental in the definition of the concept of weak hydrogen bonds. The identification of short C–H$\cdots$O and C–H$\cdots$N contacts as hydrogen bonds has been possible thanks to such analyses.[18]

A description of some case studies follows.

9.3.1.1 *Hydrogen Bond Linearity*

The tendency to linearity of O–H$\cdots$O hydrogen bonds has been assessed by the study of the distribution of angles θ in carbohydrates:[39] after the cone correction (see Section 9.5.4), θ shows a clear preference for values between 170 and 180°, showing a tendency to linearity. Angular preferences have been also assessed[17,39] by building scatter plots of angles θ against distances d for all contacts found in crystal

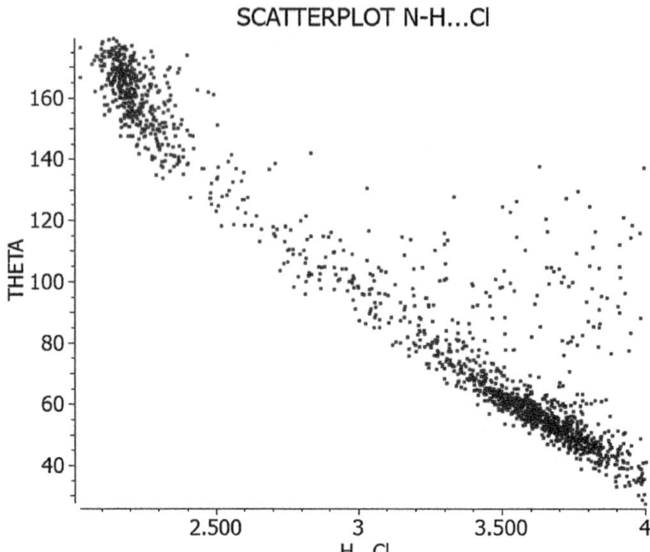

Figure 9.3   Example of angular scatterplot of N–H$\cdots$Cl angle (°) against H$\cdots$Cl distance (Å) for C–NH$_3^+$ donors and Cl$^-$ acceptors, whereby all contacts with H$\cdots$Cl$<$4.0 Å are reported, whether they represent a hydrogen bond or not. Different clusters represent strong, weak, bifurcated interactions, as explained in ref. 39.

structures (see for example Figure 9.3 for N–H$\cdots$Cl contacts). The dispersion in the scatterplots is considerable,[40] but there are usually highly populated clusters of data points at short distances and fairly linear angles, showing the expected preference for linearity of strong contacts, while other clusters indicate minor components of bifurcated bonds, non-bonding next-neighbour contacts, randomly scattered non-bonding second-neighbour contacts, and empty regions due to stereochemical hindrance in the D$\cdots$A approach.

Another study[41] has shown that the linearity of the interactions is correlated to the polarity of the donor. In fact, the propensity to linearity of O–H$\cdots$O=C, C–H$\cdots$O=C, and C–H$\cdots$H–C interactions decreases in the order O–H$>$C$\equiv$C–H$>$C=CH$_2$$>$–CH$_3$, in the same way as the polarity of the donor group is decreased.

This kind of approach has been use to investigate the preferred direction in which the acceptor is approached by the donor group (represented for instance by the C=O$\cdots$H angle Φ in the above cases). It has been shown that in general this corresponds to the orientation of electron lone pairs. For instance, angular histograms representing Φ for N/O–H$\cdots$O/S=C hydrogen bonds,[33] show that the oxygen lone pair lobes are in the R$_2$C=O plane and form angles of about 120° with

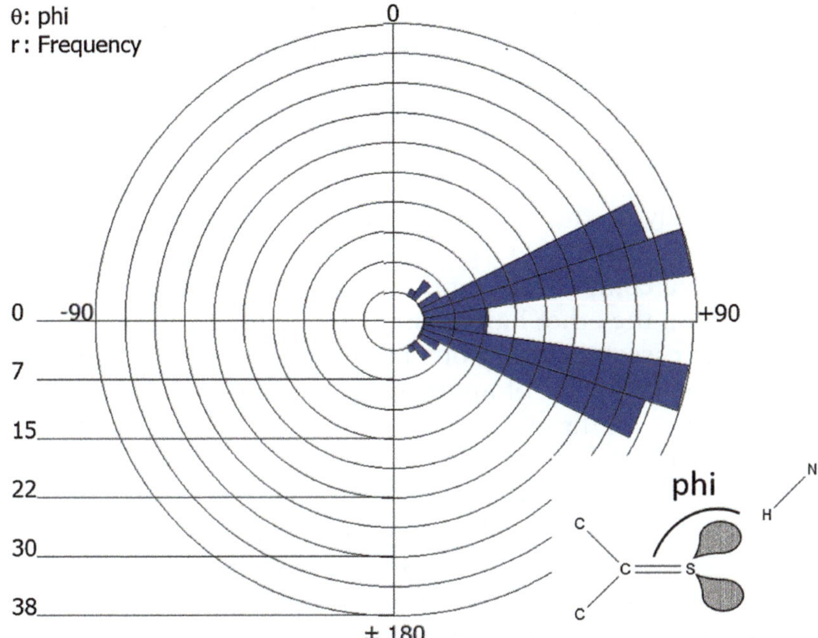

Figure 9.4 Angular histogram (°) of phi angles (C=S···H) representing the directional preferences of NH···S=C hydrogen bonds. Lone pair directionality is evident. These data have been obtained by considering interactions where N–H and C=S are coplanar (putting |sin[torsion(NHSC)]| < 0.1 as a filter in the CSD search). Compare to ref. 39.

the C=O bond. The directionality towards the lone pair lobes is much more pronounced in thiocarbonyl groups, that are located at about 105°, as shown by the higher frequency of occurrences in the angular histogram (Figure 9.4).

9.3.1.2 Hydrogen Bonds and Coordination Chemistry

The hydrogen-bond acceptors may be metal-bound halides; the efficiency in this role of the M–Cl group has been investigated[42] by comparing structures containing O/N–H and M–Cl, C–Cl, or Cl⁻. The percentage of short H···Cl interactions found in these three classes of compounds was taken as an indicator of the strength of the corresponding D–H···Cl interaction. The sequence of acceptor strength is Cl⁻ > M–Cl ≫ C–Cl.

9.3.2 Interactions Involving Aromatic Groups

Supramolecular non-bonded interactions which involve pi systems have attracted growing interest in the last decades to explain the aggregation in the solid state. Data mining on crystallographic databases has played a crucial role in identifying and characterizing such interactions. Some examples follow.

9.3.2.1   CH···pi Interactions

A database study on the role of the CH/π(C6 aromatic) in the crystal packing[43,44] has shown that a CH/π short distance (<3.05 Å) occurs in 54% of organic compounds containing at least a YCH$_3$ and an aromatic group; on the other hand, short OH/π interactions are observed only in 4% of cases. Several geometric descriptors have been defined to quantify the geometry of approach between CH and the aromatic ring. The values of these descriptors have been retrieved form the Cambridge Structural Database and analysed by mono- and bivariate statistics. Monovariate statistics on distances show that the mean CH/π distance decreases as the proton acidity increases; scatterplots of the C–H···π access angle on the CH/π plane distance[44] (Figure 9.5) show that the more acidic the proton, the more linear is the approach, as already observed for CH/n hydrogen bonds.[41]

9.3.2.2   pi···pi Interactions

Similar geometric descriptors have allowed investigation of the geometry of $\pi\cdots\pi$ interactions between metal-coordinated pyridines.[34] The lowest value for the centroid–centroid contacts between

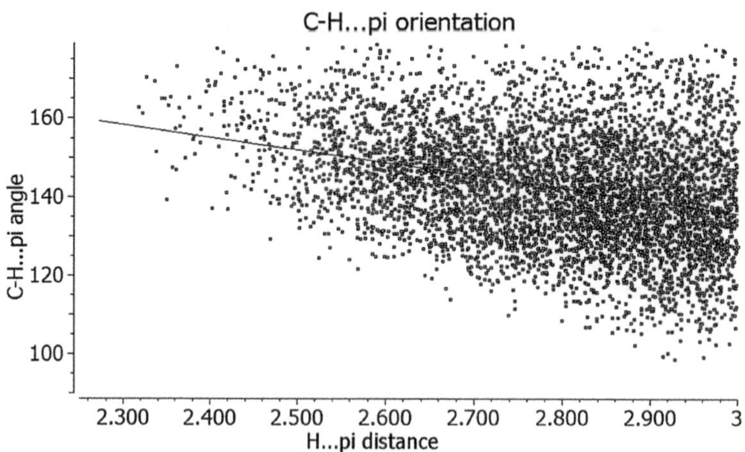

Figure 9.5   Correlation between the access angle (°) of the CH group to the aromatic plane and the H···pi distance (Å). Parameters have been calculated by a CSD search considering the centroid of the ring. A tendency to linearity for shorter distances is displayed (compare to ref. 44).

two pyridine fragments is found to be slightly below 3.4 Å, while a relative maximum in the number of occurrences is found around 3.8 Å. In the vast majority of cases the interplane angle is near zero, but the angle formed between the ring-centroid vector (CC) and the ring normal to one of the pyridine planes averages 27°. A correlation between the displacement angle and the centroid–centroid distance (Figure 9.6) reveals a slightly smaller tendency for parallel displacement at shorter distance.

9.3.3 Halogen Bonds

The marked tendency of the halogens X = Cl, Br, I to form short contacts to each other and to electronegative N and O atoms has been observed by studying the supramolecular geometry of C–X$\cdots$O=C< systems.[45] The shortest X$\cdots$O interactions have a marked preference to form along the extension of the C–X bond. By contrast C–Cl$\cdots$Cl–C interactions tend to form with C–Cl$\cdots$Cl angles close to 90°.[46] These observations have opened one of the most prolific chapters of the modern crystal engineering, paving the way for the design of new crystalline systems based on halogen bond interactions, that has been rationalized by theoretical studies and is now a well-accepted tool for the fabrication of supramolecular entities.[47]

9.4 Databases and the Study of Interaction patterns

'Crystallography needs its own Sherlock Holmes to crack the secrets of molecules as they assemble into crystals.' (G. R. Desiraju, 2002)[48]

Crystal engineering is the design and synthesis of molecular solid-state structures with desired properties based on an understanding and exploitation of intermolecular interactions. The properties of the material are a direct and predetermined consequence of the way in which arrays of molecules are assembled in the solid state.[49,50] A totally deliberate design would require an *'a priori'* complete knowledge of (i) the relationship between structure and function and (ii) the physics of self-assembly and crystallization.

This approach is rarely successful mainly because the factors which rule self-assembly in the solid state are still obscure to a large extent, and molecules cannot be easily convinced to line up according to a predetermined scheme.[51]

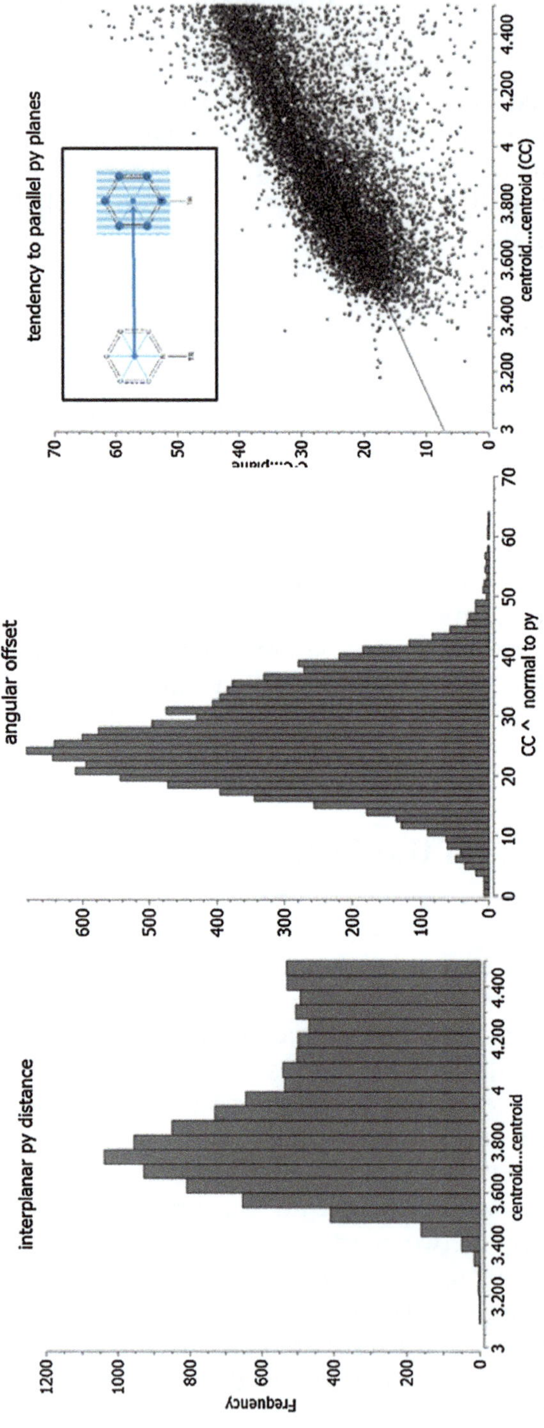

Figure 9.6 Left: for the centroid··centroid distances (Å) for metal-bound pyridine rings the relative maximum in the number of occurrences is found around 3.8 Å. Centre: histogram of the angular offset, expressed as the angle (°) between the centroid··centroid (CC) vector and the normal to one pyridine plane (see inset). Right: correlation between the centroid··centroid distance and the angle between the CC vector and the normal to one pyridine plane shows that the shortest the stacking the more parallel are the rings (compare ref. 34).

9.4.1 Crystal Structure Prediction: The Holy Grail

'This does not mean that they would have cracked the problem of predicting all organic crystal structures.' (Sally Price, 2015)[52]

Decades of struggles and debates on the possibility of predicting crystal structures[53–55] have made this subject a cutting edge aspect of crystal engineering. It is nowadays very clear that the final actual crystal structure is the outcome of a delicate balance between many elusive factors, and the differences in energy for different alternative assemblies of a given molecule are often within the uncertainty limits of the computational methods themselves. Crystal structure prediction (CSP) theoretical methods offer a variety of approaches;[55] one of the latest views is oriented to look at the molecule as a continuous distribution of charge density, and at intermolecular cohesion as the integration of many point-to-point electrostatic interactions, attractive and repulsive, plus the repulsion due to Pauli exclusion principle.[56–58] The Cambridge Crystallographic Data Centre organizes periodically a contest where all the state-of-the-art software for CSP competes in a blind test to predict the structure of new compounds, known to the organizers but unreleased to the competitors.[59] The last blind test (2015) showed that significant progress has been made in the setup of reliable methods that may predict correctly the packing of a new compound,[52] but we are far from a 100% confidence level. It must be also noted that the outcome of the crystallization process is controlled by kinetics, while CSP operates by considering thermodynamic stability. Therefore, observed structures, which may be the easiest to obtain, are not necessarily always the most stable from a thermodynamic point of view.

This picture accounts for the hard life that chemists have in designing and actually obtaining a desired three-dimensional network.

9.4.2 The Pragmatic Approach to the Study of Interaction Patterns

In fact, much bulk experimental work in CE still relies upon heuristic, trial and error, or even serendipitous protocols.[60]

The two main strategies currently in use for crystal engineering are based on hydrogen bonding/weak interactions and coordination complexation, and the key concepts used to assist the design are the supramolecular synthon and the secondary building unit, which are considered as operative tools.

9.4.2.1 Patterns of Supramolecular Interactions and the Tool of Supramolecular Synthons

Decoding the association patterns observed in actual crystal structures may be viewed as the first step to understand the rules of molecular self-assembly. With the growth of the structural data available in the literature, a pragmatic approach is based on the use of the experimental information collected in the databases.[61] The propensity of several functional groups to generate recurrent patterns whenever they associate in the solid state may be considered as a measure of the stability of that pattern.[62] These patterns are recognized by the term 'supramolecular synthons',[63] and are a practical tool used to rationalize the supramolecular synthesis of a crystal structure. According to molecular tectonics, a crystal is the convolution of molecular structure into three-dimensional networks; these are built through robust supramolecular self-association patterns capable of achieving close packing with the help of symmetry operators.[64,65] This practical approach requires ranking of the modes of association for the functional groups present in the molecular constituents and of their most probable three-dimensional distribution in the crystal. Particularly recurrent intermolecular patterns may be then exploited to design crystal structures.[63] Patterns of interactions can be obtained by manual inspection or more rigorously, with the use of crystallographic databases. Databases offer a precious mine of structural data that enables assessment of the propensity to supramolecular association through the analysis all the known occurrences. Moreover, information collected in the databases represents an experimental picture of the relative influence of thermodynamics and kinetics on the crystallization.[48]

Some examples follow.

9.4.2.1.1 Hydrogen Bonded Rings

The occurrence of hydrogen-bonded ring motifs formed between two organic molecules has been assessed by implementing a method that does not need any prior knowledge of the topology or chemical constitution of the motifs.[66] All intermolecular ring motifs comprising ≤ 20 atoms formed with $N-H\cdots N$, $N-H\cdots O$, $O-H\cdots N$ and $O-H\cdots O$ hydrogen bonds in organic structures in the Cambridge Structural Database have been classified. 75 bimolecular motifs occurring in >12 structures in the Cambridge Structural Database were ranked according to their frequency of occurrence and according to their probabilities of formation, *i.e.* their frequency relative to the number of possible motifs which could have formed. These probabilities point to the relative robustness of known and potential supramolecular synthons.

9.4.2.1.2 Carboxylic Acids

Structural motifs based on the interaction between carboxylic acids and carboxylic acids have been classified by using clustering and multivariate analysis.[67,68] All acid–acid fragments were extracted from the Cambridge Structural Database, and a matrix containing the differences between all the interatomic distances and angles within the fragments has been built. This matrix was then subjected to cluster analysis in order to group the fragments into similar motifs based on intermolecular contacts (Figure 9.7). It has been shown that the majority of the contacts in the acid fragments occur "in front" of the central contact group, group A. Hydrogen bonds to the "exo-" position occur in only 7% of cases (group C). A number of weaker hydrogen bonds are included in group E. A number of non-hydrogen bonded contacts just inside the defined van der Waals distance (particularly groups B and D) are also observed. Group F consists of structural outliers due to errors in the output coordinates. Interaction patterns for carboxylic acid dimers and catemers were also

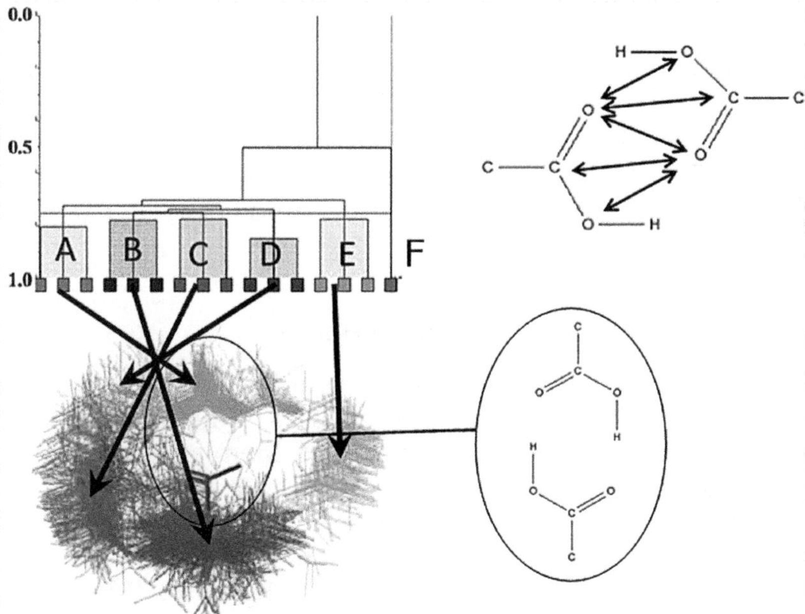

Figure 9.7 Similarity dendrogram (top left) from the cluster analysis of acid…acid contacts, built from the cluster analysis of all the interatomic distances and angles within the acid···acid fragment (top right). Bottom: cumulative plot of all the occurring interactions extracted from CSD (left) showing the spatial distribution of the clusters around the central fragment. The most populated fragment (A) corresponds to the acid···acid supramolecular dimer (left).

identified[69] by using data retrieved directly from the Cambridge Structural Database and analysed by Mercury.[70]

9.4.2.1.3 Phenyl Embraces

PPh_3 groups frequently aggregate by forming six-fold phenyl embraces (6PE) based on six concerted edge-to-face interactions between phenyl groups.[71] The relevance of this supramolecular pattern has been investigated by the analysis of all structures containing at least one $M–PPh_3$ fragment (M = transition metal) present in the Cambridge Structural Database. The aggregation pattern was characterized by the $P\cdots P$ separation and by the $M–P\cdots P–M$ angle; all interactions with $P\cdots P$ in the range 5.5–8.0 Å and angle in the range 120–180° were considered. The scatterplot of these two parameters shows a preponderance of interactions with $P\cdots P = 6.4$–7.4 Å and $\theta = 160$–180°, corresponding to regular 6PE. These values have been taken as the paradigm to analyse 6PE in successive studies.

9.4.3 Metal–organic Frameworks and the use of Databases

Metal–organic frameworks (MOFs) are a class of nanoporous crystalline materials synthesized by bonding metal nodes to multitopic organic linkers. In principle, the chemical building blocks can be rationally selected to tailor MOFs for applications. In practice, it is difficult to predict the complex relationship among the building blocks, the resulting framework structures, and the emergent physical properties prior to synthesis.[72]

9.4.3.1 The Reticular Chemistry Resource

However, the realization that MOFs could be designed and synthesized in a rational way from molecular building blocks led to the emergence of a discipline called reticular chemistry. MOFs can be represented as a special kind of graph called a periodic net, where the metal nodes and the ligands are related to the vertices and the edges of the net. A database of more than 1600 such nets has been created, the Reticular Chemistry Structure Resource (RCSR, http://rcsr.net/), that can be searched by symbol, name, keywords, and attributes. The resource also contains searchable data for polyhedra and layers.[73]

This was used in a successive work where the TOPOS program package was used to generate all subnets of 3- to 12-coordinated binodal nets taken from the database.[74]

9.4.3.2 The Geometry of SBU

The role of the metal-ligand mode of coordination may also be explored with the help of databases, in order to determine the geometry of the secondary building units (SBUs) that act as nodes in the architecture of the MOF. A comprehensive study of transition-metal carboxylate clusters which may serve as SBUs for the construction and synthesis of metal–organic frameworks (MOFs) has been performed from a search of molecules and extended structures archived in the Cambridge Structure Database. The geometries of 131 SBUs, their connectivity and composition have been described and a list of the wide variety of transition-metal carboxylate clusters which may serve as secondary building units (SBUs) in the construction and synthesis of metal–organic frameworks has been compiled.[75]

9.4.3.3 Databases and Interpenetration

MOFs are usually designed in order to exploit their porosity, generally for gas storage purposes. Interpenetration is in this case detrimental to the application of these materials.

The occurrence of interpenetration in metal–organic and inorganic networks has been investigated by a systematic analysis of the CSD and ICSD structural databases. A comprehensive list of inter-penetrating metal–organic 3D structures from CSD and ICSD, that were analyzed on the basis of their topologies, is available.[76,77]

9.4.3.4 Databases, MOFs and Material Science

With the purpose of identifying new ionic conductors for Na ions in ternary Na oxides, the Voronoi–Dirichlet approach has been applied to the Inorganic Crystal Structure Database and some new procedures have been introduced to the algorithm implemented in the program package ToposPro. The main new features are the use of data mined values, which are then used for the evaluation of void spaces, and a new method of channel size calculation.[78]

9.5 Some Cautionary Words

'Data is not information, information is not knowledge, knowledge is not understanding, understanding is not wisdom.' (attributed to Clifford Stoll and Gary Schubert).[79]

Crystallographic databases tend to contain those structures that are most interesting to chemists, or easier to solve, or easier to crystallize, as concisely expressed by the well-known sentence[80] *'the number of forms known for a given compound is proportional to the time and money spent in research on that compound'*. Therefore, some caution must be used when the statistical approach requires the use of a random collection of data. *Databases are socially biased,*[51] since the structures contained in the database are the result of human selection.

9.5.1 Thermodynamic Bias

One classical example is the apparent higher density and stability of racemic crystals compared to their chiral counterparts claimed by Wallach, and named Wallach's rule.[81] The observation on which the rule was based was affected by a bias in the data considered for the analysis: in fact when the racemic crystals are more stable than their chiral counterparts the latter can only be obtained by resolution prior to crystallization; on the other hand, when the racemic crystals are less stable than their chiral analogues, they do not crystallize from the racemic mixture, and spontaneous resolution occurs. Therefore, the cases when racemic crystals are less dense than the separate enantiomers do not appear in the database.

9.5.2 Human bias

Structure solution and refinement of crystals with many atoms in the asymmetric unit was initially hard work; therefore, it is possible that structures with $Z' > 1$ are more frequent in reality than in the Cambridge Structural Database. It is possible that a certain amount of structures with a high number of molecules in the asymmetric unit has been discarded by crystallographers since they were difficult to treat. However, the frequency of $Z' > 1$ structures in the database has been almost unchanged since 1970, suggesting that the improvements in data collection and computing resources has not significantly influenced their occurrence.[82,83] Disordered, modulated, and incommensurate structures are probably underrepresented for the same reasons.[24]

9.5.3 Errors

Crystallographic databases contain errors such as wrong space-groups assignments,[25] abnormally low densities due to missing included

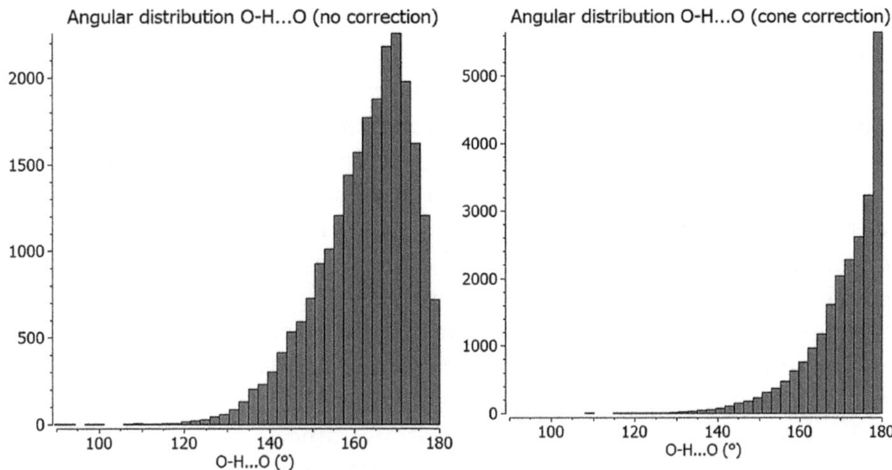

Figure 9.8　Effect of the cone correction on the angular distribution of O–H···O angles (°) for hydrogen bonds in carbohydrates.

solvent, and multiple inclusion of the same structure under different identification codes.[84] Usually the outliers in distributions of geometric parameters obtained from crystallographic databases result from wrong data rather than from novel and unexpected properties.

9.5.4　Uncertainty and Data Selection

Like all experimental data, structural data are affected by uncertainties, and some standard cautions are normally practised in database work. Crystal structures are usually considered in statistical analyses only if they respond to the criteria:[35] R factor <0.1 (although this could cause a bias against 'problematic' compounds, as discussed above),[20] atomic coordinates present, chemical/crystallographic connectivity match, no errors, no disorder. The positions of hydrogen atoms are normalised to the neutron diffraction value. For angular distributions of D–H···A systems the cone correction must be applied for the difference in solid angle sweep of the donor hydrogen as a function of θ with the weighting factor $1/\sin\theta$ (Figure 9.8).[39,40]

9.6　Statistical Methods

The tools of chemometrics, historically developed for analytical chemistry, have found many applications in structural chemistry especially in the investigations related to crystallographic databases. The information hidden in the relationships between all the data may be disclosed by statistical analysis and reveal important new chemical knowledge. An exhaustive review of statistical and numerical methods of data analysis was provided by R. Taylor and F. H. Allen,[16] and theory and applications may be found in a series of papers.[85–94] The most important concepts used in crystallographic database analysis are: (a) descriptive statistics (mean, median and standard deviation) for a distribution of, for instance, a specific bond length observed in many crystal structures; (b) parametric and non-parametric tests to assess the significance of differences between means; (c) the use of covariance, correlation and regression to determine the extent and nature of any relationship between pairs of parameters; and (d) multivariate methods, such as principal components analysis (PCA) and cluster analysis (CA). PCA pinpoints multiple correlations among all variables, and provides the linear combinations of variables that describe the greatest amount of sample variance. Its main application is the reduction of dimensionality of a multivariate sample. Cluster analysis is a technique complementary to PCA, that groups together similar points in the multidimensional data space, providing clouds of data points that are often useful in building a model. PCA and CA are appropriate for structural problems which require the analysis of three or more parameters for each substructure retrieved from the database; examples are the analysis of conformational preferences of n-membered rings,[93] where each conformer is described by n torsion angles, or the analysis of metal coordination spheres, where each sphere is characterised by the $[n(n-1)]/2$ L–M–L valence angles in an MLn species.[95] The application of factor analysis to analytical chemistry has been outlined by Malinowski and Howery,[96] while Massart and Kaufman have described the use of cluster analysis.[97]

9.7   Knowledge Based Applications

In order to allow an easier and reliable access to derived data, knowledge-based libraries have been built.[98] These libraries contain validated subsets of geometric parameters derived from the original databases (mostly the Cambridge Structural Database), which have been polished by eliminating outliers, errors, *etc.*

Mogul collects intramolecular geometry data such bond lengths, valence angles and acyclic torsions. One particular geometric parameter of interest may be selected within a complete molecule or a substructural query generated in the Mogul graphical interface, and then the distribution of the corresponding values may be obtained.

IsoStar is a library of intermolecular interactions between pairs of groups A$\cdots$B.[99] One central group A and one probe group B are selected; all the A$\cdots$B contacts observed in the CSD, PDB, and in some cases by theoretical calculations, are reported in a three-dimensional scatterplot by superimposing all the A moieties, showing the experimental distribution of B around A. The scatterplot can be converted to a contoured surface showing the density of contact groups around the central group. Although the distributions of many interaction pairs are pre-calculated, the program Isogen allows one to generate population–geometry distributions for all interactions to be generated. Examples of the use of Isostar for the optimization of docking geometry and interactions are found in medicinal chemistry.[100,101]

CSDSymmetry is a relational database constructed using Microsoft Access.[102] It contains information about space group, point group, Z, Z', and symmetry of Wyckoff positions for each entry in the CSD. Auxiliary tables, linked to the main table, list the symmetry elements which belong to 38 common point groups, the symmetry operators, and Wyckoff positions of the 230 space groups. These tables allow selection of molecules by the symmetry elements that characterise them or by their membership of a space group with a particular symmetry operator.

9.8 Conclusion

'A large part of the chemical literature nowadays reports on answers that lack a question [. . .] small molecule organic crystal structures conceal the answers to some of the most challenging and fascinating questions in modern chemistry.' (A. Gavezzotti)[10]

Databases are goldmines that enshrine a treasure of hidden knowledge. We are challenged to exercise our intelligence to extract what databases conceal.

References

1. H. M. Berman *et al.*, *IUCrJ*, 2015, **2**, 45.
2. B. MacMahon, *J. Res. Natl. Inst. Stand. Technol*, 1996, **101**, 347.
3. *Acta Crystallogr., Sect. B*, 2002, **B58**, 317–422.
4. *Acta Crystallogr., Sect. D*, 2002, **D58**, 879–920.
5. F. H. Allen and R. Taylor, *Chem. Soc. Rev.*, 2004, **33**, 463.
6. F. H. Allen and R. Taylor, *Chem. Commun.*, 2005, 5135.
7. C. R. Groom and F. H. Allen, *Angew. Chem., Int. Ed.*, 2014, **53**, 662.
8. S. Gražulis *et al.*, *Nucleic Acids Res.*, 2012, **40**, D420.
9. S. Grazulis *et al.*, *J. Appl. Crystallogr.*, 2009, **42**, 726.
10. A. Gavezzotti, *Crystallogr. Rev.*, 1998, 7, 5.

11. A. Domenicano and P. Murray-Rust, *Tetrahedron. Lett.*, 1979, 2283.
12. F. H. Allen *et al.*, *J. Chem. Soc., Perkin Trans.*, 1987, **2**, S1.
13. A. G. Orpen *et al.*, *J. Chem. Soc., Dalton Trans.*, 1989, S1.
14. R. A. Engh and R. Huber, *Acta Crystallogr. Sect. A*, 1991, **47**, 392.
15. H. B. Buergi and J. D. Dunitz, *Acc. Chem. Res.*, 1983, **16**, 153.
16. *Structure Correlation*, ed. H.-B. Buergi and J. D. Dunitz, VCH Publishers, Weinheim, 1994.
17. T. Steiner and W. Saenger, *Acta Crystallogr. Sect. B*, 1992, **B48**, 819.
18. R. Taylor and O. Kennard, *J. Am. Chem. Soc.*, 1982, **104**, 5063.
19. G. Gilli *et al.*, *J. Am. Chem. Soc.*, 1989, **111**, 1023.
20. A. Gavezzotti, *Molecular Aggregation - Structure Analysis and Molecular Simulation of Crystals and Liquids*, Oxford University Press, USA, 2007.
21. J. D. Dunitz, G. Filippini and A. Gavezzotti, *Tetrahedron*, 2000, **56**, 6595.
22. A. Gavezzotti and G. Filippini, *J. Am. Chem. Soc.*, 1995, **117**, 12299.
23. E. Pidcock and W. D. S. Motherwell, *Cryst. Growth Des.*, 2004, **4**, 611.
24. C. P. Brock, *J. Res. Natl. Inst. Stand. Technol.*, 1996, **101**, 321.
25. R. E. Marsh, *Acta Crystallogr., Sect. B*, 2005, **B61**, 359.
26. A. J. C. Wilson, *Acta Crystallogr. Sect. A*, 1988, **A44**, 715.
27. A. J. C. Wilson, *Acta Crystallogr. Sect. A*, 1990, **A46**, 742.
28. A. J. C. Wilson, *Acta Crystallogr. Sect. A*, 1993, **A49**, 795.
29. C. P. Brock and J. D. Dunitz, *Chem. Mater.*, 1994, **6**, 1118.
30. C. P. Brock and L. L. Duncan, *Chem. Mater.*, 1994, **6**, 1307.
31. M. A. Lloyd and C. P. Brock, *Acta Crystallogr. Sect. B*, 1997, **53**, 780.
32. R. S. Rowland and R. Taylor, *J. Phys. Chem.*, 1996, **100**, 7384.
33. F. H. Allen *et al.*, *Acta Crystallogr. Sect. B*, 1997, **53**, 680.
34. C. Janiak, *J. Chem. Soc., Dalton Trans.*, 2000, 3885.
35. A. Nangia, *CrystEngComm*, 2002, **4**, 93.
36. A. Bacchi, *Engineering of Crystalline Materials Properties -NATO Science for Peace and Security Series B: Physics and Biophysics* ed. J. J. Novoa, D. Braga and L. Addadi, 2007, Springer, Dordrecht, The Netherlands, pp. 33–58.
37. A. Bond, *Organic Crystal Engineering: Frontiers in Crystal Engineering*, ed. E. R. T. Tiekink, J. Vittal and M. Zaworotko, Wiley, 2010.
38. *The Impact of Electronic Publishing on the Academic Community*, ed. O. Kennard and I. Butterworth, London, Portland Press Ltd. 1997, pp. 159–166.
39. T. Steiner, *Angew. Chem., Int. Ed.*, 2002, **41**, 48.
40. G. R. Desiraju, *Acc. Chem. Res.*, 2002, **35**, 565.
41. T. Steiner and G. R. Desiraju, *Chem. Commun.*, 1998, 891.
42. G. Aullon *et al.*, *Chem. Commun.*, 1998, 653.
43. Y. Umezawa *et al.*, *Bull. Chem. Soc. Jpn.*, 1998, **71**, 1207.
44. M. Nishio, *CrystEngComm*, 2004, **6**, 130.
45. J. P. M. Lommerse *et al.*, *J. Am. Chem. Soc.*, 1996, **118**, 3108.
46. S. L. Price *et al.*, *J. Am. Chem. Soc.*, 1994, **116**, 4910.
47. P. Metrangolo *et al.*, *Acc. Chem. Res.*, 2005, **38**, 386.
48. G. R. Desiraju, *Nat. Mater.*, 2002, **1**, 77.
49. G. R. Desiraju, *Crystal Engineering: The Design of Organic Solids*, Elsevier, Amsterdam, 1989.
50. *Crystal Engineering: From Molecules and Crystals to Materials*, ed. D. Braga, F. Grepioni and G. A. Orpen, NATO ASI Series, Kluwer, Dordrecht, 1999, vol. 538.
51. A. Gavezzotti, *Molecular Aggregation - Structure Analysis and Molecular Simulation of Crystals and Liquids*, Oxford University Press, USA, 2007.
52. E. Gibney, *Nature*, 2015, **527**, 20.
53. A. Gavezzotti, *Acc. Chem. Res.*, 1994, **27**, 309.
54. R. S. Woodley and R. Catlow, *Nat. Mater.*, 2008, **7**, 937.

55. S. L. Price, *Chem. Soc. Rev.*, 2014, **43**, 2098.
56. S. L. Price, *Rev. Comput. Chem.*, 2000, **14**, 225.
57. J. D. Dunitz and A. Gavezzotti, *Angew. Chem., Int. Ed.*, 2005, **44**, 1766.
58. J. Kendrick *et al.*, *Chem. – Eur. J.*, 2011, **17**, 10736.
59. G. M. Day *et al.*, *Acta Crystallogr. Sect. B*, 2005, **B61**, 511.
60. M. Rafilovich and J. Bernstein, *J. Am. Chem. Soc.*, 2006, **128**, 12185.
61. E. Nauha and J. Bernstein, *J. Pharm. Sci.*, 2015, **104**, 2056.
62. R. Taylor, *CrystEngComm*, 2014, **16**, 6852.
63. G. R. Desiraju, *Angew. Chem., Int. Ed. Engl.*, 1995, **34**, 2311.
64. J. D. Wuest, *Chem. Commun.*, 2005, 5830.
65. M. W. Hosseini, *Acc. Chem. Res.*, 2005, **38**, 313.
66. F. H. Allen *et al.*, *New J. Chem.*, 1999, 25.
67. A. Parkin *et al.*, *CrystEngComm*, 2006, **8**, 257.
68. A. Collins *et al.*, *CrystEngComm*, 2007, **9**, 245.
69. L. D'Ascenzo and P. Auffinger, *Acta Cryst. Sect. B*, 2015, **B71**, 164.
70. C. F. Macrae, *J. Appl. Cryst.*, 2008, **41**, 466.
71. I. Dance and M. Scudder, *J. Chem. Soc., Chem. Commun.*, 1995, 1039.
72. O. Keeffe *et al.*, *J. Solid State Chem.*, 2000, **152**, 3.
73. A. Peskov *et al.*, *Acc. Chem. Res.*, 2008, **41**, 1782.
74. V. A. Blatov and D. M. Proserpio, *Acta Cryst Sect. A.*, 2009, **A65**, 202.
75. D. J. Tranchemontagne *et al.*, *Chem. Soc. Rev.*, 2009, **38**, 1257.
76. V. A. Blatov *et al.*, *CrystEngComm*, 2004, **6**, 378.
77. I. A. Baburin *et al.*, *J. Sol. State Chem.*, 2005, **178**, 2452.
78. F. Meutzner *et al.*, *Chem. – Eur. J.*, 2015, **21**, 16601.
79. M. R. Keeler, *Nothing to Hide: Privacy in the 21st Century*, Bloomington, iUniverse Inc, Indiana, USA, 2006, p. 112.
80. W. C. McCrone, *Polymorphism in Physics and Chemistry of the Organic Solid State*, ed. D. Fox, M. M. Labes and A. Weissberger, Wiley Interscience, New York, 1965, vol. II, pp. 726–767.
81. C. P. Brock, W. B. Schweizer and J. D. Dunitz, *J. Am. Chem. Soc.*, 1991, **113**, 9811.
82. T. Steiner, *Acta Crysallogr. Sect. B.*, 2000, **B56**, 673.
83. K. M. Steed and J. W. Steed, *Chem. Rev.*, 2015, **115**, 2895.
84. J. Van de Streek and S. D. Motherwell, *Acta Cryst. Sect. B*, 2005, **B61**, 504.
85. R. Taylor and O. Kennard, *J. Chem. Inf. Comput. Sci.*, 1986, **26**, 28.
86. R. Taylor and O. Kennard, *Acta Crystallogr. Sect. A.*, 1985, **A41**, 85.
87. T. Auf der Heyde, *Angew. Chem., Int. Ed. Engl.*, 1994, **33**, 823.
88. F. H. Allen, M. J. Doyle and R. Taylor, *Acta Crystallogr. Sect. B.*, 1991, **B47**, 29.
89. F. H. Allen, M. J. Doyle and R. Taylor, *Acta Crystallogr. Sect. B.*, 1991, **B47**, 41.
90. F. H. Allen, M. J. Doyle and R. Taylor, *Acta Crystallogr. Sect. B.*, 1991, **B47**, 50.
91. F. H. Allen and O. Johnson, *Acta Crystallogr. Sect. B.*, 1991, **B47**, 62.
92. F. H. Allen and R. Taylor, *Acta Crystallogr. Sect. B.*, 1991, **B47**, 404.
93. F. H. Allen, M. J. Doyle and T. P. E. Auf der Heyde, *Acta Crystallogr. Sect. B.*, 1991, **B47**, 412.
94. L. M. C. Buydens *et al.*, *Chemom. Intell. Lab. Syst.*, 1991, **49**, 121.
95. T. P. E. Auf der Heyde, Analyzing, *J. Chem. Ed.*, 1990, **67**, 461.
96. E. R. Malinowski and D. G. Howery, *Factor Analysis in Chemistry*, Wiley, New York, 1980.
97. *The Interpretation of Analytical Chemical Data by Cluster Analysis*, L. D. Massart and L. Kaufman, Wiley, New York, 1983.
98. J. Chisholm *et al.*, *CrystEngComm.*, 2006, **8**, 11.
99. I. J. Bruno *et al.*, *J. Comp.-Aided. Mol. Des.*, 1997, **11**, 525.
100. C. Bissantz, B. Kuhn and M. Stahl, *J. Med. Chem.*, 2010, **53**, 5061.
101. K. A. Brameld, B. Kuhn, D. C. Reuter and M. Stahl, *J. Chem. Inf. Model.*, 2008, **48**, 1.
102. J. W. Yao *et al.*, *Acta Crystallogr. Sect. B.*, 2002, **B58**, 640.

Section 3: Isolated Intermolecular Interactions

10 Intermolecular Interactions in Crystals

Peter Politzer,*[a,b] Jane S. Murray[a,b] and Timothy Clark[c]

[a] Department of Chemistry, University of New Orleans, New Orleans, LA 70148, USA; [b] CleveTheoComp, 1951 W. 26th Street, Cleveland, OH 44113, USA; [c] Computer-Chemie-Centrum, Department Chemie und Pharmazie, Friedrich-Alexander-Universität Erlangen-Nürnberg, Nägelsbachstrasse 25, 91052 Erlangen, Germany
*Email: ppolitze@uno.edu

10.1 Coulomb's Law

Any chemical system – an atom, ion, molecule, molecular complex, crystal, *etc.* – can be viewed as a collection of nuclei and electrons. These can be regarded as positive and negative point charges, with attractive electrostatic forces between unlike charges and repulsive forces between like charges. The magnitude of the electrostatic force $F(R)$ that a point charge Q_a exerts upon a second point charge Q_b is given by Coulomb's Law as,

$$\boldsymbol{F}(R) = k\frac{Q_a Q_b}{R^2} \qquad (10.1)$$

where R is the separation of the charges, measured from Q_a to Q_b, and k is a constant. If the charges have different signs, then $F(R)$ is negative indicating that the force on Q_b is opposite to the direction of increasing R and hence it is toward Q_a; the charges are attracting.

Intermolecular Interactions in Crystals: Fundamentals of Crystal Engineering
Edited by Juan J. Novoa
© The Royal Society of Chemistry 2018
Published by the Royal Society of Chemistry, www.rsc.org

When the charges have the same sign, a positive F shows that Q_b is being repelled in the direction of increasing R.

By classical physics (and still true in the cyber age), a force is the negative gradient of a potential energy. In the present one-dimensional situation, this means that $F(R) = -dE/dR$. Accordingly, the energy $\Delta E(R)$ of the interaction between Q_a and Q_b can be obtained by integrating $-F(R)$ from infinite separation to R:

$$\Delta E = -\int_{R=\infty}^{R} k\frac{Q_a Q_b}{R^2}\mathrm{dR} = \frac{kQ_a Q_b}{R} \tag{10.2}$$

When Q_a and Q_b have opposite signs, they attract and $\Delta E(R)$ is negative; when they have the same sign, they repel and $\Delta E(R)$ is positive.

A point charge Q_a creates both an "electric field" $\varepsilon_a(R)$ and an "electrical potential" $V_a(R)$ at any distance R in the surrounding space:

$$\varepsilon_a(R) = k\frac{Q_a}{R^2} \tag{10.3}$$

$$V_a(R) = k\frac{Q_a}{R} \tag{10.4}$$

The significance of $\varepsilon_a(R)$ and $V_a(R)$ is that another point charge Q placed at a distance R from Q_a will feel a force $F(R) = Q\varepsilon_a(R)$ and its interaction energy with Q_a will be,

$$\Delta E(R) = QV_a(R) \tag{10.5}$$

Eqn (10.4) gives the electrical potential created at a distance R by a single point charge Q_a. In the context of a molecular solid, however, our interest is in the potential created by a molecule. Since its nuclei and electrons can be viewed as point charges, the electrical potential $V(r)$ that they create at any point r can be obtained, in principle, by summing eqn (10.4) over all of the nuclei and electrons. Within the Born–Oppenheimer approximation, the nuclei can be treated as having fixed positions.[1] However, this does not apply to the electrons, which are in constant rapid motion. Accordingly, while we can simply sum eqn (10.4) over the nuclei, the effect of the electrons at r is obtained by integrating over the electronic density $\rho(r')$ at every other point r':

$$V(r) = \sum_{A} \frac{Z_A}{|\mathbf{R}_A - r|} - \int \frac{\rho(r')\mathrm{dr}'}{|r' - r|} \tag{10.6}$$

In eqn (10.6), Z_A is the charge on nucleus A, located at $\mathbf{R}_A$. The denominators are the magnitudes of the vector distances from the point of interest r to each nucleus A and each infinitesimal unit of electronic charge $\rho(r')dr'$. Eqn (10.6) gives the total electrical potential at r as the sum of the nuclear and electronic potentials at that point. The equation is written in atomic units (au), in which the Coulomb's Law constant k equals one, the charge on an electron is -1 and the charge on a proton is $+1$. A potential $V(r)$ of 1 au is equivalent to 27.21 volts.

We emphasize that $V(r)$ is a real property of a molecule (or other system). It is a physical observable, that can be determined both experimentally (by diffraction methods[2-4]) and computationally. $V(r)$ is positive or negative in a given region of a molecule depending upon whether the contribution of the nuclei or that of the electrons is dominant there.

10.2 Forces Within Nuclear/Electronic Systems

Newtonian physics dictates that in a system of nuclei and electrons at equilibrium, the resultant force felt by each nucleus must be zero. This means that the repulsions between a given nucleus and the other nuclei must be exactly balanced by the attractions between that nucleus and the electrons. Eqn (10.6) provides a means for describing the force felt by a particular nucleus X having a charge Z_X. Let the point r in eqn (10.6) be the position of nucleus X, *i.e.* $r = \mathbf{R_X}$. Then the electrical potential at nucleus X due to the electrons and other nuclei is,

$$V(\mathbf{R_X}) = \sum_{A \neq X} \frac{Z_A}{|\mathbf{R_A} - \mathbf{R_X}|} - \int \frac{\rho(r')dr'}{|r' - \mathbf{R_X}|} \tag{10.7}$$

The interaction energy between nucleus X and the electrons and other nuclei is therefore, by extension of eqn (10.5), $Z_X V(\mathbf{R_X})$. Since the negative gradient of an interaction energy is a force, it follows that the force felt by nucleus X is,

$$\mathbf{F}(\mathbf{R_X}) = -\nabla Z_X V(\mathbf{R_X}) \tag{10.8}$$

Calculating the gradient as indicated in eqn (10.8) results in,

$$\mathbf{F}(\mathbf{R_X}) = -Z_X \sum_{A \neq X} \frac{Z_A(\mathbf{R_A} - \mathbf{R_X})}{|\mathbf{R_A} - \mathbf{R_X}|^3} + Z_X \int \frac{\rho(r')(r' - \mathbf{R_X})dr'}{|r' - \mathbf{R_X}|^3} \tag{10.9}$$

Eqn (10.9) gives the resultant force exerted upon nucleus X by the electrons and the other nuclei in the molecule. At equilibrium, this must be zero.

According to eqn (10.9), the forces felt by each nucleus within a molecule, complex, molecular crystal, *etc.* – and hence its bonding interactions – are purely Coulombic: repulsions with the other nuclei and attractions with the electrons. But surely this cannot be the whole story! It appears to be purely classical physics, ignoring the exotic complexities of quantum mechanics.

So we turn now to quantum mechanics, in the form of the Hellmann–Feynman theorem. This theorem can be traced back to at least five different people,[5–9] three of whom became Nobel Laureates. Hellmann and Feynman, whose names it bears, were apparently the last two. Hellmann was a German physicist who emigrated to the Soviet Union after the Nazis took power in Germany. This had a sad ending; in 1938, he was denounced and executed. Feynman's story is happier; he derived the theorem independently as part of an undergraduate thesis!

Consider a system with an energy E, a time-independent Hamiltonian operator H and a normalized wave function Ψ. By the Schrödinger equation, $E = \langle \Psi^* | H | \Psi \rangle$. The theorem proves that the derivative of the energy with respect to any parameter λ that appears explicitly in the Hamiltonian is given by,

$$\frac{\partial E}{\partial \lambda} = \left\langle \Psi^* \left| \frac{\partial H}{\partial \lambda} \right| \Psi \right\rangle \tag{10.10}$$

This result may seem to be almost trivial,[10] but nevertheless the derivation of eqn (10.10) was at one time criticized.[11] The criticism was later refuted,[12] and the validity of eqn (10.10) is not in question.

It has most frequently been applied in the context of a system of nuclei and electrons, within the Born–Oppenheimer approximation. The Hamiltonian operator is then,

$$H = -\frac{1}{2}\sum_A \nabla_A^2 - \frac{1}{2}\sum_i \nabla_i^2 + \sum_A \sum_{B>A} \frac{Z_A Z_B}{|\mathbf{R}_B - \mathbf{R}_A|} - \sum_A \sum_i \frac{Z_A}{|\mathbf{r}_i - \mathbf{R}_A|}$$
$$+ \sum_i \sum_{j>i} \frac{1}{|\mathbf{r}_j - \mathbf{r}_i|} \tag{10.11}$$

In eqn (10.11), A and B pertain to the nuclei, i and j to the electrons. If λ is now taken to be a coordinate of nucleus X and eqn (10.10) is invoked, then the derivatives of most of the terms in eqn (10.11) give zero since the terms do not explicitly contain the coordinates of nucleus X. When the differentiation indicated by eqn (10.10) is performed in turn for each coordinate of X, then we have the gradients of

E in each coordinate direction, and by eqn (10.8) these yield the negative of the force felt by nucleus X:[9,13,14]

$$\mathbf{F}(\mathbf{R}_X) = -\sum_{B \neq X} \frac{Z_X Z_B (\mathbf{R}_B - \mathbf{R}_X)}{|\mathbf{R}_B - \mathbf{R}_X|^3} + \int \frac{Z_X (\mathbf{r}' - \mathbf{R}_X)\rho(\mathbf{r}')\mathrm{d}\mathbf{r}'}{|\mathbf{r}' - \mathbf{R}_X|^3} \tag{10.12}$$

Eqn (10.12) is exactly equivalent to eqn (10.9). We have arrived at the same result *via* quantum mechanics, starting from the Schrödinger equation, as we did *via* classical physics, starting from Coulomb's Law: the forces felt by any nucleus in a system of nuclei and electrons are due entirely to the Coulombic interactions of that nucleus with the electrons and the other nuclei. The forces depend only upon the electronic density and the positions and charges of the nuclei. Arriving at this conclusion from the Schrödinger equation should actually not be surprising; the forces are related to the electrical potential terms in the Hamiltonian, and as Reed *et al.* pointed out,[15] these are all Coulombic.

Does this mean that quantum mechanics is not needed? No. Quantum mechanics provides a means to obtain the electronic density and the positions of the nuclei that go into eqn (10.9) and (10.12). But once we have all this, Coulomb's Law suffices.

The Hellmann–Feynman theorem, in the form of eqn (10.12), can be viewed as a forerunner of the Hohenberg–Kohn theorem.[16] This showed that the electronic density alone is the fundamental determinant of the properties of a system of nuclei and electrons; even the electrical potential of the nuclei is a functional of the electronic density.

Slater described the Hellmann–Feynman theorem as one of "...the most powerful theorems applicable to molecules and solids..."[17] and there is ample evidence of this. For example, eqn (10.10) has been the basis for deriving exact atomic and molecular energy formulae;[18–21] eqn (10.12) has provided considerable insight into covalent bonding.[12,22–26] Nevertheless, the Hellmann–Feynman theorem has not received the recognition that it merits. In 1981, referring in particular to eqn (10.12), Deb observed that "...the apparent simplicity of the H–F theorem had evoked some skepticism and suspicion..."[27] Even more than twenty years later, Fernandez Rico *et al.* commented that "...the possibilities that it opens up have been scarcely exploited..."[14] This situation needs to be remedied!

10.3 Intermolecular Interactions

Our present focus is upon noncovalent intermolecular interactions, as within molecular crystals. The molecules are typically far enough apart that each can, to a good approximation, be viewed as retaining its identity. Eqn (10.6) gives the electrical potential $V(r)$ that the nuclei and electrons of each molecule create at any point r in the surrounding space; thus if we multiply the $V(r)$ of molecule A at each point in the space of a neighboring molecule B by the nuclear or electronic charge of B at that point, we will obtain the energy of interaction between A and B. In doing this, we are treating the interaction as purely Coulombic, and the Hellman–Feynman theorem provides rigorous justification for doing so.

For the interaction to be attractive, therefore, a region of positive $V(r)$ on A should be near a region of B in which its electronic charge predominates or a region of negative $V(r)$ on A should be near a region of B in which its nuclear charge predominates. To a considerable extent, therefore, favorable intermolecular interactions in crystals (or molecular complexes) can be understood, at least qualitatively, in terms of the attractive forces between regions of positive $V(r)$ on one molecule and negative $V(r)$ on the other.

However, there is an important caveat: $V(r)$ is typically computed with eqn (10.6) for an isolated molecule with a "static" electronic density, prior to any interaction. This is designated the "electrostatic" potential. As two molecules approach each other, however, the electric field of each will polarize the charge distribution of the other to some extent, and correspondingly modify its $V(r)$. The effect of this upon the interaction is stabilizing.

Consider the noncovalent interaction between a positive site on a molecule A and a negative site on a molecule B, to form the complex A$\cdots$B. The electronic densities of A and B will be polarized (*i.e.* will shift) as depicted in **1**:

A----B

← ←

1

The polarization shown in **1** has been confirmed by plots displaying the difference between the computed electronic densities of a complex A$\cdots$B and those of free A and B placed at the same separation as in the complex.[28–31] These plots show, as in **1**, that the electric field of the negative site on B polarizes the electronic density of the positive site of A away from B, while the electric field of the positive site on A polarizes the electronic density of B towards A.

Thus the electronic densities and hence the $V(r)$ of both A and B change somewhat during the course of the interaction. It is the *instantaneous* $V(r)$ of A at every stage of the interaction that is interacting at each point r with the *instantaneous* nuclear or electronic charge of B at that point.

For relatively large separations of A and B (*i.e.* greater than the sum of their van der Waals radii), these effects of polarization are usually relatively minor and the $V(r)$ of the isolated molecules generally provide good qualitative insight into at least the initial stages of noncovalent interactions.[32–34] However, there are exceptions, as shall be pointed out. It is essential to keep in mind that polarization is an integral, stabilizing part of Coulombic interactions.

The Hellmann–Feynman theorem allows intermolecular interactions to be described in a straightforward manner solely in terms of Coulomb's Law. As Levine put it (in the context of molecules), "...there are no 'mysterious quantum-mechanical forces' acting..."[13] To some theoreticians, these conclusions are distressingly simple, seemingly ignoring quantum mechanics (despite coming directly from the Schrödinger equation). Much more elaborate descriptions of noncovalent interactions are frequently advanced, an example of which will be given in the next section in terms of that venerable intermolecular interaction, the hydrogen bond.

10.4 Hydrogen Bonding

For analyzing and interpreting noncovalent interactions between molecules, as in a complex or a molecular solid, $V(r)$ is commonly computed on the "surfaces" of the molecules, since it is through these surface potentials that the molecules perceive each other at relatively large separations.[32–34] If these surfaces are beyond the van der Waals radii of the atoms, then the $V(r)$ are what the molecules "see" of each other. A molecular surface cannot be defined rigorously since the molecule's electronic density $\rho(r)$ extends, in principle, to infinity. Bader *et al.* pointed out, however, that outer contours of $\rho(r)$, such as the 0.001 au or 0.002 au, are effective as surfaces;[35] they reflect the specific features of the particular molecule (*e.g.* lone pairs, π electrons and atomic anisotropy). The 0.001 au contour in particular encompasses about 97% of the molecule's electronic charge and it is somewhat beyond atomic van der Waals radii,[36] which makes it appropriate for analyzing noncovalent interactions. When $V(r)$ is computed on a molecular surface, it is labeled $V_S(r)$. Its locally most positive and most negative values, of which there may be several, are designated as the $V_{S,max}$ and $V_{S,min}$, respectively.

Molecular electrostatic potentials are customarily expressed in energy units, *e.g.* kcal per mol. When this is done, then by eqn (10.5) the stated value of $V(\mathbf{r})$ actually corresponds to the interaction energy of the molecule with a positive point charge at the position $\mathbf{r}$. This is misleading, however, because in reality the electric field of the positive point charge would perturb the electronic density of the molecule and hence $V(\mathbf{r})$ would change from what it is for the isolated molecule, for which it was computed. To avoid this situation, we will give $V(\mathbf{r})$ in units of potential, volts. The conversion factor is: $1\text{ volt} = 23.06\text{ kcal mol}^{-1}$.

For our examples of hydrogen bonding, we will consider two molecules: 2-aminopyrimidine (**2**) and succinic acid (**3**). In Figure 10.1 are shown the electrostatic potentials on their 0.001 au molecular

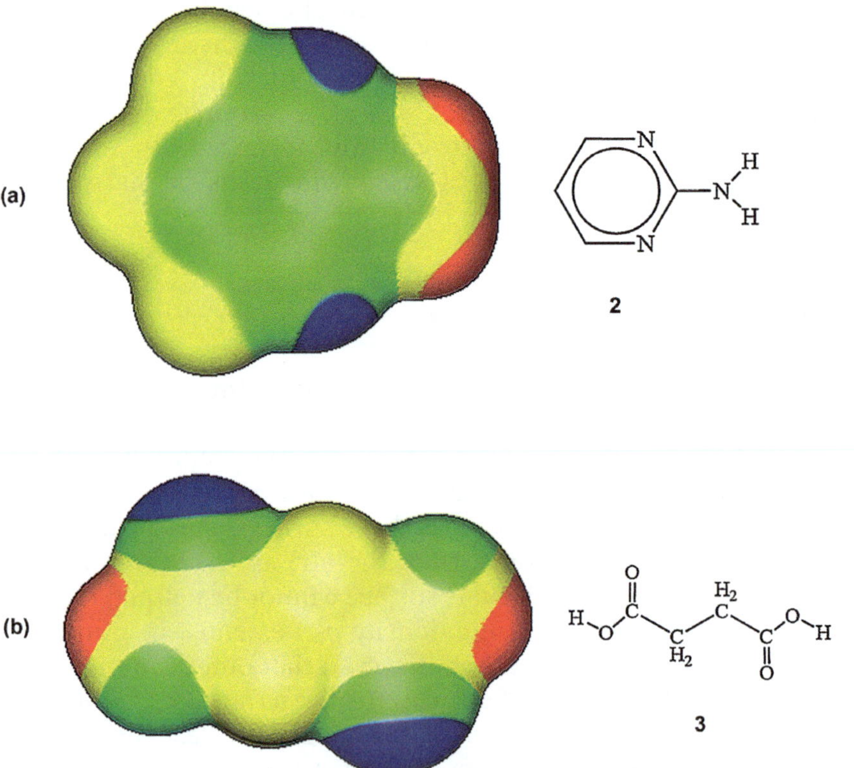

Figure 10.1   Calculated electrostatic potentials on the 0.001 au molecular surfaces of 2-aminopyridimine (**2**) and succinic acid (**3**). Color ranges, in volts: red, greater than 0.87; yellow, from 0.87 to 0; green, from 0 to −0.87; blue, more negative than −0.87. Most positive potentials on **2** are for amino hydrogens (red), $V_{S,max} = 1.6$ volts; most negative are for ring nitrogens, (blue), $V_{S,min} = -1.4$ volts. In **3**, hydroxyl hydrogens have most positive potentials, $V_{S,max} = 2.3$ volts; doubly-bonded oxygens have most negative, $V_{S,min} = -1.3$ volts.

surfaces, computed with eqn (10.6) at the density functional B3PW91/6-31G(d,p) level with Gaussian 09[37] and plotted with the WFA-SAS code.[38] (These procedures have been used for all of the electrostatic potentials to be presented.)

2

3

The most positive potentials on the surface of **2**, Figure 10.1(a), are associated with the amino hydrogens, with $V_{S,max} = 1.6$ volts; less positive are the ring hydrogens. The entire region above the ring is negative, but the strongest negative potentials are by the ring nitrogens, which have $V_{S,min}$ of -1.4 volts. Accordingly, **2** can be expected to form multiple hydrogen bonds through the amino hydrogens and the ring nitrogens. This is exactly what is observed in the 2-aminopyrimidine crystal lattice:[39] every molecule is involved in both N–H$\cdots$N and N$\cdots$H–N interactions with each of two neighboring molecules.

On the surface of **3**, Figure 10.1(b), the most positive potentials are those of the hydroxyl hydrogens, $V_{S,max} = 2.3$ volts; the CH_2 hydrogens are much less positive. There are prominent negative potentials associated with both the hydroxyl oxygens and the doubly-bonded ones, but the latter are more strongly negative, with $V_{S,min} = -1.3$ volts *vs.* -0.76 volts. In principle, all of these oxygens could be involved in hydrogen bonding, but the doubly-bonded ones are much more likely on the basis of their $V_{S,min}$. Indeed, the crystal lattice of succinic acid has chains of molecules linked by both O–H$\cdots$O=C and C=O$\cdots$H–O hydrogen bonds at each end.[40]

2 and **3** can of course also form hydrogen bonds with molecules other than themselves; an example is the **2:3** co-crystal.[41] In this, every molecule of **2** participates in N–H$\cdots$O=C and N$\cdots$H–O interactions with each of two neighboring molecules of **3**.

We have computed the interaction energies for the bimolecular hydrogen-bonded complexes **2**$\cdots$**2**, **3**$\cdots$**3** and **2**$\cdots$**3**, which have the same pairwise interactions as in the crystals of **2** and **3** and in the **2:3** co-crystal. The interaction energy of a complex A$\cdots$B is given by eqn (10.13),

$$\Delta E = E(A\cdots B) - [E(A) + E(B)] \tag{10.13}$$

in which $E(\text{A}\cdots\text{B})$, $E(\text{A})$ and $E(\text{B})$ are the energies of the complex and of its separate components, respectively.

Since the $V_{\text{S,max}}$ of the interacting hydrogens is considerably more positive on **3** than on **2**, 2.3 volts *vs.* 1.6 volts, while the ring nitrogens on **2** are just slightly more negative than the doubly-bonded oxygens on **3**, −1.4 volts *vs.* −1.3 volts, it would be anticipated from these numbers alone that the strongest interaction would be in **3**$\cdots$**3** and the weakest in **2**$\cdots$**2**. The ΔE values at 298 K, computed with the B3PW91/6-31G(d,p) procedure, are in agreement with this expectation (see below):

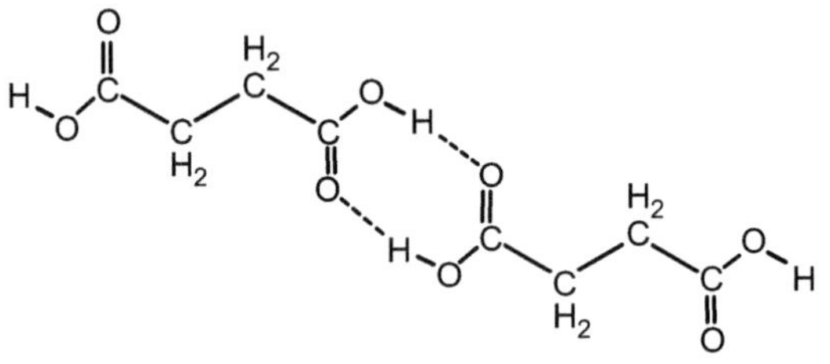

3----3, ΔE = −16 kcal/mol

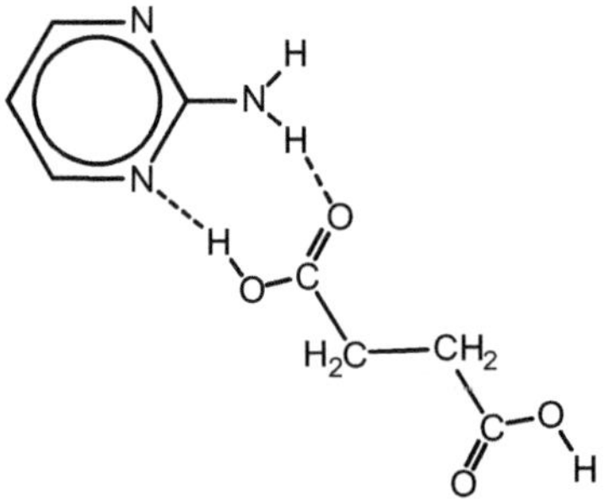

2----3, ΔE = −15 kcal/mol 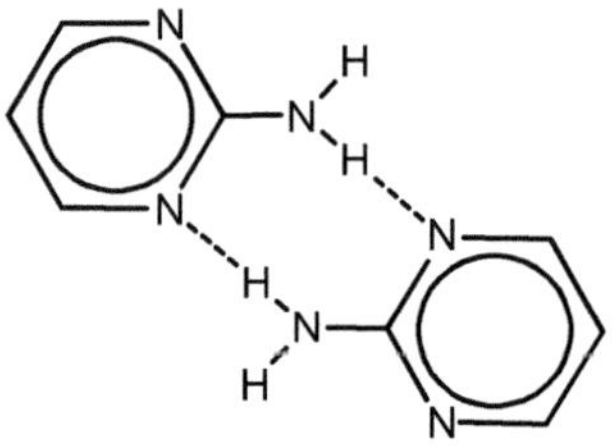   **2----2**, ΔE = −10 kcal/mol

The fact that **3**$\cdots$**3** and **2**$\cdots$**3** have nearly the same interaction energies is consistent with the fact that the co-crystal does form.[41] It should be noted that these ΔE values would be slightly more negative if the mutually polarizing effects of the interacting molecules were taken into account.

In Figure 10.2 are the computed electrostatic potentials on the molecular surfaces of the three complexes, **3**$\cdots$**3**, **2**$\cdots$**3** and **2**$\cdots$**2**. In each instance, the potential in the intermolecular region is only weakly positive or negative; considerable neutralization has occurred. However, each molecule also retains one strongly positive hydrogen and at least one strongly negative site available for further hydrogen

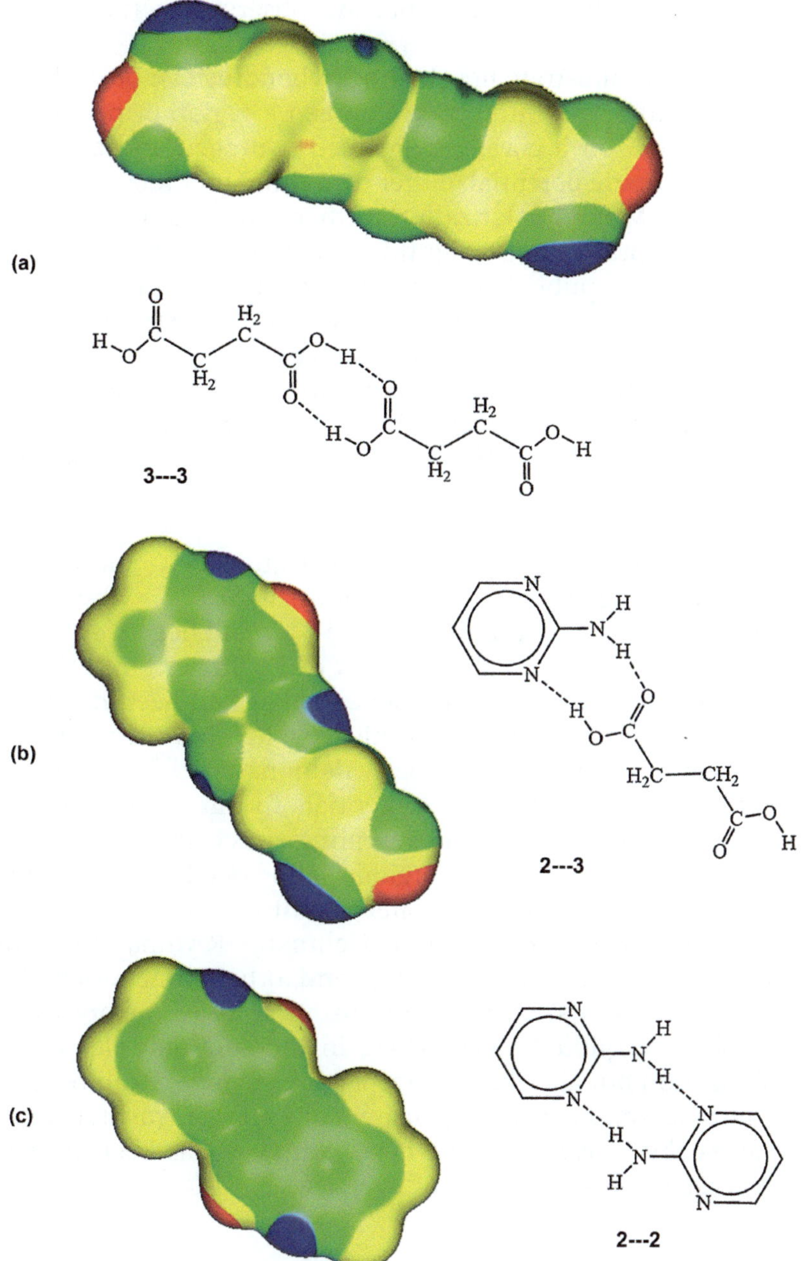

Figure 10.2 Calculated electrostatic potential on the 0.001 au molecular surface of **3···3**, **2···3** and **2···2**. Color ranges, in volts: red, greater than 0.87; yellow, from 0.87 to 0; green, from 0 to −0.87; blue, more negative than −0.87. Potentials in intermolecular regions are only weakly positive or weakly negative due to partial neutralization. However, each molecule still has a strongly positive hydrogen (red) and at least one strongly negative site (blue) available for further interactions, as in a crystal lattice.

bonding, to form the networks that are observed in the respective crystal lattices.

These examples illustrate how hydrogen bonding can be understood as simply the attractive interaction between a region of positive electrostatic potential on a covalently-bonded hydrogen and a region of negative electrostatic potential, either on a different molecule or in the same molecule. Mutual polarization further stabilizes the interaction.

Compare the preceding straightforward explanation with one that has recently been published:[42] "…hydrogen bonding has contributions from electrostatic interactions between permanent multipoles, polarization, or induction interactions between permanent and induced multipoles, dispersion arising from instantaneous multipoles-induced multipoles, charge-transfer-induced covalency, and exchange correlation effects from short-range repulsion due to overlap of the electron distribution."

Any questions?

Notwithstanding this recent "clarification," Coulombic descriptions of hydrogen bonding have a long history.[43-49] About twenty years ago, it was shown[50] that hydrogen $V_{S,max}$ and negative site $V_{S,min}$ correlate well with empirical measures of hydrogen bond donating and hydrogen bond accepting tendencies that had been determined from experimental solution properties by Kamlet, Taft, Abraham, *et al.*[51-54] Quite recently, it has been demonstrated[55] that this capability of molecular electrostatic potentials for identifying and ranking hydrogen bond donating and accepting sites complements Etter's empirical "rules" for predicting organic solid-state hydrogen bonding patterns,[56,57] as well as providing insight into what is occurring.

How does one reconcile the simple Hellmann–Feynman/Coulombic interpretation of hydrogen bonding (and other noncovalent intermolecular interactions) with the imposing array of factors that are often invoked, as quoted above. These include not only electrostatics and polarization but also such concepts as exchange, Pauli repulsion, orbital interaction, correlation, dispersion, charge transfer, *etc.* We will address this question in the next section. It has been discussed more extensively in several recent overviews.[58-60]

10.5   Mathematical Modeling and Physical Reality

The energy ΔE of an intermolecular interaction between molecules A and B is given by eqn (10.13). The more negative is ΔE, the stronger is the interaction. The quantities $E(A\cdots B)$, $E(A)$ and $E(B)$ can be obtained by solving (nearly always approximately) the respective Schrödinger equations, $H\Psi = E\Psi$, for A$\cdots$B, A and B. This produces a wave function and an associated energy for each of the three. An energy is a physical observable; a wave function is not; in itself, it has no direct physical meaning. The enormous importance of a wave function is that it allows us to determine quantities that do have physical meaning, such as the energy and the electronic density, among others.

According to the Hellmann–Feynman theorem, the magnitude of ΔE for the formation of A$\cdots$B can be fully explained in terms of simply the Coulombic interaction between A and B, which must be understood to include their mutual polarization. However, such straightforward explanations are viewed with suspicion by some theoreticians. Current analyses of noncovalent interactions typically argue that ΔE includes contributions from not only electrostatics and polarization but also exchange, Pauli repulsion, orbital inter-action, correlation, dispersion, charge transfer and sometimes others. We shall now consider these supposed contributions.

10.5.1 Electrostatics and Polarization

It is quite common to treat electrostatics and polarization as two separate contributions to a noncovalent interaction between molecules A and B. The electrostatic contribution is viewed as involving the *unperturbed* charge distributions of isolated A and isolated B but placed at their separation in the A$\cdots$B complex. This is a purely imaginary, artificial situation! As pointed out in Section **10.3**, as soon as the interaction begins, the molecules' charge distributions begin to be polarized by each other's electric fields. The reality of this is shown by density difference plots,[28–31] as discussed in relation to structure **1**.

The Coulombic interaction of A and B should properly be viewed as a multi-step process in which complementing electrostatic and po-larizing effects are iterated to self-consistency.[31] An interaction energy computed with A and B being kept in their unperturbed states, without including the stabilizing effect of polarization, is actually an upper bound to the correct value (*i.e.*, it is not sufficiently negative).

10.5.2 Exchange, Pauli Repulsion and Orbitals

A "proper" wave function must satisfy certain requirements.[13] It must reflect the fact that the electrons are indistinguishable; this is han-dled by introducing "exchange." It must be antisymmetric with respect to the interchange of any two electrons. This leads to what is called "Pauli repulsion." These are simply mathematical requirements upon the form of the wave function. Taking them into account for AB, A and B does affect their calculated energies, as does any modification of their wave functions; however, exchange and Pauli repulsion do *not* correspond to physical forces.[12,13,17,61,62]

Orbitals are a clever way to express a multi-electron wave function, but they are just mathematical constructs. A wise choice of orbitals will produce a better wave function and a better energy, but the orbitals themselves have no physical reality,[63] nor does their overlap.

10.5.3   Electronic Correlation and Dispersion

Electronic correlation refers to the fact that the actual electronic density of a system of nuclei and electrons reflects the *instantaneous* repulsive interactions between them. The movements of the electrons are correlated; "they tend to keep out of each other's way."[13] It is not sufficient to consider only the average electronic interactions. Correlation has significant energetic consequences, but it is simply a component of the overall electronic repulsions, and thus is part of the total Coulombic interactions within the system; it is not something additional.

Dispersion is the name given to the ubiquitous weak attraction that exists between any atoms, molecules, *etc.*, even inert gas atoms. It is commonly viewed as an interaction between instantaneous induced dipoles that result from the correlated movements of the electrons (see above).[61,64] A different explanation was given by Feynman, according to whom dispersion involves intramolecular (or intraatomic) attractions between a nucleus and its own electronic charge that has been polarized into the region between the molecules (or atoms).[9] Feynman's view has been confirmed by several studies,[65–68] and the $1/R^6$ dependence of the dispersion energy has been derived from the Hellmann–Feynman theorem, but nonetheless Feynman's explanation is largely ignored.

In the present context, however, the important point is that by either description, whether as dipole/dipole or nuclear/electronic attractions, dispersion is a purely Coulombic effect, and is simply part of the overall polarization that accompanies the interaction. It is not something separate.

10.5.4   Charge Transfer

We come now to charge transfer, a favorite component of noncovalent interactions for many theoreticians. We will illustrate the concept in terms of a hydrogen bond, R–H···B, between a molecule R–H and a negative site B. According to the charge transfer argument, a small fraction of an electron is transferred from an occupied orbital on B (the "charge donor") to an antibonding orbital of R–H (the "charge acceptor"). This weakens the R–H bond and explains why its stretching frequency is often lower and its length greater in the complex R–H···B than in the free molecule R–H. (In some cases, however, a rebellious R–H bond inconsiderately shifts to a higher frequency and shorter length.)

Noncovalent intermolecular interactions are accordingly frequently described as "donor–acceptor." This approach can sometimes be useful as a purely mathematical model, but it should not be taken seriously; it does not represent physical reality. Electrons are indivisible, and cannot be fractionated; orbitals – whether bonding, antibonding, or otherwise – are a very convenient way of expressing molecular wave functions, but they have no physical existence.[63]

The physical reality is polarization, as shown in **1**. Charge transfer is simply a mathematical means of modeling the redistribution (polarization) of electronic charge that accompanies a noncovalent intermolecular interaction.[30,31,69] This redistribution is shown by density-difference plots, as discussed in Section **10.3**. In the case of the hydrogen bond R–X···B, it would be represented as,

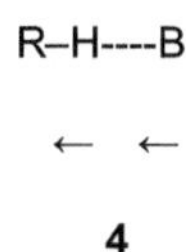

4

which has been confirmed by density-difference plots,[70–72] just as for **1**.

It was noted by Morokuma and Kitaura already in 1981 that, "If one uses a complete basis set, one cannot distinguish between the charge transfer and the polarization."[73] Others have made similar observations.[15,74–78] It is increasingly recognized that charge transfer, in noncovalent interactions, is a mathematical model (and not the only one) for the actual physical event, which is polarization.

The changes in stretching frequency and bond length (*e.g.* of the R–H bond in **4**) can readily be explained in terms of the molecule's permanent and induced dipole moments and the electric field of the negative site,[79–81] using the formalisms of Hermansson[82] or Qian and Krimm.[83] No orbital role need be invoked. The effect of an external electric field upon a stretching frequency of a neighboring molecule was demonstrated graphically by Hennemann *et al.*[81]

The artificial, nonphysical nature of charge transfer as invoked in noncovalent interactions can be illustrated by a modified version of an example due to Stone and Misquitta.[77] Consider the hydrogen-bonded complex of two water molecules, $HO-H\cdots OH_2$. A study using a wave function, expressed as usual in terms of basis orbitals centered on the atoms of both molecules, concluded that charge transfer from an occupied orbital on one molecule to an empty orbital on the other accounts for about 36% of the interaction energy.[84] However, one could equally well express the wave function of the complex in terms of orbitals centered on only one of the water molecules. If a sufficient number and variety of orbitals were used, the resulting electronic density would be very similar to that resulting from the first wave function; both would show the polarization depicted in **4**. But the charge transfer, "determined" in the same manner as before, would be zero since one of the molecules has no orbitals to give or to receive electronic charge. The physical reality, polarization, is the same by both approaches and is expressed by the electronic density, an observable; the charge transfer, a mathematical artifact, is illusory.

10.5.5   Interaction Energy Decomposition

A favorite approach to analyzing noncovalent interactions is to "decompose" the interaction energy into separate contributions from all or some subset of the factors that have been discussed in the preceding Sections, **10.5.1** to **10.5.4**. This is claimed to provide insight into the detailed nature of each particular interaction. The fact that it is meaningless is not allowed to be a deterrent.

Exchange, Pauli repulsion and orbitals are part of the mathematics that produces a proper wave function; they do not correspond to physical forces that promote or hinder an interaction. Charge transfer, as invoked in a noncovalent interaction, is simply one way of representing mathematically a particular physical observable, polarization. It is not a separate factor. Electrostatics, polarization, correlation and dispersion are all intrinsic parts of the Coulombic interaction. They are interdependent; any attempt to uncouple them is purely artificial, as has been extensively discussed.[30,59,69]

Since there is no correct way to perform interaction energy decomposition, anyone can feel free to invent his/her own procedure, secure in the knowledge that it cannot be proven to be more wrong than any other. A recent study cites 16 different decomposition schemes that have been proposed.[85] They do not even agree as to what are the "contributing" factors, and the results can be quite contradictory. For example, two studies examined the interactions of methyl halides and trifluoromethyl halides with formaldehyde. One concluded that the stabilizing effects are primarily electrostatics and dispersion;[86] the other, for exactly the same molecules, found them to be charge transfer and polarization (which are actually the same thing), with electrostatics contributing only "slightly."[87] So what insight has been obtained?

10.6   A Survey of Some Noncovalent Intermolecular Interactions

10.6.1 σ-Hole Interactions

We have demonstrated in Section **10.4** that hydrogen bonding can be understood as the interaction between a region of positive electrostatic potential on the hydrogen and a site of negative potential, whether on the same or a different molecule. The positive region on the hydrogen is roughly hemispherical and its maximum value, the $V_{S,max}$, is usually approximately along the extension of the bond to the hydrogen. (Note the hydrogens in Figure 10.1 and the noninteracting ones in Figure 10.2.)

In 1992, it was shown that covalently-bonded halogens can also have local positive electrostatic potentials on their outer surfaces, along the extensions of the bonds to them, even though other parts of the halogen atoms' surfaces may be negative.[88–90] An example is the bromine atom in 5-bromopyrimidine (**5**), Figure 10.3; note the positive region on its outer surface, with its $V_{S,max}$ on the extension of the C–Br bond.

5

This discovery was surprising, since halogen constituents of molecules are generally viewed as being negative in character. However, it explained what had seemed to be an enigma called "halogen

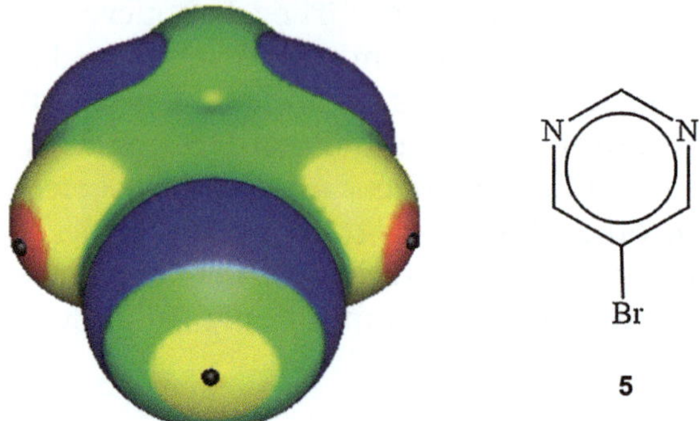

Figure 10.3   Calculated electrostatic potential on the 0.001 au molecular surface of 5-bromopyrimidine (**5**). Color ranges, in volts: red, greater than 0.87; yellow, from 0.87 to 0.43; green, from 0.43 to 0; blue, less than 0 (negative). Locations of most positive potentials, the $V_{S,max}$, are indicated by black hemispheres. The bromine has a positive region (yellow) on its outer surface, along the extension of the C–Br bond.

bonding," the fact that many covalently-bonded halogens can interact attractively (noncovalently) with negative sites. This appears to be a case of negative attracting negative! Such interactions had been known since the 19th century,[91,92] and had been extensively characterized crystallographically in the mid-20th,[93,94] but remained puzzling until it was pointed out in 1992[88–90] and subsequently[95–98] that they can be understood as simply the positive potential on the halogen interacting with the negative site. The latter can be the lone pair of a Lewis base, the π electrons of an unsaturated molecule, an anion, *etc.*

The positive outer region was labeled a "positive σ-hole."[99] It is a consequence of the redistribution of electronic density that accompanies an atom's participation in a covalent bond; the atom's electronic density becomes anisotropic, with less on its outer side (along the extension of the covalent bond) than on its lateral sides.[58,60,100,101] The focused nature of a σ-hole potential results in halogen bonding usually being close to linear; *i.e.*, in a complex R–X···B, where X is a halogen, the angle R–X–B is in the vicinity of 180°. The X···B separation is typically less than or similar to the sum of the respective van der Waals radii.

The positive outer portions of covalently-bonded hydrogens, as in Figures 10.1–10.3, can also be viewed as σ-holes. However, they are approximately hemispherical, not as narrowly focused as those of the halogens, because a hydrogen has only one electron and that is involved in the bonding. As a result, hydrogen bonding tends to be somewhat less directional than halogen bonding.[102,103]

During the years 2007–2009, it was shown that positive σ-holes can also be found on covalently-bonded atoms of Groups VI,[104] V[105] and IV.[106] They are again on the extensions of single (and sometimes multiple) bonds to the atoms, meaning that there can be two for Group VI atoms, three for Group V and four for Group IV. (There can even be more if the atoms are hypervalent.[106,107]) Examples of positive σ-holes on the Group VI molecule **6**, the Group V molecule 7 and the Group IV molecule **8** are in Figures 10.4–10.6.

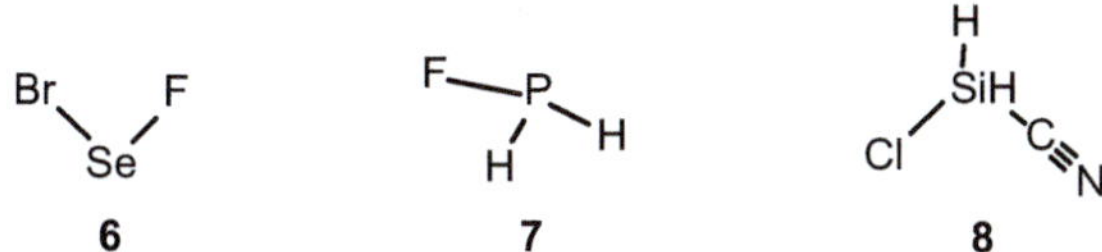

Just as for the halogens, positive σ-holes on Group VI–IV atoms can interact in a directional manner with negative sites. Recognition of this explains numerous noncovalent interactions involving these atoms that had been observed experimentally over a period of many years, as discussed in recent summaries.[101,108]

It is interesting to note that in 1992 Burling and Goldstein observed computationally that the sulfur in thiazole, **9**, has regions of positive

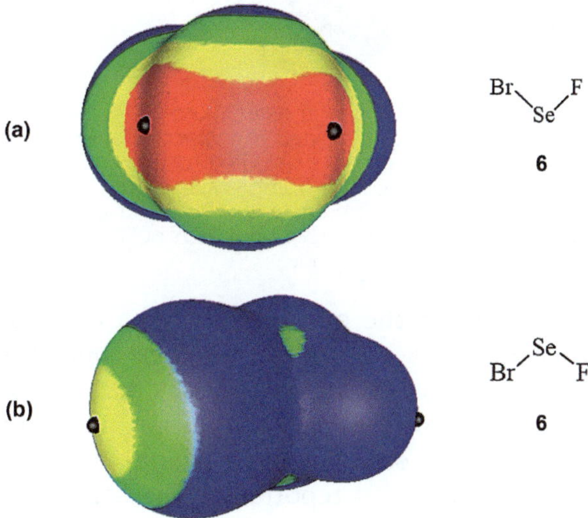

Figure 10.4 Calculated electrostatic potential on the 0.001 au molecular surface of SeFBr (**6**). Color ranges, in volts: red, greater than 0.87; yellow, from 0.87 to 0.43; green, from 0.43 to 0; blue, less than 0 (negative). Locations of most positive potentials, the $V_{S,max}$, are indicated by black hemispheres. The selenium has positive σ-holes on the extensions of the Br–Se and F–Se bonds; the bromine has one on the extension of the Se–Br bond.

Figure 10.5 Calculated electrostatic potential on the 0.001 au molecular surface of PH$_2$F (**7**). Color ranges, in volts: red, greater than 0.87; yellow, from 0.87 to 0.43; green, from 0.43 to 0; blue, less than 0 (negative). Locations of most positive potentials, the $V_{S,max}$, are indicated by black hemispheres. The phosphorus has positive σ-holes on the extensions of the F–P and both H–P bonds.

Figure 10.6 Calculated electrostatic potential on the 0.001 au molecular surface of SiH$_2$(Cl)(CN) (**8**). Color ranges, in volts: red, greater than 0.87; yellow, from 0.87 to 0.43; green, from 0.43 to 0; blue, less than 0 (negative). Locations of most positive potentials, the $V_{S,max}$, are indicated by black hemispheres. The silicon has positive σ-holes on the extensions of the Cl–S, the H–Si and the C–Si bonds. The chlorine has one on the extension of the Si–Cl bond.

electrostatic potential on the extensions of the C–S bonds.[109] They used these to account for S···O and Se···O close contacts that stabilize certain biologically-active thiazole and selenazole conformations. These positive regions correspond to what are now labeled σ-holes, although they were found on a two-dimensional plot rather than on a molecular surface. It was also reported in the same year, 1992, that Brinck *et al.* had found positive potentials on halogen surfaces.[88]

Since σ-holes result from the redistribution of electronic charge (*i.e.* polarization) that accompanies bond formation, the magnitudes of their most positive potentials, the $V_{S,max}$, depend upon the polarizabilities and relative electron-attracting powers of the participants in the bonding.[60,101,110] Thus, within the series of molecules R–X, where X is a halogen, the σ-hole $V_{S,max}$ usually increases when going from fluorine to iodine, as X becomes more polarizable and less electronegative. In Figure 10.4, the selenium σ-hole on the extension of the F–Se bond has a more positive $V_{S,max}$ than that on the extension of the Br–Se bond, 1.6 volts *vs.* 1.4 volts, due to the greater electronegativity of fluorine compared to bromine. For the same reason, the σ-hole $V_{S,max}$ due to the F–P bond in Figure 10.5 is greater than those produced by the H–P bonds, 1.5 volts *vs.* 0.70 volts.

Since the first-row atoms tend to have lower polarizabilities and greater electronegativities than the larger ones, their σ-hole $V_{S,max}$ are often weakly positive or even negative. For instance, the fluorine $V_{S,max}$ in **6**, Figure 10.4, is –0.50 volts, while that of the more polarizable and less electronegative bromine is 0.80 volts. (Even when a σ-hole is negative, its potential is still less negative than the surrounding surface; thus it still has a $V_{S,max}$ but it is negative.) At one time it was indeed believed that fluorine does not form halogen bonds. However, it is now well established that it *can* have a positive $V_{S,max}$ and halogen bond if the remainder of the molecule is sufficiently electron-attracting,[98,111,112] *e.g.* as in F_2 and FCN.[101] (In FCN, as in the other halocyanides, the entire halogen surface has a positive potential, with the σ-hole being the most positive.) Similarly, carbon,[106] nitrogen[105] and oxygen[104] can also have positive σ-holes and participate through them in interactions with negative sites.

Furthermore, the absence of a positive σ-hole does not necessarily preclude an attractive σ-hole interaction with a negative site; a positive σ-hole can sometimes be *induced* by the polarizing electric field of an external negative charge.[81] For instance, the chlorine in H_3C–Cl has a negative or near-neutral σ-hole.[101] Yet it has been found computationally to form a noncovalent complex with formaldehyde, H_3C–Cl$\cdots$O=CH_2,[86] with $\Delta E < 0$. The formation of this complex despite the absence of a positive region on the chlorine has been cited by some, *e.g.* Ding *et al.*,[113] as a failure of the Coulombic interpretation. What it actually demonstrates, however, is the inadequacy of considering only the isolated, unperturbed molecules prior to interaction. Clark *et al.* have shown that the polarization caused by an external electric field, such as that of the formaldehyde oxygen, can *induce* a positive σ-hole on the chlorine, making possible the Coulombic attraction that results in the formation of the complex.

Another example is the complex $H_3P\cdots NSH$, which is predicted computationally[114] even though the phosphorus has a negative or near-neutral σ-hole.[105] This can again be easily explained as the result of polarization, the electric field of the nitrogen inducing a positive σ-hole on the phosphorus. The Coulombic interpretation is alive and well, but polarization is an inherent part of it!

It has already been pointed out that positive σ-holes are often found in conjunction with regions of negative potential on the same atoms. This can be seen for the bromine in Figure 10.3, the selenium and bromine in Figure 10.4, the phosphorus in Figure 10.5 and the chlorine in Figure 10.6. (Our experience has been that covalently-bonded Group IV atoms have completely positive surfaces, with the σ-holes having the most positive potentials. Note the silicon in Figure 10.6.)

The presence of regions of both positive and negative electrostatic potential on the same atom suggests that it can interact attractively with both negative and positive sites. This was actually demonstrated some time ago, predating studies of molecular surface electrostatic potentials. In extensive surveys of organic halide crystal structures in the Cambridge Structural Database, Murray-Rust *et al.* found numerous highly directional halogen close contacts.[115–117] For a halogen X in a molecule R–X, close contacts with nucleophilic components of other molecules (negative sites) were approximately linear, along the extension of the R–X bond (**10**), while close contacts with electrophilic components (positive sites) involved the lateral sides of the halogen (**11**).

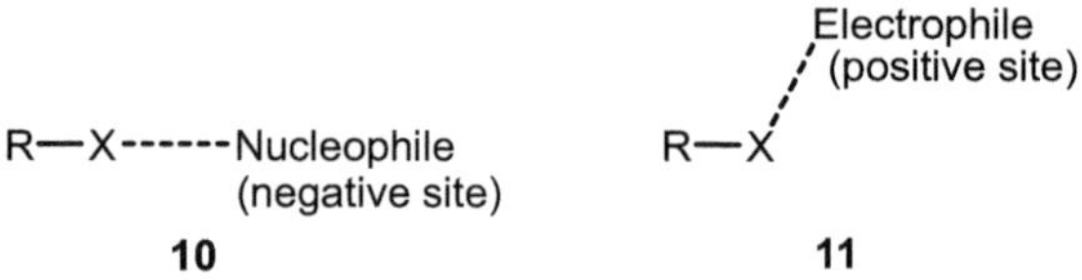

During approximately the same period, Parthasarathy *et al.* reported analogous findings for sulfides and selenides.[118–120] For compounds $R_1R_2S(Se)$, crystallographic surveys showed that close contacts with nucleophilic (negative) sites were along the extensions of the R_1–S(Se) and/or R_2–S(Se) bonds (**12**), and those with electrophilic (positive) sites were above or below the R_1–S(Se)–R_2 plane (**13**).

R_1, R_2 S(Se) ---- Nucleophile (negative site)

12

Electrophile (positive site) R_1, R_2 S(Se)

13

The close contacts with nucleophiles, **10** and **12**, can now be explained as positive σ-hole interactions with negative sites, while **11** and **13** involve the negative lateral sides of the halogen, sulfur or selenium (Figures 10.3, 10.4 and 10.6) interacting with positive sites.

Since a σ-hole interaction can be explained as a Coulombic attraction between regions of positive and negative electrostatic potentials (as modified by polarization), it can be expected to generally become stronger as the $V_{S,max}$ of the σ-hole is more positive and the $V_{S,min}$ of the negative site is more negative. It has indeed been found that, for interactions with a given negative site, the computed interaction energy ΔE, eqn (10.13), varies approximately linearly with the magnitude of the σ-hole $V_{S,max}$;[100,101,121,122] ΔE becomes more negative as $V_{S,max}$ is more positive. The correlation cannot be exact, since (a) the $V_{S,max}$ values are typically computed for the isolated molecules prior to interaction and hence do not reflect polarization, and (b) secondary interactions between other portions of the molecules are only partially taken into account by the $V_{S,max}$ of the σ-holes. For instance, in the complex $F_2Se\cdots NH_3$, weak $F\cdots H$ attractions occur that the selenium $V_{S,max}$ and the nitrogen $V_{S,min}$ only partially take into account, and that will affect the overall ΔE.[104]

For a series of complexes in which both the positive and negative sites are changing, both the σ-hole $V_{S,max}$ and the negative site $V_{S,min}$ must be taken into account, *e.g.* by a double regression analysis in which the interaction energy is expressed as,[60,101]

$$\Delta E = c_1 V_{S,max} + c_2 V_{S,min} + c_3 \tag{10.14}$$

Quite satisfactory agreement between predicted and directly computed ΔE has been obtained with eqn (10.14), despite the limiting factors mentioned above. This supports the Coulombic explanation of these interactions.[60,100,101,121,122]

10.6.2 The Fallacy of Atomic Charges

Chemical experience, as well as the concept of electronegativity, encourages us to think of some atoms in molecules as having more positive or more negative character than others. This can be useful in qualitative terms. However, there is also an urge to try to quantify this (an urge to which one of the present authors also once succumbed[123]), and this results in the unfortunate practice of assigning numerical charges to atoms in molecules, which Price described as "a travesty of bonding theory."[124] There is no physically valid way of doing this;[34] thus everyone can invent his or her own procedure. More than 30 had been proposed already 20 years ago,[125] and this number has doubtless increased. The results can vary remarkably; for instance, six different procedures have assigned charges to the carbon in nitromethane that range from -0.478 to $+0.564$![126]

Assigning a single positive or negative charge to an atom in a molecule can be very misleading, since it conceals the fact that the surfaces of many such atoms have regions of both positive and negative electrostatic potentials (Figures 10.3–10.6) and can interact attractively with both negative and positive sites. This was observed in the crystallographic surveys of Murray-Rust *et al.*,[115–117] structures **10** and **11**, and Parthasarathy *et al.*,[118–120] structures **12** and **13**, and in computational studies.[30,69,100,127,128] A special case is the phenomenon of "like attracting like," *i.e.*, the positive σ-hole on an atom in one molecule interacting with the negative potential *on the same atom* in another identical molecule. Crystalline Cl_2, **14**, is an example of this;[129] another is the complex $ClH_2P\cdots PH_2Cl$.[130] Interactions such as **10–14** could not be predicted on the basis of atomic charges.

$$Cl-Cl----\overset{\displaystyle \overset{Cl}{|}}{Cl}$$

14

It has been pointed out that the inadequacy of atomic charges for describing inter- and intramolecular interactions is reflected in force fields that assign single charges to atoms in molecules.[95,127] There have accordingly been recent efforts to make such force fields more realistic.[131–135]

10.6.3 π-Hole Interactions

Some molecules have regions of positive electrostatic potential, with local maxima $V_{S,max}$, that are perpendicular to a planar portion of the molecular framework. An example is the positive potentials above and below the cyclic ureide **15** (Figure 10.7), with $V_{S,max}$ near the carbonyl carbon. Such regions, which have been labeled "π-holes,"[136] are associated with specific atoms in a variety of molecules, *e.g.* with the

Figure 10.7 Calculated electrostatic potential on the 0.001 au molecular surface of the cyclic ureide **15**. Color ranges, in volts: red, greater than 0.87; yellow, from 0.87 to 0.43; green, from 0.43 to 0; blue, less than 0 (negative). The black hemisphere shows the location of the π-hole $V_{S,max}$; it is associated with the carbonyl carbon.

sulfur in SO_2, the nitrogen in FNO_2, the borons in the BX_3 series and the carbons in the X_2CO, where X is a halogen.

15

16

As with σ-holes, the $V_{S,max}$ of π-holes increase as the atom is more polarizable and its molecular environment more electron-withdrawing. Thus the $V_{S,max}$ of the π-hole of silicon in F_2SiO is 2.9 volts at the B3PW91/6-31G** computational level, while that of the less polarizable carbon in F_2CO is 1.8 volts.[136] However the carbon π-hole $V_{S,max}$ is greater in F_2CO than in Cl_2CO (0.99 volts), due to the greater electronegativity of fluorine.

Molecules with π-holes can interact attractively through them with negative sites. This helps to explain, for instance, the susceptibility of many carbonyl carbons to nucleophilic attack.[137] The strengths of π-hole interactions, just as those involving σ-holes, depend upon the magnitudes of the $V_{S,max}$ and $V_{S,min}$ and the respective polarizabilities.[101,136]

π-Holes can be associated with regions of molecules as well as with specific atoms. Prominent examples are the π regions of unsaturated molecules, provided that they have sufficient electron-withdrawing substituents.[138] This is shown for the 1,3,5-tricyanobenzene molecule,

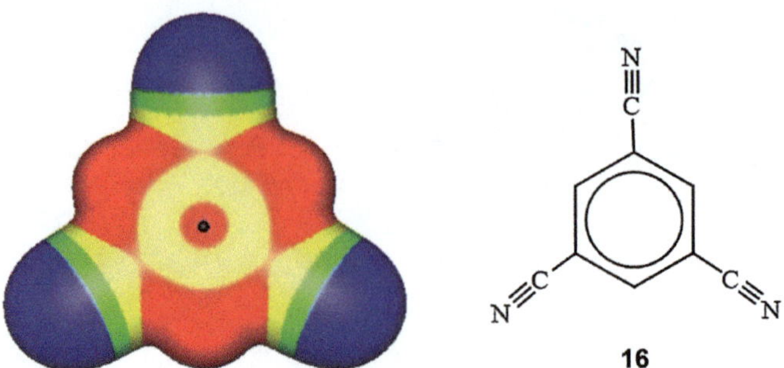

Figure 10.8   Calculated electrostatic potential on the 0.001 au molecular surface of 1,3,5-tricyanobenzene, **16**. Color ranges, in volts: red, greater than 0.87; yellow, from 0.87 to 0.43; green, from 0.43 to 0; blue, less than 0 (negative). The black hemisphere shows the location of the π-hole $V_{S,max}$, which is associated with the π region of **16**.

16, in Figure 10.8; it has a positive $V_{S,max}$ of 1.0 volts above and below the center of the ring. In cyanobenzene, on the other hand, the electrostatic potential above the ring is completely negative.

A recent study of the interactions of substituted benzenes with HCN demonstrated again the important role that can be played by polarization[138] (which includes dispersion – see Section **10.5.3**). Those benzene derivatives having negative (positive) potentials over their rings were found to interact attractively with the hydrogen (nitrogen) of the HCN, as expected. In addition, however, derivatives having *weakly* negative (positive) potentials interacted *attractively* with the nitrogen (hydrogen) of HCN. These latter interactions could not be predicted from the electrostatic potentials of the isolated, unperturbed molecules; they are the result of polarization.

10.7   Thermodynamic Stability

It is customary to describe the strength of an intermolecular interaction by means of the associated change in energy ΔE or enthalpy ΔH. These quantities tend to be quite similar; for the complex $2\cdots3$, shown earlier, $\Delta E = -15.2$ kcal mole^{-1} and $\Delta H = -15.8$ kcal mole^{-1}. From the standpoint of thermodynamic stability, however, it is the change in free energy that is important, ΔG. At a constant absolute temperature T, this is given by,

$$\Delta G = \Delta H - T\Delta S \tag{10.15}$$

ΔS being the associated change in entropy. The requirement for thermodynamic stability is that ΔG be negative.

The interactions of molecules to form larger systems, whether a gas phase complex, a liquid or a solid, is typically accompanied by a decrease in entropy, $\Delta S < 0$, because it diminishes the rotational and translational degrees of freedom of the molecules. The quantity $T\Delta S$ is then negative. If it is larger in magnitude than ΔH, then $\Delta G > 0$ even though ΔH may be negative.

This is frequently the case for weakly- or moderately-bound gas phase molecular complexes.[101,139,140] For instance the formation of the gas phase σ-hole complex $(NC)_2S \cdots NH_3$ has a computed $\Delta E = -6.3$ kcal mole^{-1}, $\Delta H = -6.9$ kcal mole^{-1} and $T\Delta S = -9.6$ kcal mole^{-1}, resulting in $\Delta G = 2.7$ kcal mole^{-1},[101] which makes the complex thermodynamically unstable. Note, however, that having $\Delta G > 0$ does not completely preclude the formation of a complex; it simply means that the equilibrium constant for its formation is less than one.

For stronger gas phase interactions, the magnitude of ΔH may be larger than that of $T\Delta S$, so that if $\Delta H < 0$, then $\Delta G < 0$. An example is the complex **2**$\cdots$**3**. Since it involves two hydrogen-bonding interactions, the ΔH for its formation is relatively large in magnitude, -15.8 kcal mole^{-1}, and ΔG is negative, -5.3 kcal mole^{-1}.

In molecular crystals, with their extensive intermolecular interactions, various factors can affect ΔH and ΔS.[139,141–143] This can lead to the overall ΔG for the formation of the crystal being negative (and the crystal being thermodynamically stable) even though ΔG may be positive for forming the corresponding gas phase complex. For instance, oxalyl chloride, **17**, and 1,4-dioxane, **18**, form a 1:1 co-crystal, the positive σ-holes of the chlorines in **17** interacting with the oxygens of **18**. This co-crystal was one of the early halogen-bonded solids to be characterized crystallographically.[144] However the bimolecular gas phase complex **19** has a positive computed ΔG.[140]

17　　　　　**18**　　　　　**19**

10.8　Discussion and Summary

Researchers sometimes have a tendency to overanalyze. For example, consider hydrogen bonding, which is a favorite prey. Hydrogen bonding has been described as: classical or nonclassical, proper or improper, blue-shifted or red-shifted, dihydrogen, anti-hydrogen (a rebellious form?), H-σ and H-π, positive or negative charge-assisted, resonance-assisted, polarization-assisted, *etc.* But all of these are fundamentally just Coulombic interactions between regions of positive electrostatic potential on the hydrogens (positive σ-holes) and negative sites.

Other σ-hole interactions have similarly been overclassified. For Group VII atoms, the term "halogen bonding" does at least have a historical basis. However, using the labels chalcogen bonding, pnicogen bonding and tetrel bonding for Groups VI, V and IV, respectively, obscures the key point that these are all fundamentally similar interactions, involving a positive σ-hole and a negative site.

We suggest that chemistry has progressed by emphasizing unifying similarities, not differences; *e.g.* Mendeleev focused upon elements having similar properties, Brønsted focused upon a unifying feature of acids and a unifying feature of bases, Lewis emphasized a general characteristic of covalent bonds (the electron pair), not the differences in detail between C–C, C–N, N–F, *etc.*

Starting from the Hellmann–Feynman theorem (and the form of the Hamiltonian itself), we have emphasized the Coulombic nature of noncovalent interactions, which must be understood to include polarization, electronic correlation and dispersion. In many instances, these interactions can be viewed as involving positive σ-holes or π-holes and can be interpreted and ranked on the basis of the $V_{S,max}$ of the σ- or π-hole and the $V_{S,min}$ of the negative site (which may be a lone pair, an anion, π electronic charge, *etc.*). The Coulombic explanation is perfectly adequate even for those noncovalent interactions that are stronger than usual (*i.e.* have more negative ΔE and shorter intermolecular separations than is typical).[58,100,101,136,145,146] These are observed when a particularly highly positive $V_{S,max}$ or highly negative $V_{S,min}$ is involved, and/or large polarizabilities. Thus these strong interactions can again be explained as Coulombic, taking polarization into account; it is not necessary to invoke any new factors. They have sometimes been described as having some degree of "dative" or "coordinate covalent" character, but these are just labels for a high level of polarization.

While the focus upon the $V_{S,max}$ and $V_{S,min}$ can be quite useful, it must be kept in mind that they normally are computed for the molecules prior to interaction. Accordingly, they do not include the effects of the polarization that accompanies the interaction, and they do not fully account for secondary attractions or repulsions between other portions of the molecules.

Furthermore, it must be recognized that the $V_{S,max}$ and $V_{S,min}$ correspond to single points on the molecular surfaces, whereas interactions actually involve regions. As was recently shown,[55] hydrogen bonds to the oxygens of nitroanilines sometimes involve a *ridge* of negative electrostatic potential between the $V_{S,min}$ of the oxygens of an NO_2 group. It is also not unusual for positive or negative regions that are in close proximity to coalesce, sometimes resulting in a single $V_{S,max}$ or $V_{S,min}$ that is associated with the coalesced region rather than with any particular atom. This can be seen, for instance, in some diaryl ureas.[55]

Throughout this discussion of noncovalent intermolecular interactions, we have sought to emphasize the distinction between mathematical modeling and physical reality. The models may be enticingly elegant, and the reality boringly straightforward. So we will conclude by quoting two quite competent theoreticians:[147] Newton: "Nature is pleased with simplicity." Einstein: "Nature is the realization of the simplest conceivable mathematical ideas."

References

1. M. Born and J. R. Oppenheimer, *Ann. Phys.*, 1927, **389**, 457.
2. R. F. Stewart, *Chem. Phys. Lett.*, 1979, **65**, 335.
3. *Chemical Applications of Atomic and Molecular Electrostatic Potentials*, ed. P. Politzer and D. G. Truhlar, Plenum, New York, 1981.
4. C. L. Klein and E. D. Stevens, *Structure and Reactivity*, ed. J. F. Liebman and A. Greenberg, VCH, New York, 1988, ch. 2, pp. 25–64.
5. E. Schrödinger, *Ann. Phys.*, 1926, **80**, 437.
6. P. Güttinger, *Z. Phys.*, 1932, **73**, 169.
7. W. Pauli, *Handbuch der Physik*, Springer, Berlin, 1933, paragraph 11.
8. H. Hellmann, *Z. Phys.*, 1933, **85**, 180.
9. R. P. Feynman, *Phys. Rev.*, 1939, **56**, 340.
10. J. I. Musher, *Am. J. Phys.*, 1966, **34**, 267.
11. C. A. Coulson and R. P. Bell, *Trans. Faraday Soc.*, 1945, **41**, 141.
12. T. Berlin, *J. Chem. Phys.*, 1951, **19**, 208.
13. I. N. Levine, *Quantum Chemistry*, Prentice-Hall, Upper Saddle River, NJ, 5th edn, 2000.
14. J. Fernandez Rico, R. López, I. Ema and G. Ramírez, *J. Chem. Theory Comput.*, 2005, **1**, 1083.
15. A. E. Reed, L. A. Curtiss and F. Weinhold, *Chem. Rev.*, 1988, **88**, 899.
16. P. Hohenberg and W. Kohn, *Phys. Rev. B*, 1964, **136**, 864.
17. J. C. Slater, *J. Chem. Phys.*, 1972, **57**, 2389.
18. L. L. Foldy, *Phys. Rev.*, 1951, **83**, 397.
19. E. B Wilson, Jr., *J. Chem. Phys.*, 1962, **36**, 2232.
20. P. Politzer, in *Fundamental World of Quantum Chemistry: A Tribute to the Memory of Per-Olov Löwdin*, ed. E. J. Brandas and E. S. Kryachko, Kluwer, Dordrecht, The Netherlands, 2003, vol. I, pp. 631–638.
21. P. Politzer, *Theor. Chem. Acc.*, 2004, **111**, 395.

22. R. F. W. Bader and G. A. Jones, *Can. J. Chem.*, 1961, **39**, 1253.
23. P. Politzer, *J. Phys. Chem.*, 1965, **69**, 2132.
24. R. F. W. Bader and A. D. Bandrauk, *J. Chem. Phys.*, 1968, **49**, 1653, 1666.
25. F. L. Hirshfeld and S. Rzotkiewicz, *Mol. Phys.*, 1974, **27**, 1319.
26. J. Autschbach and W. H. E. Schwarz, *J. Phys. Chem. A*, 2000, **104**, 6039.
27. B. M. Deb, in *The Force Concept in Chemistry*, ed. B. M. Deb, Van Nostrand Reinhold, New York, 1981, p. ix.
28. M. Solimannejad, M. Malekani and I. Alkorta, *J. Phys. Chem. A*, 2010, **114**, 12106.
29. S. Scheiner, *J. Chem. Phys.*, 2011, **134**, 164313.
30. P. Politzer, K. E. Riley, F. A. Bulat and J. S. Murray, *Comput. Theor. Chem.*, 2012, **998**, 2.
31. T. Clark, J. S. Murray and P. Politzer, *Aust. J. Chem.*, 2014, **67**, 451.
32. P. Politzer and J. S. Murray, *J. Mol. Struct.: Theochem*, 1998, **425**, 107.
33. P. Politzer and J. S. Murray, *Theor. Chem. Acc.*, 2002, **108**, 134.
34. J. S. Murray and P. Politzer, *Wiley Interdiscip. Rev. : Comput. Mol. Sci.*, 2011, **1**, 153.
35. R. F. W. Bader, M. T. Carroll, J. R. Cheeseman and C. Chang, *J. Am. Chem. Soc.*, 1987, **109**, 7968.
36. J. S. Murray and P. Politzer, *Croat. Chem. Acta*, 2009, **82**, 267.
37. M. J. Frisch, G. W. Trucks, H. B. Schlegel, G. E. Scuseria, M. A. Robb, *et al.*, *Gaussian 09*, Revision A.1, Gaussian, Inc., Wallingford, CT, 2009.
38. F. A. Bulat, A. Toro-Labbé, T. Brinck, J. S. Murray and P. Politzer, *J. Mol. Model.*, 2010, **16**, 1679.
39. J. Scheinbeim and E. Schempp, *Acta Crystallogr.*, 1976, **B32**, 607.
40. H. J. Verweel and C. H. MacGillavry, *Z. Kristallogr.*, 1940, **102**, 60.
41. M. C. Etter, D. A. Adsmond and D. Britton, *Acta Crystallogr.*, 1990, **C46**, 933.
42. E. Arunan, G. R. Desiraju, R. A. Klein, J. Sadlej, S. Scheiner, I. Alkorta, D. C. Clary, R. H. Crabtree, J. J. Dannenberg, P. Hobza, H. G. Kjaergaard, A. C. Legon, B. Mennucci and D. J. Nesbitt, *Pure Appl. Chem.*, 2011, **83**, 1619.
43. L. Pauling, *The Nature of the Chemical Bond*, Cornell University Press, Ithaca, NY, 3rd edn, 1960.
44. P. Kollman, J. McKelvey, A. Johansson and S. Rothenberg, *J. Am. Chem. Soc.*, 1975, **97**, 955.
45. H. Umeyama and K. Morokuma, *J. Am. Chem. Soc.*, 1977, **99**, 1316.
46. P. Politzer and K. C. Daiker, in *The Force Concept in Chemistry*, ed. B. M. Deb, Van Nostrand Reinhold, New York, 1981, ch. 6, pp. 294–387.
47. A. D. Buckingham and P. W. Fowler, *Can. J. Chem.*, 1985, **63**, 2018.
48. S. Lin and C. E. Dykstra, *Chem. Phys.*, 1986, **107**, 343.
49. A. C. Legon and D. J. Millen, *Acc. Chem. Res.*, 1987, **20**, 39.
50. H. Hagelin, J. S. Murray, T. Brinck, M. Berthelot and P. Politzer, *Can. J. Chem.*, 1995, **73**, 483.
51. M. J. Kamlet, J.-L. M. Abboud, M. H. Abraham and R. W. Taft, *J. Org. Chem.*, 1983, **48**, 2877.
52. M. H. Abraham, P. L. Grellier, D. V. Prior, J. J. Morris and P. J. Taylor, *J. Chem. Soc., Perkin Trans. 2*, 1990, 521.
53. R. W. Taft and J. S. Murray, in *Quantitative Treatments of Solute/Solvent Interactions*, ed. P. Politzer and J. S. Murray, Elsevier, Amsterdam, 1994, ch. 3, pp. 55–82.
54. M. H. Abraham, in *Quantitative Treatments of Solute/Solvent Interactions*, ed. P. Politzer and J. S. Murray, Elsevier, Amsterdam, 1994, ch. 4, pp. 83–134.
55. P. Politzer and J. S. Murray, *Cryst. Growth Des.*, 2015, **15**, 3767.
56. M. C. Etter, *Acc. Chem. Res.*, 1990, **23**, 120.
57. M. C. Etter, *J. Phys. Chem.*, 1991, **95**, 4601.
58. P. Politzer, J. S. Murray and T. Clark, *Top. Curr. Chem.*, 2015, **358**, 19.

59. P. Politzer, J. S. Murray and T. Clark, *J. Mol. Model.*, 2015, **21**, 52, (1–10).
60. P. Politzer and J. S. Murray, in *Noncovalent Forces, Challenges and Advances in Computational Chemistry*, ed. S. Scheiner, Springer, 2015, ch. 10, pp. 291–321.
61. J. O. Hirschfelder, C. F. Curtiss and R. B. Bird, *Molecular Theory of Gases and Liquids*, Wiley, New York, 1954.
62. R. F. W. Bader, *Chem. – Eur. J.*, 2006, **12**, 2896.
63. E. R. Scerri, *J. Chem. Ed.*, 2000, **77**, 1492.
64. F. London, *Trans. Faraday Soc.*, 1937, **33**, 8.
65. L. Salem and E. B. Wilson, Jr., *J. Chem. Phys.*, 1962, **36**, 3421.
66. J. O. Hirschfelder and M. A. Eliason, *J. Chem. Phys.*, 1967, **47**, 1164.
67. R. F. W. Bader and A. K. Chandra, *Can. J. Chem.*, 1968, **46**, 953.
68. K. L. C. Hunt, *J. Chem. Phys.*, 1990, **92**, 1180.
69. T. Clark, P. Politzer and J. S. Murray, *Wiley Interdiscip. Rev.: Comput. Mol. Sci.*, 2015, **5**, 169.
70. J. Joseph and E. D. Jemmis, *J. Am. Chem. Soc.*, 2007, **129**, 4620.
71. G. Sánchez-Sanz, C. Trujillo, I. Alkorta and J. Elguero, *Comput. Theor. Chem.*, 2012, **991**, 124.
72. J. Wang, J. Giu and J. Leszczynski, *J. Comput. Chem.*, 2012, **33**, 1587.
73. K. Morokuma and K. Kitaura, in *Chemical Applications of Atomic and Molecular Electrostatic Potentials*, ed. P. Politzer and D. G. Truhlar, Plenum, New York, 1981, ch. 10, pp. 215–242.
74. W. A. Sokalski and S. M. Roszak, *J. Mol. Struct.: Theochem*, 1991, **234**, 387.
75. S. Scheiner, *Reviews in Computational Chemistry*, ed. K. B. Lipkowitz and D. B. Boyd, VCH, Weinheim, Germany, 1991, vol. 2, ch. 5, pp. 165–218.
76. J. Chen and T. Martínez, *Chem. Phys. Lett.*, 2007, **438**, 315.
77. A. J. Stone and A. J. Misquitta, *Chem. Phys. Lett.*, 2009, **473**, 201.
78. R. J. Azar, P. R. Horn, E. J. Sundstrom and M. Head-Gordon, *J. Chem. Phys.*, 2013, **138**, 84102.
79. W. Wang, N. B. Wang, W. Zheng and A. Tian, *J. Phys. Chem. A*, 2004, **108**, 1799.
80. J. S. Murray, M. C. Concha, P. Lane, P. Hobza and P. Politzer, *J. Mol. Model.*, 2008, **14**, 699.
81. M. Hennemann, J. S. Murray, P. Politzer, K. E. Riley and T. Clark, *J. Mol. Model.*, 2012, **18**, 2461.
82. K. Hermansson, *J. Phys. Chem. A*, 2002, **106**, 4695.
83. W. Qian and S. Krimm, *J. Phys. Chem. A*, 2002, **106**, 6628.
84. R. Z. Khaliullin, A. T. Bell and M. Head-Gordon, *Chem. – Eur. J.*, 2009, **15**, 851.
85. Y. Mo, P. Bao and J. Gao, *Phys. Chem. Chem. Phys.*, 2011, **13**, 6760.
86. K. E. Riley and P. Hobza, *J. Chem. Theory Comput.*, 2008, **4**, 232.
87. M. Palusiak, *J. Mol. Struct.: Theochem*, 2010, **945**, 89.
88. T. Brinck, J. S. Murray and P. Politzer, *Int. J. Quantum Chem.*, 1992, **44**(Suppl. 19), 57.
89. T. Brinck, J. S. Murray and P. Politzer, *Int. J. Quantum Chem.*, 1993, **48**, 73.
90. J. S. Murray, K. Paulsen and P. Politzer, *Proc. – Indian Acad. Sci., Chem. Sci.*, 1994, **106**, 267.
91. F. Guthrie, *J. Chem. Soc.*, 1863, **16**, 239.
92. I. Remsen and J. F. Norris, *Am. Chem. J.*, 1896, **18**, 90.
93. O. Hassel and C. Rømming, *Q. Rev., Chem. Soc.*, 1962, **16**, 1.
94. H. A. Bent, *Chem. Rev.*, 1968, **68**, 587.
95. P. Auffinger, F. A. Hays, E. Westhof and P. Shing Ho, *Proc. Natl. Acad. Sci. U. S. A.*, 2004, **101**, 16789.
96. F. F. Awwadi, R. D. Willett, K. A. Peterson and B. Twamley, *Chem. – Eur. J.*, 2006, **12**, 8952.
97. P. Politzer, P. Lane, M. C. Concha, Y. Ma and J. S. Murray, *J. Mol. Model.*, 2007, **13**, 305.

98. P. Politzer, J. S. Murray and M. C. Concha, *J. Mol. Model.*, 2007, **13**, 643.
99. T. Clark, M. Hennemann, J. S. Murray and P. Politzer, *J. Mol. Model.*, 2007, **13**, 291.
100. P. Politzer and J. S. Murray, *Chem. Phys. Chem.*, 2013, **14**, 278.
101. P. Politzer, J. S. Murray and T. Clark, *Phys. Chem. Chem. Phys.*, 2013, **15**, 11178.
102. Z. P. Shields, J. S. Murray and P. Politzer, *Int. J. Quantum Chem.*, 2010, **110**, 2823.
103. A. C. Legon, *Phys. Chem. Chem. Phys.*, 2010, **12**, 7736.
104. J. S. Murray, P. Lane, T. Clark and P. Politzer, *J. Mol. Model.*, 2007, **13**, 1033.
105. J. S. Murray, P. Lane and P. Politzer, *Int. J. Quantum Chem.*, 2007, **107**, 2286.
106. J. S. Murray, P. Lane and P. Politzer, *J. Mol. Model.*, 2009, **15**, 723.
107. T. Clark, J. S. Murray, P. Lane and P. Politzer, *J. Mol. Model.*, 2008, **14**, 689.
108. P. Politzer, J. S. Murray, G. V. Janjić and S. D. Zarić, *Crystals*, 2014, **4**, 12.
109. F. T. Burling and B. M. Goldstein, *J. Am. Chem. Soc.*, 1992, **114**, 2313.
110. J. S. Murray, L. Macaveiu and P. Politzer, *J. Comput. Sci.*, 2014, **5**, 590.
111. D. Chopra and T. N. Guru Row, *CrystEngComm*, 2011, **13**, 2175.
112. P. Metrangolo, J. S. Murray, T. Pilati, P. Politzer, G. Resnati and G. Terraneo, *Cryst. Growth Des.*, 2011, **11**, 4238.
113. X. Ding, M. Tuikka and M. Haukka, *Recent Advances in Crystallography*, ed. B. B. Jason, InTech, 2012, ch. 7, pp. 143–168, , DOI: 10.5772/48592.
114. M. Solimannejad, M. Gharabaghi and S. Scheiner, *J. Chem. Phys.*, 2011, **134**, 24312, (1–6).
115. P. Murray-Rust and W. D. S. Motherwell, *J. Am. Chem. Soc.*, 1979, **101**, 4374.
116. P. Murray-Rust, W. C. Stallings, C. T. Monti, R. K. Preston and J. P. Glusker, *J. Am. Chem. Soc.*, 1983, **105**, 3206.
117. N. Ramasubbu, R. Parthasarathy and P. Murray-Rust, *J. Am. Chem. Soc.*, 1986, **108**, 4308.
118. R. E. Rosenfield, Jr, R. Parthasarathy and J. D. Dunitz, *J. Am. Chem. Soc.*, 1977, **99**, 4860.
119. T. N. Guru Row and R. Parthasarathy, *J. Am. Chem. Soc.*, 1981, **103**, 477.
120. N. Ramasubbu and R. Parthasarathy, *Phosphorus Sulfur*, 1987, **31**, 221.
121. K. E. Riley, J. S. Murray, P. Politzer, M. C. Concha and P. Hobza, *J. Chem. Theory Comput.*, 2009, **5**, 155.
122. A. Bundhun, P. Ramasami, J. S. Murray and P. Politzer, *J. Mol. Model.*, 2013, **19**, 2739.
123. P. Politzer and R. R. Harris, *J. Am. Chem. Soc.*, 1970, **92**, 6451.
124. S. L. Price, *J. Chem. Soc., Faraday Trans.*, 1996, **92**, 2997.
125. J. Meister and W. H. E. Schwarz, *J. Phys. Chem.*, 1994, **98**, 8245.
126. K. B. Wiberg and P. R. Rablen, *J. Comput. Chem.*, 1993, **14**, 1504.
127. P. Politzer, J. S. Murray and M. C. Concha, *J. Mol. Model.*, 2008, **14**, 659.
128. P. Politzer, J. S. Murray and T. Clark, *Phys. Chem. Chem. Phys.*, 2010, **12**, 7748.
129. V. G. Tsirelson, P. F. Zou, T.-H. Tang and R. F. W. Bader, *Acta Crystallogr.*, 1995, **A 51**, 143.
130. P. Politzer and J. S. Murray, in *Concepts and Methods in Modern Theoretical Chemistry, Vol. 1: Electronic Structure and Reactivity*, ed. K. Ghosh and P. Chattaraj, Taylor and Francis, New York, 2013, pp. 181–199.
131. M. A. A. Ibrahim, *J. Comput. Chem.*, 2011, **32**, 2564.
132. M. Kolař and P. Hobza, *J. Chem. Theory Comput.*, 2012, **8**, 1325.
133. M. Carter, A. K. Rappé and P. Shing Ho, *J. Chem. Theory Comput.*, 2012, **8**, 2461.
134. W. L. Jorgensen and P. Schyman, *J. Chem. Theory Comput.*, 2012, **8**, 3895.
135. S. Y. Liem and P. L. A. Popelier, *Phys. Chem. Chem. Phys.*, 2014, **16**, 4122.
136. J. S. Murray, P. Lane, T. Clark, K. E. Riley and P. Politzer, *J. Mol. Model.*, 2012, **18**, 541.
137. J. S. Murray, P. Lane, T. Brinck, P. Politzer and P. Sjoberg, *J. Phys. Chem.*, 1991, **95**, 844.

138. J. S. Murray, Z. P.-I. Shields, P. G. Seybold and P. Politzer, *J. Comput. Sci.*, 2015, **10**, 209.
139. X. Lu, H. Li, X. Zhu, W. Zhu and H. Liu, *J. Phys. Chem. A*, 2011, **115**, 4467.
140. P. Politzer and J. S. Murray, *CrystEngComm*, 2013, **15**, 3145.
141. M. S. Searle and D. H. Williams, *J. Am. Chem. Soc.*, 1992, **114**, 10690.
142. M. S. Searle, D. H. Williams and U. Gerhard, *J. Am. Chem. Soc.*, 1992, **114**, 10697.
143. L. Liu and Q.-X. Guo, *Chem. Rev.*, 2001, **101**, 673.
144. E. Damm, O. Hassel and C. Romming, *Acta Chem. Scand.*, 1965, **19**, 1159.
145. J. E. Del Bene, I. Alkorta and J. Elguero, *J. Phys. Chem. A*, 2010, **114**, 12958.
146. P. Politzer and J. S. Murray, *Theor. Chem. Acc.*, 2012, **131**, 1114, (1–10).
147. W. Isaacson, *Einstein: His Life and Universe*, Simon and Schuster, New York, 2007, p. 549.

11 The Nature of the Hydrogen Bond, from a Theoretical Perspective

Steve Scheiner

Utah State University, Logan, UT 84322-0300, USA
Email: steve.scheiner@usu.edu

11.1 Introduction

The impact of the hydrogen bond (HB) is vast and far-reaching. Water would not be a liquid at room temperature without its presence between individual molecules. Its influence is seen in all processes that occur within the aqueous phase, as well as in a number of other solvents. The solid state, too, is replete with examples of crystal structures controlled in part by the influence of HBs. The HB is an integral component in the three-dimensional structure of a majority of biological molecules, including proteins, carbohydrates, and nucleic acids. The functioning of most enzymes is highly dependent upon these intermolecular forces.

All chemists have some familiarity with the concept of the HB. As we were taught in our introductory chemistry class, an HB occurs when the H atom of AH acts as a bridge with another molecule (D), typically denoted by the AH··· D shorthand. The atoms to which the H atom is connected, in both the proton donor molecule AH and the acceptor D are highly electronegative F, O, or N. The strength of the HB, that is the energy required to pull the AH··· D complex apart into

Intermolecular Interactions in Crystals: Fundamentals of Crystal Engineering
Edited by Juan J. Novoa

Published by the Royal Society of Chemistry, www.rsc.org

separate AH and D molecules, is typically in the range of 3–8 kcal mol^{-1}, much weaker than a covalent bond.

Most of us were further taught that the HB owes much of its ability to hold the two molecules together to electrostatic forces. Specifically, the electronegative character of the A atom pulls electron density from the H atom, resulting in a strong bond dipole. The H with its partial positive charge is attracted to the negative end of the dipole moment of the D molecule. There is also supposed to be a certain amount of "covalent character" to the HB, wherein the two electrons in the D lone pair delocalize with the two electrons in the D–H sigma bond, forming what is sometimes termed a 3-center, 4-electron (3c-4e) bond. While much of this explanation is in fact correct, it represents a gross oversimplification, as will be expanded upon in this chapter.

How does one detect the presence of an HB? There are a number of phenomena which are connected to the formation of such a bond. With regard to spectroscopic measurements, one of the surest indicators is connected with the A–H stretching mode. Upon forming an HB, the ν(AH) frequency shifts to the red by a substantial amount, commonly in the order of several hundred cm^{-1}. The intensity of this band is strengthened, and the band is usually broadened. The NMR chemical shift of the bridging proton moves downfield, corresponding to a lowering of its chemical shielding.

Examination of crystal structures provides additional signals of the formation of an HB. The first type of evidence is structural. The close proximity of the AH and D groups is an important criterion, since the HB has only a limited range. A second issue rests on the θ(AH$\cdots$D) angle, since a strong HB will usually place the atoms in a near linear arrangement. The extraction and analysis of the detailed electron density within the crystal also enables an electronic criterion. The formation of an AH$\cdots$D HB will typically enhance the electron density in the region between H and D, much as in any chemical bond, albeit less so than in a covalent bond.

The above phenomena have more power than to act as mere signals of the presence of an HB, but can also provide information about its strength. For example, the degree of red shift of the ν(AH) stretching frequency is strongly correlated with the strength of the HB. The same is true for the downfield shift of the proton's NMR signal. Likewise, the distance found in the crystal between the proton-donating and accepting atoms is another measure of the HB strength, as is the enhancement of the electron density between the pertinent nuclei.

Despite their power to identify the presence of HBs, and estimate their strength, there exist certain complications as well. In examination of a crystal structure, for instance, the proximity of two electronegative atoms within H-bonding range does not necessarily prove that there is a HB between these atoms or indeed, even an attractive force at all. There is a myriad of intra and intermolecular forces within a crystal that can act to shove these atoms close together, even against a repulsive force, a necessary sacrifice to satisfy the gods of overall stabilization of the entire system. The analysis of the electron density can certainly supply clues of an attraction, but again such data are not conclusive. Similar issues weigh against spectroscopic data as evidence of an HB. There are factors other than an HB that can red shift the stretching frequency of an A–H bond, Fermi resonance among them. As an added complication, it has become apparent in recent years that there exists a class of HBs that do not shift to the red at all, but in the opposite direction.

Quantum chemical methods have the potential to fill these gaps. Given a particular geometry, calculations of this type can easily distinguish an attractive from a repulsive interaction. The nature of this interaction is calculated directly, so is not dependent upon the derived properties of the electron density, or the shifts of individual spectroscopic bands. A system can be devised which focuses only on the HB, circumventing other interactions that might be present in a larger system. The masking and sometimes confounding influence of solvent effects can be circumvented, again centering the focus on the particular interaction of interest. More than this, quantum chemical methods can probe the most fundamental nature of the HB. The total attractive force can be dissected into its individual components, with physical meaning. For example, it is possible to calculate separately the electrostatic attractive energy and the so-called covalent element. And each of these can be further analyzed. The electrostatic potential of each of the two partner molecules can be elucidated within the context of the isolated molecules, to make evident how the positive regions of one molecule align with the negative areas of the other. Before proceeding with an exposition of various sorts of HBs, the following represents an expanded explanation of the particular tools that are used by quantum chemists in this area.

11.2 Computational Tools

11.2.1 Energetics

One of the more important facets of any particular interaction is its "strength". Most commonly, this issue refers to the amount of energy required to break the bond in question. Taking the hydrogen bond between proton donor AH and D as an example, reaction 11.1 describes the association or binding, so is typically exothermic, with a negative ΔE, which is computed as the difference between the energy of the HB complex and the sum of the energies of the two monomers AH and D, eqn (11.2)

$$AH + D \rightarrow AH \cdots D \tag{11.1}$$

$$\Delta E = E(AH \cdots D) - \{E(AH) + E(D)\} \tag{11.2}$$

It is typically assumed that the geometry of the complex $AH \cdots D$ will be fully optimized, as will those of the isolated monomers. In this case, ΔE is usually termed the binding energy, although this is by no means universal. But an alternate prescription would leave AH and D in the geometries which they adopt within the complex. In this case ΔE is frequently denoted the interaction energy because it more directly addresses the interaction of the two monomers within the complex. In the majority of cases, any geometrical rearrangement of AH and D within the complex is minor, and has only a small effect on their energies, so that the binding and interaction energies are usually fairly similar. But this may not always be the case, particularly for strong interactions which produce substantial geometrical changes in each subunit. Even though ΔE is usually negative for reaction 1, it is not unusual to nonetheless describe the binding energy as a positive quantity.

There is a subtle complication in calculations of ΔE that arises from the variation principle. Normally, a given set of orbitals is used to describe the electronic structure of AH, and another set for D. The $AH \cdots D$ complex, however, utilizes the union of the two latter sets of orbitals. The larger basis set of the complex gives it an unfair advantage in the sense that the variation principle states that the energy of a system becomes more negative as the basis is enlarged. A more negative energy for $AH \cdots D$ will lead to an overly negative ΔE for reaction 1. This spurious inflation of the binding energy, called basis set superposition error, is usually corrected by a counterpoise procedure,[1] wherein the energy of monomer AH is computed in a basis set that includes not only its own orbitals, but also those of D (and *vice versa* for subunit D).

The interaction energy up to this point refers only to the electronic contribution to the thermodynamic ΔE. The full ΔE includes also vibrational, rotational, and translational terms. Of these, the term that introduces the largest contribution is the zero-point vibrational energy (ZPE), which takes account of the various vibrational modes of reactants and product. Following this correction, it is common to denote the new quantity as $\Delta E + \text{ZPE}$. The quantum calculation of vibrational frequencies, rotational constants, and so forth permits the ready computation of factors required to elucidate ΔH, ΔS, and ΔG.

11.2.2 Energy Dissection

A prevalent question that often arises in connection with HBs has to do with the factors that contribute to the binding. The vagueness of this question has promoted a long list of different terms in common usage: electrostatic, covalent, dipole–dipole, van der Waals, London forces, charge transfer, donor/acceptor, and on and on. Some of these quantities have clearer definition than others and it is not unusual to see these terms applied differently by different authors. There have been numerous attempts to dissect the total interaction energy into its constituent parts. Kitaura and Morokuma[2–4] defined the electrostatic component as the Coulombic attraction (or repulsion) between the charge distributions of the two monomers in their pristine state, *i.e.* before the two molecules are allowed to perturb one another's charge clouds. This definition, which seems sensible and along the lines of what most would suggest, implicitly contains within it dipole–dipole, dipole–quadrupole, and higher terms in the multipole expansion. Using the same frozen wave functions, these authors defined an exchange, or steric, repulsion that results when the charge clouds penetrate one another.

Additional attractive forces are connected with the adjustment of each molecular charge distribution resulting from the interaction, and it is here that different energy decomposition schemes diverge the most. Kitaura and Morokuma differentiated between charge that crossed a boundary from one molecule to another, which they termed charge transfer, and charge redistributions that remained on a single molecule, referred to as polarization. It must be understood however, that such a distinction is artificial, dependent upon where the boundary is drawn, as well as other factors. This fact motivated others to provide a more rigorous means to separate the two phenomena.[5–10] Other schemes avoid this distinction, leaving the charge redistribution as a single term which goes by several names, including induction, orbital interaction, or simply polarization. The last major contributor to the attractive force between molecules originates in the instantaneous fluctuations of charge of one molecule, and its effect upon its partner. This phenomenon is commonly dubbed dispersion, but it also goes by the names London or even van der Waals forces.

Perhaps the most widespread means of dissecting interaction energy at the current time is symmetry-adapted perturbation theory[11–14] or SAPT. This formalism provides electrostatic (ES), induction (IND), and dispersion (DISP) energies, but the effects of exchange reside not only in a first-order exchange energy, but also its effects upon other terms, in the form of exchange-induction and exchange-dispersion. SAPT is particularly flexible in that it can be applied at progressively higher levels of perturbation theory, leading to additional terms, and to wave functions computed at either the Hartree–Fock or higher levels.

Among other energy partitioning schemes in common usage, there is LMO-EDA which is based[15] on localized orbitals. In addition to electrostatic, polarization, and dispersion terms, LMO-EDA provides separate attractive exchange and repulsion terms. Another method that has found wide application is based on Natural Bond Orbital (NBO) treatment, and is termed NEDA.[16–19] While the preceding may be the methods in most widespread use, it is by no means an all-encompassing list; there are also other procedures that have been proposed.[20–28]

11.2.3 Electrostatic Potentials

As stated earlier, the electrostatic energy is usually defined as the Coulombic interaction between the charge distributions of the two monomers. There are various means to visualize this interaction, the most common of which presents the electrostatic potential around each isolated molecule in a color format, where for instance the most positive regions are blue and the most negative are red. On any given surface, the potential will have one or more minima and maxima, whose values can provide a quantitative means of comparing one molecule with another. It is common for these extrema to be denoted $V_{s,\min}$ and $V_{s,\max}$. Still another measure dispenses with the idea of a fixed density or atomic radius and instead presents an isopotential surface. In other words, the figure may illustrate a surface on which the potential is equal to a preselected constant value.

Other measures of the electrostatic potential forgo a visual presentation, relying instead on numerical values in tabular form. For example, a charge can be assigned to each atom. But of course a table of atomic charges provides a much cruder picture of the full potential. This approach can be improved by adding atomic dipoles, or even quadrupoles. There remains, however, the question of how to make these assignments. There are numerous prescriptions for this purpose, some of the most common of which are Mulliken, NBO, and AIM schemes. Comparison shows that different schemes often provide very different charges, so some caution is necessary in the interpretation of this data.

11.2.4 Atoms in Molecules (AIM)

The electron density contains a great deal of information about the character of the bonding within any system. One means of extracting the character of chemical bonding rests on the topology of the density ρ, and most particularly its Laplacian $\nabla^2\rho$. This "atoms-in-molecules" (AIM) analysis[29–32] leads to the concept of basins that surround each atom and separate it from the others, and the total density within this basin can then lead to the assignment of AIM atomic charges. Also of importance, there are zero-gradient curves, termed bond paths, between atoms that are viewed as noncovalent bonds in the context of intermolecular interactions. Along each path there is a bond critical point, more or less midway between the two atoms. Bond paths are not restricted to pairs of nuclei, but can also connect other areas, such as π bonds.[33–35] The numerical value of the density ρ and Laplacian $\nabla^2\rho$ at the bond critical point, has been shown in many cases to correlate nicely with other measures of the strength of the bond, *e.g.* distance or energetics. The AIM procedure has the advantage that it relies entirely on the electron density, and as such lends itself to application to experimental densities acquired *via* crystal structure analyses.

11.2.5 NBO

The natural bond orbital (NBO) method transforms the fully delocalized wave function into one more in line with chemists' conventional ideas about individual chemical bonds and lone pairs. In terms of noncovalent forces, NBO analysis of a dimeric complex provides a series of second-order perturbation energies $E(2)$ that correspond to the energetic consequence of the transfer of charge from one orbital of the first molecule, to another orbital of the second. In the case of an H-bonded complex AH$\cdots$D, for example, there is typically a sizable $E(2)$ that corresponds to the transfer of charge from the lone pair of D to the σ^* antibonding orbital of AH: $n_D \rightarrow \sigma^*(AH)$. It is this buildup of density in the latter antibond which has been implicated as the source of the weakening and stretching of the A–H covalent bond in the H-bonded complex.

The intermolecular charge transfers of the NBO prescription would best fall under the heading of induction energy. But it must be understood that induction is by definition a broader quantity, including all motions of electron density, in all orbitals, and encompasses both inter and intramolecular transfers.

11.2.6　Electron Density Redistributions

The mutual perturbations of each molecule upon its partner can be explicitly visualized as a density shift map. One starts with the total electron density of the complex and then subtracts from it the sum of densities of the monomers, prior to the interaction occurring, analogous to eqn (11.1) above.

$$\Delta\rho = \rho(\mathrm{AH}\cdots\mathrm{D}) - \{\rho(\mathrm{AH}) + \rho(\mathrm{D})\} \tag{11.3}$$

The resulting map offers a three-dimensional perspective on shifts within each monomer, as well as density transfers from one molecule to the other. Maps such as these have clearly displayed, for example, the loss of density that occurs around the bridging proton in HBs, and the increase within the region occupied by the lone pair of the electron donor atom. Quantitative encapsulations of the charge shifts are often provided by comparisons of atomic charges before and after the interaction occurs. But of course a single number assigned to an atom can miss important details that are clearly visible in the full map.

11.2.7　Perturbations of the Monomers

When two molecules engage in an interaction with one another, even a relatively weak one, each exerts a force which perturbs the internal properties of the other. One manifestation is the change in internal geometries. Perhaps the most famous example of this effect is the stretch undergone by the A–H covalent bond when it participates in a AH$\cdots$D H-bond. But other geometrical changes can occur as well, including stretches and contractions of other bonds, as well as modifications of internal bond angles. The rearrangement of electron density that accompanies formation of a noncovalent bond has repercussions on the spectral features of each monomer as well. Taking HBs as an example once again, the reduction of electron density in the vicinity of the bridging proton leads to a diminution of the NMR chemical shielding σ and consequently to a downfield shift of the signal of this proton. Another example from the H-bonding interaction is the red shift of the A–H stretching frequency in the vibrational spectrum of AH$\cdots$D, along with an intensification of this band. This reduced frequency is connected with a weakening of the covalent A–H bond, which has been attributed to the increased population of the σ^*(AH) antibonding orbital. The latter is consistent with the idea that charge transfer into this antibonding orbital takes place from the lone pair of the electron donor atom D. The magnitudes of the downfield NMR shift as well as the red shift of ν(AH) have been shown in numerous studies to correlate with energetic and geometric measures of the strength of the HB.

11.2.8 Spectroscopic Features

And indeed, in a broader sense, it is the spectroscopic analysis that provides the strongest bridge between calculations and direct experimental observations. The calculations can perform a full vibrational analysis that unambiguously establishes the nature of each and every mode, including the specific atomic motions, along with its frequency and intensity. Such an analysis can easily distinguish between two bands that might have similar frequencies, and be difficult to disentangle from a purely experimental perspective. On the other hand, most such calculations rely on a full double harmonic approximation. Without accurate anharmonic corrections, the calculated frequencies can be somewhat different than experimental quantities, which can lead to problems with the low-frequency intermolecular modes of an H-bonded complex. But despite such complications, the calculations are usually adept at reproducing changes induced in intramolecular modes by the formation of an H-bond, *e.g.* red or blue shifts, or intensification of a band.

NMR chemical shifts and coupling constants represent a second point of contact between experiment and theory. The advantage of the computational extraction of these quantities is that they are calculated for one very specific geometry, so can pin down the effect of the HB on each individual atom. Due to their relatively long time scale, experimental NMR spectra, on the other hand, yield averages over a range of different geometries. An example is the rapid rotation of methyl groups, which generally leads to a single signal for all three H atoms, even if only one proton is involved in an HB at a time.

11.2.9 Complications with Direct Comparison

The latter issues represent only some of the factors that prevent a strict correspondence between calculated data and experimental measurements. Elaborating further on the time scale issue, a crystal structure represents an average over vibrations and a large set of unit cells. In contrast, quantum calculations refer to a single configuration, usually the equilibrium geometry with all interatomic forces vanished. Zero-point motions are not included, although their energetic consequence can be added. As a second distinction, the geometry within a crystal includes all forces, both intra and intermolecular, long as well as short range. By their nature, quantum calculations are usually limited to a small segment of the crystal, perhaps only a pair of molecules, and thus exclude long-range forces, as well as even direct contacts with other surrounding molecules. The same limitations apply to comparisons with the solution phase. It is difficult for quantum calculations to accurately mimic the rapid fluctuations of surrounding solvent molecules, or to include a sufficiently large number of them. Solvation is frequently modeled by ignoring individual solvent molecules, and instead treating the entire solvent phase as merely a polarizable continuum in which the system of interest is immersed.

Another very important factor is the level of theory being applied. The accuracy of the results will usually depend sensitively on the size and flexibility of the basis set which is used to model the atomic orbitals of the various atoms in the system. *Ab initio* calculations further depend upon the method used to include the effects of electron correlation. There are presently a very large and still rapidly growing number of functionals available for density functional theory (DFT) calculations. And there is not yet widespread agreement as to which are the most suitable for studying particular chemical phenomena, such as H-bonding.

Whereas the strength of an intermolecular interaction such as an HB can be accurately and unambiguously calculated by eqn (11.2), the situation is not so simple for the case of an intramolecular interaction. Reaction 11.1 is no longer relevant as there are no separate AH and D units; that is, the two constituents of the noncovalent bond cannot be fully separated from one another. How then can the system with this bond broken be defined? Unfortunately, there is no simple answer to this thorny question. This situation has motivated a number of different definitions over the years, but all provide somewhat different answers, so the issue remains unsettled. On the positive side, methods of analysis like NBO and AIM do not require the HB be dissociated into two fully separate units, so can help to discern the relative strengths of different intramolecular HBs, and even approximate their energetic strength.

11.3   Modern Understanding of H-bonds

The early basic ideas about HBs have remained intact for the most part. An H atom acts as a bridge between two units, whether inter or intramolecular. The HB is partially electrostatic in that the bridging H atom is partially positively charged, and thus attractive to the electron donor. There is also a good degree of covalent character in the sense that electron density moves around, both within each molecule, and across from the electron donor to the acceptor. Density around the bridging H is diminished, leading to a downfield shift of its NMR signal. Some density moves into the AH σ^* antibonding orbital which usually acts to weaken the A–H covalent bond, with its attendant lengthening and red shift of its stretching frequency. However, it has recently become recognized that the latter is not universally true, as certain HBs result in the opposite trend of a bond contraction and blue shift. The geometry of the HB strives toward a linear AH$\cdots$D arrangement, although a certain amount of deformation is allowed without sacrificing much of the HB energy.

If there is one issue that has changed the most from the earlier understanding of HBs to our present view, it is the broadening[36] of the concept. Within the context of the original AH$\cdots$D HB, the proton donor and acceptor atoms were limited to the highly electronegative F, O, and N. Secondly, the bridging proton was envisioned as partially sharing a lone electron pair of the acceptor atom. The atoms that can act in this same capacity in HBs are now thought to extend well below the first row of the periodic table, encompassing nearly all of the halogens, chalcogens, and pnicogens, especially S.[37–58] Indeed, even the tetrel family of C, Si, and so on have been well established to engage in HBs.[59,60] The source of electrons has now been extended beyond a simple lone pair. The π bonds in alkenes and alkynes can serve in a similar capacity, as well as more extended π-electron systems, as in conjugated or aromatic species.[61–67] Even σ bonds can donate electrons,[35,68] as is the case in what have come to be called dihydrogen bonds.[69–79] Metals can participate as well.[71,80–91] Nor are HBs limited to closed shell systems, as radicals[92–100] are also participants in these interactions.

One could expend a great deal of effort in documenting all of the evidence that each of the extensions to various atoms and electron sources do indeed correspond to a true HB. But the aim of this particular chapter is not such an encyclopedic compendium, but rather an illustration to the reader of how theoreticians understand the underlying nature of HBs, and how they go about gathering the relevant information. For this purpose, the remainder of this chapter focuses on one particular sort of non-traditional HB wherein the CH group serves as proton donor. After a brief summary of the historical hints that such a HB could exist, the chapter progresses to a detailed examination of the CH$\cdots$O HB from a number of theoretical angles, and illustrates how the various theoretical tools outlined above have been applied to this problem.

11.4 The CH Group as Proton Donor – Historical Context

It may be surprising that the idea that the C atom could serve as proton donor in a HB appeared in the literature prior to 1940. Probably the first suggestion[101] was derived from boiling point measurements, followed soon thereafter[102] by studies of the liquid phase for systems like $CHCl_3$ and ethers, and then this concept offered an explanation of the high dissociation constant of *o*-toluic and *n*-butyric acids.[103] The ability of neighboring groups such as Cl, Br, and phenyl to activate the CH donor was emphasized in 1940.[104] It was as early as 1943 that the CH$\cdots$O concept was extended[105] to biological systems such as proteins like collagen. These ideas were not accepted universally; alternate explanations were offered. Echoing criticisms that have commonly been voiced about extension of HBs to atoms other than N, O, and F, it was suggested that the proximity of the C and O atoms could be the result of crystal packing forces, and not the product of a genuine attractive force. The change observed in the CH stretching frequency might result instead from intermolecular geometry[106] or perhaps steric crowding.[107]

But the weight of experimental evidence began to swing more heavily toward the recognition of the CH group as a proton donor in a legitimate HB. The 1950s yielded multiple confirmatory reports[106,108–113] that included not only additional heat of mixing results, but also NMR, IR, and X-ray data. More accurate data of various sorts[114–123] continued to buttress the idea in the 1960s, and greatly extended the range of systems in which they were observed to include collagen and polyglycine among others.

The momentum continued, and even accelerated as time went on. Just citing work from the last few years, such bonds have been noted and used to help explain such observations as the cyclic nature of certain amides,[124] imidazolium ionic liquid,[125] the microwave structure of the $HF/H_2C{=}CHCl$ heterodimer,[126] the stability of cucurbituril/guest assemblies,[127] the hydration of pivotal tropane alkaloids,[128] and as the driving force for self-assembly of supramolecular polymer structures.[129] Overcoming initial skepticism in the biological community, the CH···O HB now occupies an important niche, and its known implications continue to mount. It has been recognized in all classes of biomolecules from oligosaccharides and carbohydrates,[130–132] to nucleic acids,[133–145] and proteins and polypeptides.[146–164] The X-ray community was quick to catch on[115,119,120,165,166] to the idea of CH···O HBs, and there are countless such examples in the literature.[167–175]

The low electronegativity of C, roughly equal to that of H, would argue against the existence of CH bond dipole, and certainly against any appreciable magnitude of such a dipole. Why then can the CH group engage in such a wide diversity of HBs, in all sorts of systems? Answering this question brings us to a tour of the theoretician's toolbox, and how each method of analysis brings more light to the issue. Probing this question also goes right to the heart of what is meant by a HB, what is its fundamental nature. The following summary focuses on the work of the author, based on its familiarity, but it must be understood that this set of data comprises only a small fraction of the totality of work that has been accomplished by countless researchers in this field.

11.5 CH···O H-bonds in Small Systems

Early work probed the question of whether a CH···O interaction is a true HB or a different sort of attractive interaction. As one example, Novoa *et al.*[176] looked at the CH_4/H_2O pair and found the single optimized geometry directed a CH group toward the O atom, although the binding energy of 0.6 kcal mol^{-1} was rather small. (This value contradicted[177] a larger amount from an earlier calculation.[178]) The authors pointed out that a large fraction of even this small binding would be missed if electron correlation were not included. The fundamental nature of this interaction was later probed[179] by consideration of the topology of its electron density, which does indeed contain a bond critical point between the H and O atoms, a requisite of AIM formalism for classification as a bond.

In keeping with this early emphasis on small model systems, which prevent other forms of attraction from clouding the interpretation, our group expanded the range of interaction energy. Beginning with the simplest system which might contain a CH···O HB, the binding energy between CH_4 and OH_2 is very weak,[180] in the order of 0.5 kcal mol^{-1} or less. But there is a strong substituent effect in that electron withdrawing agents on the proton donor molecule polarize the CH bond, making it a more potent donor. In numerical terms, with each replacement of a H atom on CH_4 by F, the HB energy climbs by about 1 kcal mol^{-1}, to the point where the $F_3CH\cdots OH_2$ HB is very nearly as strong as that in the water dimer. As is typical of HBs, the progressive strengthening is also accompanied by a shortening; in this case $R(H\cdots O)$ contracts by about 0.14 Å with each F substitution. These findings were later confirmed by others[181,182] for very similar systems. This idea of inductive effects strengthening a $CH\cdots O$ HB is not limited to F, but has been demonstrated for a range of other electron-withdrawing substituents like Cl.[183–186] These trends are characteristic of HBs so argue for this classification of a $CH\cdots O$ interaction.

Unlike conventional HBs such as $OH\cdots N$ where the OH bond is stretched upon formation of the complex, and its stretching frequency shifted to the red, there were some indications that $CH\cdots O$ and $CH\cdots N$ HBs often behave in a contrary fashion.[187–193] Some studies[180,194] addressed the matter of C–H bond contraction and blue shift, specifically to determine whether such behavior might exclude this interaction from its HB classification. It was determined first that in all other respects, the $CH\cdots O$ interaction fits into this rubric. Electron density shifts are characteristic of typical HBs, and the NMR signal of the bridging proton shifts downfield[195] by an amount proportional to the HB strength.

Also in line with HB expectations are the components of the interaction energy arising from an energy decomposition. In fact, this sort of dissection provided some insight into the nature of this contrary behavior. All HBs, whether conventional or $CH\cdots O$, exhibit a large electrostatic attraction, which is complemented by smaller polarization, charge transfer, and dispersion terms, all balanced by a repulsive exchange energy. And in all cases, a small stretch of the CH/OH covalent bond magnifies all of these factors. The difference resides only in the *amount* of this magnification. The combined magnification of the attractive components outweighs the greater repulsion energy when the OH bond is stretched, albeit by only a small amount, so this bond is stretched when the $OH\cdots O$ bond is formed. The opposite is true for $CH\cdots O$ where the enhanced repulsion overwhelms the magnified attractive term, forcing the CH bond to contract. This central idea, that the direction of shift of the CH stretching frequency resides in a fine balance between two sets of opposing forces, has been confirmed by numerous research groups, each using a different definition of these forces.[185,196–221]

As one might expect, O is not the only atom that can accept a CH proton. Calculations[187,194,222–229] have shown that N serves an equivalent purpose. In fact, due the greater basicity of N, CH···N HBs tend to be stronger than CH···O ones.[230] Like their CH···O counterparts, the NMR signal of the bridging proton is also deshielded in CH···N HBs.[231] Despite the generic weakness of CH as a proton donor, computations have shown that it is strong enough to engage F as an acceptor.[186,210,232–239] Other proton-accepting atoms are S,[207,240–244] even as strongly as 7 kcal mol^{-1},[245,246] Cl[247–250] or I.[251] Even an atom as weakly basic as P[252,253] can serve this function, as can carbenes.[254] And there is an extensive literature concerning CH···π HBs, where the electron donor is an aromatic π system.[255–260]

Another major factor which influences the direction of C–H stretching frequency shift is the hybridization of the C atom. Calculations[194] show that whereas the sp^3 CH of alkanes commonly (but not universally) leads to a blue shift, the opposite result of a red shift occurs for the sp-hybridized alkynes, which generally form stronger CH···O HBs. Given their intermediate sp^2 hybridization, and HB strengths intermediate between alkanes and alkynes, it was not surprising to learn that alkenes manifest only very small shifts, sometimes to the red and other times to the blue. But again, whether alkane, alkene, or alkyne, all properties of these CH···O HBs, *e.g.* NMR chemical shifts or electron density shifts, mimic those of conventional HBs such as OH···O. Comparable trends were noted later with other systems,[218,240,261] some of which contained the analogous CH···N HBs.[262]

HBs are well known for their ability to reinforce one another. That is, a string of n H-bonded molecules is commonly bound together by a force which exceeds that occurring within $n-1$ simple dimers, in what is frequently referred to as cooperativity. A detailed examination[263] revealed that this same phenomenon is common to CH···O HBs as well. This cooperativity is observed not only in the energetics, but also in the HB lengths and in the electron density shifts resulting from the HB formation. On the other hand, there is little evidence of cooperativity within the context of CH covalent bond contraction or CH stretching frequency shifts. Whether conventional or CH···O, the cooperativity in either sort of HB is reduced when the system is immersed in a model solvent. The cooperative aspects of CH···O HBs have been the subject of some inquiry by others as well.[264–269] The overriding conclusion is that these HBs act much like any others in the sense that synergistic, positive cooperativity will occur if a central molecule acts simultaneously as both electron donor and acceptor, while double donor or acceptor activity will result in an overall weakening or negative cooperativity.

11.6   CH···O H-bonds in Biological Systems

While study of small model systems *in vacuo* is of course irreplaceable as a means of elucidating the fundamentals of a given interaction, it is necessary to go beyond this paradigm, to larger systems that are more representative of those commonly encountered in crystals, for example. One type of extension involves some consideration of the surrounding environment. Turning once more to the simple fluoromethanes as model proton donor, examination of how its CH···O HBs are affected by solvent provides some insight into the situation within solution or a crystal. Figure 11.1 shows how the binding energies of any of the F_nCH_{4-n} molecules with water are progressively

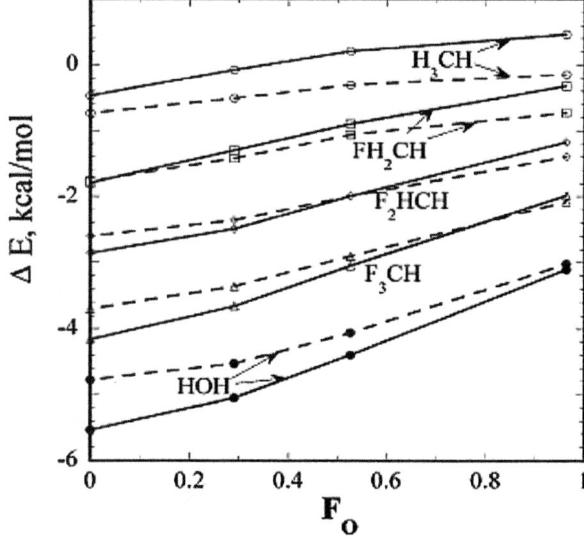

Figure 11.1   Interaction energies of each indicated proton donor molecule with H_2O as proton acceptor. Horizontal axis $F_o = (\varepsilon - 1)/(\varepsilon + 2)$ measures polarizability of surrounding medium, where ε is dielectric constant. B3LYP/6-31+G** and MP2/aug-cc-pVTZ results indicated by solid and broken lines, respectively.
Reprinted with permission from S. Scheiner and T. Kar., Effect of Solvent upon CH-O Hydrogen Bonds with implications for Protein Folding, *J. Phys. Chem. B* 2005, **109**, 3681. Copyright 2005 American Chemical Society.

weakened as the polarizability of the surrounding medium is increased,[270] and the same is true of the conventional OH$\cdots$O HB in the water dimer. Again, the behavior of the CH$\cdots$O interaction is quite similar to that of classical HBs.

Of course, a continuum homogeneous polarizable medium is only a rough approximation of the interior of a crystal or of aqueous solution with numerous specific interactions. In order to bring the model one step closer to the real situation, a number of discrete water molecules were placed[270] around the H-bonded systems, forming a first solvation sphere. HB energies in this primitive solvated system were rather close to the same quantities computed with the dielectric continuum model. A second issue with the latter solvation model is the choice as to what value of dielectric constant most correctly models the system of interest. In the case of a protein, for instance, it is not uncommon in the literature for a value of $\varepsilon = 4$ to be considered to simulate a generic protein interior but of course each protein is different, and even within a single protein, some domains will be more polarizable than others. The same is true for the interior of most crystals.

11.6.1 Amino Acids

$F_3CH + OH_2$ is only the roughest of models of a CH$\cdots$O HB within a protein, even if it does provide some fundamental insights. A first effort to study a more realistic model donor started[271] with a set of amino acids, $NH_2C^\alpha HRCOOH$, all of which contain a $C^\alpha H$ group. Gly, Ala, Val, Ser, and Cys were all paired with a water molecule as they engaged in a $C^\alpha H\cdots OH_2$ HB. There was very little sensitivity to the nature of the sidechain R, with binding energies all in the 1.9–2.5 kcal mol^{-1} range. Also rather constant was the $R(C^\alpha\cdots O)$ distance, 3.31–3.35 Å, and the NMR downfield chemical shift of the $C^\alpha H$ proton which varied between -1.35 and -1.71 ppm. There was a bit more sensitivity in the contraction of the C–H bond, from 0.3 to 3.1 mÅ; all of the C–H stretching frequencies were to the blue, in the 14–56 cm^{-1} range, so this pattern fits the idea of blue stretches accompanying sp^3 hybridization.

In order to expand the scope to charged amino acids, Lys$^+$ was modeled[271] by the R sidechain of $(CH_2)_4NH_3^+$ and Asp$^-$ by CH_2COO^-. As one might expect, the presence of a positive charge, even one that is removed from the $C^\alpha H$ by a hydrocarbon chain, enhanced the binding energy with OH_2 up to 4.9 kcal mol^{-1}. But perhaps surprisingly, this stronger bond did not result in much change in any of the other parameters: $R(C^\alpha\cdots O)$, $\Delta r(C^\alpha H)$, and $\Delta\sigma_H$ were in line with the values obtained for the neutral amino acids. Likewise, for the anionic Asp^-, even though its binding to OH_2 is much weaker.

11.6.2 Dipeptides

Within the context of the amino acid model, the $C^\alpha H$ is surrounded by a $-NH_2$ on one side and $-COOH$ on the other. An expansion of each to a full peptide group leads to a glycyl dipeptide $CHONHC^\alpha H_2CONH_2$ that better represents the setting of this central group within a protein. The ability of this $C^\alpha H$ group to participate in an HB was tested,[272] this time using the carbonyl O of formamide H_2NCHO as a more representative proton acceptor within a protein.

The dipeptide model introduces a good deal of flexibility into the donor molecule. This flexibility is represented primarily by the dihedral angles φ and ψ, that are commonly used to denote the rotation of the two peptide groups around the $C-C^\alpha$ and $C^\alpha-N$ bonds. This work[272] centered around the two conformations of dipeptides that represent minima on their potential energy surface. The C7 minimum derives its name from the presence of a seven-membered ring that contains an intramolecular $NH\cdots O$ HB, while a five-membered ring occurs in the C5 structure. Note that neither internal HB directly involves the $C^\alpha H_2$ group which is available to participate in a $CH\cdots O$ HB with the neighboring formamide.

The interaction energy of a conventional $NH\cdots O$ HB with the formamide carbonyl O atom of the C7 dipeptide is 7.5 kcal mol^{-1}, considerably larger than the 2.3 kcal mol^{-1} of the $C^\alpha H\cdots O$ HB. In the case of the C5 structure, however, the $NH\cdots O$ HB is cut by a factor of three to only 2.5 kcal mol^{-1}, whereas the $C^\alpha H\cdots O$ HB is slightly larger, at 3.8 kcal mol^{-1}. In other words, the $CH\cdots O$ HB is stronger than the $NH\cdots O$ HB for the C5 geometry of the dipeptide. This dipeptide structure with $(\varphi,\psi)=(180°, 180°)$ represents an extended structure of a polypeptide backbone, so is certainly an important segment of the Ramachandran (φ,ψ) region, not far from the β-sheet geometry (more about the β-sheet below).

What can account for this surprising and remarkable sensitivity of the $NH\cdots O$ HB to the conformation of the dipeptide? The threefold reduction of this HB strength is particularly puzzling as the geometry of this bond, including $R(NH\cdots O)$, is nearly the same in the C5 and C7 conformers. A detailed inquiry[273] expanded the (φ,ψ) conformational space of the dipeptide to cover the entire Ramachandran map, not just the C5 and C7 areas, as shown in Figure 11.2. It was found

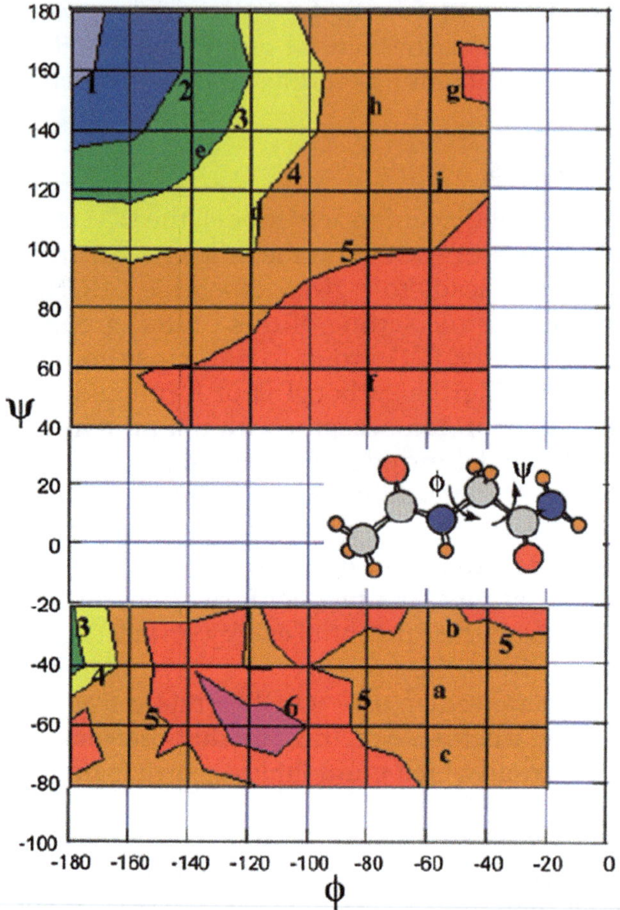

Figure 11.2   Binding energy of water to $CH_3CONHCH_2CONH_2$, as a function of internal dihedral angles ϕ and ψ of glycyl dipeptide. Numerical labels indicate binding energy (kcal per mol); Letters indicate standard locations of (a) α-helix, (b) 3_{10}-helix, (c) π-helix, (d) parallel β-sheet, (e) antiparallel β-sheet, (f) 2.2_7 ribbon, (g) collagen triple helix, (h) PPII, and (i) type II β-bend.
Reprinted with permission from S. Scheiner., The Strength with which a Peptide Group can form a Hydrogen Bond varies with the Internal Conformation of the Polypeptide Chain, *J. Phys. Chem. B* 2007, **111**, 11312. Copyright 2007 American Chemical Society.

that the NH$\cdots$O HB energy is nearly uniform over the majority of the (φ,ψ) map, as indicated by the red and orange sections of Figure 11.2, but becomes progressively weaker as one moves toward the fully extended $(-180°,180°)$ structure in the upper left corner, at which point the HB energy nearly vanishes entirely. The region of weakened NH$\cdots$O HB extends over a fairly wide area, which may be categorized as $-180° < \varphi < -100°$, and $100° < \psi < 180°$.

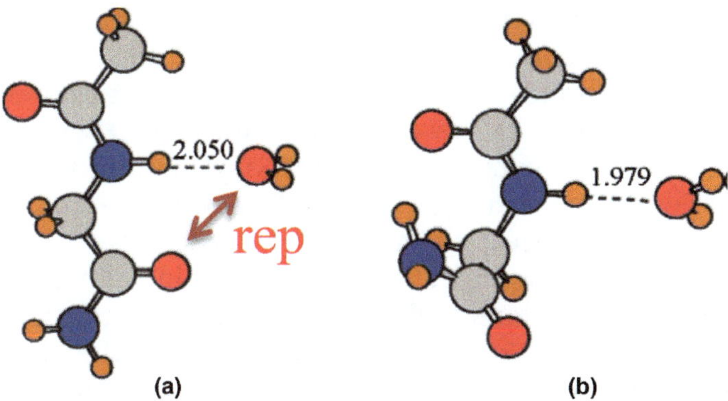

Figure 11.3 Geometries optimized for dipeptide-water system, with dipeptide in (a) $(\phi,\psi)=(-180°, -180°)$ and (b) $(-80°, 80°)$ conformations. Reprinted with permission from S. Scheiner., The Strength with which a Peptide Group can form a Hydrogen Bond varies with the Internal Conformation of the Polypeptide Chain, *J. Phys. Chem. B* 2007, **111**, 11312. Copyright 2007 American Chemical Society.

Careful scrutiny of the data[273] pointed to one particular feature as the prime culprit. In the extended structure of the dipeptide, Figure 11.3a shows that the proton donor NH is close to the carbonyl O of the neighboring peptide unit. (It is in fact this proximity which leads to the common characterization of this structure as C5.) A negative region of electrostatic potential emanates from this carbonyl O atom, which acts as a shield of sorts, pushing an approaching proton acceptor away from the NH group as indicated by the red double arrow in Figure 11.3a, and thereby weakening the incipient NH$\cdots$O HB. As the dipeptide curls away from the $(-180°, -180°)$ extended structure, the neighboring carbonyl O moves away from the NH, leaving this group exposed to the approaching proton acceptor group, and allowing the NH$\cdots$O HB to achieve its normal potential, as for example in Figure 11.3b where $(\varphi,\psi)=(-80°, +80°)$.

An important conclusion arising from this work is that it is incorrect to consider the strength of any particular HB, NH$\cdots$O or otherwise, as a given or constant. The actual binding energy can be heavily influenced by the conformation adopted by the protein, particularly if the structure places the proton donor group in the vicinity of another group with a strong electrostatic potential.

11.6.3   Longer Chains

Given the variability of NH$\cdots$O HB strengths in polypeptides, and its weakness in particular in extended conformations, one is naturally led to consider β-sheets, wherein each strand adopts a fairly extended geometry. The conventional wisdom holds that the strands are held together by interstrand NH$\cdots$O HBs, but even an idealized visualization of the structure of these sheets shows that CH groups of one strand lie in close proximity to the carbonyl O atoms of the next strand. Could not the ensuing CH$\cdots$O HBs contribute to the stability of the interchain linkages, just as the NH$\cdots$O HBs do?

This question was specifically addressed[274] in both parallel and anti-parallel configurations of β-sheets. Beginning with the geometry of a full double-strand extent of anti-parallel polyglycine, a piece was excised such that each strand contains both CH and NH donor groups, lying opposite the carbonyl O of the other strand. The HCONHCH$_2$CONH$_2$ dimer was then optimized to yield both NH$\cdots$O and CH$\cdots$O HBs. The former were shorter than the latter, with R(NH$\cdots$O) = 1.97 Å and R(CH$\cdots$O) = 2.57 Å, as displayed in Figure 11.4. But as shown above, the length can be deceiving, as a short HB is not necessarily a strong one. It was necessary to extract the energetic contribution of each sort of HB, separate from the others. The question of how to disentangle individual interactions, when both are present simultaneously, has been a thorny problem for some time.

The issue was addressed[274] by removing one proton donor group at a time, and then computing the interaction energy of the remaining dimer. As indicated in Figure 11.4a, the NH$\cdots$O HB was deleted by replacing the terminal NH$_2$ groups on each strand by a H atom, leaving a CH stump that is both a weak proton donor, and too far away from the carbonyl O to engage in a HB in any case. The interaction between the two HCONHC$^\alpha$H$_2$COH molecules consists only of C$^\alpha$H$\cdots$O HBs. The CH$\cdots$O HBs were deleted by simply removing the central CH$_2$ group of each strand, which leaves only a set of HCONH$_2$ molecules, bound solely by NH$\cdots$O HBs, as indicated in Figure 11.4b. It should be stressed that the geometry of the dimer was left unchanged by each deletion, to avoid contamination of the results by changes in HB geometry. In this manner, it was determined that the pair of NH$\cdots$O HBs contribute 14 kcal mol^{-1}, as compared to 10 kcal mol^{-1} for the CH$\cdots$O HBs.

The location where the strands are cut is an arbitrary one. If this location is shifted down a bit, the energetic contributions of the NH$\cdots$O HBs drop from 14 to only 10 kcal mol^{-1}, placing it precisely on a par with the CH$\cdots$O HBs. And it should be stressed that this result is not an artifact of the use of a pair of dipeptides; extension to tripeptides has no substantive effect on the results. The situation is

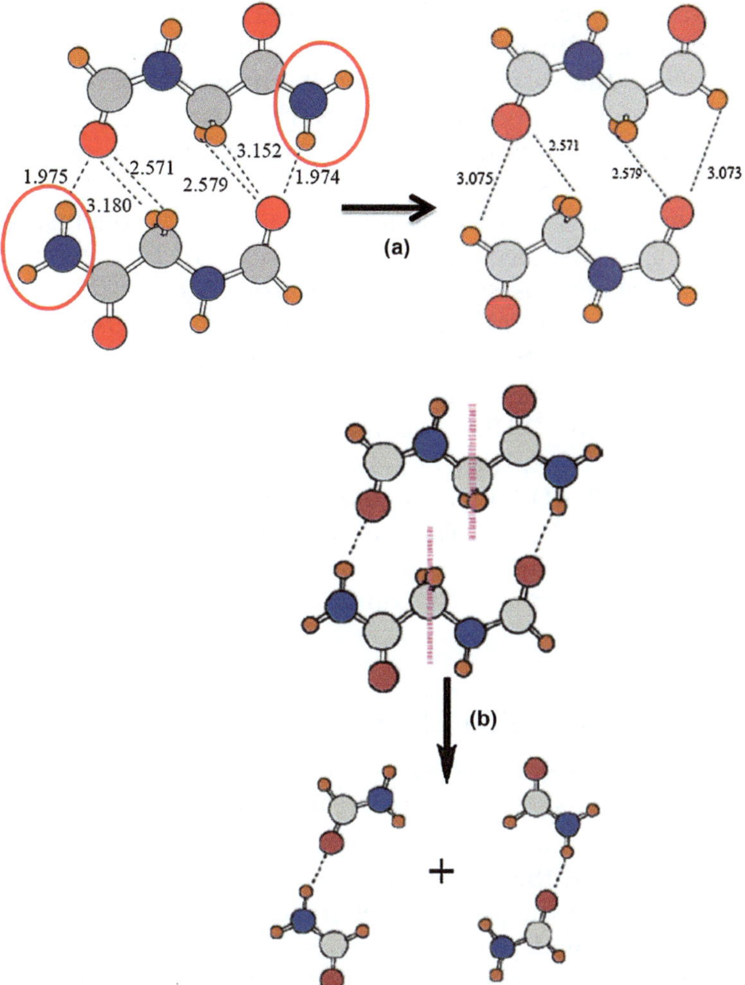

Figure 11.4　Atom excisions made in HCONHCH$_2$CONH$_2$ dimer. (a) Replacement of terminal NH$_2$ groups by H, to leave only CH$\cdots$O HBs. (b) Removal of central CH$_2$ groups, leaving behind only NH$\cdots$O.
Reprinted with permission from S. Scheiner., Contributions of NH-O and CH-O Hydrogen Bonds to the Stability of β-Sheets in Proteins, *J. Phys. Chem. B* 2006, **110**, 18670. Copyright 2006 American Chemical Society.

somewhat different in the case of a parallel β-sheet. An equivalent partitioning into separate NH$\cdots$O and CH$\cdots$O HB energetic contributions shows that the latter makes a *larger* contribution than does the former. Specifically, the 8.3 kcal mol^{-1} contribution from the pair of NH$\cdots$O HBs is superseded by a 9.5 kcal mol^{-1} contribution from the CH$\cdots$O HBs.

In summary, whichever model is adopted, whether parallel or anti-parallel, whether dipeptide or tripeptide, and wherever the cut is made to excise the system for study, it can hardly be said that the interstrand binding is solely due to NH$\cdots$O HBs. The CH$\cdots$O HBs clearly make contributions which are comparable to, and perhaps even larger than, those of NH$\cdots$O HBs.

Later calculations verified some of the primary conclusions. For example, the importance of the C^αH$\cdots$O=C HB to the interaction energy in a β-sheet was supported by examinations[275,276] of β-sheet models of (Gly)$_n$ and (Ala)$_n$ which found CH$\cdots$O HBs to be almost as strong as NH$\cdots$O in the anti-parallel structure, but stronger in parallel. Additional confirmation came from Guo *et al.* in 2009.[277] In that same year, the C^αH$\cdots$O HB was computed to be stronger than NH$\cdots$O in dipeptide and tripeptide models of the parallel β-sheet geometry Ala.[278] In fact, CH$\cdots$O HBs have shown some propensity to stabilize the α-helix as well.[155,278,279] There is also experimental support for the importance of CH$\cdots$O HBs in β-sheets;[280] the β-pleated sheet structure for β-sulfidocarbonyls[281] contains not only CH$\cdots$O, but also CH$\cdots$S HBs.

11.6.4 Amino Acid Side Chains

Of course, the C^αH groups of the polypeptide skeleton are not the only ones capable of engaging in a CH$\cdots$O HB. The amino acid sidechains also contain CH protons, some of which are situated near electron withdrawing groups that ought to impart to them an added potency. Those situated near the positively charged termini of the Lys and Arg residues come immediately to mind, as do the C^βH hydrogens of Ser that are adjacent to a hydroxyl group. A particularly interesting class are the aromatic CH atoms of Phe, His, Tyr, and Trp whose sp^2 hybridization should enhance their potency, particularly if they lie near an electron-withdrawing atom.

The ability of these aromatic CH groups to engage in a CH$\cdots$O HB was assessed,[282] and placed in the context of other HBs with which these side chains might participate. In particular, the CH$\cdots$O HBs were compared to conventional NH$\cdots$O and OH$\cdots$N HBs, as well as OH$\cdots\pi$ interactions where the π system of the aromatic group serves as electron donor; water was used as the partner molecule.

As anticipated, the conventional HBs were the strongest, with binding energies varying between 4 and 7 kcal mol^{-1}. OH$\cdots\pi$ HBs were weaker, between 2 and 4 kcal mol^{-1}, and CH$\cdots$O slightly weaker still lying in the 1–2 kcal mol^{-1} range. The HB lengths correlated with these binding energies, with $R(H\cdots O)$ varying between 2.9 and 3.0 Å for conventional HBs, up to 3.3–3.4 Å for $CH\cdots O$. Consistent with the sp^2 hybridization of the proton donor C atom, both contractions and stretches were observed for the $CH\cdots O$ HBs, and stretching frequency changes were small, and both red and blue. As in the case of other HBs, the isotropic NMR signal of the bridging $CH\cdots O$ proton shifted downfield. The $OH\cdots\pi$ proton's NMR shift went in the opposite direction, due to the usual magnetic field currents of the aromatic ring above which it lies. The CH protons lying in closer proximity to electron-withdrawing atoms, such as the N atoms of His, displayed a somewhat greater propensity to engage in $CH\cdots O$ interactions.

Of particular interest was the effect of placing a charge on the aromatic proton donor. The protonated imidazole model of His formed a very strong HB, amounting to 10–11 kcal mol^{-1}. Consonant with this greater strength was a HB length that was reduced by 0.3 Å, and a strong blue shift of its CH stretching frequency, of 78–118 cm^{-1}. The NMR shift of this CH proton was also enhanced, by a factor of 2. Other work has supported these ideas. A $CH^+\cdots O$ HB of protonated imidazole was in part responsible[283] for the self-assembly of a triple helical structure.

An early calculation of the interaction involving a methylpyridinium cation with dimethyl ether[284] also found a strong $CH^+\cdots O$ HB with a binding energy of as much as 13 kcal mol^{-1}. Regarding other amino acid side chains, the $C^\delta H$ group of proline engages in $CH\cdots O$ HBs, within the context of real protein geometries.[285]

11.7 Effect of Charge

Indeed, placing a charge on either subunit has been known for some time[286,287] to amplify the binding in HBs, so it is natural to expect that the same ought to be true for $CH\cdots O$ HBs as well. And in fact, there was some evidence this might be true in a few cases.[288,289] This idea was probed systematically[290] in a series of systems that all employed the carbonyl O of *N*-methylacetamide (NMA) as the proton acceptor. Beginning with neutral proton donors $S(CH_3)_2$ and $N(CH_3)_3$ as a point of reference, both formed optimized complexes with NMA in which a CH of each methyl group present was engaged in a $CH\cdots O$ HB. The total binding energies amounted to 4.9 and 2.1 kcal mol^{-1}, respectively for the S and N systems. The situation changed dramatically when an extra methyl group was added to each, so that the new donors were the $S(CH_3)_3{}^+$ and $N(CH_3)_4{}^+$ cations. The various $CH\cdots O$

HBs all contracted by 0.2–0.4 Å, and the binding energies rose to 20.6 and 18.8 kcal mol^{-1}, for the S and N systems, respectively, an amplification by a factor of 4–9. (The optimized geometries for the amines are illustrated in Figure 11.5a and b.) These HBs are very strong, stronger than any neutral conventional HBs, and in fact on a par with ionic HBs of the OH^{+}···O or NH^{+}···O sort.

It is understood that displacing any HB from a gas-phase situation to a solvated environment will weaken it. And such was found[290] to be the case with all HBs studied, ionic as well as neutral. Indeed, the binding strengths of the ionic systems were more dramatically reduced with increasing dielectric constant ε of the surrounding polarizable continuum model of solvation. But nevertheless, the ionic systems remained more tightly bound than the neutral complexes, even for a high ε of 78 that simulates water.

As the methyl groups on these cations are lengthened to ethyl, propyl, *etc.*, the alkyl H atoms occur at various distances from the heteroatomic center of charge. For example, the methylene CH_2

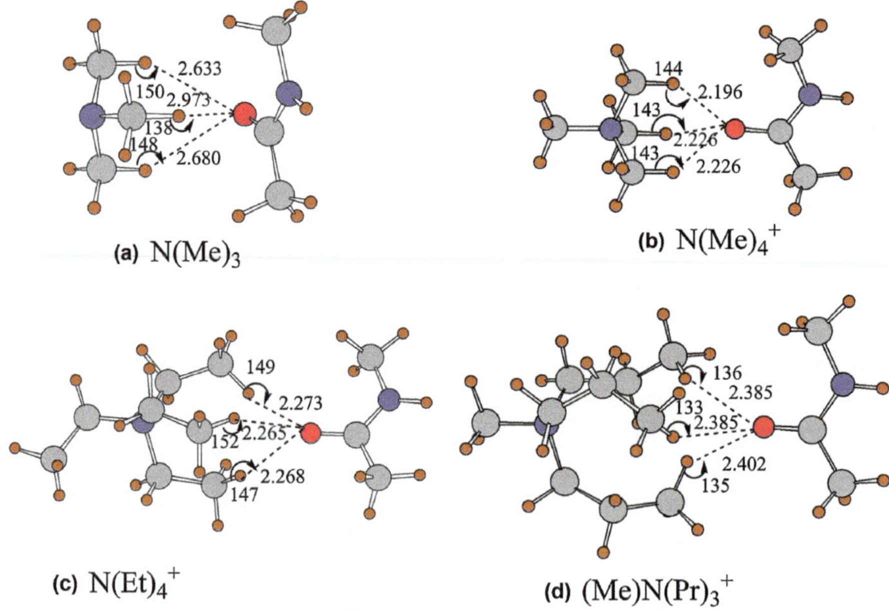

Figure 11.5   Optimized geometries of (a) N(Me)$_3$, and (b) N(Me)$_4^+$, (c) N(Et)$_4^+$ and (d) NMe(Pr)$_3^+$ complexes with NMA as H bond acceptor. Distances in Å and angles in degrees.
Reprinted with permission from U. Adhikari and S. Scheiner., Magnitude and Mechanism of Charge Enhancement of CH-O Hydrogen Bonds, *J. Phys. Chem. A* 2013, **117**, 10551. Copyright 2013 American Chemical Society.

hydrogens lie adjacent to N/S, while the terminal methyl protons are removed by one additional C–C linkage. This distinction was found[290] to make an important difference. If the CH$\cdots$O HBs involved the terminal methyl groups, then each chain lengthening, *i.e.* methyl $\rightarrow$ ethyl $\rightarrow$ propyl, very substantially reduced the binding energy with the NMA acceptor. Taking the N series as an example, the binding energy of 20.3 kcal mol^{-1} of $N(CH_3)_4{}^+$ was reduced to 14.1 kcal mol^{-1} for $N(Et)_4{}^+$, and then to 10.5 kcal mol^{-1} for $MeN(Pr)_3{}^+$. As displayed in Figure 11.5, this reduced binding energy is accompanied by elongations of the relevant $R(CH^+\cdots O)$ HBs. This decrease is much more gradual if instead of terminal methyl groups, the HBs are rather formed with the CH_2 protons lying adjacent to S/N. The elongation to ethyl and then to propyl results in binding energies of 18.2 and 17.5 kcal mol^{-1}, greatly diminished reductions. One might anticipate that the positive charge on CH protons drops as one moves along the alkyl chain further from the heteroatomic center of charge, which would help to explain this distinction. And indeed, careful scrutiny of the electrostatic potential confirmed this suspicion.

Since these are charged systems, it is tempting to presume that the two entities are bound together primarily by Coulombic attraction. And indeed a SAPT decomposition of the interaction energy reveals a strong electrostatic component, exceeding 20 kcal mol^{-1}. However, one cannot ignore other attractive forces: induction and dispersion together contribute between 13 and 16 kcal mol^{-1}. The induction signals its presence in a number of ways, that also confirm that there are indeed *bona fide* HBs present in these ionic complexes. First of all, there are large NBO values of $E(2)$ for the charge transfer from the proton-accepting O atom to the $\sigma^*(CH)$ antibonding orbitals of the donors, as much as 18 kcal mol^{-1}, characteristic of HBs. Second, maps of electron density redistribution that accompany complexation contain the characteristic trademarks of HBs: losses of density around the bridging proton and gains in the area of the lone pair of the proton acceptor atom. The identity of these interactions as true HBs is further confirmed by downfield shifts of the NMR signal of the bridging protons. These shifts are as large as 2 ppm for the ionic systems, more than twice the magnitude of the corresponding quantities in the neutral counterparts.

As indicated above, the optimal arrangements of these ionic systems contain a trifurcated HB wherein one H atom from each of three different methyl groups interacts directly with the proton acceptor O. The binding is weakened if this trifurcation involves three protons from the *same* methyl group, by as much as 35%. This reduction likely has a geometric cause in that there is a good deal of deviation of each $\theta(\text{CH}\cdots\text{O})$ angle from linearity when all H atoms come from the same methyl group. This sort of nonlinearity can be avoided if there is but a single proton involved in the $\text{CH}\cdots\text{O}$ bond. While this single linear HB is slightly superior to a trifurcated interaction with a single methyl group, it remains weaker than the optimal arrangement of three $\text{CH}\cdots\text{O}$ HBs arising from three separate methyls.

The addition of charge can also make the normally weak $\text{CH}\cdots\pi$ interaction into an HB of surprising strength. As one example, the addition of a fourth methyl group to trimethylamine, changing it into the tetramethylammonium ion magnifies[291] its $\text{CH}\cdots\pi$ binding energy with a proton acceptor π-system by a factor of 4–7. The interaction is strengthened when the π-system is conjugated, as in 1,3-butadiene, and takes another step for aromatic systems. This interaction can be quite strong, as much as 15.5 kcal mol^{-1} in the case of indole. Contrary to a common presumption that Coulombic forces ought to dominate, it is dispersion that acts as the prime contributor. Even in these very strongly bound systems, the CH bond length tends to shorten upon formation of the complex. Many of these conclusions are consistent with other work in the literature.[292–296]

Not only can a cation like $\text{S(CH}_3)_3{}^+$ engage in $\text{CH}\cdots\text{O}$ HBs, but there is also the possibility of other sorts of interactions, with which they can be compared. Figure 11.6a illustrates a standard $\text{CH}^+\cdots\text{O}$ HB[297] between $\text{S(CH}_3)_3{}^+$ and N-methylacetamide (NMA); the binding energy of this structure is 16.3 kcal mol^{-1}. The presence of the three methyl groups also presents the possibility of a trifurcated HB wherein three separate methyl groups donate a proton to the same O atom. This geometry, shown in Figure 11.6b, is more stable by some 4 kcal mol^{-1}. Another possibility has the geometrical appearance of a trifurcated HB, with all three protons connected to the same C, as in Figure 11.6c. But analysis reveals the true interaction does not involve these H atoms at all, but rather the C atom. This so-called tetrel bond involves transfer of charge from the O atom to the $\sigma^*(\text{CS})$ antibonding orbital. It is somewhat weaker than the two sorts of $\text{CH}\cdots\text{O}$ HB. The charge can be transferred to the opposite end of the C–S bond, but still into the $\sigma^*(\text{CS})$ antibonding orbital, in Figure 11.6d which is commonly denoted a chalcogen bond. In this case, this chalcogen bond is quite strong, comparable to the trifurcated $\text{CH}\cdots\text{O}$ arrangement. This chalcogen bond is further strengthened when the methyl group opposite the approaching O atom is replaced by a strongly electron-withdrawing substituent such as F. Also quite strong is the $\text{SH}^+\cdots\text{O}$ HB that can arise[297] for a cation such as $\text{CH}_3\text{SH}_2{}^+$ or

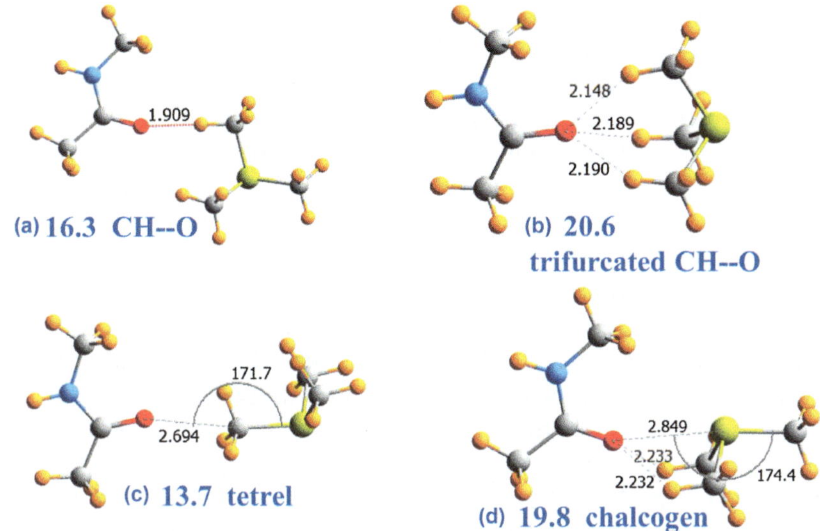

Figure 11.6 Optimized geometries of complexes of NMA with $(CH_3)_3S^+$. Counterpoise-corrected binding energies (kcal per mol) in blue, distances in Å, angles in deg.
Reprinted with permission from S. Scheiner., Comparison of CH-O, SH-O, Chalcogen, and Tetrel Bonds Formed by Neutral and Cationic Sulfur-Containing Compounds, *J. Phys. Chem. A* 2015, **119**, 9189. Copyright 2015 American Chemical Society.

CH_3SHF^+. Calculations of the analogous neutral pairs[297] revealed similar arrangements but much weaker binding.

Just as the HB can be strengthened by a positive charge on the donor, the same can be achieved by an anionic proton acceptor. Using CF_3H as a common CH proton donor, the strength of $CH\cdots X^-$ HBs was assessed[298] for a range of anions. The binding energies are sensitive to the magnitude of the charge on the anion, but are less sensitive to the nature of the anion. The various monoanions are bound to CF_3H by some 12–17 kcal mol^{-1}, with the exception of the much stronger HB with F^- which amounts to 26 kcal mol^{-1}. The binding is inversely related to the number of O atoms on which the charge can be dispersed. That is, formate and acetate with two O atoms, are most strongly bound, and the larger HSO_4^- and $H_2PO_4^-$ with four O atoms engage in a weaker HB; with three O atoms, NO_3^- forms a HB of intermediate strength. The dianions are bound much more strongly with $E_b \sim 27$ kcal mol^{-1}, and the PO_4^{3-} trianion stronger still, at 45 kcal mol^{-1}. When there are two O anionic atoms available, the $CH\cdots O^-$ HB is usually bifurcated, although one $CH\cdots O$ distance is typically shorter, and presumably stronger, than the other. This asymmetry is particularly pronounced in the phosphate series

$(H_2PO_4^-, HPO_4^{2-}$ and $PO_4^{3-})$, where the bonding pattern may perhaps better be described as non-bifurcated.

The CH bond is elongated and its stretching frequency redshifted[298] in these ionic HBs. These quantities are as large as 64 mÅ, and 1000 cm^{-1}, respectively, and are closely related to the binding energy. The degree of this shift is reduced in the bifurcated HBs where neither CH$\cdots$O configuration is close to linear. This diminution is caused by the reduction in overlap between the electron donor lone pair and the acceptor $\sigma^*(CH)$ antibonding orbital, which in turn decreases the amount of charge transferred into the latter orbital. The formation of the HB also reduces the chemical shielding around the bridging proton, by an amount between 2 and 12 ppm, and this downshift of the NMR signal is also correlated with the strength of the HB. These HBs are composed largely of a Coulombic force, with some 52–65% of the binding energy attributed to the electrostatic component. Another 20–38% is due to induction and a smaller residual to dispersion. The increase of charge (mono to di to trianion) enhances all components, especially induction. The results are reinforced by earlier work in the literature[211,299–304] that treated comparable sorts of systems, albeit with cruder levels of theory. While the binding energies of the CH$\cdots$X$^-$ HBs were comparable to their CH$^+\cdots$X analogues,[290] they differ primarily in that the former show a larger downfield shift of the bridging proton's NMR signal.

Of course, if placing a charge on one subunit can increase the binding energy, one would expect an even greater enhancement if both subunits contain opposite charges. Indeed, there is ample evidence that such cation-anion charge pairs can form strong HBs. For example D'Oria and Novoa[305] found that the interaction energies of OH$\cdots$O HBs could approach 100 kcal mol^{-1} for monocation-monoanion pairs, and were considerably greater, even exceeding 200 kcal mol^{-1} for dication-dianion pairs. The strong attractions between ions of opposite charge seem sufficiently powerful to overwhelm the repulsions of like-charge ions, at least in certain cases. Prior calculations[306] document the overall stability of clusters containing two cations and two anions, where the former consist of tetraphenyl phosphine species.

In the specific case of CH donors, the placement of a halide anion in a position where it can form a pair of ionic CH$^+\cdots$X$^-$ HBs with a bipodal receptor,[307] clearly illustrates the binding strength of systems with oppositely charged ions. As illustrated in Table 11.1, as the receptor's charge varies from 0 to +1 to +2, the interaction energy increases. However, in this case, the increase is not linearly

Table 11.1 Interaction energy (kcal per mol) of CH-bonding bistriazole-pyridine receptor with halide anions.[307]

Charge on receptor	F^-	Cl^-	Br^-	I^-
0	9.70	5.44	4.77	4.10
+1	13.18	7.35	6.42	5.47
+2	11.12	8.57	7.77	6.74

proportional to the charge, due in part to the large size of the cationic receptor which can spread the positive charge over a large group of atoms.

11.8 Geometrical Distortions

The geometrical preferences of any particular intermolecular interaction represent only one of many within the context of a crystal. The final structure is hence a compromise between all of these many factors. As a result, HBs within crystals are typically stretched and bent relative to their intrinsic optimum geometries. It is therefore important to understand how such distortions influence the energetics and other properties of these HBs. Quantum calculations are well suited to such a task, as any particular geometry can be considered, equilibrium or any distortions therefrom.

11.8.1 Stretches

As one example, a series of different interactions, not only HBs but also related halogen, chalcogen, and pnicogen bonds were examined.[308] Figure 11.7 illustrates how the energy of the system rises as each such bond is stretched, ΔR representing the stretch from the equilibrium structure. The HBs, in red, show first that HF is a stronger proton donor than is HOH, so forms a stronger HB, 12 *vs.* 6 kcal mol^{-1}. As a stronger HB, its energy climbs more quickly than does the $HOH\cdots NH_3$ system. In fact, the shapes of all the curves in Figure 11.7 are quite similar, HB or otherwise. Looking at this situation from a proportional standpoint, a stretch of any system by 1 Å reduces the binding energy by roughly half for each system. But in any case, there is no sharp cutoff as to when the HB ceases to be attractive.

What might be considered an appropriate cutoff or threshold? That is, after how much of a stretch can one say that an H or other non-covalent bond is "broken"? This is a question that has lain at the heart of many discussions of H-bonds.[169] The answer clearly depends

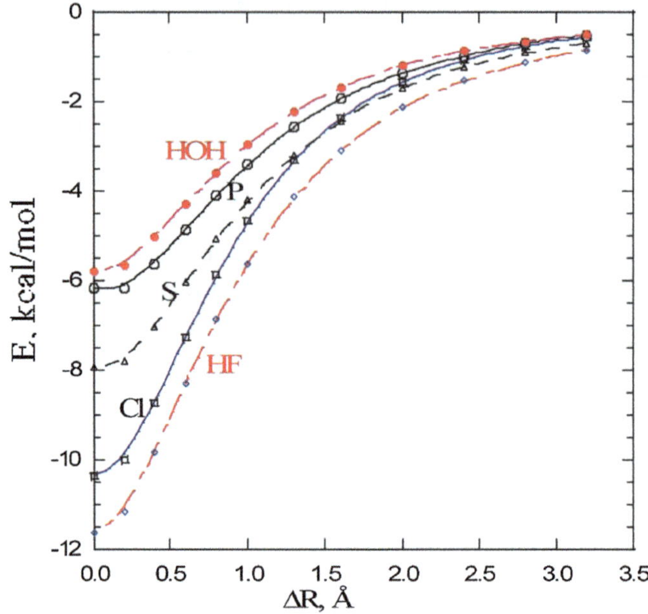

Figure 11.7 Weakening of binding energy of various complexes, all involving NH_3 as electron donor, as intermolecular distance is stretched. HF and HOH form H-bonds with NH_3. P label indicates $H_2FP\cdots NH_3$ complex, S refers to $HFS\cdots NH_3$, and Cl to $FCl\cdots NH_3$. The zero of energy in each case refers to fully separated monomers. Optimized equilibrium distance in each complex is taken as $\Delta R = 0$.
Reprinted with permission from Scheiner, S. *CrystEngComm*, 2013, **15**, 3119 with permission from the Royal Society of Chemistry.

upon one's definition of a bond. A criterion of an attractive force, even if a small one, is probably unsatisfactory as the attraction persists to even very long distances. A threshold of a particular bond energy would lead to the idea that the stronger interactions may be stretched a greater amount. Taking 4 $kcal\,mol^{-1}$ as a sample cutoff, the $FH\cdots NH_3$ H-bond may be stretched by 1.3 Å, while the allowed stretch in $FH_2P\cdots NH_3$ is only 0.8 Å. Still another view might envision the stretch required to reduce the binding energy to half its maximal value, in which case all the systems can sustain a stretch of some 1.0 ± 0.1 Å.

It might be instructive to consider how particular choices of a bond distance cutoff correlate with binding energy. In the strongly H-bonded $FH\cdots NH_3$ complex, the equilibrium $R(H\cdots N)$ distance is 1.69 Å. Adding in the $r(FH)$ bond length of 0.96 Å, the $R(F\cdots N)$ distance is 2.65 Å. If one were to assert a $R(F\cdots N)$ cutoff of 3.02 Å, the sum of $r_F + r_N$ van der Waals radii,[309] the stretch of 0.37 Å would leave a residual H-bond energy of 10 $kcal\,mol^{-1}$. Even taking the longer cutoff of 3.2 Å would correspond to a value of 8.5 kcal mol^{-1}. An R(H$\cdots$N) distance of 2.75 Å, corresponding to the sum of $r_H + r_N$ van der Waals radii, and representing a stretch of 1.06 Å, is still associated with a rather strong interaction, of some 5.5 kcal mol^{-1}. One of the longer H-bond threshold values proposed in the literature sets the R(F$\cdots$N) distance at 4 Å,[310] which would reduce the interaction energy to 1.0 kcal mol^{-1}.

Is there a particular function to which the die-off of HB energies can be fitted with respect to distance? *Ab initio* calculations[311] considered a wide range of HB interactions, both neutral and ionic, as well as pnicogen, chalcogen, and halogen-bonded dimers. It was found that the drop in interaction energy with bond stretch ΔR can be fitted to a common power n, in the functional form ΔR^{-n}. This exponent is smaller for charged H-bonds, as compared to neutral systems, where n varies in the order pnicogen < chalcogen < halogen bond. The decay is slowest for the electrostatic term, followed by induction and then by dispersion. The values of the exponent n are smaller for electrostatic energy than would be expected if it arose purely as a result of classical multipole interactions, such as dipole–dipole for the neutral systems. The exponents are larger when the fitting is done with respect to intermolecular distance R, rather than to its stretch relative to equilibrium length, although still not precisely matching what might be expected on classical grounds.

11.8.2 Bends

It is generally understood that H-bonds prefer to maintain the bridging proton directly along the line between donor and acceptor atom. But of course, since this is not always possible within the context of crystals, it is necessary to have some information about the energetic penalty of angular distortions. Again, FH and HOH were taken[312] as proton donors to NH_3 as acceptor, as well as FPH_2, FSH, and FCl as alternate electron acceptors. The rise in energy as each intermolecular interaction is bent away from linearity is illustrated in Figure 11.8. Just as for the stretching, the stronger FH$\cdots NH_3$ HB rises in energy more quickly than HOH$\cdots NH_3$ as the HB is bent. But the energetics of bending each interaction can be fitted by a simple parabolic expression $E = 1/2\ k(\Delta\theta)^2$ where $\Delta\theta$ represents the distortion from equilibrium, and k a bending force constant.

It is generally considered that a stronger interaction will likewise exhibit a larger bending force constant. For that reason, the ratio between k and ΔE was considered.[312] The four H-bonding systems all

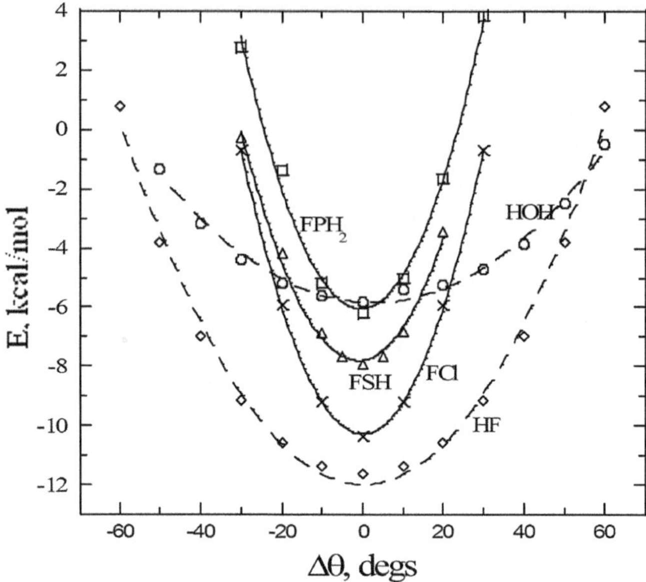

Figure 11.8 Rise in energy that accompanies angular distortion in the complex of each indicated electron acceptor with NH_3. Curves represent parabola that are fitted to the data points shown. H-bonding systems denoted by broken curves.
Reprinted from *Chem. Phys. Lett.* 532, 31, U. Adhikari, S. Scheiner., Sensitivity of pnicogen, chalcogen, halogen and H-bonds to angular distortions, 31–35, Copyright 2012 with permission from Elsevier.

have roughly equivalent $k/\Delta E$ ratios between 1.6 and 2.8 rad^{-2}. But the pnicogen, halogen, and chalcogen bonds are considerably more sensitive to angular distortions than are any of the H-bonded systems. This conclusion is based not only on k itself, but also when the ratio between k and the optimized binding energy is considered. To what can the weaker sensitivity of HBs to angular distortion be attributed? It might be tempting to attribute angular deformation effects to Coulombic forces, as is commonly assumed in the literature. However, a careful decomposition of the binding energy leads instead to the conclusion that it is steric repulsion that is responsible. The electron density of each proton-donating molecule has a sort of oblong egg shape, with the narrow end along the direction of the H-bond. Consequently, as a partner molecule is rotated away from the linearity, the density diminishes significantly, which would mitigate any rise in the steric repulsion. So the lesser sensitivity of H-bonds to angular distortion, in comparison to other related interactions, is attributed to the narrower shape of the electron density of the proton donor along the H-bond axis direction.

How are these trends affected when the HB contains a charged entity? A later work[313] expanded the field of systems to cationic proton donors, of both CH^+ and NH^+ varieties. In summary, all sorts of HBs, whether CH or NH donor, cationic, anionic, or neutral, suffer a loss of binding energy as angular deformations are introduced into the geometry of the HB. This loss of energy is very roughly proportional to the intrinsic strength of the HB, as measured by binding energy E_b. This quantity is in turn much larger for charged than for neutral HBs, and NH acts as a stronger donor than does CH. However, these distinctions are nearly washed out when the HB energy sensitivity to distortion is normalized by dividing the bending force constant by the optimal binding strength. In general, the k/E_b ratios are slightly larger for NH donors than for CH, but only slightly so. Regarding the effect of charge, the ionic $F_3CH\cdots$anion systems are characterized by a slightly larger k/E_b ratio than their neutral counterparts. On the other hand, the removal of the positive charge induces a small increase in this ratio for both CH and NH ends of the $HNCH^+$ donors.

Energy decomposition reveals that the attractive components of HBs, *i.e.* electrostatic, induction, and dispersion energies, all become more stabilizing upon angular distortion of the HB. Outweighing this effect, though, is the even larger destabilization caused by increasing exchange repulsion, which is thus identified as the major contributor to the loss of HB energy induced by bending. This phenomenon appears to be a general one, common not only to the ionic and neutral $NH\cdots D$ and $CH\cdots D$ HBs considered here, but also to the related halogen, chalcogen, and pnicogen bonds examined earlier. This predominance of exchange repulsion has been verified[314–317] by other calculations, at least for neutral dimers.

11.9 Summary

This chapter has hopefully illuminated a number of issues for the reader. Both experimental and theoretical work over the years has greatly expanded the range of what is deemed an H-bond. Proton donor and acceptor atoms extend far beyond the original very limited set of O, F, and N, and the charge being transferred to the donor can arise from sources other than lone pairs, *e.g.* π-systems. Theoretical calculations offer a number of useful tools to analyze these interactions, supplementing the information that is available from experiment.

The expanded discussion of one sort of unconventional H-bond, involving CH as a donor group, has offered a detailed illustration of just how theoretical methods can contribute to our understanding. The work described has included explanations of the applications of the various theoretical tools. More than that, it is hoped that the reader is now familiar with the way in which computational chemists look at H-bonds, the language that is used, and the principal concepts. Of course, this sort of analysis is not limited to H-bonds, but is commonly employed in increasing our understanding of all sorts of noncovalent interactions.

References

1. S. F. Boys and F. Bernardi, *Mol. Phys.*, 1970, **19**, 553.
2. K. Morokuma, *J. Chem. Phys.*, 1971, **55**, 1236.
3. K. Kitaura and K. Morokuma, *Int. J. Quantum Chem.*, 1976, **10**, 325.
4. K. Morokuma, *Acc. Chem. Res.*, 1977, **10**, 294.
5. W. J. Stevens and W. H. Fink, *Chem. Phys. Lett.*, 1987, **139**, 15.
6. W. Chen and M. S. Gordon, *J. Phys. Chem.*, 1996, **100**, 14316.
7. A. van der Vaar and K. M. Merz, *J. Phys. Chem. A*, 1999, **103**, 3321.
8. P. Salvador, M. Duran and I. Mayer, *J. Chem. Phys.*, 2001, **115**, 1153.
9. D. G. Fedorov and K. Kitaura, *J. Comput. Chem.*, 2007, **28**, 222.
10. R. Z. Khaliullin, A. T. Bell and M. Head-Gordon, *J. Chem. Phys.*, 2008, **128**, 184112.
11. K. Szalewicz, B. Jeziorski and S. Rybak, *Int. J. Quantum Chem.*, 1991, **QBS18**, 23.
12. B. Jeziorski, R. Moszynski and K. Szalewicz, *Chem. Rev.*, 1994, **94**, 1887.
13. K. Szalewicz and B. Jeziorski, Symmetry-adapted perturbation theory of inter-molecular interactions, in *Molecular Interactions. From Van der Waals to Strongly Bound Complexes*, ed. S. Scheiner, Wiley, New York, 1997, p. 3.
14. R. M. Parrish and C. D. Sherrill, *J. Chem. Phys.*, 2014, **141**, 044115.
15. P. Su and H. Li, *J. Chem. Phys.*, 2009, **131**, 014102.
16. E. D. Glendening and A. Streitwieser, *J. Chem. Phys.*, 1994, **100**, 2900.
17. E. D. Glendening, *J. Am. Chem. Soc.*, 1996, **118**, 2473.
18. G. K. Schenter and E. D. Glendening, *J. Phys. Chem.*, 1996, **100**, 17152.
19. E. D. Glendening, *J. Phys. Chem. A*, 2005, **109**, 11936.
20. T. Ziegler and A. Rauk, *Theor. Chim. Acta*, 1977, **46**, 1.
21. P. S. Bagus, K. Hermann and C. W. J. Bauschlicher, *J. Chem. Phys.*, 1984, **80**, 4378.
22. A. van der Vaart and K. M. Merz, *J. Phys. Chem. A*, 1999, **103**, 3321.
23. Y. Mo, J. Gao and S. D. Peyerimhoff, *J. Chem. Phys.*, 2000, **112**, 5530.
24. R. Z. Khaliullin, M. Head-Gordon and A. T. Bell, *J. Chem. Phys.*, 2006, **124**, 204105.
25. M. Mitoraj and A. Michalak, *J. Mol. Model.*, 2007, **13**, 347.
26. P. Reinhardt, J.-P. Piquemal and A. Savin, *J. Chem. Theory Comput.*, 2008, **4**, 2020.
27. Q. Wu, P. W. Ayers and Y. Zhang, *J. Chem. Phys.*, 2009, **131**, 164112.
28. P. R. Horn, E. J. Sundstrom, T. A. Baker and M. Head-Gordon, *J. Chem. Phys.*, 2013, **138**, 134119.
29. R. F. W. Bader, *Atoms in Molecules, A Quantum Theory*, Clarendon Press, Oxford, 1990, vol. 22, p. 438.

30. R. F. W. Bader, J. R. Cheeseman, K. E. Laidig, K. B. Wiberg and C. Breneman, *J. Am. Chem. Soc.*, 1990, **112**, 6530.
31. P. L. A. Popelier, *Atoms in Molecules. An Introduction*, Prentice Hall, Harlow, UK, 2000.
32. P. L. A. Popelier and R. F. W. Bader, *Chem. Phys. Lett.*, 1992, **189**, 542.
33. M. Domagala and S. J. Grabowski, *Chem. Phys.*, 2009, **363**, 42.
34. S. J. Grabowski and J. M. Ugalde, *J. Phys. Chem. A*, 2010, **114**, 7223.
35. S. J. Grabowski, *J. Phys. Org. Chem.*, 2013, **26**, 452.
36. E. Arunan, G. R. Desiraju, R. A. Klein, J. Sadlej, S. Scheiner, I. Alkorta, D. C. Clary, R. H. Crabtree, J. J. Dannenberg, P. Hobza, H. G. Kjaergaard, A. C. Legon, B. Mennucci and D. J. Nesbitt, *Pure Appl. Chem.*, 2011, **83**, 1637.
37. C. A. Morgado, J. P. McNamara, I. H. Hillier, N. A. Burton and M. A Vincent, *J. Chem. Theory Comput.*, 2007, **3**, 1656.
38. T. Kobayashi and M. Hirota, *Chem. Lett.*, 1972, 975.
39. G. Duan, V. H. Smith and D. F. Weaver, *Mol. Phys.*, 2001, **99**, 1689.
40. Z. Latajka and S. Scheiner, *Chem. Phys.*, 1997, **216**, 37.
41. F. Wennmohs, V. Staemmler and M. Schindler, *J. Chem. Phys.*, 2003, **119**, 3208.
42. J. M. Hermida-Ramón, E. M. Cabaleiro-Lago and J. Rodríguez-Otero, *J. Chem. Phys.*, 2005, **122**, 204315.
43. E. M. Cabaleiro-Lago, J. Rodrguez-Otero and A. Pena-Gallego, *J. Phys. Chem. A*, 2008, **112**, 6344.
44. H. S. Biswal and S. Wategaonkar, *J. Phys. Chem. A*, 2009, **113**, 12763.
45. H. S. Biswal and S. Wategaonkar, *J. Phys. Chem. A*, 2009, **113**, 12774.
46. M. Solimannejad and S. Scheiner, *Int. J. Quantum Chem.*, 2011, **111**, 3196.
47. M. Solimannejad, M. Gharabaghi and S. Scheiner, *J. Chem. Phys.*, 2011, **134**, 024312.
48. S. Scheiner, *J. Chem. Phys.*, 2011, **134**, 094315.
49. B. J. Mintz and J. M. Parks, *J. Phys. Chem. A*, 2012, **116**, 1086.
50. K. Grzechnik, K. Rutkowski and Z. Mielke, *J. Mol. Struct.*, 2012, **1009**, 96.
51. J. Nadas, S. Vukovic and B. P. Hay, *Comput. Theor. Chem.*, 2012, **988**, 75.
52. A. Bhattacherjee, Y. Matsuda, A. Fujii and S. Wategaonkar, *ChemPhysChem.*, 2013, **14**, 905.
53. S. Bhattacharyya, A. Bhattacherjee, P. R. Shirhatti and S. Wategaonkar, *J. Phys. Chem. A*, 2013, **117**, 8238.
54. V. S. Minkov and E. V. Boldyreva, *J. Phys. Chem. B*, 2013, **117**, 14247.
55. C. L. Andersen, C. S. Jensen, K. Mackeprang, L. Du, S. Jørgensen and H. G. Kjaergaard, *J. Phys. Chem. A*, 2014, **118**, 11074.
56. A. S. Hansen, L. Du and H. G. Kjaergaard, *J. Phys. Chem. Lett.*, 2014, **5**, 4225.
57. R. B. Viana and A. B. F. da Silva, *Comput. Theor. Chem.*, 2015, **1059**, 35.
58. V. R. Mundlapati, S. Ghosh, A. Bhattacherjee, P. Tiwari and H. S. Biswal, *J. Phys. Chem. Lett.*, 2015, **6**, 1385.
59. I. Alkorta and J. Elguero, *J. Phys. Chem.*, 1996, **100**, 19367.
60. X. An, H. Liu, Q. Li, B. Gong and J. Cheng, *J. Phys. Chem. A*, 2008, **112**, 5258.
61. E. Ventura, S. A. D. Monte, W. Fragoso, C. F. Braga and R. C. M. U. Araújo, *Int. J. Quantum Chem.*, 2006, **106**, 1009.
62. S. Scheiner and S. J. Grabowski, *J. Mol. Struct.*, 2002, **615**, 209.
63. K. P. Gierszal, J. G. Davis, M. D. Hands, D. S. Wilcox, L. V. Slipchenko and D. Ben-Amotz, *J. Phys. Chem. Lett.*, 2011, **2**, 2930.
64. S. Kumar, V. Pande and A. Das, *J. Phys. Chem. A*, 2012, **116**, 1368.
65. J. C. Amicangelo, D. G. Irwin, C. J. Lee, N. C. Romano and N. L. Saxton, *J. Phys. Chem. A*, 2013, **117**, 1336.
66. O. Takahashi, CH...π Interaction in Organic Molecules, in *Noncovalent Forces*, ed. S. Scheiner, Springer, Dordrecht, 2015, vol. 19, p. 47.

67. G. Aragay, D. Hernández, B. Verdejo, E. Escudero-Adán, M. Martínez and P. Ballester, *Molecules*, 2015, **20**, 16672.
68. I. Alkorta, J. Elguero and J. E. D. Bene, *Chem. Phys. Lett.*, 2010, **489**, 159.
69. G. Orlova and S. Scheiner, *J. Phys. Chem. A*, 1998, **102**, 4813.
70. G. Orlova, S. Scheiner and T. Kar, *J. Phys. Chem. A*, 1999, **103**, 514.
71. N. V. Belkova, E. S. Shubina and L. M. Epstein, *Acc. Chem. Res.*, 2005, **38**, 624.
72. T. Kar and S. Scheiner, *J. Chem. Phys.*, 2003, **119**, 1473.
73. M. Solimannejad and S. Scheiner, *J. Phys. Chem. A*, 2005, **109**, 11933.
74. P. C. Singh and G. N. Patwari, *Chem. Phys. Lett.*, 2006, **419**, 265.
75. H. Cybulski, E. Tymiska and J. Sadlej, *ChemPhysChem.*, 2006, **7**, 629.
76. M. Jablonski, *Chem. Phys. Lett.*, 2009, **477**, 374.
77. B. G. d. Oliveira and M. N. Ramos, *Int. J. Quantum Chem.*, 2009, **110**, 307.
78. D. J. Wolstenholme, J. Flogeras, F. N. Che, A. Decken and G. S. McGrady, *J. Am. Chem. Soc.*, 2013, **135**, 2439.
79. S. J. Grabowski, What is Common for Dihydrogen Bond and H...σ Interaction—Theoretical Analysis and Experimental Evidences, in *Noncovalent Forces*, ed. S. Scheiner, Springer, Dordrecht, 2015, vol. 19, p. 159.
80. H. Schmidbaur, H. G. Raubenheimer and L. Dobrzanska, *Chem. Soc. Rev.*, 2014, **43**, 345.
81. H. Nuss and M. Jansen, *Angew. Chem., Int. Ed.*, 2006, **45**, 4369.
82. E. S. Kryachko, A. Karpfen and F. Remacle, *J. Phys. Chem. A*, 2005, **109**, 7309.
83. E. S. Kryachko and F. Remacle, *Chem. Phys. Lett.*, 2005, **404**, 142.
84. L. Brammer, *Dalton Trans.*, 2003, 3145.
85. G. Orlova and S. Scheiner, *Organometallics*, 1998, **17**, 4362.
86. T. S. Thakur and G. R. Desiraju, *J. Mol. Struct.: THEOCHEM*, 2007, **810**, 143.
87. S. Rizzato, J. Bergès, S. A. Mason, A. Albinati and J. Kozelka, *Angew. Chem., Int. Ed.*, 2010, **49**, 7440.
88. L. R. Falvello, *Angew. Chem., Int. Ed.*, 2010, **49**, 10045.
89. R. Sanchez-de-Armas and M. S. G. Ahlquist, *Phys. Chem. Chem. Phys.*, 2015, **17**, 812.
90. G. Zhang, H. Yue, F. Weinhold, H. Wang, H. Li and D. Chen, *ChemPhysChem.*, 2015, **16**, 2424.
91. M. Gao, J. Cheng, X. Yang, W. Li, B. Xiao and Q. Li, *J. Chem. Phys.*, 2015, **143**, 054308.
92. I. Alkorta, I. Rozas and J. Elguero, *Ber. Bunsenges. Phys. Chem.*, 1998, **102**, 429.
93. A. Gil, M. Sodupe and J. Bertran, *Chem. Phys. Lett.*, 2004, **395**, 27.
94. M. Solimannejad, S. G. Shirazi and S. Scheiner, *J. Phys. Chem. A*, 2007, **111**, 10717.
95. M. Solimannejad, C. J. Neilsen and S. Scheiner, *Chem. Phys. Lett.*, 2008, **466**, 136.
96. M. Solimannejad and S. Scheiner, *Chem. Phys. Lett.*, 2006, **429**, 38.
97. S. Hammerum, *J. Am. Chem. Soc.*, 2009, **131**, 8627.
98. Q. Li, H. Wang, Z. Liu, W. Li, J. Cheng, B. Gong and J. Sun, *J. Phys. Chem. A*, 2009, **113**, 14156.
99. M. Solimannejad and S. Scheiner, *Mol. Phys.*, 2009, **107**, 713.
100. Q.-Z. Li and H.-B. Li, Hydrogen Bonds Involving Radical Species, in *Noncovalent Forces*, ed. S. Scheiner, Springer, Dordrecht, 2015, vol. 19, p. 107.
101. W. D. Kumler, *J. Am. Chem. Soc.*, 1935, **57**, 600.
102. S. Glasstone, *Trans. Faraday Soc.*, 1937, **33**, 200.
103. J. F. J. Dippy, *Chem. Rev.*, 1939, **25**, 151.
104. C. S. Marvel, M. J. Copley and E. Ginsberg, *J. Am. Chem. Soc.*, 1940, **62**, 3109.
105. M. L. Huggins, *Chem. Rev.*, 1943, **32**, 195.
106. W. G. Schneider and H. J. Bernstein, *Trans. Faraday Soc.*, 1956, **52**, 13.
107. W. F. Forbes, *Can. J. Chem.*, 1962, **40**, 1891.
108. G. M. Kosolapoff and J. F. McCullough, *J. Am. Chem. Soc.*, 1951, **73**, 5392.

109. M. W. Dougill and G. A. Jeffrey, *Acta Cryst.*, 1953, **6**, 831.
110. G. S. Parry, *Acta Cryst.*, 1954, **7**, 313.
111. C. M. Huggins, G. C. Pimentel and J. N. Shoolery, *J. Chem. Phys.*, 1955, **23**, 1244.
112. S. Pinchas, *Anal. Chem.*, 1955, **27**, 2.
113. S. Pinchas, *Anal. Chem.*, 1957, **29**, 334.
114. I. Brown and F. Smith, *Austr. J. Chem.*, 1960, **13**, 30.
115. D. J. Sutor, *Nature*, 1962, **195**, 68.
116. D. J. Sutor, *J. Chem. Soc.*, 1963, 1105.
117. A. Allerhand and P. v. R. Schleyer, *J. Am. Chem. Soc.*, 1963, **85**, 1715.
118. S. Pinchas, *J. Phys. Chem.*, 1963, **67**, 1862.
119. G. Ferguson and J. Tyrrell, *Chem. Commun.*, 1965, 195.
120. A. L. Bednowitz and B. Post, *Acta Cryst.*, 1966, **21**, 566.
121. S. Krimm, *Science*, 1967, **158**, 530.
122. S. Krimm and K. Kuroiwa, *Biopolymers*, 1968, **6**, 401.
123. G. N. Ramachandran and R. Chandrasekharan, *Biopolymers*, 1968, **6**, 1649.
124. J. Sandoval-Lira, L. Fuentes, L. Quintero, H. Höpfl, J. M. Hernández-Pérez, J. L. Terán and F. Sartillo-Piscil, *J. Org. Chem.*, 2015, **80**, 4481.
125. A. Khrizman, H. Y. Cheng, G. Bottini and G. Moyna, *Chem. Commun.*, 2015, **51**, 3193.
126. H. O. Leung and M. D. Marshall, *J. Phys. Chem. A*, 2014, **118**, 9783.
127. R. Joseph, A. Nkrumah, R. J. Clark and E. Masson, *J. Am. Chem. Soc.*, 2014, **136**, 6602.
128. P. Écija, M. Vallejo-López, L. Evangelisti, J. A. Fernández, A. Lesarri, W. Caminati and E. J. Cocinero, *ChemPhysChem.*, 2014, **15**, 918.
129. C. Rest, M. J. Mayoral, K. Fucke, J. Schellheimer, V. Stepanenko and G. Fernández, *Angew. Chem., Int. Ed.*, 2014, **53**, 700.
130. M. Zierke, M. Smieško, S. Rabbani, T. Aeschbacher, B. Cutting, F. H.-T. Allain, M. Schubert and B. Ernst, *J. Am. Chem. Soc.*, 2013, **135**, 13464.
131. Y. Yoneda, K. Mereiter, C. Jaeger, L. Brecker, P. Kosma, T. Rosenau and A. French, *J. Am. Chem. Soc.*, 2008, **130**, 16678.
132. M. C. Wahl and M. Sundaralingam, *Trends Biochem. Sci.*, 1997, **22**, 97.
133. O. Khakshoor, S. E. Wheeler, K. N. Houk and E. T. Kool, *J. Am. Chem. Soc.*, 2012, **134**, 3154.
134. M. Y. Anzahaee, J. K. Watts, N. R. Alla, A. W. Nicholson and M. J. Damha, *J. Am. Chem. Soc.*, 2011, **133**, 728.
135. A.-C. Uldry, J. M. Griffin, J. R. Yates, M. Perez-Torralba, M. D. S. Maria, A. L. Webber, M. L. L. Beaumont, A. Samoson, R. M. Claramunt, C. J. Pickard and S. P. Brown, *J. Am. Chem. Soc.*, 2008, **130**, 945.
136. E. Freisinger, I. B. Rother, M. S. Lüth and B. Lippert, *Proc. Natl. Acad. Sci. U. S. A.*, 2003, **100**, 3748.
137. J. Parsch and J. W. Engels, *J. Am. Chem. Soc.*, 2002, **124**, 5664.
138. M. Brandl, K. Lindauer, M. Meyer and J. Sühnel, *Theor. Chem. Acc.*, 1999, **101**, 103.
139. C. L. Kielkopf, S. White, J. W. Szewczyk, J. M. Turner, E. E. Baird, P. B. Dervan and D. C. Rees, *Science*, 1998, **282**, 111.
140. R. Biswas and M. Sundaralingam, *J. Mol. Biol.*, 1997, **270**, 511.
141. P. Auffinger and E. Westhof, *J. Mol. Biol.*, 1997, **274**, 54.
142. S. Metzger and B. Lippert, *J. Am. Chem. Soc.*, 1996, **118**, 12467.
143. M. C. Wahl, S. T. Rao and M. Sundaralingam, *Nat. Struct. Biol.*, 1996, **3**, 24.
144. I. Berger, M. Egli and A. Rich, *Proc. Natl. Acad. Sci. U. S. A.*, 1996, **93**, 12116.
145. M. Egli and R. V. Gessner, *Proc. Natl. Acad. Sci. U. S. A.*, 1995, **92**, 180.
146. P. Venugopalan and R. Kishore, *Chem. – Eur. J.*, 2013, **19**, 9908.
147. H. Yang and M. W. Wong, *J. Am. Chem. Soc.*, 2013, **135**, 5808.
148. S. Horowitz and R. C. Trievel, *J. Biol. Chem.*, 2012, **287**, 41576.

149. J. C.-H. Chen, B. L. Hanson, S. Z. Fisher, P. Langan and A. Y. Kovalevsky, *Proc. Natl. Acad. Sci. U. S. A.*, 2012, **109**, 15301.
150. D. Sheppard, D.-W. Li, R. Godoy-Ruiz, R. Brschweiler and V. Tugarinov, *J. Am. Chem. Soc.*, 2010, **132**, 7709.
151. N. Sukumar, F. S. Mathews, P. Langan and V. L. Davidson, *Proc. Natl. Acad. Sci. U. S. A.*, 2010, **107**, 6817.
152. A. Schmiedekamp and V. Nanda, *J. Inorg. Biochem.*, 2009, **103**, 1054.
153. Z. Liu, G. Wang, Z. Li and R. Wang, *J. Chem. Theory Comput.*, 2008, **4**, 1959.
154. P. W. Hildebrand, S. Günther, A. Goede, L. Forrest, C. Frömmel and R. Preissner, *Biophys. J.*, 2008, **94**, 1945.
155. H. Park, J. Yoon and C. Seok, *J. Phys. Chem. B*, 2008, **112**, 1041.
156. S. K. Panigrahi and G. R. Desiraju, *Proteins*, 2007, **67**, 128.
157. A. Senes, D. E. Engel and W. F. DeGrado, *Curr. Opin. Struct. Biol.*, 2004, **14**, 465.
158. B. S. Kang, Y. Devedjiev, U. Derewenda and Z. S. Derewenda, *J. Mol. Biol.*, 2004, **338**, 483.
159. K. Manikandan and S. Ramakumar, *Proteins Struct. Funct. Genet.*, 2004, **56**, 768.
160. S. Sarkhel and G. R. Desiraju, *Proteins*, 2004, **54**, 247.
161. R. Bhattacharyya and P. Charkabarti, *J. Mol. Biol.*, 2003, **331**, 925.
162. K. M. Lee, H.-C. Chang, J.-C. Jiang, J. C. C. Chen, H.-E. Kao, S. H. Lin and I. J. B. Lin, *J. Am. Chem. Soc.*, 2003, **125**, 12358.
163. S. K. Singh, M. M. Babu and P. Balaram, *Proteins Struct. Funct. Genet.*, 2003, **51**, 167.
164. A. Senes, I. Ubarretxena-Belandia and D. M. Engelman, *Proc. Natl. Acad. Sci. U. S. A.*, 2001, **98**, 9056.
165. I. Goldberg, *Acta Cryst.*, 1975, **B31**, 754.
166. R. Kaufmann, A. Knöchel, J. Kopf, J. Oehler and G. Rudolph, *Chem. Ber.*, 1977, **110**, 2249.
167. C. Rest, A. Martin, V. Stepanenko, N. K. Allampally, D. Schmidt and G. Fernandez, *Chem. Commun.*, 2014, **50**, 13366.
168. M. V. Sigalov, E. P. Doronina and V. F. Sidorkin, *J. Phys. Chem. A*, 2012, **116**, 7718.
169. C. H. Schwalbe, *Cryst. Rev.*, 2012, **18**, 191.
170. S. Jin, M. Guo, D. Wang, S. Wei, Y. Zhou, Y. Zhou, X. Cao and Z. Yu, *J. Mol. Struct.*, 2012, **1020**, 70.
171. I. D. Madura, J. Zachara, H. Hajmowicz and L. Synoradzki, *J. Mol. Struct.*, 2012, **1017**, 98.
172. L.-Y. You, S.-G. Chen, X. Zhao, Y. Liu, W.-X. Lan, Y. Zhang, H.-J. Lu, C.-Y. Cao and Z.-T. Li, *Angew. Chem., Int. Ed.*, 2012, **51**, 1657.
173. A. M. Vibhute, R. G. Gonnade, R. S. Swathi and K. M. Sureshan, *Chem. Commun.*, 2012, **48**, 717.
174. Y. Sonoda, M. Goto, T. Ikeda, Y. Shimoi, S. Hayashi, H. Yamawaki and M. Kanesato, *J. Mol. Struct.*, 2011, **1006**, 366.
175. T. S. Thakur, M. T. Kirchner, D. Bläser, R. Boese and G. R. Desiraju, *Phys. Chem. Chem. Phys.*, 2011, **13**, 14076.
176. J. J. Novoa, B. Tarron, M.-H. Whangbo and J. M. Williams, *J. Chem. Phys.*, 1991, **95**, 5179.
177. J. J. Novoa, P. Constans and M.-H. Whangbo, *Angew. Chem., Int. Ed. Engl.*, 1993, **32**, 588.
178. P. Seiler, G. R. Weisman, E. D. Glendening, F. Weinhold, V. B. Johnson and J. D. Dunitz, *Angew. Chem., Int. Ed. Engl.*, 1987, **26**, 1175.
179. J. J. Novoa, P. Lafuente and F. Mota, *Chem. Phys. Lett.*, 1998, **290**, 519.
180. Y. Gu, T. Kar and S. Scheiner, *J. Am. Chem. Soc.*, 1999, **121**, 9411.
181. I. Alkorta, I. Rozas and J. Elguero, *J. Fluor. Chem.*, 2000, **101**, 233.

182. J. B. L. Martins, J. R. S. Politi, A. D. Braga and R. Gargano, *Chem. Phys. Lett.*, 2006, **431**, 51.
183. E. Cubero, M. Orozco and F. J. Luque, *Chem. Phys. Lett.*, 1999, **310**, 445.
184. E. S. Kryachko and T. Zeegers-Huyskens, *J. Phys. Chem. A*, 2001, **105**, 7118.
185. S. N. Delanoye, W. A. Herrebout and B. J. van der Veken, *J. Am. Chem. Soc.*, 2002, **124**, 7490.
186. B. J. van der Veken, S. N. Delanoye, B. Michielsen and W. A. Herrebout, *J. Mol. Struct.*, 2010, **976**, 97.
187. B. L. Bedell, L. Goldfarb, E. R. Mysak, C. Samet and A. Maynard, *J. Phys. Chem. A*, 1999, **103**, 4572.
188. N. Karger, A. M. Amorim da Costa and J. A. Ribeiro-Claro, *J. Phys. Chem. A*, 1999, **103**, 8672.
189. P. Hobza and Z. Havlas, *Chem. Rev.*, 2000, **100**, 4253.
190. M. P. M. Marques, A. M. A. da Costa and P. J. A. Ribeiro-Claro, *J. Phys. Chem. A*, 2001, **105**, 5292.
191. B. Reimann, K. Buchhold, S. Vaupel, B. Brutschy, Z. Havlas, V. Spirko and P. Hobza, *J. Phys. Chem. A*, 2001, **105**, 5560.
192. P. Hobza and Z. Havlas, *Theor. Chem. Acc.*, 2002, **108**, 325.
193. R. Gopi, N. Ramanathan and K. Sundararajan, *J. Phys. Chem. A*, 2014, **118**, 5529.
194. Y. Gu, T. Kar and S. Scheiner, *J. Mol. Struct.*, 2000, **552**, 17.
195. Y. Gu, T. Kar and S. Scheiner, *J. Mol. Struct.: THEOCHEM*, 2000, **500**, 441.
196. A. Masunov, J. J. Dannenberg and R. H. Contreras, *J. Phys. Chem. A*, 2001, **105**, 4737.
197. L. Pejov and K. Hermansson, *J. Chem. Phys.*, 2003, **119**, 313.
198. W. Qian and S. Krimm, *J. Phys. Chem. A*, 2002, **106**, 6628.
199. W. Qian and S. Krimm, *J. Phys. Chem. A*, 2002, **106**, 11663.
200. K. Hermansson, *J. Phys. Chem. A*, 2002, **106**, 4695.
201. S. N. Delanoye, W. A. Herrebout and B. J. van der Veken, *J. Am. Chem. Soc.*, 2002, **124**, 11854.
202. X. Li, L. Liu and H. B. Schlegel, *J. Am. Chem. Soc.*, 2002, **124**, 9639.
203. I. V. Alabugin, M. Manoharan, S. Peabody and F. Weinhold, *J. Am. Chem. Soc.*, 2003, **125**, 5973.
204. I. V. Alabugin, M. Manoharan and F. A. Weinhold, *J. Phys. Chem. A*, 2004, **108**, 4720.
205. K. S. Rutkowski, P. Rodziewicz, S. M. Melikova, W. A. Herrebout, B. J. van der Veken and A. Koll, *Chem. Phys.*, 2005, **313**, 225.
206. Y. Yang, W.-J. Zhang and X.-M. Gao, *Chin. J. Chem.*, 2006, **24**, 887.
207. M. Jablonski, *J. Mol. Struct.: THEOCHEM*, 2007, **820**, 118.
208. A. Y. Li, *J. Chem. Phys.*, 2007, **126**, 154102.
209. J. Joseph and E. D. Jemmis, *J. Am. Chem. Soc.*, 2007, **129**, 4620.
210. B. Michielsen, W. A. Herrebout and B. J. van der Veken, *ChemPhysChem.*, 2008, **9**, 1693.
211. A. Y. Li, *J. Mol. Struct.: THEOCHEM*, 2008, **862**, 21.
212. A. Karpfen and E. S. Kryachko, *J. Phys. Chem. A*, 2009, **113**, 5217.
213. S. J. Grabowski, *J. Phys. Chem. A*, 2011, **115**, 12789.
214. A. Karpfen and E. S. Kryachko, *Chem. Phys.*, 2005, **310**, 77.
215. E. S. Kryachko and A. Karpfen, *Chem. Phys.*, 2006, **329**, 313.
216. A. Karpfen, *Phys. Chem. Chem. Phys.*, 2011, **13**, 14194.
217. A. K. Chandra and T. Zeegers-Huyskens, *Chem. Phys.*, 2013, **410**, 66.
218. O. Donoso-Tauda, P. Jaque and J. C. Santos, *Phys. Chem. Chem. Phys.*, 2011, **13**, 1552.
219. Y. Mo, C. Wang, L. Guan, B. Braïda, P. C. Hiberty and W. Wu, *Chem. – Eur. J.*, 2014, **20**, 8444.
220. M. Jabłoński, *J. Comput. Chem.*, 2014, **35**, 1739.

221. M. D. Struble, C. Kelly, M. A. Siegler and T. Lectka, *Angew. Chem,. Int. Ed.*, 2014, **53**, 8924.
222. Q. Gou, G. Feng, L. Evangelisti and W. Caminati, *J. Phys. Chem. A*, 2014, **118**, 737.
223. A. Dey, S. I. Mondal and G. N. Patwari, *ChemPhysChem.*, 2013, **14**, 746.
224. B. Kharat, V. Deshmukh and A. Chaudhari, *Struct. Chem.*, 2012, **23**, 637.
225. S. Maity, A. Dey, G. N. Patwar, S. Karthikeyan and K. S. Kim, *J. Phys. Chem. A*, 2010, **114**, 11347.
226. A. A. Howard, G. S. Tschumper and N. I. Hammer, *J. Phys. Chem. A*, 2010, **114**, 6803.
227. M. Jablonski and M. Palusiak, *J. Phys. Chem. A*, 2010, **114**, 2240.
228. G. Liu, H. Wang and W. Li, *J. Mol. Struct.: THEOCHEM*, 2006, **772**, 103.
229. M. Mons, I. Dimicoli, B. Tardivel, F. Piuzzi, V. Brenner and P. Millié, *Phys. Chem. Chem. Phys.*, 2002, **4**, 571.
230. P. R. Shirhatti, D. K. Maity and S. Wategaonkar, *J. Phys. Chem. A*, 2013, **117**, 2307.
231. A. V. Afonin, A. V. Vashchenko, T. Takagi, A. Kimura and H. Fujiwara, *Can. J. Chem.*, 1999, **77**, 416.
232. G. Zhang, W. He and D. Chen, *Mol. Phys.*, 2014, **112**, 1736.
233. C. L. Christenholz, D. A. Obenchain, R. A. Peebles and S. A. Peebles, *J. Phys. Chem. A*, 2014, **118**, 1610.
234. N. Lu, R. M. Ley, C. E. Cotton, W.-C. Chung, J. S. Francisco and E.-I. Negishi, *J. Phys. Chem. A*, 2013, **117**, 8256.
235. M. J. Biller and S. Mecozzi, *Mol. Phys.*, 2012, **110**, 377.
236. M. A. Vincent and I. H. Hillier, *Phys. Chem. Chem. Phys.*, 2011, **13**, 4388.
237. B. G. Saar, G. P. O'Donoghue, A. H. Steeves and J. W. Thoman, *Chem. Phys. Lett.*, 2006, **417**, 159.
238. L. B. Favero, B. M. Giuliano, S. Melandri, A. Maris, P. Ottaviani, B. Velino and W. Caminati, *J. Phys. Chem. A*, 2005, **109**, 7402.
239. P. Lu, G.-Q. Liu and J.-C. Li, *J. Mol. Struct.: THEOCHEM*, 2005, **723**, 95.
240. M. Domagala and S. J. Grabowski, *Chem. Phys.*, 2010, **367**, 1.
241. D. J. Wolstenholme, J. J. Weigand, E. M. Cameron and T. S. Cameron, *Phys. Chem. Chem. Phys.*, 2008, **10**, 3569.
242. M. Domagala and S. J. Grabowski, *J. Phys. Chem. A*, 2005, **109**, 5683.
243. A. V. Afonin, D. D. Toryashinova and E. Y. Schmidt, *J. Mol. Struct.: THEOCHEM*, 2004, **680**, 127.
244. S. A. C. McDowell, *Chem. Phys. Lett.*, 2006, **424**, 239.
245. W. S. Hopkins, M. Hasan, M. Burt, R. A. Marta, E. Fillion and T. B. McMahon, *J. Phys. Chem. A*, 2014, **118**, 3795.
246. G. Tamasi, F. Botta and R. Cini, *J. Mol. Struct.: THEOCHEM*, 2006, **766**, 61.
247. G. Feng, Q. Gou, L. Evangelisti, M. Vallejo-López, A. Lesarri, E. J. Cocinero and W. Caminati, *Phys. Chem. Chem. Phys.*, 2014, **16**, 12261.
248. F. Ito, *J. Mol. Struct.*, 2012, **1012**, 43.
249. M. Castro, I. Nicolas-Vazquez, J. I. Zavala, F. Sanchez-Viesca and M. Berros, *J. Chem. Theory Comput.*, 2007, **3**, 681.
250. V. Balamurugan, J. Mukherjee, M. S. Hundal and R. Mukherjee, *Struct. Chem.*, 2007, **18**, 133.
251. R. Pazout, J. Houskova and M. Dusek, *Struct. Chem.*, 2011, **22**, 1325.
252. B. Michielsen, C. Verlackt, B. J. van der Veken and W. A. Herrebout, *J. Mol. Struct.*, 2012, **1023**, 90.
253. P. Ramasami and T. A. Ford, *J. Mol. Struct.*, 2012, **1023**, 163.
254. S. T. Howard and C. D. Abernethy, *J. Comput. Chem.*, 2004, **25**, 649.
255. F. Yu, *Int. J. Quantum Chem.*, 2013, **113**, 2355.
256. S. Karthikeyan, V. Ramanathan and B. K. Mishra, *J. Phys. Chem. A*, 2013, **117**, 6687.

257. M. Majumder, B. K. Mishra and N. Sathyamurthy, *Chem. Phys.*, 2013, **557**, 59.
258. B. Michielsen, J. J. J. Dom, B. J. van der Veken, S. Hesse, Z. Xue, M. A. Suhm and W. A. Herrebout, *Phys. Chem. Chem. Phys.*, 2010, **12**, 14034.
259. P. R. Shirhatti and S. Wategaonkar, *Phys. Chem. Chem. Phys.*, 2010, **12**, 6650.
260. O. Takahashi, Y. Kohno, Y. Gondoh, K. Saito and M. Nishio, *Bull. Chem. Soc. Jpn.*, 2003, **76**, 369.
261. Y. Wang and P. B. Balbuena, *J. Phys. Chem. A*, 2001, **105**, 9972.
262. S. D. Wetmore, R. Schofield, D. M. Smith and L. Radom, *J. Phys. Chem. A*, 2001, **105**, 8718.
263. T. Kar and S. Scheiner, *J. Phys. Chem. A*, 2004, **108**, 9161.
264. D. P. Malenov, G. V. Janjic, D. Ž. Veljkovic and S. D. Zaric, *Comput. Theor. Chem.*, 2013, **1018**, 59.
265. X. Gao, Y. Liu, H. Li, J. Bian, Y. Zhao, Y. Cao, Y. Mao, X. Li, Y. Xu, Y. Ozaki and J. Wu, *J. Mol. Struct.*, 2013, **1040**, 122.
266. K.-M. Lee, J. C. C. Chen, H.-Y. Chen and I. J. B. Lin, *Chem. Commun.*, 2012, **48**, 1242.
267. M. Solimannejad, M. Malekani and I. Alkorta, *Mol. Phys.*, 2011, **109**, 1641.
268. A. K. Samanta, P. Pandey, B. Bandyopadhyay and T. Chakraborty, *J. Phys. Chem. A*, 2010, **114**, 1650.
269. Q. Li, X. An, B. Gong and J. Cheng, *J. Phys. Chem. A*, 2007, **111**, 10166.
270. S. Scheiner and T. Kar, *J. Phys. Chem. B*, 2005, **109**, 3681.
271. S. Scheiner, T. Kar and Y. Gu, *J. Biol. Chem.*, 2001, **276**, 9832.
272. S. Scheiner, *J. Phys. Chem. B*, 2005, **109**, 16132.
273. S. Scheiner, *J. Phys. Chem. B*, 2007, **111**, 11312.
274. S. Scheiner, *J. Phys. Chem. B*, 2006, **110**, 18670.
275. M. V. Vener, A. N. Egorova, D. P. Fomin and V. G. Tsirelson, *Chem. Phys. Lett.*, 2007, **440**, 279.
276. R. Parthasarathi, S. S. Raman, V. Subramanian and T. Ramasami, *J. Phys. Chem. A*, 2007, **111**, 7141.
277. H. Guo, A. Gorin and H. Guo, *Interdiscip. Sci.: Comput. Life Sci.*, 2009, **1**, 12.
278. M. V. Vener, A. N. Egorova, D. P. Fomin and V. G. Tsirel'son, *Russ. J. Phys. Chem. B*, 2009, **3**, 541.
279. S. M. LaPointe, S. Farrag, H. J. Bohórquez and R. J. Boyd, *J. Phys. Chem. B*, 2009, **113**, 10957.
280. G. Pohl, J. A. Plumley and J. J. Dannenberg, *J. Chem. Phys.*, 2013, **138**, 245102.
281. S. Hussain, G. Das and M. K. Chaudhuri, *J. Mol. Struct.*, 2007, **837**, 190.
282. S. Scheiner, T. Kar and J. Pattanayak, *J. Am. Chem. Soc.*, 2002, **124**, 13257.
283. H.-C. Chang, K. M. Lee, J.-C. Jiang, M.-S. Lin, J.-S. Chen, I. J. B. Lin and S. H. Lin, *J. Chem. Phys.*, 2002, **117**, 1723.
284. F. M. Raymo, M. D. Bartberger, K. N. Houk and J. F. Stoddart, *J. Am. Chem. Soc.*, 2001, **123**, 9264.
285. H. Guo, R. F. Beahm and H. Guo, *J. Phys. Chem. B*, 2004, **108**, 18065.
286. C. A. Deakyne and M. Meot-Ner, *J. Am. Chem. Soc.*, 1999, **121**, 1546.
287. C. E. Cannizzaro and K. N. Houk, *J. Am. Chem. Soc.*, 2002, **124**, 7163.
288. T. Liu, J. Gu, X.-J. Tan, W.-L. Zhu, X.-M. Luo, H.-L. Jiang, R.-Y. Ji, K.-X. Chen, I. Silman and J. L. Sussman, *J. Phys. Chem. A*, 2002, **106**, 157.
289. O. V. Shishkin, G. V. Palamarchuk, L. Gorb and J. Leszczynski, *Chem. Phys. Lett.*, 2008, **452**, 198.
290. U. Adhikari and S. Scheiner, *J. Phys. Chem. A*, 2013, **117**, 10551.
291. B. Nepal and S. Scheiner, *J. Phys. Chem. A*, 2014, **118**, 9575.
292. M. Meot-Ner and C. A. Deakyne, *J. Am. Chem. Soc.*, 1985, **107**, 469.
293. C. A. Deakyne and M. Meot-Ner, *J. Am. Chem. Soc.*, 1985, **107**, 474.
294. K. S. Kim, J. Y. Lee, S. J. Lee, T.-K. Ha and D. H. Kim, *J. Am. Chem. Soc.*, 1994, **116**, 7399.

295. J. Y. Lee, S. J. Lee, H. S. Cho, S. J. Cho, K. S. Kim and T.-K. Ha, *Chem. Phys. Lett.*, 1995, **232**, 67.
296. T. Liu, J. Gu, X.-J. Tan, W.-L. Zhu, X.-M. Luo, H.-L. Jiang, R.-Y. Ji, K.-X. Chen, I. Silman and J. L. Sussman, *J. Phys. Chem. A*, 2001, **105**, 5431.
297. S. Scheiner, *J. Phys. Chem. A*, 2015, **119**, 9189.
298. B. Nepal and S. Scheiner, *Chem. – Eur. J.*, 2015, **21**, 1474.
299. E. S. Kryachko and T. Zeegers-Huyskens, *J. Phys. Chem. A*, 2002, **106**, 6832.
300. A. Y. Li and X. H. Yan, *Phys. Chem. Chem. Phys.*, 2007, **9**, 6263.
301. X. Zarate, M. C. Daza and J. L. Villaveces, *J. Mol. Struct.: THEOCHEM*, 2009, **893**, 77.
302. L. Pedzisa and B. P. Hay, *J. Org. Chem.*, 2009, **74**, 2554.
303. J.-M. Fan, L. Liu and Q.-X. Guo, *Chem. Phys. Lett.*, 2002, **365**, 464.
304. S. A. C. McDowell, *Chem. Phys. Lett.*, 2007, **441**, 194.
305. E. D'Oria and J. J. Novoa, *J. Phys. Chem. A*, 2011, **115**, 13114.
306. E. D'Oria, D. Braga and J. J. Novoa, *CrystEngComm*, 2012, **14**, 792.
307. B. Nepal and S. Scheiner, *Chem. – Eur. J.*, 2015, **21**, 13330.
308. S. Scheiner, *CrystEngComm*, 2013, **15**, 3119.
309. A. Bondi, *J. Phys. Chem.*, 1964, **68**, 441.
310. G. R. Desiraju, *Acc. Chem. Res.*, 1996, **29**, 441.
311. B. Nepal and S. Scheiner, *Chem. Phys.*, 2015, **456**, 34.
312. U. Adhikari and S. Scheiner, *Chem. Phys. Lett.*, 2012, **532**, 31.
313. B. Nepal and S. Scheiner, *Chem. Phys. Lett.*, 2015, **630**, 6.
314. A. J. Stone, *J. Am. Chem. Soc.*, 2013, **135**, 7005.
315. S. M. Huber, J. D. Scanlon, E. Jimenez-Izal, J. M. Ugalde and I. Infante, *Phys. Chem. Chem. Phys.*, 2013, **15**, 10350.
316. L. Wang, J. Gao, F. Bi, B. Song and C. Liu, *J. Phys. Chem. A*, 2014, **118**, 9140.
317. M. Kolár, J. Hostaš and P. Hobza, *Phys. Chem. Chem. Phys.*, 2014, **16**, 9987.

12 The CH$\cdots\pi$ Hydrogen Bond

Osamu Takahashi*[a] and Motohiro Nishio[b]

[a] Institute for Sustainable Sciences and Development, Hiroshima University, Higashi-Hiroshima-shi, Hiroshima 739-8526, Japan; [b] The CHPI Institute, 794-7-910 Izusan, Atami-shi, Shizuoka 413-0002, Japan
*Email: shu@hiroshima-u.ac.jp

12.1 Introduction

12.1.1 Definition of the Hydrogen Bond

The hydrogen bond (H-bond) is an important molecular force that controls the shape of a molecule and arrangement of organic crystals.[1] The H-bond is categorized into four types,[2] namely the conventional (or normal, ordinary, classical, such as OH$\cdots$O), CH$\cdots n$ (such as CH$\cdots$O),[3–6] XH$\cdots\pi$ (X$=$O, N), and the CH$\cdots\pi$ H-bond. Pauling defined the conventional H-bond as the interaction between a hard acid (OH or NH) and a hard base (O or N).[7] This is the strongest among the four H-bonds considered above.

As chemistry developed, the definition of the H-bond has been extended. In 1960, Pimentel and McClellan argued that a H-bond exists between a functional group A–H and an atom or a group B in the same or a different molecule, when: (1) there is evidence of bond formation (association or chelation) and (2) there is evidence that this new bond linking A–H and B specifically involves the hydrogen atom already bonded to A.[8] Note that no restriction has been made on the chemical nature of the donors and acceptors, nor the energy and the

Intermolecular Interactions in Crystals: Fundamentals of Crystal Engineering
Edited by Juan J. Novoa
© The Royal Society of Chemistry 2018
Published by the Royal Society of Chemistry, www.rsc.org

geometry of the participants. The definition of Pimentel has proved to be very useful to the progress of modern chemistry.

A new definition of the H-bond was proposed by the IUPAC commission several years ago. The short definition is as follows: *This is an attractive interaction between a hydrogen atom from a molecule or a molecular fragment X-H in which X is more electronegative than H, and an atom or a group of atoms in the same or a different molecule, in which there is evidence of bond formation.*[9] As the basis for determination of H-bonding, six criteria and several characteristics of H-bonds have been compiled. The $CH\cdots\pi$ H-bond is the weakest limit of H-bonding and is also included in this definition. The present definition includes a broader type of H-bond than before.

12.1.2 Weak Hydrogen Bonds

According to Pearson's hard and soft acid and base principle (HSAB),[10] the $CH\cdots O$ H-bond is an interaction between a soft acid (SA: C–H) and a hard base (HB: lone pairs of O or N). The $XH\cdots\pi$ H-bond (X = O and N) is the interaction between a hard acid (HA: OH or NH) and a soft base (SB: π-electron systems).

12.2 The $CH\cdots\pi$ Hydrogen Bond. What Is It?

The $CH\cdots\pi$ H-bond is the interaction between an SA (C–H) and an SB (π-system). Evidence for the $CH\cdots\pi$ H-bond is provided by various methods: crystallographic, database (Cambridge Structural Database: CSD) analyses, spectroscopy,[11] and computational methods.[12]

12.2.1 Crystallographic Evidence

The crystallographic method provides good evidence for a weak bond, especially when effects from the electronic substituent are supplied. In 1962, Sutor performed pioneering works to introduce crystallographic evidence for weak H-bonds such as $CH\cdots O$.[3] Distance and angle parameters of the putative H-bonded atoms are used in evaluating the strength of the interaction.[4] In 1982, Taylor and Kennard inspected crystallographic data of weak H-bonds such as $CH\cdots O$, $CH\cdots N$ and $CH\cdots Cl$ by using the CSD neutron diffraction data.[5] Interested readers are also referred to Geffrey (1997),[13] and Desiraju and Steiner (1999).[6]

12.2.2 Evidence from Crystallographic Database Analysis

Crystallographic database analyses provide a firm basis for the H-bond nature of the CH$\cdots\pi$ interaction. Interested readers are referred to our reviews.[14]

Umezawa *et al.*[15] looked for short distances which could indicate intramolecular and intermolecular CH$\cdots\pi$ H-bonds in a search of the CSD. In the case of intramolecular interactions, they found that *ca.* 29% of compounds have short CH$\cdots\pi$ (aromatic) distances, suggesting that there may be stronger interactions operating than are due to van der Waals dispersion forces. The values for OH$\cdots\pi$ and NH$\cdots\pi$ interactions were 1.4% and 2.7%, respectively. Differences in these ratios for intermolecular interactions were more significant; 75.1, 5.1, and 9.3% for CH$\cdots\pi$, OH$\cdots\pi$, and NH$\cdots\pi$ interactions, respectively. These values were obtained from the CSD version 515 (181 309 entries). These results will be modified if CSD data from other versions were used. However, the number of the entries included in the database is relatively large in the version used, and a conclusion derived in their study will not be affected. Their conclusion suggested that the CH$\cdots\pi$ H-bond is ubiquitous and significant in the development of the intramolecular conformation and intermolecular configuration in crystals.

As another important point to assert the existence of the H-bond, the H-bond trends directionality, and the three atoms X–H–Y usually tend toward linearity because of dipole–dipole interaction between functional groups. This is one of the important features of the H-bond as, in contrast, the van der Waals force does not have the directionality and can be expressed as a simple function of the distance. For the CH$\cdots\pi$ H-bond, similar trends have been indicated computationally. A minimum of the potential energy surface as a function of bending angle of X–H–Y was found at 180 degrees using *ab initio* MO calculations although the binding energy was extremely small.[17] The energy expression was not described by a simple function, because its interaction is different with different types of CH$\cdots\pi$ bonds. According to the database study using the CSD, the directionality has been found to correlate with the CH$\cdots\pi$-plane distance (D_{pln}), depending on the strength (or acidity) of the proton donor.[3,16] Histograms depicting the distribution of CH hydrogen atoms against the C–H$\cdots\pi$ access angle (α) and correlation diagrams between D_{pln} and α are given in Figures 12.1 and 12.2.[17] The directionality of CH$\cdots\pi$ bonds may not be linear due to uncategorized restrictions in crystals. Although the energy components of CH$\cdots\pi$ interaction, which are

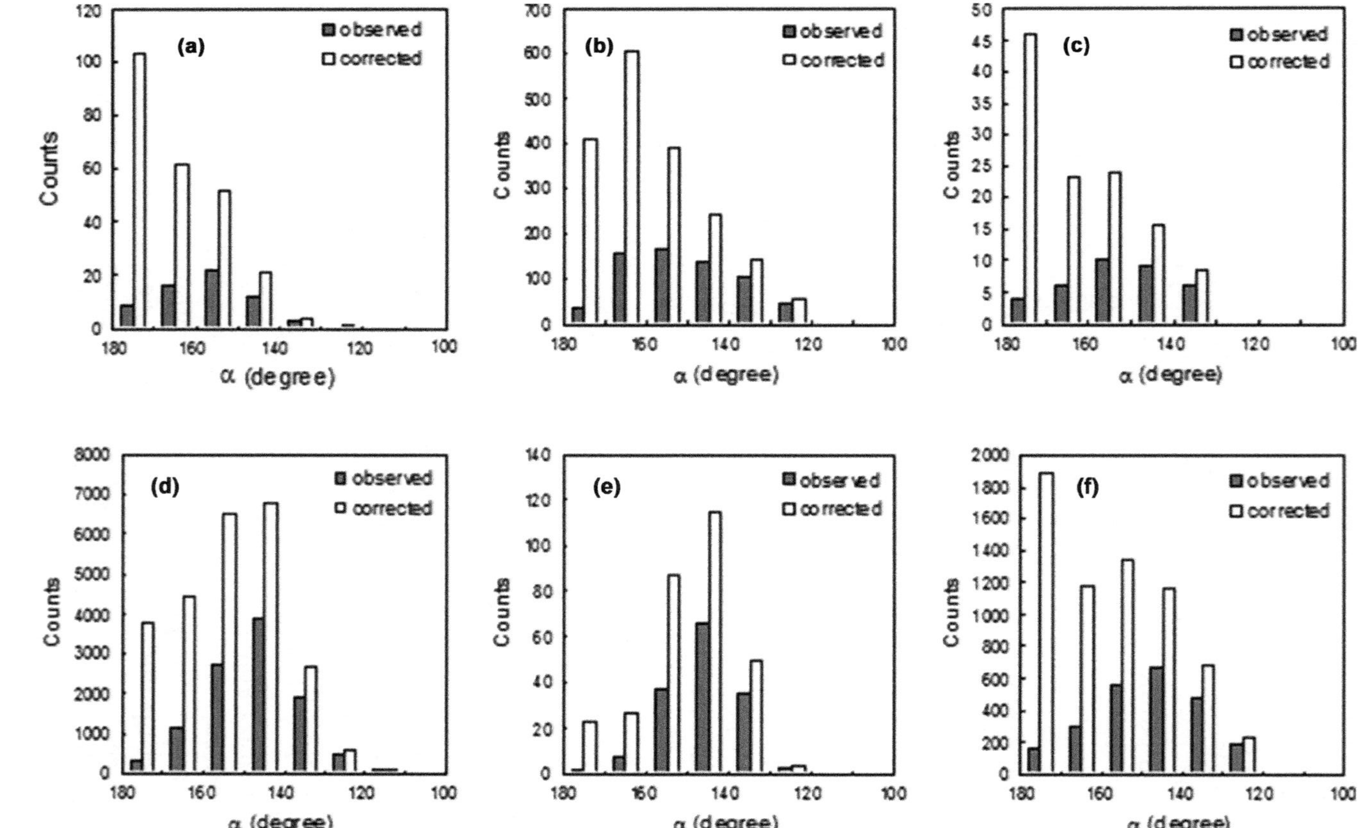

Figure 12.1 Histograms showing the distribution of CH hydrogen atoms ($D_{px1} \leq 1.4$ Å, $D_{pln} < 2.9$ Å) against the C–H$\cdots\pi$ access angle (α). (a) Cl_3CH, (b) Cl_2CH_2, (c) sp-CH, (d) sp^2-CH (aromatic CH), (e) sp^2-CH (aromatic CH, neutron data), (f) CCH_3. The data of only organic crystals with no disorder and $R \leq 5\%$ were used for (d) and (f). Open bars: observed. Shaded bars: corrected by a factor of $1/\sin\alpha$.
Reproduced from ref. 17 with permission of the Chemical Society of Japan.

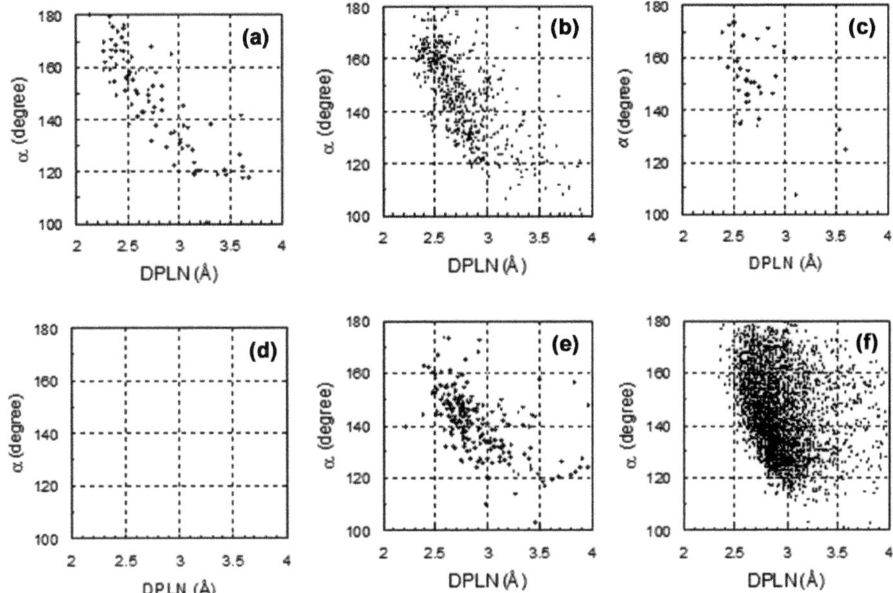

Figure 12.2 Scatter plots showing dependence of the C–H···π access angle (α) on the CH···π plane distance in the range of $D_{pxl} < 1.4$ Å. (a) Cl_3CH, (b) Cl_2CH_2, (c) sp-CH, (d) sp^2-CH (aromatic CH), (e) sp^2-CH (aromatic CH, neutron data), (f) CCH_3. The data of only organic crystals with no disorder and $R \leq 5\%$ were used for (d) and (f).
Reproduced from ref. 17 with permission of the Chemical Society of Japan.

described later, are mostly dispersion forces (no orientational dependence), it is emphatically shown that the CH···π H-bond has directionality unless the orientation between functional groups is not linear. Thus molecular recognition between molecules can be enabled through the CH···π H-bond.

12.2.3 Evidence from NMR Spectroscopy

Nuclear magnetic resonance (NMR) spectroscopy is nowadays one of the standard tools used to determine the structure in organic molecules in condensed phases. Hill *et al.* studied conformational isomerism of a ball-shaped cyclophane **1** using NMR and first principles density functional theory (DFT) calculations (Scheme 12.1).[11] The most stable conformer of **1** is shown in Figure 12.3. According to ^{13}C-NMR and ^{1}H-NMR, the most stable conformer was predicted as a highly symmetric cyclic benzene tetramer, in which four intramolecular CH···π H-bonds cooperatively operate. These analyses were consistent with the DFT calculations.

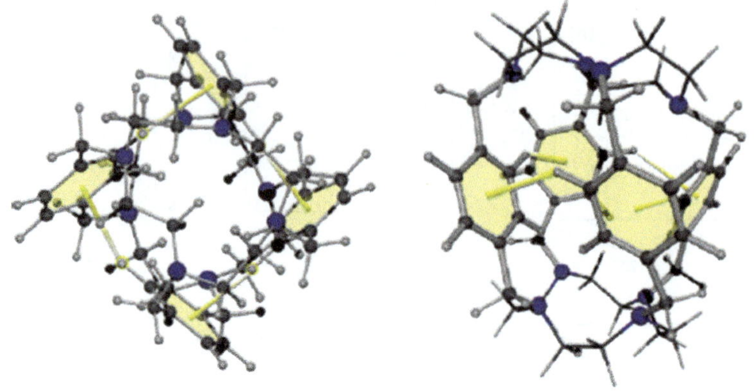

Figure 12.3   Most stable conformer of **1**. CH$\cdots\pi$ bonds shown as yellow lines. Reproduced from ref. 11 with permission from the PCCP Owner Societies.

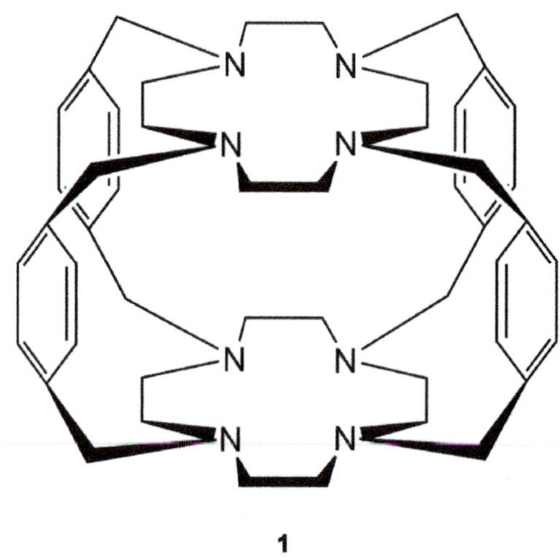

1

Scheme 12.1

12.2.4   X-Ray Photoelectron Spectroscopy

Evidence has also been provided by the X-ray photoelectron spectroscopy (XPS) of *n*-propyl cyanide, shown in Figure 12.4. The conformation of *n*-propyl cyanide substantiated that a CH$\cdots\pi$ H-bonded species prevails. Only peaks assigned as terminal methyl groups were split, and this is a direct result of the existence of gauche and anti-conformers. The interatomic distance between one of the methyl hydrogens and an acetylenic carbon was found to be appreciably shorter than the van der Waals distance, which was obtained by high level *ab initio* molecular orbital (MO) calculations.[18]

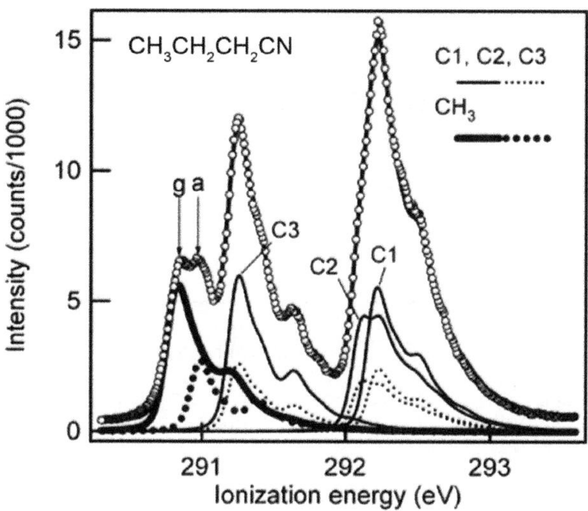

Figure 12.4 X-ray photoelectron spectra of *n*-propyl cyanide. Reprinted from ref. 18 with permission of the Royal Society of Chemistry and the PCCP Owner Societies.

12.2.5 Infrared (IR) Spectroscopy

IR spectroscopy is a powerful tool with which to study the features of H-bonding, and H-bonded systems have been investigated using this spectroscopy extensively. It is common to measure a peak shift of an XH stretching mode between an H-bonded system and a gas phase system, and the strengths and the thermodynamic properties of CH···π interaction can be determined. It can be difficult to detect a CH···π H-bond directly in the gas phase because of the small peak shift. However, Fujii *et al.*,[19] reported the direct observation of intermolecular CH···π H-bonding of benzene with acetylene using an infrared–ultraviolet double resonance spectroscopic technique in the gas phase. In Figure 12.5, the observed IR spectrum of jet-cooled benzene–acetylene complex is depicted. A peak at 3266.7 cm^{-1} was assigned as the ν_3 band, which is the anti-symmetric CH stretching mode of the acetylene molecule, and this band shows a remarkable blue shift of 22 cm^{-1} compared with a free acetylene molecule. This remarkable feature of the complex clearly demonstrated that the intermolecular interaction between acetylene and benzene consists of the CH···π H-bond in the gas phase. The observation indicated that the intermolecular interaction between the activated CH group and π-electrons is characterized as a π-hydrogen bond rather than as a van der Waals interaction.[19]

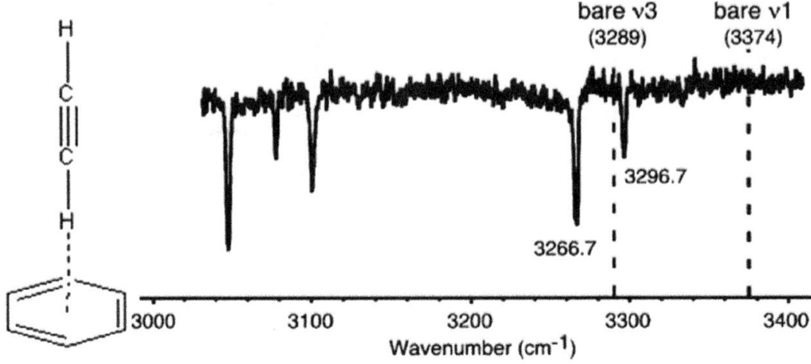

Figure 12.5   IR spectrum of benzene–acetylene complex. Unperturbed vibrational frequencies of CH stretching vibrations in bare acetylene are indicated by the dashed lines.
Reprinted with permission from A. Fujii, S. Morita, M. Miyazaki, T. Ebata and N. Mikami., A Molecular Cluster Study on Activated CH/π Interactions: Infrared Spectroscopy of Aromatic Molecule-Acetylene Clusters, *J. Phys. Chem. A*, 2004, 108, 2652. Copyright 2004 American Chemical Society.

Table 12.1   Structural parameters for $C_6H_6 \cdots C_2H_2$.[a]

	R_{cm}[b]/Å	$R_{CH\cdots\pi}$[c]/Å
$r_0(R_{cm})$	4.1546	2.4921
r_s	4.1430	2.4717
Ab initio	4.0387	2.3694

[a]Ref. 21.
[b]The center of mass distance.
[c]Perpendicular distance from H atom to the benzene plane.

12.2.6   Microwave Spectroscopy

Microwaves excite the rotational motion of a molecule, and microwave spectroscopy is used to determine accurate molecular structures through the rotational constants. Tubergen *et al.*[20] reported rotational spectra of 1-phenyl-2-propanol, methamphetamine and 1-phenyl-2-propanone molecules using a molecular beam technique. They also compared their experimental results with theoretical calculations. The conformations of these molecules shown as the lowest energy were found to be stabilized by weak intramolecular OH···π, NH···π and CH···π H-bonds, respectively. Ulrich *et al.*[21] studied the rotational spectra of benzene–acetylene complexes including the several isotopically substituted species using chirped-pulse Fourier-transform microwave spectroscopy; determined intermolecular CH···π distances and intermolecular interaction energies are consistent with the presence of CH···π H-bonds observed by other spectroscopies and previous theoretical calculations. In Table 12.1, experimentally determined structural parameters are summarized together with theoretical ones.

12.3 Characteristics of the CH· · ·π Hydrogen Bond

As in the other H-bonds, the CH· · ·π H-bond has distinct directionality.[17,22] The interaction energy depends on the nature of the molecular fragments, CH as well as π-groups. The stronger the proton donating ability of the CH group, the larger the stabilizing effect. For typical cases involving aliphatic and aromatic CH groups as the H-donor, the energy of a one unit CH· · ·π hydrogen bond is *ca.* 1.5–2.5 kcal mol^{-1}. Interactions involving aromatic CHs are stronger than aliphatic ones. For CH· · ·π bonds involving an electron-withdrawing atom or group, the energy of interaction becomes comparable to the conventional H-bond. As for the acceptor, the electron density of the π-group is relevant. A significant characteristic of the CH· · ·π H-bond is that it occurs in polar, protic solvents such as water. This is of paramount importance in the consideration of the effect in biochemistry.

To understand the nature of various types of H-bonds, the intermolecular interaction energy between functional groups in H-bonds can be decomposed into physical components such as electrostatic, exchange repulsion, induction, and dispersion energies. Several energy decomposition schemes have been proposed before. The pioneering work is the so-called Kitaura–Morokuma energy partitioning,[23] which was developed within the framework of the Hartree–Fock theory, *i.e.*, one electron orbital picture. Of the proposed schemes, symmetry adapted perturbation theory (SAPT)[24–26] is one of the reliable procedures and has been widely applied recently. This theory is based on the perturbation theory and can be applied for various computational levels including *ab initio* MO theory with electron correlation and density functional theory (DFT). Several typical weak H-bonds and their energy components are collected in Table 12.2. The electrostatic energy (E_{es}) is dominant for stronger H-bonds such as OH· · ·O. On the other hand, the dispersion energy (E_{disp}) is dominant in weaker H-bonds such as CH· · ·π.

12.4 Computational Study

12.4.1 How to Describe the CH· · ·π Hydrogen Bond

As discussed in the previous section, the main contribution to CH· · ·π H-bonding is the dispersion force. Thus high-level *ab initio*

Table 12.2 Energy decomposition for various types of H-bonds using the SAPT method (in $kcal\,mol^{-1}$).

Type of H-bond	Example	$E_{total}{}^a$	$E_{es}{}^a$	$E_{er}{}^a$	$E_{ind}{}^a$	$E_{disp}{}^a$	$E_{other}{}^a$	E_{es}/E_{total}	E_{disp}/E_{total}
$OH\cdots O^b$	$H_2O\cdots H_2O$	−4.60	−8.22	8.30	−1.38	−2.36	−0.94	1.79	0.51
$OH\cdots\pi^c$	$H_2O\cdots C_6H_6$	−2.86	−2.94	3.78	−1.05	−3.18		1.03	1.11
$NH\cdots\pi^c$	$NH_3\cdots C_6H_6$	−2.08	−2.05	3.40	−0.60	−3.32		0.98	1.60
$CH\cdots\pi^c$	$CH_4\cdots C_6H_6$	−1.27	−0.86	2.23	−0.10	−2.4	−0.16	0.68	1.89
$CH\cdots\pi^d$	$C_2H_2\cdots C_6H_6$	−2.54	−2.89	5.46	−0.55	−3.75	−0.72	1.14	1.48

[a]The total interaction (E_{total}), electrostatic (E_{es}), exchange repulsion (E_{er}), induction (E_{ind}), dispersion (E_{disp}), and other effects (E_{other}) energies, respectively.
[b]Ref. 52.
[c]Ref. 53.
[d]Ref. 54.

Table 12.3 Molecular properties of the benzene–methane complex.

Ref.	$E_e{}^a$	$R_e{}^b$	Level of theory
Brédas (1989)[55]	1.02	3.71	MP2/6-31G(d)
Sakaki (1993)[56]	0.57	4.07	MP2/MIDI-4, BSSE
Calderone (1998)[57]	0.96	3.681	MP2/3-21G//HF/3-21G
Tsuzuki (2000)[27]	1.45	3.8	CCSD(T), basis set limit
Tarakeshwar (2001)[58]	1.48	3.525	MP2/aug-cc-pVDZ
Ran (2006)[28]	1.05	2.548^c	MP2/aug(d,p)-6-311G(d,p)
Smith (2014)[59]	1.44	3.76	CCSD(T)/CBS

[a]Binding energy (in $kcal\,mol^{-1}$) between benzene and methane.
[b]Distance (in Å) between carbon atom of methane and benzene plane.
[c]Distance (in Å) between a nearest hydrogen atom of methane and benzene plane.

calculations may be needed to estimate molecular properties such as binding energy between molecules correctly. Table 12.3 lists molecular properties of the benzene–methane complex with a dimer model. Systematic studies were performed by Novoa and Mota,[12] and Tsuzuki *et al.*[27] more than ten years ago. A theoretical procedure including electron correlation is needed to describe the system, with quite large basis sets such as cc-pVQZ. Using large basis sets, the problem of basis set superposition error (BSSE) will be relatively small, which is comparable with the amount of $CH\cdots\pi$ H-bond. Several systematic computational studies followed.[28–30]

Albertí *et al.*[31] obtained analytical interaction potential functions between benzene and methane using improved Lenard–Jones potentials. A typical two-dimensional map of the probability density of methane molecules around benzene is depicted in Figure 12.6. Using their potential functions, calculated molecular properties were consistent with the previous *ab initio* results. Also they performed molecular

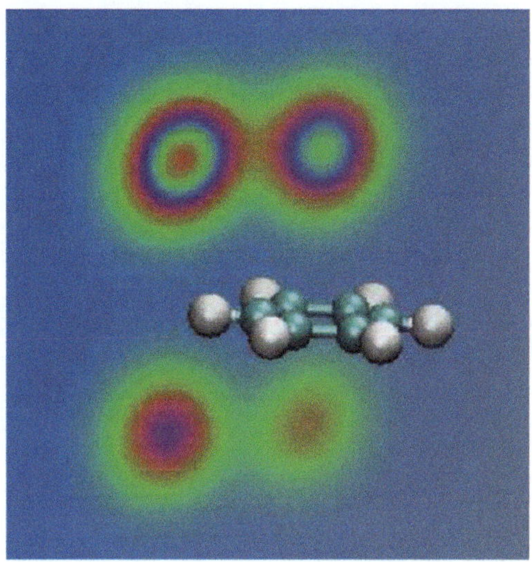

Figure 12.6 Two-dimensional map of the probability density of CH_4 molecules around benzene for benzene–$(CH_4)_3$ system.
Reprinted with permission from M. Albertí, A. Aguilar, J. M. Lucas and F. Pirani., Competitive Role of CH_4-CH_4 and CH-π Interactions in C_6H_6-$(CH_4)_n$ Aggregates: The Transition from Dimer to Cluster Features, *J. Phys. Chem. A*, 2012, 116, 5480. Copyright 2012 American Chemical Society.

dynamics simulations using their potentials and discussed macroscopic properties of benzene–$(CH_4)_n$ ($n > 1$–10) clusters. It can still be difficult to perform such simulations using *ab initio* all electron quantum chemistry calculations using current computer resources. However, their potential functions may be useful for theoretical calculations in crystals in which weak H-bond interactions are dominant.

12.4.2 Cooperative Effects in Crystals

One of the significant results of CH·· ·π H-bonding in crystals is the cooperative effect. This may be due to the existence of CH functionals everywhere in organic crystals and the dispersion force for the main contribution in the CH·· ·π H-bond. From a theoretical viewpoint, Ran and Wong[28] reported the cooperative effect of the CH·· ·π H-bond between benzene and various hydrocarbons. They used CCSD(T)/aug-cc-pVTZ//MP2/aug(d,p)-6-311G(d,p) level of theory and estimated the binding energy between benzene and modeled saturated hydrocarbons. Optimized structures are shown in Figure 12.7. Multiple CH groups were interacted with the π face of benzene, and calculated interaction energies are 2–3 times larger than that of the prototypical methane–benzene complex.

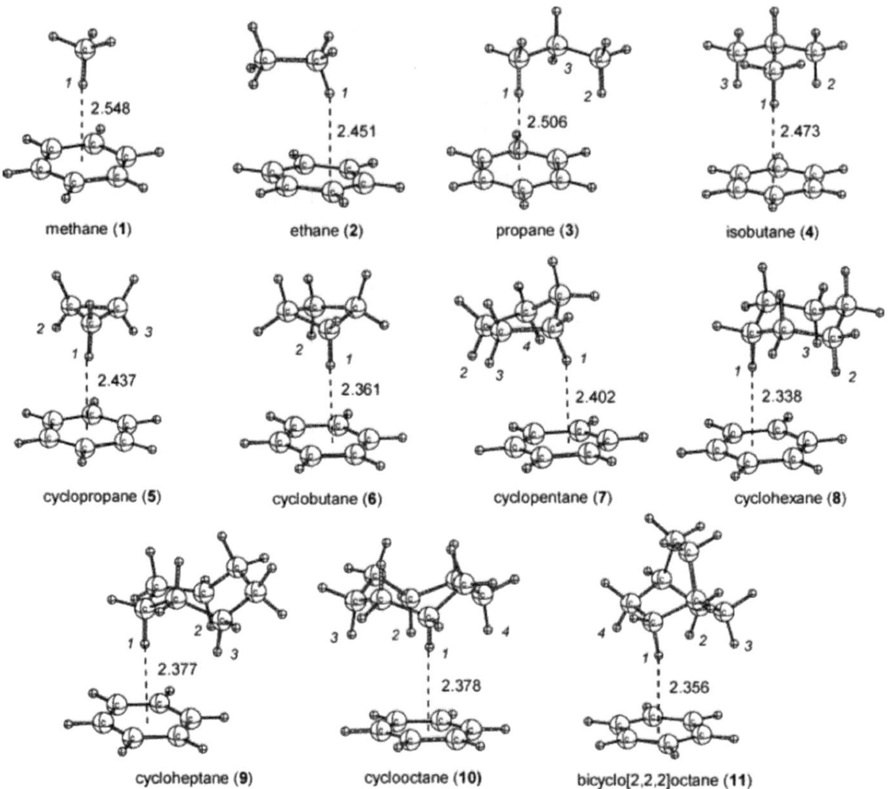

Figure 12.7   Optimized structure of benzene with saturated hydrocarbons. Reprinted with permission from J. Ran and M. W. Wong., Saturated Hydrocarbon-Benzene Complexes: Theoretical Study of Cooperative CH/π Interactions, *J. Phys. Chem. A*, 2006, 110, 9702. Copyright 2006 American Chemical Society.

Smith and Patkowski[32] systematically investigated interactions between methane and polycyclic aromatic hydrocarbons (PAHs). Their results showed that the singly coordinated minimum configuration was the global minimum but not, however, for naphthalene and larger PAHs, *i.e.*, the triply coordinated configuration was the global minimum, as displayed in Figure 12.8. It should be noted that interaction energies are not proportional to the number of coordination. The energy of the CH⋯π H-bond highly depends on C–H⋯π angle.[17] In the case that the guest molecule is fixed as methane, interaction energies per CH bond become small as the number of CH bonds increases.

As another important point, the so-called packing effect in crystals is one of the reasons for the cooperative effect. Kobayashi and Saigo[33] performed theoretical calculations at periodic boundary conditions

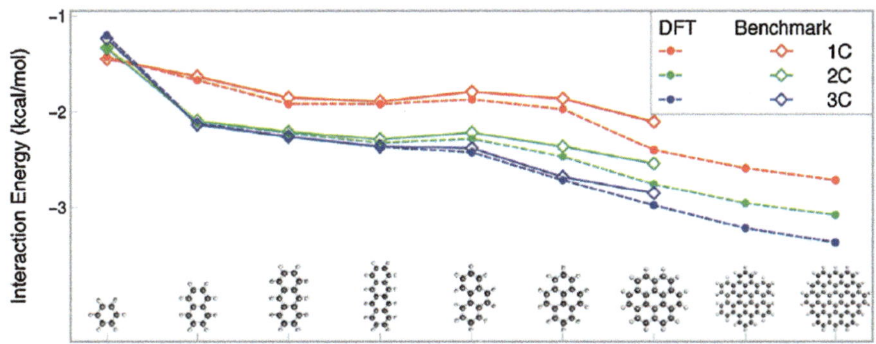

Figure 12.8 Comparison of the interaction energies calculated by different approaches for the 1C, 2C, and 3C minimum structures of all PAH. The DFT results are computed at the B3LYP-D3/aug-cc-pVDZ level. The "Benchmark" values are calculated at the MP2 + DCCSD(T) level. Reprinted with permission from D. G. A. Smith and K. Patkowski., Interactions between Methane and Polycyclic Aromatic Hydrocarbons: A High Accuracy Benchmark Study, *J. Chem. Theory Comput.*, 2013, 9, 370. Copyright 2013 American Chemical Society.

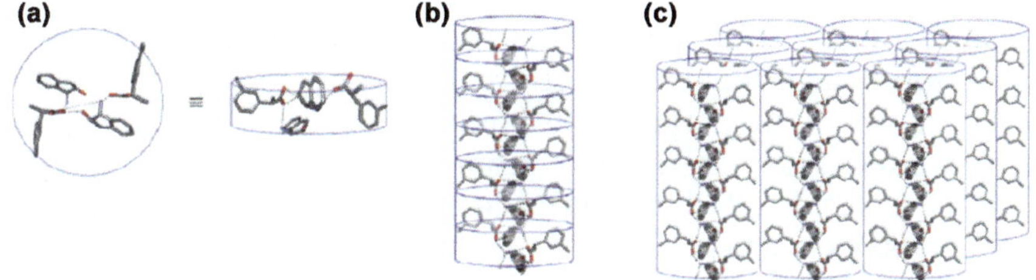

Figure 12.9 Three models. (a) A minimal molecular unit (MMU), as the smallest isolated molecular cluster, (b) a one-dimensional helical column (1D), and (c) a three-dimensional crystal (3D).
Reprinted with permission from Y. Kobayashi and K. Saigo., Periodic ab Initio Approach for the Cooperative Effect of CH/*n* Interaction in Crystals: Relative Energy of CH/n and Hydrogen-Bonding Interactions, *J. Am. Chem. Soc.*, 2005, 127, 15054. Copyright 2005 American Chemical Society.

for crystal structures of diastereomeric salts of mandelic acid derivatives with *p*-methyl-1-phenulethylamine and amino alcohols. Their calculations were carried out for three kinds of molecular arrangements. A minimal molecular unit (MMU), as the smallest isolated molecular cluster; a one-dimensional helical column (1D), and a three-dimensional crystal (3D), as shown in Figure 12.9. Their theoretical calculations were carried out at the Hartree–Fock (HF) level. Although estimation of binding energy of CH· · ·π H-bonds using the HF level is not necessarily suitable, those in the crystal model

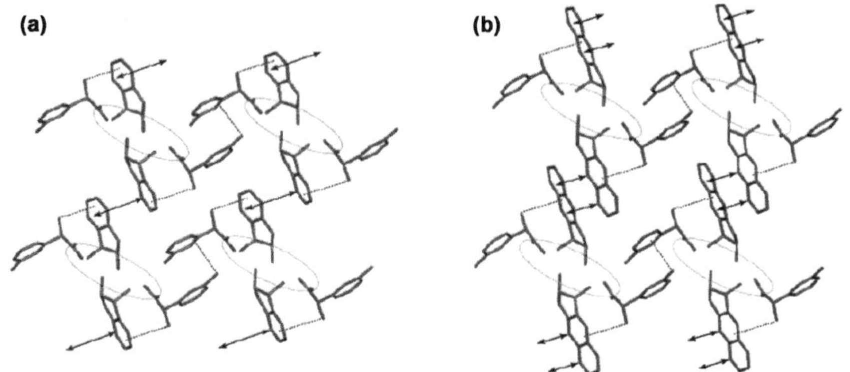

Figure 12.10 Packing modes of helical columns by CH···π H-bonds in (a) **2** (b) **3**. Arrows and the blue dotted lines show the CH···π H-bonds and the circles show OH···O hydrogen bonds.
Reprinted with permission from Y. Kobayashi and K. Saigo., Periodic ab Initio Approach for the Cooperative Effect of CH/π Interaction in Crystals: Relative Energy of CH/π and Hydrogen-Bonding Interactions, *J. Am. Chem. Soc.*, 2005, 127, 15054. Copyright 2005 American Chemical Society.

predicted an attractive interaction. They concluded that such an attractive force including CH···π H-bonding is caused by the cooperative effect originating from surrounding molecules, and suggested that the effect may largely arise from the so-called static packing, which is a result of the electrostatic contribution of long-range interactions in crystals. Figure 12.10 depicts packing modes of helical columns by CH···π H-bonds in **2** and **3** (Scheme 12.2). Substituting the phenyl group with a naphthyl group in ammonium cations in the salt crystal, leads to a different CH···π H-bonding network being constructed.

Sureshan *et al.*[34] studied the structure of (±)-3,6-di-*O*-benzyl-1,2-*O*-isopropylidene-*allo*-inositol **4** which showed evidence for the CH···π H-bond both by X-ray crystallography and NMR (Scheme 12.3). One of the two ether groups in the molecule adopts a gauche conformation stabilized by strong intramolecular CH···π H-bonding, the other adopts the anti-conformation stabilized by intermolecular CH···π H-bonds in the crystal. Both intra- and intermolecular CH···π H-bonds manifest cooperative edge-to-face CH···π networks, as shown in Figure 12.11.

Recently, Kumar *et al.*[35] examined the cooperative influence of non-covalent interactions such as CH···O and CH···π H-bonds in Ni(II)/Pd(II) heteroleptic dithio-dipyrrin complexes, illustrated in Figure 12.12. To characterize these compounds, they determined

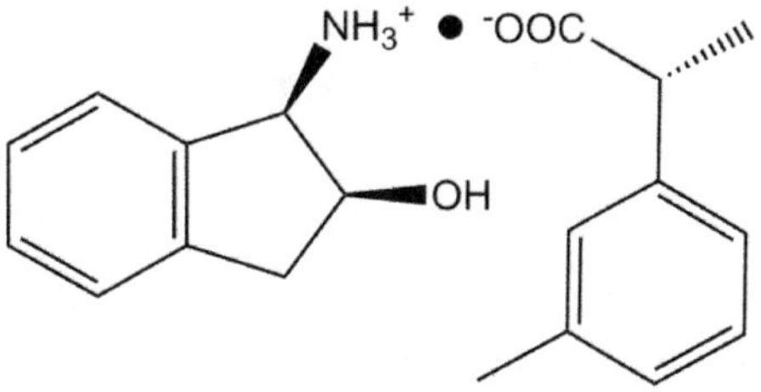

2

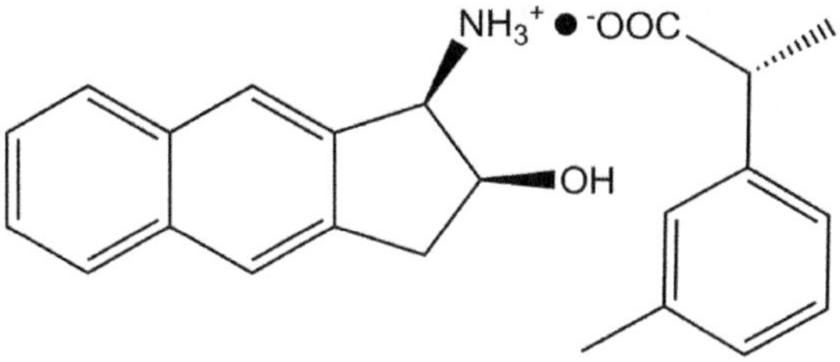

3

Scheme 12.2

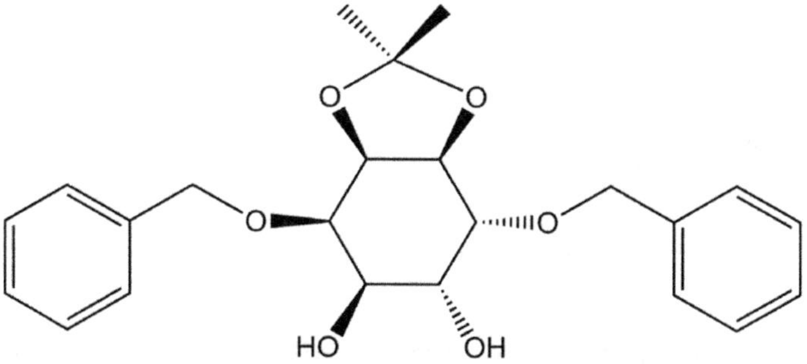

4

Scheme 12.3

NMR, IR, and UV-Vis spectra, and X-ray diffraction. Furthermore, theoretical calculations with DFT were also performed to know the positions of hydrogen atoms.

Dziubek *et al.*[36] reported a crystallographic study of phenylacetylene (PA) under high pressure. They named a new phase of this molecule as the β phase in which there are both the $\equiv$CH$\cdots\pi$ (C$\equiv$C) and $\equiv$CH$\cdots\pi$ (arene) H-bonds in operation. This phase is the most stable phase of PA from its freezing pressure at 0.4 to 1.25 GPa. They

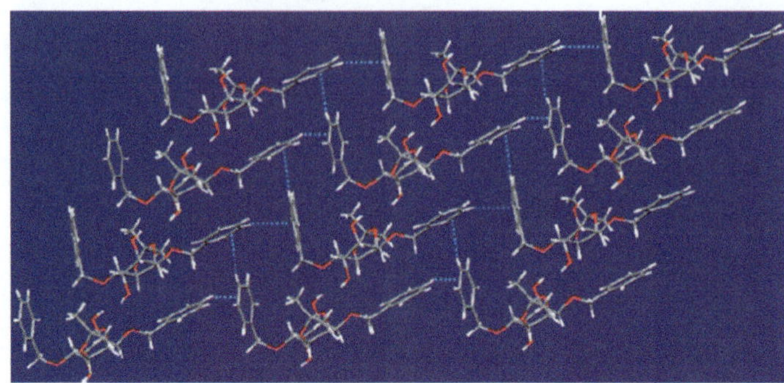

Figure 12.11   Cooperative edge-to-face CH···π H-bonding network of **4**.
Reprinted from ref. 34 with permission from the Royal Society of Chemistry.

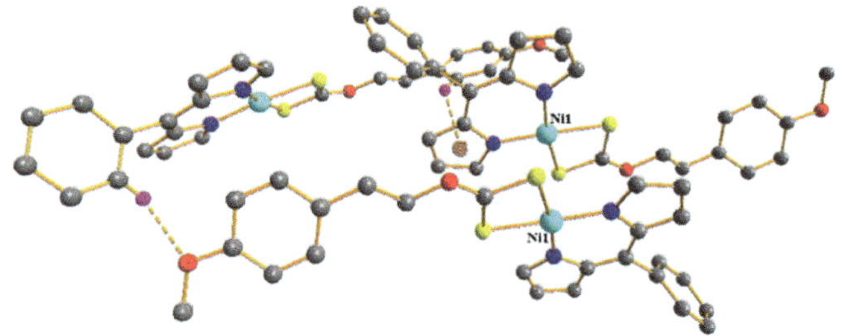

Figure 12.12   Intermolecular OH···C and CH···π H-bond in Ni(II)/Pd(II) heteroleptic dithio-dipyrrin complexes.
Reprinted from Polyhedron, 89, V. Kumar, V. Singh, A. N. Gupta, S. K. Singh, M. G.B. Drew, N. Singh., Cooperative influence of ligand frameworks in sustaining supramolecular architectures of Ni(II)/Pd(II) heteroleptic dithiodipyrrin complexes via non-covalent interactions, 304–312. Copyright 2015 with permission from Elsevier.

concluded in the paper that this illustrates the stabilizing role of CH···π contacts, particularly at high pressure, since their energy is considerably larger as compared to other interactions. Steiner already pointed out the formation of cooperative CH···π networks for compounds including terminal alkynes (Figure 12.13), in 1995.[37,38]

Zhao *et al.*[39] designed a system, shown in Scheme 12.4, to examine the strength of intramolecular CH···π H-bonds and studied cooperativity of the CH···π H-bond in the solid state and in solution. They measured folding energies ΔG as a function of alkyl carbons forming intramolecular CH···π H-bonds; this is shown in Figure 12.14. The folding energies were measured by integration of the ^{1}H NMR spectra (23 °C, CDCl$_3$). The additivity of the CH···π H-bonds in methyl and ethyl arms was shown. However, when

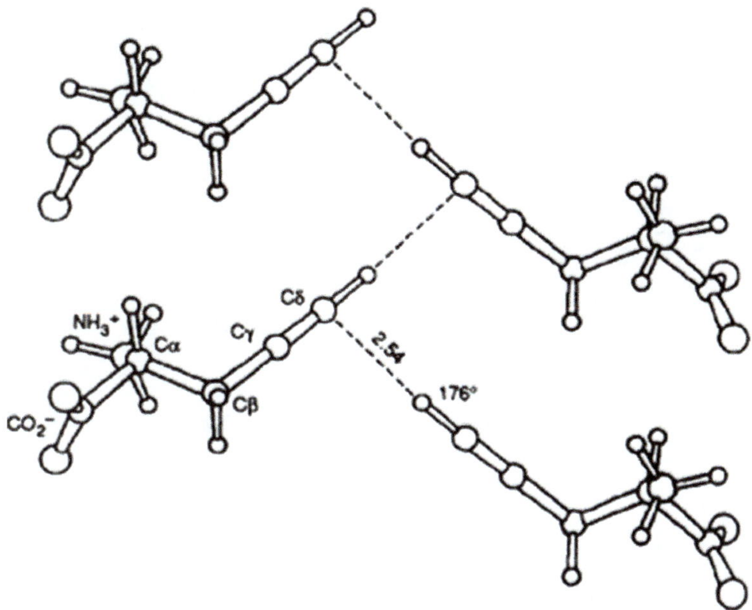

Figure 12.13 Molecular conformation and cooperative CH···π H-bonds in crystals including terminal alkynes.
Reproduced from ref. 37 with permission from the Royal Society of Chemistry.

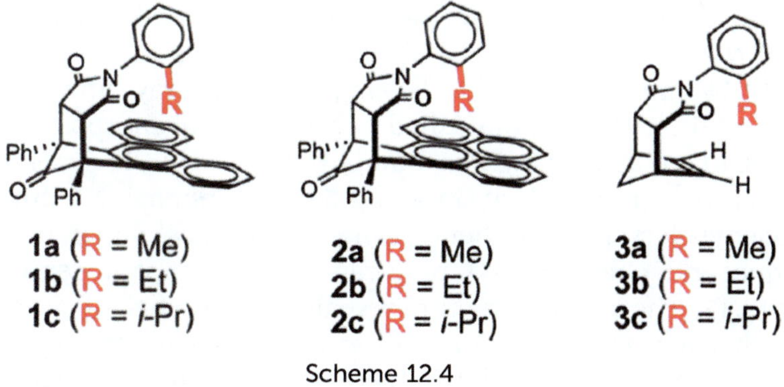

1a (R = Me)
1b (R = Et)
1c (R = *i*-Pr)

2a (R = Me)
2b (R = Et)
2c (R = *i*-Pr)

3a (R = Me)
3b (R = Et)
3c (R = *i*-Pr)

Scheme 12.4

there was an *iso*-propyl arm the interaction energy was higher than predicted. This means strong positive cooperativity is operating.

Reddy *et al.*[40] demonstrated a stabilization of ribbon-like guanosine derivatives by various H-bonds such as NH···O, NH···N, OH···N, OH···O, and CH···π. Each specific interaction was probed by means of NMR chemical shifts calculated using the DFT gauge-including projector-augmented wave (GIPAW) approach for the full crystal

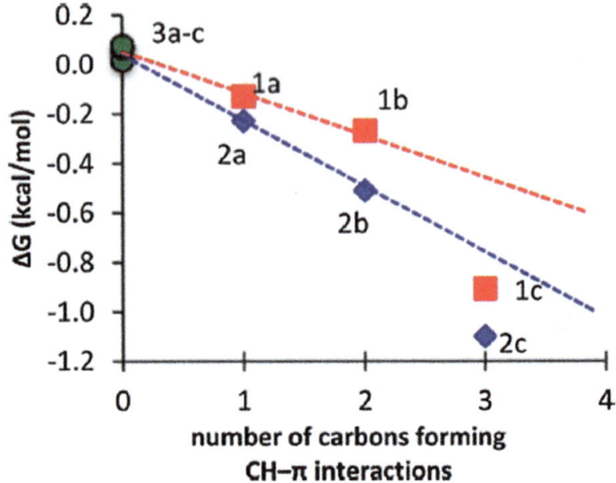

Figure 12.14   Folding energy ΔG (in kcal mol^{-1}) as a function of number of carbons forming CH$\cdots$π H-bonds.
Reprinted with permission from C. Zhao, P. Li, M. D. Smith, P. J. Pellechia and K. D. Shimizu., Experimental Study of the Cooperativity of CH-π Interactions, *Org. Lett.*, 2014, 16, 3520. Copyright 2014 American Chemical Society.

and extracted isolated single molecules. An overview of inter- and intramolecular H-bonding interactions in 3′,5′-*O*-isopropylideneguanosine hemihydrate is shown in Figure 12.15.

Tokutome and Okuno[41] prepared a novel ynamine compound, N^1,N^1,N^4,N^4-tetraphenylbuta-1,3-diyne-1,4-diamine, shown in Figure 12.16, and characterized by single crystal X-ray diffraction, ^{1}H and ^{13}C NMR. The compound had two crystal polymorphs, and the difference in crystal packing was found to originate in CH$\cdots$π H-bonds.

Karle *et al.*[42] prepared crystals of 2,5-di-(3-biphenyl)-1,1-dimethyl-3,4-diphenylsilacyclo-pentadiene using the physical vapor transport technique and characterized by X-ray crystallography, depicted in Figure 12.17. The crystal structure reveals the presence of several CH groups that act as donors for CH$\cdots$π H-bonds with nearby phenyl groups in an edge-to-face motif. Theoretical electronic structure calculations indicate that there is a decoupling between the CH$\cdots$π H-bonds that stabilizes the crystal structure and the electronic interactions that give the largest electronic couplings defining the preferred electron hopping directions.

Raffo *et al.*[43] studied crystalline structures of some alkoxy-substituted benzoic acids. The synthon is a H-bonded head-to-head dimer with CH$\cdots$π H-bonds. For 3-methoxy-benzoic acids, CH$\cdots$π H-bond involving C atoms at the 4-positions of neighboring molecules

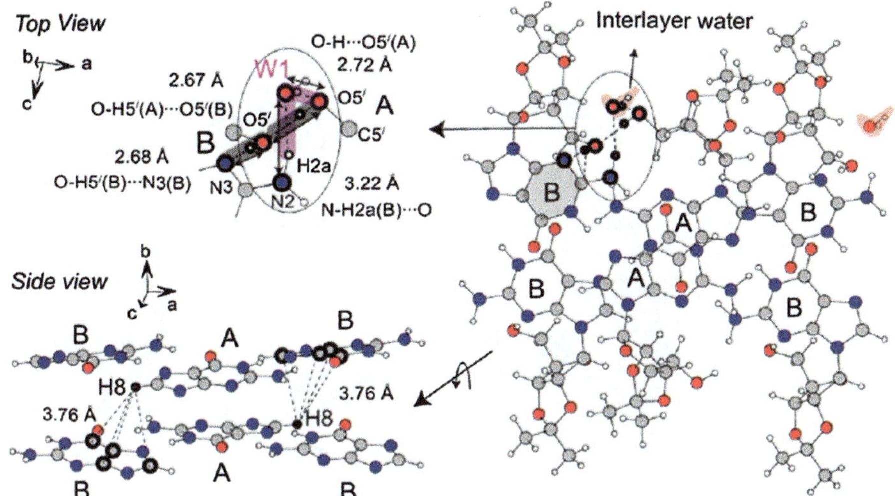

Figure 12.15 Overview of inter- and intramolecular H-bonding interactions in 3',5'-O-isopropylideneguanosine hemihydrate.
Reprinted with permission from G. N. M. Reddy, A. Marsh, J. T. Davis, S. Masiero and S. P. Brown., Interplay of Noncovalent Interactions in Ribbon-like Guanosine Self-Assembly: An NMR Crystallography Study, *Crystal Growth & Design*, 2015, 15, 5945. Copyright 2015 American Chemical Society.

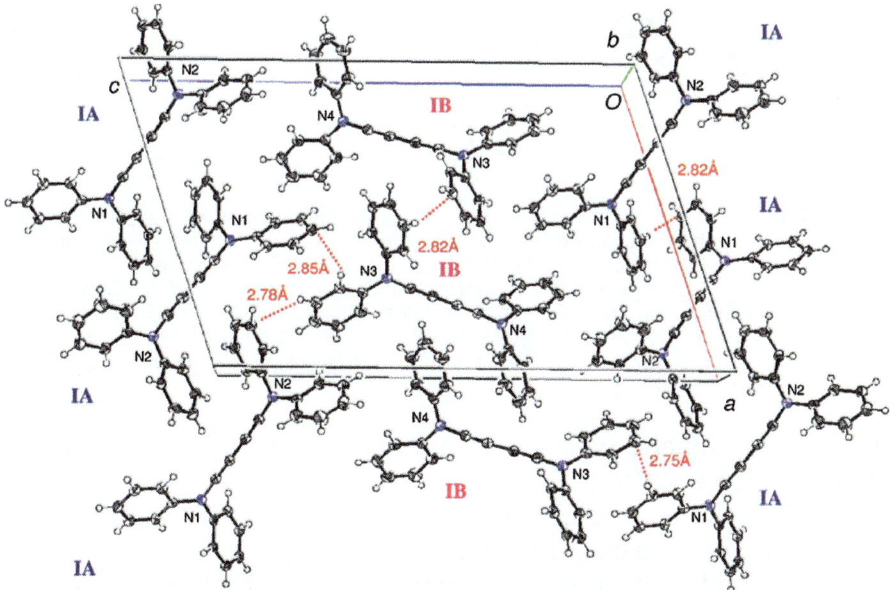

Figure 12.16 Crystal packing of N^1,N^1,N^4,N^4-tetraphenylbuta-1,3-diyne-1,4-diamine.
Reprinted from Journal of Molecular Structure, 1047, Tokutome Y. and Okune T., Preparations, crystal polymorphs and DFT calculations of N^1,N^1,N^4,N^4-tetraphenylbuta1,3-diyne-1,4-diamine, Copyright 2013 with permission from Elsevier.

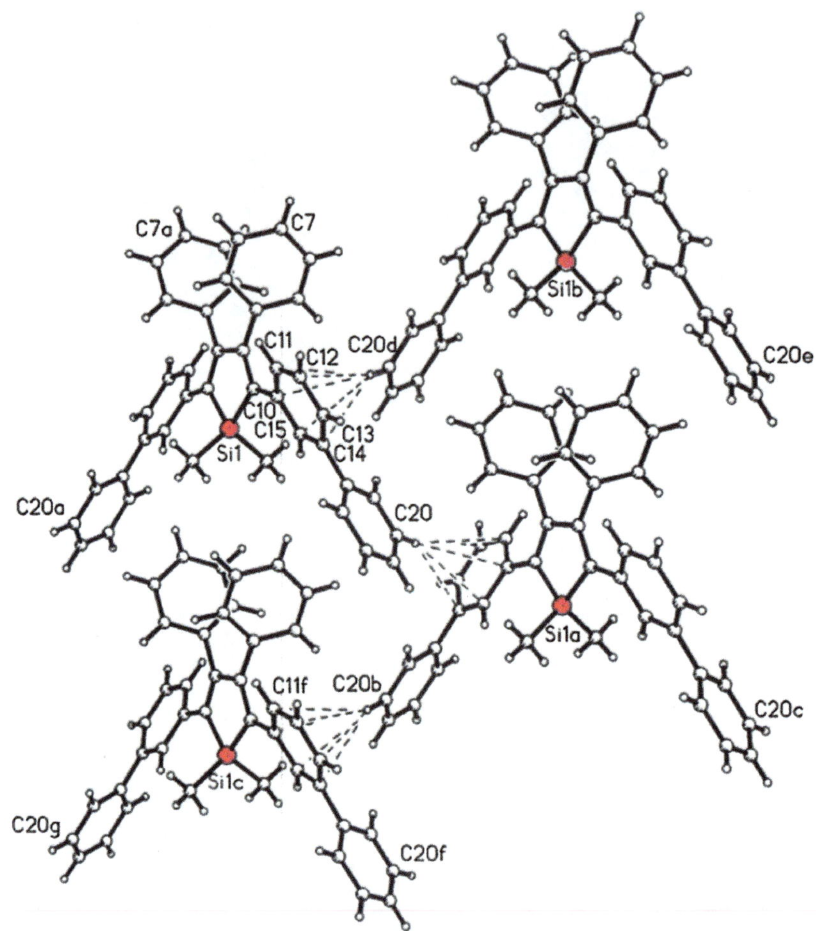

Figure 12.17　Crystal structure of 2,5-di-(3-biphenyl)-1,1-dimethyl-3,4-diphenylsila-cyclo-pentadiene (PPSPP). Dashed lines indicate CH···π H-bonds. Reprinted with permission from I. L. Karle, R. J. Butcher, M. A. Wolak, D. A. da Silva Filho, M. Uchida, J.-L. Brédas and Z. H. Kafafi., Cooperative CH···π Interactions in the Crystal Structure of 2,5-Di(3-biphenyl)-1,1-dimethyl-3,4-diphenyl-silole and its effect on its electronic properties, *J. Phys. Chem. C*, 2007, 111, 9543. Copyright 2007 American Chemical Society.

organize the dimeric synthons in a perpendicular orientation; this is shown in Figure 12.18.

On the basis of findings from weak hydrogen bonds in organic molecules and crystals, Birchall *et al.*[44] applied them to derive the self-assembly of aromatic carbohydrate amphiphiles in the formation of supramolecular hydrogels. To explore the self-assembly structure in solution, theoretical calculations were performed, and they proposed four possible dimers, shown in Figure 12.19. These

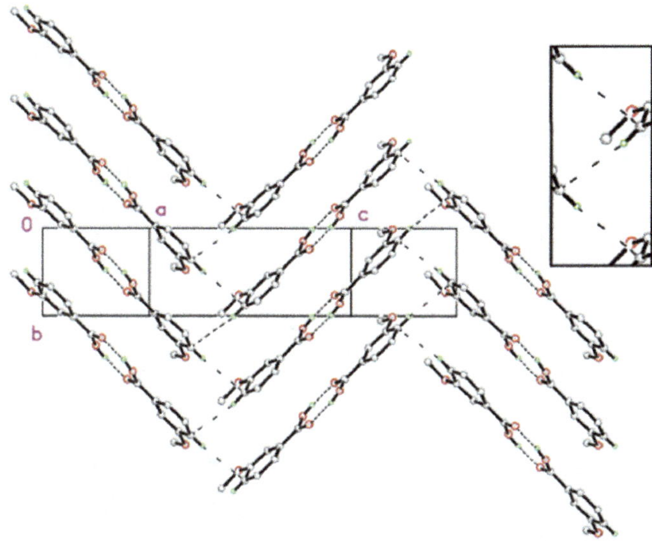

Figure 12.18 Crystal structure of 3-methoxy-benzoic acid. Dashed lines correspond to CH···π H-bonds.
Reprinted from Journal of Molecular Structure, 1070, P. A. Raffo,
l. Rossi, P. Alborés, R. F. Baggio, F. D. Cukiernik., Alkoxy-benzoic acids:
Some lacking structures and rationalization of the molecular features
governing their crystalline architectures, Copyright 2014 with permission from Elsevier.

include (i) *via* J-stacking which involves a mixture of XH···π H-bonds
between the sugar and the aromatic moiety and T-stacking between
the aromatic residues, (ii) aromatic–aromatic π–π stacking,
(iii) H-bonding between the sugars, and (iv) XH···π hydrogen bonds
between the aromatic group and the sugar, illustrated in Figure 12.19.

12.4.3 Predicting Crystal Structures of Organic Molecules

As discussed in the previous sections, the estimation of interaction
energy between molecules in crystals is now becoming possible to
discuss qualitatively. And molecular structures of moderate size can
be predicted using an *ab initio* level of theory. However, it is still
difficult to predict *a priori* the crystal structures of molecules.[45] The
difficulty in growing crystals means that the structure of the first
suitable crystal obtained becomes regarded as the crystal structure,
i.e., experimentally obtained crystal structures may not be the global
most stable structures. The development of computational algorisms
to obtain crystal structures may be needed. To overcome this difficulty, several computational strategies have been proposed. In the
1990s, an HF level code including periodic boundary conditions was

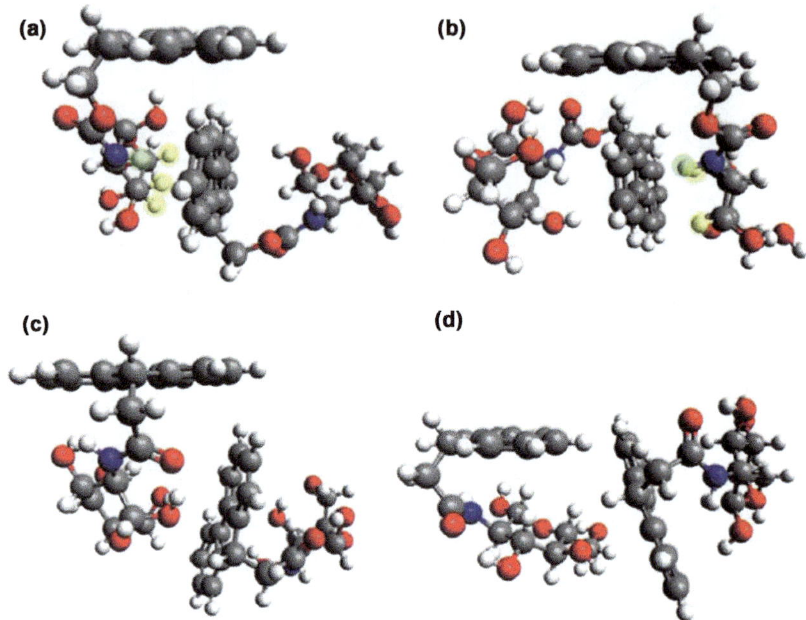

Figure 12.19 Dimer configurations of J-stacked dimers of a–d.
Reprinted from ref. 44 with permission from the Royal Society of Chemistry.

developed.[46] Later, this code was extended to the MP2 level of theory. Devarajan *et al.*[47] studied molecular interactions in cellulose Iα using an FMO-based method at the MP2 level. The fragment molecular orbital (FMO)[48] based method is a strong candidate. Beran *et al.*[49] proposed a fragment-based quantum/classical hybrid many-body interaction-model for infinite systems. Using their procedure, lattice energies for several different molecular crystals can be calculated to within a few $kJ\,mol^{-1}$ of the experimental values. As another approach, crystal structure calculations based on the DFT with dispersion force corrections are also possible. Ishikawa *et al.*[50] performed these for cellulose III$_I$ crystals to analyze intermolecular H-bonds.

For complex systems, it is still difficult to study the structure in crystals using computational methods such as *ab initio* MO theory by themselves. It is common practice to use such procedures in combination with experimental studies.

12.5   Summary and Prospect

We conclude that the CH$\cdots\pi$ H-bond plays an important role in crystal packing not only in organic but also in organometallic compounds. In an organic molecule, CH groups acting as H-donors and lone pairs in heteroatoms or π orbitals acting as H-acceptors are present in most cases. In this sense the CH··· π hydrogen bond is ubiquitous in organic systems. Although the energy of a one unit CH··· π H-bond is weak, it becomes much more significant because many interactions are possible in crystals.[51]

Computational power is becoming stronger, year-by-year, and useful codes have been provided recently. Moreover, accurate theoretical calculations at an *ab initio* level can be combined with experimental studies. Understanding of weak H-bonds in crystals will be realized. In the near future, structure predictions, by *ab initio* quantum chemical procedures without any assumptions, will become achievable.

Acknowledgements

O. T. was supported by a Grant-in-Aid for Scientific Research (C) (No.15K04755) from the Ministry of Education, Culture, Sports, Science, and Technology, Japan. The author thanks the Information Media Center at Hiroshima University for the use of a grid of high-performance PCs and the Research Center for Computational Science in Okazaki, Japan. O.T. thanks Prof. Ebata at Hiroshima University for his helpful support.

References

1. M. Nishio, M. Hirota and Y. Umezawa, *The CH/π Interaction. Evidence, Nature, and Consequences*, Wiley-VCH, NY, 1998.
2. M. Nishio and M. Hirota, *Tetrahedron*, 1989, **45**, 7201.
3. D. J. Sutor, *Nature*, 1962, **195**, 68; D. J. Sutor, *J. Chem. Soc.*, 1963, 1105.
4. N. N. L. Madhavi, A. K. Katz, H. L. Carrell, A. Nangia and G. R. Desiraju, *Chem. Commun.*, 1997, 1953.
5. R. Taylor and O. Kennard, *J. Am. Chem. Soc.*, 1982, **104**, 5063.
6. G. R. Desiraju and T. Steiner, *The Weak Hydrogen Bond in Structural Chemistry and Biology*, Oxford University Press, Oxford, 1999.
7. L. Pauling, *The Nature of the Chemical Bond*, Cornell Univ. Press, Ithaca, NY, 3rd edn, 1960, ch. 12.
8. G. C. Pimentel and A. L. McClellan, *The Hydrogen Bond*, W. H. Freeman, San Francisco, 1960, pp. 194–195.
9. E. Arunan, G. R. Desiraju, R. A. Klein, J. Sadlej, S. Scheiner, I. Alkorta, D. C. Clary, R. H. Crabtree, J. J. Dannenberg, P. Hobza, H. G. Kjaergaard, A. C. Legon, B. Mennucci and D. J. Nesbitt, *Pure Appl. Chem.*, 2011, **83**, 1619.
10. R. G. Pearson, *Science*, 1966, **151**, 172.
11. J. P. Hill, R. Scipioni, M. Boero, Y. Wakayama, M. Akada, T. Miyazaki and K. Ariga, *Phys. Chem. Chem. Phys.*, 2009, **11**, 6038.

12. J. J. Novoa and F. Mota, *Chem. Phys. Lett.*, 2000, **318**, 345; O. Takahashi, Y. Kohno and K. Saito, *Chem. Phys. Lett.*, 2003, **378**, 509; J. Oddershed and S. Larsen, *J. Phys. Chem. A*, 2004, **108**, 1057.

13. G. A. Geffrey, *An Introduction to Hydrogen Bonding*, Oxford University Press, Oxford, 1997.

14. M. Nishio, *CrystEngComm*, 2004, **6**, 130; M. Nishio, *Phys. Chem. Chem. Phys.*, 2011, **13**, 13873.

15. Y. Umezawa, S. Tsuboyama, H. Takahashi, J. Uzawa and M. Nishio, *Tetrahedron*, 1999, **55**, 10047.

16. H. Takahashi, S. Tsuboyama, Y. Umezawa, K. Honda and M. Nishio, *Tetrahedron*, 2000, **56**, 6185.

17. O. Takahashi, Y. Kohno, S. Iwasaki, K. Saito, M. Iwaoka, S. Tomoda, Y. Umezawa, S. Tsuboyama and M. Nishio, *Bull. Chem. Soc. Jpn.*, 2001, **74**, 2421.

18. T. D. Thomas, L. H. Sæthre and K. Børve, *Phys. Chem. Chem. Phys.*, 2007, **9**, 719.

19. A. Fujii, S. Morita, M. Miyazaki, T. Ebata and N. Mikami, *J. Phys. Chem. A*, 2004, **108**, 2652.

20. M. J. Tubergen, R. J. Lavrich, D. F. Plusquellic and R. D. Suenram, *J. Phys. Chem. A*, 2006, **110**, 13188.

21. N. W. Ulrich, N. A. Seifert, R. E. Dorris, R. A. Peebles, B. H. Pate and S. A. Peebles, *Phys. Chem. Chem. Phys.*, 2014, **16**, 8886.

22. S. Bracco, A. Comotti, P. Valsesia, M. Beretta and P. Sozzani, *CrystEngComm*, 2010, **12**, 2318.

23. K. Kitaura and K. Morokuma, *Int. J. Quantum Chem.*, 1976, **10**, 325.

24. B. Jeziorski, R. Moszynski and K. Szalewicz, *Chem. Rev.*, 1994, **94**, 1887.

25. W. Chen and M. S. Gordon, *J. Phys. Chem.*, 1996, **100**, 14316.

26. H. L. Williams and C. F. Chabalowski, *J. Phys. Chem. A*, 2000, **105**, 646.

27. S. Tsuzuki, K. Honda, T. Uchimaru, M. Mikami and K. Tanabe, *J. Am. Chem. Soc.*, 2000, **122**, 3746.

28. J. Ran and M. W. Wong, *J. Phys. Chem. A*, 2006, **110**, 9702.

29. K. E. Riley, M. Pitoňák, J. Černý and P. Hobza, *J. Chem. Theory Comput.*, 2009, **6**, 66.

30. J. W. G. Bloom, R. K. Raju and S. E. Wheeler, *J. Chem. Theory Comput.*, 2012, **8**, 3167.

31. M. Albertí, A. Aguilar, J. M. Lucas and F. Pirani, *J. Phys. Chem. A*, 2012, **116**, 5480.

32. D. G. A. Smith and K. Patkowski, *J. Chem. Theory Comput.*, 2013, **9**, 370.

33. Y. Kobayashi and K. Saigo, *J. Am. Chem. Soc.*, 2005, **127**, 15054.

34. K. M. Sureshan, T. Uchimaru, Y. Yao and Y. Watanabe, *CrystEngComm*, 2008, **10**, 493.

35. V. Kumar, V. Singh, A. N. Gupta, S. K. Singh, M. G. B. Drew and N. Singh, *Polyhedron*, 2015, **89**, 304.

36. K. Dziubek, M. Podsiadło and A. Katrusiak, *J. Am. Chem. Soc.*, 2007, **129**, 12620.

37. T. Steiner, *J. Chem. Soc., Chem. Commun.*, 1995, 95–96.

38. T. Steiner, E. B. Starikov, A. M. Amado and J. J. C. Teixeira-Dias, *J. Chem. Soc., Perkin Trans. 2 (1972-1999)*, 1995, 1321–1326.

39. C. Zhao, P. Li, M. D. Smith, P. J. Pellechia and K. D. Shimizu, *Org. Lett.*, 2014, **16**, 3520.

40. G. N. M. Reddy, A. Marsh, J. T. Davis, S. Masiero and S. P. Brown, *Cryst. Growth Des.*, 2015, **15**, 5945.

41. Y. Tokutome and T. Okuno, *J. Mol. Struct.*, 2013, **1047**, 136.

42. I. L. Karle, R. J. Butcher, M. A. Wolak, D. A. da Silva Filho, M. Uchida, J.-L. Brédas and Z. H. Kafafi, *J. Phys. Chem. C*, 2007, **111**, 9543.

43. P. A. Raffo, L. Rossi, P. Alborés, R. F. Baggio and F. D. Cukiernik, *J. Mol. Struct.*, 2014, **1070**, 86.

44. L. S. Birchall, S. Roy, V. Jayawarna, M. Hughes, E. Irvine, G. T. Okorogheye, N. Saudi, E. De Santis, T. Tuttle, A. A. Edwards and R. V. Ulijn, *Chem. Sci.*, 2011, **2**, 1349.
45. S. L. Price, *Acc. Chem. Res.*, 2009, **42**, 117.
46. R. Dovesi, M. Causa, R. Orlando, C. Roetti and V. R. Saunders, *J. Chem. Phys.*, 1990, **92**, 7402.
47. A. Devarajan, S. Markutsya, M. H. Lamm, X. Cheng, J. C. Smith, J. Y. Baluyut, Y. Kholod, M. S. Gordon and T. L. Windus, *J. Phys. Chem. B*, 2013, **117**, 10430.
48. *The Fragment Molecular Orbital Method. Practical Applications to Large Molecular Systems*, ed. D. G. Fedorov and K. Kitaura, CRC Press, NY, 2009.
49. G. J. O. Beran and K. Nanda, *J. Phys. Chem. Lett.*, 2010, **1**, 3480–3487.
50. T. Ishikawa, D. Hayakawa, H. Miyamoto, M. Ozawa, T. Ozawa and K. Ueda, *Carbohydr. Res.*, 2015, **417**, 72.
51. *The Importance of Pi-Interactions in Crystal Engineering*, ed. E. R. T. Tiekink and J. Zukerman-Schpector, Wiley, NY, 2012.
52. O. Takahashi, CH. . .π interaction in organic molecules, in *Noncovalent Forces*, ed. S. Scheiner, Springer International Publishing, 2015.
53. J. W. G. Bloom, R. K. Raju and S. E. Wheeler, *J. Chem. Theory Comput.*, 2012, **8**, 3167.
54. B. K. Mishra, M. M. Deshmukh and R. Venkatnarayan, *J. Org. Chem.*, 2014, **79**, 8599.
55. J. L. Brédas and G. B. Street, *J. Chem. Phys.*, 1989, **90**, 7291.
56. S. Sakaki, K. Kato, T. Miyazaki, Y. Musashi, K. Ohkubo, H. Ihara and C. Hirayama, *J. Chem. Soc., Faraday Trans.*, 1993, **89**, 659.
57. A. Calderone, R. Lazzaroni and J. L. Brédas, *Synth. Met.*, 1998, **95**, 1.
58. P. Tarakeshwar, H. S. Choi and K. S. Kim, *J. Am. Chem. Soc.*, 2001, **123**, 3323.
59. D. G. A. Smith and K. Patkowski, *J. Chem. Theory Comput.*, 2013, **9**, 370.

13 Hydrogen Bonds and Halogen Bonds – A Comparative Study

Sławomir J. Grabowski[a,b]

[a] Faculty of Chemistry, University of the Basque Country and Donostia, International Physics Center (DIPC), P.K. 1072, 20080 Donostia, Spain;
[b] IKERBASQUE, Basque Foundation for Science, 48011 Bilbao, Spain
Email: s.grabowski@ikerbasque.org

13.1 Introduction

The hydrogen bond (HB) is the subject of numerous studies, since its importance in various physical, chemical and biological processes is well known.[1,2] However in recent years other so-called noncovalent interactions are often investigated; among them the halogen bond (XB) is one of the most often analyzed.[3] In general both, HB and XB, are classified as Lewis acid–Lewis base interactions;[4] the following schemes: A–H···B and A–X···B are applied for them hereafter, respectively; where B designates the Lewis base centre, H and X are the hydrogen and halogen atoms, while A is any atom. The latter is, for HB, usually the electronegative centre while for XB it is most often the carbon atom. However other cases of A–X and A–H bonds involved in XB and HB interactions are known; for example, Sb–X and Si–X[5] or the proton donating bonds containing the A-centre not characterized by high electronegativity, like C–H and SiH.[6]

The first studies on XB interactions were performed over two hundred years ago.[7,8] More recent studies are mentioned here. O. Hassel reported X-ray crystal analyses[9] where the complexes

Intermolecular Interactions in Crystals: Fundamentals of Crystal Engineering
Edited by Juan J. Novoa
© The Royal Society of Chemistry 2018
Published by the Royal Society of Chemistry, www.rsc.org

between dihalogens or halocarbons and electron donor moieties were identified. The analysis of crystal structures containing halogen bond links was also performed by H. A. Bent;[10] important characteristics of XB interaction were described there; for example, short $X \cdots B$ distances, often shorter than the corresponding sum of van der Waals radii, and the directionality of the XB interaction. Legon carried out an analysis based on the microwave spectroscopy of a variety of halogen bonded complexes in the gas phase.[11,12] The geometric characteristics of XBs were discussed by Ramasubbu *et al.*[13] who performed a detailed statistical analysis based on crystal structures taken from the Cambridge Structural Database (CSD).[14] One can also mention numerous studies by Metrangolo, Resnati and co-workers where the halogen bond interactions are systematized and where their properties are discussed.[15,16]

Various monographs and review articles on the halogen bond have been published so far.[3,10,11,16–21] This is why this chapter does not pretend to be a review but is targeted on a comparison between the halogen bond and the hydrogen bond. Since for both HB and XB interactions H and X centres may act as electron acceptors, thus there are numerous studies where these interactions are compared and where various similarities between them are discussed. However, it seems that the detailed comparison between the HB and the XB, based on former and recent experimental and theoretical results, has not been performed yet.

13.2 The Dual Character of Halogen and Hydrogen Centres

There are numerous and various definitions of the HB and XB interactions; one of the first definitions of the hydrogen bond was presented by Pauling who stated that *"under certain conditions an atom of hydrogen is attracted by rather strong forces to two atoms, instead of only one, so that it may be considered to be acting as a bond between them. This is called the hydrogen bond".*[22] Pauling pointed out that the hydrogen atom is located only between the most electronegative atoms and that it usually interacts much more strongly with one of them; the latter interaction is simply the covalent bond (A–H); the other interaction ($H \cdots B$) is much weaker and mostly electrostatic in nature.[22] This definition does not cover numerous interactions that were analyzed later and which were classified as hydrogen bonds; for example interactions where the C–H bond plays a role of the proton donor[6] or those where the A–H bond is in contact with a π-electron system, playing the role of the Lewis base; acetylene, ethylene, aromatic moieties and other π-electron systems may be mentioned.[23] One can also mention the hydrogen bonds where the σ-electrons of molecular hydrogen are the acceptor of a proton.[24,25] Such interactions were covered early on by the definition of Pimentel and McClellan[26] where it is stated that:

"A hydrogen bond exists between a functional group A–H and an atom or a group of atoms B in the same or a different molecule when

(a) There is evidence of bond formation (association or chelation),
(b) There is evidence that this new bond linking A–H and B specifically involves the hydrogen atom already bonded to A."

However, it is not precisely defined here what does it mean "evidence of bond formation"; besides, one can see that according to this definition the hydride bond may be also classified as hydrogen bond. In the hydride bond the negatively charged hydrogen, $H^{-\delta}$, acting as the Lewis base centre, is linked with the Lewis acid center, A; one can designate this link as $B–H^{-\delta}\cdots A$; such interaction was also named in literature as the inverse hydrogen bond.[27,28]

The recent definition recommended by IUPAC seems to resolve the problem of the hydride bond that is not classified as an HB: *"The hydrogen bond is an attractive interaction between a hydrogen atom from a molecule or a molecular fragment X–H in which X is more electronegative than H, and an atom or a group of atoms in the same or a different molecule, in which there is evidence of bond formation."*[29] Similarly, the IUPAC definition of halogen bond may be cited; *"A halogen bond occurs when there is evidence of a net attractive interaction between an electrophilic region associated with a halogen atom in a molecular entity and a nucleophilic region in another, or the same, molecular entity."*[30]

A few other definitions of HBs and XBs are mentioned later in this chapter. Referring to the IUPAC definitions one can see that the key word "evidence" occurs there, similarly as in the definition of Pimentel and McClellan,[26] that does not allow precise definition of the hydrogen and halogen bonds. However, the following question concerning the halogen bond interaction arises; why such electronegative centres as halogen atoms may act as the electron acceptors; *i.e.* why they may act as the Lewis acid centres.

The dual character of halogens was observed early on; for example organic crystal structures were analyzed and it was found that the C–X bonds interact with nucleophiles approximately in the elongation of these bonds, while the X halogen atoms form links with electrophilic moieties approximately in the orthogonal direction.[13] This was explained as being connected to the electron density anisotropy of halogen atoms, expressed by the electron charge depletion in the C–X direction and with its accumulation in the orthogonal direction (Scheme 13.1).[31] It was also explained that the anisotropy of the halogens' van der Waals radii observed[32] is a consequence of such charge distribution. It means that the same halogen centre may play the role of the Lewis acid and of the Lewis base – it depends on the site of the X-centre. Figure 13.1 presents a fragment of the crystal structure of 2,5-dibromoisophthalic acid monohydrate[33] as an example where various interactions may be observed; the bromine centre's contacts are indicated (Figure 13.1) with the nucleophile (the O-centre), approximately along the C–Br direction and with the electrophile (the H-atom of the water molecule). In other words, the Br-centre participates in a C–Br···O halogen bond as well as in an O–H···Br hydrogen bond as the Lewis acid and as the Lewis base, respectively.

The source of the anisotropy of the electron density of halogen atoms as well as their dual character, especially if they are connected with carbon atoms, is explained by the σ-hole concept[5,35,36] that is described in detail in the other chapters of this monograph. Briefly speaking, for the X-atom the positive region of the electrostatic potential (EP) often appears in the elongation of A–X bond that is the consequence of the electron depletion of the outer lobe of the half-filled p-orbital of the X atom which is involved in the formation of the bond to the A-centre. The positive EP in the extension of the A–X bond

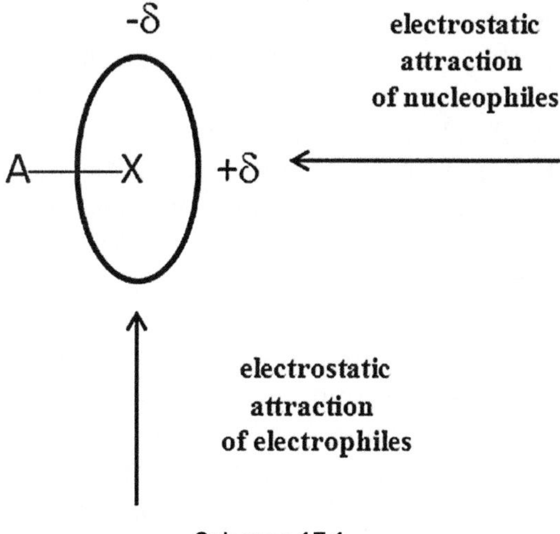

Scheme 13.1

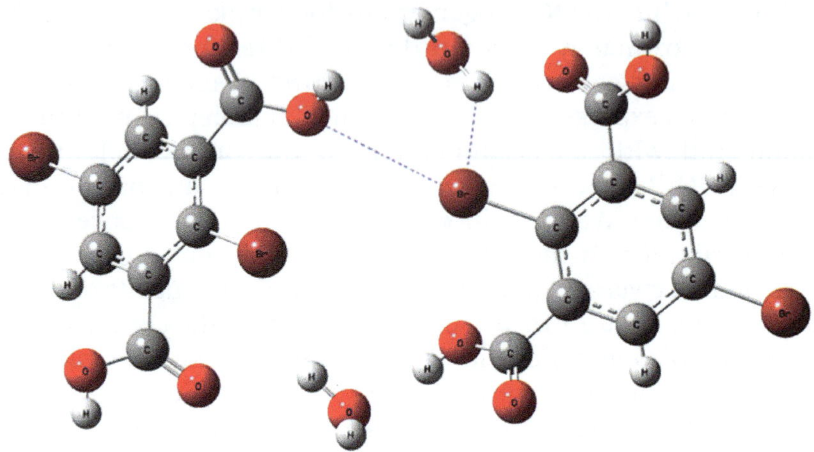

Figure 13.1 A fragment of the crystal structure of 2,5-dibromoisophthalic acid monohydrate;[33] the crystal structure taken from Cambridge Structural Database[14] and visualized by GaussView 5.0.9 program;[34] CSD refcode of the crystal structure: RELROR. All the following figures in this chapter where the crystal structure fragments are presented are taken from CSD[14] and all are visualized by GaussView program.[34]

explains the Lewis acid properties of halogen centres, thus the emphasis is put on the electrostatic nature of the XB interaction in the definition proposed by Politzer, Murray and Clark; "*A halogen bond is a highly directional, electrostatically-driven noncovalent interaction between a region of positive electrostatic potential on the outer side of the halogen X in a molecule R–X and a negative site B, such as a lone pair of a Lewis base or the π-electrons of an unsaturated system.*"[35]

The σ-hole concept explains not only the nature of halogen bonds but also of other interactions where the σ-holes, characterized by a positive EP, play the role of Lewis acid centres. One can mention chalcogen, pnicogen and tetrel bonds where the Groups 16, 15 and 14 elements, respectively, act as the electron acceptors.[5,35,36] Recently it was justified that even noble atoms (Group 18) in numerous compounds possess σ-holes with corresponding regions of the positive EP, and that they may act as Lewis acid centres.[37]

The A–H⋯B hydrogen bond is also classified as a σ-hole bond interaction[38] since the H-atom connected with the more electronegative A-centre is characterized by the enhanced positive EP. Figure 13.2 presents the EP maps for the 0.001 au electron density surfaces of the CF_3X molecules (X = F, Cl and Br) while Figure 13.3 shows the EP map of the CF_3H moiety for comparison; the X and H centres may be involved in XB and HB interactions, respectively. The increase of the EP attributed to the σ-hole is observed with the

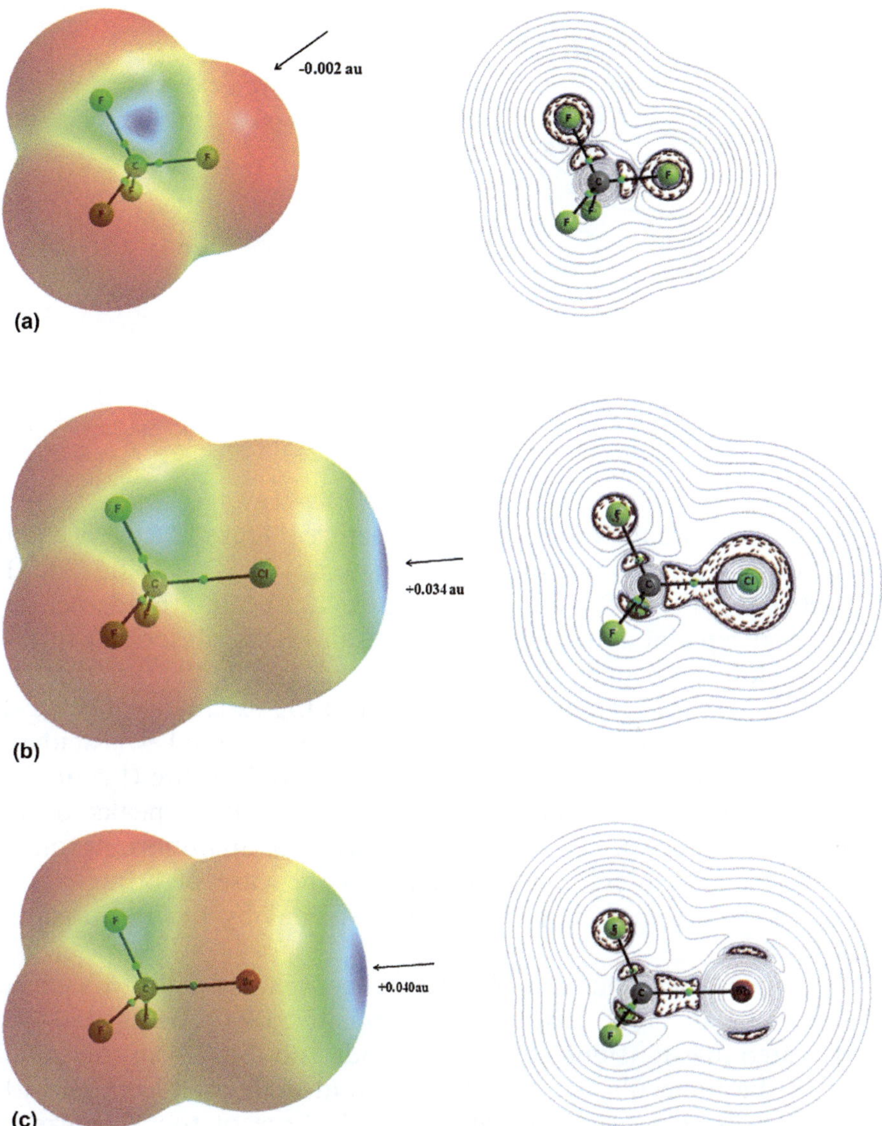

Figure 13.2   The maps of the electrostatic potential calculated at the 0.001 au molecular electron density surfaces (left side) and the corresponding molecular graphs (right side) for the following molecules (a) CF_4, (b) CF_3Cl, (c) CF_3Br. In the EP maps, red and blue colors correspond to negative and positive EP, respectively. The EP maxima are designated by arrows. For molecular graphs solid lines correspond to bond paths, big circles to attractors and small circles to bond critical points, the isolines of the Laplacian of electron density are presented; positive values are depicted in solid lines and negative values in broken lines (this is a rule for other figures of this chapter based on calculations); MP2/aug-cc-pVTZ level of calculations.

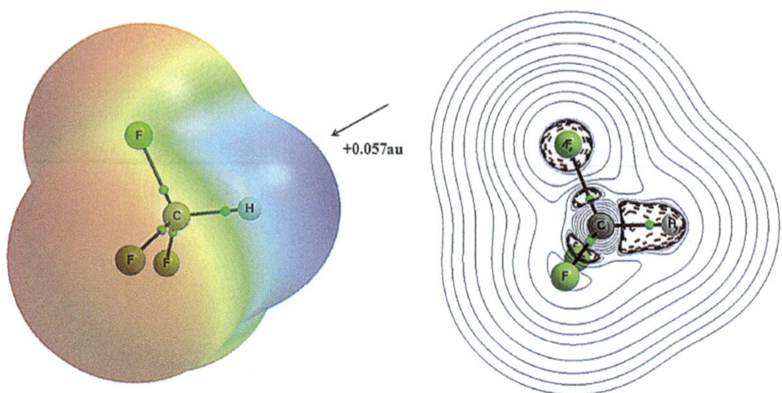

Figure 13.3 The map of the electrostatic potential (0.001 au molecular electron density surface) and the molecular graph with the isolines of the Laplacian of electron density for CF_3H molecule.

increase of the atomic number of the X-centre (Figure 13.2); it is equal to -0.002, $+0.034$, $+0.040$ and $+0.049$ au for F, Cl, Br and I, respectively (CF_3I is not shown in Figure 13.2). This trend was observed and explained in earlier studies;[39] also it was described earlier that the fluorine usually does not form halogen bonds (for the CF_4 presented here a negative EP is observed in the extension of the C–F bond). However, rare cases of the F-centre acting as the Lewis acid are known. The maximum EP of 0.057 au is observed for the H-centre of the CF_3H molecule. The positive EP at X and H centres speaks for the electrostatic nature of XBs and HBs or at least for an important role of the electrostatic interaction, similarly to the case of other σ-hole bonds, it seems that they are mainly steered by the electrostatic contribution.

Differences between the X and H centres are also observed (Figures 13.2 and 13.3). The regions of positive EP for halogen atoms are situated in extensions of the C–X bonds and they cover only small parts of the molecular surfaces, *i.e.* around their intersections with the C–X axes. However, the whole hemisphere of the H-centre is characterized by a positive EP in the case of the CF_3H moiety (Figure 13.3) and in general this is usually observed for the A–H proton donating bond. Consequently, both HB and XB are directional interactions but a narrower range is observed for the A–X···B angle (near 180°) than for the A–H···B angle.[11,35] The X-centres, even if they are characterized by positive EPs (σ-holes), also possess regions characterized by negative EP in the direction orthogonal (or nearly so) to the C–X bond (Figure 13.2); hence they may also play the role of Lewis bases. Figure 13.1 mentioned earlier presents a crystal structure where the bromine atom acts simultaneously as a nucleophilic and an electrophilic centre; such a dual role is not the case for an H-atom since its whole hemisphere is characterized by a positive EP.

Figures 13.2 and 13.3 present also contour maps of the Laplacian of the electron density for the CF_3X and CF_3H moieties mentioned above. The regions of the negative Laplacian (designated by areas with broken lines) indicate the concentration of the electron density. For the CF_3X series one can see changes of the distribution of the electron density at the X-atoms which are in agreement with observations concerning the electrostatic potential. An approximately spherical concentration of the electron density is observed for the F-centre – that is why F in CF_4 does not act as the Lewis acid. Next, for chlorine it is a slightly deformed distribution. A slightly broader region of the electron density concentration is observed in the direction orthogonal to the C–Cl bond than in the extension of this bond. This is why rather weak Lewis acid properties of chlorine centres are usually observed. For bromine, small regions of the negative Laplacian in the direction orthogonal to the C–Br bond are observed while in the extension of this bond there is an area of the positive EP; this is why Br-centres usually show strong Lewis acid properties. In a case of the CF_3H species, the electron density is shifted from the H-centre to the carbon. The electron charge depletion at the H-atom results in the total EP being strongly affected by the nucleus electrostatic potential and a totally positive EP hemisphere attributed to the hydrogen is observed.

The results concerning the electron charge distribution presented here for H and X centres are in agreement with experimental studies. For example, the halogen bond interactions and the experimental charge densities were analyzed for crystals of chlorine,[40] chlorine-fluoride[41] and hexachlorobenzene.[42,43] The nucleophilic and electrophilic sites at the X-centres in these crystals were found since maps of the Laplacian of electron density were determined. Figure 13.4 presents two examples of crystal structures of chlorine[40] and hexachlorobenzene[43] where one can see the Cl···Cl contacts between electrophilic (electron charge depletion) and nucleophilic (electron charge concentration) sites.

The above examples show that the halogen – halogen attractive Lewis acid – Lewis base interactions often formed in crystal structures result from the dual character of halogen atoms. Such arrangements were detected early on; Sakurai *et al.* analyzed in 1963 the crystal structure of 2,5-dichloroaniline and they found two types of chlorine – chlorine arrangements.[44] Desiraju and Parthasarathy have generalized that two types of attractive halogen – halogen contacts are

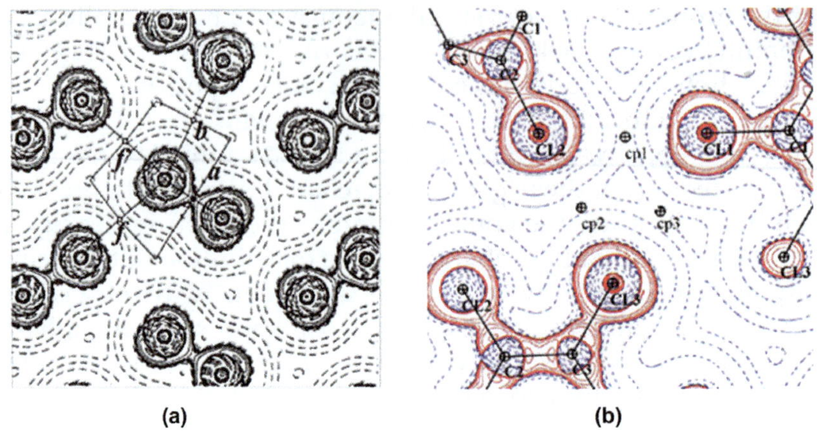

(a)   (b)

Figure 13.4   (a) The Laplacian distribution for the (100) plane of solid chlorine with solid contours corresponding to negative $\nabla^2\rho$ and broken ones to a positive Laplacian; bond paths between chlorine atoms are shown; reprinted with permission from V. G. Tsirelson, P. F. Zou, T. H. Tang and R. F. W. Bader, Acta Crystallogr. Sect. A, 1995, 51, 143. Copyright 1995 International Union of Crystallography. (b) The Laplacian distribution for the solid C_6Cl_6, bond paths within molecules are shown as well as the intermolecular critical points (cp); reprinted with permission from M. E. Brezgunova, E. Aubert, S. Dahaoui, P. Fertey, S. Lebègue, C. Jelsch, J. G. Ángyán and E. Espinosa, Charge Density Analysis and Topological Properties of Hal3-Synthons and Their Comparison with Competing Hydrogen Bonds, *Cryst. Growth Des.*, 2012, 12, 5373. Copyright 2012 American Chemical Society.

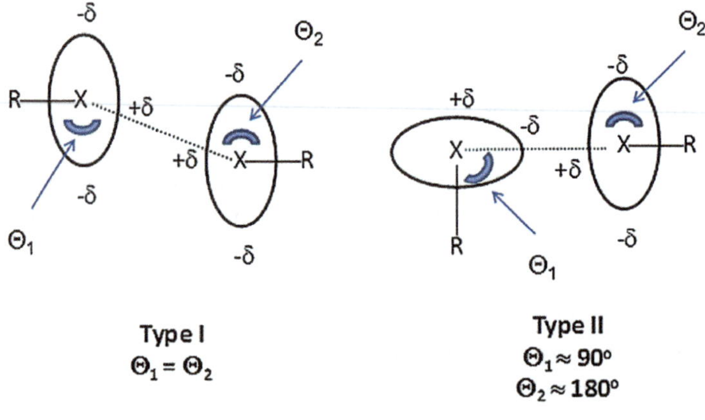

Scheme 13.2

common in crystal structures (Scheme 13.2).[45] These arrangements were later discussed and analyzed in numerous articles and reports[18,31,46,47] and the type I arrangements were further subdivided into *trans* and *cis* systems.[48] The type II arrangement fits into the IUPAC and other definitions of the halogen bond but it seems that the type I arrangement does not since it is a contact which arises from crystal packing;[3,21] this is discussed in the next sections of this chapter.

Both arrangements often occur in crystal structures; for example, C–Br···Br–C contacts were analyzed theoretically as well as through a search of the Cambridge Structural Database.[46] It was confirmed that the arrangements preferred in crystal structures correspond to the above-mentioned types I and II (Scheme 13.2). Figure 13.5 shows a fragment of the crystal structure of (1*R*(*S*),3*S*(*R*),4*S*(*R*),6*R*(*S*))-1,3,4,6-tetrabromo-1,2,3,4,5,6-hexahydropentalene[49] where the Br···Br contact of 3.587 Å is by ∼0.1 Å shorter than double the bromine van der Waals radius (3.7 Å) and the C–Br···Br angles are equal to 135.3°, meaning it corresponds to the type I contact (Scheme 13.2). Figure 13.6 presents the crystal structure of *syn*-4-bromobenzaldehyde oxime[50] where the Br···Br intermolecular distance is 3.639 Å, close to double the van der Waals Br radius, and where the C–Br···Br angles are equal to 98° and 164° – this is a type II contact. One can see for the latter structure (Figure 13.6) the evident dual character of bromine centres that form halogen bonds (with other bromine atoms), acting as Lewis acids in the extension of C–Br bonds and as Lewis bases in the direction orthogonal to the C–Br bonds. The interactions described above may be named as dihalogen bonds or dihalogen contacts; however, one can see that there are a lot of possibilities here; the X···X contacts between the same centres of the same molecules as in the structures presented in Figures 13.5 and 13.6; between different centres and/or different molecules but with the same type of X-atoms being in contact (Cl···Cl, Br···Br, *etc.*), or between different halogen atoms as in Cl···Br, Br···I *etc.*

The above-described halogen–halogen interactions concern cases where both X-centres possess dual character; however, one of the X-centres in the contact may be characterized only by Lewis base properties,[51,52] like in the case of halogen anions playing the role of

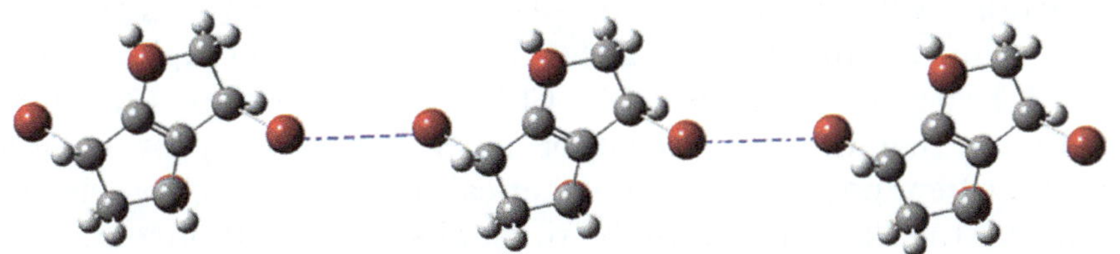

Figure 13.5   A fragment of the crystal structure of 2(1*R*(*S*),3*S*(*R*),4*S*(*R*),6*R*(*S*))-1,3,4,6-tetrabromo-1,2,3,4,5,6-hexahydropentalene;[49] CSD refcode of the crystal structure: KAWPUU.

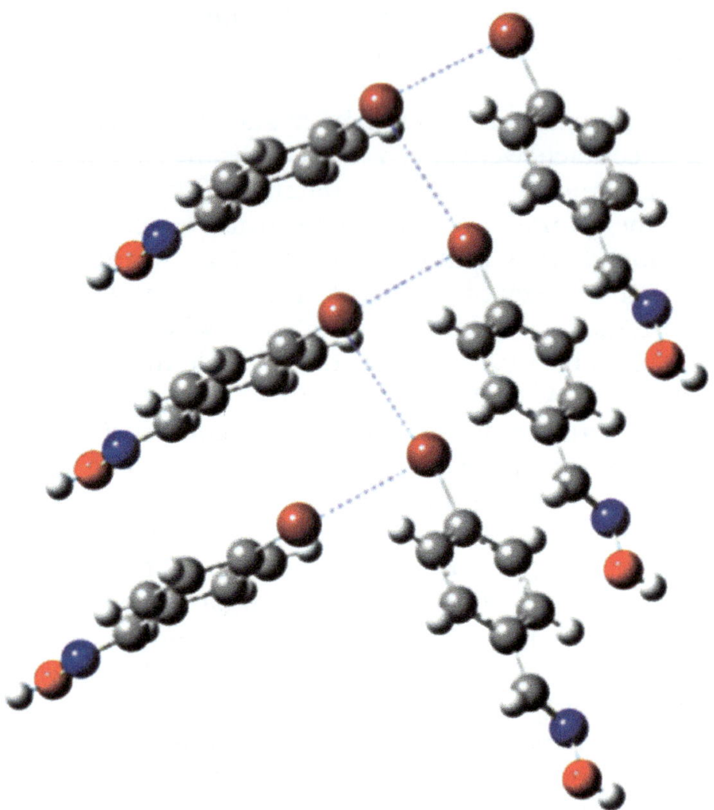

Figure 13.6   A fragment of the crystal structure of *syn*-4-bromobenzaldehyde oxime;[50] CSD refcode of the crystal structure: BAGWOW.

electron donors. One would expect that such A–X$\cdots$X$^-$ halogen bonds, assisted by a negative charge, CAXB(−)s, should be characterized by stronger interactions than typical XBs; analogously to hydrogen bonds, where the charge assisted hydrogen bonds, CAHBs, are stronger than their neutral counterparts, HBs.[53,54] For example, for the C–Br$\cdots$Br–C interactions found in the CSD the energies of interactions vary from −2.4 to −0.4 kcal mol^{-1} (MP2/aug-cc-pVTZ level)[46] while the binding energy for the CF$_3$Cl$\cdots$Cl$^-$ complex is equal to −7.3 kcal mol^{-1} (MP2/6-311++G(d,p) level).[51] Figure 13.7 shows the contour map of the Laplacian of the electron density for the latter complex where one can see the spherical distribution of the electron density for the Cl$^-$ anion, with the ring of its concentration around the nucleus; this is why the chlorine anion acts only as the Lewis base. The known ellipsoidal electron charge distribution for the Cl-centre involved in C–Cl bonds, described earlier above, is shown (Figure 13.7);

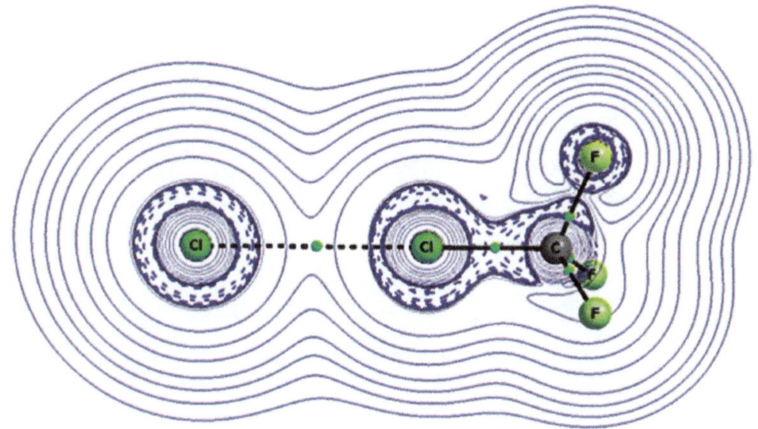

Figure 13.7　The molecular graph with the isolines of the Laplacian of electron density for $CF_3Cl\cdots Cl^-$ complex; positive values of $\nabla^2\rho$ are depicted by solid lines and negative values by broken lines.

the Cl-centre possesses dual character. There are different levels of calculations for the bromine and chlorine species mentioned above.[46,51] However, the bromine in C–Br should be characterized by stronger Lewis acid properties than chlorine due to the greater positive EP for the former centres than for the latter one (Figure 13.2). The MP2/ 6-311$++$G(d,p) calculations were also performed for the neutral $CF_3Cl\cdots ClCH_3$ complex linked by a type II interaction (Scheme 13.2) and the binding energy is equal to -0.7 kcal mol^{-1}.[51] In other words, these results prove that charge assistance enhances the halogen bond interaction.

For the CF_3H moiety (Figure 13.3), the whole hemisphere of the H-atom is characterized by a positive EP that acts as an electron acceptor in hydrogen bonded complexes. In general, the H-atom cannot act as a Lewis acid and a Lewis base simultaneously as is observed for halogens, as for the Cl, Br and I centres in the CF_3X species discussed earlier. The H-atom may play the role of a Lewis base in hydrides where it is negatively charged;[27,28] however the same H-atom centre does not possess dual character. In other words the H-centre is either an acidic or a basic one.[27] That is why, aside from the hydrogen bonds, there are hydride bonds with the central hydric H-atom (negatively charged) and dihydrogen bonds (DHBs) that are contacts between the hydrogen atoms, $H^{-\delta}\cdots^{+\delta}H$, characterized by opposite charges.[55] The latter interactions (DHBs), often classified as a sub-class of hydrogen bonds,[54,56] may be compared with the above-described $X\cdots X$ contacts, especially the type II interactions (Scheme 13.2). Figure 13.8 presents

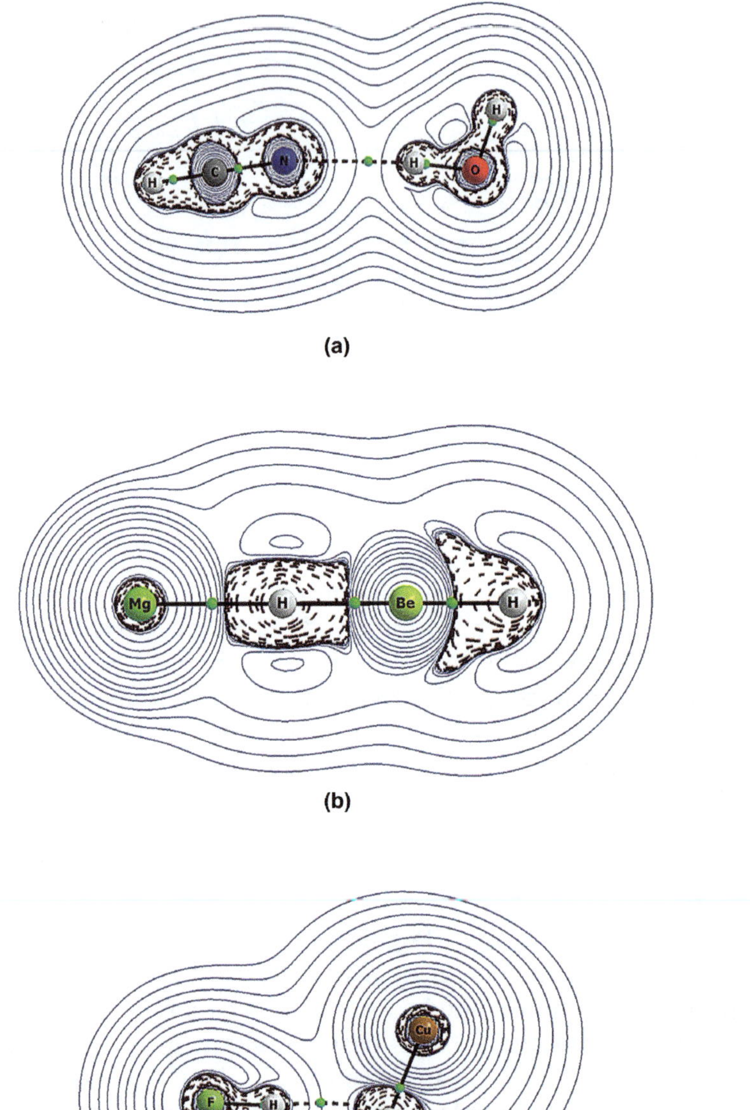

(a)

(b)

(c)

Figure 13.8   The molecular graphs with the isolines of the Laplacian of electron density for complexes of (a) HOH···NCH, (b) HBeH···Mg^{2+}, (c) FH···HCu; positive values of $\nabla^2\rho$ are depicted by solid lines and negative values by broken lines.

three cases of the above-mentioned interactions. The HOH$\cdots$NCH complex is linked through an O–H$\cdots$N hydrogen bond. One can see here the contact between the H-atom, where the depletion of the electron charge density in the extension of the O–H bond is observed, with the electron rich region of nitrogen. A very interesting situation is observed for the Be–H$\cdots$Mg^{2+} hydride bond (inverse hydrogen bonding) in the HBeH$\cdots$Mg^{2+} complex (Figure 13.8).[28] The H and Mg centres are here in intermolecular contact, the first one characterized by the broad area of the negative Laplacian of electron density, and the second one by the depletion of the electron density. Only a small ring of negative Laplacian is observed for the magnesium that is located close to the corresponding nucleus and that is attributed to the core electrons. In general, one can characterize this complex as a linearly located sequence of ions; H$^-$, Be^{2+}, H$^-$ and Mg^{2+}; the QTAIM atomic charges equal for this sequence -0.78, $+1.73$, -0.82 and $+1.88$ au, respectively; close to the $-1, +2, -1, +2$ au distribution characteristic for the idealized ionic Lewis structure. For the F–H$\cdots$H–Cu complex (Figure 13.8)[57] the protic (connected with fluorine) hydrogen is characterized by the depletion of the electron charge in the extension of the F–H bond while the hydric hydrogen connected with copper is characterized by the area of electron density concentration. In general, for these three types of interactions, hydrogen, hydride and dihydrogen bonds, where the complexes are linked through the hydrogen (or through two hydrogen atoms as in the FH$\cdots$HCu complex), contacts between the electron rich and electron poor regions are observed, similar to the observation for halogen bond interactions.

13.3　Hydrogen and Halogen Bonds Are Ruled By the Same Mechanisms

It was explained in the previous section that the halogen bond is characterized by numerous properties similar to those of the hydrogen bond. Even more, it seems that both interactions are ruled by the same mechanisms. It was pointed out that the A–H$\cdots$B hydrogen bond is a combination of two effects: the hyperconjugative A–H bond weakening and the rehybridization-promoted A–H bond strengthening.[58] The first effect is well known[59,60] and it is connected with the electron charge transfer from the lone pair of the B Lewis base centre to the antibonding σ^* orbital of the A–H bond; the corresponding $n(B) \rightarrow \sigma^*_{AH}$ donor-acceptor interaction is the most important one connected with the electron charge transfer from the Lewis base to the Lewis acid in hydrogen bonded systems. The second one, the rehybridization process connected with further electron charge density shift from the H-atom to the A-centre and the increase of the positive charge of the H-atom, leads to the increase in the s-character of the A-atom hybrid orbital in the A–H bond.[58,61]

It was found recently that the hyperconjugative and rehybridization processes, similarly as for the hydrogen bond, may be considered as steering the formation of numerous noncovalent interactions such as for example, the dihydrogen bond, halogen bond, dihalogen bond, hydride bond *etc.*[51,52,61–63] It was found that for the A–X···B halogen bond and for the other interactions there is a donor–acceptor interaction between the lone pair of the Lewis base and the antibonding orbital of the Lewis acid; in the case of the A–X···B halogen bond, it is the $n_B \rightarrow \sigma^*_{AX}$ orbital–orbital interaction between the n_B lone pair and the σ^*_{AX} antibonding orbital. The formation of XB also leads to the increase in s-character of the A-atom hybrid orbital in the A–X bond.[61,62]

The changes of the s-character may be roughly explained within the NBO (Natural Bond Orbital) approach[59,60] in the following way. The natural bond orbital for the localized σ bond, σ_{AB}, between A and B atoms is formed from the orthonormal hybrids h_A and h_B (named natural hybrid orbitals, NHOs, eqn (13.1)).

$$\sigma_{AB} = c_A h_A + c_B h_B \tag{13.1}$$

The NHOs are composed from effective valence-shell atomic orbitals optimized for the corresponding, chosen wave function. The antibonds are formed in the similar way as the bond orbitals (eqn (13.2)).

$$\sigma^*_{AB} = c_B h_A - c_A h_B \tag{13.2}$$

For example, for the water molecule (MP2/6-311++G(d,p) calculations) not involved in any interactions, the O–H natural bond orbital has the following form.[61,64]

$$\sigma_{OH} = c_H h_H + c_O h_O = 0.5203\ \text{s(H)} + 0.8540\ \text{sp}^{3.35}\text{(O)} \tag{13.3}$$

This means that the hydrogen natural hybrid orbital, h_H, possesses purely s-character while the oxygen hybrid h_O orbital is mostly of the p-type. The coefficient (superscript) 3.35 corresponds to 77% p-orbital contribution (23% of s-orbital and, as one would expect, no contributions from orbitals with higher angular momentum quantum numbers such as d or f). The c_H and c_O coefficients show that this is a strongly polarized covalent bond with an electron density shift to the oxygen atom where the percentage of the electron density amounts to 72.9% ($|c_O|^2 \times 100\%$). For the water dimer involved in the O–H$\cdots$O hydrogen bond interaction (see Figure 13.9) the following natural orbital corresponds to the proton donating OH bond.

$$\sigma_{OH} = c_H h_H + c_O h_O = 0.5033\ \text{s(H)} + 0.8641\ \text{sp}^{2.89}(O) \qquad (13.4)$$

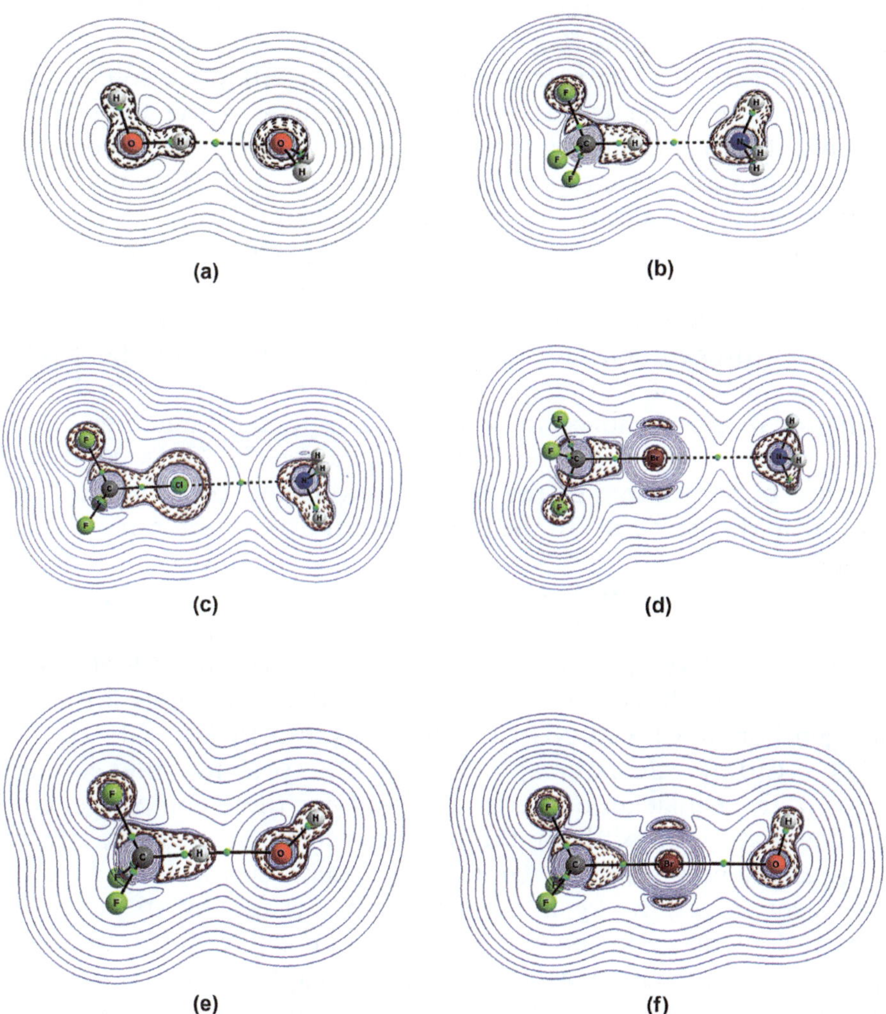

Figure 13.9 The molecular graphs with the isolines of the Laplacian of electron density for complexes of (a) HOH$\cdots$OH$_2$, (b) CF$_3$H$\cdots$NH$_3$, (c) CF$_3$Cl$\cdots$NH$_3$, (d) CF$_3$Br$\cdots$NH$_3$, (e) CF$_3$H$\cdots$OH$^-$, (f) CF$_3$Br$\cdots$OH$^-$; positive values of $\nabla^2\rho$ are depicted by solid lines and negative values by broken lines.

The most important changes connected with complexation concern here the increase of the s-character of the h_O hybrid orbital from 23.0% in the monomer to 25.7% in the dimer and the increase of the polarization of the O–H bond, from 72.9% to 74.9% (percentage of the electron density at the oxygen centre). It was mentioned above that such changes are observed not only for HBs but also for other interactions, as for example XB. This is why one can expect the increase of the electron density at the carbon centre and the increase of the s-character of the carbon hybrid orbital for C–X bonds involved in halogen bonds with the X-centre acting as the Lewis acid.

The increase of s-character as a result of the HB formation, or as a result of other interactions, is in agreement with the Bent rule[65] which states that atoms maximize their s-character in hybrid orbitals directed towards electropositive substituents and they maximize their p-character towards electronegative substituents. The formation of an HB leads to the increase of the positive charge of the H-atom[60,61,66] (in most cases although there are several exceptions[67]). Hence the hydrogen atom becomes more electropositive after the A–H···B hydrogen bond formation; thus according to the Bent rule, the s-character of the neighbouring A-atom hybrid orbital also increases in the complex, in comparison with the monomer. Similarly, the formation of the A–X···B halogen bond leads to an increase of the charge of the halogen atom and consequently, according to the Bent rule, to the increase of the s-character in A-hybrid orbital.

A few systems are presented here as examples to show changes of various parameters that result from complexation; the hydrogen bonded complexes the water dimer, $HOH \cdots OH_2$, the $CF_3H \cdots NH_3$ system and the $CF_3H \cdots OH^-$ complex as well as complexes linked by halogen bonds, $CF_3Cl \cdots NH_3$, $CF_3Br \cdots NH_3$ and $CF_3Br \cdots OH^-$ (MP2/ 6-311++G(d,p) calculations).[61] Figure 13.9 presents the contour maps of the Laplacian of electron density for these complexes. One can see that for all of them, the intermolecular interactions, HBs or XBs, exactly correspond to contacts between Lewis acid and Lewis base centres, *i.e.* between the site characterized by the electron density depletion and the site of electron density concentration, respectively. The small areas of the negative Laplacian of the electron density in the direction approximately orthogonal to the C–Br bond for the CF_3Br complexes, and the positive Laplacian in the extension of the C–Br bond allow prediction of the greater stabilization of these systems in comparison with the $CF_3Cl \cdots NH_3$ complex, where the thin area of the negative Laplacian is detected in the extension of C–Cl bond. The latter electron density distributions for the CF_3X species coincide with the greater positive EP for the σ-hole of the bromine centre than for chlorine one (see previous section). Similarly, the change of other parameters resulting from the complexation, including in the s-character as described above, should be more significant for the bromine complexes than for the chlorine one.

Table 13.1 presents selected parameters for complexes mentioned above as well as the corresponding characteristics for monomers not involved in interactions; the species presented in the table correspond to energetic minima. The negative binding energies indicate that all complexes are stable. The binding energy, E_{bin}, is calculated as the difference between the energy of the complex and the sum of energies of monomers in their energetic minima; *i.e.* not involved in any interactions. It means that the deformation energy resulting from complexation is included here, and the binding energies are corrected for BSSE errors.[68] The deformation energy is positive since the formation of the complex leads to a change of geometry of monomers that in the complex are "taken out" from their energetic minima; however, the complex as a whole corresponds to a minimum.[69,70]

Table 13.1 shows that charge assisted hydrogen and halogen bonds, CAHBs and CAXBs, are much stronger than those corresponding to neutral complexes ("more negative" E_{bin} corresponds to the stronger interaction). For neutral hydrogen and halogen bonded systems $-E_{bin}$ amounts to ~4 kcal mol^{-1} and ~2–3 kcal mol^{-1}, respectively, while for CAHBs and CAXBs this value exceeds 20 kcal mol^{-1}. The greater positive electrostatic potential at Br than at Cl for the CF_3X species results in a stronger interaction with NH_3 in the bromine complex than in the chlorine one. The halogen bonds in bromine complexes are characterized by strengths comparable with the corresponding hydrogen bonds; $-E_{bin}$ is equal to 4.0 and 3.2 kcal mol^{-1} for the $CF_3H\cdots NH_3$ and $CF_3Br\cdots NH_3$ complexes, respectively; and it amounts 23.4 and 23.8 kcal mol^{-1} for the $CF_3H\cdots OH^-$ and $CF_3Br\cdots OH^-$, respectively. These results are in agreement with other studies carried out since it was reported that the A–H$\cdots$B hydrogen bonds are in general stronger than the A–X$\cdots$B halogen bonds; however, the strength of the latter interactions increases in the order Cl < Br < I;[71] for X = Br both interactions begin to be comparable and for X = I the halogen bond is often stronger than the hydrogen bond counterpart. These theoretical results are also in agreement with other observations. For example, Metrangolo *et al.* have pointed out that often in the crystal structures halogen bonds are preferred to hydrogen bonds.[16]

Table 13.1 Comparison of parameters of complexes linked by hydrogen or halogen bond; s-character (%), polarization (%), H/X charge (au), the charge transferred to the Lewis acid unit as a result of complexation, Q (au), the hydrogen or halogen atom volume, $V(H/X)$ ($Å^3$) and binding energy corrected for BSSE, E_{bin}, (kcal per mol); the orbital–orbital interaction energy, E_{NBO} (kcal per mol), the charges and volumes calculated by QTAIM approach; the parameters of monomers optimized separately are included in parentheses; MP2/6-311++G(d,p) calculations (results from ref. 61).

Parameter	$HOH\cdots OH_2$	$CF_3H\cdots NH_3$	$CF_3H\cdots OH^-$	$CF_3Cl\cdots NH_3$	$CF_3Br\cdots NH_3$	$CF_3Br\cdots OH^-$
s-Character	25.66 (22.96)	32.50 (30.66)	37.17 (30.66)	27.60 (26.62)	27.31 (26.00)	31.13 (26.0)
Polarization	74.67 (72.93)	59.59 (57.06)	66.0 (57.06)	48.06 (47.00)	52.40 (50.77)	59.53 (50.77)
Q	−0.015	−0.018	−0.090	−0.013	−0.021	−0.255
H/X charge	+0.611 (+0.567)	+0.180 (+0.107)	+0.391 (+0.107)	−0.050(−0.086)	+0.103 (+0.056)	+0.159 (+0.056)
$V(H/X)$	2.27 (3.48)	5.17 (6.28)	3.02 (6.28)	30.03 (30.70)	36.42 (37.63)	35.87 (37.63)
E_{bin}	−4.5	−4.0	−23.4	−2.0	−3.2	−23.8
E_{NBO}	6.9	6.6	41.2	2.2	4.3	52.2

The other results presented in Table 13.1 show the same mechanisms steering the formation of hydrogen and halogen bonds. Scheme 13.3 summarizes the mutual properties of the hydrogen and halogen bonds; two complexes are presented that are included in Table 13.1, the water dimer, $HOH\cdots OH_2$, and the $F_3CCl\cdots NH_3$ complex. For both there is the electron charge transfer from the Lewis base unit, H_2O for the HB interaction and NH_3 for the XB one, to the Lewis acid moiety. Such transfer is equal to 15 and 13 millielectrons for the mentioned species, respectively. The similar orbital-orbital overlap is observed for both interactions, the $n_O \rightarrow \sigma^*_{OH}$ interaction for HB system and the $n_N \rightarrow \sigma^*_{CCl}$ for XB with the corresponding energies of 6.9 and 2.2 kcal mol^{-1}, respectively (designated as E_{NBO}s in Table 13.1). The binding energy, E_{bin}, for the water dimer is equal to -4.5 kcal mol^{-1}, while for the $F_3CCl\cdots NH_3$ complex it amounts -2.0 kcal mol^{-1}. The further electron density shift is accompanied by the above-mentioned rehybridization process that leads to the outflow of electron density from H and Cl atoms of HB and XB complexes to the parts of the molecules playing the role of the Lewis acids. It results in the increase of the positive charge of H and Cl atoms by 0.044 and 0.036 au (QTAIM calculations) in comparison with the H_2O and F_3CCl moieties not involved in HB and XB interactions, and this is connected with the corresponding decrease of the H and Cl atomic volumes (QTAIM calculations). The rehybridization process observed for both HB and XB interactions, leads consequently to the increase of the s-character of the O and C hybrid orbitals for O–H and C–Cl bonds, respectively. In a case of the O–H bond it increases from 23.0 in the water monomer to 25.7 in the water dimer; for the C–Cl bond it increases from 26.6 in the CF_3Cl monomer to 27.6 in the $CF_3Cl\cdots NH_3$ complex. It is interesting that the complexation for both interactions leads to the increase in the polarization of the Lewis acid bonds (O–H and C–Cl); the polarization is understood here as the percentage of the electron density on the O or C atom. The $H\cdots O$ and $Cl\cdots N$ intermolecular distances of 1.950 and 2.292 Å for these complexes are both shorter than the corresponding sums of van der Waals radii of 2.6 and 3.3 Å,[22] respectively.

All these changes resulting from complexation for species presented in Scheme 13.3 are observed for the other HB and XB complexes presented in Table 13.1; for the charge assisted interactions, CAHB and CAXB, the corresponding changes are much more forcible. One can see that the chlorine atom in CF_3Cl is characterized by a negative charge (Table 13.1), in spite of the fact that it may act as the Lewis acid centre through the region of positive EP. The negative

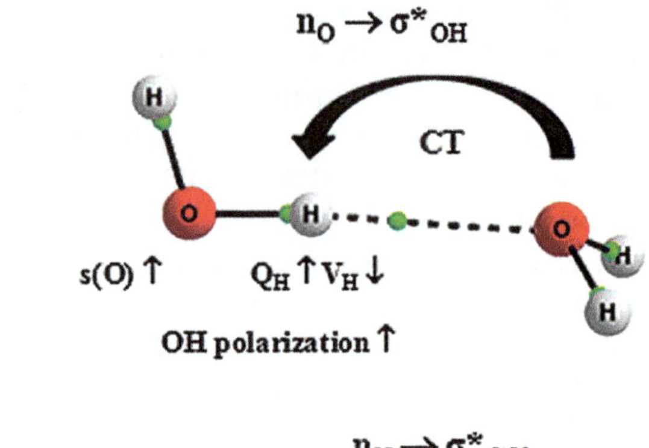

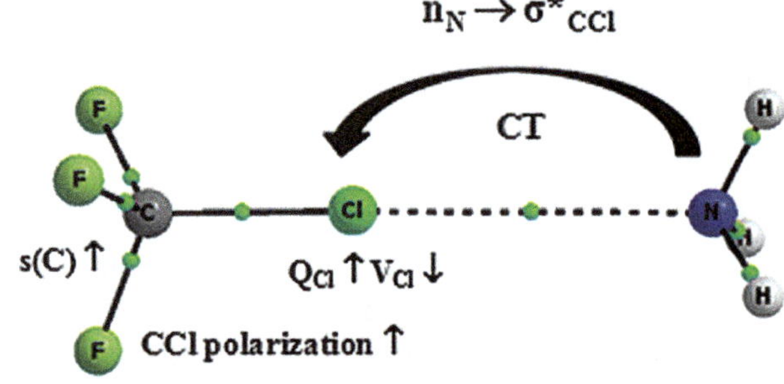

Scheme 13.3

Cl-charge is "less negative" (more positive) in the complex with NH_3 according to the mechanisms described above. For the CF_3Br species a positive charge of Br is observed that increases in complexes with NH_3 and OH^-. It is worth mentioning that the centre's charge is not a proper parameter expressing its Lewis acid or Lewis base properties; one can see from the above results that the negatively charged chlorine does not necessarily act as the Lewis base; it was explained in detail that a better parameter determining the properties of any atomic centre is the electrostatic potential.[5]

13.4 The Energetic Characteristics

The halogen bond is often competitive with the hydrogen bond; especially in cases of heavier halogen atoms which are characterized by greater positive EP values than the lighter X-centres. For example, for the $CF_3X\cdots NH_3$ complexes linked through halogen bonds, the binding energy is equal to -2.2; -3.2 and -4.8 kcal mol^{-1} for $X = Cl$, Br and I, respectively (MP2(full)/cc-pVTZ level of calculations; energies corrected for BSSE, for the iodine atom the scalar-relativistic energy-consistent small core Dirac–Fock effective-core pseudopotential was used).[72] For the hydrogen bonded $CF_3H\cdots NH_3$ complex, the binding energy is equal to -3.9 kcal mol^{-1} (MP2/6-311++G(3df,3pd) level, BSSE corrected energy).[73] In spite of different basis sets used, it seems the results are comparable since the extended basis sets were used for calculations for both complexes and these are close to the complete basis set limit. Hence one may expect a stronger interaction for the C–I$\cdots$N halogen bond than for the C–H$\cdots$N hydrogen bond.

The calculated X$\cdots$N distances for the above-mentioned XB complexes are equal to 3.048, 2.970 and 2.967 Å for Cl, Br and I species, respectively and these correspond to distances shorter than the sums of van der Waals radii by 0.30, 0.52 and 0.72 Å (the Pauling van der Waals radii were applied here; Cl (1.81 Å); Br (1.95 Å), I (2.15 Å) and N (1.54 Å)).[22] The difference between the sum of van der Waals radii and the corresponding atom-atom intermolecular distance is often considered as a rough measure of the strength of interaction, especially for hydrogen bonds.[54] The above-mentioned X$\cdots$N distances are in line with the order of the corresponding binding energies.

The question often arises in the literature of the nature of the halogen bond; similar controversies concern the hydrogen bond that is characterized, as it was presented earlier here, by numerous properties similar to the halogen bond. In the case of the hydrogen bond, a main subject of controversies and disputes concerns the question of whether it is electrostatic or covalent in nature; or at least which of them is the dominant one.[54,59,60] The σ-hole concept[35,36] shows clearly the crucial role of the electrostatic interaction for XBs. For these interactions the distribution of the electrostatic potential (EP) at the molecular surfaces determines the arrangement of species forming the complex. The directionality of the A–X$\cdots$B halogen bond especially can be explained by the location of the Lewis acid region in the extension of A–X bond; a similar situation is observed for the hydrogen bond and for other σ-hole bonds.[35,36]

The statement of the electrostatic nature of XBs and of the other σ-hole bonds is supported by numerous correlations between the maximum EP located at the σ-hole region and the total interaction energy.[36] It was stated that the σ-hole interaction may be understood "in terms of the electrostatic attraction between the positive σ-hole and a negative site – which might be a lone pair of a Lewis base, π electrons of unsaturated molecules, an anion or a hydride, the negative regions of another covalently-bonded atom, *etc.*"[36] The polarization is an intrinsic component of the electrostatic interaction and the dispersion can also have an important role in the total σ–hole interaction.[5,36,74] In other words the σ-hole bonds, including XB and

HB when classified as this type of interaction, are electrostatic/polarization interactions accompanied by dispersion.

Different decomposition schemes can be applied to determine contributions to the energy of interaction. However, it is worth mentioning that the contributions that result from different schemes differ significantly between themselves in their physical meanings and thus it is difficult to compare them; even more – often it is not possible to perform such comparison. However, more important is that often these terms resulting from mathematical approaches applied to quantum chemistry methods do not have a simple physical interpretation. The comparison and description of different decomposition schemes was carried out in several review articles.[54,75]

Eqn (13.5) presents the decomposition scheme – the variation–perturbation approach[76–78] – that seems to be successful in the analysis of numerous interactions, especially in hydrogen bonded complexes.[54,79] It is briefly described here since terms of similar meanings to those of this decomposition often appear in other approaches. The following interaction energy components can be obtained in this decomposition:

$$\Delta E = E_{EL}^{(1)} + E_{EX}^{(1)} + E_{DEL}^{(R)} + E_{CORR} \tag{13.5}$$

where $E_{EL}^{(1)}$ is the first order electrostatic term describing the Coulomb interaction of static charge distributions of both units in the complex considered; $E_{EX}^{(1)}$ is the repulsive first order exchange component resulting from the Pauli exclusion principle; and $E_{DEL}^{(R)}$ and E_{CORR} correspond to higher order delocalization and correlation terms. The delocalization term contains all classical induction, exchange-induction, *etc.* from second order up to infinity. The latter term corresponds to the part of the interaction related to the electron charge density shifts as a result of complexation; this term is often decomposed into the charge transfer contribution (CT) and the polarization term (POL). However, such partitioning leads to strongly basis set dependent components (CT and POL) while $E_{DEL}^{(R)}$ is much less sensitive to the basis sets' changes. These contributions, CT and POL, appear in the Kitaura–Morokuma decomposition scheme (K–M),[80] one of the first approaches that was proposed to analyze the character of interactions. However, the latter K–M decomposition is restricted to the Hartree–Fock level of calculations; note that the sum of CT and POL terms does not correspond to the delocalization term (eqn (13.5)) as it was explained earlier here. In a rough approximation, the CT term may be understood as the energy corresponding to the electron charge density transfer between units constituting a complex. The POL contribution is roughly connected with the electron charge density shifts as a result of complexation but within the units. However, it seems the accurate partitioning of these shifts into those concerning ranges of interacting units and those between units in the complex is not possible; similarly, the decomposition of the delocalization term into CT and POL is not accurate. It is worth mentioning that these electron density shifts are related to the induction energy term that appears in other decompositions or approaches (such as the symmetry-adapted perturbation theory approach – SAPT[81]), and that they are often identified with the covalent character of interaction.[54]

Numerous studies on the hydrogen bond, A–H$\cdots$B, and the dihydrogen bond (DHB), A–H$\cdots$H–B, were performed with the use of the above-described (eqn (13.5)) decomposition scheme.[79] It was found that the shortening of the H$\cdots$B or H$\cdots$H distance, for HB and DHB, respectively, is connected with the simultaneous increase of all interaction energy terms (in a case of negative attractive contributions, it is the increase of absolute values). It was observed that for the shorter distances, *i.e.* the stronger interactions, the first order Heitler–London energy term, $E_{\text{H-L}}^{(1)} = E_{\text{ES}}^{(1)} + E_{\text{EX}}^{(1)}$, is positive since the attractive electrostatic interaction is outweighed by the exchange repulsion term; thus the system may be energetically stable owing to the delocalization interaction energy, $E_{\text{DEL}}^{(R)}$ and/or the correlation interaction energy, E_{CORR}.[82]

Figure 13.10 presents the relation between the Heitler–London, $E_{\text{H-L}}^{(1)}$, term and the delocalization interaction energy, $E_{\text{DEL}}^{(R)}$. The hydrogen bonded systems (designated by black circles) concern centrosymmetric complexes linked by the double O–H$\cdots$O, N–H$\cdots$O or O–H$\cdots$N interactions;[83] these are the dimers of formic, acetic and pyrrole-2-carboxylic acid; in the latter case three configurations of the dimer are taken into account; additionally, the dimer of formamide is considered, its fluoro-derivative (both linked by N–H$\cdots$O HBs) and the tautomeric forms of two latter complexes (linked by O–H$\cdots$N HBs). For this sample of complexes a linear correlation between $E_{\text{H-L}}^{(1)}$ and $E_{\text{DEL}}^{(R)}$ was found ($R^2 = 0.967$). This correlation shows that if the attractive electrostatic interaction is not able to compensate the repulsive exchange interaction ($E_{\text{H-L}}^{(1)} > 0$) for stable hydrogen bonded complexes then the delocalization becomes more important; this is usually observed for strong hydrogen bonds.[79]

Figure 13.10 also presents the results for halogen bonded systems (open circles), where the F–Cl and F–Br species act as the Lewis acid (through the Cl and Br centres) while the following species play a role

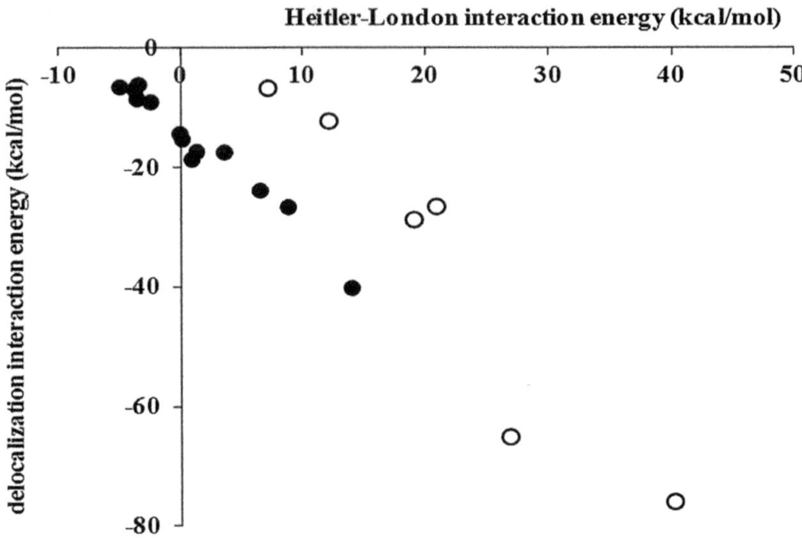

Figure 13.10   The relationship between the Heitler–London interaction energy term and the delocalization term for hydrogen and halogen bonded systems (both terms in kcal per mol); decomposition according to eqn (13.5).

Table 13.2   The decomposition interaction energy terms (eqn (13.5); kcal per mol) for the XB systems (MP2/aug-cc-pVTZ level; results from ref. 84).

Complex	$E_{EL}^{(1)}$	$E_{DEL}^{(R)}$	$E_{EX}^{(1)}$	E_{CORR}	ΔE	$E_{H-L}^{(1)}$
F–Cl$\cdots$NH$_3$	−39.1	−26.6	60.0	−7.6	−13.3	20.9
F–Br$\cdots$NH$_3$	−47.4	−28.8	66.5	−6.7	−16.4	19.1
F–Cl$\cdots$C$_2$H$_2$	−11.0	−6.8	18.2	−5.1	−4.7	7.2
F–Br$\cdots$C$_2$H$_2$	−18.5	−12.3	30.7	−6.5	−6.6	12.2
F–Cl$\cdots$Cl$^-$	−69.9	−76.0	110.2	−10.5	−46.2	40.3
F–Br$\cdots$Cl$^-$	−74.6	−65.1	101.5	−8.5	−46.7	26.9

of Lewis bases: C$_2$H$_2$ – by π-electrons; NH$_3$ – through nitrogen and the Cl$^-$ anion.[84] Table 13.2 shows interaction energy terms for these halogen bonded systems. The XB complexes are characterized by strong interactions and only for the acetylene complexes, where weaker Lewis base properties of π-electrons are revealed, is $-\Delta E$ less than 10 kcal mol^{-1}. For XB complexes assisted by a negative charge, CAXB(−)s, $-\Delta E$ is greater than 46 kcal mol^{-1}. Similarly, very strong interactions are usually observed for charge assisted HBs.[53,54,79] For the neutral complexes presented in Table 13.2, strong interactions partly result from a large positive EP at the Cl and Br sites corresponding to σ-holes; for the F–Cl and F–Br species the electrostatic potential is equal to +40 and +48 kcal mol^{-1}, respectively (B3PW91/6-311G* level).[5]

The other halogen σ-holes that may be involved in halogen bonds are usually characterized by lower values of EP. For example, for the CF_3Cl and CF_3Br species the EP at Cl and Br is equal to $+20$ and $+25$ kcal mol^{-1}, respectively.[5]

Table 13.2 shows that the delocalization interaction is comparable to the electrostatic interaction and that, for all XB complexes considered, the latter interaction is not sufficient to compensate the Pauli repulsion. However, the XB complexes characterized by rather strong interactions are considered here to show the role of the polarization interaction. These results show that HBs and XBs are steered by the same mechanisms; however, this comparison could only be performed since the same decomposition approach was applied for both HB and XB complexes. Besides, the latter conclusions should be treated carefully due to reservations concerning the decomposition approaches that were mentioned at the beginning of this section. These results are in line with the σ-hole bond definition that it is an electrostatic/polarization interaction that is often supported by a dispersive interaction.[36]

Other decomposition schemes were also applied to analyze XB complexes. For example, the SAPT approach[81] was applied for the $CH_3Cl\cdots OCH_2$, $CH_3Br\cdots OCH_2$ and $H_3Cl\cdots OCH_2$ complexes characterized by XB interactions of 1–3 kcal mol^{-1} $(-E_{\text{bin}})$. It was found that for these species the halogen bond is largely dependent on both electrostatic and dispersion type interactions while the induction interaction is negligible.[85]

13.5 The QTAIM characteristics

The Quantum Theory of Atoms in Molecules (QTAIM)[66,86–88] is a powerful tool that may provide additional information on the nature of interactions. It was stated that the bond paths correspond to stabilizing interactions;[89,90] the bond path is a line of maximum electron density linking attractors (attractors are the local maxima of electron density attributed to positions of atoms). The bond critical point (BCP) is a point on the bond path of minimum electron density; the characteristics of the BCP inform of the type of interaction; these are, for example: electron density at BCP (ρ_C), its Laplacian $(\nabla^2\rho_C)$ and the total electron energy density at the BCP (H_C).[87] The negative Laplacian of the electron density at a BCP informs of the concentration of the electron density in the interatomic region that is typical of a covalent bond. However, sometimes the $\nabla^2\rho_C$ value is positive but H_C is negative and such atom–atom contacts are classified as interactions with partially covalent character.[91–93] If $\nabla^2\rho_C$ and H_C are both positive, contacts are usually attributed to closed-shell interactions.[87]

It is worth mentioning that for A–H$\cdots$B hydrogen bonds, for the H$\cdots$B bond path, the corresponding BCP often possesses a negative value of H_C that occurs for strong and partially covalent nature HBs.[54,92] Sometimes even if the $\nabla^2\rho_C$ value is negative for an H$\cdots$B BCP such HBs are very strong and covalent in nature, like for the [FHF]$^-$ anion.[94]

Figure 13.11 presents the relation between the H$\cdots$B distance and the ratio of delocalization and electrostatic interaction energy terms (decomposition corresponding to eqn (13.5) based on the MP2/6-311++G(d,p) results was applied) for hydrogen and dihydrogen bonded systems (in a case of DHBs the H$\cdots$H intermolecular distance is considered).[79] One can see that for the H$\cdots$B/H distances greater than 1.8 Å both $\nabla^2\rho_C$ and H_C values are positive which means that the corresponding HBs are not so strong, the dimer of water is an example of such a situation. In the H$\cdots$B/H distance region between 1.2 and 1.8 Å $\nabla^2\rho_C>0$ and $H_C<0$ occurs for strong HBs with partially covalent character; the formic acid dimer is an example of such a

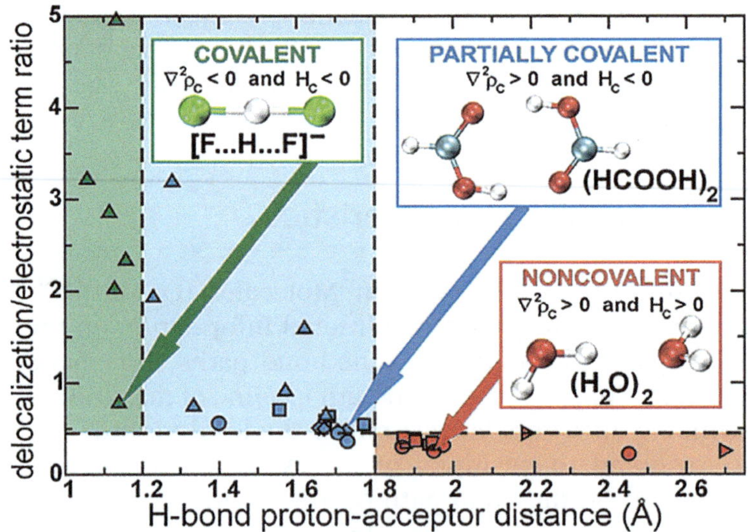

Figure 13.11 The relationship between the H$\cdots$B/H distance (in Å) and the ratio of delocalization and electrostatic interaction energy terms; decomposition according to eqn (13.5); reprinted with permission from S.J. Grabowski, W.A. Sokalski, E. Dyguda and J. Leszczynski, Quantitative Classification of Covalent and Noncovalent H-Bonds, J. Phys. Chem. B, 2006, 110, 6444, Copyright 2006 American Chemical Society.

complex. HBs may be classified as covalent in nature for $H\cdots B/H$ distances lower than 1.2 Å since $\nabla^2\rho_C < 0$ (and consequently $H_C < 0$). Such a situation occurs for the above-mentioned $[FHF]^-$ anion where both $H\cdots F$ contacts possess characteristics of covalent bonds. One can see (Figure 13.11) that with the shortening of the $H\cdots B/H$ distance the importance of the electron density shift effects increase – the delocalization interaction becomes more important than the electrostatic one.

It seems that for XB complexes, a similar situation occurs as for HBs; the results presented earlier show the importance of delocalization interaction for strong halogen bonds (Table 13.2) while for weaker interactions dispersive and electrostatic terms are more important.[85] Table 13.3 presents QTAIM characteristics for the strong XBs analyzed earlier (Table 13.2). The F–Cl and F–Br molecules act as Lewis acids through Cl and Br centres, respectively. The electron density at the BCP, ρ_C, often correlates with the binding energy;[95,96] such a rough correlation is also observed here since for the weakest interactions with acetylene (Table 13.2), the least values of ρ_C are observed, $\sim$0.02–0.03 au (Table 13.3). For complexes with NH_3, ρ_C amounts to $\sim$0.06 au and for the strongest charge assisted interactions, ρ_C is equal to $\sim$0.08–0.09 au; note that the ρ_C values of $\sim$0.1 au and more are typical for covalent bonds.[87] The H_C values are in line with the above trend; H_C is positive for complexes with acetylene while for other systems it is negative, "more negative" for the strongest CAXBs – complexes with Cl^-. It is worth mentioning that a similar situation was observed for other complexes linked through halogen bonds;[61] for the strongest CAXB interactions, *i.e.* for $Br\cdots O$ links in the $CF_3Br\cdots OH^-$ and $HCCBr\cdots OH^-$ complexes, the corresponding BCPs are characterized by negative H_C values, while for other species they are positive. Even for CAXBs in the chlorine analogues

Table **13.3** The QTAIM parameters (in au) for the halogen bonded complexes; the atomic charges of FCl and FBr molecules are given as well as the electron charge density shifts from the Lewis base to the Lewis acid, Q_{trans}; QTAIM calculations based on MP2/aug-cc-pVTZ wave functions (results from ref. 84).

Complex	ρ_C	$\nabla^2\rho_C$	H_C	$Q_{Cl/Br}$	Q_F	Q_{trans}
$F-Cl\cdots NH_3$	0.065	0.138	−0.014	+0.316	−0.498	−0.182
$F-Br\cdots NH_3$	0.064	0.124	−0.016	+0.427	−0.598	−0.171
$F-Cl\cdots C_2H_2$	0.022	0.072	0.001	+0.377	−0.441	−0.064
$F-Br\cdots C_2H_2$	0.029	0.080	0.001	+0.462	−0.547	−0.085
$F-Cl\cdots Cl^-$	0.088	0.096	−0.028	+0.132	−0.616	−0.484
$F-Br\cdots Cl^-$	0.076	0.080	−0.023	+0.281	−0.679	−0.398

CF$_3$Cl$\cdots$OH$^-$ and HCCCl$\cdots$OH$^-$ the H_C values are positive, since the Cl-centre is characterized by a lower positive EP value than the Br-centre.

Figure 13.12 shows examples of the XB complexes analyzed here (Tables 13.2 and 13.3); the FCl complexes are presented. One can see that for the Cl-centre in contact with the Lewis base (C$_2$H$_2$, NH$_3$ and Cl$^-$), the region of the σ-hole characterized by a large positive value of EP (as was mentioned earlier here, much greater than for CF$_3$Cl) is characterized by a positive Laplacian of electron density in the extension of F–Cl bond for two complexes; with NH$_3$ and C$_2$H$_2$. In a case of the F–Cl$\cdots$Cl$^-$ complex there is the large electron charge density shift from the Cl$^-$ Lewis base to the F–Cl species of -0.484 au (Table 13.3). This is why there is a greater accumulation of electron density at the Cl-atom acting as a Lewis acid; its charge amounts to $+0.132$, less than for the two remaining FCl complexes (Table 13.3). Hence the central chlorine atom in the latter complex is characterized by "the ring of negative Laplacian" – the positive Laplacian in the elongation of F–Cl bond is cancelled (Figure 13.12). Similar tendecies are observed for FBr complexes; *i.e.* the greatest electron charge density transfer for the complex with Cl$^-$, next with NH$_3$ and the lowest one for the FBr$\cdots$C$_2$H$_2$ species.

The halogen-halogen contacts described in the previous sections should be addressed here. There are two types of R$_1$–X$\cdots$X–R$_2$ interactions (Scheme 13.2),[31,45] where type I is usually not classified as an XB interaction. It was stated recently that "type I is a geometry based contact that arises from close packing and is found for all halogens. It is not a halogen bond."[21] It is true that the type II interaction is exactly a contact between the Lewis acid and Lewis base regions of halogens and that this is not the case for the type I. However, recent studies on halogen–halogen stabilizing interactions provide results that allow us to verify the view of these interactions. A theoretical study has been performed on the halogen–halogen contacts in R–X$\cdots$X–R complexes (R = H, Cl, F and X = Cl, Br, I). Different theoretical approaches were applied such as QTAIM, NBO and the analysis of the electrostatic potentials. It was found that the electrostatic interactions, as well as interactions attributed to electron charge density shifts, play a substantial role in determining the geometry of the complexes.[97] Additionally, the recent hole-lump approach[98] was applied to deepen understanding of the nature of interactions for these complexes.[97] Eskandari and Zariny introduced this approach to analyze the XB interaction;[98] they showed that the halogen bond is an interaction between a hole (region of a charge depletion and excess

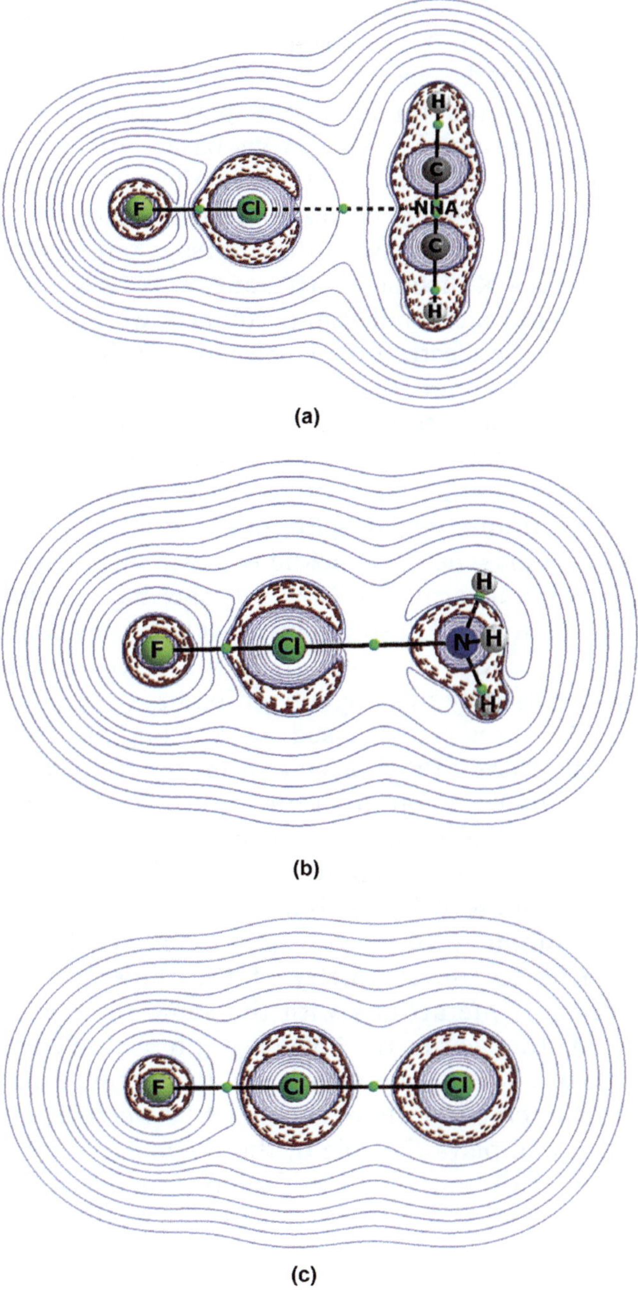

Figure 13.12 The molecular graphs with the isolines of laplacian of electron density for complexes of; (a) FCl···C$_2$H$_2$, (b) FCl···NH$_3$, (c) FCl···Cl$^-$; positive values of $\nabla^2\rho$ are depicted in solid lines and negative values in broken lines; for the FCl···C$_2$H$_2$ complex one of bond paths (designated by broken line) links the Cl-attractor with the non nuclear attractor (NNA) of acetylene.

kinetic energy) and a lump (region of a charge concentration and excess potential energy). It was also shown that the P$\cdots$P, P$\cdots$N and N$\cdots$N pnicogen bonds can be categorized as hole-lump interactions.[99] In other words, the hole corresponds to the Lewis acid site while the lump corresponds to the Lewis base site.

In the case of R–X$\cdots$X–R dihalogen interactions (classified as type I – see Scheme 13.2), the dimers corresponding to energetic minima are characterized by C_i symmetry (with the inversion centre situated at the mid-point of the X$\cdots$X distance – MP2/aug-cc-pVTZ level).[97] For such complexes, two equivalent hole-lump contacts were detected corresponding to the single electron density bond path linking halogen attractors. For the FBr dimer ($E_{bin} = -3.3$ kcal mol^{-1}) there is the bond path linking two Br-attractors that corresponds to stabilizing interactions; the ρ_C value at the corresponding BCP is equal to 0.015 au, and H_C is positive. This path does not connect exactly the sites of maximum (positive) and minimum (negative) EPs of the bromine surfaces since it goes through regions where the acidic or basic properties of the surfaces are not clearly determined. However, one may conclude that the FBr molecules are oriented in such a way that two equivalent hole-lump links between Br-centres correspond to two Lewis acid–Lewis base interactions expressed by the single bond path.

For the F-Br dimer discussed here the $n\text{Br}^1[\text{molecule1}] \rightarrow \sigma(\text{FBr})^*[\text{molecule2}]$ and $n\text{Br}^2[\text{molecule1}] \rightarrow \sigma(\text{FBr})^*[\text{molecule2}]$ orbital-orbital interactions were found within the NBO approach, with the corresponding energies of interaction of 1.1 and 7.7 kcal mol^{-1};[97] superscripts at Br designate here two different lone pairs of the same bromine centre. These orbital-orbital overlaps (bromine lone pair orbital – F–Br antibond orbital) correspond to molecule1 $\rightarrow$ molecule2 interactions. Due to the symmetry of the dimer, the same molecule2 $\rightarrow$ molecule1 interactions with the same interaction energies are detected. In other words, the I type X–X interaction corresponds to the contact between sites where each site realizes its dual character. For II type X–X interaction, there is contact between sites possessing clear character, one plays the role of the Lewis base and the second one of the Lewis acid – that is why this interaction is better fixed to the halogen bond definitions mentioned in this chapter.

13.6   Summary – Similarities and Differences

Similarities and differences between halogen bonds, XBs, and hydrogen bonds, HBs, are described in this chapter. It seems that both may be classified as σ-hole bonds; for both, the electrostatic interactions are very important since they are responsible for the arrangement of molecules in complexes or in greater aggregates.[35,36]

The electron charge density shifts as a result of complexation are also very important for halogen and hydrogen bonds; the interaction related to these effects may be named as polarization and it is an intrinsic component of the electrostatic interaction.[36] This is why HBs and XBs, and other σ-hole bonds, are defined as electrostatic/polarization interactions accompanied by dispersion.[36] The latter qualification of the σ-hole bonds is confirmed by various interaction energy decomposition schemes and approaches. One can see that for both interactions, XB and HB, the electrostatic interaction is more important at larger distances between interacting species and with the shortening of this distance the electron charge density shift effects start to play a more important role. For short distances, halogen and hydrogen bonds possess partly covalent character. However, the decomposition approaches differ between themselves significantly in the physical meanings of the interaction energy terms; this is why one should be careful in the interpretation and discussion of the interaction energy decomposition results.

The Quantum Theory of Atoms in Molecules (QTAIM)[86] is a powerful tool for analyzing the nature of interactions; the QTAIM results are in agreement with the energetic ones as concerns the properties of the A–X$\cdots$B halogen bond and the A–H$\cdots$B hydrogen bond. The characteristics of the X$\cdots$B and H$\cdots$B bond critical points follow the same changes for both interactions. The increase of the strength of interaction is connected with the increase of the electron density at BCP up to large values, accompanied by the negative total electron energy density at BCP (H_C); the latter value is positive for weaker interactions.

It is also important that two processes occuring for hydrogen bonds, the hyperconjugation and the rehybridization,[58] are common for the other noncovalent interactions (σ-hole bonds),[51,52] as was explained for HBs and XBs in this chapter.

Desiraju has pointed out that the HB and XB interactions are similar because halogen and hydrogen are both electrophilic in nature.[21] And that they are distinctive because the size of halogen atom is comparable to the sizes of other atoms in the A–X$\cdots$B systems, which is not the case for hydrogen in the A–H$\cdots$B arrangement.[21] The halogen bond is intermediate in energy terms between strong and weak hydrogen bonds and both interactions are directional ones.[21] The directionality of both interactions is nicely explained by the

σ-hole concept;[5] A–X$\cdots$B arrangements are usually closer to linearity than A–H$\cdots$B ones since, for the former interactions, the σ-hole characterized by positive EP is closely related to a restricted region located in the extension of the A–X bond. In the latter case, the whole H-atom hemisphere possesses positive EP thus there are "more places" for nucleophilic attack.

Numerous analogies and dissimilarities between HB and XB were presented here; however numerous were omitted. For example, the HB is usually characterized by an A–H bond elongation resulting from complexation, with a corresponding shift of the A–H stretching mode to lower frequencies (to the red);[1,2] however, there are so-called blue-shifting hydrogen bonds characterized by the A–H bond shortening after HB formation, with the A–H stretching mode shifting to higher frequencies (to the blue).[100] Similar effects are observed for A–X$\cdots$B halogen bonds with the blue shifts for A–X bond; however, such XBs are more frequent than HBs.[62]

The similarities and differences between dihydrogen and dihalogen bonds were briefly described. In the case of the latter interactions much greater diversity is observed than for DHBs. One can also compare the A–H$\cdots$H–B DHB, where the role of the Lewis base centre is played by the hydric H-atom (connected with B), with the halogen-hydride bond, A–X$\cdots$H–B,[4] at first look a rather strange interaction where for the X$\cdots$H contact the electronegative halogen is the Lewis acid centre while the H-atom possesses the Lewis base centre properties.

The unique phenomenon that does not occur for H-centres while it is observed for X-ones is that the halogen may be characterized by multivalency; several studies were performed on XBs where a multivalent halogen acts as a Lewis acid; these interactions are a sub-class of XBs, characterized by a few distinct properties if compared with XBs with monovalent halogen.[101,102] Finally, it seems that the proton transfer process characteristic for numerous hydrogen bonded systems[54,103] does not occur for halogen bonded species; if only because of the large sizes of halogens compared with hydrogen. However, there are a few studies where the possibility of halogen cation transfers are considered; for example, the crystal structures of complexes of pyridines with N-iodoimide derivatives show different positions of iodine atom in A–I$\cdots$B $\leftrightarrow$ [A]$^{-}\cdots$[I–B]$^{+}$ systems.[104]

One may mention other numerous examples of analogies and dissimilarities between HB and XB; the main ones were described in this chapter. The cooperativity effect that is a phenomenon occurring for both HB and XB interactions (and for numerous other types of interactions) is described in another chapter of this monograph.

Table 13.4 summarizes the characteristics of both interactions. One can see that the main differences are connected with the central H/X atom; hydrogen is less electronegative than any halogen atom, it is smaller than any halogen, it is positively charged while halogens are usually negatively charged *etc.* However, the nature of HB and XB interactions is the same (Table 13.4) and both are classified as the σ-hole bonds in spite of differences between the central atoms.

It seems that the same nature and the same numerous properties of HB and XB are a result of the σ-hole of the halogen centre that is characterized a positive electrostatic potential. The σ-hole is responsible for the properties of the halogen bond, not the whole X-atom; for

Table 13.4 The comparison of the hydrogen bond with the halogen bond.

Parameters of interaction	A–H$\cdots$B hydrogen bond	A–X$\cdots$B halogen bond
Nature	Electrostatic/covalent accompanied by dispersion	Electrostatic/covalent accompanied by dispersion
Nature	Importance of covalency increases with the strength of interaction	Importance of covalency increases with the strength of interaction
Type	σ-Hole bond	σ-Hole bond
Mechanism	Hyperconjugation and rehybridization	Hyperconjugation and rehybridization
Orbital–orbital interaction	$n(B) \rightarrow \sigma^*_{AH}$	$n(B) \rightarrow \sigma^*_{AX}$
Energy of interaction (kcal per mol)	2–40	The strength increases in the order: $Cl < Br < I$, comparable with HBs
Directionality	Directional	Directional, more than HB
Influence of interaction on the A–H/X bond length	Usually increases – red shift, sometimes decreases – blue shift	Increases or decreases, blue shift occurs more often than in HBs
H/X electronegativity ref. 22	H – 2.1	F – 4.0; Cl – 3.2; Br – 3.0; I – 2.7
H/X atomic radius (Å) ref. 105	H – 0.25	F – 0.5; Cl – 1; Br – 1.15; I – 1.4
H/X charge	Positive	Usually negative
H/X electrostatic potential	Positive for the whole hemisphere	Usually positive in the extension of A–X bond; negative in the orthogonal direction to A–X
Acidity/basicity of H/X	Acidic	Usually dual acidic/basic character of the same X centre
H/X valency	Monovalent	Monovalent or multivalent
Cooperativity	Occurs	Occurs
Proton (or X^+) transfer	Occurs very often	Few cases of the preliminary stage of this process are known

the hydrogen it is not so important since its whole hemisphere possesses a positive electrostatic potential.

Acknowledgements

Financial support comes from Eusko Jaurlaritza (GIC IT-588-13) and the Spanish Office for Scientific Research (CTQ2012-38496-C05-04). Technical and human support provided by Informatikako Zerbitzu Orokora – Servicio General de Informática de la Universidad del País Vasco (SGI/IZO-SGIker UPV/EHU), Ministerio de Ciencia e Innovación (MICINN), Gobierno Vasco Eusko Jaurlanitza (GV/EJ), European Social Fund (ESF) is gratefully acknowledged.

References and Notes

1. G. A. Jeffrey and W. Saenger, *Hydrogen Bonding in Biological Structures*, Springer, Berlin, 1991.
2. G. J. Jeffrey, *An Introduction to Hydrogen Bonding*, Oxford University Press, New York, 1997.
3. G. Cavallo, P. Metrangolo, R. Milani, T. Pilati, A. Priimagi, G. Resnati and G. Terraneo, *Chem. Rev.*, 2016, **116**, 2478.
4. P. Lipkowski, S. J. Grabowski and J. Leszczynski, *J. Phys. Chem. A*, 2006, **110**, 10296.
5. P. Politzer and J. S. Murray, *ChemPhysChem*, 2013, **14**, 278.
6. G. R. Desiraju and T. Steiner, *The Weak Hydrogen Bond in Structural Chemistry and Biology*, Oxford University Press, New York, 1999.
7. J. J. Colin and H. F. Gaultier de Claubry, *Ann. Chim.*, 1814, **90**, 87.
8. J. J. Colin, *Ann. Chim.*, 1814, **91**, 252.
9. O. Hassel and J. Hvoslef, *Acta Chem. Scand.*, 1954, **8**, 873.
10. H. Bent, *Chem. Rev.*, 1968, **68**, 587.
11. A. C. Legon, *Angew. Chem., Int. Ed.*, 1999, **38**, 2686.
12. A. C. Legon, *Chem. – Eur. J.*, 1998, **4**, 1890.
13. N. Ramasubbu, R. Parthasarathy and P. Murray-Rust, *J. Am. Chem. Soc.*, 1986, **108**, 4308.
14. R. Wong, F. H. Allen and P. Willett, *J. Appl. Crystallogr.*, 2010, **43**, 811.
15. P. Metrangolo and G. Resnati, *Chem. – Eur. J.*, 2001, 7, 2511.
16. P. Metrangolo, H. Neukirch, T. Pilati and G. Resnati, *Acc. Chem. Res.*, 2005, **38**, 386.
17. J. M. Dumas, L. Gomel and M. Guérin, *Molecular interactions involving organic halides,* The Chemistry of Functional Groups, Supplement D, Wiley, New York, 1983, pp. 983–1020.
18. M. Formigué and P. Batail, *Chem. Rev.*, 2004, **104**, 5379.
19. *Halogen Bonding. Fundamentals and Applications*, ed. P. Metrangolo and G. Resnati, Springer-Verlag, Berlin, Heidelberg, 2008.
20. E. Parisini, P. Metrangolo, T. Pilati, G. Resnati and G. Terraneo, *Chem. Soc. Rev.*, 2011, **40**, 2267.
21. A. Mukherjee, S. Tothadi and G. R. Desiraju, *Acc. Chem. Res.*, 2014, **47**, 2514.

22. L. Pauling, *The Nature of the Chemical Bond*, Cornell University Press, New York, 3rd edn, 1960.
23. M. Nishio, M. Hirota and Y. Umezawa, *The CH/π Interaction: Evidence, Nature and Consequences*, Wiley-VCH, Inc., New York, 1998.
24. J. J. Szymczak, S. J. Grabowski, S. Roszak and J. Leszczynski, *Chem. Phys. Lett.*, 2004, **393**, 81.
25. S. J. Grabowski, W. A. Sokalski and J. Leszczynski, *Chem. Phys. Lett.*, 2006, **432**, 33.
26. G. C. Pimentel and A. L. McClellan, *The Hydrogen Bond*, W. H. Freeman and Company, San Fransisco and London, 1960.
27. I. Alkorta, I. Rozas and J. Elguero, *Chem. Soc. Rev.*, 1998, **27**, 163.
28. S. J. Grabowski, W. A. Sokalski and J. Leszczynski, *Chem. Phys. Lett.*, 2006, **422**, 334.
29. E. Arunan, G. R. Desiraju, R. A. Klein, J. Sadlej, S. Scheiner, I. Alkorta, D. C. Clary, R. H. Crabtree, J. J. Dannenberg, P. Hobza, H. G. Kjaergaard, A. C. Legon, B. Mennucci and D. Nesbitt, *Pure Appl. Chem.*, 2011, **83**, 1637.
30. G. R. Desiraju, P. S. Ho, L. Kloo, A. C. Legon, R. Marquardt, P. Metrangolo, P. Politzer, G. Resnati and K. Rissanen, *Pure Appl. Chem.*, 2013, **85**, 1711.
31. F. Zordan, L. Brammer and P. Sherwood, *J. Am. Chem. Soc.*, 2005, **127**, 5979.
32. S. C. Nyburg and W. Wong-Ng, *Proc. R. Soc. London, Ser. A*, 1979, **367**, 29.
33. S. K. Surampudi, G. Nagarjuna, D. Okamoto, P. D. Chaudhuri and D. Venkataraman, *J. Org. Chem.*, 2012, **77**, 2074.
34. GaussView, Version 5, R. Dennington, T. Keith and J. Millam, Semichem Inc., Shawnee Mission, KS, 2009.
35. P. Politzer, J. S. Murray and T. Clark, *Phys. Chem. Chem. Phys.*, 2010, **12**, 7748.
36. P. Politzer, J. S. Murray and T. Clark, *Phys. Chem. Chem. Phys.*, 2013, **15**, 11178.
37. A. Bauzá and A. Frontera, *Angew. Chem., Int. Ed.*, 2015, **54**, 7340.
38. J. S. Murray, K. E. Riley, P. Politzer and T. Clark, *Aust. J. Chem.*, 2010, **63**, 1598.
39. T. Clark, M. Hennemann, J. S. Murray and P. Politzer, *J. Mol. Model.*, 2007, **13**, 291.
40. V. G. Tsirelson, P. F. Zou, T. H. Tang and R. F. W. Bader, *Acta Crystallogr., Sect. A*, 1995, **51**, 143.
41. R. Boese, A. D. Boese, D. Blaser, M. Yu. Antipin, A. Ellern and K. Seppelt, *Angew. Chem., Int. Ed. Engl.*, 1997, **36**, 1489.
42. T. T. T. Bui, S. Dahaoui, C. Lecomte, G. R. Desiraju and E. Espinosa, *Angew. Chem., Int. Ed.*, 2009, **48**, 3838.
43. M. E. Brezgunova, E. Aubert, S. Dahaoui, P. Fertey, S. Lebègue, C. Jelsch, J. G. Ángyán and E. Espinosa, *Cryst. Growth Des.*, 2012, **12**, 5373.
44. T. Sakurai, M. Sundaralingam and G. A. Jeffrey, *Acta Crystallogr.*, 1963, **16**, 354.
45. G. R. Desiraju and R. Parthasarathy, *J. Am. Chem. Soc.*, 1989, **111**, 8725.
46. M. Capdevila-Cortada and J. J. Novoa, *CrystEngComm*, 2015, **17**, 3354.
47. M. Capdevila-Cortada, J. Castelló and J. J. Novoa, *CrystEngComm*, 2014, **16**, 8232.
48. D. Chopra, *Cryst. Growth Des.*, 2012, **12**, 541.
49. D. D. Gunbas, F. Algy, T. Hokelek, W. H. Watson and M. Balcy, *Tetrahedron*, 2005, **61**, 11177.
50. A. D. Ward, V. R. Ward and E. R. T. Tiekink, *Z. Kristallogr. – New Cryst. Struct.*, 2001, **216**, 563.
51. S. J. Grabowski, *J. Phys. Chem. A*, 2011, **115**, 12340.
52. S. J. Grabowski, *J. Phys. Chem. A*, 2012, **116**, 1838.
53. P. Gilli, V. Bertolasi, V. Ferretti and G. Gilli, *J. Am. Chem. Soc.*, 1994, **116**, 909.
54. S. J. Grabowski, *Chem. Rev.*, 2011, **111**, 2597.
55. R. H. Crabtree, P. E. M. Siegbahn, O. Eisenstein, A. L. Rheingold and T. F. A. Koetzle, *Acc. Chem. Res.*, 1996, **29**, 348.
56. H. Cybulski, M. Pecul and J. Sadlej, *J. Chem. Phys.*, 2003, **119**, 5094.

57. S. J. Grabowski and F. Ruipérez, *Phys. Chem. Chem. Phys.*, 2016, **18**, 12810.
58. I. V. Alabugin, M. Manoharan, S. Peabody and F. Weinhold, *J. Am. Chem. Soc.*, 2003, **125**, 5973.
59. E. Reed, L. A. Curtiss and F. Weinhold, *Chem. Rev.*, 1988, **88**, 899.
60. F. Weinhold and C. Landis, *Valency and Bonding, A Natural Bond Orbital Donor – Acceptor Perspective*, Cambridge University Press, Cambridge, 2005.
61. S. J. Grabowski, *Phys. Chem. Chem. Phys.*, 2013, **15**, 7249.
62. W. Wang and P. Hobza, *J. Phys. Chem. A*, 2008, **112**, 4114.
63. S. J. Grabowski, *J. Mol. Model.*, 2013, **19**, 4713.
64. The results presented were taken from ref. 61; however the natural bond orbitals presented here (eqn (13.3) and (13.4)) were not discussed there in detail.
65. H. Bent, *Chem. Rev.*, 1961, **61**, 275.
66. P. Popelier, *Atoms in Molecules. An Introduction*, Prentice Hall, Pearson Education Limited, New York, 2000.
67. J. A. Platts, S. T. Howard and B. R. F. Bracke, *J. Am. Chem. Soc.*, 1996, **118**, 2726.
68. S. F. Boys and F. Bernardi, *Mol. Phys.*, 1970, **19**, 553.
69. S. J. Grabowski, *Annu. Rep. Prog. Chem., Sect. C.*, 2006, **102**, 131.
70. S. J. Grabowski, A. J. Sadlej, W. A. Sokalski and J. Leszczynski, *Chem. Phys.*, 2006, **327**, 151.
71. P. Politzer, P. Lane, M. C. Concha, Y. Ma and J. S. Murray, *J. Mol. Model.*, 2007, **13**, 305.
72. M. Tawfik and K. J. Donald, *J. Phys. Chem. A*, 2014, **118**, 10090.
73. S. J. Grabowski, *J. Phys. Chem. A*, 2011, **115**, 12789.
74. P. Politzer, K. E. Riley, F. A. Bulat and J. S. Murray, *Comput. Theor. Chem.*, 2012, **998**, 2.
75. A. M. Pendas, M. A. Blanco and E. Francisco, *J. Chem. Phys.*, 2006, **125**, 184112.
76. W. A. Sokalski, S. Roszak and K. Pecul, *Chem. Phys. Lett.*, 1988, **153**, 153.
77. W. A. Sokalski and S. Roszak, *J. Mol. Struct.: (THEOCHEM)*, 1991, **234**, 387.
78. R. W. Gora, W. A. Sokalski, J. Leszczynski and V. B. Pett, *J. Phys. Chem. B*, 2005, **109**, 2017.
79. S. J. Grabowski, W. A. Sokalski, E. Dyguda and J. Leszczynski, *J. Phys. Chem. B*, 2006, **110**, 6444.
80. K. Kitaura and K. Morokuma, *Int. J. Quantum Chem.*, 1976, **10**, 325.
81. B. Jeziorski, R. Moszyński and K. Szalewicz, *Chem. Rev.*, 1994, **94**, 1887.
82. R. W. Gora, S. J. Grabowski and J. Leszczynski, *J. Phys. Chem. A*, 2005, **109**, 6397.
83. S. J. Grabowski, *Croat. Chem. Acta*, 2009, **82**, 185.
84. S. J. Grabowski and W. A. Sokalski, *ChemPhysChem*, 2017, DOI: 10.1002/cphc.201700224.
85. K. E. Riley and P. Hobza, *J. Chem. Theory Comput.*, 2008, **4**, 232.
86. R. F. W. Bader, *Acc. Chem. Res.*, 1985, **18**, 9.
87. R. F. W. Bader, *Atoms in Molecules, A Quantum Theory*, Oxford University Press, Oxford, 1990.
88. *Quantum Theory of Atoms in Molecules: Recent Progress in Theory and Application*, ed. C. J. Matta and R. J. Boyd, Wiley-VCH, New York, 2007.
89. R. F. W. Bader, *J. Phys. Chem. A*, 1998, **102**, 7314.
90. R. F. W. Bader, *J. Phys. Chem. A*, 2009, **113**, 10391.
91. S. Jenkins and I. Morrison, *Chem. Phys. Lett.*, 2000, **317**, 97.
92. I. Rozas, I. Alkorta and J. Elguero, *J. Am. Chem. Soc.*, 2000, **122**, 11154.
93. W. D. Arnold and E. Oldfield, *J. Am. Chem. Soc.*, 2000, **122**, 12835.
94. For a recent review on [FHF]⁻ anion look at: S. J. Grabowski, *Crystals*, 2016, **6**, 3
95. O. Mó, M. Yánez and J. Elguero, *J. Chem. Phys.*, 1992, **97**, 6628.
96. E. Espinosa, E. Molins and C. Lecomte, *Chem. Phys. Lett.*, 1998, **285**, 170.
97. D. J. R. Duarte, N. M. Peruchena and I. Alkorta, *J. Phys. Chem. A*, 2015, **119**, 3746.
98. K. Eskandari and H. Zariny, *Chem. Phys. Lett.*, 2010, **492**, 9.

99. K. Eskandari and N. Mahmoodabadi, *J. Phys. Chem. A*, 2013, **117**, 13018.
100. P. Hobza and Z. Havlas, *Chem. Rev.*, 2000, **100**, 4253.
101. W. Wang, *J. Phys. Chem. A*, 2011, **115**, 9294.
102. S. J. Grabowski, *Chem. Phys. Lett.*, 2014, **605–606**, 131.
103. G. Zundel, in *Advances in Chemical Physics*, ed. I. Prigogine and S. A. Rice, Wiley, New York, 2000, vol. 111.
104. O. Makhotkina, J. Lieffrig, O. Jeannin, M. Fourmigue, E. Aubert and E. Espinosa, *Cryst. Growth Des.*, 2015, **15**, 3464.
105. J. C. *Slater, J. Chem. Phys.*, 1964, **41**, 3199.

14   The Cation–π Interaction

Dennis A. Dougherty

Division of Chemistry and Chemical Engineering, California Institute
of Technology, Pasadena, CA 91125, USA
Email: dad@caltech.edu

14.1   Introduction

The cation–π interaction is now recognized as a powerful, broadly
utilized noncovalent binding interaction. Fundamental studies in the
gas phase have established the potential magnitude of the effect and,
when combined with computation, the physical origins of the binding. Early studies of cyclophanes and other model receptors showed
the potential of the cation–π interaction as an important force for
molecular recognition in solution. Perhaps the most compelling
evidence for the importance of cation–π interactions comes from
studies of biological systems, where cation–π interactions play an important role in both determining protein structure and in impacting
intermolecular interactions, both small molecule to macromolecule
and macromolecule to macromolecule (Figure 14.1).

Molecular crystal structures have also played an important role in
establishing the significance and nature of the cation–π interaction,
but it is fair to say the majority of studies of the interaction have
explored different contexts. It is clear, however, that the cation–π
interaction can be a powerful contributor to intermolecular interactions in molecular crystals. In the present work we will briefly
review the foundational studies and essential features of the cation–π
interaction. We will then provide an overview of prototype examples

Intermolecular Interactions in Crystals: Fundamentals of Crystal Engineering
Edited by Juan J. Novoa
© The Royal Society of Chemistry 2018
Published by the Royal Society of Chemistry, www.rsc.org

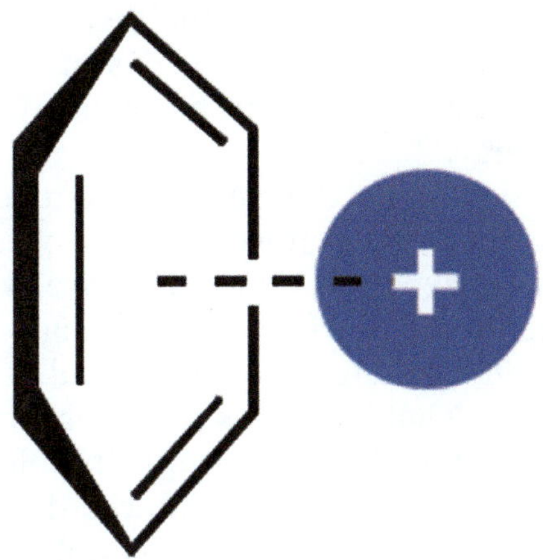

Figure 14.1 The cation–π interaction.

of the role these interactions can play in solid state structures and packing. Several extensive reviews of the cation–π interaction have appeared, and we refer the reader to those for more detailed presentations.[1–11]

14.2 The Gas Phase and Theoretical Studies

It has been over 30 years since high pressure mass spectrometry studies established that K^+ binds to benzene with a $-\Delta H^{\circ}$ of 19 kcal mol^{-1} in the gas phase.[12] This is a large number, especially when compared to the $K^+\cdots$water interaction, which is 18 kcal mol^{-1}. Subsequent studies established other key binding energies, including: $Na^+\cdots$benzene, 28 kcal mol^{-1}; $Li^+\cdots$benzene, 38 kcal mol^{-1}; $NH_4^+\cdots$benzene, 19 kcal mol^{-1}; $NMe_4^+\cdots$benzene, 9 kcal mol^{-1}; $Na^+\cdots$ethylene, 12 kcal mol^{-1}; and so on.[13–18] These substantial binding energies established the potential of the cation–π interaction as a potent noncovalent binding force, easily comparable in magnitude to, for example, a hydrogen bond.

From the beginning a predominantly electrostatic model for the cation–π interaction has been proposed. Benzene and ethylene both have a substantial quadrupole moment (benzene's is larger), and a good starting point for thinking about a cation–π interaction is to align the cation with that quadrupole. Certainly other factors such as ion$\cdots$induced dipole and dispersion forces contribute to varying degrees, depending on the nature of the cation and the π system. However, to first order, it is the buildup of negative electrostatic potential on the face of simple π systems that provides an attractive binding site for cations. Indeed, for a series of related systems, the variation in this electrostatic interaction explains the variation in the cation–π binding energy.[19,20]

A simple way to think about the electrostatic model is to recognize that sp^2 carbon is more electronegative than hydrogen. This produces 6 (4) bond dipoles in benzene (ethylene) that give rise to a molecular quadrupole. Importantly, sp^3 carbon has essentially the same electronegativity as hydrogen, and so cyclohexane and ethane are not significant cation binders. That said, it is not appropriate to term these binding interactions as merely an ion···quadrupole interaction. That would ignore contributions from ion···induced dipole and dispersion forces. Also, an ion···quadrupole interaction implies a specific distance dependence that is not universally followed in cation–π interactions. In general, the distance dependence of the cation–π interaction is not steep,[21] and so even ions that are not at their optimal separation from the π system (which is essentially van der Waals contact) can still experience an energetically meaningful interaction (Figure 14.2).

Based on this analysis, it should be evident that there is a strong directional component to the cation–π interaction. For a simple π system like benzene, the optimal binding arrangement is directly over the center of the ring, along the six-fold axis. High level calculations establish that there is no meaningful distortion of, for example, benzene when an ion such as Na^+ binds to it. Also, as mentioned above, the distance dependence of the cation–π interaction is not overly steep. An ion can move a full Å away from the π system and still retain almost half its optimal binding energy.[21]

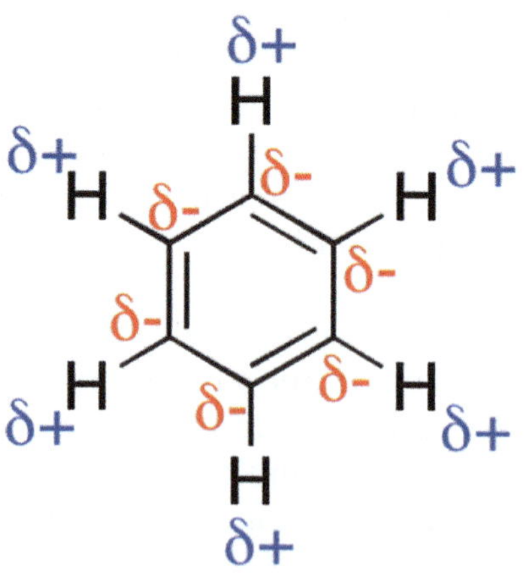

Figure 14.2   Bond dipoles that lead to a molecular quadrupole.

Moving significantly off the center of the ring is deleterious to the cation–π interaction. In the limit, of course, placing the cation along the edge of the π system produces a repulsive interaction. This electrostatic potential pattern for simple π systems such as benzene and ethylene – negative over the face of the π system and positive on the edge – is essential to π–π interactions, which are discussed elsewhere in this volume.

Most modern computational methods reproduce experimental cation–π binding energies well for prototype systems such as $Na^+\cdots$benzene. However, when more complex cations are involved, not all methods perform equally well. A recent study indicates that the M06 density functional with the relatively modest 6-31G** basis set performs well.[21] The situation is much less clear with the various "modeling" packages of the sort used to probe protein structures. These molecular mechanics-based methods often underestimate the cation–π interaction substantially, and caution should be exercised when applying them to the kinds of systems discussed below.

14.3 The Magnitude of the Cation–π Interaction and the Aromatic Box

With any noncovalent interaction, it is natural to ask about the potential energetic contribution it can make to a binding energy or a crystal packing arrangement. And, as with any noncovalent interaction, this is a complex question that depends on the context of the interaction and the reference chosen. The gas phase data cited above establish that, intrinsically, the cation–π interaction can be very strong indeed – more powerful than most other noncovalent interactions. More relevant, perhaps, are studies in condensed media – solutions and the solid state. While many cation–π interactions have been observed in the solid state, it is always difficult to assign an energetic value to any one interaction in a crystal.

In contrast, a large number of efforts to quantify the cation–π interaction have been undertaken in solution. Different systems produce contrasting results, and we provide a brief overview here. More extensive discussions have been presented in the reviews noted above. Several studies have involved a small molecule model system in which an organic cation (ammonium, guanidinium...) interacts with a single aromatic ring, typically in aqueous media. An attraction is generally seen, but the energetic consequences are typically small, on the order of 0.5 kcal mol^{-1}.[8] As always, it is difficult to know to what extent the model system allows an optimal cation–π interaction.

The first model system to establish the viability of the cation–π interaction in aqueous media was a cyclophane system developed by Dougherty, typified by structure **1**.[4] The cyclophane is a *general* receptor for organic cations such as alkylated ammoniums, quinoliniums, guanidiniums, sulfoniums … These studies were the first to show that a binding site comprised of simple π systems can pull a soluble, organic cation out of water and into a "hydrophobic" cavity. An interesting comparison is for the guests **2** and **3**, which are nearly identical in size, shape, and hydrophobic surface area. Remarkably, the cationic **3** is bound more tightly than **2** ($\Delta\Delta G^\circ = 2.5$ kcal mol^{-1}), despite the fact the **3** is better solvated than **2** in water by ~46 kcal mol^{-1}. This and many other cyclophane studies suggest that the cation–π interaction can contribute several kcal mol^{-1} to small molecule binding. Cyclophane **1**, of course, contains six π systems that could contribute to cation binding. While it seems unreasonable to think that all six could contribute equally, it may be that several cation–π interactions are contributing to the binding. As discussed below, in some ways cyclophane **1** presaged a pattern often seen in biological binding sites.

Cation–π interactions have also been established to be quite effective at contributing to catalysis in a wide range of reactions. A recent review cites dozens of reactions in which stabilizing cation–π interactions in the transition state contribute to catalysis.[7] In many cases, computational studies and product ratios support contributions of several kcal mol^{-1} for cation–π interactions stabilizing a particular transition state.

There have also been a number of studies of cation–π interactions in naturally occurring, protein binding sites, and these have often produced strong cation–π effects. Several studies have compared the binding of cationic drugs or similar compounds to that of an analogous neutral molecule.[22,23] Effects on the order of 2–3 kcal mol^{-1} are seen, favoring the cationic ligand. Based on such studies, Diederich declared that "the cation–π interaction is one of the strongest driving forces in biological complexation processes".[24]

A number of studies have probed the binding of a natural ligand to its protein binding site, choosing to subtly alter the protein rather than the small molecule.[25] For example, the binding of acetylcholine (ACh) to its native ACh receptor has been extensively investigated. In this case, the protein has evolved to bind the ligand, and so, in contrast to model systems, one can anticipate that binding interactions will have highly favorable if not perfectly optimal geometries. Progressively fluorinating the aromatic ring of the protein that contributes to binding a cationic small molecule provides a subtle, precise probe of cation–π binding. The high electronegativity of fluorine diminishes the negative electrostatic potential that dominates the cation–π interaction, and so binding is diminished. This approach has been applied to over two dozen small molecule···protein interactions, and the results suggest a range of 2–5 kcal mol^{-1} for the contribution of a single aromatic residue making a cation–π interaction to the native ligand. Other small molecules, such as pharmaceuticals targeting the receptor, have also been probed, and again strong cation–π interactions are seen.

Studies of binding sites from a range of proteins that vary considerably in function and structure reveal an intriguing feature of the cation–π interaction. Often, perhaps usually, these naturally evolved cation–π binding sites do not possess just one π system, but rather a cluster of 3–5 rings that embraces the cation, suggesting the possibility of multiple cation–π interactions. In some cases, the cation is almost completely encapsulated by what has come to be known as the "aromatic box" (in proteins, all the relevant π systems are aromatic – the side chains of Phe, Tyr, and Trp).[4,23,26–28] A prototype box from the acetylcholine binding protein (AChBP) is shown in Figure 14.3 (pdb file 1UW6); it is comprised of three Tyr and two Trp residues.[26] Computational studies establish that biological cations such as $R-NH_3^+$ and $R-N(CH_3)_3^+$ can indeed benefit from multiple cation–π interactions when surrounded by an aromatic box.[21] Interestingly, however, functional studies using the fluorination strategy mentioned above generally indicate that one particular ring of an aromatic box makes the dominant cation–π interaction.[25] The cluster of π systems seen in an aromatic box is reminiscent of the situation with cyclophane **1**, and we will see similar clusters in molecular crystals.

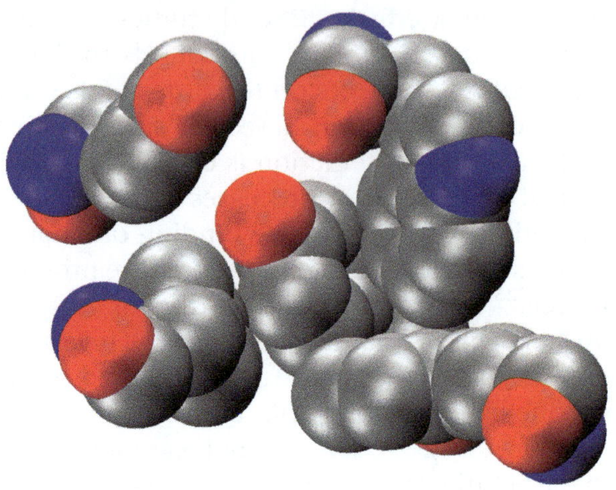

Figure 14.3 The aromatic box of AChBP.

A different manifestation of cation–π interactions in biological systems is seen in the intrinsic folding of proteins, perhaps a system more relevant to crystal packing. The collection of natural amino acid contains two cations – Lys and Arg – and three π systems – Phe, Tyr, and Trp. Analysis of the Protein Data Bank shows that cation–π interactions – (Lys/Arg···Phe/Tyr/Trp) – are common in proteins. On average, there is one such interaction for every 77 residues in the PDB, and so over 500 000 cation–π interactions can be found in these crystal structures.[29] A simple program to identify such interactions is available on the web (capture.caltech.edu). To the extent that protein folding resembles crystal packing, it can be seen that the cation–π interaction is a potent driving force for organizing complex systems.

14.4 Cation–π Interactions in Molecular Crystals

It should be clear from the above examples that the cation–π interaction is powerful enough to strongly influence intermolecular interactions, and many examples of well-documented cation–π interactions in molecular crystals have appeared. Indeed, in a recent review Biradha and Santra have noted that "cation–π interactions are very strong and reliable intermolecular interactions in the solid state".[30] Here we provide an overview of some recurring motifs. Several recent reviews provide a much more comprehensive analysis.[5,30,31]

Many studies have focused on the cation–π interaction between alkali metals and aromatic π systems. One of the earliest and most-cited examples involves alkali metal salts of tetraphenylborates and related compounds.[32] There are many such structures, and inevitably, the metal ion is sandwiched between multiple phenyl rings, providing inter-molecular connections between adjacent tetraphenylborates. For example, in $K \cdot BPh_4$ each K^+ is surrounded by four phenyl groups, two each from two different BPh_4^- units (**4**) (in this and other images, cation–π interactions are denoted by purple, dashed lines; some substituents are deleted for clarity). Recent work has used solid state NMR (^{23}Na and ^{39}K) to further characterize these complexes.[33] Geometrically, these structures look like perfect cation–π interactions, and are in some ways reminiscent of the aromatic box discussed above. There is a complication, however, in that each BPh_4 unit carries a negative charge. It is difficult to parse out the relative contribution to metal binding provide by this charge *vs.* the intrinsic electrostatic potential of the arene.

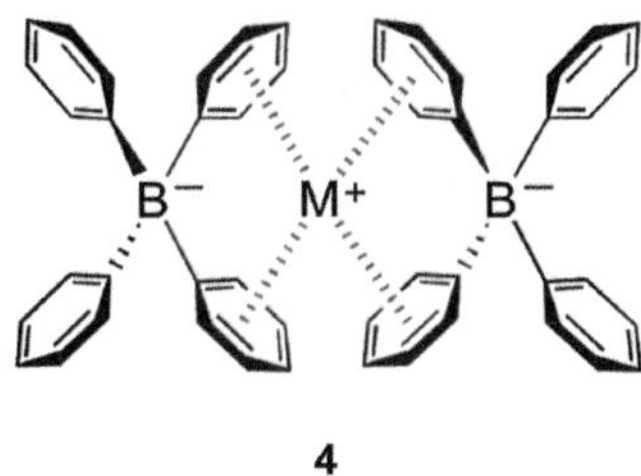

4

Cation–π interactions have been seen in a number of other anionic systems in which an extended arene such as 9,9′-bianthracene,[34] rubrene,[35] or tetraphenylallene[36] is reduced by Na metal. The Na^+ is found over the center of aromatic rings, typically also complexed by amines or ethers. Similarly, deprotonation of 1-phenylcyclohexene by *n*-butylsodium gave an allyl anion in which the Na^+ lies over the face of the phenyl ring, not the allyl anion.[37] As with the tetra-phenylborates, detailed analysis of the role of cation–π interactions *vs.* the direct interaction of the Na^+ with the negative charge of the π system is difficult to sort out.

Ferrocene is a neutral molecule with cyclopentadienide rings that bear formal negative charge, and several examples of cation–π interactions involving ferrocene have appeared. Mulvey described a molecular complex involving hexamethyldisilazide (HMDS): $[K \cdot (Cp_2Fe)_2 \cdot toluene_2]^+[Mg(HMDS)_3]^-$ (**5**).[38] In this elegant structure, a K^+ is coordinated by two toluene molecules and by two ferrocene Cp rings, again making 4 cation–π interactions to a single metal ion. This concept was expanded to create polymeric solid state chains, in which, for example, a Na^+ ion coordinates two HMDS nitrogens and one ferrocene (**6**).[39] The second Cp of the ferrocene then binds another comparable Na^+, creating the chain. Thus, the cation–π interaction was use to rationally design a well ordered, solid state chain. Another example that used ferrocene, but to a different end, was described by Gokel, who prepared a ferrocene that had a CH_2Ph on one Cp ring and a $CH_2NMe_3^+$ on the other.[40] The structure crystallized such that an intermolecular cation–π interaction forms between the quaternary ammonium of one molecule and the Ph of another, in a chain-like geometry (7).

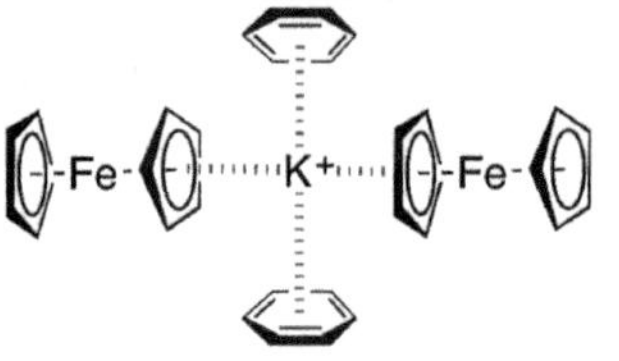

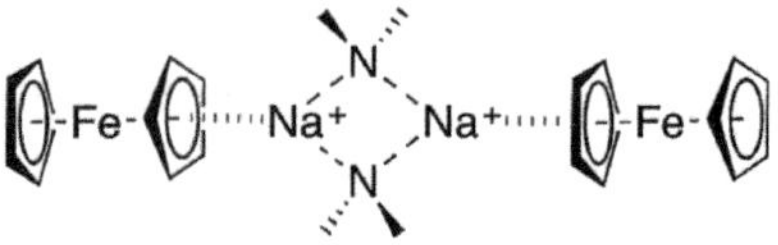

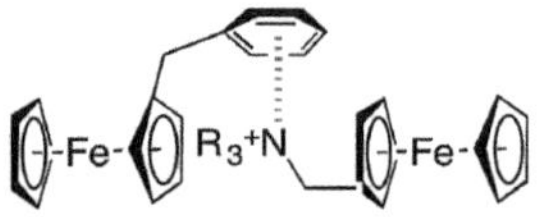

Other systems involving solid state cation–π interactions and alkali metals have been reported. A clever series of compounds developed by Gokel combines a crown ether with a pendant aromatic group.[31] In the absence of an ion, the aryl rings are directed away from the crown (8). When the crown binds the metal, the side chains wrap around, "lariat style", to cap the ion in a compelling cation–π interaction (9). A variety of complexes with Na^+ or K^+ ions, and with side chains such as benzene, phenol, or indole (mimicking the amino acids Phe, Tyr, and Trp) made strong cation–π interactions. An earlier structure of K^+ complexed to dibenzo-18-crown-6 shows a benzene molecule from solvent making an ideal cation–π interaction to the K^+ ion (10).[41] In another case where a cation–π interaction augments an additional metal binding motif, a tris(aryloxy)silane thiol is reacted with Na metal. In the resulting thiolate, the Na^+ binds to the S^- but also sits over the face of one of the aryl rings of the silane (11).[42]

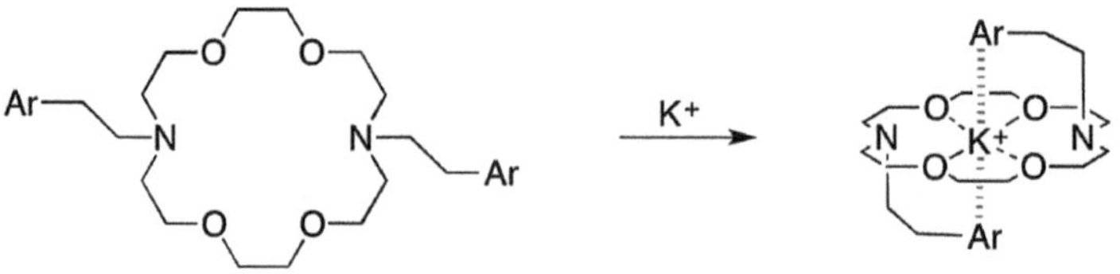

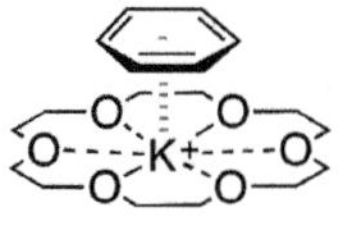

10

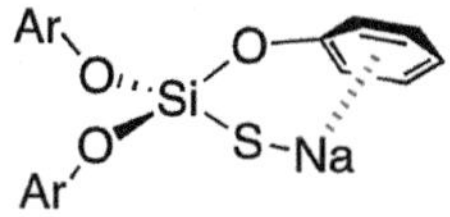

11

We noted above the interaction of a quaternary ammonium group with a ferrocene, and "quats" have been found to make intermolecular cation–π interactions to arenes in several structures.[40] A classic case is the choline ester of indole-3-acetic acid (**12**).[43] In the solid state, the quaternary trimethylammonium group folds back onto the indole ring to make a cation–π interaction that has also been seen in solution. This system is of special interest because it mimics the interaction seen when ACh binds to its natural protein binding sites, including the nicotinic ACh receptor, the muscarinic ACh receptor, and ACh esterase.[4] In each case, the quaternary ammonium moiety lies directly over the face of a π system provided by a Trp (indole side chain) or a Tyr (phenol side chain).

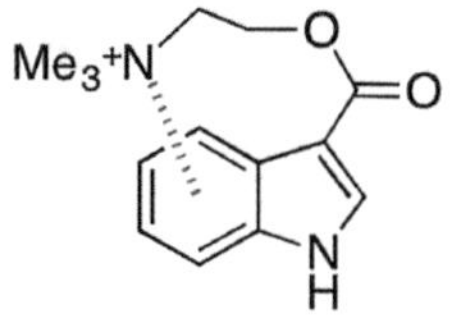

12

A fascinating molecule presents what has been argued to be the first case of light alkali metals (Li^+, Na^+) binding to a molecule exclusively through cation–π interactions.[44] Mixing the diboryne shown with Li^+ or Na^+ salts of noncoordinating ions produces the structure shown (**13**). A clear cation–π interaction in a neutral molecule is involved.

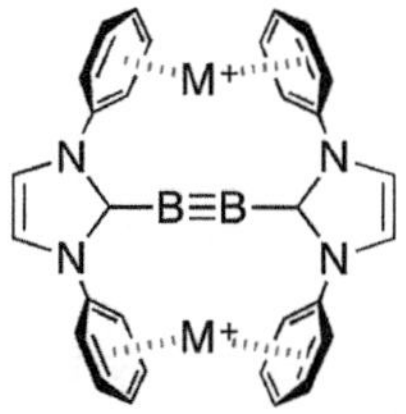

13

A remarkable example of the use of cation–π interactions to control intermolecular geometries in the solid state has been provided by a series of studies by Yamada and co-workers.[45] A classic example of solid state chemistry is the photochemical dimerization of styrene derivatives. A tetrasubstituted cyclobutane results, but product stereochemistry is essentially determined by crystal packing, and its prediction/control is quite challenging. Building on earlier solution phase work,[46] Yamada reasoned that a cation–π interaction could be used to overcome this problem. The protonated pyridine chalcones shown (**14**), on solid state photodimerization, give a single pure product with complete control of regiochemistry and stereochemistry (**15**). This is because the pyridinium hydrochloride salts crystallize parallel to each other in a head-to-tail orientation, due to a cation–π interaction between the pyridinium of one molecule and the benzene or anisole ring of another (**16**). In sharp contrast, photolysis of the unprotonated, neutral pyridine chalcone gives a complex mixture of products. This approach has been extended to include many other types of alkenes with cationic and aromatic substituents,[47] establishing the generality of the cation–π interaction in controlling crystal packing patterns. This is a rare example of rationally designing a molecule so as to encourage a particular packing pattern. To some extent this work was presaged by studies of diaryldiacetylenes, in which a phenyl–perfluorophenyl stacking interaction was used to control crystal packing and thus solid state photochemistry.[48]

Another example of a protonated pyridine ring stacking on a neutral arene arises in the aryl-substituted terpyridine shown (**17**).[49] In the solid state, a protonated pyridine stacks perfectly over the methoxyphenyl group, in a prototype cation–π interaction.

17

While perhaps not perfectly relevant to the topic of molecular crystals, we note that cation–π interactions have been employed to develop nanoscale structures. One example uses calixarenes, cyclic tetramers with an aromatic cavity (**18**). The aromatic walls on the interior of a calixarene are well documented to bind cations such as NMe_4^+ through cation–π interactions, and many structures with a cationic species embedded in a calixarene have appeared.[50] In the particular case of a hydroquinone-derived calixarene $(R=OH)$, the calixarene forms extended tubes in the solid state, mediated by hydrogen bonds. Added Ag^+ ions are attracted to the interior of the calixarene tubes through cation–π interactions. On reduction, the neutral Ag is no longer attracted to the calixarene, and so it assembles into very well defined Ag nanowires.[51] Many other examples of cation–π interactions in nanosystems such as nanotubes, nanocapsules, nanogel-composites and so on are given in the recent review by Mahadevi and Sastry.[5]

18

14.5   Conclusions

It is now well established that the cation–π interaction is a strong and universally employed noncovalent interaction. In the present work we have summarized the key features of the cation–π interaction, as revealed by a variety of techniques and systems. Not surprisingly, this interaction is common in molecular crystals. In this brief overview, we have tried to show prototypical examples where cation–π interactions are especially relevant and that are representative of the kinds of structures seen. We anticipate many more examples of using cation–π interactions to understand and control molecular crystals.

References

1. D. A. Dougherty and D. A. Stauffer, *Science*, 1990, **250**, 1558–1560.
2. D. A. Dougherty, *Science*, 1996, **271**, 163–168.
3. J. C. Ma and D. A. Dougherty, *Chem. Rev.*, 1997, **97**, 1303–1324.
4. D. A. Dougherty, *Acc. Chem. Res.*, 2013, **46**, 885–893.
5. A. S. Mahadevi and G. N. Sastry, *Chem. Rev.*, 2013, **113**, 2100–2138.
6. A. S. Mahadevi and G. N. Sastry, *Chem. Rev.*, 2016, **116**, 2775–2825.
7. C. R. Kennedy, S. Lin and E. N. Jacobsen, *Angew. Chem., Int. Ed.*, 2016, DOI: 10.1002/anie.201600547.
8. F. Biedermann and H.-J. Schneider, *Chem. Rev.*, 2016, **116**, 5216–5300.
9. Y. Zhao, Y. Cotelle, N. Sakai and S. Matile, *J. Am. Chem. Soc.*, 2016, **138**, 4270–4277.
10. C. Bissantz, B. Kuhn and M. Stahl, *J. Med. Chem.*, 2010, **53**, 6241.
11. N. S. Scrutton and A. R. C. Raine, *Biochem. J.*, 1996, **319**, 1–8.
12. J. Sunner, K. Nishizawa and P. Kebarle, *J. Phys. Chem.*, 1981, **85**, 1814–1820.
13. O. M. Cabarcos, C. J. Weinheimer and J. M. Lisy, *J. Chem. Phys.*, 1998, **108**, 5151–5154.
14. O. M. Cabarcos, C. J. Weinheimer and J. M. Lisy, *J. Chem. Phys.*, 1999, **110**, 8429–8435.
15. J. C. Amicangelo and P. B. Armentrout, *J. Phys. Chem. A*, 2000, **104**, 11420–11432.
16. D. Kim, S. Hu, P. Tarakeshwar and K. S. Kim, *J. Phys. Chem. A*, 2003, **107**, 1228–1238.
17. C. A. Deakyne and M. Meotner, *J. Am. Chem. Soc.*, 1985, **107**, 474–479.
18. M. Meotner and C. A. Deakyne, *J. Am. Chem. Soc.*, 1985, **107**, 469–474.
19. S. Mecozzi, A. P. West Jr. and D. A. Dougherty, *Proc. Natl. Acad. Sci. U. S. A.*, 1996, **93**, 10566–10571.
20. S. Mecozzi, A. P. West Jr. and D. A. Dougherty, *J. Am. Chem. Soc.*, 1996, **118**, 2307–2308.
21. M. R. Davis and D. A. Dougherty, *Phys. Chem. Chem. Phys.*, 2015, **17**, 29262–29270.
22. A. Y. Ting, I. Shin, C. Lucero and P. G. Schultz, *J. Am. Chem. Soc.*, 1998, **120**, 7135–7136.
23. K. Scharer, M. Morgenthaler, R. Paulini, U. Obst-Sander, D. W. Banner, D. Schlatter, J. Benz, M. Stihle and F. Diederich, *Angew. Chem., Int. Ed.*, 2005, **44**, 4400–4404.
24. L. M. Salonen, M. Ellermann and F. Diederich, *Angew. Chem., Int. Ed.*, 2011, **50**, 4808–4842.
25. E. B. Van Arnam and D. A. Dougherty, *J. Med. Chem.*, 2014, **57**, 6289–6300.
26. K. Brejc, W. J. van Dijk, R. V. Klaassen, M. Schuurmans, J. van Der Oost, A. B. Smit and T. K. Sixma, *Nature*, 2001, **411**, 269–276.
27. E. A. Meyer, R. K. Castellano and F. Diederich, *Angew. Chem., Int. Ed.*, 2003, **42**, 1210–1250.
28. L. M. Salonen, M. Ellermann and F. Diederich, *Angew. Chem., Int. Ed.*, 2011, **50**, 4808–4842.

29. J. P. Gallivan and D. A. Dougherty, *Proc. Natl. Acad. Sci. U. S. A.*, 1999, **96**, 9459–9464.
30. K. Biradha and R. Santra, *Chem. Soc. Rev.*, 2013, **42**, 950–967.
31. G. W. Gokel, L. J. Barbour, R. Ferdani and J. X. Hu, *Acc. Chem. Res.*, 2002, **35**, 878–886.
32. R. Ferdani, J. X. Hu, W. M. Leevy, J. Pajewska, R. Pajewski, V. Villalobos, L. J. Barbour and G. W. Gokel, *J. Inclusion Phenom. Macrocyclic Chem.*, 2001, **41**, 7–12.
33. A. Wong, R. D. Whitehead, Z. H. Gan and G. Wu, *J. Phys. Chem. A*, 2004, **108**, 10551–10559.
34. H. Bock, Z. Havlas, D. Hess and C. Nather, *Angew. Chem., Int. Ed.*, 1998, **37**, 502–504.
35. H. Bock, K. GharagozlooHubmann, C. Nather, N. Nagel and Z. Havlas, *Angew. Chem., Int. Ed. Engl.*, 1996, **35**, 631–632.
36. H. Bock, K. Ruppert, Z. Havlas and D. Fenske, *Angew. Chem., Int. Ed. Engl.*, 1990, **29**, 1042–1044.
37. S. Corbelin, J. Kopf, N. P. Lorenzen and E. Weiss, *Angew. Chem., Int. Ed. Engl.*, 1991, **30**, 825–827.
38. G. W. Honeyman, A. R. Kennedy, R. E. Mulvey and D. C. Sherrington, *Organometallics*, 2004, **23**, 1197–1199.
39. J. J. Morris, B. C. Noll, G. W. Honeyman, C. T. O'Hara, A. R. Kennedy, R. E. Mulvey and K. W. Henderson, *Chem. – Eur. J.*, 2007, **13**, 4418–4432.
40. J. X. Hu, L. J. Barbour and G. W. Gokel, *New J. Chem.*, 2004, **28**, 907–911.
41. D. C. Hrncir, R. D. Rogers and J. L. Atwood, *J. Am. Chem. Soc.*, 1981, **103**, 4277–4278.
42. A. Dolega, W. Marynowski, K. Baranowska, M. Smiechowski and J. Stangret, *Inorg. Chem.*, 2012, **51**, 836–843.
43. K. Aoki, K. Murayama and H. Nishiyama, *J. Chem. Soc., Chem. Commun.*, 1995, 2221–2222.
44. R. Bertermann, H. Braunschweig, P. Constantinidis, T. Dellermann, R. D. Dewhurst, W. C. Ewing, I. Fischer, T. Kramer, J. Mies, A. K. Phukan and A. Vargas, *Angew. Chem., Int. Ed.*, 2015, **54**, 13090–13094.
45. S. Yamada and Y. Tokugawa, *J. Am. Chem. Soc.*, 2009, **131**, 2098–2099.
46. S. Yamada, N. Uematsu and K. Yamashita, *J. Am. Chem. Soc.*, 2007, **129**, 12100–12101.
47. S. Yamada, K. Yamagami and S. Oaku, *Tetrahedron Lett.*, 2016, **57**, 2451–2454.
48. G. W. Coates, A. R. Dunn, L. M. Henling, D. A. Dougherty and R. H. Grubbs, *Angew. Chem., Int. Ed. Engl.*, 1997, **36**, 248–251.
49. A. Das, A. D. Jana, S. K. Seth, B. Dey, S. R. Choudhury, T. Kar, S. Mukhopadhyay, N. J. Singh, I.-C. Hwang and K. S. Kim, *J. Phys. Chem. B*, 2010, **114**, 4166–4170.
50. A. T. Macias, J. E. Norton and J. D. Evanseck, *J. Am. Chem. Soc.*, 2003, **125**, 2351–2360.
51. B. H. Hong, S. C. Bae, C. W. Lee, S. Jeong and K. S. Kim, *Science*, 2001, **294**, 348–351.

15 Intramolecular Beryllium Bonds. Further Insights into Resonance Assistance Phenomena

O. Brea,[a] I. Alkorta,[b] I. Corral,[a] O. Mó,[a] M. Yáñez*[a] and J. Elguero[b]

[a] Universidad Autónoma de Madrid, Departamento de Química, Facultad de Ciencias, Módulo 13 and Institute of Advanced Chemical Sciences (IadChem), Campus de Excelencia UAM-CSIC, Cantoblanco, Madrid 28049, Spain;
[b] Instituto de Química Médica, CSIC, Juan de la Cierva, 3, Madrid 28006, Spain
*Email: manuel.yanez@uam.es

15.1 Introduction

A significant number of phenomena in chemistry are related to the interaction between closed-shell systems. To this category belong most of the non-covalent interactions known up to date, such as hydrogen bonds,[1–3] halogen bonds,[4–6] di-hydrogen bonds,[7,8] pnicogen bonds,[9–12] tetrel,[13,14] chalcogen interactions,[15–17] and beryllium bonds.[18,19] All the aforementioned linkages can be classified as acid–base interactions in the Lewis sense, because in all cases one of the interacting subunits behaves as a Lewis base able to transfer some electronic charge to the second interacting subunit acting as a Lewis acid. For the particular case of the well known hydrogen bonds, $X-H\cdots A$, the proton acceptor, A, is the system which behaves as a Lewis base by donating charge from its electron lone-pairs to the antibonding X–H orbitals of the proton

Intermolecular Interactions in Crystals: Fundamentals of Crystal Engineering
Edited by Juan J. Novoa

donor. This leads to a lengthening of the corresponding X–H bond and to a red-shifting of its X–H stretching frequency, both effects being signatures of these kinds of non-covalent interactions. In beryllium bonds, the Lewis acid is a BeXY derivative,[18] whereas the Lewis base is any compound with some electron donor capacity, namely systems having sites with lone-pairs of electrons. In this case the charge donation from the base goes not only to the Be–X or Be–Y antibonding orbitals, resulting in a lengthening of the corresponding bonds, but also to the empty p orbitals of the Be atom, leading to a hybridization change of this atomic center reflected in a significant bending of the X–Be–Y moiety.[18]

Particularly interesting situations arise when the center acting as a Lewis base and that behaving as a Lewis acid form part of the same molecular compound, usually known as ditopic systems.[20–23] When the electron affinity of the acidic center is significant and the electron donor capacity of the basic center is also high, the aforementioned coincidence opens the possibility of self-assembling, and the formation of polymers becomes an important issue. Paradigmatic examples can be found in the carboxylic acid dimers[24–27] stabilized through two intermolecular hydrogen bonds (HBs), in which the carbonyl group of one of the monomers behaves as a basic center with respect to the hydroxyl group acting as the Lewis acid of the second monomer (see Scheme 15.1a). This is also the case, for instance, in (iminomethyl)beryllium hydride which is able to form *n*-mers which may be cyclic (see Scheme 15.1b) or linear (see Scheme 15.1c),[28] because the terminal imino group is an excellent electron donor and the terminal Be–X (X = H, F) is a very good electron acceptor, so the formation of quite strong Be···N beryllium bonds gives a substantial stability to the dimers, trimers, tetramers and, in general, *n*-mers. In Scheme 15.1 the structures of the cyclic and linear decamers are shown.[28]

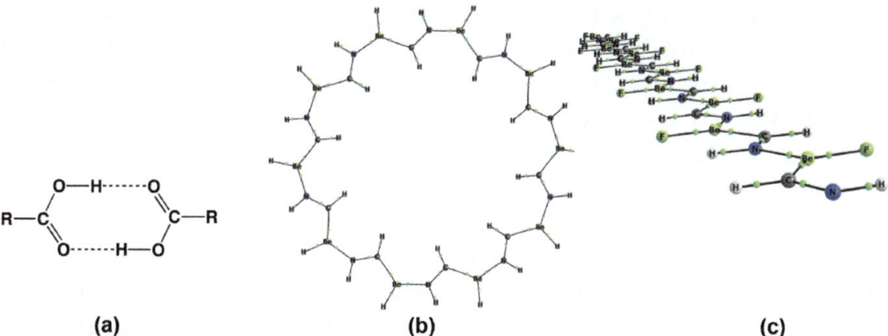

(a) (b) (c)

Scheme 15.1 Self-assembled structures triggered by the formation of: (a) hydrogen bonds, (b) and (c) beryllium bonds.

Many other ditopic ligands are the building blocks of a large series of coordination polymers[29] and linear coordination supramolecules[30,31] or lead to rather interesting molecular aggregates,[32,33] or play a role in supramolecular gel chemistry.[34] Ditopic ligands have been also employed for solvent extraction of cations and anions profiting from their ability to behave simultaneously as a base and as an acid.[35]

However, in more flexible ditopic systems it is also possible to observe intramolecular interactions between the two active sites. These kinds of interactions are not possible, for instance, in normal carboxylic acids where the planarity and the rigidity of the acidic function prevents the hydroxyl group from approaching sufficiently to the carbonyl group to form an intramolecular HB, but it is possible in many other compounds, such as malonaldehyde[36,37] and its derivatives, tropolone,[38,39] diols, sugars and a plethora of other organic and inorganic compounds. These intramolecular linkages are not exclusively HBs, but can also involve chalcogen–chalcogen interactions[15,40] or intramolecular halogen bonds.[41–43] One of the signatures of intramolecular interactions is the difficulty in establishing quantitatively their energetic contribution to the stability of the molecule. For this reason, information about the strength of intramolecular bonds is usually obtained through indirect magnitudes, such as the internuclear distances between the interacting centers, the electron density within the bonding region, or the shifting of some specific vibrational frequencies. This difficulty in accurately evaluating the interaction energy of an intramolecular interaction is closely related with the vagueness or ambiguity on the factors that may contribute to the strength of the interaction. A typical example of the foggy basis behind some useful concepts is found in the so-called resonance assisted hydrogen bonds (RAHBs).[44–50] The strength of intramolecular HBs in unsaturated systems is clearly larger than in the saturated analogues. This evidence is usually explained by claiming that these linkages are stronger because they enhance the resonance in the unsaturated compound,[44] which is however not so convincing an argument.[51,52] As a matter of fact, several papers, among them several from our group,[51–60] question not the effect but its interpretation as a resonance enhancement. This new viewpoint could be summarized by asserting that the unsaturated nature of the compound is the factor behind the larger strength of the intramolecular linkage. In other words, the intramolecular HB is not stronger due to a resonance assisted effect, but due to the fact that the interaction is assisted by the resonance of the system, due to several concomitant features. On the one hand, because the rigidity of the skeleton of an unsaturated compound strongly facilitates the interaction between the HB donor and the HB acceptor, and this is well reflected by the magnetic properties of the system, which are very sensitive to structural details. On the other hand, because the presence of unsaturation in the system modulates both the intrinsic basicity and the intrinsic acidity of the two active centers involved, enhancing the acid–base interaction.

The aim of the present chapter is three-fold: (a) to investigate, for the first time, the nature and characteristics of intramolecular beryllium bonds in molecular environments similar to those in which different intramolecular hydrogen bonds have been found and characterized, (b) to analyze these new intramolecular linkages to gain further insight into possible resonance assisted phenomena, and (c) to explore whether these compounds are able to dimerize forming Be–Be bonds, and what the characteristics of these bonds are, taking into account that Be_2 is only bound by a very weak interaction arising only from electron correlation.

For this purpose, and in order to ensure that we are going to have a similar molecular environment to directly compare intramolecular beryllium bonds with intramolecular HBs, we have used as suitable model systems the Be derivatives which can be formed by making replacements in the acidic group (the hydroxyl group) of malonaldehyde (**1**), some of its derivatives (**2, 3**) and tropolone (**4**), as well as in the corresponding saturated analogues (**5–8**) (see Scheme 15.2). Two approaches were used to build up the compounds with intramolecular beryllium bonds: the substitution of the H atom of the hydroxyl group by a BeH group (compounds ***n*-OBeH**, *n* = 1–8) and the substitution of the whole hydroxyl group by a BeH group (compounds ***n*-BeH**, *n* = 1–8), as shown in Scheme 15.2.

15.2 Computational Details

The geometries of the different compounds and complexes investigated in this study have been optimized by using the B3LYP functional associated with a 6-31+G(d,p) basis set expansion. The B3LYP approach combines the Becke's three-parameter nonlocal hybrid exchange potential[61] with the nonlocal correlation functional of Lee, Yang and Parr[62] and it has been found to be rather reliable for the description of intermolecular beryllium bonds.[18] Nevertheless, in order to ensure the reliability of this model for the description of intramolecular beryllium bonds, we have assessed the results obtained at

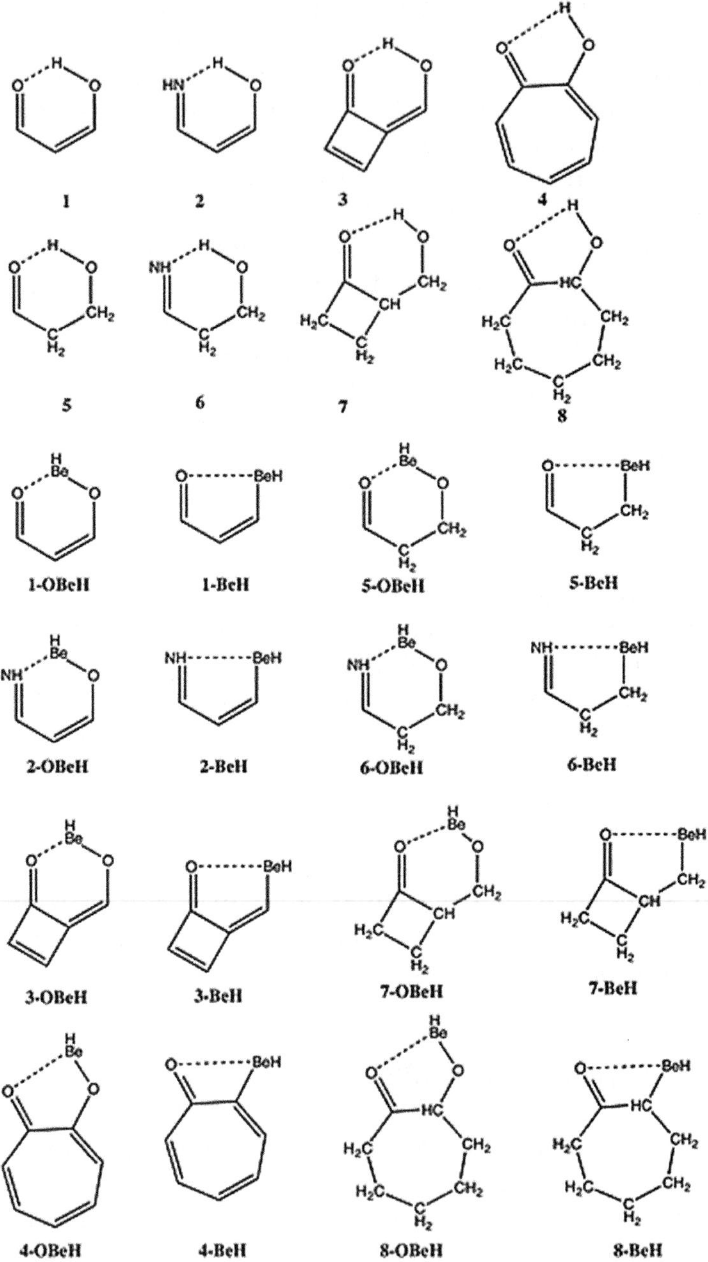

Scheme 15.2   Systems characterized by the existence of intramolecular hydrogen bonds (**1–8**) and the related compounds able to present intramolecular beryllium bonds, which are obtained from the former, by replacing the H atom of the OH group by a –BeH substituent (compounds *n*-**OBeH**, *n* = 1–8) or replacing the whole –OH group by a –BeH group (compounds *n*-**BeH**, *n* = 1–8).

the B3LYP/6-31+G(d,p) by using as a reference the G4 theory[63] for two derivatives, namely **1-OBeH** and **1-BeH**, as suitable examples. The G4 approach is a high-level *ab initio* composite method which uses B3LYP/6-31G(2df,p) optimized geometries and thermal corrections, and high-level empirical corrections, to yield final energies at an effective CCSD(T,full)/G3LargeXP+HF limit level.[63] This method is very well suited for the calculation of different thermodynamic magnitudes. Indeed, an assessment of the method on 454 experimental energies yielded an average absolute deviation from experiment of 3.47 kJ mol^{-1} (0.83 kcal mol^{-1}). However, since the G4 composite method, as mentioned above, is based also on the use of B3LYP optimized structures, we have assessed the reliability of these geometries by comparing them with those obtained at the MP2/6-31+G(d,p) level for the same two reference compounds. The agreement between both B3LYP and MP2 optimized geometries is rather good, although, in general, the B3LYP method tends to overbind the Be atom to the C=O basic site, but this effect does not compromise the conclusions of our analysis, based on general trends.

One signature of beryllium bonds is that they produce a significant distortion on the electron density distribution of the Lewis base participating in the interaction.[64] In order to quantitatively analyze these electron density perturbations we have used two different, but somehow complementary methods, namely the atoms in molecules (AIM) theory[65] and the Natural Bond Orbital (NBO) approach.[66] The AIM theory is based on the topological analysis of the electron density, $\rho(r)$ and its Laplacian, $\nabla^2\rho(r)$. The topology of $\rho(r)$ of any chemical system is characterized by the presence of maxima, associated with the position of the nuclei, saddle points located between two maxima or inside cyclic structures, usually named bond critical points (BCPs) and ring critical points (RCPs), respectively, and minima associated with the existence of cage structures. On top of that, the so called molecular graphs, which are the representation of the bond paths, defined as the lines of maximum density connecting two maxima and containing a BCP, provide an accurate description of the molecular structure since these molecular graphs are able to account for bent bonds, usually associated with ring-strain phenomena and which cannot be identified from the optimized geometries. Also importantly, the changes observed in the values of the electron density at the BCPs permit a quantitative measure of bond strengthening or bond weakening effects. These AIM calculations have been carried out by using the AIMAll program package.[67]

The NBO approach is very well suited to analyzing the formation of the new linkages between a basic site and an acidic site, because it permits quantification of charge transfers between occupied and empty orbitals within a given molecular system, through the calculation of the second-order orbital interaction energies involved in the interaction between these orbitals. Furthermore, this method, based on the use of localized natural orbitals, permits the description of the bonding of a molecule in terms of hybrid orbitals and core and lone pairs, obtained as local block eigenvectors of the one-particle density matrix. The NBO approach also allows the calculation of the Wiberg bond order (WBO)[68] which is a good index by which to measure the strength of a chemical bond. All the NBO calculations have been done by using the NBO-3.1 program package.[69]

15.3 Results and Discussion

15.3.1 Intramolecular Beryllium Bonds in Malonaldehyde-like Systems

The optimized structures of compounds **1-OBeH** and **1-BeH** are shown in Figure 15.1. The first conspicuous fact is that in all these compounds a rather strong O–Be intramolecular bond is formed, since the O–Be distances are much smaller than the sum of the van der Waals radii of both atoms. In all cases a BCP is located within the O–Be bonding region with electron densities of the same order of magnitude as the normal covalent bonds between Be and H atoms. An idea of the relative strength of these intramolecular O–Be bonds can be achieved by comparing the relative stability of the cyclic forms shown in Figure 15.1 with the conformers in which the alkyl chain is extended and no contact between the BeH group and the oxygen atom of the carbonyl group can be established. From now on these open structures will be named adding (**t**) to the name of the corresponding cyclic conformer.

It can be seen that both **1-OBeH** and **1-BeH** cyclic structures are more stable than the corresponding open forms, indicating that the new intramolecular O···Be bond significantly contributes to the stabilization of the former with respect to the latter. A comparison between the molecular graphs shows that on going from the open to the cyclic structure a significant decrease of the electron density at the O–BeH and the C–BeH bonds takes place. This bond weakening is reflected in a sizable lengthening of these bonds in the cyclic structure, but this destabilizing feature is counterbalanced by the formation of the new Be···O intramolecular beryllium bond. A comparison of the

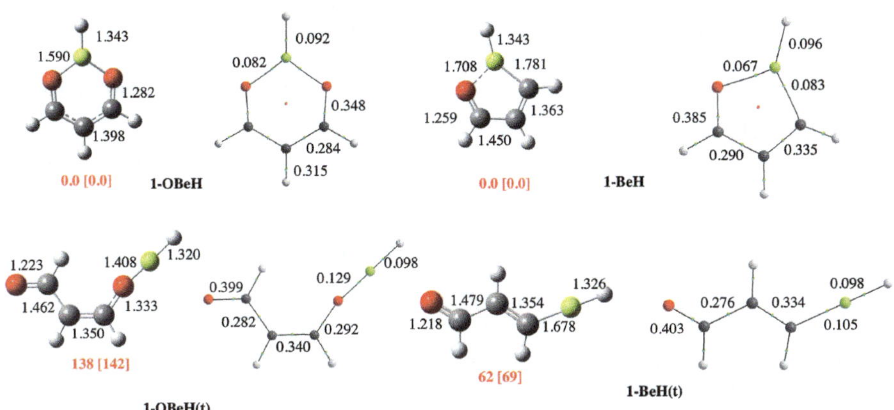

Figure 15.1 B3LYP/6-31+G(d,p) optimized structures and molecular graphs of unsaturated compounds showing O···Be intramolecular beryllium bonds and their corresponding open conformers. In red, the relative stabilities of the open forms with respect to the cyclic ones are given in kJ per mol. Values in square brackets were calculated at the G4 level of theory. Internuclear distances are in Å. In the molecular graphs, green and red dots denote BCPs and RCPs respectively. Electron densities are in a.u.

Be···O internuclear distance, the value of the electron density at the corresponding BCP, and the WBO (0.27 *vs.* 0.11) shows that the intramolecular beryllium bond is stronger in **1-OBeH** than in **1-BeH**. Consistently, the energy gap between the cyclic structure and the open one is also significantly larger for **1-OBeH** than for **1-BeH**. It is worth noting that the agreement between the relative energies obtained using the high-level G4 approach and the less computationally demanding B3LYP/6-31+G(d,p) theoretical model is rather good.

The stronger Be···O beryllium bond in **1-OBeH** is a consequence of two concomitant facts: the larger intrinsic basicity of the carbonyl group and the larger intrinsic acidity of the BeH group of this structure with respect to those of the **1-BeH** compound. It is rather obvious that the BeH group of **1-OBeH** should be a better Lewis acid than the same group in the **1-BeH** compound, because in the former the Be atom is attached to an electron withdrawing group (C–O) whereas in the latter it is attached to a carbon atom. Indeed, this prediction is corroborated by the G4 calculated hydride affinity of the corresponding open forms (see Scheme 15.3).

The enhanced basicity of the C=O group of **1-OBeH** is also confirmed by the G4 calculated proton affinity of compound **1-OBeH(t)** (853 kJ mol^{-1}) which is 45 kJ mol^{-1} larger than that of compound **1BeH(t)** (808 kJ mol^{-1}). All these findings are consistent with the results of the NBO analysis. According to this approach, the

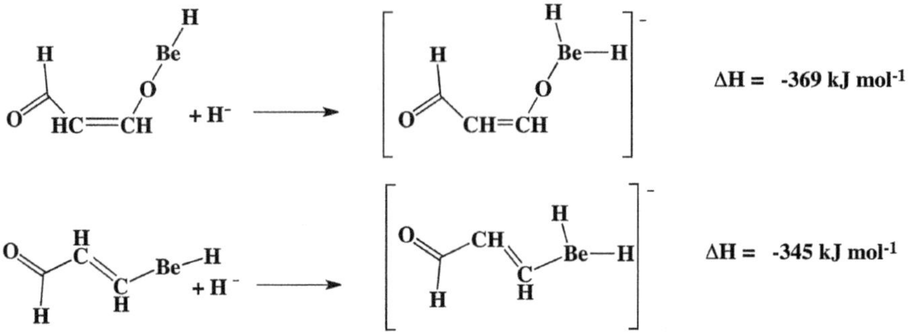

$$\Delta H = -369 \text{ kJ mol}^{-1}$$

$$\Delta H = -345 \text{ kJ mol}^{-1}$$

Scheme 15.3 Hydride affinities of compounds **1-OBeH(t)** and **1-BeH(t)** calculated at the G4 level of theory.

intramolecular beryllium bond in **1-OBeH** molecule is a consequence of a strong dative bond (with a second-order interaction energy of 99 kJ mol^{-1}) from one of the oxygen lone-pairs of the carbonyl group into the empty p orbital (actually a sp^2 hybrid) of the Be atom. It is the population of the p orbital of the metal which is responsible for the bending of the O–Be–H, because the hybridization of the Be atom changes from sp in the open forms, where no beryllium bond is possible, to sp^2 in the cyclic structures. A similar charge transfer into the σ^*_{BeH} antibonding orbital, with a second order interaction energy of 133 kJ mol^{-1}, is responsible for the lengthening of the Be–H bond that is observed on going from the open form to the cyclic one. For **1-BeH**, the aforementioned charge donation from the CO group to Be is smaller, due as mentioned above to a smaller electron donor capacity of the former and a smaller electron acceptor ability of the latter. Hence, the second order interaction energy between the oxygen lone-pair and the empty p orbital of Be is only 67 kJ mol^{-1}, and the one corresponding to the interaction with the σ^*_{BeH} antibonding orbital is 110 kJ mol^{-1}.

The strength of the C=O$\cdots$Be interaction in **1-OBeH** is responsible for the symmetric arrangement of the O–Be–O bridge, which differs from the behavior observed for malonaldehyde, where the O–H–O bridge is clearly non symmetric with an O–H distance much larger for the intramolecular hydrogen bond than for the O–H group acting as proton donor. We will come back later on to this question.

The situation is different when dealing with the corresponding saturated analogues, whose optimized structures and molecular graphs are shown in Figure 15.2.

A comparison of Figures 15.1 and 15.2 clearly shows that the O$\cdots$Be intramolecular beryllium bond in both **5-OBeH** and **5-BeH** compounds

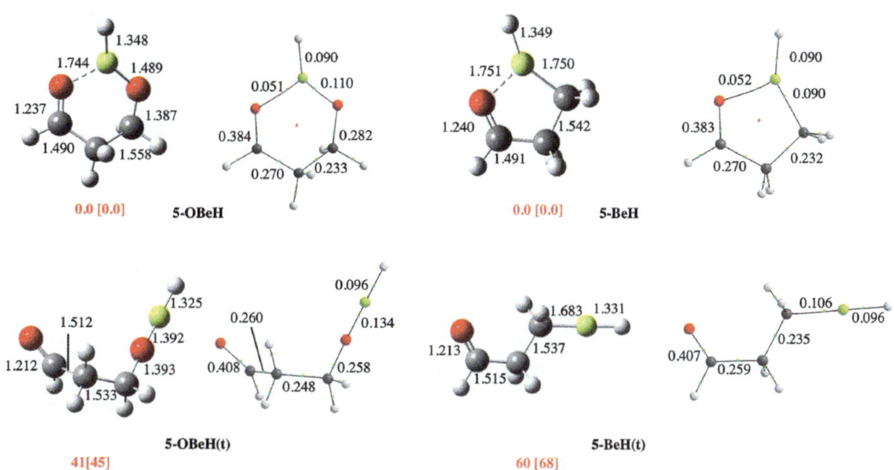

Figure 15.2 B3LYP/6-31+G(d,p) optimized structures and molecular graphs of saturated compounds showing a O···Be intramolecular beryllium bonds and their corresponding open conformers. Same conventions as in Figure 15.1.

is weaker than in the corresponding unsaturated counterparts, **1-OBeH** and **1-BeH**. The same was observed when the intramolecular hydrogen bond of malonaldehyde was compared with that in its saturated analogue, which was frequently used as an argument in favor of resonance assistance in malonaldehyde. However, the main reason for the substantial decrease in the strength of the O···Be beryllium bond in **5-OBeH** and **5-BeH,** as compared with the one in **1-OBeH** and **1-BeH**, is again twofold: (i) the decrease in the electron donor ability of the carbonyl group in **5-OBeH** and **5-BeH** as compared with those of **1-OBeH** and **1-BeH**, respectively; (ii) the concomitant decrease of the Lewis acidity of the corresponding BeH group. Indeed, the G4 calculated proton affinities for **5-OBeH(t)** and **5-BeH(t)** (833 and 793 kJ mol^{-1}, respectively) are clearly lower than those for the unsaturated counterparts **1-OBeH(t)**, **1-BeH(t)** (853 and 808 kJ mol^{-1}, respectively), as well as the corresponding hydride affinities (-312 and -319 kJ mol^{-1}, to be compared with the values in Scheme 15.3).

However, in contrast with the unsaturated analogues, in both **5-OBeH** and **5-BeH** the intramolecular beryllium bond is of similar strength when considering not only the O···Be distances but also the densities at the BCPs and the WBOs (0.1 *vs.* 0.095), and is only slightly stronger in the **5-OBeH** molecule. Still, the basicity of the **5-OBeH** is 40 kJ mol^{-1} greater than that of **5-BeH**, but, as indicated above, the acidity of **5-BeH** is 7 kJ mol^{-1} larger than that of **5-OBeH**. These opposite effects partially counterbalance each other, with the result that the O···Be intramolecular bond is only slightly shorter in **5-OBeH** than in **5-BeH**. In this respect, it is important to note that whereas the energy gap between the **5-BeH** (cyclic) and the **5-BeH(t)** (open) forms is very similar to that between the corresponding unsaturated analogues, **1-BeH** and **1-BeH(t)** (see Figures 15.1 and 15.2), the energy gap between **5-OBeH** and the **5-OBeH(t)** is much smaller than the one between **1-OBeH** and **1-OBeH(t)**. This seems to suggest an extra stabilization of the unsaturated **1-OBeH** cyclic structure, likely due to some resonant effects. Indeed, as we have discussed in previous paragraphs, the formation of the intramolecular O···Be beryllium bond implies a rather large charge transfer from the oxygen lone-pairs of the carbonyl group to the empty orbitals of Be. This results in a deep distortion of the electron density distribution of the rest of the molecule, which is apparent by comparing the molecular graphs of the **1-OBeH** (cyclic) and the **1-OBeH(t)** (open) structures. Upon cyclization of the system, a significant weakening of the covalent O–Be linkage and accordingly a reinforcement of the corresponding C–O bond is observed. Concomitantly, the densities at the C–C bonds are also affected, and the two clearly localized C–C bonds in the open **1-OBeH(t)** structure (one single and the other double) become identical by symmetry in the cyclic **1-OBeH**, with a density slightly higher than that associated with a C–C single bond (see Figure 15.1).

This electron density redistribution is in agreement with the picture that is obtained through the use of the Natural Resonance Theory (NRT)[70] in the framework of the NBO approach, which shows that the unsaturated **1-OBeH** cyclic structure is stabilized by the participation of the three different resonant structures depicted in Scheme 15.4.

We can then conclude that the higher relative stability of the cyclic **1-OBeH** with respect to the saturated counterpart **5-OBeH** is due to resonance, as is the case for the stability of malonaldehyde and its

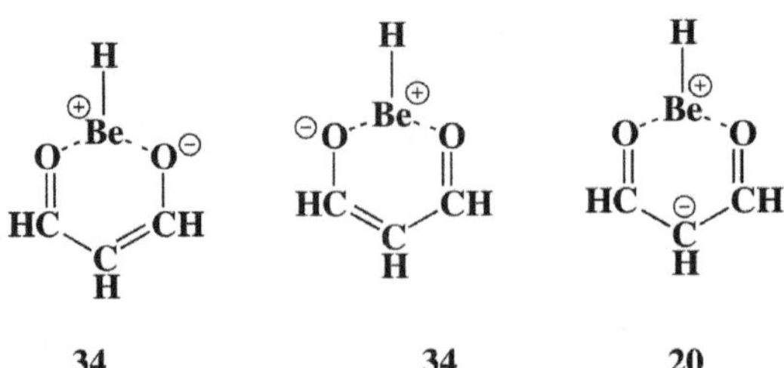

| 34 | 34 | 20 |

Scheme 15.4   Natural Resonance Theory weights (%) of the most important resonant structures contributing to the stability of **1-OBeH** derivative. Note that the NBO describes the BeH group as an independent unit.

derivatives.[71] Note, however, that as we have been claiming for similar systems showing intramolecular hydrogen bonds, the existence of this resonance does not result in a stronger intramolecular beryllium bond, but the other way around. It is the much stronger beryllium bond, resulting from a larger intrinsic basicity of the basic site (C=O group) and from a larger Lewis acidity of the –BeH group, which leads to a symmetric structure with two identical Be$\cdots$O linkages resulting necessarily in an enhancement of the resonance within the system. In summary, as indicated in ref. 60 concerning hydrogen bonds, the quasi-aromaticity observed in the unsaturated compounds is consistent with hydrogen-bonding-assisted resonance rather than with RAHBs. This argument is also consistent with the fact that the beryllium bond does not participate as such in any of the resonant structures stabilizing the system, since any of the resonant forms present O–Be bonds.

In order to gain some insight on the effect of changing the nature of the basic center involved in the formation of the intramolecular beryllium bond, we have included in our survey the derivatives that could be obtained from the set discussed above by replacing their carbonyl group by a C=NH group. These will be stabilized through the formation of Be$\cdots$NH intramolecular beryllium bonds. These compounds are **2-OBeH**, **2-BeH**, **6-OBeH** and **6-BeH**. The optimized structures and their molecular graphs are shown in Figure 15.3. The trends observed are similar to those discussed above for Be$\cdots$O bonds and for similar reasons. The intramolecular beryllium bond in **2-OBeH** is stronger than in **2-BeH** due to a larger basicity of the =NH group (956 *vs.* 923 kJ mol^{-1}) and to a larger Lewis acidity of the –BeH group in the former, reflected in a larger hydride affinity (-355 *vs.* -325 kJ mol^{-1}). Again, the differences between these intramolecular beryllium bonds are much smaller between **6-OBeH** and **6-BeH**, because the intrinsic basicity of the imino groups in the extended conformers, **6-OBeH(t)** and **6-BeH(t)** are equal (933 kJ mol^{-1}) and only the hydride affinity of **6-OBeH(t)** (299 kJ mol^{-1}) is slightly larger than that of **6-BeH(t)** (295 kJ mol^{-1}). It is worth noting that the energy gaps between the cyclic and the open forms of these series of compounds are larger than the ones calculated for the carbonyl containing analogues discussed above. This is an indication that the Be$\cdots$NH intramolecular bonds are stronger than the Be$\cdots$O bonds, due to a much larger intrinsic basicity of the imino group with respect to the carbonyl one (956 *vs.* 853 kJ mol^{-1} for **2-OBeH(t)** and **1-OBeH(t)**, respectively, and 923 *vs.* 808 kJ mol^{-1} for **2-BeH(t)** and **1-BeH(t)**, respectively). Nevertheless, this effect is partially counterbalanced by a

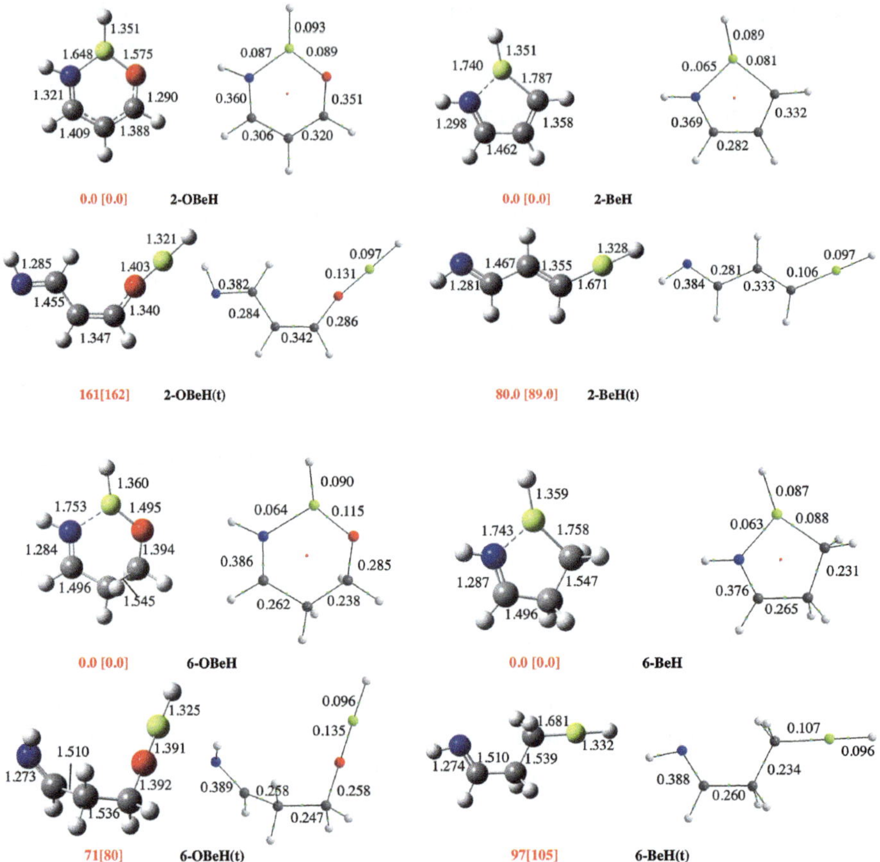

Figure 15.3 B3LYP/6-31+G(d,p) optimized structures and molecular graphs of unsaturated and saturated compounds showing a N···Be intramolecular beryllium bond and their corresponding open conformers. Same conventions as in Figure 15.1.

concomitant decrease of the hydride affinity of the BeH group in the nitrogen containing compounds with respect to the oxygen containing analogues.

Again, as it was found above for the couple **1-OBeH/1-BeH**, the gap between the cyclic **2-OBeH** and the open structures **2-OBeH(t)**, is larger than the one between **2-BeH** and **2-BeH(t)**, because the **2-OBeH** is stabilized through resonance effects associated with the participation of the five dominant resonant forms depicted in Scheme 15.5.

As in the case of the compound **1-OBeH**, the enhanced stability of the cyclic **2-OBeH** compound is a direct consequence of the resonance triggered by the large strength of the Be···NH intramolecular bond. The relative stabilities of the other three cyclic compounds, namely

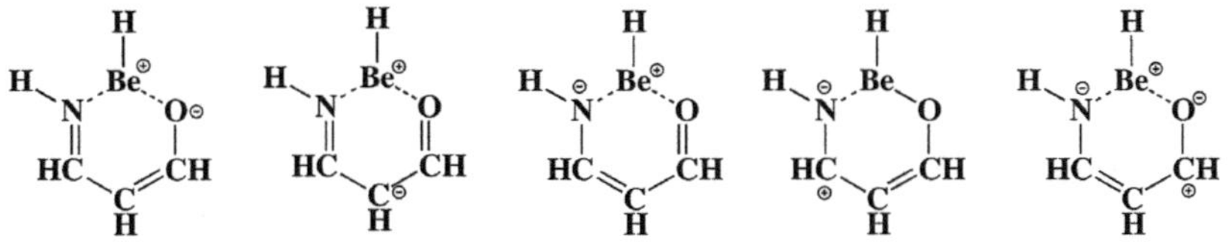

Scheme 15.5 Natural Resonance Theory weights (%) of the most important resonant structures contributing to the stability of **2-OBeH** derivative. Note that, with the only exception of the fourth resonant structure, the NBO describes the BeH group as an independent unit.

6-OBeH, **2-BeH** and **6-BeH** with respect to the corresponding open forms, are rather similar. In these cases, where resonance effects have a negligible role, the differences between the cyclic and the open structures are mainly due, although not exclusively, to the formation of the intramolecular beryllium bond.

15.3.2 Effects of the Rigidity of the Molecular Skeleton

At this point we consider it of interest to investigate what the effect would be of imposing some rigidity on the system exhibiting an intramolecular beryllium bond. For this purpose, we consider a system in which the malonaldehyde-like moiety is fused to a four membered ring, as in species **3-OBeH** and **3-BeH** and in their saturated counterparts **7-OBeH** and **7-BeH**.

From the structures and molecular graphs in Figure 15.4, the effect of the geometrical constraints imposed by the four membered ring on the strength of the beryllium intramolecular bond is apparent. Indeed, whereas in compound **1-OBeH** the beryllium atom is symmetrically located between the two oxygen atoms, in compound **3-OBeH** this symmetric arrangement is not possible because the C1–C2 bond which belongs to the four membered ring is forced to be essentially a single bond, and as a consequence the C2–C5 is in effect a much shorter double bond. The result is that the O···Be intramolecular beryllium bond is necessarily much longer and much weaker in compound **3-OBeH** than in compound **1-OBeH**. It is then the rigidity of the sigma skeleton that is the factor that strongly influences the strength of this intramolecular linkage, very much in the same way as it was suggested to occur when dealing with the intramolecular hydrogen bond in (*Z*)-4-(hydroxymethylene)cyclobut-2-enone (**3** in Scheme 15.2) as compared with that in malonaldehyde (**1** in Scheme 15.2).[54]

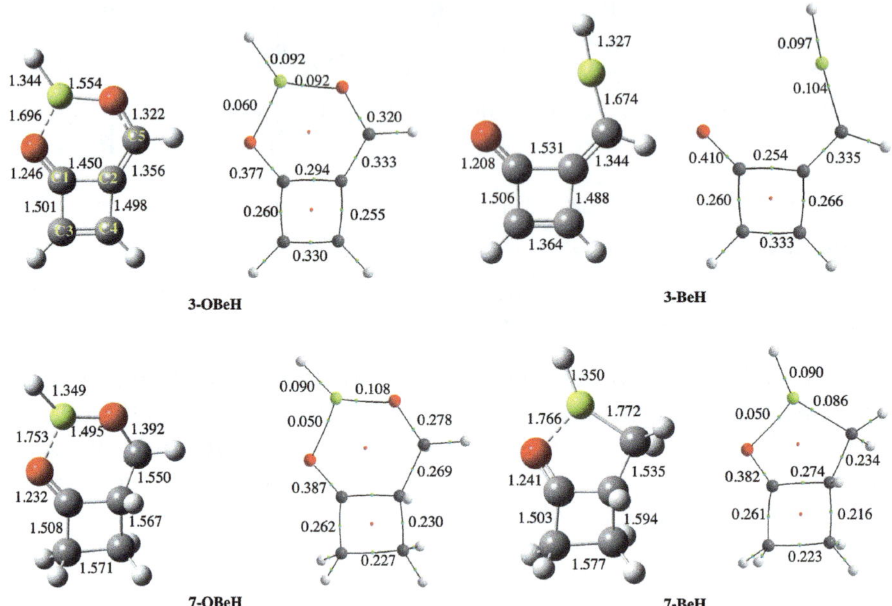

Figure 15.4 B3LYP/6-31+G(d,p) optimized structures and molecular graphs of compounds related to (*Z*)-4-(hydroxymethylene)cyclobut-2-enone (**3**) and 2-(hydroxymethyl)cyclobutanone (**7**). Compounds **3-OBeH** and **7-OBeH** are obtained by replacing the H atom of the OH group in compounds **3** and **7** by a BeH group. Compounds **3-BeH** and **7-BeH** are obtained by replacing the OH group in compounds **3** and **7** by a BeH group. Same conventions as in Figure 15.1.

The effect of the rigidity of the four membered ring is even more dramatic in compound **3-BeH**, whose equilibrium conformation, differently to what has been found above for the **1-BeH** analogue, does not exhibit any intramolecular beryllium bond. In this case, the rigidity imposed by the four membered cycle forces the distance between the Be atom and the carbonyl oxygen to be too large as to interact in an effective way. We found, however, a structure for this system that does present a beryllium bond (see Figure 15.5). However, the formation of this cyclic structure requires a significant lengthening of the C–Be bond, and consequently this local minimum lies 7 kJ mol^{-1} above the non-cyclic equilibrium conformation, in terms of Gibbs free energies.

As it was found for the previous two series of compounds, the intramolecular beryllium bond in the **7-OBeH** saturated species (see Figure 15.4) is weaker than in the corresponding unsaturated one, **3-OBeH**. Once more this is a consequence of a higher intrinsic basicity of the carbonyl group of the **3-OBeH** unsaturated compound

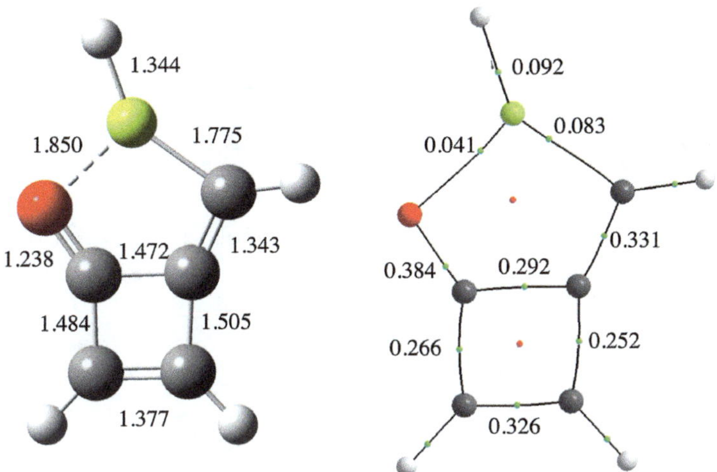

Figure 15.5 Local minimum of the **3-BeH** compound exhibiting a Be···O intramolecular bond. This minimum lies 7 kJ mol^{-1} above the global minimum. Same conventions as in Figure 15.1.

(823 kJ mol^{-1}) and of a concomitant higher hydride affinity of its BeH group (-342 kJ mol^{-1}) with respect to those of the corresponding saturated derivative (801 and -300 kJ mol^{-1}, respectively). However, whereas the rigidity of the σ-skeleton in **3-BeH** hinders the formation of the beryllium intramolecular bond, the larger flexibility of the aliphatic chain in the corresponding saturated analogue, **7-BeH**, facilitates the formation of this non-covalent linkage.

15.3.3 Intramolecular Beryllium Bonds in Tropolone-like Systems

The third set of compounds considered in our survey could be considered as derivatives of tropolone (**4** in Scheme 15.2) and 2-hydroxycycloheptanone (**8** in Scheme 15.2), in which the H atom of the OH group has been replaced by a BeH group to yield compounds **4-OBeH** and **8-OBeH**, or the whole OH group has been replaced by a BeH group, to yield compounds **4-BeH** and **8-BeH**. This set represents a different molecular environment, where the basic sites, as in the compounds studied in the previous section, are attached to a cyclic structure, the difference being not only the size of the ring, a seven-membered ring instead of a four-membered ring, but the possibility of having an aromatization of the seven-membered cyclic system, triggered by the formation of the intramolecular beryllium bond.

The structures and molecular graphs of these four compounds are shown in Figure 15.6. The first conspicuous fact is that whereas

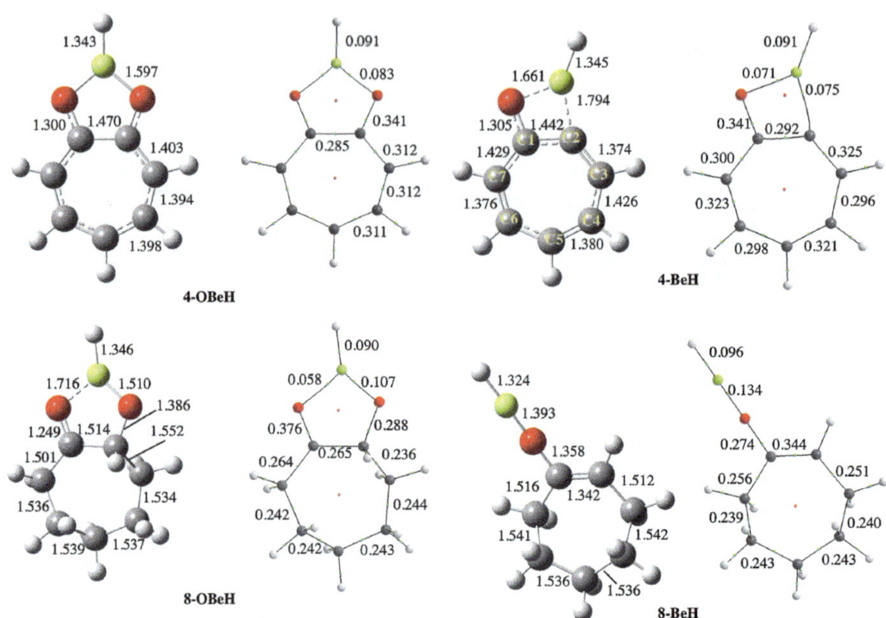

Figure 15.6 B3LYP/6-31+G(d,p) optimized structures and molecular graphs of compounds related to tropolone (**4**) and 2-hydroxycycloheptanone (**8**). Compounds **4-OBeH** and **8-OBeH** are obtained by replacing the H atom of the OH group in compounds **3** and **8** by a BeH group. Compounds **4-BeH** and **8-BeH** are obtained by replacing the OH group in compounds **4** and **8** by a BeH group. Same conventions as in Figure 15.1.

tropolone, which is stabilized by an intramolecular O–H···O hydrogen bond, presents a cycle with a clear alternation of single and double bonds, the **4-OBeH** compound presents an almost totally aromatic ring where the only typically single C–C bond is the one to which the two oxygen atoms of the system are attached. The remaining C–C bonds of the seven membered ring exhibit almost identical bond lengths, in agreement with the fact that the electron densities at the BCPs are also practically equal. As it was discussed above for **1-OBeH**, the large basicity of tropolone (894 kJ mol^{-1}), 40 kJ mol^{-1} larger than that of **1-OBeH(t)** system, results in a very strong Be···O intramolecular beryllium bond, in which, as it was found above for **1-OBeH**, the Be atom sits symmetrically between both oxygen atoms.

The great strength of the beryllium bond formed leads therefore to a symmetrization of the **4-OBeH** system, which is not observed in tropolone, ratifying that it is the strength of the intramolecular interaction that is the key factor in the resonance stabilization of the system, and not the other way around. As illustrated in Figure 15.6, this effect is almost totally destroyed on going to the **4-BeH**, where the formation of the O···Be intramolecular bond requires to almost cleave the C2–Be bond. This forces a much higher localization within the seven-membered ring, because the weakening of the C2–Be bond necessarily results in a strengthening of the C2–C3 bond, which becomes a clear double bond, forcing an alternation around the cycle.

As in previous cases, the intramolecular beryllium bond is weaker in the saturated derivative **8-OBeH**, which is a direct consequence of the significant decrease of the intrinsic basicity of the carbonyl group and the concomitant decrease in the Lewis acidity of the –BeH group on going from the unsaturated to the saturated compound. In fact, the intrinsic basicity of tropolone (891 kJ mol^{-1}), which can be considered as a good model system to estimate the intrinsic basicity of the carbonyl group of **4-OBeH,** is significantly higher than that of 2-hydroxycycloheptanone (815 kJ mol^{-1}), which is a good model for the basicity of this group in compound **8-OBeH**. On the other hand, our calculated intrinsic acidity of cyclohepta-2,4,6-trienol (1525 kJ mol^{-1}), which would be a good measure of the intrinsic acidity of **4-OBeH**, is found to be 35 kJ mol^{-1} larger than that of cycloheptanol, which is a good model for the acidity of compound **8-OBeH**.

Our attempts to locate a local minimum for **8-BeH** featuring an intramolecular beryllium bond, similar to the one found for the unsaturated analogue, **4-BeH**, failed, because they collapsed to the structure depicted in Figure 15.6, in which the C–Be bond cleavage initiated in compound **4-BeH** becomes complete in the saturated analogue, **8-BeH**. Consequently, Be is covalently attached to the carbonyl oxygen atom only. This is reflected not only in an electron density at the O–Be BCP almost twice as large in **8-BeH** than in **4-BeH**, but also in the fact that for **4-BeH** only a dative bond from the lone-pairs of the oxygen to beryllium is found. In contrast, for compound **8-BeH**, a very polar covalent bond, with 91% participation of the sp hybrids at the oxygen atom and 9% of the sp^2 orbitals at the beryllium atom, is found. This possibility is very unlikely when dealing with the unsaturated analogue **4-BeH**, because the complete breaking of the C2–Be bond and the eventual migration of the BeH group towards the oxygen atom would lead a carbene-like structure which would be rather unstable.

15.3.4 Beryllium Bridges. Dimerization of Malonaldehyde-like Systems

In this section we investigate the dimerization of Be-containing malonaldehyde-like systems, namely **1-OBeH** and **1-BeH** compounds, to see whether the self-assembly of these systems leads to the formation of Be–Be bonds. The Be_2 dimer itself is a very weakly bound species, the most recent experimental value for its dissociation energy being 935 cm^{-1},[72] and for which only 12 vibrational states have been resolved.[73,74] Be_2 has been also a challenge for chemical theory,[72–85] although the most recent studies indicate that the bonding arises from electron correlation effects.[84] In other words, in the absence of electron correlation effects, the Be_2 system would be unbound, which is the prediction of basic molecular orbital theory. We have shown for many different cases the ability of beryllium compounds to form very stable complexes in which they behave as very good electron acceptors, reflecting the electron-deficient character of beryllium derivatives. This nature is also behind the strong intramolecular beryllium bonds that we have analyzed and discussed in preceding sections. The question we want to address now is whether these compounds, in which the propensity of Be to accept electrons is at least partially satisfied through the formation of an intramolecular beryllium bond, are able to self-assemble through the direct interaction between the beryllium atoms. For this purpose we have optimized the structures of the dimers of **1-OBeH** and **1-BeH**. As illustrated in Figure 15.7, the dimerization leads to the formation of Be–H–Be bridges, very much as the ones that stabilize diborane, so the two monomers lie in the same plane, whereas the two hydrogens of the bridge lie in a perpendicular one. These bridges are characterized by BCPs whose electron densities are about 40% lower than in normal Be–H bonds as in beryllium dihydride ($\rho_{BCP} = 0.102$ a.u.).

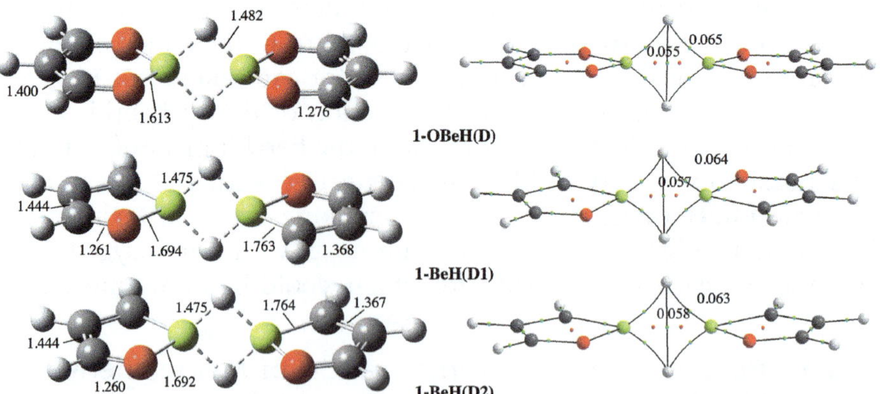

Figure 15.7 B3LYP/6-31+G(d,p) optimized structures and molecular graphs of the dimers of **1-OBeH** and **1-BeH** compounds. Note that for the latter, two different isomers which are practically degenerate, namely **1-BeH(D1)** and **1-BeH(D2)**, are local minima on the potential energy surface. Same conventions as in Figure 15.1.

The immediate consequence is that the length of the Be–H bond increases by about 0.14 Å with respect to the isolated monomers. Rather interestingly, the effects on the other two bonds in which the Be atom participates is the opposite in the **1-OBeH(D)** dimer to those in the **1-BeH(D1)** and **1-BeH(D2)** ones. As shown in Figure 15.7, in the former the O–Be distances increase by 0.023 Å with respect to the isolated monomers, whereas in the latter the O–Be and C–Be distances shrink by 0.014 and 0.017 Å, respectively. It is also worth noting that the Be–H–Be bridges are rather similar for both **1-OBeH** and **1-BeH** dimers. Coherently the WBOs for all these Be–H linkages are equal (0.446). No BCP is found between the two beryllium atoms, although one is observed between the two hydrogen atoms involved in the bridge. This AIM result is in contrast with the values of the WBOs, which are negligibly small between the two hydrogens of the bridge, but sizably large (0.26 and 0.29, respectively) between the two Be atoms. The bonding within the bridge is viewed as two electron three-center bonds, shown in Figure 15.8, which involve 17% participation of sp^3 orbitals of both Be atoms and 65% participation of the 1s orbital of the H atom. Each of these three-center bonds is populated by 1.97 e^-. The nature of these bridges is not substantially different from those responsible for the stability of diborane, the only significant difference being the relative participation of the orbitals of boron in the bonding, which amounts to 28%, reflecting the larger electronegativity of the boron atom with respect to the Be atom. The calculated dimerization enthalpy of these Be-containing dimers is significantly high (-105 kJ mol^{-1}) and of the same order of magnitude as that of diborane (-141 kJ mol^{-1}).[86]

Two different dimers are possible for the **1-BeH** compound, depending on the relative orientation of the two monomers (see Figure 15.7). Quite interestingly however, both structures, **1-BeH(D1)**

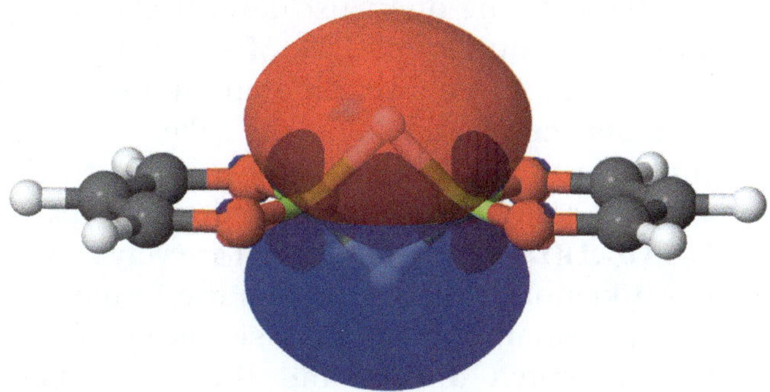

Figure 15.8   Three-center localized natural orbitals responsible for the formation of two Be–H–Be bridges in dimer **1-OBeH(D1)**.

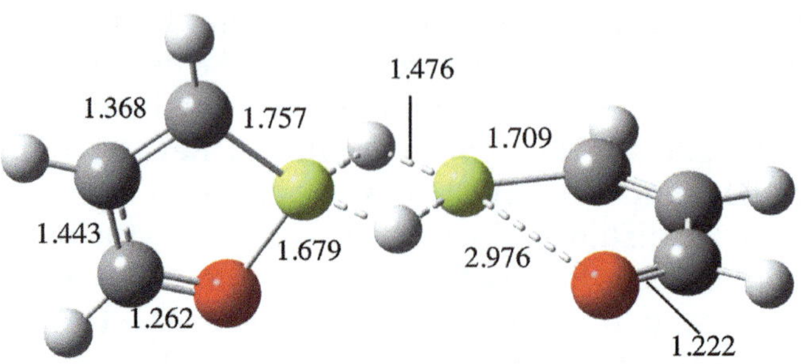

Figure 15.9 B3LYP/6-31+G(d,p) optimized structures for the transition state connecting **1-BeH(D1)** and **1-BeH(D2)** dimers. Bond distances are in Å.

and **1-BeH(D2)**, are practically degenerate in terms of Gibbs free energies, the latter being only 1 kJ mol^{-1} lower in energy than the former, with binding energies higher (-164 kJ mol^{-1}) than that of **1-OBeH(D)**. In view of the identical stability of both conformers we considered it of interest to calculate the transition state (TS) between both structures (see Figure 15.9).

The calculated barrier height for the **1-BeH(D1)/1-BeH(D2)** isomerization is 82 kJ mol^{-1}. One could naively expect a very low activation barrier associated with the rotation of one of the monomers to be perpendicular to the plane of the other monomer. However, although this is indeed the displacement connecting the minima with the TS, the rotation forces the oxygen of the rotating moiety to lie in the same plane as the Be–H–Be bridge and too much close to one of the hydrogens. This strong repulsion leads to a significant distortion of the rotating monomer and a significant lengthening of the Be–O bond, which explains the rather high activation barrier. The main conclusion then is that in the gas phase the **1-BeH** dimerization should lead to a 50/50 mixture of both conformers in equilibrium, because no interconversion between them should be expected.

The dimerization of the corresponding saturated systems obeys rather similar patterns as the ones just described in the preceding paragraphs. In this case, due to the lack of symmetry in the corresponding monomers, there are two different dimers for both the **5-OBeH** and the **5-BeH** compounds. For the sake of conciseness, we present in Figure 15.10 only the ones associated with the **5-OBeH** compound. As in the case of the unsaturated counterparts, the two dimers, namely **5-OBeH(D1)** and **5-OBeH(D2)** are very close in energy, the latter being 2.4 kJ mol^{-1} less stable than the former.

It is interesting to remark on the great similarity of the Be–H–Be bridges in these saturated dimers and their corresponding unsaturated analogues (see Figure 15.7), as far as internuclear distances

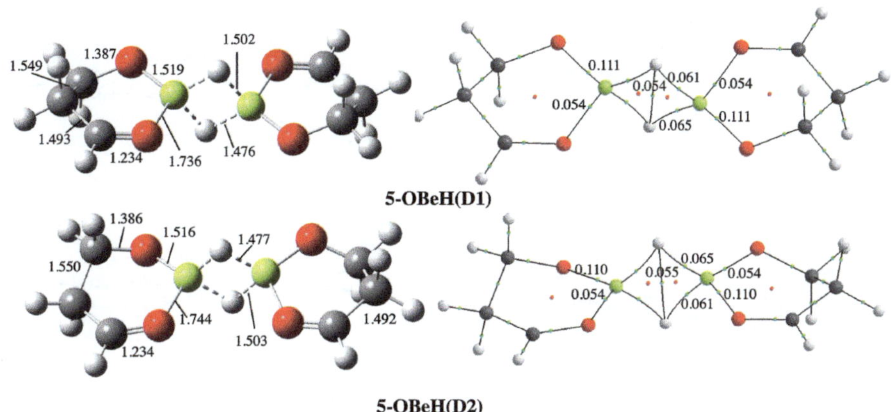

Figure 15.10 B3LYP/6-31+G(d,p) optimized structures and molecular graphs of the dimers of **5-OBeH** compound. Two different isomers which are practically degenerate, namely **5-OBeH(D1)** and **5-OBeH(D2)**, are local minima on the potential energy surface. Same conventions as in Figure 15.7.

and electron densities are concerned. Also the three-center bonds obtained through NBO analyses are essentially identical, with negligible differences in the participation of the Be orbitals, which for the unsaturated derivative, as mentioned above, is 17%, whereas for the saturated analogue is 18%. As a matter of fact, their binding energies are almost identical to that of the unsaturated analogue $(-102 \text{ kJ mol}^{-1})$. The question that arises quite naturally is whether these bridge linkers are also formed when the hydrogen atoms are replaced by other atoms or functional groups. To answer this question, we have considered three different cases, (i) the substitution of the H atoms by more electronegative chlorine atoms, (ii) the substitution of the H atom by a group with a clear inductive effect, such as the methyl group, and (iii) the substitution by a bulkier and aromatic group, such as the phenyl group.

The structures and molecular graphs of chlorine containing complexes have been plotted in Figure 15.11 for the unsaturated systems. The results are similar for the corresponding saturated analogues and therefore they are not going to be discussed in detail.

As illustrated in Figure 15.11, the replacement of H by chlorine atoms does not alter significantly the bonding pattern of the corresponding dimers, in which the two monomers are held together through Be–Cl–Be bridges. As expected, the length of the Be–Cl links are larger than the Be–H ones due to the larger size of the chlorine atoms, and, for the same reason the electron densities are necessarily

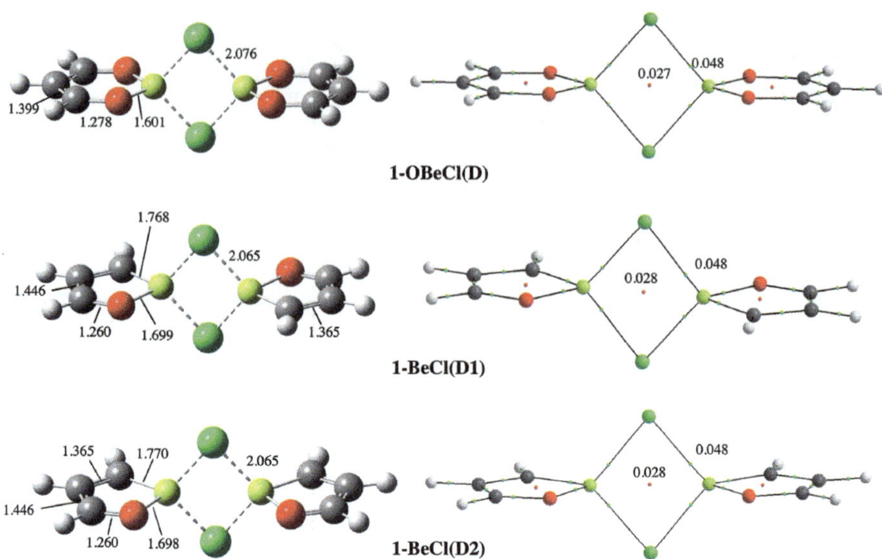

Figure 15.11   B3LYP/6-31+G(d,p) optimized structures and molecular graphs of the dimers obtained when the H atoms of **1-OBeH** and **1-BeH** derivatives are substituted by chlorine atoms. Two different isomers which are practically degenerate, namely **1-BeCl(D1)** and **1-BeCl(D2)**, are local minima on the potential energy surface. Bond distances are in Å. Same conventions as in Figure 15.7.

smaller. However, the NBO analysis indicates that the interaction between Be and Cl in these bridges is stronger than between Be and H in the bridges of the **1-OBeH(D)**, as reflected in a larger WBO (0.517 *vs.* 0.446). In fact, whereas in **1-OBeH(D)** dimers, as indicated above, the interaction is a dative bond from the H electron density into the empty Be orbitals, in the **1-OBeCl(D)**, **1-BeCl(D1)** and **1-BeCl(D2)**, the NBO analysis finds a very polar covalent bond between a sp^3 hybrid in the Cl atom, with a contribution of 91%, and a sp^3 hybrid in Be with a contribution of 9%. Consistently, the electron densities at the BCPs of the bridges are equal in all these dimers. Also, the Gibbs free energy gap between the two conformers **1-BeCl(D1)** and **1-BeCl(D2)** is negligibly small, the latter being 0.9 kJ mol^{-1} lower in free energy than the former.

Upon methyl substitution the changes in the bonding of the dimers are not significant with respect to the unsubstituted parent dimer. In Figure 15.12, as a suitable example, we present the structure and the molecular graph of **1-BeMe(D1)**. The electron densities at the BCPs are rather similar to the unsubstituted dimer, whereas the WBO of the Me–H bonds within the bridge is 0.390. As expected, the localized Be–CH$_3$ bond is less polar than for the chlorine derivative with a participation of the sp^3 orbitals of Be of 14%.

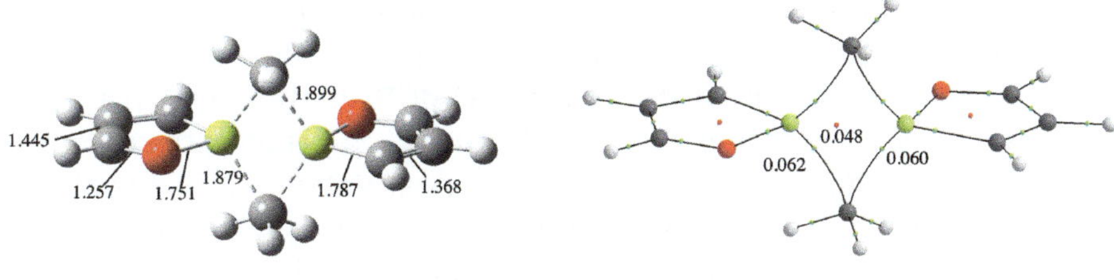

1-BeMe(D1)

Figure 15.12 B3LYP/6-31+G(d,p) optimized structures and molecular graphs of the dimer obtained when the H atoms of **1-BeH** derivative are substituted by methyl groups. Same conventions as in Figure 15.7.

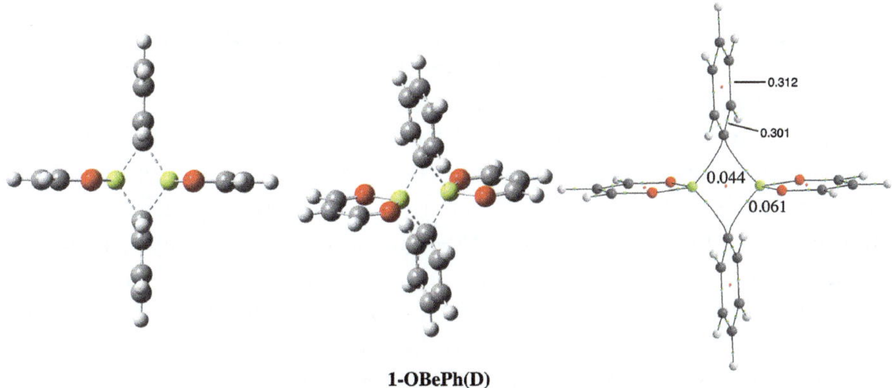

1-OBePh(D)

Figure 15.13 Two different perspectives of the B3LYP/6-31+G(d,p) optimized structures of the dimer obtained when the H atoms of **1-BeH** derivative are substituted by phenyl groups. The third graphic corresponds to the molecular graph. Same conventions as in Figure 15.7.

The substitution by phenyl groups does not alter the bonding patterns observed in the dimers discussed so far either. The compound **1-OBePh(D)**, chosen as a suitable example to illustrate the characteristics of the dimers with phenyl substituents, has D_{2h} symmetry like both the unsubstituted **1-OBeH(D)** and the chlorine containing analogue **1-OBeCl(D)**. The Be–C_6H_5 bonds are similar to those seen in the methyl derivative, with WBOs of 0.399. The bonding of the aromatic ring to both beryllium atoms, as in previous cases, occurs through Be–C–Be three-center bonds. These bonds have a significant contribution from the sp^2 hybrid of the phenyl carbon atom which amounts to 77%, reducing the participation of the beryllium bonds to 11%. This significant participation of the carbon orbital has a non-negligible effect on the aromaticity of the phenyl derivative. This is reflected in a substantial negative charge (−0.49) of both phenyl groups. On the other hand, as shown in Figure 15.13, the electron density at the BCPs between the carbon atom participating in the bridge and its two neighbors in the aromatic ring (0.301 a.u.) is smaller than the density at the other C–C BCPs within the ring. This bonding pattern resembles that of a Wheland intermediate, with the difference that these intermediates were postulated for electrophilic aromatic substitutions and therefore they are cationic, whereas in the present case they are electron rich, with a negative charge close to half of an electron, as mentioned above.

15.4　Concluding Remarks

Intramolecular beryllium bonds have the same intrinsic characteristics as intermolecular ones. These characteristics are a direct consequence of the charge transfer from the lone pairs of the basic site to the $\sigma_{BeH}{}^{*}$ antibonding orbital and to the empty p orbitals of the Be atom. As a consequence, and similarly to what has been observed for the intermolecular case, the BeX_2 moiety distorts significantly, departing from linearity, while the Be–X bonds become longer. In some cases, this charge donation is so large that a new covalent linkage between the Be atom and the basic site of the molecule is formed. In general, intramolecular beryllium bonds share many similarities with hydrogen bonds within the same molecular environment, though the former are in general stronger and therefore produce stronger distortions of the electron density distribution within the system. This electron density redistribution for unsaturated compounds leads to a significant enhancement of the resonance within the π-system, because the formation of the Be bond necessarily implies a weakening of the bonds in which the basic site participates, and therefore leads to a complete rearrangement of the remaining bonds of the system. It is important then to emphasize that the resonance enhancement is larger in systems that exhibit an intramolecular beryllium bond, than in similar systems which are stabilized through an intramolecular hydrogen bond, just because the former interactions are stronger than the latter. It is also worth noting that there is a reasonably good linear correlation between the strength of these two kinds of interactions, as measured by the electron density at the corresponding BCPs. This is shown in Figure 15.14, for some of the systems included in this study.

More importantly, the greater strength of the intramolecular beryllium bonds in unsaturated derivatives with respect to the saturated

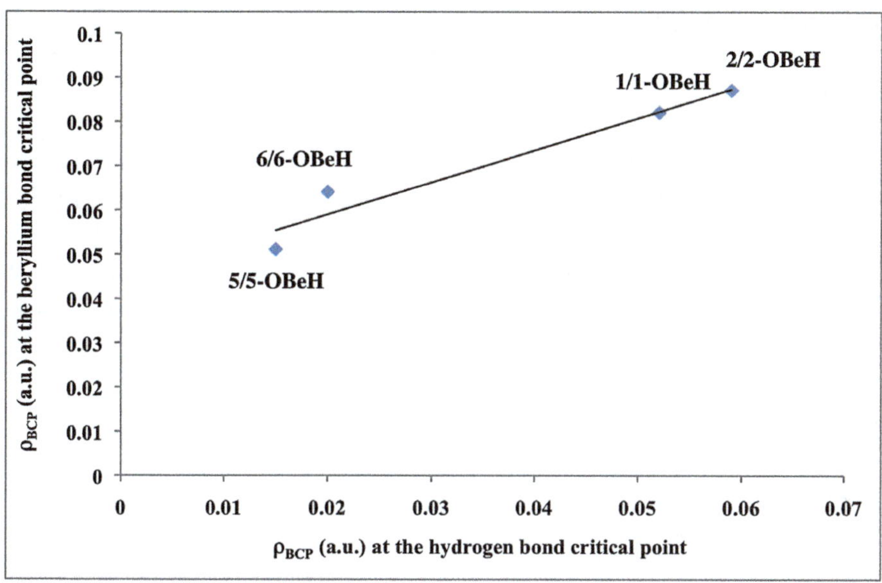

Figure 15.14 Linear correlation between the electron density at the beryllium bond critical point and at the hydrogen bond critical point [ρ_{BCP} (Be) = 0.7265 ρ_{BCP} (HB) + 0.0445; r^2 = 0.946)] for compounds in which the hydrogen bond has been replaced by a beryllium bond (see Scheme 15.1 for nomenclature).

analogues does not arise from a resonance assisted mechanism but it is essentially due to two concomitant intrinsic effects: a larger basicity of the basic site and a larger acidity of the –BeH group in the unsaturated derivatives. The immediate consequence is that the larger the strength of the intramolecular interaction the larger the stabilization of the system by resonance. This is rather well illustrated by the fact that the equilibrium structure in malonaldehyde is not symmetric, whereas that of the BeH containing analogue, **1-OBeH**, is. Therefore, the resonance stabilization of the latter is much larger than the resonance stabilization of the former, because the intramolecular beryllium bond is stronger than the hydrogen bond.

The self-assembly of Be-containing malonaldehyde-like structures takes place through the formation of Be–H–Be bridges, but non-direct Be–Be interactions are observed. These bridges are rather similar, and only slightly weaker than those stabilizing diborane. The substitution of the H atom in these bridges by either halogen atoms, such as Cl, alkyl groups or phenyl groups does not alter this arrangement and the two moieties appear still connected through Be–X–Be (X = Cl, CH_3, C_6H_5) bridges.

Acknowledgements

This work has been partially supported by the Ministerio de Economía y Competitividad (Projects No. CTQ2015-63997-C2-1-P and CTQ2013-43698-P), by the STSM COST Action CM1204, and by the Project FOTOCARBON-CM S2013/MIT-2841 of the Comunidad Autónoma de Madrid. OB acknowledges the Ministerio de Ciencia, Cultura y Deportes of Spain by a FPU grant. Computational time at Centro de Computación Científica (CCC) of Universidad Autónoma de Madrid is also acknowledged.

References

1. G. C. Pimentel and A. L. McClelland, *The Hydrogen Bond*, W.H. Freeman and Co., San Francisco, 1960.
2. G. A. Jeffrey, *An Introduction to Hydrogen Bonding*, Oxford University Press New York, 1997.
3. S. J. Grabowski, *Hydrogen Bonding. New Insights*, 2006.
4. P. Metrangolo and G. Resnati, *Chem. – Eur. J.*, 2001, **7**, 2511–2519.
5. T. Clark, M. Hennemann, J. S. Murray and P. Politzer, *J. Mol. Model.*, 2007, **13**, 291–296.
6. P. Politzer, J. S. Murray and T. Clark, *Phys. Chem. Chem. Phys.*, 2013, **15**, 11178–11189.
7. I. Alkorta, J. Elguero and C. Foces-Foces, *Chem. Commun.*, 1996, 1633–1634.
8. V. I. Bakhmutov, *Dihydrogen Bond: Principles, Experiments, and Applications*, John Wiley & Sons, 2008.
9. S. Zahn, R. Frank, E. Hey-Hawkins and B. Kirchner, *Chem. – Eur. J.*, 2011, **17**, 6034–6038.
10. J. E. Del Bene, I. Alkorta, G. Sánchez-Sanz and J. Elguero, *J. Phys. Chem. A*, 2011, **115**, 13724–13731.
11. S. Scheiner, *Acc. Chem. Res.*, 2013, **46**, 280–288.
12. S. Scheiner, *Int. J. Quantum Chem.*, 2013, **113**, 1609–1620.
13. A. Bauzá, T. J. Mooibroek and A. Frontera, *Angew. Chem., Int. Ed.*, 2013, **52**, 12317–12321.
14. I. Alkorta, I. Rozas and J. Elguero, *J. Phys. Chem. A*, 2001, **105**, 743–749.
15. P. Sanz, O. Mó and M. Yáñez, *J. Phys. Chem. A*, 2002, **106**, 4661–4668.
16. P. Sanz, O. Mó and M. Yáñez, *New J. Chem.*, 2002, **26**, 1747–1752.
17. W. Z. Wang, B. M. Ji and Y. Zhang, *J. Phys. Chem. A*, 2009, **113**, 8132–8135.
18. M. Yáñez, P. Sanz, O. Mó, I. Alkorta and J. Elguero, *J. Chem. Theory Comput.*, 2009, **5**, 2763–2771.
19. E. Fernández Villanueva, O. Mó and M. Yáñez, *Phys. Chem. Chem. Phys.*, 2014, **16**, 17531–17536.
20. D. Parker, *J. Chem. Soc., Chem. Commun.*, 1985, 1129–1131.
21. F. Zhang, T. Morawitz, S. Bieller, M. Bolte, H. W. Lerner and M. Wagner, *Dalton Trans.*, 2007, 4594–4598.
22. F. Tancini and E. Dalcanale, Polymerization with Ditopic Cavitand Monomers, in *Supramolecular Polymer Chemistry*, Wiley-VCH Verlag GmbH & Co, 2011.
23. T. K. Kim, K. J. Lee, M. Choi, N. Park, D. Moon and H. R. Moon, *New J. Chem.*, 2013, **37**, 4130–4139.

24. R. Mualem, E. Sominska, V. Kelner and A. Gedanken, *J. Chem. Phys.*, 1992, **97**, 8813–8814.
25. A. Witkowski, *J. Chem. Phys.*, 1970, **52**, 4403–4407.
26. A. Winkler and P. Hess, *J. Am. Chem. Soc.*, 1994, **116**, 9233–9240.
27. F. O. Koller, M. Huber, T. E. Schrader, W. J. Schreier and W. Zinth, *Chem. Phys.*, 2007, **341**, 200–206.
28. I. Alkorta, J. Elguero, M. Yáñez and O. Mó, *Phys. Chem. Chem. Phys.*, 2014, **16**, 4305–4312.
29. R. Dobrawa and F. Wurthner, *J. Polym. Sci., Part A: Polym. Chem.*, 2005, **43**, 4981–4995.
30. Y. Yan and J. B. Huang, *Coord. Chem. Rev.*, 2010, **254**, 1072–1080.
31. A. D. Burrows, M. F. Mahon, P. R. Raithby, A. J. Warren, S. J. Teat and J. E. Warren, *CrystEngComm*, 2012, **14**, 3658–3666.
32. F. Wurthner, C. C. You and C. R. Saha-Moller, *Chem. Soc. Rev.*, 2004, **33**, 133–146.
33. B. Lippert and P. J. S. Miguel, *Chem. Soc. Rev.*, 2011, **40**, 4475–4487.
34. J. W. Steed, *Chem. Commun.*, 2011, **47**, 1379–1383.
35. D. J. White, N. Laing, H. Miller, S. Parsons, S. Coles and P. A. Tasker, *Chem. Commun.*, 1999, 2077–2078.
36. W. F. Rowe, R. W. Duerst and E. B. Wilson, *J. Am. Chem. Soc.*, 1976, **98**, 4021–4023.
37. R. S. Brown, *J. Am. Chem. Soc.*, 1977, **99**, 5497–5499.
38. N. Sanna, F. Ramondo and L. Bencivenni, *J. Mol. Struct.*, 1994, **318**, 217–235.
39. M. A. Ríos and J. Rodriguez, *Can. J. Chem.*, 1991, **69**, 201–204.
40. P. Sanz, O. Mó and M. Yáñez, *Phys. Chem. Chem. Phys.*, 2003, **5**, 2942–2947.
41. D. L. Widner, Q. R. Knauf, M. T. Merucci, T. R. Fritz, J. S. Sauer, E. D. Speetzen, E. Bosch and N. P. Bowling, *J. Org. Chem.*, 2014, **79**, 6269–6278.
42. M. Yahia-Ouahmed, V. Tognetti and L. Joubert, *Comput. Theor. Chem.*, 2015, **1053**, 254–262.
43. B. V. Pandiyan, P. Deepa and P. Kolandaivel, *Phys. Chem. Chem. Phys.*, 2015, **17**, 27496–27508.
44. G. Gilli, F. Bellucci, V. Ferretti and V. Bertolasi, *J. Am. Chem. Soc.*, 1989, **111**, 1023–1028.
45. V. Bertolasi, L. Nanni, P. Gilli, V. Ferretti, G. Gilli, Y. M. Issa and O. E. Sherif, *New J. Chem.*, 1994, **18**, 251–261.
46. V. Bertolasi, P. Gilli, V. Ferretti and G. Gilli, *J. Chem. Soc., Perkin Trans. 2*, 1997, 945–952.
47. P. Gilli, V. Bertolasi, V. Ferretti and G. Gilli, *J. Am. Chem. Soc.*, 2000, **122**, 10405–10417.
48. B. K. Paul, N. Ghosh, R. Mondal and S. Mukherjee, *Photochem. Photobiol. Sci.*, 2015, **14**, 1147–1162.
49. M. Dracinsky, L. Cechova, P. Hodgkinson, E. Prochazkova and Z. Janeba, *Chem. Commun.*, 2015, **51**, 13986–13989.
50. D. Rusinska-Roszak and G. Sowinski, *J. Phys. Chem. A*, 2015, **119**, 3674–3687.
51. I. Alkorta, J. Elguero, O. Mó, M. Yáñez and J. D. Bene, *Mol. Phys.*, 2004, **102**, 2563–2574.
52. P. Sanz, O. Mó, M. Yáñez and J. Elguero, *J. Phys. Chem. A*, 2007, **111**, 3585–3591.
53. Y. R. Mo, *J. Mol. Model.*, 2006, **12**, 221–228.
54. P. Sanz, O. Mó, M. Yáñez and J. Elguero, *ChemPhysChem*, 2007, **8**, 1950–1958.
55. J. F. Beck and Y. R. Mo, *J. Comput. Chem.*, 2007, **28**, 455–466.
56. Y. R. Mo, *J. Phys. Chem. A*, 2012, **116**, 5240–5246.
57. R. W. Gora, M. Maj and S. J. Grabowski, *Phys. Chem. Chem. Phys.*, 2013, **15**, 2514–2522.
58. A. R. Nekoei and M. Vatanparast, *New J. Chem.*, 2014, **38**, 5886–5891.
59. K. Sutter, G. A. Aucar and J. Autschbach, *Chem. – Eur. J.*, 2015, **21**, 1–19.

60. M. P. Romero-Fernández, M. Avalos, R. Babiano, P. Cintas, J. L. Jimenez and J. C. Palacios, *Tetrahedron*, 2016, **72**, 95–104.
61. A. D. Becke, *J. Chem. Phys.*, 1993, **98**, 5648–5652.
62. C. Lee, W. Yang and R. G. Parr, *Phys. Rev. B: Condens. Matter Mater. Phys.*, 1988, **37**, 785–789.
63. L. A. Curtiss, P. C. Redfern and K. Raghavachari, *J. Chem. Phys.*, 2007, **126**, 12.
64. O. Brea, O. Mó, M. Yáñez, I. Alkorta and J. Elguero, *Chem. – Eur. J.*, 2015, **21**, 12676–12682.
65. R. F. W. Bader, *Atoms in Molecules. A Quantum Theory*, Clarendon Press, Oxford, 1990.
66. A. E. Reed, L. A. Curtiss and F. Weinhold, *Chem. Rev.*, 1988, **88**, 899–926.
67. T. A. Keith, *AIMAll (Version 13.05.06)*, (2013) AIMAll (Version 13.05.06) Gristmill Software, Overland Park, KS, 2013; aim.tkgristmill.com.
68. K. B. Wiberg, *Tetrahedron*, 1968, **24**, 1083–1088.
69. F. Weinhold, *NBO Program*, (2001) University of Wisconsin System, Madison.
70. E. D. Glendening and F. Weinhold, *J. Comput. Chem.*, 1998, **19**, 593–609.
71. M. Palusiak, S. Simon and M. Sola, *Chem. Phys.*, 2007, **342**, 43–54.
72. V. V. Meshkov, A. V. Stolyarov, M. C. Heaven, C. Haugen and R. J. LeRoy, *J. Chem. Phys.*, 2014, **140**, 064315.
73. J. M. Merritt, V. E. Bondybey and M. C. Heaven, *Science*, 2009, **324**, 1548–1551.
74. K. Patkowski, V. Spirko and K. Szalewicz, *Science*, 2009, **326**, 1382–1384.
75. M. R. A. Blomberg and P. E. M. Siegbahn, *Int. J. Quantum Chem.*, 1978, **14**, 583–592.
76. S. Evangelisti, G. L. Bendazzoli and L. Gagliardi, *Chem. Phys.*, 1994, **185**, 47–56.
77. J. Starck and W. Meyer, *Chem. Phys. Lett.*, 1996, **258**, 421–426.
78. R. J. Gdanitz, *Chem. Phys. Lett.*, 1999, **312**, 578–584.
79. J. M. L. Martin, *Chem. Phys. Lett.*, 1999, **303**, 399–407.
80. M. W. Schmidt, J. Ivanic and K. Ruedenberg, *J. Phys. Chem. A*, 2010, **114**, 8687–8696.
81. J. Koput, *Phys. Chem. Chem. Phys.*, 2011, **13**, 20311–20317.
82. W. Helal, S. Evangelisti, T. Leininger and A. Monari, *Chem. Phys. Lett.*, 2013, **568**, 49–54.
83. J. M. Matxain, F. Ruiperez and M. Piris, *J. Mol. Model.*, 2013, **19**, 1967–1972.
84. M. El Khatib, G. L. Bendazzoli, S. Evangelisti, W. Helal, T. Leininger, L. Tenti and C. Angeli, *J. Phys. Chem. A*, 2014, **118**, 6664–6673.
85. M. J. Deible, M. Kessler, K. E. Gasperich and K. D. Jordan, *J. Chem. Phys.*, 2015, **143**, 084116.
86. L. T. Redmon, G. D. Purvis and R. J. Bartlett, *J. Am. Chem. Soc.*, 1979, **101**, 2856–2862.

16 On the Nature of Hydrogen–Hydrogen Bonding[†]

Juan C. García-Ramos,[a] Fernando Cortés-Guzmán*[a] and Chérif F. Matta*[b,c]

[a] Instituto de Química, Universidad Nacional Autónoma de México. Av. Universidad 3000, Ciudad Universitaria 4510 Ciudad de México (México); [b] Dept. of Chemistry & Physics, Mount Saint Vincent University, Halifax, Nova Scotia, Canada B3M 2J6; [c] Dept. of Chemistry, Dalhousie University, Halifax, Nova Scotia, Canada, B3H 4R2
*Email: Cherif.Matta@msvu.ca; fercor@unam.mx

16.1 Introduction

Experimental[1,2] and computational evidence accrued for more than a decade has enlarged the realm of weak hydrogen–hydrogen interactions.[3–7] A new type of weak hydrogen–hydrogen interaction was identified and characterized in 2003[4] and was termed "hydrogen–hydrogen bonding (or H···H bonding)."[‡] H···H bonding is neither a type of hydrogen bonding nor even a variant of the dihydrogen

---

[†] This chapter is dedicated to the memory of **Professor Richard F. W. Bader (1931–2012)** and dedicated to **Professor Jesús Hernández-Trujillo** for their key contributions in the discovery and characterization of hydrogen–hydrogen bonding.

[‡] In this chapter, *"bonding"* (a verb) and *"bond"* (a noun) are used interchangeably to designate the same phenomenon. As emphasized by Richard F. W. Bader[133] on a number of occasions, it may be more precise to use the verb to indicate that chemical bonding is a dynamic *interaction* and not an *object*. The universality of the usage of the word "bond" is such that we use it as well in this chapter even when we really mean "bonding". It may be of interest to study the etymology of the words describing chemical bonding in different languages and whether they indicate an interaction or a passive object with independent "existence".

Intermolecular Interactions in Crystals: Fundamentals of Crystal Engineering
Edited by Juan J. Novoa
© The Royal Society of Chemistry 2018
Published by the Royal Society of Chemistry, www.rsc.org

bond[8–11] with a hydridic and an acidic hydrogen, but rather it is a new class of interactions within or between molecules where a pair of closed-shell bonded hydrogen atoms are either identical or very similar. Generally, but not always, this pair of hydrogen atoms are electrically neutral (or close to electrical neutrality).[4–7] This new category of weak $H\cdots H$ bonding has been enlarged by the pioneering work of Wolstenholme and McGrady and their coworkers to include bonding in which the two hydrogen atoms bear significant net electric charges of the same sign, hence the designation of "homopolar hydrogen–hydrogen bonding".[12–14]

Since precise language is one of the hallmarks of science, it is important to define the different types of interactions involving hydrogens to avoid ambiguous discourse:

(1) Hydrogen bonding ($^{\delta-}X–H^{\delta+}\cdots^{\delta-}Y$): A bonding that involves one protonic hydrogen atom (a proton at the limit) shared or sandwiched between two electronegative atoms, a proton donor (X) and a proton acceptor (Y);

(2) Dihydrogen (or heteropolar hydrogen–hydrogen) bonding ($^{\delta-}X–H^{\delta+}\cdots^{\delta-}H–^{\delta+}Y$): A hydrogen bonding whereby the proton acceptor happens to be an electron rich hydrogen atom (a hydridic proton) as found in metal hydrides for example. In this mode of bonding the two hydrogen atoms bear charges of opposite signs and usually of significant magnitude;

(3) Hydrogen–hydrogen (or homopolar hydrogen–hydrogen) bonding ($X–H\cdots H–Y$): This is *not* a variant of hydrogen bonding since it does not include a proton donor and an acceptor but rather either two neutral (or close to neutral) hydrogen atoms or two hydrogen atoms bearing charges of the same sign (whether identical, similar, or differing in magnitude). It may also include bonding where one of the two hydrogen atoms bear a significant net electrical charge (of either sign) while the other is neutral or close to neutrality. It is thus advantageous to distinguish these (classically) non-electrostatic types of bonding, as a group, with a different designation.[6] We propose to adopt "hydrogen–hydrogen bonding" for this general class and reserve the term "dihydrogen bonding" for heteropolar bonding of two hydrogen atoms.

(4) Tri-hydrogen bonding ($X–H\cdots H–H$): A hydrogen bond formed between a hydrogen molecule and either a protonic or hydridic hydrogen atom whereby the latter polarizes the hydrogen

molecule with a resulting primarily electrostatic interaction just as in hydrogen bonding.

Surviving critical scrutiny is often a rite of passage of new ideas and findings in science before their incorporation into the body of knowledge. The discovery and characterization of hydrogen–hydrogen bonding in 2003[4] sparked heated debates in the literature and stimulated numerous groups to contribute to expand the definition of this interaction. We present here a short, balanced, and objective review of the computational and experimental evidence supporting the validity of this concept.

Weak interactions have gained considerable attention from experimentalists and theoreticians in recent times for their role in supramolecular structural stability and the self-assembly of molecular aggregates.[15–21] The cooperativity of large numbers of such weak interactions is responsible, for example, of the stability of the secondary, tertiary and quaternary structures of proteins, the growth of molecular crystals, and monomer–monomer recognition in self-assembled monolayers on surfaces (SAMs).[15]

Under the broad umbrella of weak interactions are those involving one or more hydrogen atom(s) as mediator(s) of the interaction. The hydrogen bond is probably the best known example and consists, as briefly mentioned in the above definitions, of a proton shared between two electronegative atoms, often denoted symbolically as $^{\delta-}X–H^{\delta+}\cdots^{\delta-}Y$.[22–27] A variant of the hydrogen bond, ubiquitous in organometallic chemistry, is the dihydrogen bond.[8–11,28–40] This special type of hydrogen bond involves not one, but two, hydrogen atoms. One of this pair of hydrogen atoms is positively charged, formally the proton of the hydrogen bond, and the second is a hydridic hydrogen atom, which gains its negative charge from a metal atom, and which acts as the proton acceptor.[8–11,28–40] Thus, and recapping the definitions advanced earlier, the dihydrogen bond can be thought of, classically, as a primarily electrostatic interaction and could be denoted as $^{\delta-}X–H^{\delta+}\cdots^{\delta-}H–M^{\delta+}$.

16.2 Observational Evidence for Hydrogen–Hydrogen (H···H) Bonding

Hydrogen–hydrogen (H···H) interaction between electrically neutral hydrogen atoms, as in C–H···H–C interactions, is ubiquitous in nature and is associated[§] with the stable structures of macromolecules and hydrocarbon crystals, for example.[1,4–6,8,17,41–44] A statistical analysis of high quality crystallographic structures ($R_f < 0.05$) deposited in the Cambridge Structural Database (CSD) reveals that "short H$\cdots$H contacts", *i.e.*, those with inter-nuclear separation smaller than twice the van der Waals (vdW) radius of a free hydrogen atom ($2 \times 1.2 = 2.4$ Å),[27] are quite common.[17] This search of the CSD was limited to interactions of the type $R_2C \Big\langle {}^{H\cdots H}_{H\cdots H} \Big\rangle CR_2$, and did not include interactions involving methyl hydrogen atoms (expected to be more common) since the uncertainty in their positions is considerably larger due to free rotation. Even with these stringent restrictions, the authors have found 23 hits with H$\cdots$H distances smaller than 2.4 Å.[17] Close H$\cdots$H distances between closed-shell neutral hydrogen atoms are, thus, experimentally observed and not infrequently. However, most experimental data are related to dihydrogen bonding and several reviews emphasize the importance of molecular complexes of the type M–H$\cdots$H–A as an intermediate in proton transfer reactions,[45,46] as well as in catalytic reduction,[47,48] and hydrogen activation.[49]

Despite the X-ray diffraction evidence for its ubiquitous occurrence, other experimental identification of H$\cdots$H interactions, *e.g.* spectroscopic, are difficult due to the small energies associated with this bonding. However, the recent technical developments in spectroscopy allow the study of this weak unconventional hydrogen interaction. Particularly, infrared (IR) and nuclear magnetic resonance (NMR) spectroscopies provide valuable information for systems in the gas phase and in solution. For the solid state, X-ray diffraction spectroscopy remains the most widely-used technique.[50]

[§]Whether chemical bonding is the cause of a system's stability or whether it is a manifestation of this stability is a meaningless *petitio principii* (asking for the starting point) informal fallacy, a common example is begging the chicken-or-egg question. What we can say is that chemical bonding emerges from the solution of the Schrödinger equation at stationary points on the potential energy surface. At a stable (energy minimized) structure, bonding is often considered to occur between atoms that are closer in space than a specified distance, say the sum of their van der Waals radii. The multiplicity or strength of that bonding is traditionally inferred and assigned on the basis of the inter-nuclear distance. More recently, and since the advent of Bader's Quantum Theory of Atoms in Molecules (QTAIM),[101] bonding is considered to coincide with the presence of bond paths in the electron density (calculated or experimental) within a stable structure. In this chapter, we say: *Bonding (no matter how defined) is "associated" with the system's stability rather than causing this stability or be caused by it* and, further, we rely on the QTAIM definition of chemical bonding.

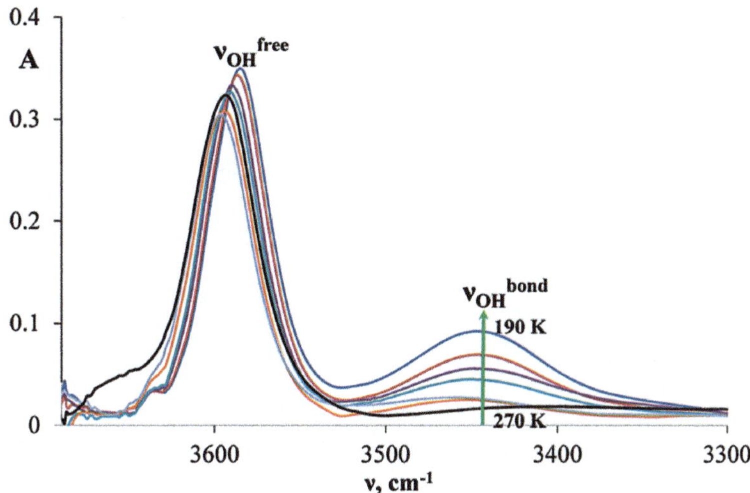

Figure 16.1 FT-IR spectra in the νOH region of trifluoroethanol (black line) in the presence of $(Ph_3P)_2Cu(\eta^2\text{-}BH_4)$ in CH_2Cl_2 in a temperature range of 190 to 270 K, showing the dihydrogen bond formation.
(Reprinted with permission from I. E. Golub, O. A. Filippov, E. I. Gutsul, N. V. Belkova, L. M. Epstein, A. Rossin, M. Peruzzini, E. S. Shubina; Dimerization mechanism of bis(triphenylphosphine)copper(I) tetrahydroborate: Proton Transfer via a dihydrogen bond. *Inorg. Chem.* 2012, 51, 6486–6497. Copyright 2012 American Chemical Society).

Infrared spectroscopy is one of the most employed techniques to study $H\cdots H$ bonding as the interaction can be followed by IR spectral changes as has been established by Epstein's group.[45] This group has found that the decrease in intensity of the νOH(free) band and the appearance of a low frequency ($\Delta\nu OH > 100$ cm⁻¹) broad intense νOH(bonded) band are a consequence of the presence of $H\cdots H$ interaction.[51] A temperature decrease also improves the information obtained by this technique as can be seen in Figure 16.1, where the $H\cdots H$ interaction between $(Ph_3P)_2Cu(\eta^2\text{-}BH_4)$ and trifluoroethanol (TFE) was identified.[52]

Another important technique employed to identify $H\cdots H$ interactions is ${}^1H$ NMR. It has shown that dihydrogen bonding affects two NMR parameters: the hydride chemical shift and the $T1$ relaxation time. Upon complexation with a proton donor, the hydride signal shifts to high field, this is typical for a hydride occupying a bridging position, whereas the $T1$ relaxation time diminishes because a proton near a hydride provides an additional source of relaxation. One example of above observations can be seen in Figure 16.2 that shows the ${}^1H$ NMR spectra in the hydride region of

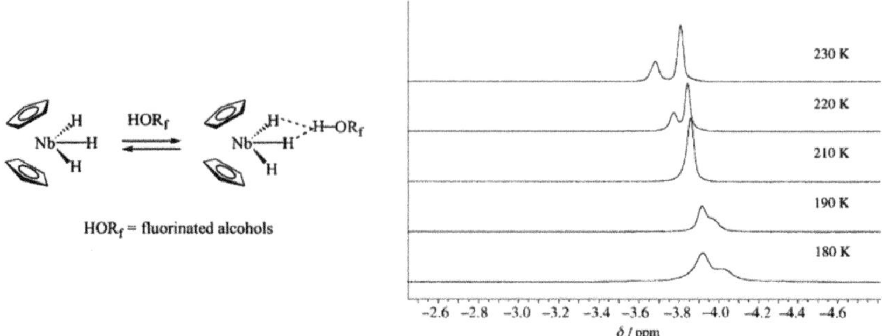

Figure 16.2   Hydrogen-bond formation in [NbCp$_2$H$_3$]/HOR$_f$ system followed by ^{1}H NMR.
(Reprinted from E. V. Bakhmutova, V. I. Bakhmutov, N. V. Belkova, M. Besora, L. M. Epstein, A. Lledós, G. I. Nikonov, E. S. Shubina, J. Tomàs, E. V. Vorontsov; First investigation of non-classical dihydrogen bonding between an early transition-metal hydride and alcohols: IR, NMR, and DFT approach. *Chem. Eur. J.* **2004**, *10*, 661–671. Copyright 2004 with permission from John Wiley and Sons).

[NbCp$_2$H$_3$] in the presence of hexafluoroisopropanol (HFIP). Changes in the hydride chemical shifts also permit the identification of the strong interaction of the fluorinated alcohol with the central hydride, due to the shift of almost 1.7 ppm compared with the coordination compound spectrum in the absence of the alcohol. The small shift in the second hydride allows the proposal of the bifurcated interaction with the alcohol hydrogen also confirmed by the DFT calculations.[53]

Recently, more sophisticated spectroscopic techniques have been employed in the identification of H···H interactions such as Laser Induced Fluorescence (LIF) employed by Ishikawa's group.[54] In this work four isomers of the type Si–H···H–O were identified for the interaction between neutral phenol and triethylsilane employing LIF, IR, IR–UV and theoretical studies. The isomers identified in the figure as A–D were proposed as Si–H···H–O due to the new bands observed on both sides of the band associated with the Ph–OH monomer in the LIF spectra (Figure 16.3, left side) and red-shift wavelengths observed in the IR spectra compared with the free phenol (Figure 16.3, right side). Theoretical study helped in the determination that the intrinsic strength of the Si–H···H–O is somewhat stronger than those of the π-HO hydrogen bond and with this the authors concluded that the relative large proton affinity of TES may be related to the fact that the hydrogen atom bonded to the silicon atom is very reactive.[54]

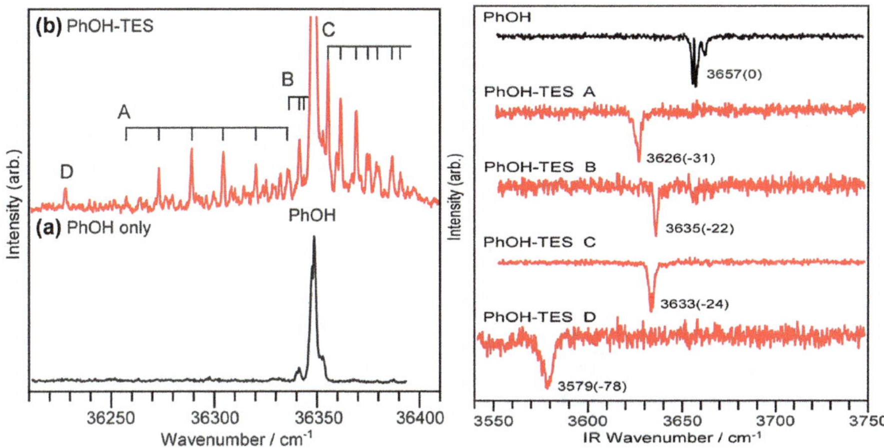

Figure 16.3 Right, LIF spectra of PhOH alone (a) and in presence of TES. Left, IR spectra showing PhOH and the PhOH/TES systems.
(Reprinted with permission from H. Ishikawa, T. Kawasaki, R. Inomata; Infrared spectroscopy of phenol-triethylsilane dihydrogen-bonded cluster and its cationic analogues: Intrinsic strength of the Si-H···H-O dihydrogen bond. *J. Phys. Chem. A* **2015**, *119*, 601–609. Copyright 2015 American Chemical Society).

16.3 Occurrence of Short H···H Distances and Their Role in the Packing of a Crystal

The existence of H···H interactions was observed experimentally and theoretically.[1–7,23,55–65] Intra- and inter-molecular H···H interactions have been found in organic, inorganic and organometallic systems in the solid state (Figure 16.4). A short H···H distance (1.62 Å) in norbornene derivatives was reported in 1999, as a candidate for the shortest observed H···H intramolecular interaction in crowded molecules.[66] Pattabhi reported a database analysis on H···H interactions of the type CH_2···H_2C and NH···HN in organic crystals in the absence of any metal atom. CH_2···H_2C interactions are observed at distances from 2.0 to 2.4 Å and within an angle range of 99–150°. NH···HN interactions were presented between 2.2–2.4 Å and 60–170°.[17]

Many aliphatic and aromatic systems have been identified in the CSD involving organic and inorganic molecules. A crystallographic search with defined parameters such as intermolecular C–H···H–C distance interaction in the range 1.6–2.0 Å, $R_f \le 3.5\%$ and acquisition data temperature ≤ 200 K gave as a result almost 3100 crystal structures, from which practically 80% were reported after 2000.

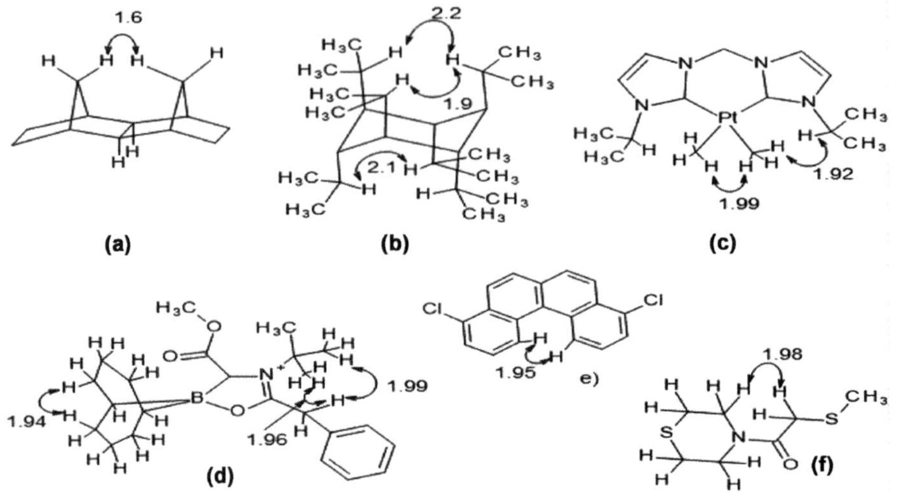

Figure 16.4 Some examples of intermolecular H···H interactions with distances of less than 2.0 Å. (a) norborane, (b) all-*trans*-1,2,3,4,5,6-hexaisopropylcyclohexane, (c) 1,1′-methylenebis(3-isopropyl-2,3-dihydro-1*H*-imidazol-2-ylidene))-dimethyl-platinum(II), (d) methyl-2-(benzyl(*t*-butoxycarbonyl)amino)-2-(9-borabicyclo(3.3.1)nonan-9-yl-acetate, (e) 4,9-dichlorobenzo[c]phenanthrene and (f) 2-(methylsulfanyl)-1-(thiomorpholin-4-yl)ethanone.
(Reprinted from *J. Chem. Cryst.*, Short H···H distances in norbornene derivatives, **29**, 1999, 523–530, S. G. Bodige, D. Sun, A. P. Marchand, N. N. Namboothiri, R. Shukla, W. H. Watson, © 1999 *Springer*, with permission of Springer.)

Some interesting examples given in Table 16.1 included H···H interactions longer than 2.0 Å besides the shorter ones.

It is possible that the H–H interactions have an important function during a chemical reaction. Schreiner *et al.* found the extraordinary stability of C–C coupling products is due to the overall attractive dispersion interactions between intramolecular H–H contact surfaces.[77] Although the H···H bonding interactions are generally weak, they can, through cooperativity, play an important role in the formation and stability of a particular crystal packing. Database analysis on H···H contacts of the type $CH_2···H_2C$ and NH···HN in organic crystals substantiate the occurrence of H···H interactions as a contributing force to the packing of the molecules in the crystal.[17] Analysis of the data on the XH···HX (X = B, C) intermolecular contacts in the crystals of organic carborane derivatives retrieved from the CSD revealed that the H···H interactions are important within the crystal.[78] The crystallographic analysis of the tricyclic derivative of [1,2,4,5]-tetrazine, showed that the packing of molecules within the

Table 16.1 Representative examples found in the CSD of compounds that present intramolecular H$\cdots$H interactions with distance shorter than 2.0 Å.

Compound	CSD identifier	C–H$\cdots$H–C (Å)	R_f^2	Ref.
Methyl-2-(benzyl(*t*-butoxycarbonyl)amino)-2-(9-borabicyclo(3.3.1)nonan-9-yl-acetate	SUSFAP	1.940 RHC–H$\cdots$H–CHR 1.965 RHC–H$\cdots$H–CH$_2$ 1.993 RHC–H$\cdots$H–CH$_2$	3.31	67
(μ-2,2′-Bipyridine-4-ylethynyl)-(μ-methylenbis(diphenylphosphine))-(μ-hydrido)-heptacarbonyl-tri-ruthenium	TUSFOE	1.915 RHC–H$\cdots$H–Ph	2.82	68
1,1′-Methylenebis(3-isopropyl-2,3-dihydro-1*H*-imidazol-2-ylidene))-dimethyl-platinum THF	BUPJIH	1.924 RHC–H$\cdots$H–CH$_2$ 1.993 RH$_2$C–H$\cdots$H–CH$_2$	3.34	69
2,6-Dibromo-9-selenobicyclo[3.3.1]nonane	JUQXEA	1.945 RHC–H$\cdots$H–CHR	2.6	70
2-(Methylsulfanyl)-1-(thiomorpholin-4-yl)ethanone	PUNJIT	1.984 RHC–H$\cdots$H–CHR	2.94	71
4,9-Dichlorobenzo[*c*]phenanthrene	RUNZIL	1.954 Ph–H$\cdots$H–Ph	2.7	72
1,1′,1″,1‴,1⁗,1‴″-(Benzene-1,2,3,4,5,6-hexaylhexakis(methylene))hexakis(4-bromopyrazole)	TUSNOM	1.928 RHC–H$\cdots$H–CHR 2.025 RHC–H$\cdots$H–CHR 2.035 RHC–H$\cdots$H–CHR	2.58	73
Bis(μ$_2$-piperidino-bis(μ$_2$-2,2,6,6-tetramethylpiperidino)-di-copper-di-lithium	TUSWEL	1.999 RH$_2$CH$\cdots$H–CH$_2$R 2.384 RHC–H$\cdots$H–CHR 2.382 RHC–H$\cdots$H–CHR 2.293 RHC–H$\cdots$HCH$_2$R 2.345 RH$_2$C–H$\cdots$HCH$_2$R	3.03	74
Bis(*t*-butylimino)-(*t*-butylaminato)-(2-((4-phenylpiperazin-1-yl)methyl)-1*H*-pyrrol-1-yl)tungsten CH$_2$Cl$_2$	TUSWOV	1.978 RHC–H$\cdots$H–Ph 1.932 RHC–H$\cdots$H–Ph 2.347 RHC–H$\cdots$H–CHR 2.307 RHC–H$\cdots$H–CHR 2.324 RH$_2$CH$\cdots$H–CH$_2$R	3.07	75
3β-Acetoxyolean-11,12-aziridin-28,13-β-olide	URENOV	1.924 RHCH$\cdots$H–CH$_2$R 2.281 RC–H$\cdots$H–CH$_2$R 2.086 RHC–H$\cdots$HCH$_2$R 2.126 RC–H$\cdots$H–CH$_2$R 2.123 RC–H$\cdots$H–CH$_2$R 2.042 RHC–H$\cdots$HCH$_2$R 2.339 RHC–H$\cdots$H–CR 2.371 RHCH$\cdots$H–CH$_2$R 2.394 RHC–H$\cdots$H–CHR	3.25	76

crystal resulted from weak interactions, where 71.6% are H$\cdots$H interactions.[79] Homopolar H$\cdots$H interactions, BH$\cdots$HB and NH$\cdots$HN exist in the solid-state structures of $LiNH_2BH_3$ and $NaNH_2BH_3$, which are surprisingly strong and play a substantial role in stabilizing the solid-state structures.[12] In homopolar H$\cdots$H interactions, electron density is shared symmetrically between the participating H atoms. Furthermore, the BH$\cdots$HB interactions mediate the release of H_2, as revealed by a ^{1}H NMR study of the gaseous decomposition products of $LiNH_2BH_3$.[12]

Considering the importance of H$\cdots$H interactions in the tridimensional crystal growth, a crystallographic search employing similar search criteria to those used for the intramolecular H$\cdots$H interactions gave the following results. For the homopolar N–H$\cdots$H–N interaction only 14 structures were found when the search was restricted to the H$\cdots$H distance range of 1.6–2.0 Å but on increasing the interaction distance to the sum of van der Waals hydrogen radii (2.4 Å), the number of structures found increases to 632, all of them with $R_f \leq 3.5$ % and acquisition data temperatures ≤ 200 K. Similar results were found for the homopolar system B–H$\cdots$H–B with 74 structures found for the shortest range and 284 structures for the longest interaction distance. The same exercise with C–H$\cdots$H–C system leads to the identification of 2460 structures when the distance range is ≤ 2.0 Å, $R_f \leq 3.5\%$ and $T \leq 200$ K. The histograms shown in Figure 16.5 give a clear idea of the importance of H$\cdots$H interactions for the crystal growth.

The H$\cdots$H interaction has a central role in the crystallization of non-polar molecules, where the CH$\cdots$HC interaction can favor a given crystal packing. The classical system in this situation is the CH_4 dimer. A study on $(CH_4)_2$ shows that the $CH_4\cdots CH_4$ dimer is bound in all possible H$\cdots$H contact arrangements, and the arrangements with more than one H$\cdots$H contact give rise to a greater amount of stabilization than does the arrangement with one H$\cdots$H contact. These interactions are found to be weakly attractive and anisotropic at short vdW distances, in good agreement with the experimental data,[80] and explain the solid state of methane under 90 K. There are very good linear relationships between H$\cdots$H bonding in complex alkanes and their stabilities and between H$\cdots$H bonding in alkanes and their boiling points. The hydrogen–hydrogen bond plays an important role on the boiling point of linear alkanes. Intermolecular interactions in alkane complexes are related to the H$\cdots$H bond, not to non-directional induced dipole–induced dipoles. Shaik *et al.*[43] studied intermolecular interactions in dimers of *n*-alkanes and polyhedranes, such as tetrahedrane, cubane, octahedrane or dodecahedrane, and

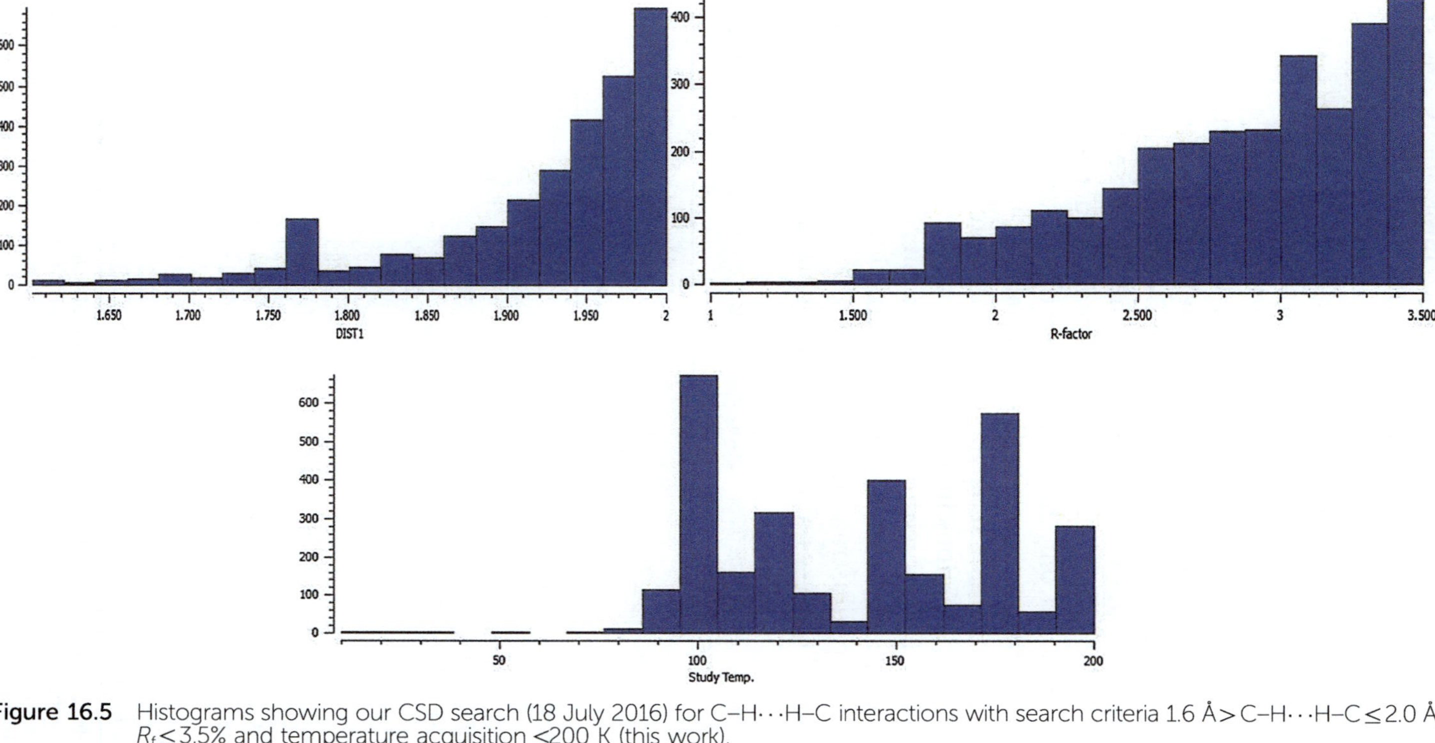

Figure 16.5 Histograms showing our CSD search (18 July 2016) for C–H···H–C interactions with search criteria 1.6 Å > C–H···H–C ≤ 2.0 Å, R_f ≤ 3.5% and temperature acquisition ≤200 K (this work).

demonstrated that attractive C–H$\cdots$H–C interactions are stronger than usually thought.[43] The authors' MP2 calculations show several (globally) stabilizing short C–H$\cdots$H–C contacts between hydrocarbon molecules irrespective of their relative orientations. They emphasized the role of the cooperativity of these weak interactions in affecting trends in the melting points of hydrocarbons.

Other examples where the C–H$\cdots$H–C interaction plays an important role are in the crystallization of steroids and sterols. Two structures of cholesterol with different crystallization solvents and two derivatives have been found in the database. Even when the solvents and substituents exert a great influence on the crystal packing, this can only help to explain the crystal growth in certain directions. Thus, to reach tridimensionality, C–H$\cdots$H–C interactions have to be involved. It is interesting to note that the number of C–H$\cdots$H–C interactions increase as well, as the solvent or heteroatom in the cholesterol structure participate in intermolecular interactions. Some selected distances of cholesterol and its derivatives are listed in Table 16.2 to illustrate the above discussion. Figure 16.6 shows the principal C–H$\cdots$H–C interactions found in the cholesterol crystal

Table 16.2   H$\cdots$H interactions found in cholesterol crystals involved in the three-dimensional growth of the crystal.

Compound	CSD identifier[a]	C–H$\cdots$H–C (Å) RHC–H$\cdots$H–CHR interactions	R_f^{2b}	Ref.
Cholesterol phenol solvate	YARXAS	2.309	4.66	81
		2.214		
		2.379		
Cholesterol phenylmethanol solvate hemihydrate	YARXEW	2.304	3.91	81
		2.397		
		2.279		
		2.347		
		2.378		
		2.359		
		2.338		
		2.359		
Bis((3S,8S,9S,10R,13R,14S,17R) cholesterol-4-iodophenol)	WOMHAI	2.342	4.38	82
		2.229		
		2.244		
		2.365		
		2.389		
24(S), 25-Epoxycholesterol	HIFKIQ	2.382	4.7	83

[a]Nomenclature employed in the Cambridge Structural Database.
[b]Reliability factor $R_\mathrm{f}^2 = \Sigma\,||F_\mathrm{obs}| - |F_\mathrm{calc}||/\Sigma\,|F_\mathrm{obs}|$, where F is the so-called structure factor and the sum extends over all the reflections measured and their calculated counterparts, respectively.

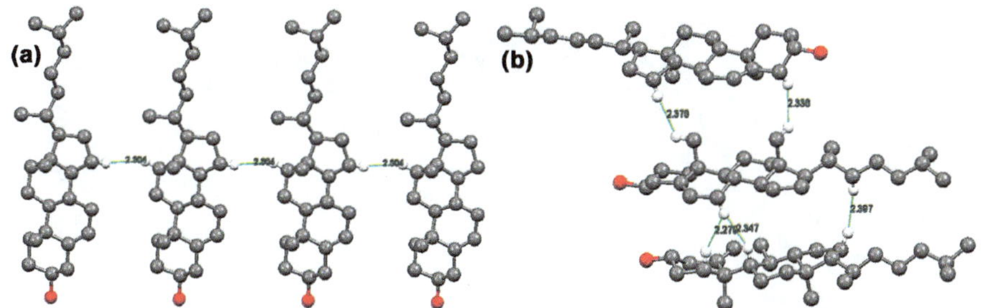

Figure 16.6   C–H···H–C interactions identified along the *b* axis (a) and *c* axis (b) of a cholesterol derivative which are involved in the tridimensional growth of the crystal.
(Image prepared using the structural data obtained from ref. 81).

that helps to explain the growth of the crystal over the axis **b** and **c** of the unit cell.

Another system that leaves no doubt as to the importance of C–H···H–C interactions for crystal growth is *all-trans*-1,2,3,4,5,6-hexaisopropylcyclohexane. This compound is the perfect example in which to identify the attractive character between the hydrogen atoms, that confirms both the presence of C–H···H–C interactions and the importance of their cooperative character to generate the crystals, because there is no other option to produce it. Figure 16.7 shows the C–H···H–C interactions involved in the crystal stabilization. Similar interactions are observed in octacyclopropyl-penta-cyclo[4.2.0.02,5.03,8.O4,7]octane,[84] *meso*-D3-trishomocubylidene-D3-trishomocubane,[85](pentacyclo[4.2.0.02,5.03,8.04,7]octane-1,4-diyldiethyne-2,1-diyl)bis(triphenylsilane),[86] 9-bromo-9-(9′-bromo-9′-homocubyl)-homo-cubane[87] and [2.2.0]*p,m,m*-cyclophane,[88] where H···H interactions range from 2.25 to 2.39 Å.

16.4   Nature of the Hydrogen–Hydrogen Interaction

The neutrality of the H···H interaction prohibits its characterization as primarily electrostatic in nature, in sharp contrast with the hydrogen bond[91,92] or its variant, the dihydrogen bond.[89] Such interaction between two closed-shell, essentially neutral (or homo-polar) hydrogen atoms is, hence, not a type of hydrogen bond since it does not consist of a proton shared between two electronegative atoms. This interaction has been termed "hydrogen–hydrogen (or H···H) bonding" to distinguish it from the qualitatively different dihydrogen bonding.

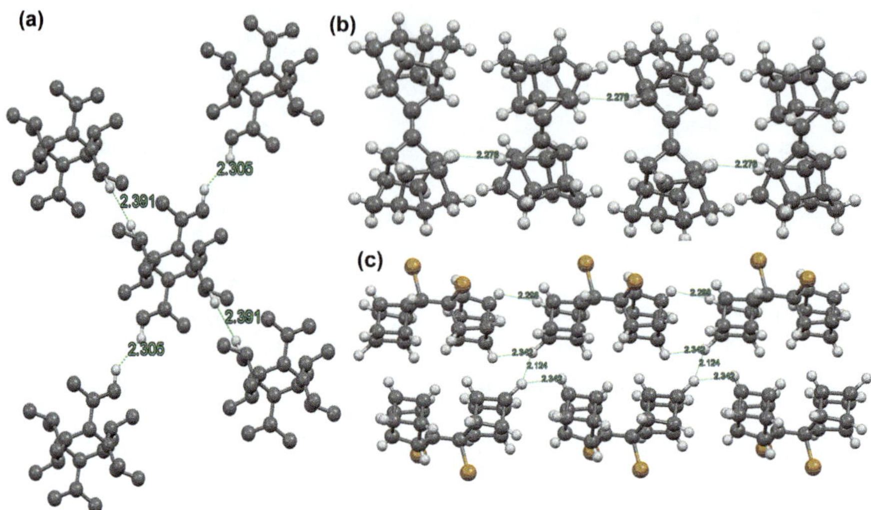

Figure 16.7 C–H···H–C interactions found in the crystal structure of (a) *all-trans*-1,2,3,4,5,6-hexaisopropylcyclohexane (other hydrogen atoms different from the interacting ones were omitted for clarity); (b) *meso*-D$_3$-trishomocubylidene-D$_3$-trishomocubane and (c) 9-bromo-9-(9′-bromo-9′-homocubyl)homocubane. H···H interaction distances are between 2.5 and 3.9 Å. (Image prepared using the structural data obtained from ref. 84–88).

Theoretical and experimental observations suggest that these short H···H closed-shell contacts are generally stabilizing, justifying the usage of the term "bond or bonding" in their designation. Figure 16.8 presents a classification intended to summarize the inter-relations between the various types of multi-hydrogen interaction discussed above.

Alkane molecules are held together in the crystal state by purportedly weak homonuclear R–H···H–R interactions. In an apparent contradiction, the high melting points and vaporization enthalpies of polyhedranes in condensed phases require quite strong inter-molecular interactions. Two questions arise: 'how strong can a weak C–H···H–C bond be?' and 'how do the size and topology of the carbon skeleton affect these bonding interactions?'[43]

In some instances, the H···H bonding is *locally and globally* stabilizing, *i.e.*, it is accompanied by a reduction in the energy of the two hydrogen atoms involved in the interaction as well as by a reduction of the total energy of the system. Examples of locally and globally stabilizing H···H interactions are those occurring in the bay region of kinked polycyclic condensed acenes[4,44] and between the hydrogen atoms attached to C1 and C11 in steroids.[92]

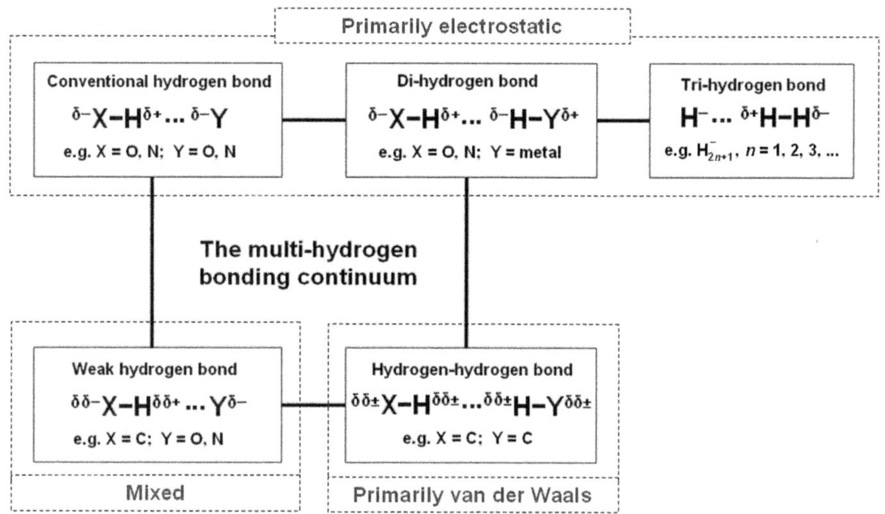

Figure 16.8 A classification showing the relationship between different types of multi-hydrogen bonding interactions.
(Reprinted with permission from C. F. Matta, L. Huang, L. Massa; Characterization of a trihydrogen bond on the basis of the topology of the electron density. *J. Phys. Chem. A* **2011**, *115*, 12445–12450. Copyright © 2011 *American Chemical Society*).

In other cases, the interaction is locally stabilizing (the energies of the two hydrogen atoms are lowered by the interaction), yet the interaction is accompanied with changes elsewhere in the molecule resulting into a net destabilization of the system. An example of the latter case is planar biphenyl (transition state) whereby the two *ortho*-hydrogen atoms are each stabilized by $\sim$8 kcal mol^{-1} with respect to the twisted global minimum, yet, simultaneously, each of the two junction carbons linking the two rings are destabilized by $\sim$22 kcal mol^{-1}.[5] The sum of the energy changes of the junction carbons, the *ortho*-hydrogens, and the (much less significant) energy changes of all the remaining atoms in biphenyl, is equal to the barrier of the planar transition state.[5]

It is worth mentioning that a recent theoretical study on biphenyl[93] shows that two other approaches, namely, the interacting quantum atoms (IQAs) and Hirshfeld atomic energy partitioning schemes, also indicate this local stabilization in the bay region hydrogen atoms. These findings demonstrate that this local stabilization is fundamentally rooted in the molecular electron density properties (see ref. 94 and 95).

The lack of distinction between *global* and *local* stabilization has led to some misunderstandings.[96–99] Experimental observation,[17]

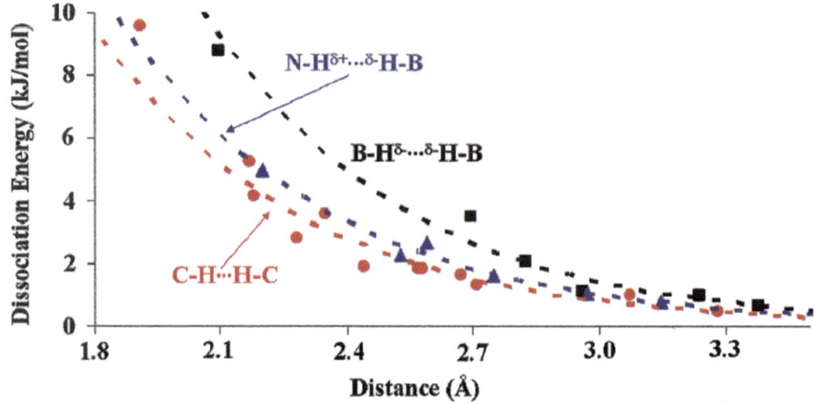

Figure 16.9 Dependence of the dissociation energy (kJ per mol), $\Delta E = 1/2\ V(\mathbf{r})$, on the bond distance (Å) for intermolecular C–H···H–C ($R^2 = 0.95$), N–H$^{\delta+}$···$^{\delta-}$H–B ($R^2 = 0.96$), and B–H$^{\delta-}$···$^{\delta-}$H–B ($R^2 = 0.98$) interactions. (Reprinted with permission from D. J. Wolstenholme, J. T. Titah, F. N. Che, K. T. Traboulsee, J. Flogeras, G. S. McGrady; Homopolar dihydrogen bonding in alkali-metal amidoboranes and its implications for hydrogen storage. *J. Am. Chem. Soc.* **2011**, *133*, 16598–16604. Copyright © 2011 *American Chemical Society*).

examination of the molecular electrostatic potential at the position of the nuclei as a function of the rotation of the two rings in biphenyl,[41] and the discovery that bond paths are *"privileged exchange channels"*[42] have all lent further support to the stabilizing nature of C–H···H–C interactions.

The energy of H···H contacts, estimated with the Espinosa–Molins–Lecomte (EML) approximation,[100] is around 2 kcal mol^{-1}, not dissimilar to the energy of a weak C–H···O interaction.[78]

Figure 16.9 displays the exponential relationship between the strength and bond length for CH···HC, NH···HB, and BH···HB interactions. These types of interactions gradually move from predominantly van der Waals at longer distances to resemble classical hydrogen bonds at the shorter end of the spectrum. The strength of these interactions increases in the order CH···HC < NH···HB < BH···HB.[12]

16.5 Theoretical Indicators of Hydrogen–Hydrogen Bonding

16.5.1 Hydrogen–Hydrogen Bonding Revealed in Molecular Orbitals

Chemists often explicate the nature of a chemical interaction on the basis of molecular orbitals. In molecular systems where an H···H interaction is observed, it is possible to find molecular orbitals that

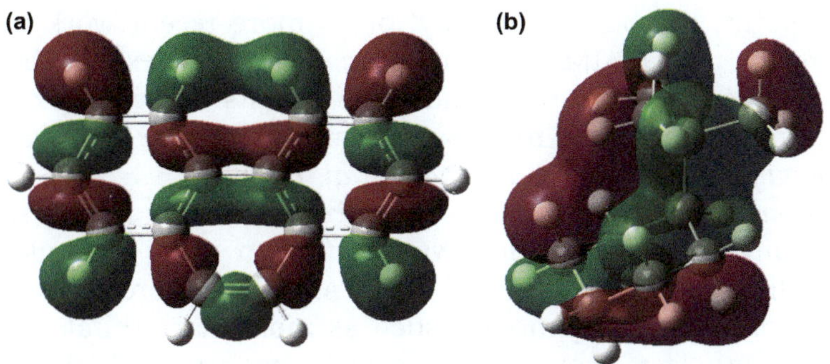

Figure 16.10   (a) Molecular orbital 39 of phenanthrene and (b) molecular orbital 23 of axial *tert*-butylcyclohexane (this work).

result from constructive overlap of the hydrogen atom orbitals. Phenathrene and the axial conformer of *tert*-butylcyclohexane present molecular orbitals that are related to the presence of intramolecular H···H interactions. The observable (total) electron density, the scalar field where the bond path is traced, is the result of the contribution of every occupied molecular orbital at every point in space. The presence of the bond path is, hence, a result of the distribution of all molecular orbitals, save those with nodes at the point of evaluation (Figure 16.10).

16.5.2   Hydrogen–Hydrogen Bonding Revealed in the Topology of the Electron Density

The presence of chemical bonding is typically assigned when specific criteria are fulfilled, whether these are interatomic distances, properties of molecular orbitals, or the properties of scalar fields such as the presence of a bond path. In this regard, the electron density is perhaps the most used scalar field to define molecular structure, on the basis of the presence of a bond path defined by the quantum theory of atoms in molecules (QTAIM).[101] We recall that the bond path is a line of locally maximum electron density that links the nuclei of atoms that are considered chemically-bonded.[42,99,102,103] It is worth noting that the 1977 discovery of the bond path[102] allowed chemists, for the first time since Loschmidt's 1861 book[104] to which the symbol of the chemical bond as a multivalent link between bonded atoms can be traced, to assign lines that connect the nuclei of bonded atoms in real physical space. There exists a unique line linking the nuclei of bonded atoms no matter what the classically-assigned bond multiplicity, the multiplicity being reflected in the properties of the bond path itself such as the density at the bond critical point, the bond ellipticity, or as more recent work has demonstrated, the QTAIM electron delocalization index[42,105] between the two atoms sharing the bond path and its associated zero-flux interatomic surface. One may assert that *the chemical bond itself was made manifest in real space by Bader et al. in 1977.*

The bond path concept has been utilized by IUPAC to characterize and classify hydrogen-bonded systems.[90,91] In most circumstances, the bonding description offered by QTAIM is concordant with that given by other bonding models such as molecular orbital (MO) analysis. Nonetheless, there are some instances in which the QTAIM results are not in agreement with those of different chemical-bonding schemes, and these are the most interesting cases (usually borderline). These discrepancies have led to reassessments of the ideas that underlie QTAIM. Most of the criticisms of the bonding interpretation of QTAIM come from the existence (or absence) of unexpected (or presumed) bond paths (BP) that are not in accordance with classic bonding models (*e.g.*, Lewis structures) or with other bond definitions like that derived from the electron localization function. Distinct scalar fields—for example, the density of the Fermi hole and the source function[106]—have been employed in an attempt to explain these apparent anomalies.

In 2003, Bader and coworkers presented $H \cdots H$ bonding as distinct from "dihydrogen bonding" on the basis that the two bonded hydrogen atoms bear identical or similar charges.[4,7] $H \cdots H$ bonds are found between the *ortho*-hydrogen atoms in planar biphenyl, between the hydrogen atoms bonded to the bay region's C(4) and C(5) carbon atoms in phenanthrene and other angular polybenzenoids, and between the methyl hydrogen atoms in the cyclobutadiene, tetrahedrane and indacene molecules corseted with tertiary-*tetra*-butyl groups. It is shown that each such $H \cdots H$ interaction, rather than denoting the presence of "nonbonded steric repulsions", makes a stabilizing contribution of up to 10 kcal mol^{-1} to the energy of the molecule in which it occurs.[4,7]

Koch and Popelier proposed several criteria to characterize weak hydrogen bonds,[107] which are also satisfied not only by heteropolar $C-H^{\delta-} \cdots ^{\delta+}H-C$ interactions but also by homopolar $C-H^{\delta+} \cdots ^{\delta+}H-C$ ones found within the crystal structures after a multipole refinement.[108] The first of the Popelier–Koch criteria requires that $H \cdots H$ bonds present a bond path and bond critical point (bcp). The second and third of their criteria are related to the ranges of values of $\rho(\mathbf{r}_{bcp})$ and $\nabla^2 \rho(\mathbf{r}_{bcp})$. The former requires values between 0.013 and 0.27 eÅ^{-3} and the latter must be positive and fall between 0.35 and

4.0 eÅ^{-5}. There must also be a relationship between $\rho(\mathbf{r}_{bcp})$ and the energy densities. The fourth Popelier criterion involves the mutual penetration of the both atoms, which is measured by the $\Delta rH_a + \Delta rH_b$ values of each interaction, where ΔrH is simply the nonbonding radius of the atom minus its bonding radius. The nonbonding radius is taken to be equivalent to the gas-phase van der Waals radius for any particular atom. The static deformation maps of two hydrogen atoms involved in an H$\cdots$H bond show a small change in the electron density between the two atoms upon formation of the H$\cdots$H bonded complex. This is not the case for two hydrogen atoms not involved in H$\cdots$H bonding, where no change was observed in the electron distribution between the two atoms.

Several chemical systems with intra or intermolecular bond paths have been observed, including in discrete systems such as the 10-vertex *arachno*-borane[109] (where extremely short intramolecular H$\cdots$H interactions ($\sim$2.1 Å) were experimentally found by X-ray diffraction determinations[110,111]), and in the study of host–guest interactions in the case of fatty acids and heart-type fatty acid binding protein (H-FABP, Figure 16.11),[112] going by the study of interactions in SAM.[113]

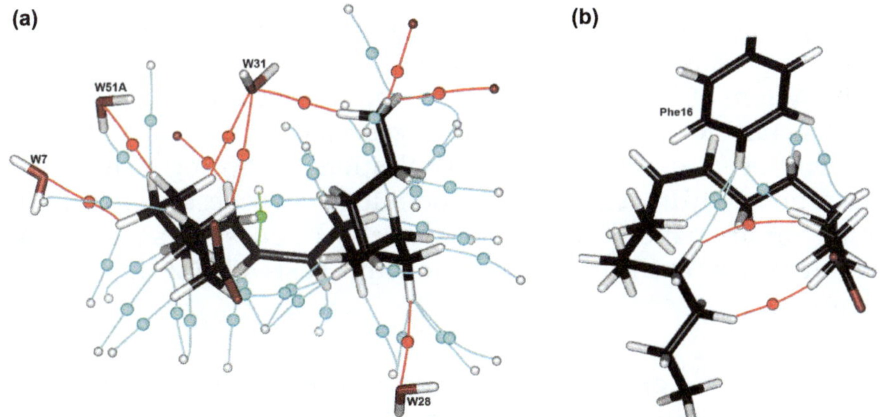

Figure 16.11 H$\cdots$H interactions found in the interaction of oleic acid and heart-type fatty acid binding protein.
(Figure taken from E. I. Howard, B. Guillot, M. P. Blakeley, M. Haertlein, M. Moulin, A. Mitschler, A. Cousido-Siah, F. Fadel, W. M. Valsecchi, T. Tomizaki, T. Petrova, J. Claudot, A. Podjarny; High-resolution neutron and X-ray diffraction room-temperature studies of an H-FABP oleic acid complex: study of the internal water cluster and ligand binding by a transferred multipolar electron density distribution. Int. Union Cryst. J. 2016, 3, 115–126 with permission. Published under a Creative Commons Attribution License.)

Other examples include the female sex-hormone molecule estrone, which exhibits several stable bond paths linking the hydrogen nuclei in experimentally and theoretically calculated densities. The analysis of atomic energies reveals that this $H\cdots H$ closed-shell bonding interaction is accompanied with an estimated local stabilization of 8–11 kcal mol^{-1}.[92] The hydrogen–hydrogen bond path has also been found to be associated with the stability of a powerful rhodium catalyst,[114] strained chelating rings of metal complexes,[115] and with increasing the "stickiness" of crowded systems whence reducing bond rotation flexibility,[116] to cite a few examples.

16.5.3 Hydrogen–Hydrogen Bonding Revealed in the Topology of ELF

Several research groups have investigated $H\cdots H$ interactions, their origin, and their nature on the basis of the topology and topography of several other scalar fields. As a prominent example, we mention the (dimensionless) electron localization function (ELF, $\eta(\mathbf{r})$),[117,118] defined as:

$$\eta(\mathbf{r}) = \frac{1}{1 + \chi_\sigma^2(\mathbf{r})}, \qquad (16.1)$$

where $\chi_\sigma^2(\mathbf{r})$ is the ratio of a localization index (the probability density for finding a same-spin electron in the vicinity of a reference electron) in the system of interest at position $\mathbf{r}$ and the corresponding value of this function for a uniform electron gas that has the same density. ELF has been used for the identification of hydrogen–hydrogen interactions. Hydrogen–hydrogen bonding has been assigned in alkane[119] or aromatic[120] complexes whose $\eta(\mathbf{r})$ values are smaller than that from a weak hydrogen bond on the basis of ELF topography (Figure 16.12).

16.5.4 Hydrogen–Hydrogen Bonding Revealed in the Topology of the Current Density

Current density plots of closed-shell intermolecular $H\cdots H$ interactions characterized by a bond critical point (BCP) show two vortices separated by a saddle, a pattern which allows for a clear definition of a pair current strength. In the case of the hydrogen atoms of the bay region of polycyclic aromatic hydrocarbons, the current of the $H\cdots H$ moiety is delocalized into a diatropic vortex which indicates stabilization (Figure 16.13).[121]

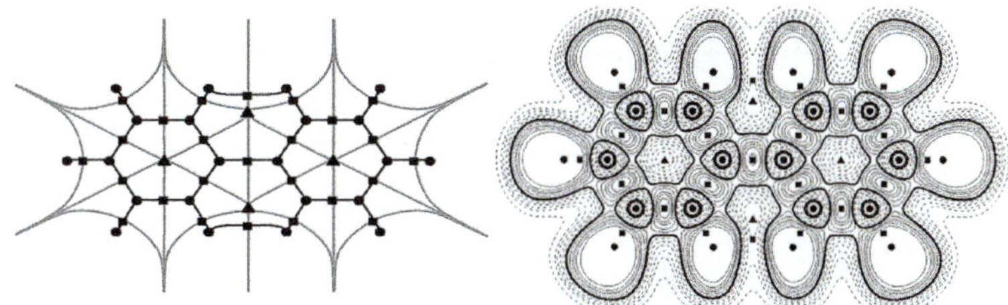

Figure 16.12 Left. Molecular graph (black lines) of planar biphenyl and interatomic surfaces (gray lines) defined by the critical points (CPs) of $\rho(\mathbf{r})$ at the plane containing the nuclei (circles). Squares are bond critical points and triangles are ring critical points. Right. ELF map showing the following isocontours of $G(\mathbf{r})$ (starting at the outermost one): 0.01, 0.05, 0.1, 0.2, 0.3, and 0.4 (dashed), 0.5 (bold), 0.6, 0.7, 0.8, 0.85, 0.9, and 0.95 (solid). Nuclei (circles), BCPs (squares), and RCPs (triangles) are superimposed.
Reprinted from *Struct. Chem.*, A theoretical study of the intramolecular interaction between proximal atoms in planar conformations of biphenyl and related systems, *18*, **2007**, 785–795, L. F. Pacios., Copyright 2007 Springer, with permission of Springer.

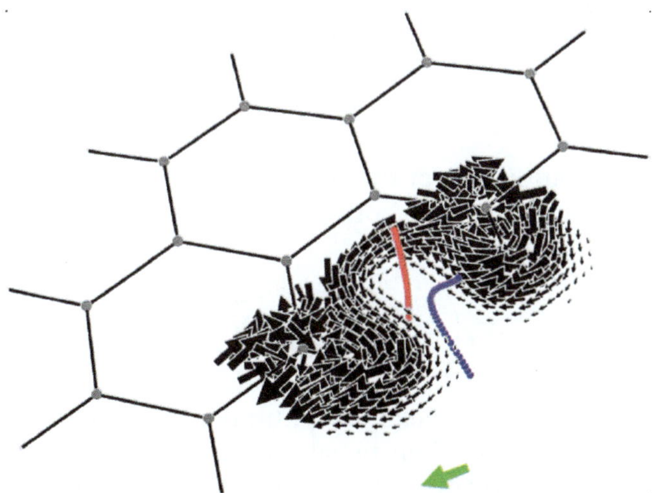

Figure 16.13 A portion of the current density ($\mathbf{J}^B$) plot for phenanthrene for a magnetic field perpendicular to the molecular plane. The saddle line is in blue and the line of paratropic centers in red.
(Reprinted with permission from P. Della Porta, R. Zanasi, G. Monaco; Hydrogen-hydrogen bonding: The current density perspective. *J. Comput. Chem.* 2015, 36, 707–716. © 2015 *John Wiley and Sons*).

16.5.5 Hydrogen–Hydrogen Bonding Revealed in the Topology of the Ehrenfest Force

The Ehrenfest force (**EF**) is the force acting on the electrons in a molecule due to the presence of the other electrons and the nuclei,

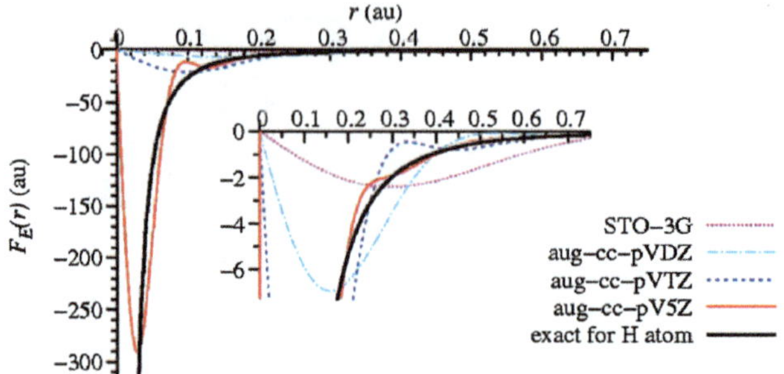

Figure 16.14 Full configuration interaction (full CI) Ehrenfest force field for the ground state hydrogen atom.
(Reprinted from A. Martin Pendas, J. Hernindez-Trujillo; The Ehrenfest force field: Topology and consequences for the definition of an atom in a molecule. J. Chem. Phys. 2012, 137, 134101. © 2012 American Institute of Physics, with permission of AIP Publishing).

defining a force density, $\mathbf{F}(\mathbf{r})$, as the electrostatic force exerted on the electrons in a generic volume element $\mathbf{dr}$ by the average distribution of the remaining electrons and the nuclei (Figure 16.14). It is defined as:[101]

$$\mathbf{F}(\mathbf{r}) = N \int d\tau' \Psi^*(-\nabla \hat{V})\Psi, \qquad (16.2)$$

where the integration is over the coordinates of all electrons, but one followed by summation over both spins. Explicitly, in Dirac notation and atomic units we write:

$$\mathbf{F}(\mathbf{r}) = N \left\langle \Psi^* \left| -\nabla_\mathbf{r} \left(-\sum_{\alpha=1}^{M} \frac{Z_\alpha}{|\mathbf{r} - \mathbf{R}_\alpha|} + \sum_{i=2}^{N} \frac{1}{|\mathbf{r} - \mathbf{r}_i|} \right) \right| \Psi \right\rangle, \qquad (16.3)$$

where $M =$ number of nuclei, $N =$ number of electrons, and the remaining symbols have their usual meanings.

Alternatively, by applying Gauss' theorem, $\mathbf{F}(\mathbf{r})$ can be expressed in terms of the divergence of the quantum stress tensor $\overleftrightarrow{\sigma}$:[101]

$$\mathbf{F}(\mathbf{r}) = N \int d\tau' \, \Psi^*(-\nabla \hat{V})\Psi = -\nabla \cdot \overleftrightarrow{\sigma}(\mathbf{r}). \qquad (16.4)$$

The integration of $\mathbf{F}(\mathbf{r})$ over the basin of a topological atom A in a given molecule yields the average EF exerted on the electron density of the atom considered, $\mathbf{F}(\Omega)$:[101]

$$\mathbf{F}(\Omega) = \int_\Omega d\mathbf{r}\, \mathbf{F}(\mathbf{r}) = N \int_\Omega d\mathbf{r} \int d\tau' \, \Psi^*(-\nabla \hat{V})\Psi, \qquad (16.5)$$

which can be obtained by means of the surface integral of the stress tensor $\overleftrightarrow{\sigma}$ in terms of the forces acting on the zero-flux boundary of Ω that this atom shares with all other atoms bonded atoms Ω':[101]

$$\mathbf{F}(\Omega) = -\sum_{\Omega' \neq \Omega} \oint \mathrm{d}S(\Omega'|\Omega, \mathbf{r})\, \mathbf{n}(\mathbf{r}) \cdot \overleftrightarrow{\sigma}(\mathbf{r}). \tag{16.6}$$

The Ehrenfest force between the *ortho*-hydrogen atoms in biphenyl, when they share an interatomic surface, is attractive with a magnitude of 0.011 a.u.[5] Martín Pendás and Hernández-Trujillo reported the topology of $\mathbf{F}(\mathbf{r})$ *using Gaussian basis sets* where they did not find any interatomic line (IAL) between the hydrogen atoms in the bay region of phenanthrene.[122] However, *using Slater basis set*, Dillen found the $H\cdots H$ IAL within $\mathbf{F}(\mathbf{r})$ in phenanthrene and planar biphenyl and also an atomic stabilization energy of 5.9 kcal mol^{-1}, when the integrations were performed with the Ehrenfest force basins.[123] The integration over the electron density basin gives a stabilization of 6.9 kcal mol^{-1}. It is worth noting that the theoretical level of the calculation was found to have a significant impact in the topology of $\mathbf{F}(\mathbf{r})$ when studying weak interactions.

16.5.6 Hydrogen–Hydrogen Bonding Revealed in the Non-covalent Interaction Indicator (NCI)

The non-covalent interactions approach (NCI) identifies inter- and intra-molecular interactions by localizing low-density and low-gradient regions between atoms where the reduced density gradient (RDG, s) tends to zero, where the RDG is defined as a dimensionless quantity:[124]

$$s(\mathbf{r}) = \frac{1}{2(3\pi^2)^{1/3}} \frac{|\nabla\rho(\mathbf{r})|}{\rho(\mathbf{r})^{4/3}}. \tag{16.7}$$

These regions are then classified by the sign of the second eigenvalue of the electron density Hessian (λ_2).

The trace of the (diagonalized) Hessian of the density, the Laplacian, is:

$$\nabla^2\rho(\mathbf{r}) = \nabla \cdot \nabla\rho(\mathbf{r}) = \underbrace{\frac{\partial^2\rho(\mathbf{r})}{\partial x^2}}_{\lambda_1} + \underbrace{\frac{\partial^2\rho(\mathbf{r})}{\partial y^2}}_{\lambda_2} + \underbrace{\frac{\partial^2\rho(\mathbf{r})}{\partial z^2}}_{\lambda_3}, \tag{16.8}$$

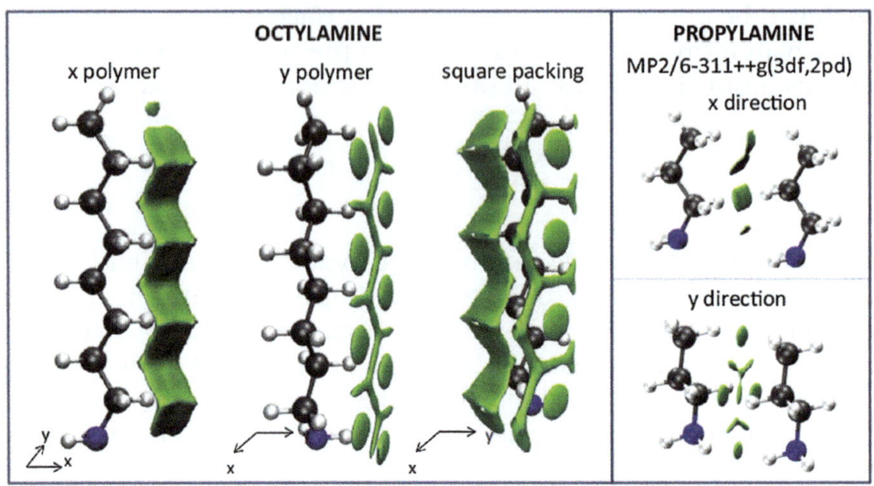

Figure 16.15   NCI surfaces showing the attractive H$\cdots$H interaction within octyla-
mine, and between units of propylamine.
(Reprinted from *Comput. Theor. Chem.* **2015**, *1053*, R. A. Boto,
J. Contreras-Garc̨ka, M. Calatayud; The role of dispersion forces in
metal-supported self-assembled monolayers., 322–327. Copyright
2015 with permission from *Elsevier*).

where λ_i ($i = 1$, 2, 3) are the three eigenvalues (curvatures). Bonding
interactions can be identified by the negative sign of λ_2, while, on the
other hand, two atoms are in non-bonded contact if $\lambda_2 > 0$.

NCI is usually interpreted as stabilizing, when electron density is
concentrated along the bonding axis ($\lambda_2 < 0$), or destabilizing, when
electron density is depleted in the same direction ($\lambda_2 > 0$).

NCI identifies electron density concentration, despite the
absence of an atomic interaction line (AIL), that is, a line of maximal
electron density in a structure which is no longer constrained to be at
its equilibrium (optimized) geometry (which is the condition to term
this line a bond path). NCI always shows a region of electron density
concentration wherever an AIL is present (Figure 16.15 and 16.16).

Cukrowski *et al.* found that H$\cdots$H interactions within different systems such as water dimers and bipyridine are characterized by
the presence of AILs and increased density in the interatomic region and
they correlated this property with interacting quantum atom energies.[125]

16.5.7   Hydrogen–Hydrogen Bonding Revealed in the Interacting Quantum Atom (IQA) Energies

Martín-Pendás *et al.* used QTAIM partitioning of the physical 3D
space to partition the total energy of a many-electron system into

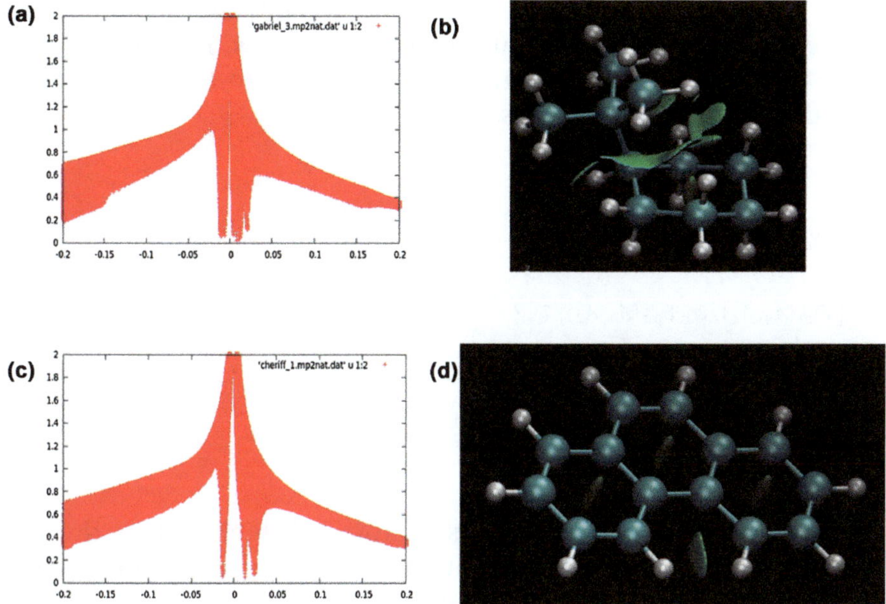

Figure 16.16 NCI surfaces that show the attractive H···H interaction in (a,b) *tert*-butylcyclohexane and (c,d) phenanthrene (this work). Figure prepared by Manuel Guevara.

intra- and interatomic terms by explicitly computing both the one- and two-electron contributions over atomic basins and every pair of basins in the molecule:[126–129]

$$E = \sum_A E_{\text{self}}^A + \frac{1}{2} \sum_A \sum_{B \neq A} E_{\text{int}}^{AB}, \tag{16.9}$$

where E is the total energy of the molecule, and where the additive atomic and diatomic energy terms are defined as:

$$E_{\text{self}}^A = T^A + V_{en}^{AA} + V_{ee}^{AA} \tag{16.10}$$

This is the part of the energy of an atom A in the molecule due to the kinetic energy of the electrons within its basin, the interaction of the electrons in its basin with its own nucleus, and the electron–electron repulsion (both classical and including all quantum and correlation effects) within its own basin, and finally, the diatomic interaction energies defined as:

$$E_{\text{int}}^{AB} = V_{nn}^{AB} + V_{en}^{AB} + V_{ne}^{AB} + V_{ee}^{AB} \quad (A \neq B), \tag{16.11}$$

where in each term the first superscript refers to atom A and the second to another atom B in the molecule.

The energy of a molecule can be partitioned exhaustively in this manner. Because of their similar summation rules, IQA energy components can be cast in a matrix format similar to the localization–delocalization matrices (LDMs):[130]

$$
\boldsymbol{\eta} \equiv
\begin{bmatrix}
E_{\text{self}}(A_1) & E_{\text{int}}(A_1,A_2)/2 & \cdots & E_{\text{int}}(A_1,A_n)/2 \\
E_{\text{int}}(A_2,A_1)/2 & E_{\text{self}}(A_2) & \cdots & E_{\text{int}}(A_2,A_n)/2 \\
\vdots & \vdots & \ddots & \vdots \\
E_{\text{int}}(A_n,A_1)/2 & E_{\text{int}}(A_n,A_2)/2 & \cdots & E_{\text{self}}(A_n)
\end{bmatrix}_{n\times n}
\begin{aligned}
\sum_{\text{row}} &= E_{\text{add}}(A_1) \\
&= E_{\text{add}}(A_1) \\
&\;\vdots \\
&= E_{\text{add}}(A_1)
\end{aligned}
\left.\vphantom{\begin{aligned}1\\2\\3\\4\end{aligned}}\right\}\sum_{i=1}^{n} E_{\text{add}}(A_i)=E
$$

$$
\underbrace{\sum_{\text{column}} \quad = E_{\text{add}}(A_1) \quad = E_{\text{add}}(A_2) \quad\quad = E_{\text{add}}(A_n)}_{\sum_{i=1}^{n} E_{\text{add}}(A_i)=E} \quad tr(\zeta) \quad = E_{\text{self}}^{\text{total}}
$$

$$
\tag{16.12}
$$

where the additive atomic energies of atom A_i (which include the self-energy plus one half of the total interaction energy terms of that atom with all other atoms in the molecule) is defined as:

$$
E_{\text{add}}(A_i) = E_{\text{self}}^{A_i} + \frac{1}{2}\sum_{j\neq i} E_{\text{int}}^{A_iA_j}. \tag{16.13}
$$

Several groups have studied the H$\cdots$H interactions within the framework of the IQA approach. Eskandari *et al.* found that the nature of H$\cdots$H interaction between *ortho*-hydrogen atoms in planar biphenyl is attractive, which disproves the classical view of steric repulsion between the hydrogens.[93]

In addition, in contrast to the traditional virial QTAIM energy analysis, IQA shows that the additive atomic energy of the *ortho*-hydrogens of biphenyl remains almost constant as a function of the central dihedral angle rotation as they form and break the H$\cdots$H bond path. Although the total *interaction contribution* to the additive atomic energy of the hydrogens plays a stabilizing role during the formation of the H$\cdots$H bond path, it is almost compensated by the destabilizing effect of the self-energy contribution. As a consequence of this cancellation of energetic contributions, the additive energy of the hydrogens remains constant as one rotates biphenyl around its central dihedral angle.[93]

In the case of zinc complexes with bipyridine (L), the formation of C–H$\cdots$H–C contacts within the ligands is provoked by a local stabilization. From Hartree–Fock IQA calculations, it was found that the local diatomic interaction energy, $E_{\text{int}}^{\text{HH}}$, amounts to -2.5, -2.7, and -2.9 kcal mol^{-1} for ZnL, ZnL$_2$, and ZnL$_3$, respectively. This change was accompanied by a decrease of the ^1J(C–H) coupling constants, from 177.06 Hz (ZnL) to 173.87 Hz (ZnL$_3$).[131]

Cukrowski *et al.* compared three descriptors of the presence of chemical interaction, (*i*) interatomic line, (*ii*) the sign of interaction energy $E_{int}^{A_iA_j}$, and (*iii*) the sign of λ_2. The ideal situation is to have an agreement between the three descriptors, *i.e.*, the presence of the bond path, $E_{int}^{A_iA_j} < 0$, and $\lambda_2 < 0$. In most cases the ideal situation occurs but there are some cases where one of descriptors contradicts the other two. In all the cases, C–H$\cdots$H–C interactions are dominated by the exchange contribution.[125]

16.6 From Explanation to Prediction: Recent Examples

Sabirov *et al.* found linear correlations between the mean polarizability and the number of H$\cdots$H bond critical points in the series of planar isomeric PAHs. In this way, BCPs can be related to the measurable molecular properties.[132] Monteiro and Firme were able to construct predictive statistical models for the boiling points of alkanes using as predictor simply the number of hydrogen–hydrogen bonding interactions in the corresponding dimers.[119] Furthermore, these authors also show that the problem of the relative stability of branched alkanes compared to lesser branched isomers may be partly explained by the stability associated with an increasing number of H$\cdots$H interactions. These findings demonstrate that a strong statistical association exists between the existence of these H$\cdots$H bond paths, as seen in the observable topography of the electron density, and measurable physicochemical properties.

16.7 Closing Remarks

Hydrogen–hydrogen bonding, proposed more than a decade ago, has been confirmed by complementary approaches, theoretical and experimental. It is no longer a subject of debate and it is time for this concept to be incorporated into the standard body of chemical knowledge. The phenomenon of hydrogen–hydrogen bonding "*is subtle but not faint*" to quote Shaik *et al.*[43] The importance of H$\cdots$H bonding stems from its ubiquity and cooperative nature. It role in stabilizing crystals is evident while its role in stabilizing biomolecules, such as proteins and nucleic acids, remains to be elucidated in fuller detail.

Note Added in Proof

An article has recently appeared that emphasized the importance of C–H$\cdots$H–C contacts in stabilizing hydrocarbon crystals (and their Si, Ge, Sn, and Pb counterparts).[135]

Acknowledgements

The authors are indebted to the late Professor Richard F. W. Bader (1931–2012)[133] and to Professor Jesús Hernández-Trijillo (of the *Universidad Nacional Autónoma de México* (UNAM)) for their authorship of the original work discussed in this chapter (along with Professor Ting-Hua Tang) and for numerous discussions over the years. The authors thank Dr. Todd A. Keith for his AIMAll programme that allowed this work to have the reach it now has,[134] and to Mr Manuel Guevara for preparing Figure 16.16 of this chapter. Professor Lou Massa is thanked for numerous valuable discussions. J.C. thanks DGAPA-UNAM for posdoctoral financial support in the form of a fellowship. F.C.G. thanks PAPIIT-UNAM, and C.F.M. the *Natural Sciences and Engineering Research Council of Canada* (NSERC), *Canada Foundation for Innovation* (CFI), and *Mount Saint Vincent University*, for financial support.

References

1. K. N. Robertson, O. Knop and T. S. Cameron, C-H$\cdots$H-C interactions in organoammonium tetraphenylborates: another look at dihydrogen bonds, *Can. J. Chem.*, 2003, **81**, 727–743.

2. K. N. Robertson, Intermolecular *Interactions in a Series of Organoammonium Tetraphenylborates*, PhD Thesis, Dalhousie University: Halifax, Canada, 2001.

3. J. A. Platts and S. T. Howard, C-H...C Hydrogen bonding involving ylides, *J. Chem. Soc., Perkin Trans. 2*, 1997, 2241–2248.

4. C. F. Matta, J. Hernández-Trujillo, T. H. Tang and R. F. W. Bader Hydrogen-hydrogen, bonding: a stabilizing interaction in molecules and crystals, *Chem. – Eur. J.*, 2003, **9**, 1940–1951.

5. J. Hernández-Trujillo and C. F. Matta, Hydrogen-hydrogen bonding in biphenyl revisited, *Struct. Chem.*, 2007, **18**, 849–857.

6. C. F. Matta, Hydrogen-hydrogen bonding: The non-electrostatic limit of closed-shell interaction between two hydrogen atoms. A critical review, *Hydrogen*

Bonding - New Insight, (Challenges and Advances in Computational Chemistry and Physics Series), Springer, 2006, pp. 337–376.

7. C. F. Matta, *Applications of the Quantum Theory of Atoms in Molecules to Chemical and Biochemical Problems*, PhD Thesis, McMaster University, Hamilton, Canada, 2002.

8. V. I. Bakhmutov, *Dihydrogen Bonds: Principles, Experiments, and Applications*, Wiley-Interscience, New Jersey, 2008.

9. R. H. Crabtree, Dihydrogen complexes: Some structural and chemical studies, *Acc. Chem. Res.*, 1990, **23**, 95–101.

10. I. Alkorta, I. Rozas and J. Elguero, Non-conventional hydrogen bonds, *Chem. Soc. Rev.*, 1998, **27**, 163–170.

11. I. Rozas, I. Alkorta and J. Elguero, Field effects on dihydrogen bonded systems, *Chem. Phys. Lett.*, 1997, **275**, 423–428.

12. D. J. Wolstenholme, J. T. Titah, F. N. Che, K. T. Traboulsee, J. Flogeras and G. S. McGrady, Homopolar dihydrogen bonding in alkali-metal amidoboranes and its implications for hydrogen storage, *J. Am. Chem. Soc.*, 2011, **133**, 16598–16604.

13. D. Wolstenholme, J. Flogeras, F. N. Che, A. Decken and G. S. McGrady, Homopolar dihydrogen bonding in alkali metal amidoboranes: Crystal engineering of low-dimensional molecular materials, *J. Am. Chem. Soc.*, 2013, **135**, 2439–2442.

14. D. J. Wolstenholme, M. M. D. Roy, M. E. Thomas and G. S. McGrady, Desorption of hydrogen from light metal hydrides: Concerted electronic rearrangement and role of H...H interactions, *Chem. Commun.*, 2014, **50**, 3820–3823.

15. S. K. Burley and G. A. Petsko, Weakly polar interactions in proteins, *Adv. Protein Chem.*, 1988, **39**, 125–189.

16. A. R. Choudhury, U. K. Urs, T. N. Guru Row and K. Nagarajan, Weak interactions involving organic fluorine: A comparative study of the crystal packing in substituted isoquinolines, *J. Mol. Struct.*, 2002, **605**, 71–77.

17. L. Damodharan and V. Pattabhi, Weak dihydrogen bond interactions in organic crystals, *Tetrahedron Lett.*, 2004, **45**, 9427–9429.

18. D. M. Bassani, L. Jonusauskaite, A. Lavie-Cambot, N. D. McClenaghan, J.-L. Pozzo, D. Ray and G. Vives, Harnessing supramolecular interactions in organic solid-state devices: Current status and future potential, *Coord. Chem. Rev.*, 2010, **245**, 2429–2445.

19. A. Datta and S. K. Pati, Dipolar interactions and hydrogen bonding in supramolecular aggregates: understanding cooperative phenomena for 1st hyperpolarizability, *Chem. Soc. Rev.*, 2006, **35**, 1305–1323.

20. M. Nishio, CH/π hydrogen bonds in crystals, *CrystEngComm*, 2004, **6**, 130–158.

21. G. T. Stewart, Liquid crystals in biology II. Origins and processes of life, *Liq. Cryst.*, 2004, **31**, 443–471.

22. G. Gilli and P. Gilli, *The Nature of the Hydrogen Bond: Outline of a Comprehensive Hydrogen Bond Theory*, Oxford University Press, Oxford, 2009.

23. *Hydrogen Bonding - New Insight, (Challenges and Advances in Computational Chemistry and Physics Series)*, ed. S. Grabowski, Springer, Dordrecht, 2006.

24. G. A. Jeffrey, *An Introduction to Hydrogen Bonding*, Oxford University Press, Oxford, 1997.

25. T. Steiner, The hydrogen bond in the solid state, *Angew. Chem., Int. Ed.*, 2002, **41**, 48–76.

26. S. N. Vinogradov and R. H. Linnell, *Hydrogen Bonding*, Van Nostrand Reinhold Co., New York, 1971.

27. L. Pauling, *The Nature of the Chemical Bond*, Cornell University Press, Ithaca, N.Y., 3rd edn, 1960.

28. M. Solimannejad and A. Boutalib, Theoretical investigation of the weakly dihydrogen bonded dimers $H_{2-n}X_nAlH\ldots HArF$ and $H_{2-n}X_nAlH\ldots HKrF$ ($n = 0$–2; $X = F$, Cl), *Chem. Phys.*, 2006, **320**, 275–280.

29. S. J. Grabowski, W. A. Sokalski and J. Leszczynski, Nature of $X\text{-}H^{\delta+}\cdots^{-\delta}H\text{-}Y$ dihydrogen bonds and X-H$\ldots\sigma$ interactions, *J. Phys. Chem. A*, 2004, **108**, 5823–5830.

30. S. Wojtulewski and S. J. Grabowski, Ab initio and AIM studies on intramolecular dihydrogen bonds, *J. Mol. Struct.*, 2003, **645**, 287–294.

31. I. Alkorta, J. Elguero, O. Mo, M. Yanez and J. E. Del Bene, Ab initio study of the structural, energetic, bonding, and ir spectroscopic properties of complexes with dihydrogen bonds, *J. Phys. Chem. A*, 2002, **106**, 9325–9330.

32. S. J. Grabowski, Ab initio calculations on conventional and unconventional hydrogen bonds - study of the hydrogen bond strength, *J. Phys. Chem. A*, 2001, **105**, 10739–10746.

33. S. J. Grabowski, High-level ab initio calculations of dihydrogen-bonded complexes, *J. Phys. Chem. A*, 2000, **104**, 5551–5557.

34. W. T. Klooster, T. F. Koetzle, P. E. M. Siegbahn, T. B. Richardson and R. H. Crabtree, Study of the N-H$\ldots$H-B dihydrogen bond including the crystal structure of BH_3NH_3 by neutron diffraction, *J. Am. Chem. Soc.*, 1999, **121**, 6337–6343.

35. P. L. A. Popelier, Characterization of a dihydrogen bond on the basis of the electron density, *J. Phys. Chem. A*, 1998, **102**, 1873–1878.

36. R. H. Crabtree, A new type of hydrogen bond, *Science*, 1998, **282**, 2000–2001.

37. R. H. Crabtree, P. E. M. Siegbahn, O. Eisenstein, A. L. Rheingold and T. F. Koetzle, A new intermolecular interaction: Unconventional hydrogen bonds with element-hydride bonds as proton acceptor, *Acc. Chem. Res.*, 1996, **29**, 348–354.

38. T. B. Richardson, S. de Gala and R. H. Crabtree, Unconventional hydrogen bonds: Intermolecular B-H$\ldots$H-N interactions, *J. Am. Chem. Soc.*, 1995, **117**, 12875–12876.

39. J. C. Jr. Lee, E. Peris, A. L. Rheingold and R. H. Crabtree, An unusual type of H$\ldots$H interaction: Ir-H$\ldots$H-O and Ir-H$\ldots$H-N hydrogen bonding and its involvement in σ − metathesis, *J. Am. Chem. Soc.*, 1994, **116**, 11014–11019.

40. R. C. Stevens, R. Bau, D. Milstein, O. Blum and T. F. Koetzle, Concept of the $H(\delta+)\ldots H(\delta\text{-})$ interaction. A low-temperature neutron diffraction study of cis-[IrH(OH)(PMe3)4]PF6, *J. Chem. Soc., Dalton Trans.*, 1990, 1429–1432.

41. L. F. Pacios and L. Gómez, Conformational changes of the electrostatic potential of biphenyl: A theoretical study, *Chem. Phys. Lett.*, 2006, **432**, 414–420.

42. A. Martin Pendas, E. Francisco, M. A. Blanco and C. Gatti, Bond paths as privileged exchange channels, *Chem. − Eur. J.*, 2007, **13**, 9362–9371.

43. J. Echeverría, G. Aullón, D. Danovich, S. Shaik and S. Alvarez, Dihydrogen contacts in alkanes are subtle but not faint, *Nat. Chem.*, 2011, **3**, 323–330.

44. D. Wolstenholme, C. F. Matta and T. S. Cameron, Experimental and theoretical electron density study of a highly twisted polycyclic aromatic hydrocarbon: 4 Methyl-[4]helicene, *J. Phys. Chem. A*, 2007, **111**, 8803–8813.

45. N. V. Belkova, E. S. Shubina and L. M. Epstein, Diverse world of unconventional hydrogen bonds, *Acc. Chem. Res.*, 2005, **38**, 624–631.

46. M. Besora, A. Lledós and F. Maseras, Protonation of transition-metal hydrides: A not so simple process, *Chem. Soc. Rev.*, 2009, **38**, 957–966.

47. W. K. Fung, X. Huang, M. L. Man, S. M. Ng, M. Y. Hung, Z. Lin and C. P. Lau, Dihydrogen-bond-promoted catalysis: catalytic hydration of nitriles with the indenylruthenium hydride complex $(\eta^5\text{-}C_9H_7)Ru(dppm)H$ (dppm = Bis(diphenylphosphino)methane), *J. Am. Chem. Soc.*, 2003, **125**, 11539–11544.

48. S. C. Gatling and J. E. Jackson, Reactivity control via dihydrogen bonding: diastereoselection in borohydride reductions of β-hydroxyketones, *J. Am. Chem. Soc.*, 1999, **121**, 8655–8656.

49. X. Yang and M. B. Hall, Monoiron hydrogenase catalysis: Hydrogen activation with the formation of a dihydrogen, Fe-H$^{δ?}$···H$^{δ+}$-O, bond and methenyl-H$_4$MPT$^+$ triggered hydride transfer, *J. Am. Chem. Soc.*, 2009, **131**, 10901–10908.

50. T. Ebata, A. Fujii and N. Mikami, Vibrational spectroscopy of small-sized hydrogen-bonded clusters and their ions, *Int. Rev. Phys. Chem.*, 1998, **17**, 331–361.

51. N. V. Belkova, E. S. Shubina and L. M. Epstein, Dihydrogen bonding, proton transfer and beyond: What we can learn from kinetics and thermodynamics, *Eur. J. Inorg. Chem.*, 2010, **2010**, 3555–3565.

52. I. E. Golub, O. A. Filippov, E. I. Gutsul, N. V. Belkova, L. M. Epstein, A. Rossin, M. Peruzzini and E. S. Shubina, Dimerization mechanism of bis(triphenylphosphine)copper(I) tetrahydroborate: Proton Transfer via a dihydrogen bond, *Inorg. Chem.*, 2012, **51**, 6486–6497.

53. E. V. Bakhmutova, V. I. Bakhmutov, N. V. Belkova, M. Besora, L. M. Epstein, A. Lledós, G. I. Nikonov, E. S. Shubina, J. Tomàs and E. V. Vorontsov, First investigation of non-classical dihydrogen bonding between an early transition-metal hydride and alcohols: IR, NMR, and DFT approach, *Chem. – Eur. J.*, 2004, **10**, 661–671.

54. H. Ishikawa, T. Kawasaki and R. Inomata, Infrared spectroscopy of phenol-triethylsilane dihydrogen-bonded cluster and its cationic analogues: Intrinsic strength of the Si-H···H-O dihydrogen bond, *J. Phys. Chem. A*, 2015, **119**, 601–609.

55. T. Peńa Ruiz, A. Navarro, G. J. Kearley and M. Fernández Gómez, Hydrogen bonds in 1-indanone: Charge density analysis and simulation of the inelastic neutron scattering spectrum in solid phase, *Chem. Phys.*, 2005, **317**, 159–170.

56. C. J. O'Brien, E. A. B. Kantchev, G. A. Chass, N. Hadei, A. C. Hopkinson, M. G. Organ, D. H. Setiadi, T.-H. Tang and D.-C. Fang, Towards the rational design of palladium-*N*-heterocyclic carbene catalysts by a combined experimental and computational approach, *Tetrahedron*, 2005, **61**, 9723–9735.

57. P. C. Singh and G. N. Patwari, Theoretical investigation of C-H...H-B dihydrogen bonded complexes of acetylenes with borane-trimethylamine, *Chem. Phys. Lett.*, 2005, **419**, 5–9.

58. P. C. Singh and G. N. Patwari, The C-H...H-B dihydrogen bonded borane-trimethylamine dimer: A computational study, *Chem. Phys. Lett.*, 2005, **419**, 265–268.

59. E. A. Zhurova, V. G. Tsirelson, V. V. Zhurov, A. I. Stash and A. A. Pinkerton, Chemical bonding in pentaerythritol at very low temperature or at high pressure: An experimental and theoretical study, *Acta Crystallogr., Sect. B: Struct. Sci.*, 2006, **62**, 513–520.

60. L. J. Farrugia, C. S. Frampton, J. A. K. Howard, P. R. Mallinson, R. D. Peacock, G. T. Smith and B. Stewart, Experimental charge-density study on the nickel(II) coordination complex [Ni(H$_3$L)] [NO$_3$][PF$_6$][H$_3$L = N,N ',N ''-tris(2-hydroxy-3-methylbutyl)-1,4,7-triaza-cyclononane]: A reappraisal, *Acta Crystallogr., Sect. B: Struct. Sci.*, 2006, **62**, 236–244.

61. R. F. Freitas and S. E. Galembeck, Computational study of the interaction between TIBO inhibitors and Y181(C181), K101, and Y188 amino acids, *J. Phys. Chem. B*, 2006, **110**, 21287–21298.

62. N. Coskun, A. Parlar, H. Karabiyik, M. Aygun and O. Buyukgungor, Supramolecular architecture of phenylcarbamoylated acetone oxime [1,1-diisopropyl-3-phenylurea] complex, *Struct. Chem.*, 2006, **17**, 431–438.

63. A. Paul, M. Kubicki, C. Jelsch, P. Durand and C. Lecomte, R-free factor and experimental charge-density analysis of 1-(2′-aminophenyl)-2-methyl-4-nitroimidazole: a crystal structure with Z′ = 2, *Acta Crystallogr., Sect. B: Struct. Sci.*, 2011, **67**, 365–378.

64. E.-E. Bendeif, C. F. Matta, M. Stradiotto, P. Fertey and C. Lecomte, Can a formally zwitterionic rhodium(I) complex emulate the charge density of a cationic rhodium(I) complex? A combined synchrotron X-ray and theoretical charge density study, *Inorg. Chem.*, 2012, **51**, 3754–3769.

65. S. J. Grabowski, A. Pfitzner, M. Zabel, A. T. Dubis and M. Palusiak, Intramolecular H···H Interactions for the Crystal Structures of [4-((E)-But-1-enyl)-2,6-dimethoxyphenyl]pyridine-3-carboxylate and [4-((E)-Pent-1-enyl)-2,6-dimethoxyphenyl]pyridine-3-carboxylate; DFT Calculations on Modeled Styrene Derivatives, *J. Phys. Chem. B*, 2004, **108**, 1831–1837.

66. S. G. Bodige, D. Sun, A. P. Marchand, N. N. Namboothiri, R. Shukla and W. H. Watson, Short H...H distances in norbornene derivatives, *J. Chem. Crystallogr.*, 1999, **29**, 523–530.

67. R. K. Everrett and J. P. Wolfe, *Tetrahedron Lett.*, 201, **56**, 3393.

68. S. Bock, C. F. Mackenzie, B. W. Skelton, L. T. Byrne, G. A. Koutsantonis and P. J. Low, Clusters as ligands: Synthesis, structure and coordination chemistry of ruthenium clusters derived from 4- and 5-ethynyl-2,2′-bipyridine, *J. Organomet. Chem.*, 2016, **812**, 190–196.

69. M. Brendel, R. Engelke, V. G. Desai, F. Rominger and P. Hoffman, Synthesis and reactivity of platinum(II) cis-dialkyl, cis-alkyl chloro, and cis-alkyl hydrido bis-N-heterocyclic carbene chelate complexes, *Organometallics*, 2015, **34**, 2870–2878.

70. V. A. Potapov, S. V. Amosova, E. V. Abramova, M. V. Musalov, K. A. Lyssenko and M. G. Finn, 2,6-Dihalo-9-selenabicyclo[3.3.1]nonanes and their complexes with selenium dihalides: Synthesis and structural characterisation, *New J. Chem.*, 2015, **39**, 8055–8059.

71. G. Kang, J. Kim, E. Kwon and T. H. Kim, Crystal structure of 2-methylsulfanyl-1-(thio-morpholin-4-yl)ethanone, *Acta Crystallogr., Sect. E: Crystallogr. Commun.*, 2015, **71**, o679.

72. B. T. Haire, K. W. J. Heard, M. S. Little, A. V. S. Parry, J. Raftery, P. Quayle and S. G. Yeates, Non-linear, cata-condensed, polycyclic aromatic hydrocarbon materials: A generic approach and physical properties, *Chem. – Eur. J.*, 2015, **21**, 9970–9974.

73. N. Koch, W. Seichter and N. Mazik, Hexapodal pyrazole-based receptors: Complexes with ammonium ions and solvent molecules in the solid state, *Tetrahedron*, 2015, **71**, 8965–8974.

74. A. J. Peel, J. Slaughter and A. E. H. Wheatley, New options in directed cupration: Studies in heteroleptic bis(amido)cuprate formation, *J. Organomet. Chem.*, 2016, **812**, 259–267.

75. G. X. Chen, A. Datta, H. C. Hsiao, C. H. Lin and J. H. Huang, Structural elucidation of tungsten compounds containing arylamine, piperazine and morpholine fragments of pyrrole and keto-amine ligands, *Polyhedron*, 2015, **101**, 299–305.

76. W. N. Tan, K. C. Wong, M. Khairuddean, M. Hemamalini and H. K. Fun, Absolute configuration of 3β-acet-oxy-olean-11,12-aziridin-28,13-β-olide, *Acta Crystallogr., Sect. E: Struct. Rep. Online*, 2011, **67**, o1220.

77. P. R. Schreiner, L. V. Chernish, P. A. Gunchenko, E. Y. Tikhonchuk, H. Hausmann, M. Serafin, S. Schlecht, J. E. P. Dahl, R. M. K. Carlson and A. A. Fokin, Overcoming lability of extremely long alkane carboncarbon bonds through dispersion forces, *Nature*, 2011, **477**, 308–311.

78. I. V. Glukhov, K. A. Lyssenko, A. A. Korlyukov and M. Y. Antipin, Nature of weak inter- and intramolecular contacts in crystals. 2. Character of electron

delocalization and the nature of X-H...H-X (X = C,B) contacts in the crystal of 1-phenyl-*o*-carborane, *Russ. Chem. Bull. Int. Ed.*, 2005, **54**, 547–559.

79. M. Owczarek, I. Majerz and R. Jakubas, Weak hydrogen and dihydrogen bonds instead of strong NHO bonds of a tricyclic [1, 2, 4, 5]-tetrazine derivative. Single-crystal X-ray diffraction, theoretical calculations and Hirshfeld surface analysis, *CrystEngComm*, 2014, **16**, 7638–7648.

80. J. J. Novoa, M. H. Whangboo and J. M. Williams, Interactions energies associated with short intermolecular contacts of C-H bonds. II. Ab initio computational study of the C-H...H-C interactions in methane dimers, *J. Chem. Phys.*, 1991, **94**, 4835–4841.

81. R. J. Galloway, S. A. Raza, R. D. Young and I. D. H. Oswald, Tracking the structural changes in a series of cholesterol solvates, *Cryst. Growth Des.*, 2012, **12**, 231–239.

82. G. Desiraju and P. M. Bhatt, Co-crystal formation and the determination of absolute configuration, *CrystEngComm*, 2008, **10**, 1747–1749.

83. T. A. Spencer, N. C. O. Tomkinson, T. M. Willson, A. A. Spencer and J. S. Russel, Efficient, stereoselective synthesis of 24(S),25-epoxycholesterol, *J. Org. Chem.*, 1998, **63**, 9919–9923.

84. A. de Meijere, S. Redlich, D. Frank, J. Magull, A. Hofmeister, H. Menzel, B. Konig and J. Svoboda, Octacyclopropylcubane and some of its isomers, *Angew. Chem., Int. Ed.*, 2007, **46**, 4574–4576.

85. A. P. Marchand, G. M. Reddy, M. N. Deshpande, W. H. Watson, A. Nagl, O. S. Lee and E. Osawa, Synthesis and reactions of meso- and dl-D3-trishomocubylidene-D3-trishomocubane, *J. Am. Chem. Soc.*, 1990, **112**, 3521–3529.

86. S. D. Karlen, H. Reyes, R. E. Taylor, S. I. Khan, F. M. Hawthorne and M. A. Garcia-Garibay, Symmetry and dynamics of molecular rotors in amphidynamic molecular crystals, *Proc. Natl. Acad. Sci. U. S. A.*, 2014, **107**, 14973–14977.

87. A. P. Marchand, V. Vidyasagar, W. H. Watson, A. Nagl and R. P. Kashyap, Pinacol condensation of homocubanone. Synthesis and chemistry of homocubylidenehomocubane, *J. Org. Chem.*, 1991, **56**, 282–286.

88. T. Lahtinen, E. Wegelius, J. Linnanto and K. Rissanen, Small hydrocarbon cyclophanes: Synthesis, X-ray analysis and molecular modelling, *Eur. J. Org. Chem.*, 2002, **2002**, 2935–2941.

89. C. F. Matta, L. Huang and L. Massa, Characterization of a trihydrogen bond on the basis of the topology of the electron density, *J. Phys. Chem. A*, 2011, **115**, 12445–12450.

90. E. Arunan, G. R. Desiraju, R. A. Klein, J. Sadlej, S. Scheiner, I. Alkorta, D. C. Clary, R. H. Crabtree, J. J. Dannenberg, P. Hobza, H. G. Kjaergaard, A. C. Legon, B. Mennucci and D. J. Nesbitt, Defining the hydrogen bond: An account (IUPAC Technical Report), *Pure Appl. Chem.*, 2011, **83**, 1619–1636.

91. E. Arunan, G. R. Desiraju, R. A. Klein, J. Sadlej, S. Scheiner, I. Alkorta, D. C. Clary, R. H. Crabtree, J. J. Dannenberg, P. Hobza, H. G. Kjaergaard, A. C. Legon, B. Mennucci and D. J. Nesbitt, Definition of the hydrogen bond (IUPAC Recommendations 2011), *Pure Appl. Chem.*, 2011, **83**, 1637–1641.

92. E. A. Zhurova, C. F. Matta, N. Wu, V. V. Zhurov and A. A. Pinkerton, Experimental and theoretical electron density study of estrone, *J. Am. Chem. Soc.*, 2006, **128**, 8849–8861.

93. K. Eskandari and C. Van Alsenoy, Hydrogen-hydrogen interaction in planar biphenyl: A theoretical study based on the interacting quantum atoms and Hirshfeld atomic energy partitioning methods, *J. Comput. Chem.*, 2014, **35**, 1883–1889.

94. F. Weinhold, P. v. R. Schleyer and W. C. McKee, Bay-type H · · · H "bonding" in cis-2-butene and related species: QTAIM versus NBO description, *J. Comput. Chem.*, 2014, **35**, 1499–1508.

95. C. F. Matta, S. Sadjadi, D. A. Braden and G. Frenking, The barrier to the methyl rotation in cis-2-butene and its isomerization energy to trans-2-butene revisited, *J. Comput. Chem.*, 2016, **37**, 143–154.

96. J. Poater, M. Solà and F. M. Bickelhaupt, A model of the chemical bond must be rooted in quantum mechanics, provide insight, and possess predictive power, *Chem. – Eur. J.*, 2006, **12**, 2902–2905.

97. J. Poater, M. Solà and F. M. Bickelhaupt, Hydrogen-hydrogen bonding in planar biphenyl, predicted by atoms-in-molecules theory, does not exist, *Chem. – Eur. J.*, 2006, **12**, 2889–2895.

98. R. F. W. Bader, Pauli repulsions exist only in the eye of the beholder, *Chem. – Eur. J.*, 2006, **12**, 2896–2901.

99. R. F. W. Bader, Bond paths anre not chemical bond, *J. Phys. Chem. A*, 2009, **113**, 10391–10396.

100. E. Espinosa, E. Molins and C. Lecomte, Hydrogen bond strengths related by topological analyses of experimentally observed electron densities, *Chem. Phys. Lett.*, 1998, **285**, 170–173.

101. R. F. W. Bader, *Atoms in Molecules: A Quantum Theory*, Oxford University Press, Oxford, U.K., 1990.

102. G. R. Runtz, R. F. W. Bader and R. R. Messer, Definition of bond paths and bond directions in terms of the molecular charge distribution, *Can. J. Chem.*, 1977, **55**, 3040–3045.

103. R. F. W. Bader, A bond path: A universal indicator of bonded interactions, *J. Phys. Chem. A*, 1998, **102**, 7314–7323.

104. J. Loschmidt, *Chemische Studien. I. A. Constitutions - Formeln der organischen Chemie in graphischer Darstellung. B. Das Mariott'sche Gesetz.*, Druck von Carl Gerold's Shon, Vienna, Vienna, 1861.

105. X. Fradera, M. A. Austen and R. F. W. Bader, The Lewis model and beyond, *J. Phys. Chem. A*, 1999, **103**, 304–314.

106. C. Gatti, F. Cargnoni and L. Bertini, Chemical information from the source function, *J. Comput. Chem.*, 2003, **24**, 422–436.

107. U. Koch and P. L. A. Popelier, Characterization of C-H-O hydrogen bonds on the basis of the charge density, *J. Phys. Chem.*, 1995, **99**, 9747–9754.

108. D. J. Wolstenholme and T. S. Cameron, Comparative study of weak interactions in molecular crystals: H-H bonds vs hydrogen bonds, *J. Phys. Chem. A*, 2006, **110**, 8970–8978.

109. E. G. Kononova, Electronic structure of 10-vertex arachno-borane and -carborane clusters, *Comput. Theor. Chem.*, 2013, **1026**, 17–23.

110. D. S. Kendall and W. N. Lipscomb, Crystal structure of tetramethylammonium tetradecahydrodecaborate. Structure of the $B_{10}H_{14}^{2-}$ ion, *Inorg. Chem.*, 1973, **12**, 546–551.

111. J. R. Wermer, N. S. Hosmane, J. J. Alexander, U. Siriwardane and S. G. Shore, Synthesis and X-ray crystal structure of arachno-6-((CH$_3$)$_3$Si)-6,9-C$_2$B$_8$H$_{13}$ through a cage-expansion reaction of nido-2,3-((CH$_3$)$_3$Si)$_2$-2,3-C $_2$B$_4$H$_6$, *Inorg. Chem.*, 1986, **25**, 4351–4354.

112. E. I. Howard, B. Guillot, M. P. Blakeley, M. Haertlein, M. Moulin, A. Mitschler, A. Cousido-Siah, F. Fadel, W. M. Valsecchi, T. Tomizaki, T. Petrova, J. Claudot and A. Podjarny, High-resolution neutron and X-ray diffraction room-temperature studies of an H-FABPoleic acid complex: study of the internal water cluster and ligand binding by a transferred multipolar electron density distribution, *IUCrJ.*, 2016, **3**, 115–126.

113. R. A. Boto, J. Contreras-Garça and M. Calatayud, The role of dispersion forces in metal-supported self-assembled monolayers, *Comput. Theor. Chem.*, 2015, **1053**, 322–327.
114. E.-E. Bendeif, C. F. Matta, M. Stradiotto, P. Fertey and C. Lecomte, Can a formally zwitterionic rhodium(I) complex emulate the charge density of a cationic rhodium(I) complex? A combined synchrotron X-ray and theoretical charge density study, *Inorg. Chem.*, 2012, **51**, 3754–3769.
115. I. Cukrowski and C. F. Matta, Hydrogen-hydrogen bonding: A stabilizing interaction in strained chelating rings of metal complexes in aqueous phase, *Chem. Phys. Lett.*, 2010, **499**, 66–69.
116. J. A. Harrison, M. A. Sajjad, P. Schwerdtfeger and A. J. Nielson, Multiple weak C-H intramolecular hydrogen bonding as an aid to minimizing bond rotation flexibility, *Cryst. Growth Des.*, 2016, **16**, 4934–4942.
117. A. D. Becke and K. E. Edgecombe, A simple measure of electron localization in atomic and molecular systems, *J. Chem. Phys.*, 1990, **92**, 5397–5403.
118. B. Silvi and A. Savin, Classification of chemical bonds based on topological analysis of electron localization functions, *Nature*, 1994, **371**, 683–686.
119. N. K. V. Monteiro and C. L. Firme, Hydrogen-hydrogen bonds in highly branched alkanes and in alkane complexes: A DFT, ab initio, QTAIM, and ELF study, *J. Phys. Chem. A*, 2014, **118**, 1730–1740.
120. L. F. Pacios, A theoretical study of the intramolecular interaction between proximal atoms in planar conformations of biphenyl and related systems, *Struct. Chem.*, 2007, **18**, 785–795.
121. P. Della Porta, R. Zanasi and G. Monaco, Hydrogen-hydrogen bonding: The current density perspective, *J. Comput. Chem.*, 2015, **36**, 707–716.
122. A. Martin Pendas and J. Hernindez-Trujillo, The Ehrenfest force field: Topology and consequences for the definition of an atom in a molecule, *J. Chem. Phys.*, 2012, **137**, 134101.
123. J. Dillen, The topology of the Ehrenfest force density revisited. A different perspective based on Slater-type orbitals, *J. Comput. Chem.*, 2015, **36**, 883–890.
124. E. R. Johnson, S. Keinan, P. Mori-Sanchez, J. Contreras-Garcia, A. J. Cohen and W. Yang, Revealing noncovalent interactions, *J. Am. Chem. Soc.*, 2010, **132**, 6498–6506.
125. I. Cukrowski, J. H. de Lange, A. S. Adeyinka and P. Mangondo, Evaluating common QTAIM and NCI interpretations of the electron density concentration through IQA interaction energies and 1D cross-sections of the electron and deformation density distributions, *Comput. Theor. Chem.*, 2015, **1053**, 60–76.
126. M. A. Blanco, A. Martin Pendas and E. Francisco, Interacting quantum atoms: A correlated energy decomposition scheme based on the quantum theory of atoms in molecules, *J. Chem. Theory Comput.*, 2005, **1**, 1096–1109.
127. E. Francisco, A. Martin Pendas and M. A. Blanco, A molecular energy decomposition scheme for atoms in molecules, *J. Chem. Theory Comput.*, 2006, **2**, 90–102.
128. A. Martin Pendas, E. Francisco and M. A. Blanco, Binding energies of first row diatomics in the light of the interacting quantum atoms approach, *J. Phys. Chem. A*, 2006, **110**, 12864–12869.
129. A. Martin Pendas, M. A. Blanco and E. Francisco, The nature of the hydrogen bond: A synthesis from the interacting quantum atoms picture, *J. Chem. Phys.*, 2006, **125**, 184112.
130. C. F. Matta, Modeling biophysical and biological properties from the characteristics of the molecular electron density, electron localization and delocalization matrices, and the electrostatic potential, *J. Comput. Chem.*, 2014, **35**, 1165–1198.

131. I. Cukrowski, J. H. de Lange and M. Mitoraj, Physical nature of interactions in ZnII complexes with 2, 2′-bipyridyl: quantum theory of atoms in molecules (QTAIM), interacting quantum atoms (IQA), noncovalent interactions (NCI), and extended transition state coupled with natural orbitals for chemical valence (ETS-NOCV) comparative studies, *J. Phys. Chem. A*, 2014, **118**, 623–637.

132. D. Sh. Sabirov, A correlation between the mean polarizability of the "kinked" polycyclic aromatic hydrocarbons and the number of H...H bond critical points predicted by Atoms-in-Molecules theory, *Comput. Theor. Chem.*, 2014, **1030**, 81–86.

133. C. F. Matta, L. Massa and T. A. Keith, Richard F. W. Bader: A True Pioneer, *J. Phys. Chem. A*, 2011, **115**, 12427–12431.

134. T. A. Keith, AIMAll/AIMStudio. http://aim.tkgristmill.com/ 2016.

135. J. Echeverría, G. Aullón and S. Alvarez, Intermolecular interactions in group 14 hydrides: Beyond C–H···H–C contacts, *Int. J. Quantum Chem.*, 2017, in press, DOI: 10.1002/qua.25432.

17 Long, Multicenter Bonds in Radical Anion π-dimers

Fernando Mota,[a] Juan J. Novoa*[a] and Joel S. Miller*[b]

[a] Dept. de Química Física and IQTCUB, Fac. de Química, Univ. Barcelona, Av. Diagonal, 645, 08028-Barcelona, Spain; [b] Department of Chemistry, University of Utah, Salt Lake City UT 84112-0850, USA
*Email: juan.novoa@ub.edu; jsmiller@chem.utah.edu

17.1 Introduction

Long, multicenter C–C bonds (LMBs) are a new sub-class of bonds that can occur between a pair of neutral or charged radicals, and, similar to covalent bonds, result in a diamagnetic compound, *i.e.*, $2\,[A]^{\bullet n\pm} \rightarrow [A]_2^{2n\pm}$ $(n=0,\,1)$. LMBs were first established for the eclipsed, cofacial π-$[TCNE]_2^{2-}$ (TCNE = tetracyanoethylene) (**1**) where the $[TCNE]^{\bullet-}$s are separated by $\sim$2.9 Å.[1–5] In addition, σ-$[A]_2^{x}$ $(x=0,2^-)$ dimers can form.[6–8]

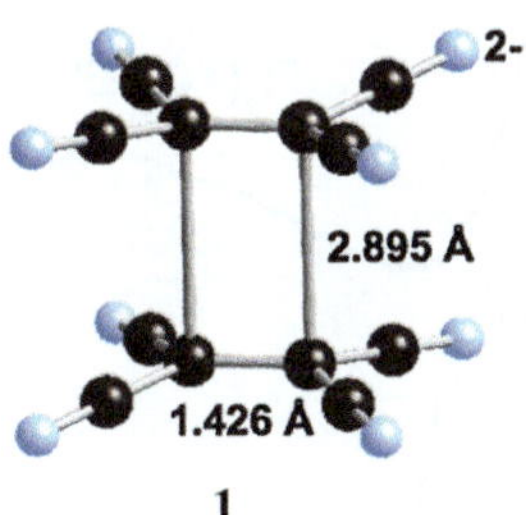

1

Experimental evidence for the intradimer bond for π-$[TCNE]_2^{2-}$ is based on structural, spectroscopic, and magnetic data. The co-facial,

Intermolecular Interactions in Crystals: Fundamentals of Crystal Engineering
Edited by Juan J. Novoa
© The Royal Society of Chemistry 2018
Published by the Royal Society of Chemistry, www.rsc.org

eclipsed, π-[TCNE]$_2^{2-}$ structure is observed for *ca.* 30 cations,[9-26] and the intradimer separation is 2.895 Å, and the central CC distance average is 1.426 Å (**1**) which is that same as observed for isolated [TCNE]$^{\bullet-}$ [1.415 Å].[27] Furthermore, unlike for either TCNE or [TCNE]$^{\bullet-}$, the nitriles bend away from the nominal TCNE plane on average by 5.0°, indicative of some sp^3 mixing with the formal sp^2-C.[27]

The spectroscopic evidence for the intradimer bond includes electronic absorption,[2,5] vibrational (IR[2,5] and Raman[24]), and NMR data.[28] π-[TCNE]$_2^{2-}$ exhibits characteristic IR ν_{CN}, ν_{CC}, and δ_{CCN},[2,5] as well as intradimer Raman[24] absorptions, that differ from those for [TCNE]$^{\bullet-}$.

The electronic absorption spectrum of π-[TCNE]$_2^{2-}$ differs from that of [TCNE]$^{\bullet-}$ as it has a new lower energy absorption at 583 nm that is red-shifted with respect to the 428 nm absorption observed for [TCNE]$^{\bullet-}$. The [TCNE]$^{\bullet-}$ and π-[TCNE]$_2^{2-}$ spectra are shown in Figure 17.1 for (NBu$_4$)$_2$[TCNE]$_2$ dissolved in acetonitrile, and Tl$_2$[TCNE]$_2$ in a KBr pellet, respectively. The maximum for the low energy absorption for π-[TCNE]$_2^{2-}$ exhibits some variation depending on the cation. As a consequence, solids possessing [TCNE]$^{\bullet-}$ are yellow-orange, while those possessing [TCNE]$_2^{2-}$ are deep purple in appearance. Solutions of [TCNE]$^{\bullet-}$ or [TCNE]$_2^{2-}$ are identical, due to the complete dissociation of [TCNE]$_2^{2-}$ to 2 [TCNE]$^{\bullet-}$ at room temperature, however, [TCNE]$_2^{2-}$ is observed in solution at low temperature.[4] The equilibrium constant in CH$_2$Cl$_2$ at 25 °C was claimed to be $\sim 7 \times 10^{-4}$ L mol^{-1}.[12] The $-\Delta H$ of dimerization is essentially the bond dissociation energy and it and the entropy for the dimer in solution

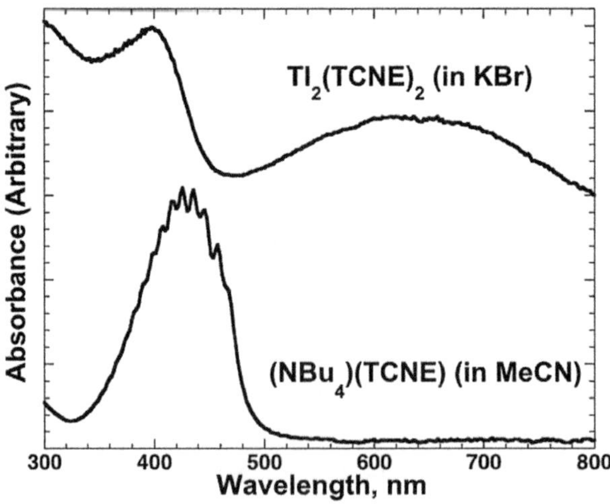

Figure 17.1 Electronic absorption spectra of (NBu$_4$)[TCNE] dissolved in acetonitrile and Tl$_2$[TCNE]$_2$ in a KBr pellet.

were reported to be 8.8 ± 1 kcal mol^{-1} and -41 eu,[12] respectively (this estimation includes the sum of the radical-anion$\cdots$radical-anion, radical-anion$\cdots$solvent and solvent$\cdots$solvent interactions, if the first solvation shell is considered only). Note that the optimum structure of π-[TCNE]$_2{}^{2-}$ in solution is thought to be similar to that in the crystal, except that the dimer is surrounded by the solvent in its first solvation shell, instead of being surrounded by counterions.[29]

[TCNE]$^{\bullet-}$ and π-[TCNE]$_2{}^{2-}$ exhibit different vibrational (IR and Raman) spectra. Planar [TCNE]^n compounds ($n = 0$, 1−) have two ν_{CN} and one δ_{CCN} IR absorptions, however, due to inversion symmetry the ν_{CC} absorption is not allowed in the IR.[30] The major ν_{CN} bands occur at 2262 and 2228, and 2183 and 2144 cm^{-1} for TCNE and [TCNE]$^{\bullet-}$, respectively,[30,31] while the δ_{CCN} absorption occurs at 522 cm^{-1} for both TCNE and [TCNE]$^{\bullet-}$.[5] [TCNE]$_2{}^{2-}$ differs from [TCNE]$^{\bullet-}$ as it should exhibit three ν_{CN}, one ν_{CC}, and three δ_{CCN} IR-allowed absorptions,[4] as respectively observed at 2191, 2170, 2163, 1365, 550, 530, and 515 cm^{-1} (Figure 17.2 as the [NEt$_4$]$^+$ salt). In addition, [TCNE]$^{\bullet-}$ and π-[TCNE]$_2{}^{2-}$ exhibit different Raman vibrational spectra. The Raman spectrum of K$_2$[TCNE]$_2$ also exhibits three ν_{CN}, one ν_{CC}, and three δ_{CCN} Raman active absorptions that respectively occur at 2217, 2178, 2149, 1431, 530, 466, and 262 cm^{-1}.[24] In addition, an intradimer bending mode, Figure 17.3a, is observed at 530 cm^{-1}. Furthermore, six low energy (<200 cm^{-1}) Raman active vibrations, not observed for [TCNE]$^{\bullet-}$, are observed at 198, 173, 155, 131, 107, and 85 cm^{-1}, Figure 17.3b. From the aforementioned correlation between stretching force constants and interaction energy, the

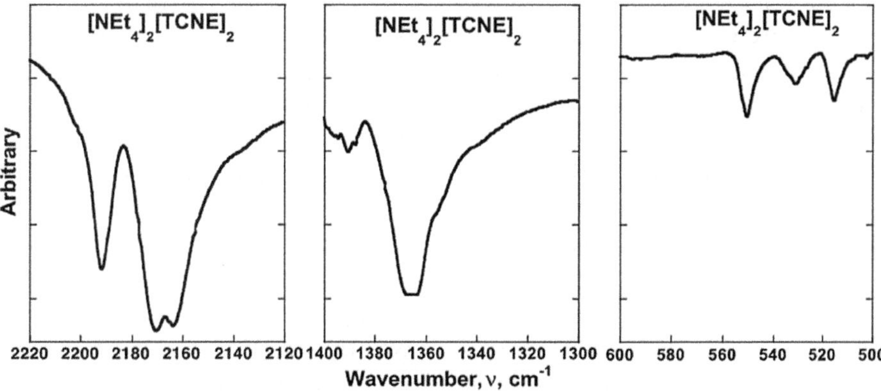

Figure 17.2 Solid state (KBr) IR spectra of π-[TCNE]$_2{}^{2-}$ in [NEt$_4$]$_2$[TCNE]$_2$ in the ν_{CN}, ν_{CC}, and δ_{CCN} absorption regions.

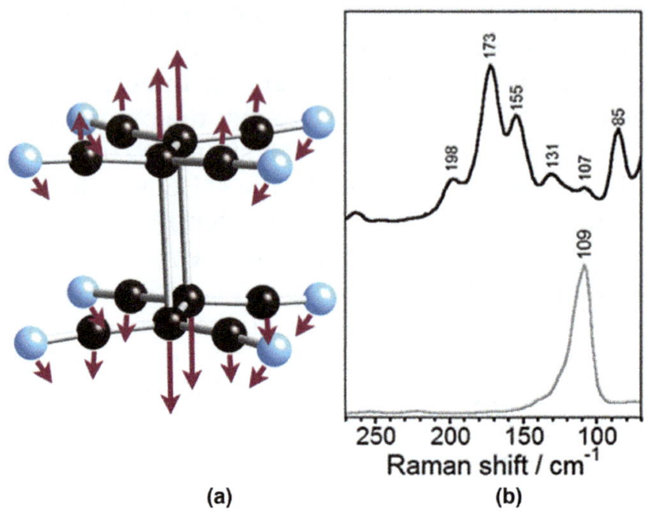

Figure 17.3   Raman active intradimer bending mode that is observed at 530 cm^{-1} (a), and low frequency Raman spectra for π-[TCNE]$_2^{2-}$ (top spectrum) and [TCNE]$^{\bullet-}$ (bottom spectrum) (b).

estimated bond energy from the computed 0.45 mdyn/Å stretching force constant is $\sim$15 kcal mol^{-1}.[24]

[TCNE]$_2^{2-}$ unlike [TCNE]$^{\bullet-}$ [32] is diamagnetic.[4] The diamagnetic nature of [TCNE]$_2^{2-}$ is validated from the magic angle spinning ^{13}C NMR spectra.[33] The $\delta_\perp$, $\delta_\perp{}'$, $\delta_\parallel$ and average δ_{iso} tensors for π-[TCNE]$_2^{2-}$ with a nominal central C atom charge of 1/2$-$ lie between the nominal central-C atom charges of zero and 1$-$ for [TCNE]^n ($n = 0$, 2$-$) respectively, Figure 17.4.

The structural (intradimer separation of 2.895 Å, 5.0° bending away of the nitriles from the nominal plane of the central sp^2-like carbons atoms), spectroscopic (new, lower energy electronic absorption; new ν_{CC} and new and shifted ν_{CN} and δ_{CCN} vibrational absorptions as well as Raman absorptions, and characteristic ^{13}C NMR $\delta_\parallel$, $\delta_\perp$, $\delta_\perp{}'$, and average, δ_{iso}, chemical shifts), and magnetic (diamagnetic ground state) data collectively indicate the formation of a 20-atom π-[TCNE]$_2^{2-}$ species. This is in accord with Pauling's definition "…there is a chemical bond between two atoms or groups of atoms in the case that the forces acting between them are such as to lead to an aggregate with sufficient stability to make it convenient for the chemist to consider it as an independent molecular species."[34,35] Thus, a bond is present between the two fragments. Due to its structural characteristics it is termed a long, multicenter CC bond (LMB).

The previous results on the LMB can be rationalized according to the following qualitative theoretical analysis: (a) all LMBs have a

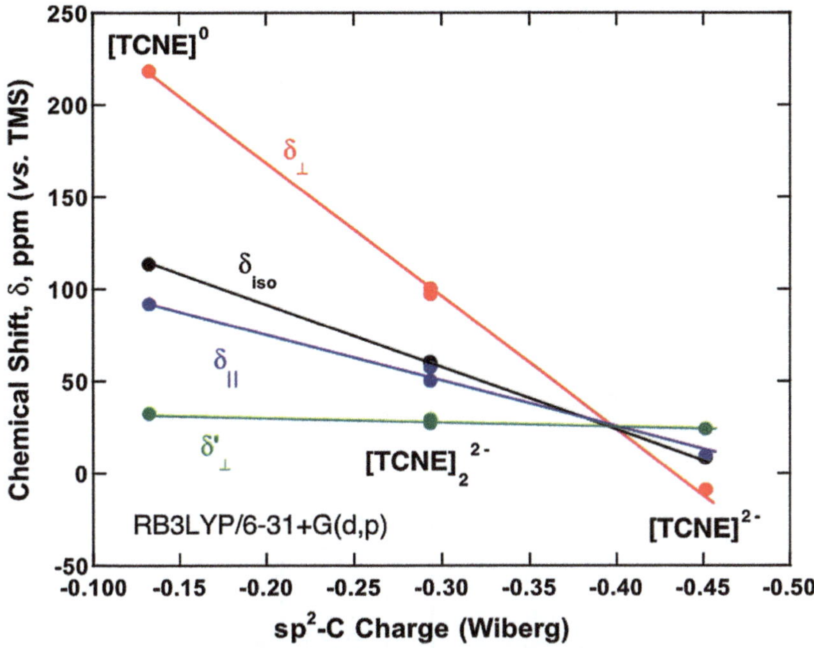

Figure 17.4 Chemical shifts of $\delta_\parallel$, $\delta_\perp$, $\delta_\perp'$, and average δ_iso tensors for TCNE, [TCNE]$^{2-}$, and [TCNE]$_2^{2-}$ as a function of Wiberg charge on the central C atoms.[33]

shared pair of electrons, one from each radical fragment, hosted in the singly occupied MO (SOMO) of such fragment; (b) due to the crystal packing, the SOMOs of both interacting fragments, the [TCNE]$^{•-}$ radical anions in π-[TCNE]$_2^{2-}$, present a non-negligible overlap; (c) similarly to what occurs for all covalent bonds (*e.g.*, $2\,H^{•} \rightarrow H_2$ or $2\,H_3C^{•} \rightarrow C_2H_6$), as a result of the overlap of the two SOMOs, this leads to the formation of a bonding (SOMO + SOMO) and an anti-bonding (SOMO − SOMO) combination in [TCNE]$_2^{2-}$, and the two electrons spin pair (*i.e.*, the SOMO + SOMO orbital becomes doubly occupied and the SOMO − SOMO orbital becomes empty). Such electronic configuration is of the closed shell type, and is associated with a closed shell singlet ground state. As for conventional covalent bonds, the electron pairing in π-[A]$_2^{2-}$ is associated with an energy gain, as experimentally observed (*vide supra*). However, the results of accurate quantitative computational analyses indicate that LMBs differ from conventional 2e$^-$/2c covalent bonds in the lack of stability of their isolated dimers (Table 17.1), due to the presence of repulsive electro-static and dispersion components, whose sum is stronger than the bonding component. It is also worth pointing out that LMBs are longer

Table 17.1 Main energetic energy components of the LMB found in the π-dimers of $[A]^{\bullet-}$, computed at the CAS(2,2)/MRMP2 level: $E_{ST} = E(S) - E(T)$ (negative values indicate that the singlet is more stable); n_a and n_{ab} are the electron orbital occupation numbers for the dimer SOMO+SOMO and SOMO−SOMO; E_{er}, E_{el}, E_{disp}, and E_{bond} are respectively the exchange-repulsion, electrostatic, dispersion and bonding components of the interaction energy; $E_{int}(2)$ is the interaction energy for the dissociation $[A_2]^{2-}$ into 2 $[A]^{\bullet-}$; and $E_{int}(4)$ is the interaction energy for (cation)$_2$A$_2$. The value of the shortest intradimer C···C distance is also given (in Å; all energies in kcal mol^{-1}). For comparison, the components were also computed for the interaction between two $^{\bullet}CH_3$ radicals producing ethane.

Dimer	$r_{c\cdots c}$	Basis	E_{ST}	n_b	n_{ab}	E_{er}	E_{el}	E_{disp}	E_{bond}	$E_{int}(2)$	$E_{int}(4)$
$^{\bullet}CH_3\cdots CH_3$	1.512[a]	6–31g(d)	−299.6	1.99	0.01	215.7	3.4	−12.2	−315.3	−108.4	n. a.
$\pi\text{-}[TCNE]_2^{2-}$	2.890[b]	cc-pVTZ	−14.1	1.63	0.37	20.1	60.8	−25.8	−12.1	+43.0	−200.6
$\pi\text{-}[TCNQ]_2^{2-}$	3.311[c]	6–31g(d)	−7.8	1.61	0.39	17.3	48.6	−26.3	−7.5	+32.1	−115.0

[a]From the optimum value for the C–C distance in ethane, computed at the B3LYP/aug-ccpVTZ level.
[b]From the 20 K structure of K$_2$[TCNE]$_2$.[24]
[c]From the structure low temperature polymorph of K$_2$[TCNQ]$_2$.[60]

and weaker than conventional covalent bonds. Typical single CC covalent bonds have average bond energies of ~ 83 kcal mol^{-1} [36] and are ~ 1.54 Å in length,[37] while a CC LMB has a bond energy of $\lesssim 15$ kcal mol^{-1} (*vide supra*), and a length of ~ 2.9 Å, *i.e.*, twice as long as a single CC covalent bond, but substantially shorter than the 3.5 Å expected for van der Waals interactions.[38] LMBs are also different to ionic bonds, where the dispersion and bonding components do not make any physically meaningful contribution.

In addition to the $\pi\text{-}[TCNE]_2{}^{2-}$ LMB, they have been observed for $\pi\text{-}[TCNQ]_2{}^{2-}$ (TCNQ = 7,7,8,8-tetracyanoquino-*p*-dimethane, Figure 17.5),[39] $\pi\text{-}[\text{cyanil}]_2{}^{2-}$,[40] $\pi\text{-}[TCNB]_2{}^{2-}$ (TCNB = 1,2,4,5-tetracyanobenzene, Figure 17.5),[41] and $\pi\text{-}[TCNP]_2{}^{2-}$ (TCNP = 1,2,4,5-tetracyanopyrazine, Figure 17.5).[41,42] LMBs have also been reported for radical cations, *e.g.*, $\pi\text{-}[TTF]_2{}^{2+}$,[20,43,44] neutral radicals, *e.g.*, [2,5,8-tri-*t*-butylphenalenyl]$_2$,[45,46] as well as for zwitterionic [TTF$^{\delta+}\cdots$TCNE$^{\delta-}$].[20] The LMBs in neutral radical dimers differ from that in anionic and cationic π-dimers, as they lack significant electrostatic interactions. Note that the bonding component of the LMB arises from π*–π*, π–π, π*–π, and nb-nb (nb = nonbonding) overlap for the anionic, cation, neutral, and zwitterionic dimers, respectively.

Herein, the properties and general trends of LMBs obtained from computational studies are compared for $\pi\text{-}[TCNE]_2{}^{2-}$ and $\pi\text{-}[TCNQ]_2{}^{2-}$, the two most commonly reported $\pi\text{-}[A]_2{}^{2-}$ (A = TCNE, TCNQ) dimers. These theoretical studies combined the results from first principles computations and a qualitative analysis on the nature of bonds. The

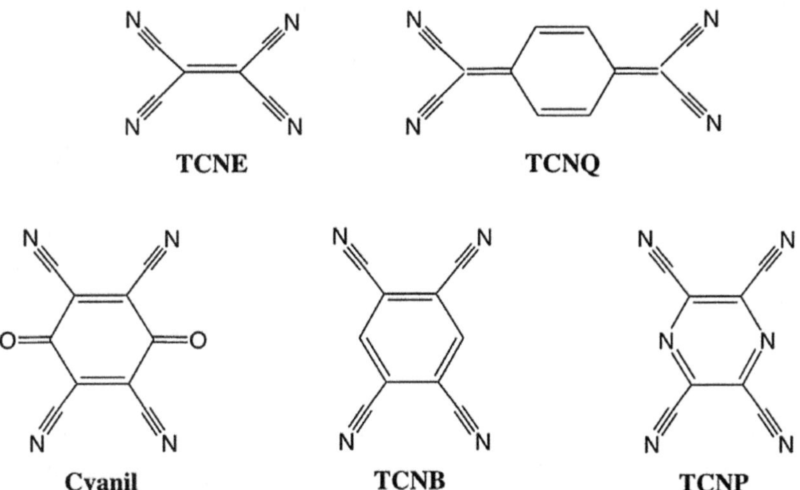

Figure 17.5 Structure of TCNE, TCNQ, cyanil, TCNB, and TCNP.

computations were done with various DFT functionals[47] and also at the CAS(n,m)/MRMP2 level, a high-level *ab initio* method,[48] capable of reproducing the experimental results, and those from CAS(n,m) calculations using all orbitals and electrons in the active space. The CAS(n,m)/MRMP2 method is a computationally demanding multireference perturbative method, similar to the CAS(n,m)/CASPT2 method,[49] and was used in all single-point calculations.[50] Geometry optimizations were done with the less demanding CASSCF(n,m)/NEVPT2 multireference perturbative method.[51] The results of all of these calculations revealed the multireferent character (*i.e.*, more than one determinant is required to represent the electronic structure, as in the case of ozone) of the closed shell singlet state for π-[TCNE]$_2^{2-}$ and π-[TCNQ]$_2^{2-}$.

17.2 A Computational Analysis of the Energetic Components of an Intermolecular Bond

The interaction energy, E_{int}, or bond energy, in principle, can be experimentally measured or theoretically computed. Extreme types of bonds are: covalent, ionic, and hydrogen bonds, and van der Waals interactions, and they can be distinguished *via* an analysis of the components of E_{int}. Furthermore, to gain deeper insights into the differences between covalent and LMB bonds, an analysis of the components of E_{int} for $[\text{A}]_2^{2n\pm}$ formed as a consequence of the creation of the new bond formally linking two $[\text{A}]^{\bullet n\pm}$ radical ions, σ or π, was undertaken. Using the basic principles of the Hayes–Stone InterMolecular Perturbation Theory (IMPT) energy decomposition analysis,[52] E_{int} between a pair of radicals ($\text{A}^\bullet + \text{B}^\bullet$) can be written as:

$$E_{\text{int}} = E_{\text{er}} + E_{\text{el}} + E_{\text{pol}} + E_{\text{ct}} + E_{\text{disp}} + E_{\text{bond}}, \tag{17.1}$$

where (a) E_{er} is the *exchange repulsion energetic component* (originating from the Pauli Exclusion Principle, which dominates at short distances and is the cause of the so-called repulsive wall); (b) E_{el} is the *electrostatic interaction component* between $\text{A}^\bullet$ and $\text{B}^\bullet$ whose multipolar distribution is taken as those for the isolated fragments, (c) E_{pol} is the *polarization energetic* component [it accounts for the change in electrostatic component that $\text{A}^\bullet$ (or $\text{B}^\bullet$) caused by the multipolar distribution of $\text{B}^\bullet$ (or $\text{A}^\bullet$)]; (d) E_{ct} is the *charge transfer* component (only relevant in donor–acceptor complexes); (e) E_{disp} is the *dispersion component* (which accounts for the difference in the correlation energy of AB and the $\text{A}^\bullet$ and $\text{B}^\bullet$), and the E_{bond} component (it accounts for the gain in energy due to the electron pairing, in AB, of the unpaired electrons of the $\text{A}^\bullet$

and $B^\bullet$; it is an energetic component only present when the interacting fragments are radicals. Except for highly polarizable molecules, E_{pol} is much smaller than E_{el} and it can be disregarded for a qualitative analysis of the dominant energetic components, and, thus,

$$E_{int} \approx E_{er} + E_{el} + E_{disp} + E_{bond}. \tag{17.2}$$

E_{disp} can be estimated from eqn (17.3) based on Löwdin's definition,[53]

$$E_{disp} = E_{int}[\text{Full-CI}] - E_{int}[\text{HF/CASSCF(min)}], \tag{17.3}$$

where $E_{int}(\text{HF/CASSCF(min)})$ is the energy obtained by the smallest CASSCF(min) active space needed to describe the electronic structure of the dimers and their corresponding fragments and their known properties (for the dimers of interest herein, it is a CAS(2,2) space that includes the bonding and antibonding combination of the monomer SOMOs and their two electrons). Note that when only one determinant is needed to describe the dimer wavefunction, as is the case in most closed shell systems, it can be estimated from a Hartree–Fock (HF) calculation, and for open shell systems it can be estimated from either an Unrestricted Hartree–Fock (UHF) or Restricted-Open Hartree–Fock (ROHF) computation. $E_{int}[\text{Full-CI}]$ is the energy computed when doing a Full-CI (a computation where all possible determinants are considered). However, as Full-CI computations are too demanding for extended monoelectronic basis sets, $E_{int}[\text{full-CI}]$ is usually taken as the total energy from either a CASPT2 or RASPT2 computation using the largest active space that one can afford. Thus,

$$E_{disp} = E_{int}[(\text{AB})\text{CASPT2(large)}] - E_{int}[\text{HF/CASSCF(min)}]. \tag{17.4}$$

17.3   Long, Multicenter Bond in π-$[\text{TCNE}]_2{}^{2-}$ Dimers

A bond implies a minimum energy structure and CAS(n,m)/MRMP2 is used herein. It is a high-level *ab initio* method designed to reproduce the results of a full-CI calculation[54] – by definition, the exact calculation for a given electronic basis set. If a minimum energy structure cannot be found at the CAS(n,m)/MRMP2 level or using its approximate variation the CAS(n,m)/NEVPT2 method, from a computational perspective a bond does not exist.[55]

The results from a CAS(n,m)/NEVPT2 geometry optimization of π-$[\text{TCNE}]_2{}^{2-}$, using the structure of $K_2[\text{TCNE}]_2$ as the basis, reveals a metastable minimum in the potential energy surface with a short intradimer $C\cdots C$ distance and with a D_{2h} orientation (Figure 17.6a,b)

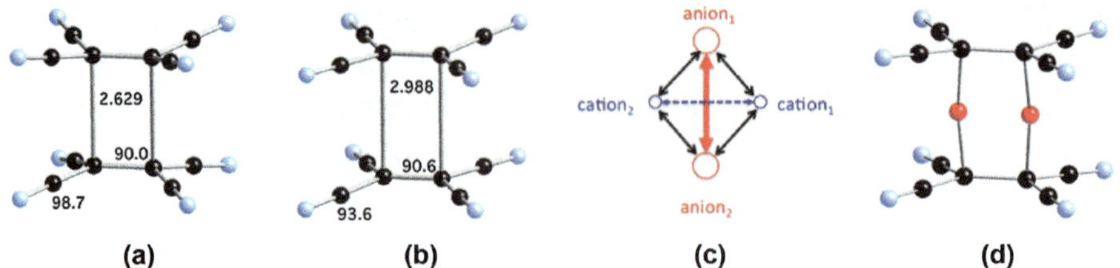

Figure 17.6 (a) Fully optimized structure of an isolated $\pi\text{-}[\text{TCNE}]_2^{2-}$ dimer computed at the CAS(2,2)/NEVPT2/aug-cc-pVTZ level; (b) structure of $\pi\text{-}[\text{TCNE}]_2^{2-}$ observed for $K_2[\text{TCNE}]_2$,[24] taken as initial structure for the geometry optimization; (c) cation₂anion₂ interactions: one anion···anion (thick solid arrow), one cation···cation (broken line arrow), and four cation···anion (thin solid arrows) interactions; (d) bond critical points found in an atoms-in-molecules analysis of the CAS(2,2)/NEVPT2/aug-cc-pVTZ density for an isolated $\pi\text{-}[\text{TCNE}]_2^{2-}$ dimer. To compare the structures in (a) and (b), the shortest intradimer C···C distances are equal in (a) and (b) the $\angle$C···C–C angle, and the average value of the $\angle$NC)···C–C···C dihedral angle are noted (distances in Å, angles in degrees).

as observed. This does not exclude that other minimum energy structure(s) could exist for this dimer. Also, due to the computed barrier with respect to dissociation (Figure 17.7a), the rupture of the D_{2h} dimer into fragments requires energy, in accord with the presence of an intradimer bond. These computational results explain the aforementioned spectroscopic and magnetic data (as in conventional covalent bonds, the dimer is diamagnetic while the interacting fragments are paramagnetic) and an intradimer bond exits for $\pi\text{-}[\text{TCNE}]_2^{2-}$, which is termed LMB due to its characteristic properties.

The computed metastability of $\pi\text{-}[\text{TCNE}]_2^{2-}$ E_{int} is $+43$ kcal mol^{-1} while the computed energy to break $\pi\text{-}[\text{TCNE}]_2^{2-}$ in $(K^+)_2(\pi\text{-}[\text{TCNE}]_2^{2-})$ is -200.6 kcal mol^{-1} and it lacks a barrier towards dissociation, *i.e.*, it is thermodynamically and kinetically stable. The computed origin of the stability of $\pi\text{-}[\text{TCNE}]_2^{2-}$ in $(K^+)_2(\pi\text{-}[\text{TCNE}]_2^{2-})$ is identified when considering all of the pairwise interactions (Figure 17.6c), *i.e.*, the E_{int} approximately results from the sum of the four cation···anion interactions (each -75.4 kcal mol^{-1}) that exceed the sum of the cation···cation (58.0 kcal mol^{-1}) plus anion···anion repulsions (43.0 kcal mol^{-1}), with the small difference attributed to polarization effects. Thermodynamic and kinetic stability was also computationally noted for $(\pi\text{-}[\text{TCNE}]_2^{2-})$ solvent$_n$ for $n>4$ (solvent $=$ dichoromethane), in accord with the observation in dichoromethane solution at low temperature.[4,56] Based on the similarity of spectra found for $\pi\text{-}[\text{TCNE}]_2^{2-}$ in solution at low temperature and in the solid, their geometry is presumed to be similar.[4]

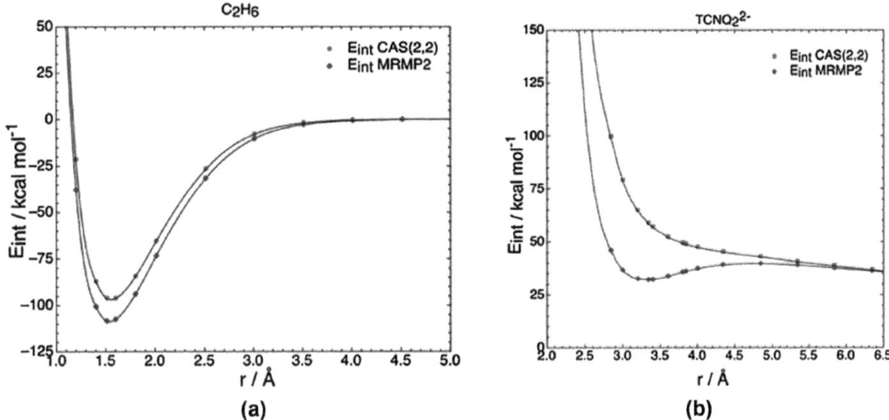

Figure 17.7 E_{int} (in kcal mol^{-1}) as a function of the shortest intradimer C–C distance (r in Å for the formation of: (a) the C–C bond in ethane by addition of two $^{\bullet}CH_3$ radicals and (b) the LMB in π-[TCNE]$_2^{2-}$ by addition of two [TCNE]$^{\bullet-}$ radicals (note the metastable minimum); The two curves are computed at the CASSCF(2,2) and CAS(2,2)/MRMP2 levels (the second one results from the first after adding the dynamical correlation computed using the MRMP2 method).

The analysis of the optimized geometry of π-[TCNE]$_2^{2-}$ (Figure 17.6a) shows (1) its D_{2h} symmetry, which allows the simultaneous formation of two $C\cdots C$ bonding components, where the $\angle C–C\cdots C–C$ is 0° and all atoms are in an eclipsed (or "on top") conformation; (2) a shortest $C\cdots C$ interfragment distance, 2.629 Å, much smaller than the sum of van der Waals radii (3.6 Å)^{38} and that structurally observed; (3) a pyramidalization of the $C_{central}$ atoms of each [TCNE]$^{\bullet-}$, while the C and N atoms of the four cyano groups are collinearly arranged with the $C_{central}$ atoms to which they are bonded; (4) an increase in the $\angle C\cdots C–C–C$(cyano) to 95° leading to pyramidalization of the $C_{central}$ atoms. These are due to the presence of LMBs and, can be taken as signatures of a bond. Except for the interdimer separation being 9%, shorter than observed, the remaining parameters have similar values to those found in the experimental structure of π-[TCNE]$_2^{2-}$.

The results from numerical simulations on the cation$_2$anion$_2$ suggest that the 9% difference between the optimized and experimental structures of π-[TCNE]$_2^{2-}$ is not due to the poor basis set or approximate computational method, but is due to the polarization of the [TCNE]$^{\bullet-}$ electron distribution caused by the presence of cations, which cannot be found when the anion and cations are both monoatomic. Polarization increases the four dominating $K^+\cdots$[TCNE]$^{\bullet-}$ attractions, which for π-[TCNE]$_2^{2-}$ also increase the anion$\cdots$anion repulsion. As a result, the anion$\cdots$anion separation also increases when cations are also present.

The interaction energy, E_{int}, is a measure of the energy difference between dimer minus that of its two dissociated fragments, and it distinguishes the interaction between the two species, which can be called $E_{\text{int}}(2)$. Similarly, $E_{\text{int}}(4)$ measures the stability of the cation$_2$(π-[TCNE]$_2{}^{2-}$) aggregate, relative to its four fragments. An analysis of the components of $E_{\text{int}}(2)$ (E_{er}, E_{el}, E_{disp}, and E_{bond}) for an isolated π-[TCNE]$_2{}^{2-}$ dimer was done to understand the nature of LMBs. The results for the relevant terms to define the stabilizing character of the $E_{\text{int}}(2)$ (E_{er} is always destabilizing for any dimer and relative orientations) indicate that (a) $E_{\text{er}} = 20.1$ kcal mol^{-1} or $0.47|E_{\text{int}}(2)|$, (b) $E_{\text{el}} = 60.8$ kcal mol^{-1} or $1.41\ |E_{\text{int}}(2)|$, (c), $E_{\text{disp}} = -25.8$ kcal mol^{-1} or $-0.6|E_{\text{int}}(2)|$, and (d) $E_{\text{bond}} = -0.12$ kcal mol^{-1} or $-0.28|E_{\text{int}}(2)|$, Table 17.1. Thus, the value of $E_{\text{int}}(2)$ is dominated by the repulsive electrostatic component originating from the monoanionic character of the interacting fragments. E_{bond} is significantly smaller, but it is the main factor that governs the preference of these dimers for a closed shell electronic ground state. The addition of two cations, as in $(\text{K}^+)_2(\pi$-[TCNE]$_2{}^{2-})$, creates four additional cation$\cdots$anion attractive interactions, which is computed to be thermodynamically stable. The orbitals from each [TCNE]$^{\bullet-}$ interact forming bonding and antibonding combinations. Except for the combinations originating from the SOMO, the bonding and antibonding combinations are doubly occupied, and do not contribute to the total energy of $(\text{K}^+)_2(\pi$-[TCNE]$_2{}^{2-})$. However, there are four ways of placing the two electrons from the two SOMOs (Figure 17.8a): (1) in the closed shell singlet state, CSS$_1$, where the two electrons occupy the bonding combination (SOMO + SOMO); (2) in the closed shell singlet state, CSS$_2$, where the two electrons are in the antibonding combination (SOMO − SOMO); (3) in the open shell singlet state, OSS, where one electron is in the SOMO + SOMO and the other in the SOMO − SOMO combination, one with spin α and the other with spin β; and (4) a triplet state, T, similar to the OSS but where both electrons have spin α. Due to the energy gain of the SOMO + SOMO relative to the SOMO (an energetic gain of Δ relative to the non-interacting anion radicals), when the two electrons in the two SOMO orbitals are in the closed shell singlet configuration (CSS$_1$ in Figure 17.8a) it results in a total energy gain of 2Δ relative to the non-interacting fragments energy. In contrast, the OSS and the T configurations have no net gain, while CSS$_2$ is energetically disfavored. Thus, despite its small weight in the interaction energy of the bonding component, it dictates the ground state as being the diamagnetic CSS$_1$.

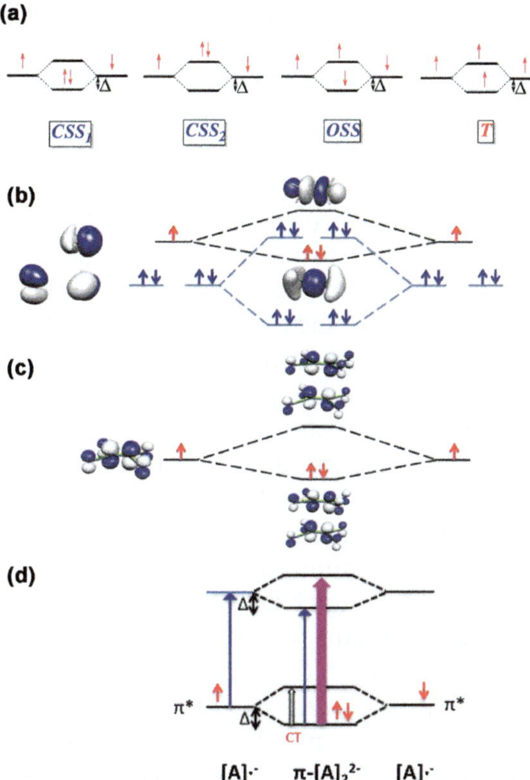

Figure 17.8 (a) Electron distribution among the HOMO, SOMO ± SOMO HOMO in the CSS$_1$ and CSS$_2$, OSS, and T states (b) MO diagrams for the interaction of two [TCNE]$^{\bullet -}$ forming π-[TCNE]$_2^{2-}$ and (c), two $^{\bullet}$CH$_3$ radicals forming ethane (d), computed lowest energy transitions expected in the UV-vis spectra of the [A]$^{\bullet -}$ and π-[A]$_2^{2-}$ originating from the SOMO orbital, in the monomer, and the dimer SOMO ± SOMO combinations.

A qualitative rationalization of the UV-vis spectra in Figure 17.1 for [TCNE]$^{\bullet -}$ and π-[TCNE]$_2^{2-}$ is given in Figure 17.8d. The low energy absorption for [TCNE]$^{\bullet -}$, possessing 17 vibrational overtones, occurs at 428 nm and is assigned to the e$_u$ SOMO → a$_g$ LUMO transition (see Figure 17.8d).[30] The lowest energy transition observed in the dimer is assigned to a dimer b$_{2u}$ HOMO (SOMO + SOMO) to b$_{1g}$ LUMO (SOMO − SOMO) transition (see Figure 17.8d) and is in accord with the observed 583 nm absorption (Figure 17.1).[4] π-[TCNE]$_2^{2-}$ exhibits a second absorption at 400 nm that is assigned to the allowed dimer HOMO b$_{2u}$ → LUMO + 1 b$_{1g}$ transition.

Due to the 1− net charge on each [TCNE]$^{\bullet -}$ fragment, E_{el} for π-[TCNE]$_2^{2-}$ is repulsive, and as a consequence the computed total energy is greater than that for two dissociated [TCNE]$^{\bullet -}$ fragments (Figure 17.7(b)). Thus, due to the $1/r$ dependence with the interdimer distance (r), at large distances, $E_{int} = E_{el}$; however, at short distance the $E_{bond} + E_{disp}$ also contributes to E_{int}. This leads a local energy minimum, separated by a barrier in the pathway from the minimum of the potential energy curve towards dissociation into two [TCNE]$^{\bullet-}$ fragments (Figure 17.7a). This barrier originates from the different sign of the $E_{bond} + E_{disp}$ and E_{el} components.[57]

The observed diamagnetic ground state of π-[TCNE]$_2^{2-}$ is replicated at all levels of computation tested (Hartree–Fock, B3LYP, or M06L density functionals, or CAS(2,2), CAS(2,2)/CASPT2, CAS(20,22)/MRMP2 computations, the latter of near-full CI quality on the space[58]). The computed ground state electronic structure for π-[TCNE]$_2^{2-}$ using any of the CAS(n,m)/MRPT2 methods is a singlet. However, rather than being described by one determinant as in the closed shell singlet determinant used in the Restricted Hartree–Fock (RHF) method (where the SOMO + SOMO and SOMO − SOMOs have orbital occupation numbers of 2.0 and 0.0, respectively), it is a mixture of determinants dominated by the lowest energy CSS$_1$ and the lowest energy open shell singlet OSS determinant.[59] with a relative weight of the CSS determinant of $\sim$80% in near-full CI calculations (this result is reproduced by a CAS(2,2) calculation, or by a CAS(6,4) calculation). This multireferent character is manifested in the fractional orbital occupation numbers for the SOMO + SOMO and SOMO − SOMOs shown in Figure 17.8b. Nonetheless, the basic features of the SOMO/SOMO overlap model of Figure 17.8b remain the same, independent of the multireferent character of the ground state.

MO calculations identify the long bond as a 2e$^-$ bond. The number of its centers in these MO computations was determined by doing an atoms-in-molecules (AIM) analysis[61] of the total density of π-[TCNE]$_2^{2-}$. When such analysis is done using the wavefunction computed in CAS(2,2)/MRMP2 and CAS(6,6)/MRMP2 computations (Figure 17.6d) two bond critical points are found, each connecting the two C$_{central}$ atoms. Identical AIM results are obtained when the analysis is performed at the HF, B3LYP and M06L levels. That is equivalent to the results from MO computations *i.e.* the long bond for π-[TCNE]$_2^{2-}$ has two C$\cdots$C bonding components, each involving two atoms. Therefore, the long bond is a 2e$^-$/4c bond from an MO analysis.

The MO diagrams for π-[TCNE]$_2^{2-}$, Figure 17.8b, and ethane from two $^{\bullet}$CH$_3$ radicals (Figure 17.8c) are similar. However, a comparison of $E_{int}(r)$ for the interacting radicals (Figure 17.7a and b) reveals that they have different types of interactions. This is also observed from an analysis of the components of the E_{int} (Table 17.1). Therefore, as

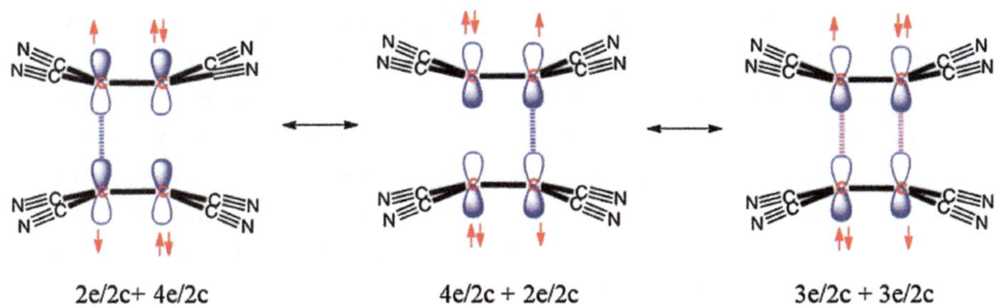

Figure 17.9 VB analysis of the electronic structure of an isolated π-[TCNE]$_2^{2-}$ dimer.

noted above, LMBs in π-[TCNE]$_2^{2-}$ differ from conventional covalent bonds such as in H_3C–CH_3.

The nature of the LMB in π-[TCNE]$_2^{2-}$ can also be analyzed from a Valence Bond (VB) perspective.[62] In a VB analysis the structure of π-[TCNE]$_2^{2-}$ can be written in two equivalent forms each involving the formation of a $2e^-/2c$ $C\cdots C$ bond and a $4e^-/2c$ non-bonding interaction (Figure 17.9). Thus, the long bond in a π-[TCNE]$_2^{2-}$ is a multicenter $6e^-/4c$ bond from the combination of two equivalent $2e^-/2c$ $C\cdots C$ components' resonant form, and it is thus an example of multicenter bonding. The D_{2h} symmetry is concomitant with the two equivalent resonant forms. The previous VB representation is equivalent to the results from a CAS(6,4) calculation (*i.e.*, $6e^-$ and 4 orbitals) when the four orbitals in the active space are the bonding and antibonding combinations of the SOMO and HOMO of each [TCNE]$^{\bullet-}$ (the MO space that results from the bonding and anti-bonding combinations of the two p orbitals perpendicular to the plane in each [TCNE]$^{\bullet-}$, which contain a total of 6 electrons). The MO diagram resulting from these CAS(6,4) calculations reproduce the MO diagram of Figure 17.8, with occupation numbers 1.98, 1.95, *1.70*, *0.32*, 0.03, and 0.02 electrons, with the middle two values corresponding to the SOMO $\pm$ SOMO orbitals (highlighted in italics, results consistent with the results from a CAS(2,2) calculation, see the values of n_a and n_b in Table 17.1).[63] Note that at the CAS(6,4) level the MO and VB descriptions are equivalent as CAS(6,4) computations discount the $2e^-\cdots 2e^-$ interactions in the description of the LMB.

In summary, the properties a LMB were described for the π-[TCNE]$_2^{2-}$ dimer, where they were first identified, and this is a model system for other examples where this type of bond occurs. The evidence for LMBs is provided by structural, spectroscopic, and magnetic data. The results for this computational analysis replicates the geometrical, spectroscopic, and magnetic features indicative of an LMB. The bond originates from the pairing of the unpaired electrons present in the fragments, as a consequence of the non-negligible overlap of the fragment orbitals. The bond involves $2e^-$ and 4 centers (2e/4c) whose nature is dominated by the electrostatic, dispersion, and bonding components. This is different to a conventional covalent, a purely ionic, or a van der Waals interaction, and cannot be a hydrogen bond in $\pi\text{-}[TCNE]_2^{2-}$ dimers, due to its lack of hydrogen atoms. The LMB name is indicative of the long intermolecular separation between the radical anions, and of multicenter character. Additionally, the interaction of $\pi\text{-}[TCNE]_2^{2-}$ with adjacent cations was computationally identified as being important for the thermodynamic and kinetic stability (isolated $\pi\text{-}[TCNE]_2^{2-}$ dimers are minimum energy but metastable species). Although the strength of each cation $\cdots$ $(\pi\text{-}[TCNE]_2^{2-})$ interaction is smaller than the (radical anion) $\cdots$ (radical anion) repulsion, their combined strength is larger, and $(\text{cation}^+)_2 \cdots$ $(\pi\text{-}[TCNE]_2^{2-})$ is stable.

17.4 Conclusions

A systematic analysis of the properties of the long multicenter bond in the prototypical $\pi\text{-}[TCNE]_2^{2-}$ dimer, where the long, multicenter bond connects two radical anions, has been carried out using high level *ab initio* methods (CAS(n,m), CAS(n,m)/MRMP2, CAS(n,m)/ NEVPT2 methods). The results obtained show that long, multicenter bonds (LMBs) share the most fundamental properties of intermolecular bonds (namely, their strength and equilibrium distance, their dependence on electrostatic and dispersion components), and also have a weak bonding component that is characteristic of covalent bonds (namely, their origin results from the overlap of the orbitals of the interacting fragments, as well as their dependence on the spin multiplicity of the aggregate). As a result, it is possible to talk about n-electron m-centered intermolecular bonds in a given spin state multiplicity. LMBs are not van der Waals bonds as they lack a bonding component. Also, they cannot be electrostatic bonds due to the presence of important bonding and dispersion components.

Acknowledgements

F. M. and J. J. N. thank MINECO (MAT2011-25972 and MAT2014-54025-P) and the Catalan Autonomous Government (2014SGR1422) for funding, and also the allocation of computer time on their machines made by CSUC and BSC.

References

1. J. J. Novoa, P. Lafuente, R. E. Del Sesto and J. S. Miller, *Angew. Chem., Int. Ed.*, 2001, **40**, 2540.
2. R. E. Del Sesto, J. S. Miller, P. Lafuente and J. J. Novoa, *Chem. – Eur. J.*, 2002, **8**, 4894.
3. J. J. Novoa, P. Lafuente, R. E. Del Sesto and J. S. Miller, *CrystEngComm*, 2002, **4**, 373.
4. I. Garcia-Yoldi, F. Mota and J. J. Novoa, *J. Comput. Chem.*, 2007, **28**, 326.
5. J. S. Miller and J. J. Novoa, *Acc. Chem. Res.*, 2007, **40**, 189.
6. σ-[TCNE]$_2^{2-}$ dimers: (a) K. I. Pokhodnya, M. Bonner, A. G. DiPasquale, A. L. Rheingold and J. S. Miller, *Chem. – Eur. J.*, 2008, **14**, 714–720; (b) J. Zhang, L. M. Liable-Sands, A. L. Rheingold, R. E. Del Sesto, D. C. Gordon, B. M. Burkhart and J. S. Miller, *J. Chem. Soc., Chem. Commun.*, 1998, 1385–1386; (c) G. Wang, H. Zhu, J. Fan, C. Slebodnick and G. T. Yee, *Inorg. Chem.*, 2007, **47**, 9461; (d) E. Shurdha, S. H. Lapidus, P. W. Stephens, C. E. Moore, A. L. Rheingold and J. S. Miller, *Inorg. Chem.*, 2012, **51**, 9655–9665; (e) J.-H. Her, P. W. Stephens, K. I. Pokhodnya, M. Bonner and J. S. Miller, *Angew. Chem. Int. Ed.*, 2007, **46**, 1521–1524; (f) K. H. Stone, P. W. Stephens, A. C. McConnell, E. Shurdha, K. I. Pokhodnya and J. S. Miller, *Adv. Mater.*, 2010, **22**, 2514–2519.
7. σ-[TCNQ]$_2^{2-}$ dimers: (a) W. E. Buschmann, A. M. Arif and J. S. Miller, *Angew. Chem. Int. Ed.*, 1998, **37**, 783; (b) S. Mikami, K. Sugiura, J. S. Miller and Y. Sakata, *Chem. Lett.*, 1999, 413; (c) S. K. Hoffmann, P. J. Corvan, P. Singh, C. N. Sethulekshmi, R. M. Metzger and W. E. Hatfield, *J. Am. Chem. Soc.*, 1983, **105**, 2612; (d) V. Dong, H. Endres, H. J. Keller, W. Moroni and D. Nöthe, *Acta Cryst.*, 1977, **B33**, 2428; (e) H. Zhao, R. A. Heintz, X. Ouyang, K. R. Dunbar, C. F. Campana and R. D. Rogers, *Chem. Mater.*, 1999, **11**, 736; (f) B. Morosin, H. J. Plastas, L. B. Coleman and J. M. Stewart, *Acta Cryst. B*, 1978, **34**, 540; (g) T. P. Radhakrishnan, D. Van Engen and Z. G. Soos, *Mol. Cryst. Liq. Cryst. B*, 1987, **150**, 473; (h) K. J. Nelson, A. R. Arif, J. S. Miller (unpublishedresults).
8. V. Zaitsev, S. V. Rosokha, M. Head-Gordon and J. K. Kochi, *J. Org. Chem.*, 2006, **71**, 52e.
9. G. Wang, C. Slebodnick and G. T. Yee, *Inorg. Chim. Acta*, 2008, **361**, 3593.
10. J. S. Miller, D. M. O'Hare, A. Chakraborty and A. J. Epstein, *J. Am. Chem. Soc.*, 1989, **111**, 7853.
11. S. V. Rosokha, B. Lorenz, T. Y. Rosokha and J. K. Kochi, *Polyhedron*, 2009, **28**, 4136.
12. J.-M. Lu, S. V. Rosokha and J. K. Kochi, *J. Am. Chem. Soc.*, 2003, **125**, 12161.
13. H. Bock, C. Nather and K. Ruppert, *Z. Anorg. Allg. Chem.*, 1992, **614**, 109.
14. G. Wang, H. Zhu, J. Fan, C. Slebodnick and G. T. Yee, *Inorg. Chem.*, 2006, **45**, 1406.
15. R. E. Del Sesto, M. Botoshansky, M. Kaftory and J. S. Miller, *CrystEngComm*, 2002, **4**, 106.
16. R. E. Del Sesto, R. D. Sommer and J. S. Miller, *CrystEngComm.*, 2001, 47.
17. J. S. Miller, D. T. Glatzhofer, C. Vazquez, R. S. McLean, J. C. Calabrese, W. J. Marshall and J. W. Raebiger, *Inorg. Chem.*, 2001, **40**, 2058.
18. H. Bock and K. Ruppert, *Inorg. Chem.*, 1992, **31**, 5094.
19. M. M. Olmstead, G. Speier and L. Szabo, *Chem. Commun.*, 1994, 541.

20. J. D. Bell, C. Moore, A. L. Rheingold and J. S. Miller, *Chem. – Eur. J.*, 2011, **17**, 9326.
21. D. Sun, S. V. Rosokha and J. K. Kochi, *J. Phys. Chem. B*, 2007, **111**, 6655.
22. D. A. Lemenovskii, R. A. Stukan, B. N. Tarasevich, Yu. L. Slovokhotov, M. Yu. Antipin, A. E. Kalinin and Yu. T. Struchkov, *Koord. Khim.*, 1981, **7**, 240.
23. H. Bock, K. Ruppert, D. Fenske and H. Goesmann, *Z. Anorg. Allg. Chem.*, 1995, **595**, 275.
24. J. Casado, S. R. González, F. J. Ramírez, J. T. L. Navarrete, S. H. Lapidus, P. W. Stephens, H.-L. Vo, J. S. Miller, F. Mota and J. J. Novoa, *Angew. Chem., Int. Ed.*, 2013, **52**, 6421.
25. G. Wang, C. Slebodnick and G. T. Yee, *Inorg. Chem.*, 2007, **46**, 9641.
26. M. T. Johnson, C. F. Campana, B. M. Foxman, W. Desmarais, M. J. Vela and J. S. Miller, *Chem. – Eur. J.*, 2000, **6**, 1805.
27. Based upon an analysis of 24 structures possessing isolated π-[TCNE]$_2^{2-}$ and 6 structures possessing isolated [TCNE]$^{\bullet-}$ with an esd ≤0.006 Å.
28. M. Strohmeier, D. H. Barich, D. M. Grant, J. S. Miller, R. J. Pugmire and J. Simons, *J. Phys. Chem. A*, 2006, **110**, 7962–7969.
29. M. Capdevila-Cortada, J. Ribas-Arino, A. Chaumont, G. Wipff and J. J. Novoa, *Chem. – Eur. J.*, 2016, **22**, 17037.
30. D. A. Dixon and J. S. Miller, *J. Am. Chem. Soc.*, 1987, **109**, 3656.
31. J. S. Miller, *Angew. Chem. Int. Ed.*, 2006, **45**, 2508.
32. W. D. Phillips, J. C. Rowell and S. I. Weissman, *J. Chem. Phys.*, 1960, **33**, 626.
33. M. Strohmeier, D. H. Barich, D. M. Grant, J. S. Miller, R. J. Pugmire, J. Simons, *J. Phys. Chem.*, 2006, **110A**, 7962.
34. L. Pauling, *The Nature of the Chemical Bond*, Cornell University Press, Ithaca, 3rd edn, 1963, p. 6.
35. Recognizing that a bond is not a force, but a (stabilizing) energy, a modified definition has been reported: "stabilizing, or attractive, interaction between atoms that significantly alters the properties of the atoms, leading to an independent or new species with new properties. ... In a more general way, a chemical bond is a construct, frequently a pictogram, used by chemists to understand the stronger attractive interactions among atoms that enable the organization and understanding of the structure, properties, reactivities, and interrelations for the growing myriad of substances. The concept of chemical bonding impacts our daily lives because it is a language enabling chemists to communicate among each other and facilitate the design and synthesis of improved substances that have benefited mankind. A chemical bond enables us to glean the order and complexity of how the basic building blocks, atoms, interact to form all substances. This language has enabled the broad enterprise of chemistry to rapidly develop and flourish into the central science." J. S. Miller, http://cen.acs.org/articles/91/i36/Trying-Explain-Bond.html.
36. R. T. Sanderson, *Chemical Bonds and Bond Energy*, Academic Press, New York, 1976.
37. L. Pauling and L. O. Brockway, *J. Am. Chem. Soc.*, 1937, **59**, 1223.
38. van der Waals and A. Bondi, *J. Phys. Chem*, 1964, **68**, 441.
39. I. Garcia-Yoldi, J. S. Miller and J. J. Novoa, *J. Phys. Chem. A*, 2009, **113**, 7124.
40. I. Garcia-Yoldi, J. S. Miller and J. J. Novoa, *PCCP*, 2008, **10**, 4106.
41. M. Capdevila-Cortada, J. S. Miller and J. J. Novoa, *Chem. – Eur. J.*, 2015, **21**, 6420.
42. J. J. Novoa, P. W. Stephens, M. Weerasekare, W. W. Shum and J. S. Miller, *J. Am. Chem. Soc.*, 2009, **131**, 9070.
43. J. B. Torrance, B. A. Scott, B. Welber, F. B. Kaufman and P. E. Seiden, *Phys. Rev. B*, 1979, **19**, 730; K. Yakushi, S. Nishimura, T. Sugano and H. Kuroda, *Acta Cryst. B*, 1980, **36**, 358; T. Sugano, H. Kuroda and K. Yakushi, *Bull. Chem. Soc. Jpn.*, 1978, **51**, 1041; A. Yamashita, H. Akutsu, J. Yamada and S. Nakatsuji, *Polyhedron*, 2005,

24, 2796; G. Matsubayashi, K. Ueyama and T. Tanaka, *J. Chem. Soc., Dalton Trans.*, 1985, 465; K. Furuta, H. Akutsu, J. Yamada and S. Nakatsuji, *Synth. Method*, 2005, **152**, 381; J. Morgado, I. C. Santos, L. F. Veiros, C. Rodrigues, R. T. Henriques, M. T. Duarte, L. Alcacer and M. Almeida, *J. Mater. Chem.*, 2001, **11**, 2108; C. Libage, M. Fourmigue, P. Batail, E. Canadell and C. Coulon, *Bull. Soc. Chim. Fr.*, 1993, **130**, 761; H. Yamochi, A. Konsha, G. Saito, K. Matsumoto, M. Kusunoki and K. Sakaguchi, *Mol. Cryst. Liq. Cryst. Sci. Technol., Sect. A*, 2001, **350**, 265; H. Akutsu, M. Masaki, K. Mori, J. Yamada and S. Nakatsuji, *Polyhedron*, 2005, **24**, 2126; M. Giffard, G. Mabon, E. Leclair, N. Mercier, M. Allain, A. Gorgues, P. Molinie, O. Neilands, P. Krief and V. Khodorkowsky, *J. Am. Chem. Soc.*, 2001, **123**, 3852; A. Slougui, L. Ouahab, C. Perrin, D. Grandjean and P. Batail, *Acta Crystallogr., Sect. C*, 1989, **45**, 388; R. C. Teitelbaum, T. J. Marks and C. K. Johnson, *J. Am. Chem. Soc.*, 1980, **102**, 2986; E. Coronado, J. R. Galan-Mascaros, C. Gimenez-Sais, C. J. Gomez-Garcia, C. Ruiz-Perez and S. Triki, *Adv. Mater.*, 1996, **8**, 737; M. D. Halling, J. D. Bell, R. J. Pugmire, D. M. Grant and J. S. Miller, J. Phys. *Chem. A*, 2010, **114**, 6622–6629.

44. I. Garcia-Yoldi, J. S. Miller and J. J. Novoa, *J. Phys. Chem. A*, 2009, **113**, 484.

45. K. Goto, T. Kubo, K. Yamamoto, K. Nakasuji, K. Sato, D. Shiomi, T. Takui, M. Kubota, T. Kobayashi, K. Yakusi, J. Ouyang and K. Nakasuji, *J. Am. Chem. Soc.*, 1999, **121**, 1619.

46. F. Mota, J. S. Miller and J. J. Novoa, *J. Am. Chem. Soc.*, 2009, **131**, 7699.

47. A. D. Becke, *J. Chem. Phys.*, 2014, **140**, 18A301.

48. K. Hirao, *Chem. Phys. Lett.*, 1992, **190**, 374.

49. B. O. Roos, K. Andersson, M. P. Fülscher, P.-O. Malmqvist, L. Serrano-Andrés, K. Pierloot, M. Merchán, *Adv. Chem. Phys.*, John Wiley & Sons, Inc., 1996, vol. 93, p. 219.

50. CAS(n,m) = Complete Active Space method where all the determinants resulting from placing n electrons in m orbitals are considered and the total energy is variationally optimized with respect to the orbitals and determinant coefficients; MRMP2 = Multireferent Second Order Møller–Plesset perturbational calculation using a Second Order Møller–Plesset method.

51. R. W. A. Havenith, P. R. Taylor, C. Angeli, R. Cimiraglia and K. Ruud, *J. Chem. Phys.*, 2004, **120**, 4619.

52. I. C. Hayes and A. Stone, *J. Mol. Phys.*, 1984, **53**, 83.

53. P.-O. Löwdin, *Phys. Rev.*, 1955, **97**, 1474; P.-O. Löwdin, *Phys. Rev.*, 1955, **97**, 1474.

54. J. Olsen, P. Jorgensen, H. Koch, A. Balkova and R. J. Bartlett, *J. Chem. Phys.*, 1996, **104**, 8007.

55. B. O. Roos, K. Andersson, M. P. Fülscher, P.-A. Malmqvist, J. L. Serrano-Andrés, K. Pierloot and M. Merchán, *Adv. Chem. Phys.*, 1996, **93**, 219; Z. Azizi, B. O. Roos and V. Veryazov, *PCCP*, 2006, **8**, 2727.

56. J.-M. Lü, S. V. Rosokha, S. V. Lindeman, I. S. Neretin and J. K. Kochi, *J. Am. Chem. Soc.*, 1797, **127**, 2005.

57. J. Jakowski and J. Simons, *J. Am. Chem. Soc.*, 2003, **125**, 16089.

58. M. Capdevila-Cortada, J. Ribas-Arino and J. J. Novoa, *J. Chem. Theor. Comput.*, 2014, **10**, 650.

59. F. Mota, J. S. Miller and J. J. Novoa, *J. Am. Chem. Soc.*, 2009, **131**, 7699.

60. M. Konno, T. Ishii and Y. Saito, *Acta Cryst.*, 1977, **B33**, 763–770.

61. R. F. W. Bader, *Atoms in Molecules*, Oxford University Press, Oxford, 1994.

62. B. Braida, K. Hendrickx, D. Domin, J. P. Dinnocenzo and P. C. Hiberty, *J. Chem. Theor. Comput.*, 2013, **9**, 2276.

63. The values of n_a and n_b respectively become 1.81 and 0.23 in a CASSCF(22,20) calculation.

Section 4: Intermolecular Interactions in Crystals

18 Revealing the Intermolecular Bonds in Molecular Crystals Through Charge Density Methods

C. Gatti* and A. Forni

CNR-ISTM, Istituto di Scienze e Tecnologie Molecolari, c/o Dept. of Chemistry, Università degli Studi di Milano, via Golgi 19, Milano 20134, Italy
*Email: c.gatti@istm.cnr.it

18.1 Introduction

Intermolecular bonds play a relevant role in molecular and supramolecular chemistry, one of the most rapidly evolving fields of chemistry, having large implications from life to materials science.[1] Molecular crystals are held together by intermolecular bonds and by the overall electrostatic and exchange interactions among composing molecules. Though the cohesive energy of a molecular crystal cannot be explained, in general, only in terms of its intermolecular interactions,[2] the latter clearly effect such energy in some way and drive, in most cases, the molecular recognition processes which eventually lead to crystallization.[3,4]

There has been in the past, and is still alive in the recent literature,[2-4] a vibrant debate on whether the observation of short distances between pairs of atoms on the peripheries of different molecules in crystals can be regarded as "structure determining" and as evidence of specific intermolecular bonding between the atoms

Intermolecular Interactions in Crystals: Fundamentals of Crystal Engineering
Edited by Juan J. Novoa
© The Royal Society of Chemistry 2018
Published by the Royal Society of Chemistry, www.rsc.org

concerned, or whether such short distances should rather be associated with an increase in potential energy and to the presence of repulsive forces. When the latter picture is embraced,[2] attractive forces among other molecular fragments, like for instance those of electrostatic nature between anions and cations, are invoked to counteract the destabilizing effects and so to ensure an overall favorable energy balance. Besides, though individual atom–atom pair interactions in crystals are still considered of some interest in such a view, they are also thought to be seldom structure determining and rather emerge from the overall packing of the molecules in the crystal as a whole.[2,5] From an *electron density perspective*, the former viewpoint focuses on the electron distribution features between the peripheral and interacting atom pairs as the dominant fact,[3,4] while the latter vision considers the interaction between the entire molecular electron distributions of proximal molecules as the determining event.[2,5] As we will illustrate in this chapter, intermolecular interactions induce distinct, and often more evident, changes to the electron distribution of the intramolecular than to that of the intermolecular linkages, so that, regardless of the adopted perspective, not only the whole molecular distribution participates into the process, but it is also to some extent affected and perturbed in all its parts upon packing.[6–9] As shown in Chapter 5, the relaxation of the electron density of the isolated molecular distribution in a molecular complex or a crystal, though not univocally determinable, has an important effect on the interaction energy balance. It is also worth recalling that crystal geometry represents a time-averaged equilibrium configuration of atoms in space where nuclei oscillate around their equilibrium positions. No net forces are acting on nuclei when they are at these locations, hence any displacement from the equilibrium geometry will simply imply a restoring force bringing all nuclei back to the equilibrium positions, with no atom in a crystal (in thermodynamic equilibrium) experiencing net repulsion or attracting forces.[3,9] The existence of a possible potential energy increase and of a repulsive force between (closed-shell) atoms in short contact is thus *to be confined* to a model of structural stability based on pair-wise, isolated, atom–atom interaction potentials,[3] like those adopted in the force field approaches. Energy Decomposition Analyses (EDA) (Chapter 5) models may also *dissect stabilizing or destabilizing* atom–atom interaction energies, but within energy partitioning frameworks quite different in nature. For instance, in the Interacting Quantum Atom Approach (IQA)[10] (see also Chapter 5), one may write the energy of a system as a sum of atomic self-energies plus a term of pair-wise additive interaction energies between atoms, made up of a classical electrostatic component and of a stabilizing quantum-mechanical correction. Binding energies relative to a given reference state (*e.g.* the isolated neutral atoms or molecules) may then be expressed in terms of the changes in the self-energies of the interacting fragments with respect to their values in the reference state, a contribution which is generally positive and destabilizing, and in terms of the pair-wise interaction energies between fragments, which are instead generally negative and stabilizing, around equilibrium.[10] The global interaction energy between two fragments (atoms or molecules) is then the result of a balance between their pair-wise interaction energy and their *deformation energies*, *i.e.* the changes to their self-energies by the interaction. Within this approach it is found, for instance, that in the head-to-tail approach of two H_2 molecules, the facing H atoms experience a *local* stabilizing interaction, as their pair-wise interaction energy is largely negative for a wide range of $H_2 \cdots H_2$ distances.[11] This occurs, despite the overall $H_2 \cdots H_2$ interaction being destabilizing, due to its basically closed-shell nature and to the unfavorable deformation energy of the interacting H_2 molecules, overriding the local $H \cdots H$ interaction energy lowering.[11] In a molecular crystal, the energy decrease due to packing will then be the result of a stabilizing compromise between molecular deformation and pair-wise interaction energies (possibly decomposed in atom–atom contributions), where, in principle, both the former and the latter may be individually stabilizing or destabilizing as a function of charge transfer, contact geometries, kind of involved atom–atom pairs, *etc.* One also has to consider that crystals mostly form under kinetic rather than under thermodynamic control, as often evidenced by the difficulty to obtain the thermodynamic polymorph in polymorphic systems.[4] Therefore, the observed molecular crystal geometry is frequently that of a local minimum on the potential energy surface containing all potential periodic molecular arrangements, rather than that corresponding to the absolute energy minimum on the surface.[4] This state of affairs further hints to the structure-determining role which is potentially played by the pair-wise contacts between peripheral atoms during the molecular recognition processes triggering crystallization,[4] and also to the importance of studying these interactions and the way they affect the interacting molecule(s) as a whole, in some detail.

Having properly set the energetic scene (and the ongoing debate it provokes), this chapter is essentially aimed at revealing *visible* features associated with intermolecular interactions, in particular in crystals, rather than dealing with the implied energy factors. The latter are amply treated, under different perspectives, in Chapters 3 and 5 and occasionally in many other chapters. The focus is also narrowed to those features that may be generically obtained through *electron-density-based* descriptors, though those based on the full first order density matrix or the pair density[12] will also be mentioned occasionally, especially for the sake of comparison. Features based on geometric factors, on kinetically derived basic structural units (synthons) and on various spectroscopic data are discussed in Chapters 11, 9 and 7–8, respectively.

Being defined in terms of a quantum observable, electron-density-based descriptors may be applied, on the same ground, to both *ab initio* and experimentally derived electron densities (EDs).[8,9,12] This property enables one to make the least unbiased comparison between experimental and theoretical outcomes, although it should not be overlooked that the ED is, rather than measured, reconstructed from the diffraction data by using a sequence of diffraction-, density-, and thermal motion models, all of which contain several simplifications (see Section 18.2 and ref. 12). Moreover, the ED obtained from first principles calculations may also suffer from various limitations such as the incomplete treatment of static and dynamic electron correlation effects (including the approximate treatment of dispersion); the basis set incompleteness and the energy cutoff in the local basis and in the plane waves approaches, respectively; the usual lack of a proper treatment of relativistic effects, *etc.*[12,13]

The chapter is organized as follows. Section 18.2 briefly reviews how the ED is reconstructed from the X-ray data, although no information is provided in this chapter about its evaluation from theory, given the plethora of sources from which such information may be easily retrieved (for crystals, see for instance ref. 12 and 13). Section 18.3 illustrates a number of electron-density-based methods able to reveal intermolecular interactions in molecular crystals, such as those applying the Quantum Theory of Atoms in Molecules,[14] the Source Function[8,15,16] and the Reduced Density Gradient descriptors.[17,18] Section 18.4 then discusses their application to a few selected paradigmatic cases.

18.2   Electron Density from Experimental Data

The ED in both position and momentum space ($\rho(\mathbf{r})$ and $\pi(\mathbf{p})$, respectively) can be experimentally obtained by scattering techniques, *i.e.*, X-ray, γ-ray or electron diffraction for $\rho(\mathbf{r})$ and Compton scattering for $\pi(\mathbf{p})$. The scattering intensity is in fact proportional to $|F|^2$, where F, the structure factor, is given by the Fourier transform of the thermally averaged density of the scattering atoms. The term 'electron density' will be here generally used, though diffraction experiments allow determination of both the electron and nuclear charge density distributions simultaneously. The possibility of obtaining the full charge density (*i.e.* electron plus nuclear) is indeed highly attractive for an accurate treatment of intermolecular interactions (with the due precautions connected with the determination of the hydrogen atom positions and thermal motion, *vide infra*), in particular when investigating dispersion-dominated interactions between medium- and large-size systems, for which quantum chemical approaches are notoriously still defective. A thorough treatment of these aspects, as well as of the methods used to obtain the charge density experimentally, is given elsewhere.[19,20]

The density distributions $\rho(\mathbf{r})$ and $\pi(\mathbf{p})$ provide different, complementary information on the behaviour of electrons in molecules as well as in condensed matter. In fact, while $\rho(\mathbf{r})$ is connected to the diagonal terms of the one-particle reduced density matrix in the position-space representation, $\pi(\mathbf{p})$ is related to the off-diagonal parts of this same matrix, so reconstruction of electron density from experimental data in both domains is desirable to get a more complete understanding of the chemical bond and intermolecular interactions.[21] At present, however, consolidated models for a quantitative determination of charge density have been developed only in the position space representation.

Such models, first introduced by Kurki–Suonio[22] and Hirshfeld[23] and later developed by Stewart[24] and Hansen and Coppens,[25] are based on the concept of deformation or asphericity of the atoms, in contrast with the conventional Independent Atom Model (IAM),[19] which assumes the molecular electron density is a sum over spherical, non-interacting atoms. The total electron density of the atoms within the asymmetric unit is expressed as a sum over finite multipolar expansions of charge density about each atomic centre, $\rho_i(\mathbf{r})$. Such atomic expansions, called pseudoatoms, depend on the actual position of the nuclei in the crystal and are assumed to be rigid, *i.e.*, they follow the motion of the nuclei upon which they are centred. Moreover, it is assumed that each pseudoatom (except hydrogen) is given by a sum of three terms: a fixed core, a valence spherical part and a valence deformable part. According to the Hansen and Coppens formalism,[25] the charge density of the i-th pseudoatom is written as:

$$\rho_i(\mathbf{r}) = P_{i,\text{core}}\rho_{i,\text{core}}(\mathbf{r}) + k_i^3 P_{i,\text{valence}}\rho_{i,\text{valence}}(k_i\mathbf{r})$$

$$+ \sum_{l=0,l_{\max}} \sum_{m=0,l} P_{i,lm\pm} y_{lm\pm}(\mathbf{r}/r) k_{i,lm\pm}^{\prime 3} R_{i,lm\pm}(k_{i,lm\pm}^\prime \mathbf{r}) \qquad (18.1)$$

where P, the electron population parameters, and k, the radial scaling factors, are determined by the method of least squares. $\rho_{\text{core}}(\mathbf{r})$ and $\rho_{\text{valence}}(\mathbf{r})$ are spherically averaged density functions, $R(\mathbf{r})$ are radial density functions, approximated by single exponential type functions, and $y_{lm\pm}(\mathbf{r}/r)$ are spherical harmonics.

Multipolar models have been implemented in several software programs[26] for their refinement against single-crystal X-ray diffracted intensities and determination of a variety of static electrostatic properties from the refined multipolar coefficients, working in both the Fourier and the real space. In the latter case, a common procedure consists in deriving properties (*e.g.*, the dipole moment) for a molecule "extracted" from the crystal. Such properties reflect the matrix effect due to the crystalline environment, which can significantly modify the intrinsic properties of the molecule, in particular in the presence of strong intermolecular interactions.

A critical comparison with results from other experiments and with theoretical results is then mandatory. In this connection, it is important to underline that, while the X-ray scattering of atoms is a very fast process ($\sim 10^{-17}$–10^{-18} s), the time-scale of a scattering measurement (10^2–10^0 s) is much longer than the period of atomic thermal vibrations ($\sim 10^{-13}$ s). The observed intensities are therefore statistical averages over the states accessible for the atomic displacements with respect to the equilibrium positions. The mean thermal electron density of an atom can be expressed as a convolution of the static density (eqn (18.1)) with the probability of having the atom displaced from its equilibrium position.[19,20] Both harmonic and, if required, anharmonic approximations can be used to represent the atomic motion. Importantly, it is exactly such hypothetical static density which can be compared with the theoretically-derived one, which excludes thermal motion.

The convolution of $\rho_i(\mathbf{r})$ and thermal motion implies that the parameters describing the static electron density and the atomic coordinates and thermal displacements are correlated during the refinement against X-ray scattering intensities. Taking into account that, owing to the Fourier transform relationship, most of radiation scattered at high angles comes from the sharp electron density distribution of the inner-shells electrons, the high-order data provide the most accurate information on both nuclear positions and thermal displacements. Only a careful determination of such parameters allows a satisfactory deconvolution of the static electron density from nuclear motion to be obtained and a guarantee that population parameters properly accommodate just the charge deformation due to bond formation, deriving from medium-order data. Only in this case, reliable static-charge properties will be obtained.

This task is even more challenging when studying intermolecular forces within the crystal, because even tinier deviations from the ideally spherical atomic distribution need to be detected and properly described. Hydrogen-bonding interactions present a further difficulty due to the additional problem of the absence of a core structure for hydrogen atoms, which prevents detailed information about their positions and thermal motion being obtained from X-ray diffraction data. Resorting to independent determination of nuclear coordinates and amplitudes of vibration by single-crystal neutron diffraction experiments is generally hampered by the difficulty of growing crystals of sufficient size as required by these experiments. While the approach of 'polarized hydrogen atoms',[27] based on the use of scattering factors obtained for terminally bonded H atoms from calculations on model systems, as implemented in VALRAY[26b] and VALTOPO[26c] codes, allows accurate H atom positions to be determined, a proper treatment of thermal motion requires independent information, which can be extracted from either a library built up on the basis of neutron diffraction experiments[28] or spectroscopy[29] or *ab initio* calculations.[30,31] In particular, a highly promising procedure has been recently developed by Woińska *et al.*[32] Simply introducing anisotropic displacement parameters of the H atoms allows inclusion in the multipole expansion of the quadrupole components, which often play an important role in determining the topological properties of hydrogen bonding.[33]

X-ray diffraction experiments must be performed at low temperature, such as that generated by a stream of cold nitrogen or helium, to attain high-order data. Operating at low temperature offers many further benefits, all pointing towards a more reliable determination of the static $\rho(\mathbf{r})$. Among them we mention (i) a reduction of thermal vibrations which makes the rigid pseudoatom approximation more reliable; (ii) a decrease of the omnipresent thermal diffuse scattering, whose peaks overlap with the elastic ones making the intensity profile broader; (iii) an increased number of data, which is of paramount importance owing to the large number of population parameters to be refined within the multipolar model, in addition to the positional and thermal parameters, in particular when anharmonic expansions are included; (iv) a reduction of the magnitude of the correlation coefficients, allowing determination of all model parameters with almost independent precision estimates.

While the originally developed multipolar models proved to be in most cases adequate to retrieve high quality experimental charge densities, the introduction, in more recent years, of area detectors is progressively changing the scene. Use of Charge-Coupled Device (CCD) and Image-Plate (IP) detectors, allowing a high redundancy of the data to be obtained and hence improved statistics, revealed some deficiencies which have stimulated the development of more flexible multipolar models. The main limitations have been individuated as (i) the use of fixed core electron densities, which yield in some cases erroneous thermal displacements, suggesting inclusion of the refinement of contraction/expansion and population parameters of the innermost shells,[34] and (ii) the rather poor description of the radial part, especially for the valence density, which calls for the adoption of more flexible radial functions.[21,35]

18.3 Electron-density-based Methods to Reveal Intermolecular Interactions in Crystals

Molecular crystals represent an enormous and ever growing source of information on intermolecular interactions. Though traditionally detected and classified in terms of close contact distances and interaction geometries, they are now increasingly being studied in terms of their effects on crystal ED distributions.[9,12,20] Depending on the kind of investigative *lens* one is using, different "fingerprints" of the intermolecular interactions are recovered. This section introduces a number of such lenses and analyses the information on the nature of intermolecular interactions they may provide, when taken alone or in combination. While, in their original formulation, both the Hirschfeld Surface (HS) analysis[36–38] and the Reduced Density Gradient (RDG) approach[17] make use of the crude IAM electron density (see Section 18.2), the Quantum Theory of Atoms in Molecules (QTAIM)[14] and the Source Function (SF)[15,16] analyses both employ the (crystal) ED as their primary ingredient. HS analysis provides extremely interesting insight on intermolecular interactions, but it is not treated in the present chapter, for the sake of space.

The emerging of QTAIM as a fascinating theoretical framework to study chemical bonding, along with its intimate link with the ED observable has, in the last 20 years, largely contributed to shift the focus of experimental charge density studies from the *deformation* to the total ED.[8,9,12,20,39] Also decisive was the role played by the ever increasing precision of the X-ray data as well as of the methodological and computational progresses in their refinement, enabling great improvement in both the precision and accuracy of the experimentally derived EDs.[12] The parallel development of *ab initio* periodic codes and of DFT functionals for the solid state[12,13] has greatly contributed to the adoption of ED-based topological tools as a common rigorous framework with which to compare and mutually validate the experimentally and theoretically derived EDs.

18.3.1 Quantum Theory of Atoms in Molecules

Searching for bond critical points (bcps, see Chapter 4) between peripheral atom pairs in molecular crystals enables one to detect intermolecular interactions and classify their nature in terms of the ED properties at bcps. According to Bader,[40] the presence of a bcp and thus of a line linking two atoms, along which the ED is a maximum relative to any lateral displacement, is a necessary condition for two atoms to be bonded to one another when the system is in a stationary state, *i.e.* in an energy minimum at a given nuclear configuration. If such a line persists when no net forces are acting on any system's nucleus, hence in a condition of stable electrostatic equilibrium, it is called a *bond path* and it provides both a necessary and a sufficient condition for bonding in the "usual chemical sense of the word".[40] The network of bond paths defines the *molecular* or, more precisely, the *crystal graph*, isolating all the pair-wise interactions present in a crystal.[41] They include the strong bonds related to the molecular backbone, the (weak) intramolecular interactions, if any, between atoms of the molecule not directly linked through such a backbone, and the intermolecular ones. The latter include several classes of bond types,[9] hydrogen bonds (HBs), the σ-hole bonds (like halogen bonds) and their π-hole bond counterparts, the $XH\cdots\pi$ (X=C, N, O, *etc.*) and the $\pi\cdots\pi$ interactions, the cation– and anion–π interactions, the van der Waals (vdW) interactions, *etc.*.. As already mentioned (see Section 18.1), intermolecular and intramolecular interactions are not separate entities and the formation of the former may induce distinct signatures on the latter and on the whole molecular distribution. This may be revealed by a comparison of a number of properties, evaluated at bcps, for the molecule in the crystal against those for a cluster of molecules or for the molecule in isolation (see the example of urea in Section 18.4.2). The ED, ρ_b, its Laplacian, $\nabla^2\rho_b$, the curvatures of the ED along the bond path, λ_3, and along directions perpendicular to such path, λ_1 and λ_2, the values of the positive definite kinetic energy density, G_b, and of the potential energy density, V_b, are all among the bcp properties normally examined.[9] Intermolecular interactions are characterized in terms of their closed-shell nature,[9,14] implying low ρ_b values, low and positive $\nabla^2\rho_b$ values, low density curvatures and with the parallel curvature λ_3 largely dominating in magnitude the perpendicular ones, small G_b and V_b values and $2G_b > |V_b|$. Moderately strong or strong HBs may, yet, exhibit properties denoting their partial shared shell character (see ref. 9 for a thorough discussion on the various classification schemes of chemical interactions based on QTAIM, and on the *caveats* one should carefully bear in mind when making use of them).

The bond path criterion was, for instance, applied to the 27 CH$\cdots$O contacts (23 inter- and 4 intra-molecular), for H$\cdots$O distances below 3.0 Å, characterising the 3,4-bi(dimethylamino)-3-cyclobutene-1,2-dione (DMACB) molecular crystal.[42] The latter represents an exemplary test case for the study of the *nature* and *function* of weak CH$\cdots$O interactions, because no other type of stronger and potentially prevailing HBs is present in the crystal. It was found that 23 out of the 27 contacts were associated to a bcp, hence *bonded*, and characterised by a large and approximately constant (120–140°) $\alpha_{CH\cdots O}$ angle value. On the other hand, the four contacts lacking a bcp, hence *non-bonded*, had instead a highly bent geometry, with $\alpha_{CH\cdots O}$ close to or even below 90°. The bond path criterion could so distinguish the contacts having an important electrostatic contribution from those of van der Waals-like nature, since electrostatic interactions are known to favour linear or close to linear geometries over bent ones, while van der Waals-type contacts do not have an energy angular dependence and the $\alpha_{CH\cdots O}$ angle may be thus even bent below 90°.

Using a *universal indicator*[40] to single out chemical interactions in a crystal is clearly an advantage as it provides the possibility of performing the analysis and comparison of their properties on the same footing, as the bcp is defined only in terms of an observable. However, the topological nature of the bcp criterion implies its inherent *yes/no*, discontinuous character, which may give rise to interpretation problems,[8,9] and, occasionally, also to harsh controversies.[43] Though staying purposely apart from the latter, it is well known that bond paths have been found where a chemist will not predict bonds[43a–c,44] and *vice versa*[45] and that significant and comparable electron sharing, as measured through the delocalization indices, may be also observed for pair of atoms not linked through a bond path.[8,11,12,45,46] This is the typical situation in systems characterized by soft potential energy surfaces and extremely flat EDs in their bonding regions, where several pairs of atoms may enter in competition for a bond path and where continuous bonding indicators like the delocalization indices or the SF contribution patterns would be probably more appropriate.[8,12] For instance, in the transition metal π-hydrocarbyl complexes, characterized by high fluxional mobility of ligands, the two-center view of bonding tied to the bond path criterion is challenged and fewer metal–carbon bond paths are observed than expected from the formal hapticity of the complex.[45] However, values of the delocalization indices do not seem to support such a discrimination. In section 18.4, we discuss a similar situation for the CH$\cdots\pi$ intermolecular interactions in the benzene crystal, where the bcp criterion and the shape of the RDG isosurfaces seem to highlight distinct bonding schemes. Decisive progress in unveiling the physical meaning of the bond path was made some time ago by Martín Pendás *et al.*[11] who showed that the bond paths correspond to *privileged exchange energy channels*. By examining a number of paradigmatic cases, it was observed that whenever two atomic pairs are competing for a bond path, the latter is found to link the pair of atoms having the larger exchange energy (evaluated through IQA, Chapter 5). As a matter of fact, systems evolving through conflict mechanisms feature exchange energy curves for the two competing pairs of atoms which almost exactly cross at the conflict catastrophe point, *i.e.* where the bond path switches from one pair to the other.[11] Interatomic exchange energies are related to the associated delocalisation indices (see Chapter 4) and so it comes as no surprise that bond paths may or may not link pairs of atoms having comparable electron sharing. Moreover, one should not be seriously bothered by the apparently contradictory descriptions obtained through the bond path criterion or from continuous descriptors, like the delocalization indices.[8,12] Apart from the distinction between their discontinuous or continuous nature, they make use of information from quite different spaces, bond paths being manifest in the position space and delocalisation indices being defined through the six-dimensional pair density. Though both are evaluated within QTAIM, these two bonding descriptors may still provide complementary insight, because of their different origin and physical meaning.[12] Clearly, this dual and complementary view also applies to the realm of intermolecular interactions. Lack of an (expected) intermolecular bcp between two peripheral atoms does not necessarily imply a lack of a (significant) interaction between these atoms.[18]

Energy properties at bcps are increasingly being used to estimate the energies associated with the corresponding interatomic interactions between molecules in crystals.[20] This practice, which is nowadays routinely adopted in experimental charge density studies and, for the sake of comparison or by its own, quite often used also in theoretical ones, deserves significant caution. It originates from an early and interesting study by Espinosa, Molins and Lecomte (EML),[47] based on 83 experimentally observed bcps for X–H$\cdots$O (X=O, C, N) interactions in crystals, spanning a $d_{\text{H}\cdots\text{O}}$ distance range of 1.6 to 3.0 Å. EML found that both G_{b} and $-V_{\text{b}}$ follow a negative exponential dependence *vs.* $d_{\text{H}\cdots\text{O}}$ over this range of distances and so do the hydrogen bond dissociation energies D_{e} computed by *ab initio* methods for a series of *in vacuo* hydrogen bonded systems in the same $d_{\text{H}\cdots\text{O}}$ interval. By assuming a common exponential factor for the V_{b} and D_{e} relationships *vs.* $d_{\text{H}\cdots\text{O}}$, EML proposed that the interaction energy E_{int} of a H$\cdots$O intermolecular interaction in a crystal may be roughly estimated through the value of the potential energy density at the corresponding bcp, $E_{\text{int}} = 0.5\ V_{\text{b}}$, where the proportionality factor has units of volume. Using such an expression, H$\cdots$O intermolecular interactions are customarily ranked as for their importance and, allegedly, also for their ability to be structure determining (see Section 18.1). This same expression has also often been used to estimate the energies of closed-shell intermolecular interactions other than the X–H$\cdots$O ones, despite the lack of any real justification for its extension.[48] Even for the X–H$\cdots$O interactions, it was shown to be rather crude, in some circumstances. For instance, when applied to the 23 unique C–H$\cdots$O contacts of DMACB crystal, the resulting bond energy estimates turned out to be clearly untenable and new relationships, fitted on C–H$\cdots$O bonds only and yielding reasonable energy estimates, have been proposed.[42] The main reason for such a failure was ascribed to the much broader range spanned by the C–H$\cdots$O distances (2.21–2.97 Å) in the DMACB crystal relative to that spanned by the C–H$\cdots$O contacts (2.22–2.59 Å) in the EML reference data set. Recently, Spackman[49] analysed an enlarged data set of 166 separate atom$\cdots$atom interactions, based on results of experimental charge density studies for 23 molecular crystals, published in the past decade. For this data set, the log-linear plot of the EML estimates of interaction energies *vs.* the interatomic separation is dominated by the H$\cdots$O hydrogen bonds and shows, as expected, a reasonably good linear behaviour ($E_{\text{int}} = -4.69 \times 10^3 \exp(-2.669\ d_{\text{H}\cdots\text{O}}\ \text{Å}^{-1})$ kJ mol^{-1}). Still, this fit greatly differs from that originally proposed by EML ($E_{\text{int}} = -25.0 \times 10^3 \exp(-3.6\ d_{\text{H}\cdots\text{O}}\ \text{Å}^{-1})$ kJ mol^{-1}), demonstrating that these relationships are not at all universal, but heavily depend on the adopted reference data set and in particular on the relative abundance of data for the various separation distance ranges (indeed, Spackman's data set contains many more interactions at separations greater than 2.6 Å, while the original EML fit was dominated by data points below 2.0 Å). About ten years ago, Gatti[9] listed a number of important *caveats* associated to the use of the EML relationship(s). Here we briefly mention two of them, while full details may be found in the original publication.[9] EML fitting relationships are characterized by large uncertainties in their parameters, implying at best only very qualitative estimates for the H-bond energies. For instance, when applied to the C–H$\cdots$O contacts in the DMACB crystal, it was found that energy estimates obtained through the exponential relationship $E_{\mathrm{int}} = -25.0 \times 10^3 \exp(-3.6\ d_{\mathrm{H}\cdots\mathrm{O}} \mathrm{\AA}^{-1})$ were about half as large as those calculated through the expression $E_{\mathrm{int}} = 0.5\ V_{\mathrm{b}}$ for those interactions falling in the range of distances of the EML data set, and even smaller for longer H$\cdots$O distances.[42] A quite discouraging result, when remembering that the second expression was obtained by comparing the former with the corresponding V_{b} exponential fit against $d_{\mathrm{H}\cdots\mathrm{O}}$. Another problematic issue concerns the dissociation and hydrogen bond energies which were evaluated by EML *in vacuo,* assuming the H$\cdots$O separation as the reaction coordinate for dissociation. While this may be considered as a reasonable approximation *in vacuo*, it might be inappropriate and quite far from reality for an HB in the crystal where many HB interactions occur simultaneously and where the H or the acceptor atoms may often take part in more than one interaction at a time. Furthermore, since the HB formation involves more important geometrical and electronic rearrangements in the molecular backbone than in the intermolecular, peripheral regions (see Section 18.4), the assumption of a single reaction coordinate for estimating the HB energy in a crystal from V_{b} is *a fortiori* quite crude. It also opens the problem of whether HB energies evaluated through the EML relationship are additive or not in a crystal, *i.e.* of whether interaction energy in the crystal may or may not be obtained by summing up EML atom$\cdots$atom energies over all relevant intermolecular bcps. A test case was made by Spackman on complexes of halogen bonded systems[50] and more recently[49] on molecular pairs extracted from the 166 atom$\cdots$atom interactions mentioned earlier. When compared with accurate molecule$\cdots$molecule interaction energies, it was found that, in general, EML energy sums largely underestimate the intermolecular energies, but in few cases the opposite seems true.[49] Spackman argues that "EML estimates of atom$\cdots$atom energies, based on modern experimental charge density analyses, provide little more than a very crude estimate of an interaction energy" and that their use for assessing the "energetic importance of specific intermolecular interactions, in the context of crystal packing and crystal engineering, can often be quite misleading and should be so discouraged".[49] In our view, the observation that "EML atom···atom energies are seldom reliable, typically underestimating and sometimes overestimating more reliable values" suggests that property values at the intermolecular bcp, like V_b, are not related to the total interaction energy, but only to some component of it, like, possibly, the quantum-mechanical component of the IQA pairwise interaction energy. When other important interaction energy components, like the classical part of the IQA pairwise interaction energy or the deformation energy of the fragments, play different energetic roles in complexes with similar separation distances, interaction energies may hardly be predictable in terms of EML energies only. Deciphering the energy component(s), if any, directly related to the local bcp energies V_b and G_b, will be object of future investigations by one of us (CG).

18.3.2 Source Function (SF) Analysis

Introduced back in 1998 by Bader and Gatti,[15] the SF permits viewing of the ED of a system at any point using a quite novel perspective and one rich in chemical insight. When the selected point is a bcp, taken as the most representative ED location for a given chemical interaction, the SF visibly reveals how local or non-local this interaction is in nature.[16,51] It does so based on the smaller or larger magnitude of contributions to the bcp density coming from all the atoms of the system other than those directly linked through the bcp. The SF analysis may be clearly applied to the study of any chemical interaction, and so also to intermolecular interactions, where, for instance, it has proved particularly useful to classify and rank HBs,[51] from the short and largely covalent charge assisted HBs, to the long and weak ones, according to their smaller or larger SF locality degree. Examples of SF applications to hydrogen-bonded systems are illustrated in Section 18.4.3, while the basic tenets of the SF tool are briefly summarized below. A comprehensive illustration of the SF is reported in ref. 16 and, in more concise ways, in ref. 8 and 52.

Bader and Gatti have shown[15] that the ED at a point **r** may be envisaged as determined by contributions from a local source LS(**r**,**r**′), operating at all other points **r**′ in the space

$$\rho(\mathbf{r}) = \int LS(\mathbf{r},\mathbf{r}') \, d\mathbf{r}' \tag{18.2}$$

where LS(**r**,**r**′),

$$LS(\mathbf{r},\mathbf{r}') = -(4\pi|\mathbf{r}-\mathbf{r}'|)^{-1}\nabla^2\rho(\mathbf{r}') \tag{18.3}$$

is given in terms of the Laplacian of the ED and of a Green's function, $(4\pi \times |\mathbf{r} - \mathbf{r}'|)^{-1}$, expressing the *influence*[53] or effectiveness of $\nabla^2 \rho(\mathbf{r}')\mathrm{d}\mathbf{r}'$ in contributing to cause the effect $\rho(\mathbf{r})$. The *cause* for the *effect, i.e.* the value of the ED at $\mathbf{r}$, is thereby related to the local behaviour of this same scalar (in terms of its Laplacian, $\nabla^2 \rho$) at all other points of space $\mathbf{r}'$.

It is then convenient to replace the operation of the local source over the whole space (eqn (18.2)) with separate LS integrations over the QTAIM atomic basins Ω

$$\rho(\mathbf{r}) = \int \mathrm{LS}(\mathbf{r}, \mathbf{r}')\,\mathrm{d}\mathbf{r}' = \sum_{\Omega} \int_{\Omega} \mathrm{LS}(\mathbf{r}, \mathbf{r}')\,\mathrm{d}\mathbf{r}' = \sum_{\Omega} \mathrm{SF}(\mathbf{r}, \Omega) \qquad (18.4)$$

enabling the ED at $\mathbf{r}$ (the *reference point, rp*) to be seen as determined by a sum of atomic contributions $\mathrm{SF}(\mathbf{r}; \Omega)$, each of which is called the *Source Function* from the atom Ω to the ED at the *rp*. The SF is therefore an interpretive tool which is deeply tied to one of the main operative notions of chemistry, namely that any local property and chemical behaviour of a system is to some extent always influenced by the remaining parts of the system.[52] Whether such an influence is small or large, it can be quantified through the SF. It is worth noting that the $\mathrm{SF}(\mathbf{r}, \Omega)$ atomic contributions in eqn (18.4) do not represent a *direct* ED donation from these atoms to $\rho(\mathbf{r})$, but simply their own capability to influence or determine such density values, similarly to the role that their atomic EDs have in determining the value of the electrostatic potential at $\mathbf{r}$.[16,52] SF contributions may either be compared as absolute values or in terms of percentage values $\mathrm{SF\%}(rp, \Omega)$,

$$\mathrm{SF}\,\%\,(rp, \Omega) = \frac{\mathrm{SF}(rp, \Omega)}{\rho(rp)} \times 100 \qquad (18.5)$$

expressing the relative ability of an atom Ω to determine ρ at the *rp*. Any exhaustive (or even fuzzy boundary) space partitioning scheme could formally be used in eqn (18.4),[16] yet adopting the QTAIM zero-flux surface recipe allows for associating $\mathrm{SF}(\mathbf{r}, \Omega)$ sources to contributions from atoms or groups of atoms defined through quantum mechanics.[14]

The SF tool provides a natural bridge for studying intermolecular interactions using experimentally and theoretically derived EDs and comparing their outcomes on the same grounds. In fact, eqn (18.2) through (18.4) imply that the $\mathrm{SF}(\mathbf{r}; \Omega)$ values are amenable also to experimental determination, provided an accurate $\nabla^2 \rho$ distribution derived from high-quality single-crystal X-ray diffraction intensity data is available.[8,16,52,54]

18.3.3 Reduced Electron Density Gradient (RDG) Analysis for Non-covalent Interactions

Non-covalent interactions (NCI) comprise a wide range of inter-molecular and intramolecular bonding types, like the hydrogen bonds, the halogen bonds and, in general, the whole class of σ- and π-hole (see 18.4.1), $XH\cdots\pi$ ($X=C$, N, O, *etc.*) and $\pi\cdots\pi$ interactions, *etc.* They are also characterized by several binding (or anti-binding) forces, such as those due to dispersion, electrostatics or Pauli's principle, which make their study particularly intriguing. NCI represent the way through which distinct molecules 'recognize' themselves and approach each other, so that decoding the interplay of inter-molecular NCI in setting up supramolecular assemblies is clearly an important step in allowing substantial progress in structural prediction capabilities to be made.[55–57] It thus comes as no surprise that research on this topic has seen the emergence of a plethora of diverse investigative tools, among which those based on the ED observable are of particular relevance, as this scalar field inherently bears a vast amount of information on how molecules influence each other when they are in close contact.[9,20,39]

A few years ago, Johnson *et al.*[17] introduced a novel NCI descriptor, based on the Reduced Electron Density Gradient (RDG),[58] enabling an easy-to-catch pictorial visualization of various kinds of NCI and a classification of their allegedly attractive or repulsive nature. The RDG is a dimensionless quantity, measuring how the ED locally deviates from that of a homogeneous electron gas, which by definition has a RDG, $s(\mathbf{r})$, equal to zero everywhere,[58]

$$s(\mathbf{r}) = \frac{1}{2(3\pi^2)^{1/3}} \frac{|\nabla\rho(\mathbf{r})|}{\rho(\mathbf{r})^{4/3}} \tag{18.6}$$

According to eqn (18.6), the RDG absolute minima have a value of zero and occur where the ED gradient vanishes, while $s(\mathbf{r})$ attains large values in regions far from the various nuclei of a system, where the total ED decays exponentially to zero and the $\rho(\mathbf{r})^{4/3}$ term approaches zero faster than $|\nabla\rho(\mathbf{r})|$. The RDG is largely used in Density Functional Theory (DFT) for developing gradient-corrected functionals of increasing quality,[58–60] but Zupan *et al.*[60] were the first to show that the exchange-energy weighted RDG (see ref. 60 for definition) varies with chemical association. It was found to decrease on passing from the isolated atoms to the molecules they form and from the latter to their molecular crystals, suggesting that lower and lower energy-weighted RDG values could be a sign of increasing chemical association within a collection of atoms. Johnson *et al.*[17] could then show that the NCI supposedly present in a system may be in fact revealed by using the simpler $s(\mathbf{r})$ distribution and by associating them to the low s-value isosurfaces in the low ED regions. Moreover, they found that the nature and strength of each interaction may be also disclosed by mapping the $\rho(\mathbf{r})^*\text{sign}(\lambda_2)$ quantity onto these RDG iso-surfaces. In fact, according to Johnson *et al.*,[17] the sign of the λ_2 curvature of the ED at each isosurface point allows for distinguishing between, locally, allegedly attractive ($\lambda_2<0$) or repulsive ($\lambda_2>0$) interactions, while the $\rho(\mathbf{r})$ value is related to their strength (for $\lambda_2<0$, the larger the magnitude of $\rho(\mathbf{r})^*\text{sign}(\lambda_2)$, the stronger and more attractive is the interaction). RDG isosurfaces coloured with $\rho(\mathbf{r})^*\text{sign}(\lambda_2)$ will be hereinafter denoted as *ρ-signed* RDG isosurfaces. Eqn (18.6) also highlights that the NCI descriptor and the QTAIM ED topology are closely related.[18] A low-RDG isosurface will clearly appear close to any critical point (CP), with its associated domain including such a point or even more CPs if reducible[18] (but there will certainly be an appropriate s isovalue below which the domain will contain only one CP and will so become irreducible). In particular, one may select an appropriate s value to yield a $1:1$ mapping between the ED bcps of a system and their associated RDG isosurfaces. It is also evident that the $\rho(\mathbf{r})^*\text{sign}(\lambda_2)$ quantity will reflect to some extent the nature of the CP included in the RDG irreducible domain. The λ_2 curvature is negative at a bcp and associated to an eigenvector perpendicular to the bond path, while it is positive and associated to an eigenvector directed in the ring plane at a ring CP.[14] The *ρ-signed* RDG isosurfaces extend in space and their *sign* may change over the isosurface, being possibly different in some surface regions from that "dictated" by the nature of the included CP. However, experience shows that a *ρ-signed* RDG isosurface enclosing a bcp is in general mostly negative, while it is almost everywhere positive when it encloses a ring CP.[17,18,61]

Applications of the RDG-NCI descriptor customarily use the IAM ED and RDG, rather than the "true" ED and RDG, as their main ingredient. Indeed, in their seminal papers, Johnson *et al.*[17] and Contreras-García *et al.*[62] showed that the IAM ED and RDG are in general able to recover the main NCI features revealed by the corresponding "true" (*ab initio*) quantities. Such a result is not fully unexpected as it is well known that molecular or crystalline EDs are, on average, only slightly modified relative to the IAM distribution[63] and that the amount of local reconstruction, $\Delta\rho(\mathbf{r}) = |\rho(\mathbf{r}) - \rho_{\text{IAM}}(\mathbf{r})|$ is everywhere much smaller than either $\rho(\mathbf{r})$ or $\rho_{\text{IAM}}(\mathbf{r})$. However, the largest *relative* changes are known to take place in the bonding (intra- and intermolecular) regions, namely those where bcps occur and where the associated RDG isosurfaces are located. Were the IAM RDG isosurfaces almost indistinguishable from those obtained with the true ED and RDG, one could conclude that the RDG isosurfaces are not really informative about NCI, since neither the atomic deform- ation due to bonding (the so called *deformation density*), nor that due to intermolecular bonding (the so called *interaction density*) are taken into account by the IAM density (and clearly the same holds true for the IAM RDG). Indeed, the IAM density should just retain that amount of information on NCI which is implicitly contained in the change of geometry due to molecular interaction and in the cumulative frozen distribution of the system's composing atoms. Saleh *et al.*[61] have therefore systematically analysed the role of the *missing information* on the RDG-NCI analysis when the IAM ED and RDG replace the "true" values, finding an interesting rationale for the observed qualitative and quantitative changes. In particular, they show that the IAM RDG isosurfaces, for a given value of s, differ systematically in size and are generally much less *structured* than those obtained with the "true" ED. In another study, Saleh *et al.*[18] have extended the NCI RDG approach to experimental crystalline EDs obtained from the multipole model refinement of charge density quality X-ray diffraction data, demonstrating that the multipolar RDG- NCI picture is in general reliable and of comparable quality to that obtained from a fully periodic *ab initio* approach. Examples taken from both such studies will be briefly highlighted in Section 18.4.4, along with the comparison of the different and complementary NCI pictures obtained by RDG and QTAIM analyses in a single case (the benzene molecular crystal). Saleh *et al.*[18,61] have discussed in great detail such differences and complementarities of the NCI pictures in three molecular crystals characterized by several NCI interactions (H-bonds, CH$\cdots\pi$ and $\pi\cdots\pi$ interactions, van der Waals contacts, steric interactions, *etc.*) and have also explored how these various kind of interactions are affected when the IAM ED and RDG replace their true (*ab initio* and experimental) counterparts. Saleh *et al.*[64] have then developed a very useful code, NCImilano, implementing the NCI RDG analysis and specifically designed for the X-ray charge density com- munity. NCImilano works both on *in vacuo* and on solid-state EDs, as evaluated by popular multipolar (XD2006)[26a] and Gaussian-based quantum mechanical (GAUSSIAN 09,[65] CRYSTAL09[66]) packages.

18.4 Selected Applications

A number of examples of the application of the concepts and methods illustrated in the previous section are presented in the following. Most of them make use of experimentally derived EDs (Section 18.2) and/or of *ab initio* periodic EDs.

The selection we made here does not reflect a single style, purposely. Subsection 18.4.1 illustrates in some detail the recent studies on σ- and π-hole interactions, which represent an area of very active research for crystal engineering and one where the charge density perspective seems to offer an interesting unifying view. The following subsections are instead aimed at presenting a few didactic examples on the use of the methods outlined in Section 18.3. Namely, QTAIM is applied to reveal and quantify a number of crystal field effects in urea and in thiazete-1,1 dioxide molecular crystals (Section 18.4.2), while the Source Function tool is used in Section 18.4.3 to characterize the nature of hydrogen bonded interactions in a series of prototypical cases. Finally, the capability of the Reduced Density Gradient approach to detect and classify NCI in the benzene molecular crystal is analysed and compared to that offered by QTAIM, in Section 18.4.4.

18.4.1 σ- and π-Hole Interactions as Viewed Through Topological Methods

σ- and π-hole bonds are electrostatically driven interactions between negative sites (*e.g.*, lone pairs of Lewis bases, π-systems, anions) and positive regions of the electrostatic potential (EP) localized respectively on the extension of a covalent bond to an atom of groups IV–VII (σ-holes)[67] or along directions perpendicular to a planar molecular fragment containing atoms of groups III–VI (π-holes).[68] According to the atom where the positive region is located, σ-hole interactions are referred to as tetrel, pnicogen, chalcogen and halogen bonds for the atoms of groups IV, V, VI and VII, respectively. Owing to their monovalent character, halogens possess only one σ-hole, while the other elements can have more σ-holes according to their valence. Among the π-hole interactions, those involving atoms of group III, the so-called triel bonds,[69] such as those formed by BX_3 and AlX_3 (X=H, halogen atoms or electron-withdrawing groups), have been known for a long time but only recently were characterized on the basis of their EP anisotropy.[70] Other examples include those originated by the EP positive regions located above and below atoms of group IV (X_2CO, XSiO), V (FNO_2, $H_5C_6PO_2$, H_3CPO_2, FPO_2) and VI (O_2O, SO_2 and SeO_2).[71] Several other chapters in the present book include a treatment of experimental and theoretical aspects of σ- and π-hole bonds in molecules and solids and their role in crystal engineering (see Chapters 10, 13 and 19). Our focus here is in discussing the results and insight that can be retrieved from charge density studies on molecular crystals, with particular emphasis on the specific role of the crystalline environment on such interactions. Macroscopic evidence of the non-negligibility of crystal packing effects on halogen bonding has been recently reported for a series of complexes based on *N*-iodoimide derivatives as halogen bond donors and pyridines as halogen-bond acceptors. These are characterized by a variable degree of ionicity going from the cocrystal, A–I$\cdots$B, to the salt form, $[A]^-\cdots[I–B]^+$.[72] Theoretical calculations on isolated dimers with externally applied a homogeneous electric field, mimicking the polarization effect induced by the local crystal electric field, have demonstrated that the iodine position, or, in other words, the stabilization of one form with respect to the other, along with the consequent implications on the charge distribution, strongly depends on the crystal environment.

Some evidence suggests that in the solid state the concepts of σ-/π-hole, which are essential to identify electrophilic sites for the isolated monomers, should be integrated within a more comprehensive approach able to fully characterize the present interactions when other packing effects come into play. A powerful tool is provided by the topological analysis of the Laplacian of the ED, $\nabla^2\rho(\mathbf{r})$, in the region of the interacting atoms.[14,73] The Laplacian for an isolated atom or for an atom in chemical combination averages to zero when integrated over its atomic basin.[14] Atomic regions with negative (positive) Laplacian are regions of charge concentration (depletion)[14] and the function $L = -\nabla^2\rho(\mathbf{r})$ is introduced and customarily topologically analysed to associate positive (negative) values to charge concentration (depletion) (CC (CD)) regions. From an energetic point of view, regions with $L > 0$ are characterized by a local excess of potential energy density, while those with $L < 0$ have a local dominance of the kinetic energy density, relative to their average virial ratio of two. Following Bader[14] and quoting Koritsanszky and Coppens,[39] "If two reactants approach each other in a Lewis acid–base-type reaction, their relative orientation can be predicted by the Laplacian functions of their electron density.

Charge concentrations/depletions of one molecule can be considered to be complementary to depletions/concentrations of the other". Hence, the topology of the Laplacian in the solid state summarizes all the main features of σ- and π-hole bonds, namely, their electrophilic–nucleophilic character, their strong directionality and their energetic basis, *i.e.* the tendency of aligning and (eventually) merging regions of excess potential energy density with regions of excess kinetic energy density. The origin of the σ-/π-hole(s) in the EP of the donor atom can be in fact associated with the CD region in its atomic valence shell, while negative sites on the acceptor correspond to the CC region. The CD and CC sites face each other in the intermolecular regions, according to a "key and lock" arrangement, indicating the dominantly electrostatic electrophilic$\cdots$nucleophilic nature of the interaction. The relative location of the involved CD and CC sites specifies its directionality. It is then evident that the Laplacian provides a criterion for the existence of σ- and π-hole intermolecular interactions in the solid state, based on the observation of the complementarity of the CC/CD sites on the interacting pair of atoms. The Laplacian distribution, in addition, gives a clear visualization of the polar flattening of the halogens,[74] otherwise indirectly inferred by inspection of intermolecular contacts. It is however worth recalling that the topological analysis of $\nabla^2\rho(\mathbf{r})$ is not always straightforward, because the correspondence between atomic shell structure and Laplacian, generally observed for second- and third-row atoms, is not similarly well defined for heavier atoms such as Br and I.[75] Moreover, the topological analysis of L in the atomic valence shell charge concentration (VSCC)[14] of light atoms provides the location of atomic electrophilic and nucleophilic sites from the observation of its $(3, +1)$ and $(3, -3)$ critical points (CPs), respectively. For heavier atoms, the whole valence shell of the atom (including its charge depletion region, VSCD) is needed and additional $(3, +3)$ and $(3, -1)$ CPs must be considered. One needs also to mention that, compared to the Laplacian distributions of gas-phase monomers and dimers, the topological analysis of L in the crystal phase (at both experimental and theoretical levels) exhibits a smaller number of critical points in the valence shell of the heavier atoms.[76] This quite general observation can be ascribed to a smoothing effect of the inherent molecular electron density distribution from the crystal environment.

Deformation electron density $\Delta\rho(\mathbf{r})$, here considered as the difference between the atom-centered multipole ED and the IAM ED, represents a complementary tool often used to provide evidence of the electrostatic nature of σ-/π-hole interactions. Electron-excess and electron-deficient regions on either sides of the interacting atoms can in fact be associated respectively with CC and CD sites of the L distribution, indicating directional $\delta^- \cdots \delta^+$ interaction. However, $\Delta\rho(\mathbf{r})$ maps are known to suffer from the unphysical and arbitrary nature of the reference function, *i.e.*, superposed free-atom charge densities, and shortcomings of using deformation densities have been discussed.[77] Such arbitrariness of the reference function could be the origin of the frequently observed significant differences between experimental and theoretical deformation densities, even when the latter are obtained by periodic calculations, at the experimental geometry. On the other hand, experimental and theoretical Laplacian distributions are generally quite similar, with only a few well known exceptions such as those observed when studying polar bonds,[39] indicating that the intrinsic experimental limitations (such as quality of data, absorption corrections, nature of scattering factors) on one side and the adopted computational protocol (theoretical approach, basis set) on the other one do not invalidate the significance of the topological analysis.

In the following, results for the various subclasses of σ- and π-hole bonds are examined in some detail. As is customarily done in the literature, we list bcp topological data for the various cases, though a preliminary note of warning here is opportune.[9,78] A comparison between these data, especially for the ρ_b, $\nabla^2\rho_b$, G_b, V_b values and much less so for the $|V_b|/G_b$ and the $H_b = G_b + V_b$ values (the latter are either a ratio or a sum between positive and negative quantities of comparable magnitude), is at most only qualitative, if the pairs of interacting atoms are not the same. In fact, these values largely depend on the row(s) of the periodic table to which the interacting atoms belong, besides the kind of chemical bond they are forming. A comparison with the corresponding data for the IAM or, better, for the non-interacting molecules superposition ED, allows a rough estimate of the values of the bcp properties one may expect for a given pair of atoms, at a certain interaction distance, to be obtained.

18.4.1.1 X···B Halogen Bonding

The first experimental charge density studies on halogen bonding (XB) interactions involving halogen atoms (X) and heteroatoms (B) were performed on a series of complexes of dihalotetrafluorobenzenes with (*E*)-1,2-bis(4-pyridyl)ethylene,[79,80] 4,4'-dipyridyl-*N,N'*-dioxide[81] and 4,4'-dipyridyl.[80] These works focused on the topological properties of electron density at the X···N(O) bcp (X=Br, I), evidencing a strict parallelism with those derived for moderate hydrogen bonds: the values of ρ_b were relatively high for $I \cdots N$ and $I \cdots O$ (0.236(2) and 0.201(4) e Å^{-3}, respectively) and slightly lower for $Br \cdots N$ (0.183(2), 0.156(2) and 0.123(1) e Å^{-3}); those of $\nabla^2 \rho_b$ were positive and low (around 2.0 e Å^{-5}); the magnitudes of the local kinetic (G_b) and potential (V_b) energy densities were comparable, leading to $|V_b|/G_b \cong 1$ and slightly negative (X=I) or positive (X=Br) $H_b = G_b + V_b$ values. All the topological features of electron density are indicative of the closed-shell nature of halogen bonding. A comparison between structural and topological parameters, as obtained from X-ray charge density analysis and theoretical calculations on the optimized gas-phase dimer, was reported for the $Br \cdots N$ interaction.[80] A significantly shorter and stronger halogen bond was found in the former case, evidencing the involvement of cooperativity effects in the crystal phase, which reinforce and shorten the dominating intermolecular interaction on an otherwise fairly flat potential energy surface separating the halogen-bonded partners.

A topological analysis of the Laplacian showed a torus of charge concentration around the halogen, perpendicular to the C–X bond, where CC CPs were located far off the direction of the intermolecular interactions. Furthermore, the CC CPs in the VSCC of N and O atoms were located close to the $X \cdots N(O)$ connecting line, and pointed toward the halogen valence-shell charge-depletion region which exhibits, in turn, CD critical points. This CP pattern is a clear signature of the previously discussed "key and lock" arrangement of the local CC and CD of electron density.

Similar topological properties of electron density were found for $Br \cdots O$ in 2,2-dibromo-2,3-dihydroinden-1-one ($\rho_b = 0.117(2)$ e Å^{-3}, $\nabla^2 \rho_b = 1.330(4)$ e Å^{-5}, $H_b = 0.001$ a.u.)[82] and $Cl \cdots O$ in 2,5-dichloro-1,4-benzoquinone ($\rho_b = 0.054(1)$ e Å^{-3}, $\nabla^2 \rho_b = 0.795(1)$ e Å^{-5}, $H_b = 0.002$ a.u.)[83] interactions, the latter approaching weak $C–H \cdots O$ hydrogen bonds from a topological point of view. Such $X \cdots B$ interactions were also characterized by means of experimental and theoretical electrostatic potential isosurface maps, as determined on fictitious, isolated molecules extracted from the crystal lattice but containing the polarization effects induced by intermolecular interactions. Similar to the EP maps computed for gas-phase isolated molecules, they clearly show, also in the solid state, the anisotropic distribution around the halogen atom, even in the case of the less polarizable chlorine atom. Such features of EP maps further support the electrostatic and directional nature of halogen bonding as an attractive interaction between the electrophilic (δ^+) region on the halogen and the nucleophilic (δ^-) region on the Lewis base.

A significantly stronger XB has been investigated by Nelyubina *et al.*[84] in the stable polymorph of iodic acid, α-HIO$_3$, where an I$\cdots$O interaction, whose contact distance closely approaches the I–O bond length (2.4830(10) *vs.* 1.8984(10) Å, respectively), approximates a hypercoordinate bond. The topological properties of $\rho(\mathbf{r})$ at the corresponding bcp ($\rho_b = 0.37$ e Å^{-3}, $\nabla^2\rho_b = 3.08$ e Å^{-5}, $H_b = -0.012$ a.u.) were in fact intermediate between those of the covalent I–O bonds on one side and of the other weaker I$\cdots$O interactions on the other side, as found in the structure. As evidenced by ELF plots, the iodine atom behaves as both donor and acceptor of ED with the neighboring oxygen atoms. A moderate O–H$\cdots$O hydrogen bond (HB) is also present in the α-HIO$_3$ structure, where the same oxygen atom acts as XB and HB acceptor. Interestingly, DFT optimization of the gas-phase halogen and hydrogen bonded dimer leads to a significant weakening of the XB interaction, while the HB is almost unchanged, implying a different influence of crystal packing effects for these two kinds of interaction. Unfortunately, no further investigation on this intriguing feature was performed.

Halogen bonding with heteroatoms should not exclude fluorine acting as the electrophilic species, though it may appear implausible owing to its high electronegativity and scarce polarizability. Some theoretical studies, however, have revealed the presence of a positive σ-hole on the fluorine atom when it is covalently bonded to groups more electron-withdrawing than fluorine or even to fluorine itself.[85] Moreover, a careful analysis of R–F$\cdots$B close contacts in the Cambridge Structural Database (CSD)[86] revealed small differences in specific populations, giving a clear indication of an anisotropic electronic distribution around covalently bound fluorines.[85] For example, interactions of fluorine atoms with oxyanions are more linear than with neutral oxygen atoms (the median values of the interaction angle were 130 *vs.* 139°, respectively), while metal cations are preferentially attracted to the negative lateral sides of the halogen (median value 123°). F$\cdots$O intramolecular interactions have been reported and characterized in the experimental charge density analysis of pentafluorobenzoic acid,[87] but the geometrical arrangement of the C–F$\cdots$O groups, with angles just below 90°, excludes any possible halogen bonding interaction between the two atoms. Moreover, QTAIM analysis on the optimized geometry, where F$\cdots$O distances are longer than those resulting from diffraction data, did not provide any (3, −1) critical point for these intramolecular contacts. On the other hand, the F$\cdots\pi$ intermolecular interactions found in tetrafluorophthalonitrile,[88] where the π-system was either a C$\equiv$N bond or a CC bond of the aromatic ring, could really be characterized as halogen bonds with $r(\text{F}\cdots c) = 3.097$,

3.051 Å and C–F$\cdots c = 163.5$, 148.1°, respectively, where c is the centroid of the C$\equiv$N or the CC bond, and, in both cases, $\rho_b = 0.05$ e Å^{-3} and $\nabla^2 \rho_b \cong 0.6$ e Å^{-5}. While a full topological analysis of the Laplacian around the fluorine atoms was not reported, maps of this function in the molecular plane provide discernible evidence of lone pairs on the sides of the fluorine atoms, perpendicular to the C–F bonding directions and pointing away from the C–F$\cdots\pi$ direction.

18.4.1.2 X$\cdots$X Halogen Bonding

One of the essential issues in charge density analyses is to discriminate between electrostatic interactions, that is, in the present context, 'true' σ- and π-hole bonds, and van der Waals, non-directional interactions, which are generally ascribable to close-packing effects. This task is particularly demanding when we are faced with weak non-covalent bonds. A case example comes from the charge density analysis of X-ray diffraction data in hexachlorobenzene, C_6Cl_6.[89] In this structure, six independent bcps associated with Cl$\cdots$Cl bonding interactions were observed, characterized by similar and small values of ρ_b and $\nabla^2 \rho_b$ ($0.03 < \rho_b < 0.06$ e Å^{-3} and $0.3 < \nabla^2 \rho_b < 0.6$ e Å^{-5}), whose magnitude falls within the range of very weak hydrogen bonds. Halogen$\cdots$halogen (C–X$_1\cdots$X$_2$–C) short contacts are ubiquitously found in the packing of organic molecules[90] and can be geometrically classified according to the values of the two $\theta_1 = $ C–X$_1\cdots$X$_2$ and $\theta_2 = $ X$_1\cdots$X$_2$–C angles. Contacts with $\theta_1 \cong \theta_2$, including both *cis* and *trans* geometry when θ_1 and θ_2 differ from 180°, are referred to as type-I, whereas contacts with $\theta_1 \cong 180°$ and $\theta_2 \cong 90°$ are referred to as type-II interactions.[90b] Type-I interactions, generally recognized as van der Waals interactions (but see also below for the I$\cdots$I$^-$ interaction) are (when ideally taken alone) deemed to be destabilizing for the crystal structure, while type-II interactions are understood as providing a stabilizing electrostatic contribution. On the basis of geometrical considerations, three of the six Cl$\cdots$Cl bonding interactions in C_6Cl_6 were classified as type-II interactions, giving rise to the 'halogen trimer synthon', well documented in crystal engineering for Cl$_3$, Br$_3$ and I$_3$,[91] while the others fall within the type-I range. Geometrical considerations aside, a possible way to discriminate between the two kinds of interactions comes from the topological analysis of the Laplacian in the VSCC region of the interacting atoms. Whereas CC regions appear on both sides of each Cl atom, perpendicular to the covalent bond axis, CD regions are observed along the bond axis, behind the nucleus (see Figure 18.1, left). The same features were observed in solid Cl$_2$[92]

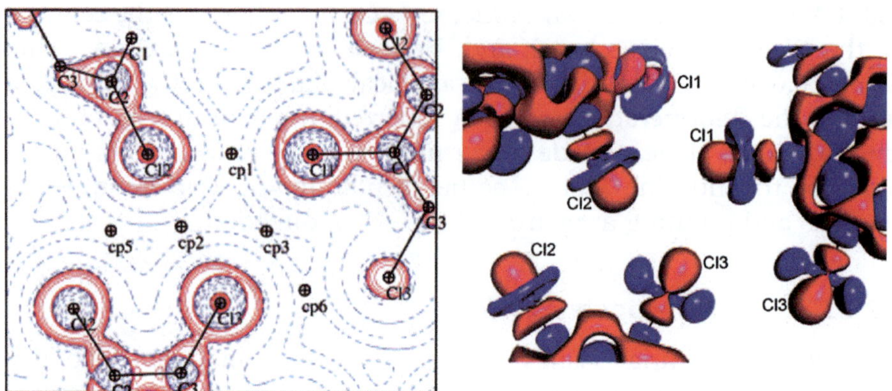

Figure 18.1 Experimental $\nabla^2\rho(\mathbf{r})$ map (left) and $\Delta\rho(\mathbf{r})$ 3D-plot (right) in the Cl_3 synthon plane of the C_6Cl_6 structure, showing the three type-II (cp1–cp3) and two type-I (cp5, cp6) Cl$\cdots$Cl interactions. Positive and negative values are represented in blue and red colours. $\nabla^2\rho(\mathbf{r})$ contours (e Å^{-5}) are drawn in logarithmic scale and $\Delta\rho(\mathbf{r})$ isosurfaces at ±0.05 e Å^{-3}. Reproduced from T.T.T. Bui, S. Dahaoui, C. Lecomte, G.R. Desiraju and E. Espinosa., The Nature of Halogen$\cdots$Halogen Interactions: A Model Derived from Experimental Charge-Density Analysis, *Angew. Chem. Int. Ed.*, 2009, **48**, 3838. Copyright Wiley-VCH Verlag GmbH & Co. KGaA. Reproduced with permission.

and I_2,[93] whose 2D layered structure is exactly related to the electron density distribution around the halogen atoms. Accordingly, the topology of $\nabla^2\rho(\mathbf{r})$, with the CC and CD regions facing each other along the three directions of the synthon, permits the geometry of the X_3 synthon to be predicted. These conclusions, confirmed by periodic calculations[94] and further strengthened by similar results obtained on the Br_3 synthon,[95] point to a description of the type-II interactions in the X_3-synthon as electrophilic – nucleophilic in nature, involving oppositely polarized regions $(X^{\delta+}\cdots X^{\delta-})$ in front of each other. This picture nicely matches the maps of deformation of the static electron density (Figure 18.1, right). Electron-excess and electron-deficient regions, plotted in blue and red, respectively, face each other along the three synthon edges, indicating directional $\delta^-\cdots\delta^+$ interactions.

The type-I (*cis* and *trans* geometry) and type-II Cl$\cdots$Cl interactions were later systematically investigated by charge density analysis in three model compounds, namely, 2-chloro-3-quinoylmethanol, 2-chloro-3-hydroxypyridine and 2-chloro-3-chloromethyl-8-methylquinoline, using high-resolution X-ray data and periodic calculations.[96] The Laplacian maps provided an unambiguous description of the nature of the Cl$\cdots$Cl interactions in the three geometrical arrangements. In type-I (both *cis* and *trans*) interactions, the lumps (CC region) on each Cl atom face away from the interaction region, while the hole on each Cl atom (CD region) faces the hole on the Cl atom participating in the Cl$\cdots$Cl interaction. In type-II interactions, the CD region hole was found to face the CC region lump, resulting in a $\delta^+\cdots\delta^-$ type of interaction.

Other homohalogen (Cl$\cdots$Cl in ClF[97] and in [ZnCl$_2$(3,4,5-trichloropyridine)$_2$][98]) and heterohalogen (Cl$\cdots$F in 2-chloro-4-fluoro-benzoic acid,[99] 4-fluorobenzoyl chloride and 2,3-difluorobenzoyl chloride,[100] with chlorine and fluorine acting as δ^+ and δ^- sites, respectively, and vice versa) intermolecular interactions have also been investigated by experimental and theoretical charge density studies, confirming in all cases the electrophilic$\cdots$nucleophilic nature of the type-II halogen$\cdots$halogen bond, with CD and CC sites on either halogen facing each other in the intermolecular region. A recent ED investigation on anion$\cdots$anion interactions between polyiodide subunits extracted from tyrosinium polyiodide hydrate[101] illustrated the unique role of the Laplacian topological analysis in discriminating between iodide anions and iodine atoms, which were otherwise indistinguishable on the basis of interatomic distances and net charge analyses.

It was also recognized that a different mechanism of XB formation takes place in the solid state when stronger interactions, such those involving charged XB donors, are expected. In the crystal of the N-methylpyrazine salt,[102] a linear I$^-\cdots$I–I$\cdots$I$^-$ motif is present with a relatively weak and closed shell I$\cdots$I$^-$ interaction ($r_{\text{I}\cdots\text{I}-} = 3.350$ Å, $\rho_b = 0.160$ eÅ^{-3}, $\nabla^2\rho_b = 1.23$ eÅ^{-5}, $H_b = -0.00135$ a.u.). Quantum chemical calculations on an optimized model I$_4^{2-}$ moiety provided a weak I$\cdots$I$^-$ interaction associated with a significant lengthening of the interatomic distance as well ($r_{\text{I}\cdots\text{I}-} = 3.591$ and 3.637 Å at MP2 and M052X levels of theory, respectively). Inspection of the theoretical and experimental Laplacian maps revealed however a remarkable difference: while the former showed a typical CC/CD arrangement, with the CC region (on the iodide anion) facing the CD region (on the molecular iodine), in the latter, CC and CD domains were located on both iodine atoms far off the I$\cdots$I$^-$ direction (a deviation from it by about 30° was found for the line connecting the CC maximum on each atom with the corresponding nucleus), similar to what observed between chloride anions in the crystal of hydroxylammonium chloride.[103] This arrangement gives rise to XBs with two CC/CD interactions, where both iodine species act at the same time as XB donor and acceptor. The different XB mechanism acting on this amphoteric interaction observed in the solid state has been ascribed to the presence of other interactions which interfere into the formation of the I$\cdots$I$^-$ halogen bonding.

A special mention should be made about charge density investigations of F$\cdots$F interactions,[79–81,87,88,99,104] because the low polarizability of the fluorine atoms could make questionable their effective role in stabilizing the crystal structure. It is however to be recalled[11] that, in the presence of a bond path (BP) connecting two atoms, the underlying pairwise interaction energy is in general negative and stabilizing even in the presence of dominating electrostatic (*i.e.*, classic) repulsive contributions, because the BP is associated with a privileged electron-exchange (*i.e.*, quantum mechanical) channel which contributes to lowering the mutual interatomic interaction energy (see Section 18.1). Though F$\cdots$F contacts below or just above the sum of van der Waals radii of the fluorine atoms are ubiquitous in fluorinated organic structures, their nature in the solid state has not yet been thoroughly explored, although some studies are appearing at intramolecular level in the gas phase.[105] A CSD[86] survey on contacts involving organic fluorine indicates a dominance of type-I over type-II contacts, in contrast to what is observed for the heavier halogens.[106] This observation suggests that they are generally to be understood as a consequence of close packing rather than a directional interaction acting as recognition unit governing – or contributing to governing – the crystal packing. In all cases, charge density studies reported on type-I F$\cdots$F interactions, with only one exception,[106] where a short type-II F$\cdots$F interaction has been intentionally engineered to infer the presence of XB with fluorine acting as electrophilic species. A structural motif was actually generated where the same fluorine (F3 in Figure 18.2a) shows amphoteric behaviour in F$\cdots$S and F$\cdots$F halogen bonds. The deformation density plot (Figure 18.2a) clearly shows that the δ^+ and δ^- regions on the involved atoms are properly oriented to give rise to the expected interactions. However, the Laplacian map (Figure 18.2b), while depicting the VSCC of sulfur pointing towards F3, does not reveal a clear signature of polar flattening on the fluorine atoms. Unfortunately, neither topological analysis of the Laplacian on these atoms nor IQA energy decomposition on the theoretically derived wavefunction, which would provide information on the relative electrostatic/exchange contributions to the F$\cdots$S and F$\cdots$F interactions, have been reported.

18.4.1.3 Chalcogen Bonding

A strong Se$\cdots$O chalcogen bonding has been characterized by charge density studies in a polymorph of Ebselen (2-phenyl-1,2-benzoselenazol-3-one),[107] revealing topological properties at bcp ($\rho_\text{b} = 0.251(2)$ e Å^{-3},

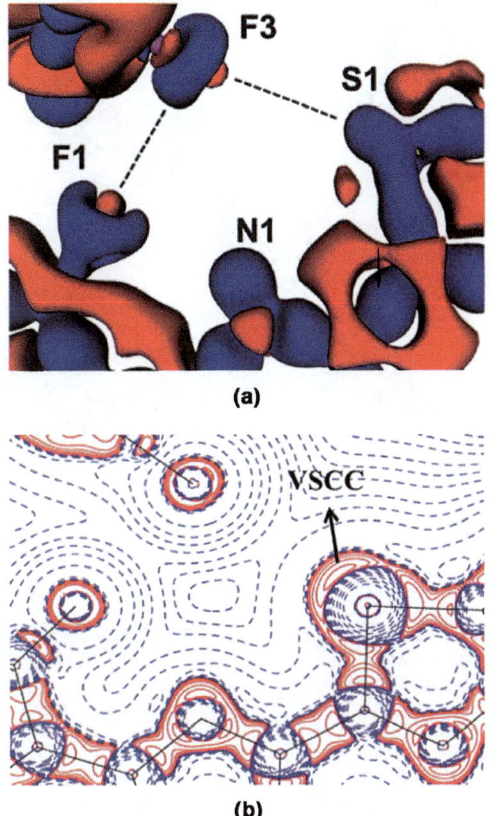

Figure 18.2 Experimental (a) $\Delta\rho(\mathbf{r})$ and (b) $\nabla^2\rho(\mathbf{r})$ maps in the F$\cdots$F and S$\cdots$F interaction regions in the pentafluorophenyl 2,2′-bithiazole structure, with positive and negative values represented by blue and red colours. $\Delta\rho(\mathbf{r})$ isosurfaces are drawn at ± 0.05 e Å^{-3}. $\nabla^2\rho(\mathbf{r})$ (e Å^{-5}) drawn on the logarithmic scale (Reproduced from ref. 106 with permission from The Royal Society of Chemistry).

$\nabla^2\rho_{\mathrm{b}} = 2.452(7)$ e Å^{-5}, $|V_{\mathrm{b}}|/G_{\mathrm{b}} = 1.11$ and $H_{\mathrm{b}} = -0.003$ a.u.) quite similar to those obtained for the I$\cdots$N halogen bonding,[79] though the full topology of charge density around the Se atom was found to differ significantly from that of the halogens. Two CD regions and the associated σ-holes have been experimentally detected in the 3D Laplacian (see Figure 18.3) and EP plots, respectively, located along the extension of the N–Se and the C–Se covalent bonds. The former was found to be more dominant owing to the greater electronegativity of the nitrogen atom compared to that of the carbon atom. Accordingly, the chalcogen bonding acceptor was found along the extension of the N–Se bond in almost linear arrangement (N–Se$\cdots$O angle, 173.54(1)°). As observed for

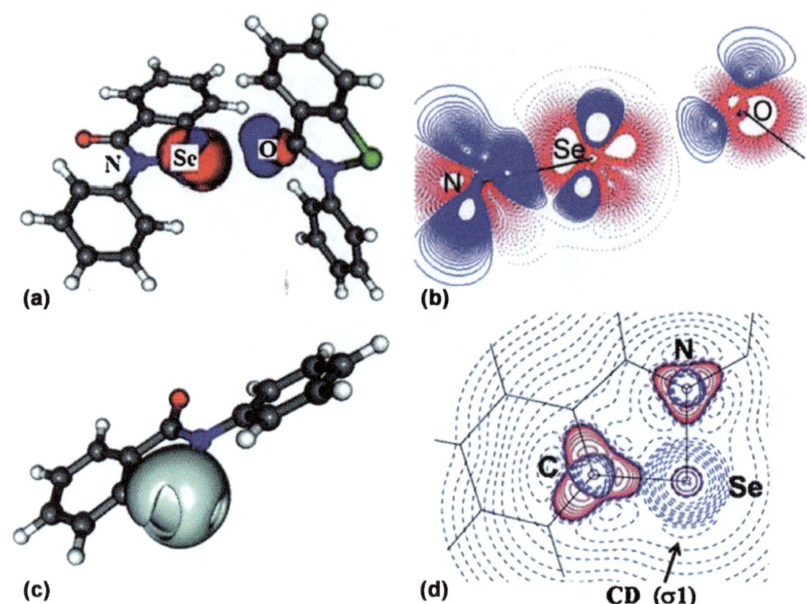

Figure 18.3   (a) 3D plot and (b) 2D maps of $\Delta\rho(\mathbf{r})$ (isosurfaces and contour levels drawn at ± 0.05 e Å^{-3} and ± 0.03 e Å^{-3}, respectively); (c) 3D plot and (d) 2D map of the Laplacian (drawn in logarithmic scale) in Ebselen, 2-phenyl-1,2-benzoselenazol-3-one (From S.P. Thomas, K. Satheeshkumar, G. Mugesh and T.N. Guru Row., Unusually Short Chalcogen Bonds Involving Organoselenium: Insights into the Se–N Bond Cleavage Mechanism of the Antioxidant Ebselen and Analogues, *Chem. Eur. J.*, 2015, **21**, 6793. Copyright Wiley-VCH Verlag GmbH & Co. KGaA. Reproduced with permission.)

halogen bonded systems, optimization of the dimer in the gas phase provides a longer Se···O distance and a more linear N–Se···O angle, a clear signature of both the stability of the identified synthon motif and the cooperativity effects present in the crystal phase.

A series of weak S···O chalcogen bonds have been characterized by X-ray diffraction studies of two polymorphs of 4,7-dibromo-5,6-dinitro-2,1,3-benzothiadiazole ($\rho_b = 0.046$–0.091 e Å^{-3}, $\nabla^2\rho_b = 0.61$–0.99 e Å^{-5}),[108] while a weak Se···O interaction has been investigated in selenophthalic anhydride (SePA) ($\rho_b = 0.049$ e Å^{-3}, $\nabla^2\rho_b = 0.62$ e Å^{-5}, $|V_b|/G_b = 0.73$ and $H_b = 0.001$ a.u.).[76] In the latter structure, the Se atom, hybridized sp^3, is bonded to two carbon atoms and gives rise to Se···O bonding with the carbonyl oxygen atom of an adjacent molecule (see Figure 18.4). Calculations on the isolated monomer showed the presence of two (3, +1) CD sites along the extension of the C – Chal bonds, where two σ-holes were located as well, and one (3, +3) CD site along the prolongation of the bisecting C$_1$–Chal–C$_2$ direction. Structural

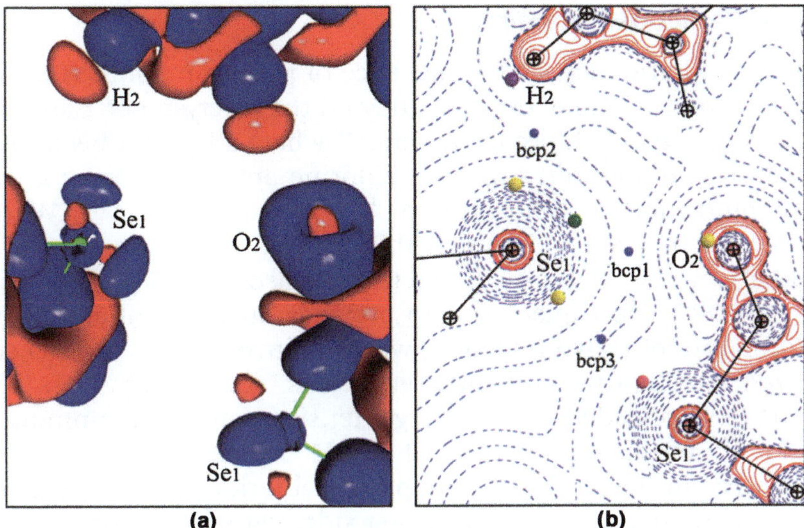

Figure 18.4 Experimental (a) $\Delta\rho(\mathbf{r})$ 3D plot and (b) $L(\mathbf{r}) = -\nabla^2\rho(\mathbf{r})$ map in the Se–O2′–Se′ plane of selenophthalic anhydride. The $\Delta\rho(\mathbf{r})$ iso-surfaces are drawn at the $\pm\,0.05$ e Å^{-3} level, positive (blue) and negative (red). $L(\mathbf{r})$ contours (e Å^{-5}) are in logarithmic scale, positive (red) and negative (blue). In (b), intermolecular bond critical points (blue) and CC (yellow) and CD (pink and green) sites involved in intermolecular interactions with Se are depicted (Reprinted with permission from M.E. Brezgunova, J. Lieffrig, E. Aubert, S. Dahaoui, P. Fertey, S. Lebègue, J.G. Ángyán, M. Fourmigué and E. Espinosa., Chalcogen Bonding: Experimental and Theoretical Determinations from Electron Density Analysis. Geometrical Preferences Driven by Electrophilic–Nucleophilic Interactions, *Cryst. Growth Des.*, 2013, **13**, 3283. Copyright (2013) American Chemical Society).

analysis of Chal$\cdots$O bonding in the TPA, SePA and TePA series where Chal$=$S, Se and Te, respectively, showed that the chalcogen bonded oxygen atom lies close to the bisecting $C_1 -$ Chal $- C_2$ direction, along which the topological analysis of L for the SePA structure indicates the presence of a $(3, -1)$ CP. These results further suggest that the Laplacian function is more appropriate to identify electrophilic sites with respect to the EP analysis, as obtained on the isolated molecule.[73] Analogously, the deformation density map, revealing positive sites on Se and O facing each other, also appears inadequate to rationalize this interaction. The same study reveals the presence of Se$\cdots$Se bonding with CD and CC sites facing each other, similar to what is observed in the experimental Laplacian maps of L-cysteine for the S$\cdots$S interaction,[109] indicating that chalcogens can interact with the molecular environment by involving simultaneously electrophilic and nucleophilic sites, exactly as observed for halogens in the X_3 synthon.[89]

18.4.1.4 Pnicogen Bonding

Experimental evidence for the presence of pnicogen bonds (PBs) has been gained by a charge density study on the co-crystal of 2-amino-5-nitropyridine and 2-chloroacetic acid,[110] where a bcp has been found between the aminic nitrogen and chlorine atoms, denoting a weak N···Cl interaction ($\rho_b = 0.054(6)$ e Å^{-3}, $\nabla^2\rho_b = 0.790(2)$ e Å^{-5}). While a possible XB occurrence between this pair of atoms can be *a priori* excluded on the basis of geometrical reasoning (C–Cl···N angle measures 98°), the Laplacian map provides unequivocal evidence of the PB nature of this interaction, with the involved CC and CD sites localized on chlorine and nitrogen atoms, respectively. Less clear evidence was provided by the charge density analysis on ammonium chloride.[111] In this structure, ···Cl···N–H··· repeating units are present where, on the basis of EP maps, the chloride anion is described as an electron density donor on either side, towards the nitrogen and the hydrogen atoms, giving rise to weak PBs and relatively stronger HBs, respectively. In this case, no Laplacian maps have been reported to enable a clear signature of PBs on the basis of the relative arrangement of the CC and CD sites to be obtained, in order to unequivocally discriminate between the supposed PB and a possible XB where the chloride anion should rather act as electron density acceptor.

18.4.1.5 Tetrel Bonding

The binding features and structural properties of tetrel bonding (TB), in selected structures extracted from the CSD,[86] have been reported by Bauzá *et al.*,[112] demonstrating the ability of heavier group-IV elements to act as Lewis acids interacting with electron donor species. Calculations on model compounds indicated that TB has comparable strength to hydrogen bonding, spanning from weak to very large interaction energies according to the tetrel species (higher binding energies were computed for heavier tetrels), and shares with the other σ-hole interactions the feature of high-directionality, suggesting its potential importance in crystal engineering. Charge density investigation of the weak C···O and C···N carbon-bonds in crystals of dimethylammonium 4-hydroxybenzoate[113] and 1,1,2,2-tetracyanocyclopropane,[114] respectively, demonstrated unequivocally the σ-hole nature of these interactions, with values of electron density and Laplacian at the bcp comparable to those reported for weak hydrogen bonds (on average, $\rho_b = 0.05$ e Å^{-3}, $\nabla^2\rho_b = 0.7$ e Å^{-5}, $|V_b|/G_b = 0.7$, and $H_b = 0.002$ a.u.).

18.4.1.6 π-Hole Bonding

Clear evidence of π-hole interaction was given *ante litteram* by Hibbs *et al.*[88] in their experimental and theoretical charge density studies of tetrafluorophthalonitrile and tetrafluoroisophthalonitrile. Several N$\cdots\pi$ interactions have been found (on average, $\rho_b = 0.06$ e Å^{-3}, $\nabla^2\rho_b = 0.7$ e Å^{-5}) as a consequence of the strongly electropositive character of the aromatic ring, in particular in the former molecule, due to the presence of several electron-withdrawing groups. Further evidence was recently provided by a charge density study on Fmoc-Leu-ψ[CH$_2$–NCS],[115] showing in its structure at 100 K the presence of short N=C=S$\cdots$N=C=S contacts, with the two interacting moieties perpendicular to each other. The authors ascribe this interaction to concomitant occurrence of both σ-hole and π-hole interactions, with the σ-hole localized on the S atom along the extension of the N=C=S axis and the π-hole localized on the N=C bond. The two positive regions interact cooperatively with the negative sites of the adjacent N=C=S moiety, so explaining the origin of the observed high directionality of this weak interaction.

18.4.2 Effects of Intermolecular Interactions on Molecular Geometry and Electron Density Distribution Explained Through Electron Density Topology

Packing effects in molecular crystals may have significant impact on the geometry and electron distribution of the composing molecules.[6–9] In the crystal, the strong intramolecular bonds coexist with the generally weak but often numerous intermolecular contacts. Their mutual presence may influence the nature of both, relative to the isolated molecule for intramolecular bonds, and/or relative to small molecular aggregates, like dimers, trimers, *etc.*, for both intramolecular and intermolecular interactions.[6–9,42] Exploitation of electrostatic forces in molecular crystals may also lead to significant geometrical and electronic perturbations of the composing molecules, even when no relevant intermolecular bonds are formed.[116] In both cases, the changes of bonding features are related to and may affect the way molecules pack together as well as their global molecular properties, like for instance their dipole moment. Can we get some qualitative and possibly quantitative insight on such changes and try to rationalize them from a chemical point of view, using QTAIM? We consider in the following two paradigmatic molecular crystal examples, one where intermolecular interactions play a relevant role (urea crystal)[6,7] and one where packing electrostatics effects dominate, while intermolecular interactions presumably have a quite limited function (a thiazete-1,1 dioxide crystal).[116]

18.4.2.1 Urea Crystal

In the crystal (space group $P\bar{4}2_1m$), urea molecules are linked to each other through hydrogen bonds (HBs) to form infinite planar tapes, with adjacent tapes being mutually orthogonal and oriented in opposite directions.[117] Cohesion between tapes realizes through another set of HBs, so that each oxygen atom turns out to be involved in four HBs, two within its own tape (O$\cdots$H$''$ length 2.06 Å, see Table 18.1 and the crystal graph, along with the gradient paths and the basin boundaries, in the inset of the table), and two shorter HBs with neighboring tapes (O$\cdots$H$'$ length 1.99 Å, also visible in the inset of Table 18.1; for more details on the crystal structure and on the complete crystal graph of urea bulk see ref. 7 and 9). Table 18.1 reports bcp properties for intramolecular and intermolecular interactions, using a number of increasingly accurate model densities, namely the IAM, the molecular, the non-interacting molecules superposition EDs and the "true" periodic crystalline ED, all obtained[6–8] using CRYSTAL 09 code, the 6-31G** basis set, the crystal geometry and, for self-consistent EDs, the Restricted Hartree–Fock method. Data shown in Table 18.1 quantify the changes in bcp properties due to the deformation density and the interaction density, as schematized in the left top inset of the table (for the effect of geometry change due to crystallization see ref. 6). The changes due to the interaction density, that is those arising from the molecular ED relaxation in the crystal, are roughly one order of magnitude smaller than those due to molecular formation. The molecular and the non-interacting molecules superposition EDs yield almost indistinguishable bcp properties for the intramolecular bonds. Relative to these two model densities, changes in the intramolecular bond properties upon crystallization range from 0 up to 3 per cent for ρ_b, but they are much larger, and as big as up to 70 and 40 per cent, for $\nabla^2\rho_b$ and λ_{3b} (Table 18.1). These latter properties are both related to second derivatives of the ED, hence more prone to respond to the environmental perturbation due to the presence of the surrounding molecules. Indeed, their changes show they are extremely *sensitive indicators* of crystal field effects and are also susceptible to an easy chemical interpretation. Such an insight may be obtained by using the dichotomous classification of bonds into shared-shell ($\nabla^2\rho_b < 0$)

Table 18.1 Intramolecular and intermolecular bond critical point properties in urea as a function of the adopted model electron density (all quantities in a.u.).

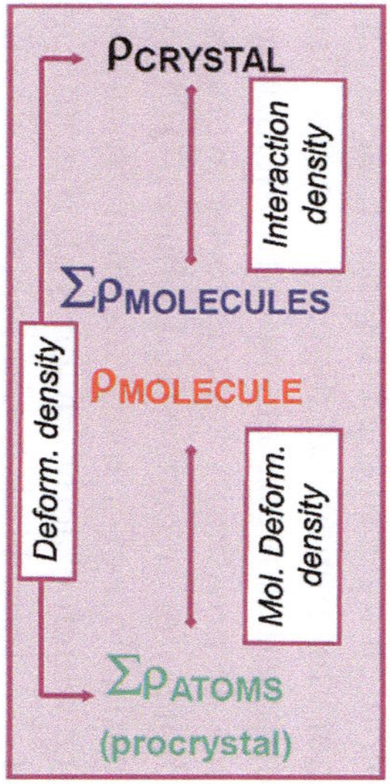

Bond X–Y	ρ_b $(\Delta\%)^a$	$\nabla^2\rho_b$ $(\Delta\%)^a$	$(\lambda_3)_b$ $(\Delta\%)^a$	ε_b $(\Delta\%)^{ab}$
C–O	0.299 (+22)	0.17 (−152)	1.02 (+42)	0.03 (−930)
	0.392 (−3)	− 0.55 (−67)	1.49 (+15)	0.07 (−2370)
	0.392 (−3)	− 0.55 (−67)	1.49 (+15)	0.07 (−2400)
	0.381	− 0.33	1.75	0.00
C–N	0.264 (+24)	− 0.20 (−83)	0.52 (+2)	0.04 (+61)
	0.341 (+2)	− 0.94 (−18)	0.73 (−38)	0.05 (+53)
	0.341 (+2)	− 0.94 (−18)	0.72 (−38)	0.05 (+53)
	0.349	− 1.15	0.53	0.10
N–H′, 1.099 Å	0.235 (+32)	− 0.39 (−80)	1.05 (−19)	0.01 (+87)
	0.345 (−0)	− 1.92 (−2)	0.78 (+11)	0.06 (−26)
	0.345 (−0)	− 1.91 (−2)	0.77 (+12)	0.06 (−28)
	0.344	− 1.95	0.88	0.05
N–H″, 1.005 Å	0.237 (+32)	− 0.40 (−80)	1.07 (−23)	0.01 (+88)
	0.351 (−1)	− 1.92 (−3)	0.71 (+18)	0.06 (−33)
	0.351 (−1)	− 1.92 (−3)	0.71 (+18)	0.06 (−33)
	0.349	− 1.97	0.87	0.05
O···H′, 1.992 Å	0.028 (−25)	0.08 (−10)	0.14 (−13)	0.07 (+7)
	0.022 (+3)	0.07 (−1)	0.12 (+1)	0.05 (+27)
	0.023	0.07	0.12	0.07
O···H″, 2.058 Å	0.025 (−34)	0.07 (−10)	0.12 (−15)	0.05 (−50)
	0.019 (−3)	0.07 (−1)	0.11 (−2)	0.01 (−78)
	0.019	0.07	0.11	0.04

^{a}Data elaborated from ref. 6 and 7. For any bond critical point property P and density model M, the property percentage change in the bulk relative to the model M, $[(P_{bulk} - P_M)/|P_{bulk}|]*100$, is reported. For each bond X–Y, property values are reported for M equal to: (i) the Independent Atom Model, IAM, electron density (first row); (ii) the isolated molecule electron density at the crystal geometry (second row); (iii) the non-interacting molecules superposition electron density, with molecules arranged as in the crystal (third row); (iv) the crystal density (fourth row). In the upper left panel, the definition of deformation, molecular deformation and interaction electron densities is illustrated in terms of the involved electron densities (EDs). $\Sigma\rho_{ATOMS}$ is the IAM ED, $\rho_{MOLECULE}$ is the isolated molecule ED, $\Sigma\rho_{MOLECULES}$ is the non-interacting molecules superposition ED and $\rho_{CRYSTAL}$ is the crystal ED. In the lower left panel of the table, $\nabla\rho$ trajectories in the plane of the molecules and of the longer O···H″ hydrogen bonds are shown for the crystal density (left) and for the non-interacting molecules superposition electron density (right).
$^b\varepsilon_b = [(\lambda_{1b}/\lambda_{2b}) - 1]$ is the bond ellipticity evaluated at the bond critical point (adimensional).

and closed-shell ($\nabla^2\rho_b > 0$) interactions, proposed a long time ago by Bader and Essén.[118] If applied on a relative scale, it provides a set of indices, whose quantitative variations along a series of chemically related compounds or, like in the present case, upon a change of phase of a given compound, neatly identify the effect these chemical perturbations have on the nature of a bond.[8,9] In particular, the decrease of ρ_b and of $\nabla^2\rho_b$ magnitudes, along with the increase of the λ_{3b} curvature, all point to the occurrence of more polar and weaker C–O and N–H bonds in the bulk, while for the C–N bond reversed changes are observed, indicating that the covalency and strength of this bond both increase upon HB formation in the crystal. The net negative charge on both O and N atoms[6] is augmented by the onset of HBs, making the nitrogen atom more disposed to share its lone electron pair with the C atom, so that the C–N bond shortens in the bulk[6] (from 1.360 to 1.345 Å) and the bond acquires a partial double character; conversely, the O atom enhances its ability to withdraw the p electron of C atom, causing a significant C=O bond lengthening in the bulk (from 1.202 to 1.261 Å), a decrease of the electron sharing between C and O atoms and, as a consequence, a shift towards a partial single-bond character for the C=O bond. Indeed, the bond ellipticity at bcp, $\varepsilon_b = [(\lambda_{1b}/\lambda_{2b}) - 1]$, which should be equal to zero for a bond with cylindrical symmetry electron distribution (*e.g.* single or triple CC bond) and larger than zero for π-double bonding,[14] increases for C–N, while it largely decreases for the C=O bond, approaching almost zero in the crystal. Note that, at variance with the intramolecular bonds, the bcp properties of the HBs in the bulk are almost matched by the simple model of the non-interacting molecular densities. Surprising as it might be, such a result *suggests that the most relevant changes induced by packing occur for intramolecular, rather than for intermolecular bonds*, that is more within the molecules themselves rather than in regions between the packed molecules (more details, examples and implications of such behaviour are thoroughly discussed in ref. 7–9).

Given the enhanced ionicity of O–H and C–O bonds and the magnitude enhancement of all atomic charges in the bulk but those of the C atom,[6] one wonders whether the molecular dipole moment also increases in the bulk. Computing such moments requires definition of the region where the electronic position vector is to be averaged out, which is not a trivial choice for a condensed phase. QTAIM, however, provides a natural definition of a molecule in a crystal as that portion of the crystal space which exactly encloses all the atomic basins of the molecule. It was so found[6] that the molecular dipole moment magnitude $|\boldsymbol{\mu}|$ increases in the crystal by 37% and by 53% relative to the isolated molecules at crystal and optimized geometry, respectively. The total molecular dipole $\boldsymbol{\mu}$ may then be rigorously partitioned in terms of a charge transfer (CT), $\boldsymbol{\mu}_{CT}$, and an atomic polarization, $\boldsymbol{\mu}_A$, component,[6,119] both quantities being origin independent for a neutral system (as is the urea molecule in the crystal defined through QTAIM). It was found that the large dipole moment increase in the bulk primarily arises from the large magnitude enhancement of $\boldsymbol{\mu}_{CT}$ which accounts for about 88% and 73% of the reported $|\boldsymbol{\mu}|$ augmentations. The magnitude of the atomic polarization component $|\boldsymbol{\mu}_A|$ instead decreases, due to the general reduction of the individual atomic polarizations upon crystallization. However, such a change concurs to slightly enhance the dipole moment enhancement due to the CT term because the atomic polarization contribution, as usually occurs, is almost collinear but oppositely directed to the CT contribution. The dipole moment enhancement in the crystal complies with a generally more polarized molecule in the bulk, where, relative to the isolated molecule, 0.066 e$^-$ has moved from each amino-group hydrogen donor to the carbonyl group acceptor. It is worth noting that the charge redistribution described by the interaction density, though leading to significant changes in some of the bcp properties (Table 18.1) and to a large dipole moment enhancement, may be difficult to visualize. Indeed, in the graphical inset of Table 18.1, the gradient paths and the atomic and molecular boundaries of the non-interacting molecules and of the crystal density look very much alike, despite the 37% enhancement of $|\boldsymbol{\mu}|$ in the crystal. Differences become evident only by plotting the interaction density itself, though the highest difference contour levels are at $\pm 2\times10^{-3}$ a.u.[8] Significant molecular dipole moment enhancements have been estimated through QTAIM in several other hydrogen bonded molecular crystals and also when only weak CH$\cdots$O interactions are operative. For instance, in the DMACB crystal (see Section 18.3.1) a percentage increase twice as big as that found in urea has been observed.[120] An extensive comparative list of molecular dipole enhancements in the crystal phase is reported in ref. 120.

Structural information and ED topology tell us that the O atom in urea is able to form four N–H$\cdots$O hydrogen bonds, though we know that in the isolated molecule it has just two lone pairs, both located in the molecular plane. A careful analysis of the density Laplacian CPs, in the Valence Shell Charge Concentration (VSCC) of the O atom, provides a chemical rationale for its enhanced H-acceptor capability in the bulk.[6,7] When, *in vacuo*, the two $(3, -3)$ L non-bonded maxima

(NBMs), related to the O atom lone pairs in urea, are topologically connected through a $(3, -1)$ L saddle point located in the molecular plane. HBs may be described as Lewis acid–base adducts, which form through the alignment of the $(3, +3)$ L minimum in the VSCC of the acidic hydrogen with a $(3, -3)$ L non-bonded maximum on the base.[121] By aligning its two non-bonded maxima with the $(3, +3)$ charge depletions (CDs) of neighboring H atoms, the O atom may form two HBs, as found in the urea dimer. However, in the bulk the noticeable elongation of the CO bond (see earlier) induces a splitting of the in-plane single saddle point into two saddle points, one above and the other below the plane of the molecule (see ref. 6 and Figure 7.3 of ref. 7). Therefore, H atoms may form HBs in the urea bulk either by approaching one of the O $(3, -3)$ NBMs in the molecular plane or one of their interconnecting saddles, which are viewed as CC maxima by H atoms approaching the O atom from above or below the molecular plane. The saddle-point splitting mechanism observed in the bulk, is replicated in the isolated urea molecule when forced at crystal geometry. The increase of the C–O bond length in the bulk, the concomitant decrease of its π-bond character and the accompanying remarkable changes in the ED distribution of the O atom in its non-bonding region, thus appears to be the key mechanism in establishing the 3D HB network in the bulk.

18.4.2.2 *Thiazete-1,1 Dioxide Crystal*

As a further example of QTAIM study of crystal field effects, selected results from an X-ray and an *ab initio* charge density study[116] on a molecular crystal of a heavily-functionalized four-membered 1,2-thiazete-1,1 dioxide ring molecule, bearing a N-bonded sulfonyl group, are briefly illustrated below. This system (hereinafter called DTC molecule or crystal) is quite uncommon: out of the >600 000 entries within the 2013 CSD release, just 16 (0.03%) contain a four-membered ring (4MR) bearing a N–SO$_2$ group and, differently from DTC, most of them have their heterocyclic core completely saturated. However, the most challenging and unusual structural feature of the DTC crystal is that in the –N–C=N–SO$_2$– moiety [see resonance structure a) in Scheme 18.1], the formally single N–C bond is found to be 0.018(3) Å shorter than the formally double N=C bond. To shed light on such an unexpected bond length inversion, a number of structurally related systems, also including the DTC molecule, were investigated, *in vacuo*, by Lo Presti[116] (see systems **1** through **4** in Figure 18.5, where **4** is DTC and where its labels ρ_{VQM}, $\rho_{\text{VQM-FROZEN}}$

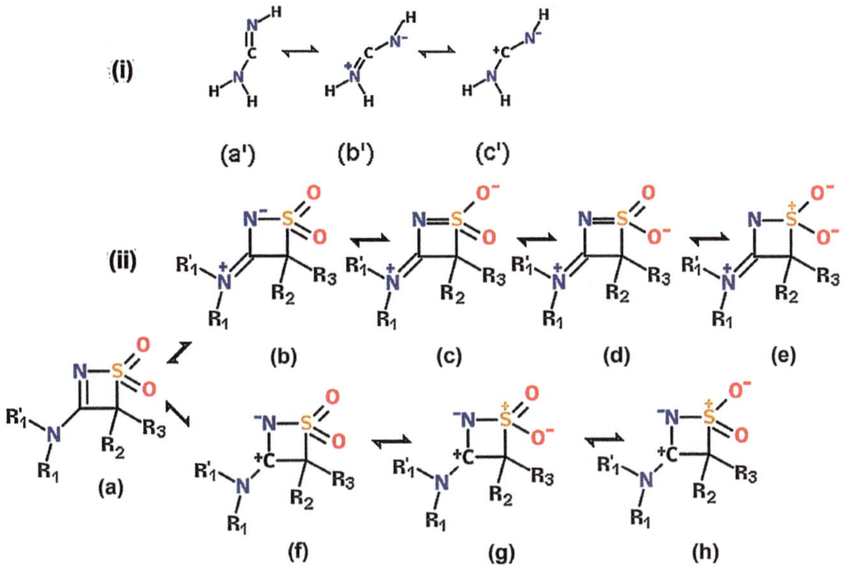

Scheme 18.1 Hypothetical and chemically conventional resonance forms for systems **1–4**.
Reprinted with permission from ref. 116. Copyright (2014) American Chemical Society.

and ρ_{PQM} further distinguish ED results obtained for DTC *in vacuo* at optimized geometry, or *in vacuo* but at the crystal geometry or in the molecular crystal, respectively). The C–N and C=N bond length evolution in these systems represents a challenging case of subtle interplay between electron delocalization and crystal field effects, with the former being also related to chemical composition. In particular, from **1** to **2**, the sulfonyl group is introduced, then from **2** to **3** 4MR cyclization occurs, while from **3** to **4** chemical substitution takes place at the C linked to S. At variance with the solid-state X-ray results, all systems **1–4** *in vacuo* (optimized geometry) show no bond length inversion and display a bond length alternation pattern in accord to the supposedly dominant resonance structure *a* (or *a′*) (Figure 18.5 and Scheme 18.1). Indeed, the C9=N2 bond distance undergoes a monotonic lengthening on going from system **1** to **4**, while its conjugated C9–N1 bond monotonically shortens by almost the same percentage amount. The bond length difference parameter, $\text{BDP} = d_{C9-N1} - d_{C9=N2}$, amounts to 0.11 Å in the conjugated methanimidamide **1** and reduces to ≈ 0.02 Å in DTC. The largest decrease (-0.037 Å) and largest increase ($+0.018$ Å) of, respectively, the formally single and formally double C–N bonds distances occur on going from **1** to **2**, upon introduction of the electron–attractor sulfonyl group. This may be easily understood in

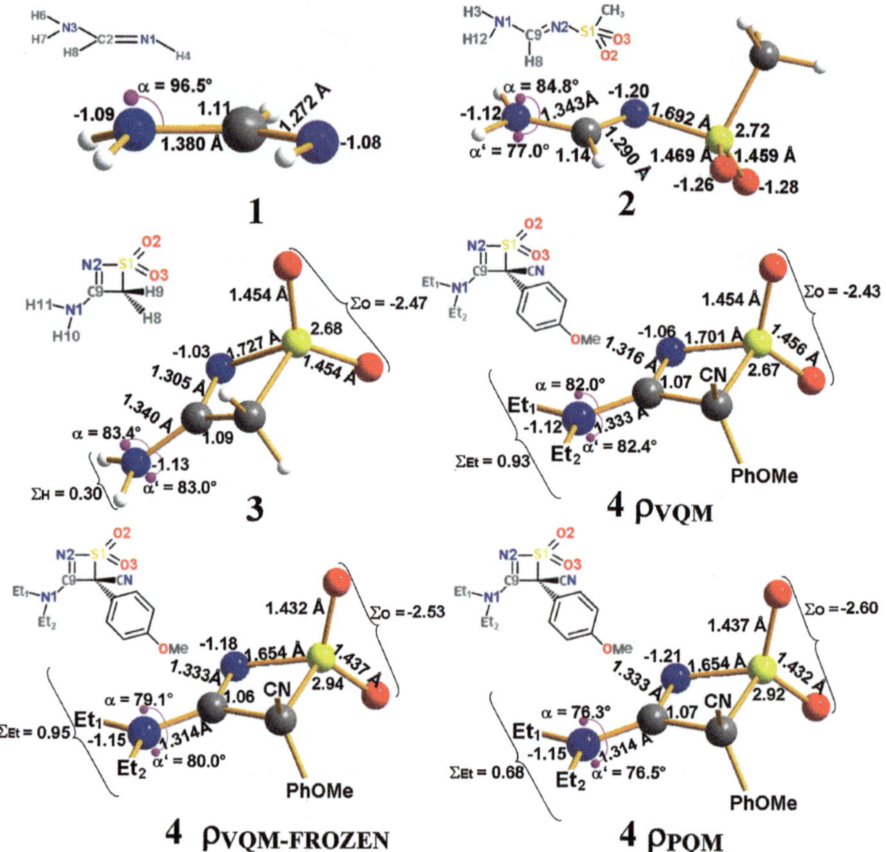

Figure 18.5 Location of $L = -\nabla^2\rho$ non bonded maxima around N1 (purple dots), bond lengths (Å) and QTAIM net charge evolution along the **1–4** system series. System **4** (DTC) is further labeled as ρ_{VQM}, $\rho_{VQM\text{-}FROZEN}$ and ρ_{PQM} to denote ED results for DTC *in vacuo* at optimized geometry, *in vacuo* at the crystal geometry and DTC molecular crystal, respectively. "Et" stands for ethyl group (Reprinted with permission from L. Lo Presti, A. Orlando, L. Loconte, R. Destro, E. Ortoleva, R. Soave and C. Gatti., Single N–C Bond Becomes Shorter than a Formally Double N=C Bond in a Thiazete-1,1-dioxide Crystal: An Experimental and Theoretical Study of Strong Crystal Field Effects, *Crystal Growth and Design*, 2014, **14**, 4418. Copyright (2014) American Chemical Society).

terms of the acquired availability of resonance forms *c–e* (and *g–h*, Scheme 18.1) following insertion of such group.

Bond length evolution and changes in the relative weight of resonance forms should also reflect in a progressive modification of the hybridization of the involved atoms. This was once more analysed in terms of the Laplacian of the electron density. Shortening of the

N1–C9 single bond and its evolution towards a double bond nature, should indeed imply a progressive change of N1 hybridization from sp^3 to sp^2. As shown in Figure 18.5, this N atom is characterized in **1** by a significant pyramidalization, with an N–H–C–H′ torsion of about 20°, and it has just *one* non bonded maximum (NBM) in its VSCC, forming an NBM–N–C α angle of about 97°. The corresponding N atom in **2** has lost most of its pyramidalization (N–H–C–H′ torsion lowered to 3°) and *two* NBMs have now appeared above and below the N atom, almost perpendicular to the H–N–C plane, and characterized by much lower α angles. The abrupt change of hybridization, the concomitant decrease of pyramidalization and the α angle(s) decrease from **1** to **2** all point to an enhanced involvement of the lone pair p-electrons on N1 into the covalent N–C bond and to its evolution towards a double C=N bond (also evidenced by a large increase of the bond ellipticity). The N1 pyramidalization seems to be completely lost when the 4MR is formed (**2** to **3** transition) and the α (and α') angles keep lowering along the series **2–4**, leading to the monotonic decrease in C–N bond length along the series. By forcing the DTC molecule to the crystal geometry, the angles α further decrease and even more decisively decrease, reaching a value of 76.3° in the crystal, where the DTC molecule is surrounded by other DTC molecules. The N–S bond weakens from **2** to **3** because of the introduction of the 4-membered ring constraint, but then it systematically shortens from 1.723 Å in **3**, to 1.701 Å in **4** and markedly further down to 1.654 Å in the DTC crystal. This trend may be interpreted in terms of an increasing importance of resonance structures *c–e*, which is corroborated, for the DTC system, by the progressive, large increase in the global negative charge of the oxygen atoms, from 2.427 e$^-$ in the geometry optimized molecule, up to a value as large as 2.596 e$^-$ in the crystal. Structures *g–h* may also play a role since charge separation does not necessarily imply bond lengthening for heteropolar bonds. The whole set of data would suggest that CN bond length inversion takes place in the crystal because the resonance forms with larger charge separation become stabilized and thus more relevant in the crystal. Evaluation of the dipole moment enhancement in the crystal confirms such a hypothesis. In the bulk, the DTC dipole moment increases by 35% (from 9.6 to 13.0 D), a percentage enhancement similar to that found in urea. In this case it is due only to the matrix effect of the crystal rather than to the geometrical change upon crystallization which, if taken alone, induces only a negligible $|\mu|$ increase (from 9.6 to 9.7 D). The general increase of net charges in the thiazete ring, and in particular in the sulfonyl groups upon crystallization augments $|\mu_{CT}|$, while the constraints of packing slightly reduce $|\mu_A|$. Both the $|\mu_A|$ decrease and the $|\mu_{CT}|$ increase concur with the large $|\mu|$ enhancement, as μ_A and μ_{CT} are almost collinear, but are oppositely directed, as in the urea crystal. One might wonder why the large dipole enhancement occurs only in the bulk, when surrounding molecules are really present, while it does not take place when their presence is considered only indirectly, through the geometrical changes they induce. This system, differently from urea, lacks any strong directional HB network.[116] Higher first electrostatic moment magnitudes, that is, an enhanced dipole moment, is required for the DTC molecule to exploit more favourable interactions with neighbouring molecules in the crystal. The driving force of bond length inversion is therefore the electrostatic contribution to the total cohesive energy of DTC. It preferentially stabilizes those DTC molecule resonance forms leading to bond length inversion, forms which become energetically convenient and manifest only in the crystal where the $|\mu|$ enhancement may be favourably exploited.[116]

18.4.3 Source Function Description of Hydrogen Bonded Systems

Hydrogen bonds (HBs), hereinafter denoted as D–H$\cdots$A interactions, are characterized by an extraordinary variety of geometries and of dominating energetic contributions, according to the nature of the H-donor, D, and that of the H-acceptor atoms, A. HB energies range from about 15–40 kcal mole^{-1} for the very short, strong hydrogen bonds (SSHB) having partly covalent nature, down to 1–4 kcal mole^{-1} for the weaker bonds, where electrostatic and/or dispersion forces play the major role.[122] Relationships among the geometrical, energetic, electronic and reactivity features of the various classes of HBs have been largely investigated by combining a variety of techniques,[122] partly also discussed in the present book, and including structural determinations through X-ray and neutron diffraction, thermochemical measures, infrared, Raman and NMR spectra, *ab initio* computations, topological studies of the ED, *etc.* . . It is shown here, through two didactic examples,[51] how the SF contributions from the various atoms involved in the H-bonding are also remarkably able to reflect the evolution of the HB nature, when a series of systems with increasing or decreasing HB lengths are compared. Furthermore, the SF HB descriptor provides interesting clues on the extent of the local/non local character of the HB interaction, as a function of both the HB length and the possible onset of cooperativity and/or electron delocalization effects.[16,51] Being applicable to both experimentally and theoretically derived EDs, the SF HB descriptor is of value also for comparative purposes.[16] In the first example, two H_2O molecules, approaching each other within the linear dimer C_s constraint, were considered by Gatti *et al.*[51] as a *gedanken* experiment to study how the atomic sources reconstructing the HB ρ_b value evolve when the $H \cdots O$ and $O \cdots O$ distances change from the values typical of weak isolated HBs to those of SSHBs. As shown in Figure 18.6 for a selected set of geometries, the atomic SF% contributions at the HB bcp were found to change radically along the reaction path, those of atoms most directly involved in the H-bond ($D–H \cdots A$) increasing and those of the remaining atoms decreasing with increasing energy and covalency of the HB.

By drawing atomic volumes proportional to the SF% contributions, this evolution becomes pictorially evident, being mirrored by a change from sources spread almost over the entire dimer to sources mostly localized on the $D–H \cdots A$ triad of atoms (Figure 18.6a). The SF% from the H involved in the HB was shown to be a characteristic *marker* of the HB nature, as it is largely negative at the equilibrium distance ($O \cdots O = 3.020$ Å), while it progressively augments with decreasing $H \cdots O$ distances and it becomes eventually positive only for the very short $O \cdots O$ and $H \cdots O$ distances.[51] Finding a negative SF contribution to the ED, which is an intrinsically positive quantity, might seem at a first sight a bit disturbing. However, the capability of the H atom, in given circumstances, to determine a negative electron density contribution at its HB bcp has instead a profound chemical meaning, as explained in the following. An atomic source function value will always be the result of a sum of local positive and negative contributions and may thus, in principle, be either positive or negative. When an atom Ω is in isolation, $SF(\mathbf{r}; \Omega)$ will be positive at any $\mathbf{r}$, because only its own basin determines the ED. But, for an atom in a polyatomic system, the situation may change. Though the local sources are usually found, on average, to dominate the local sinks, resulting in positive atomic SF contributions at bcps, the reverse may also occur in specific cases, like, for instance, for the H atom involved in weak to moderate $OH \cdots O$ HBs and for a reference point *rp* taken at the HB bcp. Inspection of the $L = -\nabla^2\rho$ maps in the plane containing the H-donor molecule and the HB bond path (Figure 18.6b), attaches a chemical meaning to such a specific behaviour. At equilibrium distance the H atom basin is notably asymmetric in shape and exhibits a quite inhomogeneous $L(\mathbf{r})$ distribution, with a large region of positive Laplacian surrounding the HB bcp. As a consequence of this combined

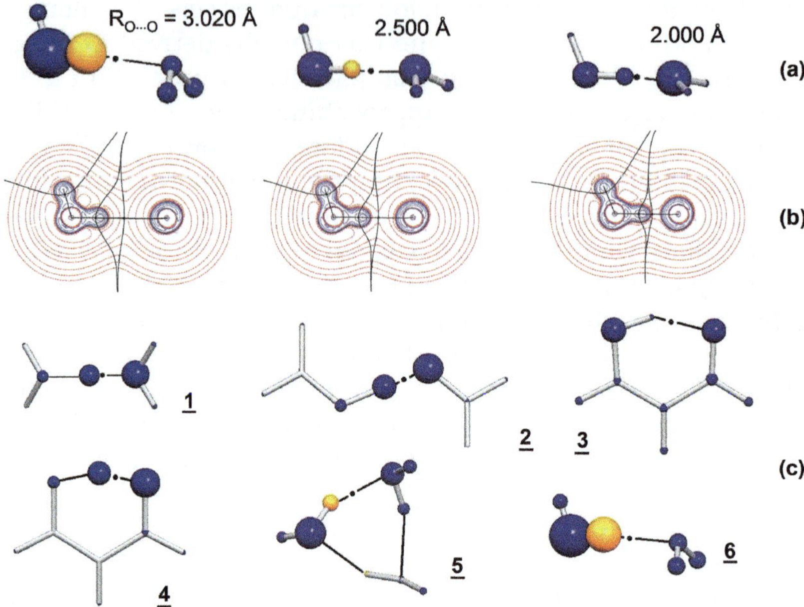

Figure 18.6   Source Function (SF) contributions at the hydrogen-bond bcp EDs, for the *gedanken* experiment for the approach of two H_2O molecules within the linear dimer C_s constraint and for various prototypical hydrogen bonded systems, spanning a large range of OH···O distances and comprising the main hydrogen-bond classes. The position of the reference point is denoted by a dot. In (a) evolution of the SF atomic contributions as the two H_2O molecules approach to each other (3.020 Å, equilibrium distance). Only three representative points along the reaction path are shown. Atoms are drawn with volumes proportional to their SF% contributions (blue: positive; yellow: negative); in (b) Isocontour plots of $L(r) = -\nabla^2\rho$, in a plane containing the H-donor molecule for the H_2O dimers shown in the first row and in (c) SF atomic contributions in prototypical hydrogen-bonded complexes. Reproduced from figure 4 *Electron Density and Chemical Bonding II*, The Source Function Descriptor as a Tool to Extract Chemical Information from Theoretical and Experimental Electron Densities, 147, 2011, 193–285, C. Gatti, Copyright Springer–Verlag Berlin Heidelberg 2011, with permission of Springer.

situation (see also eqn (18.3)) a source as negative as −72.3% results from the H at the HB bcp, while a standard positive SF contribution occurs from this same H at its O–H bond bcp. However, when the O···O distance decreases and the O–H and H···O distances become progressively alike, both the shape of the H basin and its $L(\mathbf{r})$ distribution become more and more symmetric; hence the SF% contribution from the H to its HB bcp density increases with decreasing donor to acceptor separation until it becomes eventually positive at very short O···O distances. The variation of the SF% contribution from the H to its HB bcp density, along the reaction path of approach of two water molecules, *summarizes in a single sequence of numbers* the corresponding complex changes of shape, size and Laplacian distribution within the H basin involved in the HB. These changes bear an easy chemical interpretation, in terms of an enhanced equalization of the initial O–H and H$\cdots$O bonds and of an acquired partial covalent character for the latter with decreasing O$\cdots$O distance.

Also interesting is the trend in the sum of the SF% contributions from the H and the acceptor O atom. It has a very negative value at equilibrium, it approaches zero at distances (O$\cdots$O = 2.750 Å) typical of the long chains of O–H$\cdots$O bonds in water and alcohols, and eventually reaches positive values, but no larger than 50%, at the O$\cdots$O = 2.250 Å distance typical of the charge assisted H-bonds.[122,124] The sum of SF% contributions from two covalently bonded atoms at their bcp usually exceeds 80–90%.[16] To get to SF% values similar to these, O$\cdots$O distances need to be lowered below 2.5 Å and the SF% contribution from the donor O atom has also to be included in the sum. This result implies that even at such low O$\cdots$O distances, the HB retains at least a three-center nature,[51] a vision more recently suggested also by Popelier in terms of an IQA analysis of the water dimer.[123] At equilibrium, the H atoms not directly involved in the HB are able to determine almost half of the HB bcp electron density, disclosing a fairly delocalized pattern of sources for such a bond at equilibrium. It seems reasonable to relate such delocalization of sources to the dominant electrostatic nature of the H-bond at equilibrium, in contrast to its partial covalency when the sum of SF% contributions from the H + A + D triad becomes dominant.

Gatti *et al.*[51] have tested whether their results from the study of the water dimer model could also apply to a series of prototypical "true" HB complexes (Figure 18.6c). Using the classification proposed by Gilli and Gilli,[124] they included in their study two Charge-Assisted Hydrogen Bonds, [+CAHB, *1*: $(H_2O\cdots H\cdots OH_2)^+$; −CAHB, *2*: the open form of the formic acid–formate anion complex], two Resonance Assisted Hydrogen Bonds [RAHB, malonaldehyde, in its C_s equilibrium form, *3*, and in its C_{2v} transition state, TS, *4*, for the H-atom transfer between the two oxygen atoms], a Polarization Assisted Hydrogen Bond [PAHB, *5* : cyclic homodromic water trimer] and an Isolated Hydrogen Bond [IHB, *6* : water dimer at equilibrium geometry]. The results not only confirmed most of the conclusions from the model study on the water dimer, as may be seen from the patterns of sources reported in Figure 18.6c, but also highlighted additional interesting and subtle features.[16,51] The SF% contribution from the H atom involved in the HB was confirmed as an excellent marker of the H-bond nature, but its variation with the donor to acceptor distance is enhanced relative to the water dimer model. More importantly, the SF% contributions for the RAHB in malonaldehyde, *3*, and in malonaldehyde TS, *4* exhibit a significantly decreased value from the $(H + D + A)$ triad at the HB bcp, relative to that expected from the $O{\cdots}O$ separation distance. The apparent anomaly could be ascribed to the ability of SFs to reflect the peculiar electron conjugation mechanisms operating in RAHBs. Such capability to reveal electron delocalization effects has also been corroborated by two recent dedicated studies[125,126] using theoretically[125] and either theoretically or experimentally derived EDs.[126] The set of results obtained through the study of various hydrogen-bonded prototypical complexes has led to a classification of $OH{\cdots}O$ interactions, based only on the values of the SF contributions to the HB bcp ED. For further details, the reader is directed to ref. 8, 16 and, in particular, 51.

18.4.4　Non-covalent Interactions in Molecular Crystals Viewed Through the Reduced Density Gradient Descriptor and QTAIM

As mentioned in Section 18.3.3, Saleh *et al.*[18,61] have extended the application of the NCI RDG analysis to multipolar model EDs derived from charge density quality X-ray diffraction studies. They have also investigated how these RDG isosurfaces differ either from those obtained using *ab initio* periodic approaches or from those where the multipolar model RDG is replaced by the more customarily adopted IAM electron density RDG. Saleh *et al.* have considered in their study four systems spanning various kinds of NCIs, including $\pi{\cdots}\pi$, $CH{\cdots}\pi$, $CH{\cdots}O$, $OH{\cdots}O$, $NH{\cdots}N$, $S{\cdots}H$ and $S{\cdots}S$ interactions. Here, for the sake of space and for didactic reasons, we discuss just few results involving the benzene crystal, as a prototypical example of crystal packing dominated by stacking and weak $CH{\cdots}\pi$ interactions. Details on the X-ray data quality and on the adopted multipolar model may be found in ref. 18.

Figure 18.7b shows the RDG *vs.* ρ plot relative to ED data in a cubic volume containing the asymmetric unit of the benzene crystal (3 C–H bonds, two C–C bonds; Figure 18.7a), so using only pseudo-atoms (see Section 18.2) of such a unit. It corresponds to study an isolated benzene which however feels the effect of crystal field and whose ED has been polarized by the crystalline environment. As expected, just the spikes corresponding to the intramolecular bonds (CC and CH) are seen and at large ρ values. In the Figure 18.7c, a RDG *vs.* ρ plot for

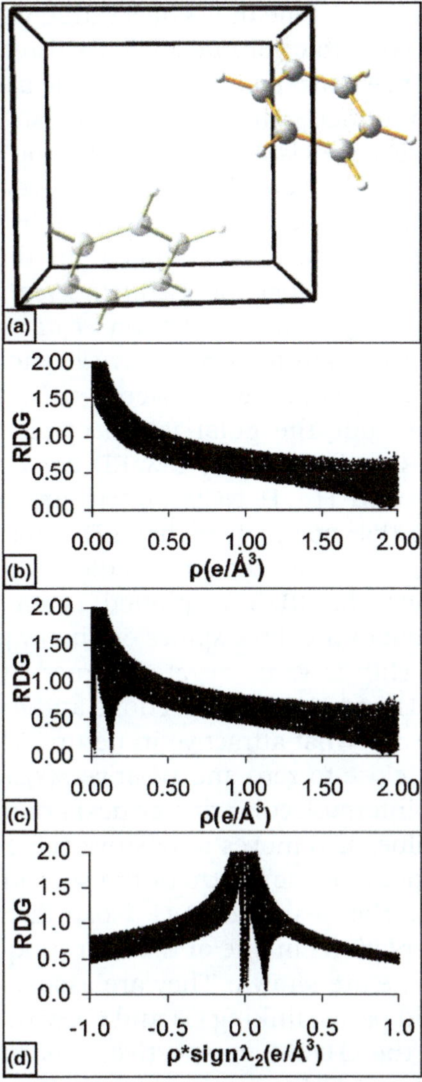

Figure 18.7 Non-Covalent Interactions (NCI) in the benzene crystal, studied through the Reduced Density Gradient (RDG) obtained from X-ray multipolar model ED. (a) Representation of the crystal region considered to build the plots; (b) RDG *versus* ρ plot for an isolated molecule in the crystal, but with its ED polarized by the crystalline environment; (c) the same as in (b) but for two benzene molecules; (d) same as (c), but multiplying the ED by the sign of λ_2.
Adapted (unpublished material has been included) from G. Saleh, C. Gatti, L. Lo Presti and J. Contreras-García, Revealing non-covalent interactions in molecular crystals through their experimental electron densities, *Chem. Eur. J.*, 2012, **18**, 15523 Copyright Wiley-VCH Verlag GmbH&Co. KGaA. Reproduced with permission.

a *pair* of neighbouring benzene molecules extracted from the crystal is shown (the considered cubic volume encloses three C–H and two C–C bonds of one reference benzene molecule, while a second molecule is almost entirely contained within the cube so that some inter-molecular contacts can be visualised). In addition to the RDG spikes at large ED values ($\rho(\mathbf{r}) > 1.8$ eÅ^{-3}), *i.e.* at ED values typical of covalent bonds in benzene, four spikes (three of which, at the lowest ED values, are nearly superimposed) also appear in the low-ED/low-RDG region of the plot. This is just the *distinctive feature of non-covalent interactions.* Analogously to what Johnson *et al.*[17] have observed by applying the RDG approach to some *in vacuo* molecular dimers, an NCI fingerprint is here similarly recovered using multipole-derived ED's, which also include the polarization effect due to the crystal field. The three RDG spikes at very low ED correspond to C···H (at slightly higher ED) and H···H NCIs and to an intermolecular ring critical point, while that at much higher ED value is the sign of the ring critical point of the benzene molecule fully contained in the cubic volume. When the RDG is plotted against the λ_2-signed ρ (Figure 18.7d), two out of the three spikes originally placed at the lowest absolute ED values shift towards negative signed-ED values. They correspond to the C···H and H···H interactions and have a local negative λ_2 curvature being somewhat attractive in nature. The remaining third spike has instead a close to zero and positive $\rho(\mathbf{r})\text{sign}(\lambda_2)$ value, being associated with the intermolecular ring critical point. The RDG spike at much higher ED value, also moves to positive values since it is related to the ring critical point at the centre of the second aromatic ring.

In Figure 18.8a, the ρ-signed-RDG isosurfaces relative to two intermolecular interactions in one of the four unique molecular pairs in crystalline benzene are shown. They are associated to two QTAIM intermolecular bond paths, linking C3 and C1 with H2 and H3 atoms, respectively. From the QTAIM perspective, this situation should in-dicate two well-defined CH···C interactions. However, the H2···C3 bond path is visibly bent and the H2 atom is located in a position roughly equidistant from all the carbon atoms of the six-carbon membered ring (6-CMR). All these features, typical of systems show-ing CH···π interactions, point to the fact that H2 should in fact interact in a similar way with all the C atoms of the symmetry-related molecule, giving rise to a CH···π attractive contact involving the whole π-electron cloud of the facing aromatic ring.

In fact, the RDG-based NCI descriptor gives rise to a donut-like-shaped, large iso-surface that covers almost entirely the 6-CMR of the other molecule. *This picture dramatically contrasts that offered by the*

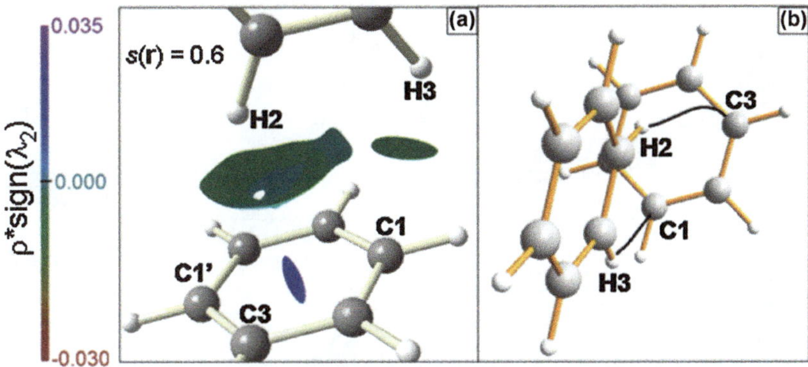

Figure 18.8 (a) RDG-based NCI isosurfaces for a molecular pair in the benzene crystal; (b) Intermolecular QTAIM bond paths for the same molecular pair. The colour scale for ρ-signed RDG isosurfaces (e bohr^{-3}) is shown on the left.
Adapted from G. Saleh, C. Gatti, L. Lo Presti and J. Contreras-García., Revealing non-covalent interactions in molecular crystals through their experimental electron densities, *Chem. Eur. J.*, 2012, **18**, 15523. Copyright Wiley-VCH Verlag GmbH&Co. KGaA. Reproduced with permission.

QTAIM bond path analysis, which is found to privilege the interaction of the H with just one single atom of the ring (Figure 18.8b). The ρ-signed RDG iso-surface of the other CH$\cdots$C interaction (H3$\cdots$C1) looks instead much smaller, disc-shaped, and only slightly negative in value. All such features comply with a conventional, very weak HB, as anticipated by the largely asymmetric location of H3 relative to the ring atoms of the other interacting molecule in the pair and by the almost straight bond path linking H3 and C1. The RDG-based NCI and the QTAIM bond path pictures nicely agree in this case. Such different behaviour from case to case is not surprising. As discussed in Section 18.3.1, when the intermolecular interactions are examined through the properties at intermolecular bcps, one unavoidably relies on an atom–atom pair description of the chemical bond, often localised and possibly discontinuous. However, many significant non-covalent interactions may have an inherently "delocalized" character and all chemical interactions are by nature characterized by continuous energy contribution changes upon lengthening or shortening.[8] The RDG index, differently from the bond path, is able to depict these inherently delocalized interactions in terms of extended and flat RDG iso-surfaces and so may differ from the QTAIM picture, in given circumstances. Yet, Saleh *et al.*[18,61] also showed that by combining the RDG-NCI analysis with QTAIM, one may often take a synergic advantage from both, as the QTAIM topological tool allows for a quantitative insight into the qualitative NCI picture offered by the RDG index.

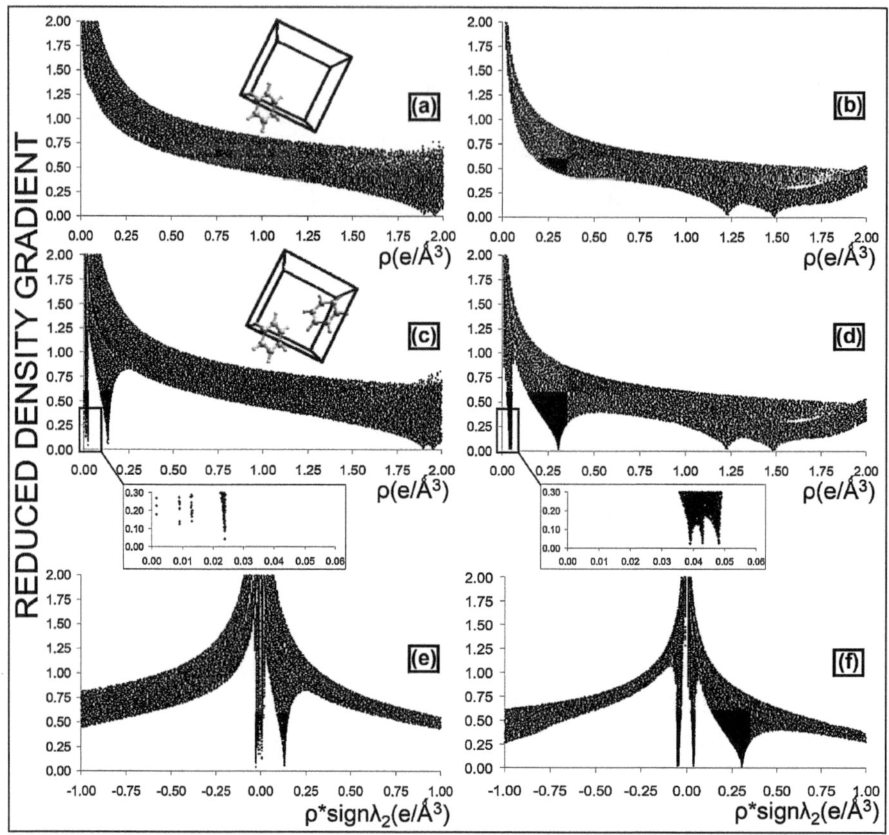

Figure 18.9 RDG analysis using the X-ray multipolar model ED (left panels) or the IAM ED (right panels). ρ *vs.* RDG for: (a) and (b) the isolated benzene (part of one molecule extracted from crystal, as shown in the inset); (c) and (d) two molecules of benzene at crystal geometry (see inset). The enlargements at low ρ values below panels *c* and *d* allow the RDG spikes to be distinguished relative to the different intermolecular interactions and to the intermolecular ring critical point; (e) and (f) are the same as in *c* and *d* but using a λ_2-signed ED.
Reprinted from *Computational and Theoretical Chemistry*, **998**, G. Saleh, C. Gatti and L. Lo Presti, Non-covalent interaction via the reduced density gradient: Independent atom model vs experimental multipolar electron densities, 148–163, Copyright 2012 with permission from Elsevier.

We conclude this chapter by examining Figure 18.9, where the RDG analysis using the X-ray multipolar model ED (left panels) for the benzene crystal is contrasted to that using the IAM ED (right panels). Comparison of panels *a* and *b* (single benzene molecule extracted from the crystal) shows that the bcps occur at lower ED values for the IAM density and that the spikes for C–H and C–C interactions are definitely more separated for that density. The spikes of the IAM-RDG are also sharper, with RDG varying much faster relative to the ED. The RDG spikes of the multipolar ED are less defined since the RDG varies much less rapidly than for the IAM density and is much less "structured" around the bcp density values. Comparison of c and d panels (relative to a pair of benzene molecules extracted from the crystal) shows a fully reversed situation for the intermolecular interactions. All four RDG spikes occur at markedly lower ED values for the multipolar ED, at about half the value of the corresponding IAM density values. In addition, the spike of the benzene molecule RCP is clearly sharper for the multipolar ED. It is a bit more difficult to compare, from panels c and d, the shape of spikes associated to the intermolecular NCI and to the ring critical point. Enlargements for such interactions at low ED values also reveal much sharper peaks for the multipolar model ED than for the IAM ED. In summary, for covalent interactions, due to the density rearrangement caused by covalent bonding, the multipolar ED is larger in value around the bcps (ED is accumulated along bonds) but its curvatures and gradient are comparatively smaller because the ED is shared, resulting in a more flat ED distribution around the bcp. The situation is reversed for the intermolecular interactions, as they are not shared in nature. The IAM density is therefore more isotropic in such case, more uniformly distributed and not contracted towards the interacting nuclei. Hence the RDG is smaller and its derivative also smaller leading to more rounded RDG peaks. Analysis of Figure 18.9 indeed shows that the "true" RDG may significantly differ from that obtained through the crude IAM, even if the related RDG isosurfaces often look qualitatively similar. Implications of such differences are thoroughly discussed in ref. 61. Relationships between the stabilization energy of molecular adducts and integrated energy densities (G, V, *etc.*) within RDG isosurface basins associated to intermolecular bonds have also been recently explored.[127]

Acknowledgements

One of us (CG) wishes to thank the Danish National Research Foundation for partial funding of this work through the Center for Materials Crystallography (DNRF93).

References

1. J. W. Steed and J. L. Atwood, *Supramolecular Chemistry*, Wiley, Chichester, 2000.
2. J. D. Dunitz, *IUCrJ*, 2015, **2**, 157.

3. C. Lecomte, E. Espinosa and C. F. Matta, *IUCrJ*, 2015, **2**, 161.
4. T. S. Thakur, R. Dubey and G. R. Desiraju, *IUCrJ*, 2015, **2**, 159.
5. J. D. Dunitz and A. Gavezzotti, *Angew. Chem., Int. Ed.*, 2005, **44**, 1766.
6. C. Gatti, V. R. Saunders and C. Roetti, *J. Chem. Phys.*, 1994, **101**, 10686.
7. C. Gatti, in *The Quantum Theory of Atoms in Molecules: From Solid State to DNA and Drug Design*, ed. C. F. Matta and R. J. Boyd, Wiley-VCH, 2007, ch. 7, p. 163.
8. C. Gatti, *Phys. Scr.*, 2013, **87**, 048102, (38pp).
9. C. Gatti, *Z. Kristallogr.*, 2005, **220**, 399.
10. M. A. Blanco, A. Martín Pendás and E. Francisco, *J. Chem. Theory. Comput.*, 2005, **1**, 1096.
11. A. Martín Pendás, E. Francisco, M. A. Blanco and C. Gatti, *Chem. – Eur. J.*, 2007, **12**, 9362.
12. C. Gatti and P. Macchi, in *Modern Charge Density Analysis*, ed. C. Gatti and P. Macchi, Springer, Dordrecht, Heidelberg, London, New York, 2012, ch. 1, p. 1.
13. C. Pisani, R. Dovesi, A. Erba and P. Giannozzi, in *Modern Charge Density Analysis*, ed. C. Gatti and P. Macchi, Springer, Dordrecht, Heidelberg, London, New York, 2012, ch. 2, p. 79.
14. R. F. W. Bader, *Atoms in Molecules: A Quantum Theory*, International Series of Monographs on Chemistry 22, Oxford University Press, Oxford, 1990.
15. R. F. W. Bader and C. Gatti, *Chem. Phys. Lett.*, 1998, **287**, 233.
16. C. Gatti, *Electron Density and Chemical Bonding II: Theoretical Charge Density Studies, Structure and Bonding*, 2012, **vol. 147**, 193.
17. E. R. Johnson, S. Keinan, P. Mori-Sanchez, J. Contreras-García, A. J. Cohen and W. Yang, *J. Am. Chem. Soc.*, 2010, **132**, 6498.
18. G. Saleh, C. Gatti, L. Lo Presti and J. Contreras-García, *Chem. – Eur. J.*, 2012, **18**, 15523.
19. P. Coppens, *X-Ray Charge Densities and Chemical Bonding*, Oxford University Press, Oxford, 1997.
20. *Modern Charge Density Analysis*, ed. C. Gatti and P. Macchi, Springer, Dordrecht, Heidelberg, London, New York, 2012.
21. J.-M. Gillet, T. Koritsanszky, in *Modern Charge Density Analysis*, ed. C. Gatti and P. Macchi, Springer, Dordrecht, Heidelberg, London, New York, 2012, ch. 5, p. 181.
22. K. Kurki-Suonio, *Acta Crystallogr., Sect. A: Found. Crystallogr.*, 1968, **24**, 379.
23. F. L. Hirshfeld, *Acta Crystallogr., Sect. B: Struct. Sci.*, 1971, **27**, 769.
24. R. F. Stewart, *Acta Crystallogr., Sect. A: Found. Crystallogr.*, 1976, **32**, 565.
25. N. K. Hansen and P. Coppens, *Acta Crystallogr., Sect. A: Found. Crystallogr.*, 1978, **34**, 909.
26. (a) A. Volkov, P. Macchi, L. J. Farrugia, C. Gatti, P. Mallinson, T. Richter and T. Koritsanszky, XD2006 – a computer program package for multipole refinement, topological analysis of charge densities and evaluation of intermolecular energies from experimental and theoretical structure factors, 2006; (b) R. F. Stewart, M. A. Spackman and C. Flensburg, *VALRAY – User's Manual*, 2.1 edn, Carnegie Mellon University and University of Copenhagen, Pittsburgh/Denmark, 2000; (c) R. Bianchi and A. Forni, *VALTOPO*: a program for the determination of atomic and molecular properties from experimental electron densities, *J. Appl. Crystallogr.*, 2005, **38**, 232; (d) V. Petricek, M. Dusek and L. Palatinus, *JANA2006, Structure Determination Software Programs*, Institute of Physics, Praha, 2006; (e) C. Jelsch, B. Guillot, L. Lagoutte and C. Lecomte, *J. Appl. Crystallogr.*, 2005, **38**, 38.
27. R. F. Stewart, J. Bentley and B. Goodman, *J. Chem. Phys.*, 1975, **63**, 3786.
28. A. Ø. Madsen, S. Mason and S. Larsen, *Acta Crystallogr., Sect. B: Struct. Sci.*, 2003, **59**, 653.
29. P. Roversi and R. Destro, *Chem. Phys. Lett.*, 2004, **386**, 472.

30. A. E. Whitten and M. A. Spackman, *Acta Crystallogr., Sect. B: Struct. Sci.*, 2006, **62**, 875.
31. D. Jayatilaka and B. Dittrich, *Acta Crystallogr., Sect. A: Found. Crystallogr.*, 2008, **64**, 383.
32. M. Woińska, S. Grabowsky, P. M. Dominiak, K. Woźniak and D. Jayatilaka, *Sci. Adv.*, 2016, **2**, e1600192.
33. I. Mata, E. Espinosa, E. Molins, S. Veintemillas, W. Maniukiewicz, C. Lecomte, A. Cousson and W. Paulus, *Acta Crystallogr., Sect. A: Found. Crystallogr.*, 2006, **62**, 365.
34. (a) A. Fischer, D. Tiana, W. Scherer, K. Batke, G. Eickerling, H. Svendsen, N. Bindzus and B. B. Iversen, *J. Phys. Chem. A*, 2011, **115**, 13061; (b) N. Bindzus, T. Straasø, N. Wahlberg, J. Becker, L. Bjerg, N. Lock, A.-C. Dippel and B. B. Iversen, *Acta Crystallogr., Sect. A: Found. Crystallogr.*, 2014, **70**, 39.
35. (a) A. Volkov, Y. Abramov, P. Coppens and C. Gatti, *Acta Crystallogr., Sect. A: Found. Crystallogr.*, 2000, **56**, 332; (b) A. Volkov, C. Gatti, Y. Abramov and P. Coppens, *Acta Crystallogr., Sect. A: Found. Crystallogr.*, 2000, **56**, 252; (c) A. Volkov, Y. Abramov and P. Coppens, *Acta Crystallogr., Sect. A: Found. Crystallogr.*, 2001, **57**, 272.
36. M. A. Spackman and P. G. Byrom, *Chem. Phys. Lett.*, 1997, **267**, 215.
37. J. J. McKinnon, A. S. Mitchell and M. A. Spackman, *Chem. – Eur. J.*, 1998, **4**, 2136.
38. M. A. Spackman and J. J. McKinnon, *CrystEngComm.*, 2002, 378.
39. T. S. Koritsanszky and P. Coppens, *Chem. Rev.*, 2001, **101**, 1583.
40. R. F. W. Bader, *J. Phys. Chem. A*, 1998, **102**, 7314.
41. V. G. Tsirelson, P. F. Zou, T.-H. Tang and R. F. W. Bader, *Acta Crystallogr., Sect. A: Found. Crystallogr.*, 1995, **51**, 143.
42. C. Gatti, E. May, R. Destro and F. Cargnoni, *J. Phys. Chem. A*, 2002, **106**, 2707.
43. (a) J. Poater, M. Sola and F. M. Bickelhaupt, *Chem. – Eur. J.*, 2006, **12**, 2889; (b) J. Poater, M. Sola and F. M. Bickelhaupt, *Chem. – Eur. J.*, 2006, **12**, 2902; (c) A. Haaland, D. J. Shorokhov and N. V. Tverdova, *Chem. – Eur. J.*, 2004, **10**, 4416; (d) S. Grimme, C. Mück-Lichtenfeld, G. Erker, G. Kehr, H. Wang, H. Beckers and H. Willner, *Angew. Chem., Int. Ed.*, 2009, **48**, 2592; (e) R. F. W. Bader, *Chem. – Eur. J.*, 2006, **12**, 2896.
44. J. Cioslowski and S. T. Mixon, *J. Am. Chem. Soc.*, 1992, **114**, 4382.
45. L. J. Farrugia, C. Evans and M. Tegel, *J. Phys. Chem. A*, 2006, **110**, 7952.
46. R. Ponec and C. Gatti, *Inorg. Chem.*, 2009, **48**, 11024.
47. E. Espinosa, E. Molins and C. Lecomte, *Chem. Phys. Lett.*, 1998, **285**, 170.
48. (a) Y. S. Chen, A. I. Stash and A. A. Pinkerton, *Acta Crystallogr., Sect. B: Struct. Sci.*, 2007, **63**, 309; (b) M. S. Pavan, R. Pal, K. Nagarajan and T. N. Guru-Row, *Cryst. Growth Des.*, 2014, **14**, 5477; (c) M. Bai, S. P. Thomas, R. Kottokkaran, S. K. Nayak, P. C. Ramamurthy and T. N. Guru-Row, *Cryst. Growth Des.*, 2014, **14**, 459.
49. M. A. Spackman, *Cryst. Growth Des.*, 2015, **15**, 5624.
50. M. A. Spackman, in *Modern Charge Density Analysis*, ed. C. Gatti and P. Macchi, Springer, Dordrecht, Heidelberg, London, New York, 2012, ch. 16, p. 553.
51. C. Gatti, F. Cargnoni and L. Bertini, *J. Comput. Chem.*, 2003, **24**, 422.
52. C. Gatti, G. Saleh and L. Lo Presti, *Acta Crystallogr., Sect. B: Struct. Sci.*, 2016, **72**, 180.
53. G. Arfken, *Mathematical Methods for Physicists*, Academic, Orlando FL, 1985.
54. L. Lo Presti and C. Gatti, *Chem. Phys. Lett.*, 2009, **476**, 308.
55. A. Gavezzotti, *Molecular Aggregation. Structure Analysis and Molecular Simulation of Crystals and Liquids*, IUCr Monographs on Crystallography 19, Oxford University Press, Oxford (UK), 2007.
56. J. Černý and P. Hobza, *Phys. Chem. Chem. Phys.*, 2007, **9**, 5291.

57. G. R. Desiraju, *Crystal Engineering. The Design of Organic Solids*, Elsevier, Amsterdam, 1989.

58. A. D. Becke, in *Modern Electronic Structure Theory*, ed. D. R. Yarkony, Advanced Series in Physical Chemistry 2, World Scientific, River Edge, NJ, 1995, p. 1022.

59. A. Zupan, J. P. Perdew, K. Burke and M. Causà, *Int. J. Quantum Chem.*, 1997, **61**, 835.

60. A. Zupan, K. Burke, M. Ernzerhof and J. P. Perdew, *J. Chem. Phys.*, 1997, **106**, 10184.

61. G. Saleh, C. Gatti and L. Lo Presti, *Comput. Theor. Chem.*, 2012, **998**, 148.

62. J. Contreras-García, W. Yang and E. R. Johnson, *J. Phys. Chem. A*, 2011, **115**, 12983.

63. R. F. Nalewajski, in *Perspectives in Electronic Structure Theory*, ed. R. F. Nalewajski, Springer, Dordrecht, Heidelberg, London, New York, 2012, p. 415.

64. G. Saleh, L. Lo Presti, C. Gatti and D. Ceresoli, *J. Appl. Cryst.*, 2013, **46**, 1513.

65. Gaussian 09, Revision E.01, M. J. Frisch, G. W. Trucks, H. B. Schlegel, G. E. Scuseria, M. A. Robb, J. R. Cheeseman, G. Scalmani, V. Barone, B. Mennucci, G. A. Petersson, H. Nakatsuji, M. Caricato, X. Li, H. P. Hratchian, A. F. Izmaylov, J. Bloino, G. Zheng, J. L. Sonnenberg, M. Hada, M. Ehara, K. Toyota, R. Fukuda, J. Hasegawa, M. Ishida, T. Nakajima, Y. Honda, O. Kitao, H. Nakai, T. Vreven, J. A. Montgomery, Jr., J. E. Peralta, F. Ogliaro, M. Bearpark, J. J. Heyd, E. Brothers, K. N. Kudin, V. N. Staroverov, R. Kobayashi, J. Normand, K. Raghavachari, A. Rendell, J. C. Burant, S. S. Iyengar, J. Tomasi, M. Cossi, N. Rega, J. M. Millam, M. Klene, J. E. Knox, J. B. Cross, V. Bakken, C. Adamo, J. Jaramillo, R. Gomperts, R. E. Stratmann, O. Yazyev, A. J. Austin, R. Cammi, C. Pomelli, J. W. Ochterski, R. L. Martin, K. Morokuma, V. G. Zakrzewski, G. A. Voth, P. Salvador, J. J. Dannenberg, S. Dapprich, A. D. Daniels, Ö. Farkas, J. B. Foresman, J. V. Ortiz, J. Cioslowski, and D. J. Fox, Gaussian, Inc., Wallingford CT, 2009.

66. CRYSTAL09, R. Dovesi, V. R. Saunders, C. Roetti, R. Orlando, C. M. Zicovich-Wilson, F. Pascale, B. Civalleri, K. Doll, N. M. Harrison, I. J. Bush, P. D'Arco, M. Llunell, User's Manual, University of Torino, Torino, 2009.

67. P. Politzer, J. S. Murray and T. Clark, *Phys. Chem. Chem. Phys.*, 2013, **15**, 11178.

68. P. Sjoberg and P. Politzer, *J. Phys. Chem.*, 1990, **94**, 3959.

69. (a) S. J. Grabowski, *ChemPhysChem*, 2014, **15**, 2985; (b) S. J. Grabowski, *ChemPhysChem*, 2015, **16**, 1470.

70. (a) E. L. Smith, D. Sadowsky, C. J. Cramer and J. A. Phillips, *J. Phys. Chem. A*, 2011, **115**, 1955; (b) M. D. Esrafili, *J. Mol. Model.*, 2012, **18**, 2003; (c) A. R. Buchberger, S. J. Danforth, K. M. Bloomgren, J. A. Rohde, E. L. Smith, C. C. Gardener and J. A. Phillips, *J. Phys. Chem. B*, 2013, **117**, 11687; (d) M. D. Esrafili and F. Mohammadian-Sabet, *Struct. Chem.*, 2016, **27**, 1157.

71. J. S. Murray, P. Lane, T. Clark, K. E. Riley and P. Politzer, *J. Mol. Model.*, 2012, **18**, 541.

72. O. Makhotkina, J. Lieffrig, O. Jeannin, M. Fourmigué, E. Aubert and E. Espinosa, *Cryst. Growth Des.*, 2015, **15**, 3464.

73. R. F. W. Bader and P. J. MacDougall, *J. Am. Chem. Soc.*, 1985, **107**, 6788.

74. S. C. Nyburg, *Acta Crystallogr., Sect. A: Found. Crystallogr.*, 1979, **35**, 641.

75. (a) R. P. Sagar, A. C. T. Ku, V. H. Smith and A. M. Simas, *J. Chem. Phys.*, 1988, **88**, 4367; (b) Z. Shi and R. J. Boyd, *J. Chem. Phys.*, 1988, **88**, 4375.

76. M. E. Brezgunova, J. Lieffrig, E. Aubert, S. Dahaoui, P. Fertey, S. Lebègue, J. G. Ángyán, M. Fourmigué and E. Espinosa, *Cryst. Growth Des.*, 2013, **13**, 3283.

77. K. B. Wiberg, R. F. W. Bader and C. D. H. Lau, *J. Am. Chem. Soc.*, 1987, **109**, 985.
78. C. Gatti and D. Lasi, *Faraday Discuss.*, 2007, **135**, 55.
79. R. Bianchi, A. Forni and T. Pilati, *Chem. – Eur. J.*, 2003, **9**, 1631.
80. A. Forni, *J. Phys. Chem. A*, 2009, **113**, 3403.
81. R. Bianchi, A. Forni and T. Pilati, *Acta Crystallogr., Sect. B: Struct. Sci.*, 2004, **60**, 559.
82. M. S. Pavan, R. Pal, K. Nagarajan and T. N. Guru Row, *Cryst. Growth Des.*, 2014, **14**, 5477.
83. V. R. Hathwar, R. G. Gonnade, P. Munshi, M. M. Bhadbhade and T. N. Guru Row, *Cryst. Growth Des.*, 2011, **11**, 1855.
84. Y. V. Nelyubina, M. Yu. Antipin and K. A. Lyssenko, *Mendeleev Commun.*, 2011, **21**, 250.
85. P. Metrangolo, J. S. Murray, T. Pilati, P. Politzer, G. Resnati and G. Terraneo, *Cryst. Growth Des.*, 2011, **11**, 4238.
86. C. R. Groom, I. J. Bruno, M. P. Lightfoot and S. C. Ward, *Acta Crystallogr., Sect. B: Struct. Sci.*, 2016, **72**, 171.
87. A. Bach, D. Lentz and P. Luger, *J. Phys. Chem. A*, 2001, **105**, 7405.
88. D. E. Hibbs, J. Overgaard, J. A. Platts, M. P. Waller and M. B. Hursthouse, *J. Phys. Chem. B*, 2004, **108**, 3663.
89. T. T. T. Bui, S. Dahaoui, C. Lecomte, G. R. Desiraju and E. Espinosa, *Angew. Chem., Int. Ed.*, 2009, **48**, 3838.
90. (a) J. A. R. P. Sarma and G. R. Desiraju, *Acc. Chem. Res.*, 1986, **19**, 222; (b) G. R. Desiraju and R. Parthasarathy, *J. Am. Chem. Soc.*, 1989, **111**, 8725; (c) V. R. Pedireddi, D. S. Reddy, B. S. Goud, D. C. Craig, A. D. Rae and G. R. Desiraju, *J. Chem. Soc., Perkin Trans.*, 1993, **2**, 2353; (d) S. L. Price, A. J. Stone, J. Lucas, R. S. Rowland and A. E. Thornley, *J. Am. Chem. Soc.*, 1994, **116**, 4910; (e) B. K. Saha, A. Nangia and J. F. Nicoud, *Cryst. Growth Des.*, 2006, **6**, 1278; (f) F. F. Awwadi, R. D. Willett, K. A. Peterson and B. Twamley, *Chem. – Eur. J.*, 2006, **12**, 8952.
91. (a) A. Anthony, G. R. Desiraju, R. K. R. Jetti, S. S. Kuduva, N. N. L. Madhavi, A. Nangia, R. Thaimattam and V. R. Thalladi, *Cryst. Eng.*, 1998, **1**, 1; (b) B. K. Saha, R. K. R. Jetti, L. S. Reddy, S. Aitipamula and A. Nangia, *Cryst. Growth Des.*, 2005, **5**, 887; (c) R. K. R. Jetti, F. Xue, T. C. W. Mal and A. Nangia, *Cryst. Eng.*, 1999, **2**, 215; (d) E. Bosch and C. L. Barnes, *Cryst. Growth Des.*, 2002, **2**, 299.
92. (a) E. D. Stevens, *Mol. Phys.*, 1979, **37**, 27; (b) V. G. Tsirelson, P. F. Zou, T. H. Tang and R. F. W. Bader, *Acta Crystallogr., Sect. A: Found. Crystallogr.*, 1995, **51**, 143.
93. F. Bertolotti, A. V. Shishkina, A. Forni, G. Gervasio, A. I. Stash and V. G. Tsirelson, *Cryst. Growth Des.*, 2014, **14**, 3587.
94. E. Aubert, S. Lebègue, M. Marsman, T. T. T. Bui, C. Jelsch, S. Dahaoui, E. Espinosa and J. G. Ángyán, *J. Phys. Chem. A*, 2011, **115**, 14484.
95. M. E. Brezgunova, E. Aubert, S. Dahaoui, P. Fertey, S. Lebègue, C. Jelsch, J. G. Ángyán and E. Espinosa, *Cryst. Growth Des.*, 2012, **12**, 5373.
96. V. R. Hathwar and T. N. Guru Row, *J. Phys. Chem. A*, 2010, **114**, 13434.
97. R. Boese, A. D. Boese, D. Bläser, M. Yu. Antipin, A. Ellern and K. Seppelt, *Angew. Chem., Int. Ed. Engl.*, 1997, **36**, 1489.
98. R. Wang, T. S. Dols, C. W. Lehmann and U. Englert, *Chem. Commun.*, 2012, **48**, 6830.
99. V. R. Hathwar and T. N. Guru Row, *Cryst. Growth Des.*, 2011, **11**, 1338.
100. A. G. Dikundwar and T. N. Guru Row, *Cryst. Growth Des.*, 2012, **12**, 1713.
101. K. Lamberts, P. Handels, U. Englert, E. Aubert and E. Espinosa, *CrystEngComm*, 2016, **18**, 3832.
102. Y. V. Nelyubina, M. Y. Antipin, D. S. Dunin, V. Y. Kotov and K. A. Lyssenko, *Chem. Commun.*, 2010, **46**, 5325.

103. Y. V. Nelyubina, M. Yu. Antipin and K. A. Lyssenko, *J. Phys. Chem. A*, 2007, **111**, 1091.
104. D. Chopra, T. S. Cameron, J. D. Ferrara and T. N. Guru Row, *J. Phys. Chem. A*, 2006, **110**, 10465.
105. (a) M. P. Johansson and M. Swart, *Phys. Chem. Chem. Phys.*, 2013, **15**, 11543; (b) R. A. Cormanich, R. Rittner, D. O'Hagan and M. Bühl, *J. Phys. Chem. A*, 2014, **118**, 7901; (c) V. Tognetti, M. Yahia-Ouahmed and L. Joubert, *J. Phys. Chem. A*, 2014, **118**, 9791; (d) M. Yahia-Ouahmed, V. Tognetti and L. Joubert, *Comp. Theor. Chem.*, 2015, **1053**, 254.
106. M. S. Pavan, K. Durga Prasad and T. N. Guru Row, *Chem. Commun.*, 2013, **49**, 7558.
107. S. P. Thomas, K. Satheeshkumar, G. Mugesh and T. N. Guru Row, *Chem. – Eur. J.*, 2015, **21**, 6793.
108. M. S. Pavan, A. K. Jana, S. Natarajan and T. N. Guru Row, *J. Phys. Chem. B*, 2015, **119**, 11382.
109. S. Dahaoui, V. Pichon-Pesme, J. A. K. Howard and C. Lecomte, *J. Phys. Chem. A*, 1999, **103**, 6240.
110. S. Sarkar, M. S. Pavan and T. N. Guru Row, *Phys. Chem. Chem. Phys.*, 2015, **17**, 2330.
111. Y. V. Nelyubina, A. A. Korlyukov and K. A. Lyssenko, *ChemPhysChem*, 2015, **16**, 676.
112. A. Bauzá, T. J. Mooibroek and A. Frontera, *Angew. Chem., Int. Ed.*, 2013, **52**, 12317.
113. S. P. Thomas, M. S. Pavan and T. N. Guru Row, *Chem. Commun.*, 2014, **50**, 49.
114. E. C. Escudero-Adán, A. Bauzá, A. Frontera and P. Ballester, *ChemPhysChem*, 2015, **16**, 2530.
115. R. Pal, G. Nagendra, M. Samarasimhareddy, V. V. Sureshbabub and T. N. Guru Row, *Chem. Commun.*, 2015, **51**, 933.
116. L. Lo Presti, A. Orlando, L. Loconte, R. Destro, E. Ortoleva, R. Soave and C. Gatti, *Cryst. Growth Des.*, 2014, **14**, 4418.
117. S. Swaminathan, B. M. Craven and R. K. McMullan, *Acta Crystallogr., Sect. B: Struct. Sci.*, 1984, **40**, 300.
118. R. F. W. Bader and H. Essén, *J. Chem. Phys.*, 1984, **80**, 1943.
119. R. F. W. Bader, A. Larouche, C. Gatti, M. T. Carroll, P. J. MacDougall and K. Wiberg, *J. Chem. Phys.*, 1987, **87**, 1142.
120. E. May, R. Destro and C. Gatti, *J. Am. Chem. Soc.*, 2001, **123**, 12248.
121. M. T. Carroll, C. Chang and R. F. W. Bader, *Mol. Phys.*, 1988, **63**, 387.
122. G. A. Jeffrey, *An Introduction to Hydrogen Bonding*, Oxford University Press, New York, 1997.
123. P. L. A. Popelier, The Chemical Bond I - 100 years old and getting stronger, *Struct. Bond.*, 2015, **169**, 71.
124. G. Gilli and P. Gilli, *J. Mol. Struct.*, 2000, **552**, 1.
125. E. Monza, C. Gatti, L. Lo Presti and E. Ortoleva, *J. Phys. Chem. A*, 2012, **115**, 12864.
126. C. Gatti, G. Saleh and L. Lo Presti, *Acta Crystallogr., Sect. B: Struct. Sci.*, 2016, **72**, 180.
127. G. Saleh, C. Gatti and L. Lo Presti, *Comput. Theor. Chem.*, 2015, **1053**, 53.

19 Noncovalent Interactions in Crystal Structures: Quantifying Cooperativity in Hydrogen and Halogen Bonds

Sławomir J. Grabowski*[a,b]

[a] Faculty of Chemistry, University of the Basque Country and Donostia, International Physics Center (DIPC), P.K. 1072, 20080 Donostia, Spain;
[b] IKERBASQUE, Basque Foundation for Science, 48011 Bilbao, Spain
Email: s.grabowski@ikerbasque.org

19.1 Introduction

The term cooperativity is usually applied to hydrogen bond (HB) interactions and it is often understood as the enhancement of the HB strength in the complex when an additional interaction is formed between one of the components of the complex and the next species.[1] There were numerous early studies on this phenomenon; it seems that the term cooperativity was firstly applied by Frank and Wen to describe interactions between water molecules in aqueous solutions.[2] They claimed that the formation of water clusters is a cooperative phenomenon where the jumping of any proton in a cluster is accompanied by a group of concerted jumps and that the cluster formation and its relaxation is related to the partially covalent character of the hydrogen bonds. The systematic theoretical study on the cooperativity effect for water was performed by Del Bene and Pople[3] who found that the chain configuration of the water trimer, linked through

Intermolecular Interactions in Crystals: Fundamentals of Crystal Engineering
Edited by Juan J. Novoa

Published by the Royal Society of Chemistry, www.rsc.org

the sequence of hydrogen bonds, has stronger intermolecular binding than two water dimers. This was one of first studies where the term cooperativity was used and it was attributed to the enhancement of the HB strength in the dimer by the additional component if the trimer is formed.[3] The above-mentioned nonadditivity of interactions was also analyzed by Kołos *et al.* in 29 geometrical configurations of water trimers with the use of the SCF LCAO MO method.[4]

Cooperativity is observed in crystal structures where symmetry relations often impose this effect. Translational symmetry is attributed to crystals from the definition, thus it enforces the existence of chains of molecules or ions[5,6] that are linked by intermolecular interactions as, for example, by hydrogen bonds.[7] The other symmetry relations in crystals may also enforce the cooperativity effects in molecular aggregates. That is why the cooperativity effect in crystal structures has been analyzed in numerous studies. For example, an analysis of the geometries of 100 O–H$\cdots$O hydrogen bonds in 24 crystal structures, resolved by the neutron diffraction method, was performed.[8] It was found that hydrogen bonds engaged in cooperative arrangements have a mean H$\cdots$O distance of 1.805(9) Å while those where the cooperative phenomenon is not detected have a mean H$\cdots$O distance of 1.869(23) Å.[9] In general, for A–H$\cdots$B hydrogen bonds, the rough dependence between the strength of HB and the H$\cdots$B distance is observed, *i.e.* the shortening of the H$\cdots$B distance is connected with the enhancement of the hydrogen bond strength.[10,11] The latter relation was contested and discussed[10,12,13] but it seems that it is often fulfilled for samples of related complexes linked through the same type interactions.[14] The clusters of water molecules linked through the chains of OH$\cdots$OH$\cdots$OH$\cdots$ hydrogen bonds often occur in crystal structures analyzed by the X-ray and neutron diffraction experiments. Figure 19.1 shows a fragment of the 1,3-bis(2-imidazolyl)benzene dihydrate structure[15] where such chains of water molecules are observed.

Studies related to hydrogen bond cooperativity were performed very early in spite of the fact this term was not applied. For example, for liquid hydrogen fluoride[18] and hydrogen cyanide[19] the high dielectric constants observed were explained by the existence of chains of HF or HCN molecules linked through F–H$\cdots$F and C–H$\cdots$N hydrogen bonds, respectively. The case of the HCN chains observed experimentally[19] was analyzed in detail by Pauling[20] who has pointed out that the enthalpy of the HB in the dimer is evaluated to be equal to 3.28 kcal mol^{-1} while the two enthalpies of HBs for the trimer amount to 8.72 kcal mol^{-1} – more than two times the value for the dimer (stable complexes or large clusters are characterized by negative

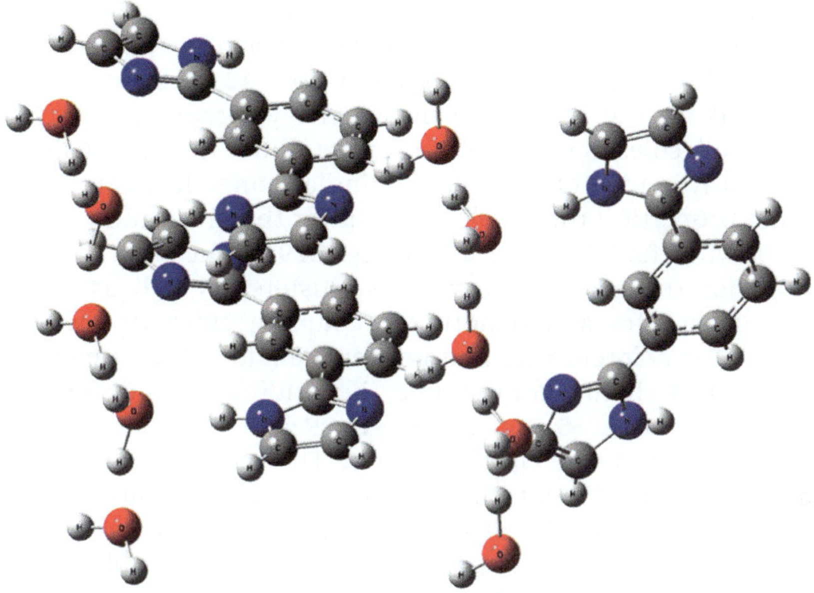

Figure 19.1 A fragment of the crystal structure of 1,3-bis(2-imidazolyl)benzene dihydrate;[15] the crystal structure taken from Cambridge Structural Database[16] and visualized by GaussView 5.0.9 program;[17] CSD refcode of the crystal structure: CIWKOJ. All the following figures in this chapter where crystal structure fragments are presented are taken from CSD[16] and all are visualized by GaussView program.[17]

values of interaction and binding energies, or enthalpies; the absolute values of enthalpies are given by Pauling[20]). Pauling claimed that "the increase in the hydrogen-bond strength with increasing degree of polymerization is interesting, and can be given a simple interpretation in terms of resonance."[20] It is interesting that the hydrogen cyanide crystal structure was determined by the X-ray diffraction technique[21] and the high symmetry of the structure was observed (*I4 mm* space group). The crystal consists of the parallel linear chains of HCN molecules linked through the C–H···N hydrogen bonds.

These two cases of the hydrogen fluoride and hydrogen cyanide are often analyzed both experimentally and theoretically as prototypes of aggregates where cooperativity effects exist. For example, the Raman spectra of the dimer and trimer of hydrogen cyanide, $(HCN)_2$ and $(HCN)_3$, in the gas phase were obtained by photoacoustic Raman spectroscopy (PARS)[22] and, following these experimental observations, *ab initio* calculations were performed on the linear and cyclic $(HCN)_n$ clusters (n up to 5 for linear aggregates and up to 4 for cyclic ones; Hartree–Fock method applied).[23] The detailed discussion on the latter results was also performed by Scheiner.[24] For HCN clusters, the nonadditivity of physical properties was detected. For example, the binding energy for the linear HCN dimer is equal to -5.60 kcal mol^{-1}, while for the linear trimer it amounts to -12.52 kcal mol^{-1}; this means that for the trimer the mean energy for the pair of neighbouring molecules is greater than for the dimer (it concerns the absolute positive values of energy). The similar nonadditivity is observed for other properties; the dipole moment for the linear dimer is equal to 3.20 D while for the trimer it amounts 7.29 D. The other interesting findings for the geometries of HCN clusters, as well as for their vibrational spectra and quadrupole coupling constants, were discussed.[23,24] It is interesting that greater nonadditivities are observed for larger HCN chains. For example, the binding energies for $(HCN)_n$ clusters are equal to -5.60, -12.52, -19.89 and -27.47 kcal mol^{-1} for n equal to 2, 3, 4 and 5, respectively. This means that the mean binding energy for the single HB amounts for those clusters: -5.60, -6.26, -6.63 and -6.87 kcal mol^{-1}, respectively. One can see that the mean absolute energy increase is lower as the number of HCN molecules in the cluster increases. One may say that the convergence of the mean binding energy is observed. Similar tendencies attributed to cooperativity effects were observed for the HF and HCl dimers and trimers analyzed theoretically.[25] Another interesting study concerns the $(H_2O)_2\cdots HF$ and $H_2O\cdots(HF)_2$ complexes analyzed theoretically (MP2/6-31++G(2d,2p) level), where the cooperative effect, defined as the difference between the value of the interaction energy of the trimer and the sum of energies of dimers, is discussed.[26]

The few examples presented here show that the cooperativity phenomenon is not a simple effect related only to the enhancement of the interaction between the pair of components of the complex if an additional unit is attached to the complex. It is connected with the interdependent changes of numerous parameters in the system considered. The situation is even more complicated since the term cooperativity is often used to describe numerous distinct physical and chemical phenomena; this is explained hereafter.

19.2 Different Meanings and Classifications of Cooperativity Effects

Scheiner has systematized cooperativity effects for hydrogen bonded systems;[24] an A–H$\cdots$B hydrogen bond may be enhanced if the AH molecule is additionally involved in an interaction with the C–H unit forming the following sequence structure; C–H$\cdots$A–H$\cdots$B (C does not refer here to the carbon atom, it designates a part of any CH unit, similarly AH and B refer to any units and only H refers to the hydrogen atom). The HB strength enhancement, if the central A–H unit acts simultaneously as the proton donor and as the proton acceptor, is named as positive cooperativity. However, the A–H unit may act as the double proton donor or the B unit may play the role of the double proton acceptor.[24] The two latter types of hydrogen bonds were named by Desiraju and Steiner[27] as bifurcated or three-centre hydrogen bonds, where the bifurcated donor and the bifurcated acceptor correspond to the double proton donor and the double proton acceptor, respectively. Scheiner explains that for those two cases "negative cooperativity" is often observed that may be explained by the electron density redistribution as a result of complexation.[24]

For example, for the A–H$\cdots$B hydrogen bond there is an electron density shift from the proton acceptor (B) to the A–H unit which results in the negative and positive charges of A–H and B units, respectively, in the complex formed. Usually for stronger hydrogen bonds, greater electron density shifts are observed.[28]

If the B unit is additionally involved in the C–H$\cdots$B interaction, then B becomes a double proton acceptor and lower electron density shift to the A–H molecule is observed since part of the electron density is transferred to the C–H one. It results in the weakening of the A–H$\cdots$B hydrogen bond. Such partial A–H$\cdots$B and C–H$\cdots$B hydrogen bonds with a common B proton acceptor often exist in crystal structures. For example, they were detected and described in the crystal structures of thioureidoalkylphosphonates.[29]

Similarly, if the A–H unit of an A–H$\cdots$B link forms a new hydrogen bond with an additional proton acceptor (say C), *i.e.* if the additional A–H$\cdots$C hydrogen bond is formed, thus the A–H molecule becomes the double proton donor. This results in electron density shifts to the A–H unit from both B and C centres. Consequently, for the triad formed, a lower shift from the B moiety to the A–H unit is observed than in a case of the corresponding dyad linked through the single A–H$\cdots$B interaction. The latter also results in the weaker A–H$\cdots$B hydrogen bond in the triad than in the dyad.

In crystal structures, where molecules or ions are linked through various types of interactions, among them through hydrogen bonds, bifurcated connections are often observed. Figure 19.2 presents fragments of two crystal structures as examples; for the structure of 1*H*-indazole-7-carboxylic acid hemihydrate,[30] the water molecule plays a role of double (bifurcated) proton acceptor, while for the structure of (3,4-dihydroxy-5-oxotetrahydrofuran-2-yl)methyl acetate,[31] the C–H bond is a double (bifurcated) proton donor.

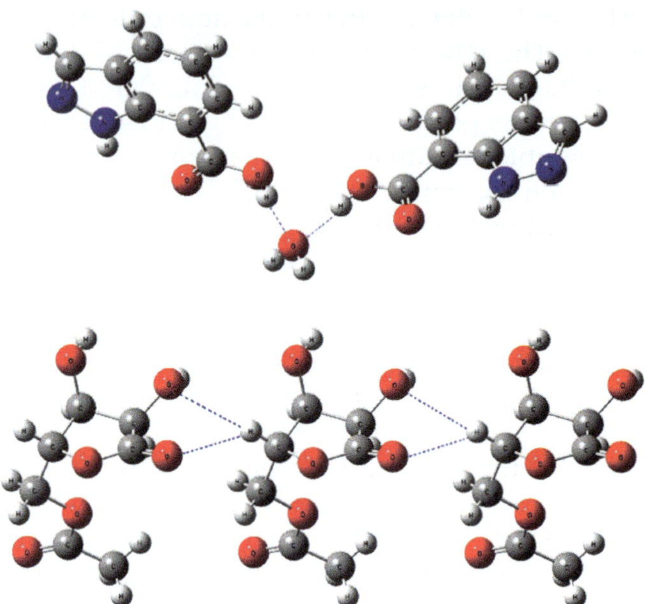

Figure 19.2 Fragments of the crystal structures of 1*H*-indazole-7-carboxylic acid hemihydrate[30] (top, CSD refcode: DONNIF) and (3,4-dihydroxy-5-oxotetrahydrofuran-2-yl)methyl acetate[31] (bottom, CSD refcode: ABA-CEN); bifurcated interactions are designated by blue broken lines.

In the case of sequence structures, the electron density shifts may be explained in a similar way as for bifurcated hydrogen bonded systems.[32] For example, for the linear HF$\cdots$HF dimer (Scheme 19.1) there is an electron density shift from the fluorine proton acceptor to the attached HF and a further shift of the density from the H-atom to the terminal fluorine. These shifts are connected with the balance between two effects accompanying the A–H$\cdots$B HB formation; the hyperconjugative weakening of the A–H bond and the rehybridization process.[33–35] In the case of sequential trimers, the central HF molecule losses electron density due to its further transfer to the additional attached unit (terminal HF molecule at the left side of Scheme 19.1). The latter electron density shift implies the further shift from the proton acceptor (HF unit at the right side of Scheme 19.1); hence the additional molecule in the trimer implies the enhancement of the strength of the primary HB interaction (Scheme 19.1).

Scheiner has pointed out that the qualitative evaluation of the cooperativity is often referred in literature to the nonadditivity which is usually defined as the difference between the total interaction energy of the cluster considered and the sum of all the pairwise interactions.[24] The latter term, corresponding to that one applied earlier

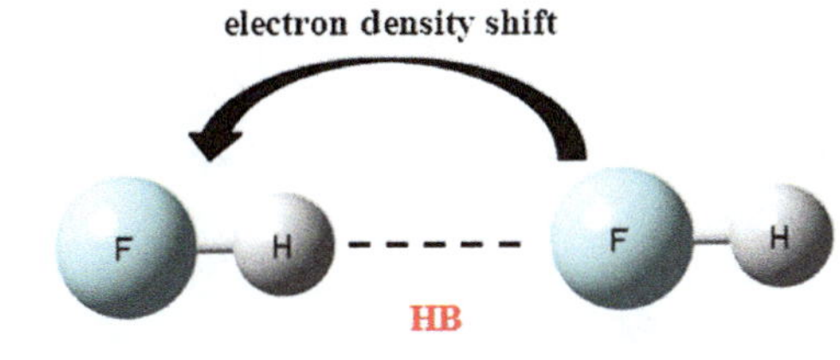

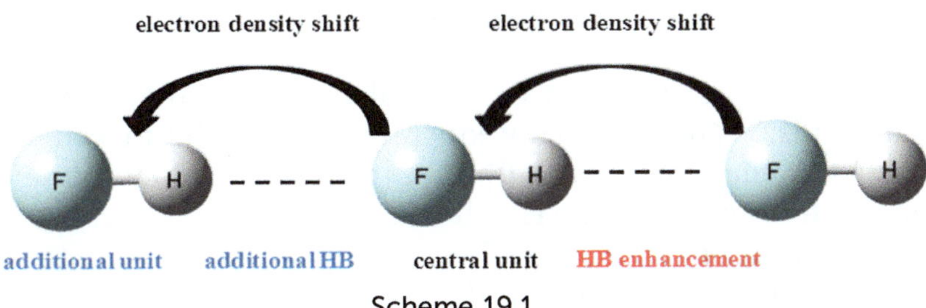

Scheme 19.1

by Rovira *et al.*[26] and described briefly in the introduction, is discussed in detail in the next section of this chapter.

Referring to the hydrogen bonds in crystals Jeffrey distinguishes two types of cooperativity;[10] the first one is named as the σ-bond cooperativity and it occurs when functional groups possessing proton donor and proton acceptor properties form continuous chains or cycles linked through hydrogen bonds. For example, this effect was detected and described in crystal structures where links between hydroxyl groups in carbohydrate molecules[36] and in cyclodextrins are observed.[37] This is the same type of cooperativity as that described earlier here and occurring for HF linear chains (see Scheme 19.1). The second type specified by Jeffrey concerns hydrogen bonds between molecules with conjugated multiple π-bonds.[10]

Numerous such intermolecular and intramolecular hydrogen bonds were classified as resonance assisted hydrogen bonds (RAHBs), where the enhancement of the strength of HB interaction is mainly a result of the resonance.[38,39] The RAHB model was first introduced for systems with conjugated double π-bonds linked through the intra-molecular HB, such as in malonaldehyde and its derivatives (Scheme 19.2, arrows show the electron density shifts). According to the RAHB model the real structure is a mixture of resonance structures (Scheme 19.2 presents resonance structures) which leads to the equalization of the formal single and double carbon–carbon bonds as

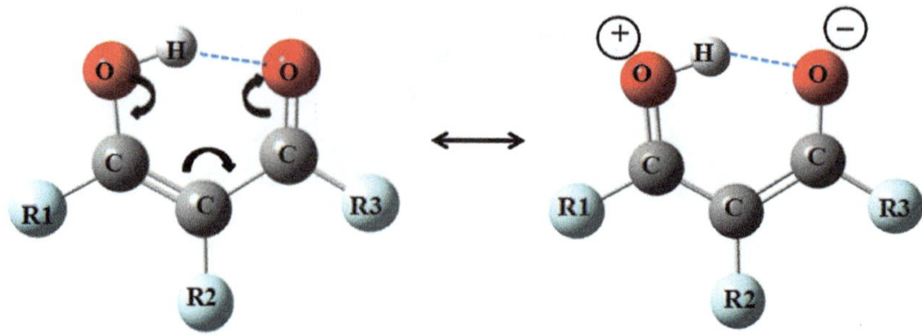

Scheme 19.2

well as to the equalization of the formal single and double carbon–oxygen bonds.[39] The latter consequently leads to the movement of the H-atom to the centre of an H-bridge (at the mid-point of the $O\cdots O$ distance); the strength of the HB should correlate with "the degree of bonds' equalization".[39]

The RAHB model was criticized in numerous studies.[40] This was justified by the finding that the existence of conjugated double bonds does not strengthen the hydrogen bond interaction, since the corresponding systems with single bonds are often characterized by stronger HBs. Additionally it was shown[41] using modern valence-bond theory,[42,43] the decomposition of the interaction energy[44,45] and quantum theory of atoms in molecules (QTAIM)[46,47] that stabilization of systems with intra- and intermolecular hydrogen bonds, which are often referred to the RAHB model, originate from charge delocalization.[41] The resonance stabilization energies for those systems are negligible and they diminish with an increasing strength of the hydrogen bond.[41] Hence it seems that the systems with conjugated double bonds linked by hydrogen bonds may be specified as a special kind of interaction and of the cooperativity effect where electron charge delocalization of the electron density is observed; however, the assumptions of the RAHB model are discussible.

It was pointed out that not only the enolones mentioned earlier here may be classified as RAHB systems but also other moieties with intramolecular hydrogen bonds (enaminones, enamino-imines, enol-imines *etc.*), as well as systems linked through intermolecular HBs (carboxylic acid dimers, amide dimers, amide–amidine couplings, DNA base pairs *etc.*).[38] It is worth mentioning that the existence of $O–H\cdots O–C$ chains was postulated early on in structures of carboxylic acids,[48] which is related to the π-bond cooperativity term introduced by Jeffrey.[10] Coulson pointed out that for β-oxalic acid the molecules are linked through the double HBs between carboxylic groups and that the decrease of the O··· O distance from 2.8 Å in the gas phase to 2.5 Å in crystal is observed which is accompanied by a doubled bond energy.[49] Recent more precise X-ray results on the crystal structures of β-oxalic acid at different temperatures[50] show the O···O distances between carboxylic groups linked through double equivalent HBs in the range of 2.665–2.680 Å in the 156–298 K temperature range. Figure 19.3 shows the layer of the H-bonded *β*-oxalic acid molecules where on one hand the cooperativity comes from the existence of the hydrogen bonded chain and on the other hand cooperativity comes from the existence of eight-member ring of two linked carboxylic groups.

Numerous crystal structures are known where the symmetry-related species are linked through carboxylic groups by two equivalent O–H··· O hydrogen bonds; these are centrosymmetric dimers with the inversion centre situated in the middle of the eight-member ring.[7] The above-mentioned motif was designated as $R_2^2(8)$ [51,52] where R refers to the ring structure, number 8 means that eight atoms form the ring while subscript 2 and superscript 2 mean that there are two proton donors and two proton acceptors here. Figure 19.4 presents an example of the crystal structure of 2,5-dinitrobenzoic acid[53] where centrosymmetric dimers are observed, similarly as in the structure of β-oxalic acid.

Jeffrey tried to generalize the concept of the cooperativity writing that "The concept that a particular configuration of single and

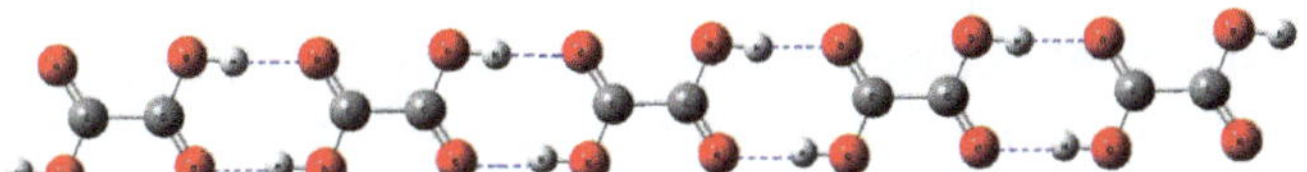

Figure 19.3 Fragment of the crystal structure of β-oxalic acid;[50] CSD refcode: OXALAC11; HBs are designated by blue broken lines.

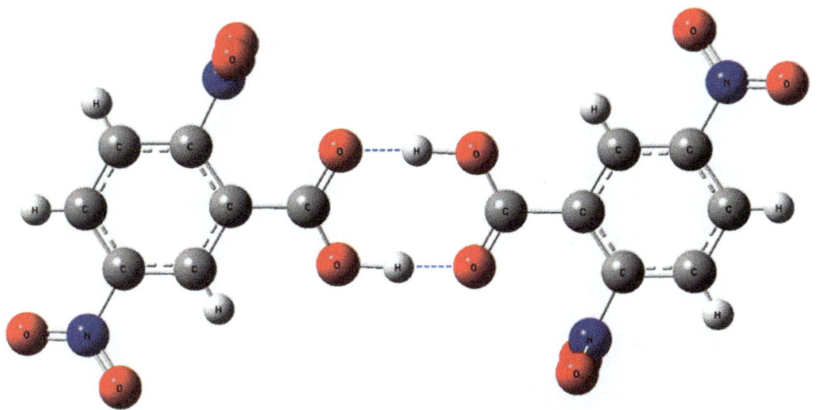

Figure 19.4 A fragment of the crystal structures of 2,5-dinitrobenzoic acid;[53] CSD refcode: DAJXUH; HBs are designated by blue broken lines.

multiple covalent bonds has a total energy greater than the sum of the energies of the individual bonds is familiar in chemistry. In valence bond theory, this extra energy is called resonance energy. In molecular orbital theory, it is called delocalization energy."[10] It was mentioned earlier that Pauling explained the increase of the strength of hydrogen bonds in the HCN chains in terms of resonance.[20] Taking into account findings that the enhancement of the hydrogen bond results from electron density shifts and electron charge delocalization (not from resonance) this generalization of Jeffrey may be extended to both σ-bond and π-bond cooperativity effects.

The cooperativity effect was presented so far for hydrogen bonded systems; however, in numerous recent studies this phenomenon is discussed for other so-called noncovalent interactions,[54] especially in studies on halogen bonding performed in recent years.[55–57] This is why the halogen bond in terms of cooperativity is also analyzed here, while only short references to the other interactions are included.

19.3 Quantifying Cooperativity – Measures and Indices

19.3.1 Energetic Measures

The cooperativity effect may be quantified by partitioning the energy of interaction into n-body terms (eqn (19.1)).[58,59]

$$\Delta E_n = E(1, 2, 3, 4 \ldots, n) - \sum_{i=1}^{n} E_i$$

$$\equiv \sum_{i=1}^{n} E(i) - \sum_{i=1}^{n} E_i \text{ relaxation (deformation energy)}$$

$$+ \sum_{i=1}^{n-1} \sum_{j>1}^{n} \Delta^2 E(ij) \text{ two-body}$$

$$+ \sum_{i=1}^{n-2} \sum_{j>i}^{n-1} \sum_{k>j}^{n} \Delta^3 E(ijk) \text{ three-body}$$

$$+ \sum_{i=1}^{n-3} \sum_{j>i}^{n-2} \sum_{k>j}^{n-1} \sum_{l>k}^{n} \Delta^4 E(ijkl) \text{ four-body}$$

$$+ \cdots \Delta^n E(1, 2, 3, 4 \cdots, n) n\text{-body}$$

$$(19.1)$$

ΔE_n is the difference between the energy of the whole system (cluster) and the sum of energies of all units (n) within. However, the energies of units refer to their energetic minima (E_i's) if they are not involved in any interactions, meaning that the energies of the separately optimized monomers (units) are considered. On the other hand, $E(i)$ expresses the energy of the i-th unit related to its geometry in the cluster. Hence $E(i) - E_i$ is the deformation energy (relaxation) for the i-th unit in the cluster, which is positive since the energy of the separate i-th unit in the cluster, $E(i)$, does not correspond to its energetic minimum; E_i does and it is "more negative" than $E(i)$. The sum of the latter $E(i) - E_i$ contributions for all units in the cluster expresses the total relaxation (deformation) energy (eqn (19.1)). It is worth mentioning that the energy of interaction is often referred to the difference between the energy of the cluster and the sum of energies of its units with geometries taken from the cluster (such energy is designated here as $\Delta E_n'$, eqn (19.2)).

$$\Delta E_n' = E(1, 2, 3, 4 \cdots, n) - \sum_{i=1}^{n} E(i) \tag{19.2}$$

The further decomposition of $\Delta E_n'$ energy leads to the same result as that one expressed by eqn (19.1) but the relaxation energy is not included. The two-body and three-body terms of eqn (19.1) (and of eqn (19.2)) are defined by eqn (19.3) and (19.4), respectively.

$$\Delta^2 E(ij) = E(ij) - [E(i) + E(j)] \tag{19.3}$$

$$\Delta^3 E(ijk) = E(ijk) - [E(i) + E(j) + E(k)] - [\Delta^2 E(ij) + \Delta^2 E(ik) + \Delta^2 E(jk)] \tag{19.4}$$

In the similar way, the $\Delta^4 E(ijkl)$ four-body and the remaining many-body terms are defined. One can see that for the complex consisting of two units (1 and 2), the ΔE_n energy (eqn (19.1)) is reduced to the $E(1,2) - (E_1 + E_2)$ difference, *i.e.* to the sum of the relaxation term and the single two-body contribution (eqn (19.5)).

$$\Delta E_n = E(1, 2) - [E_1 + E_2]$$

$$\equiv E(1) - E_1 + E(2) - E_2 \quad \text{relaxation energy} \tag{19.5}$$

$$+ E(1, 2) - [E(1) + E(2)] \quad \text{two-body term}$$

It is worth mentioning that different schemes were applied to include the basis set superposition error (BSSE) corrections to the terms of the above mentioned eqn (19.1) and (19.2), especially as such corrections are more complicated for the many-body terms.[60]

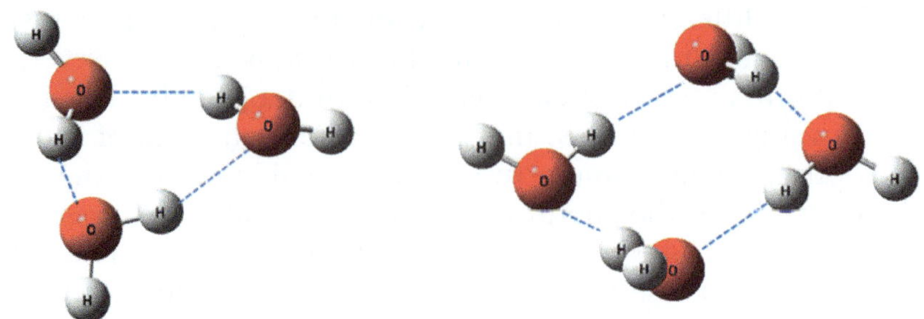

Figure 19.5   Cyclic water trimer and tetramer linked by HB interactions; HBs are designated by blue broken lines.

The relations presented above were applied to analyze different water clusters.[59] For example, the MP2/aug-cc-pVDZ calculations performed for the cyclic water trimer and tetramer, *i.e.* where water molecules are linked by O–H··· O–H··· hydrogen bonds closed within the ring (Figure 19.5), show that two-body terms dominate HB forces, accounting for 70–80% of the interaction energies. Three-body interactions account for 18% and 26% of the total interaction energies for the trimer and tetramer, respectively. The four-body contribution for tetramer accounts only 2% of the energy of interaction.[59]

Glendening has applied the DFT/NEDA approach[61] – the extension of the natural energy decomposition analysis (NEDA),[62–64] to analyze the cooperative effects in water clusters.[61] The DFT/NEDA results led to the conclusion that electrostatic, charge transfer and polarization are the most important attractive contributions to the energy of interactions. However, for the cyclic trimer and tetramer, cooperative stabilization results mostly from charge transfer interactions and slightly from polarization; in other words, the contributions related to the electron density shifts (see Scheme 19.1) are mostly connected with cooperative effects.

One can see that the equations presented in this section corres-pond to the term "nonadditivity" often applied to describe systems containing more than two units (molecules or ions).[4,24,26] For ex-ample, for the water trimers,[4] the nonadditive interaction energy was defined as the difference between the total interaction energy, ex-pressed here by eqn (19.2), and the two-body interaction energies, eqn (19.6) (1, 2 and 3 designate here water molecules).

$$\Delta E_{\text{nadd}} = \Delta E'_n - \Delta^2 E(1,2) - \Delta^2 E(2,3) - \Delta^2 E(1,3) = E(1,2,3)$$

$$- \left[E(1,2) + E(2,3) + E(1,3) \right] + E(1) + E(2) + E(3)$$

$$(19.6)$$

In general, the nonadditive energy often related to the cooperativity effect is the energy of interaction for the cluster considered without two-body contributions. Hence the cooperativity, no matter positive or negative, appears for systems containing more than two units and it is precisely defined by terms possessing a physical meaning. One can see that eqn (19.6) related to triads is exactly the three-body interaction energy term (see eqn (19.3) and (19.4))! It is worth mentioning that the slight modification of eqn (19.6) is sometimes applied to include partly the relaxation energy.[54] This means that the difference between the interaction energy of the triad and the sum of three interaction energies of dyads is considered; however, for two pairs of units in contact the interaction energy of the isolated dyad within its corresponding minimum configuration is taken into account, while for a pair of terminal units separated by the central species the interaction energy with its geometry in the triad is considered.[54]

Other descriptors of cooperativity are often applied, particularly if the clusters are composed from the same type units. For example, the bifurcated H-bond pattern in linear *N*-formylformamide chains, $[HN(CHO)_2]_n$ was analyzed[65] with the number of units (n) up to 12 (Figure 19.6).

The author[65] has considered theoretically the Z,Z conformer of *N*-formylformamide; however, it was calculated (B3LYP/6-31G(d,p) level) that the E,E conformer is more stable than Z,E and Z,Z conformers by 0.2 and 6.0 kcal mol^{-1}, respectively.[66] These theoretical findings are in agreement with the experimental observations.[66,67] Figure 19.7 presents a fragment of the crystal structure of *N*-formylformamide (E,E conformer), [66] where one can see that each terminal atom of the *N*-formylformamide molecule is involved in hydrogen bond interactions. C–H and N–H bonds play the role of proton donors while O-atoms are proton acceptors; since one of the O-centres

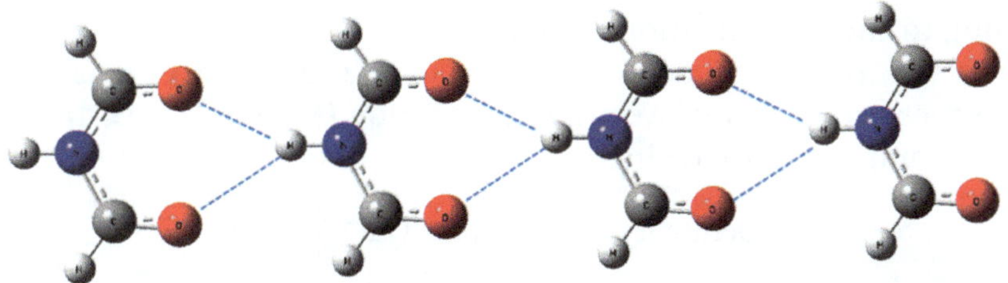

Figure 19.6 *N*-formylformamide units linked through bifurcated hydrogen bonds (based on the scheme of ref. 65); HBs are designated by blue broken lines.

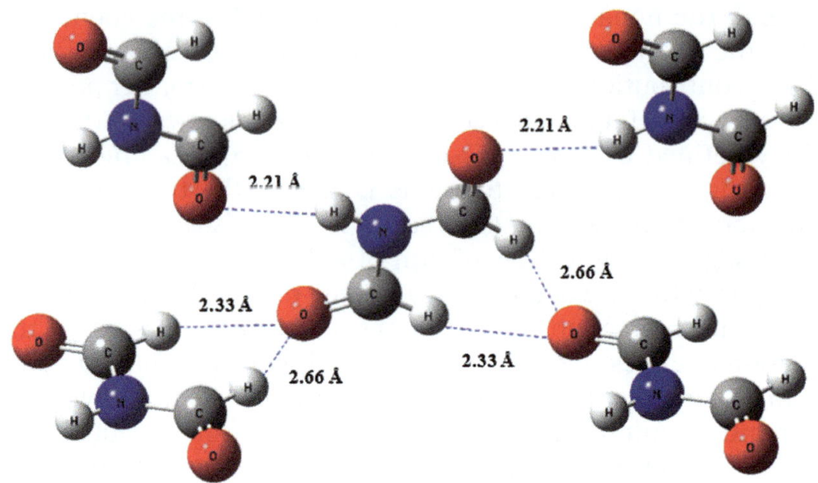

Figure 19.7 A fragment of the crystal structure of *N*-formylformamide;[66] HBs are designated by blue broken lines, the hydrogen bond lengths are indicated; CSD refcode: LUHSIR.

forms H-bonds with two C–H proton donors, thus it is a bifurcated proton acceptor (Figure 19.7).

The existence of numerous hydrogen bonds in the crystal structure in connection with the translational symmetry, and in connection with other symmetry relations, leads to numerous cooperative effects in the crystal structure of *N*-formylformamide. One can select here at least two patterns. Figure 19.8 shows the chain of molecules linked through bifurcated H-bonds; note that for the Z,Z conformer analyzed theoretically there are bifurcated NH proton donors while for the experimental E,E crystal structure there is an oxygen bifurcated proton acceptor; the (C)H$\cdots$O distances are equal here 2.33 and 2.66 Å (Figure 19.7). Figure 19.8 also shows the chain of molecules linked through N–H$\cdots$O hydrogen bonds; the H$\cdots$O distances amount 2.21 Å (Figure 19.7). One can see that both chains presented in Figure 19.8 may be classified as exhibiting π-bond cooperativity according to the classification of Jeffrey.[10]

As it was described before, Parra has analyzed the chains of Z,Z conformer molecules of *N*-formylformamide (MP2/6-31 + G(d) level); he has applied the following measure of cooperativity (eqn (19.7)).[65]

$$\text{cooperativity} = [\Delta Q_n - (n-1)\Delta Q_2]/(n-2) \qquad (19.7)$$

Q is the property of interest (energy, frequency, bond length *etc.*) while *n* designates the number of units in the cluster; for example, in the case of the dissociation energy, ΔQ_n is the dissociation energy of

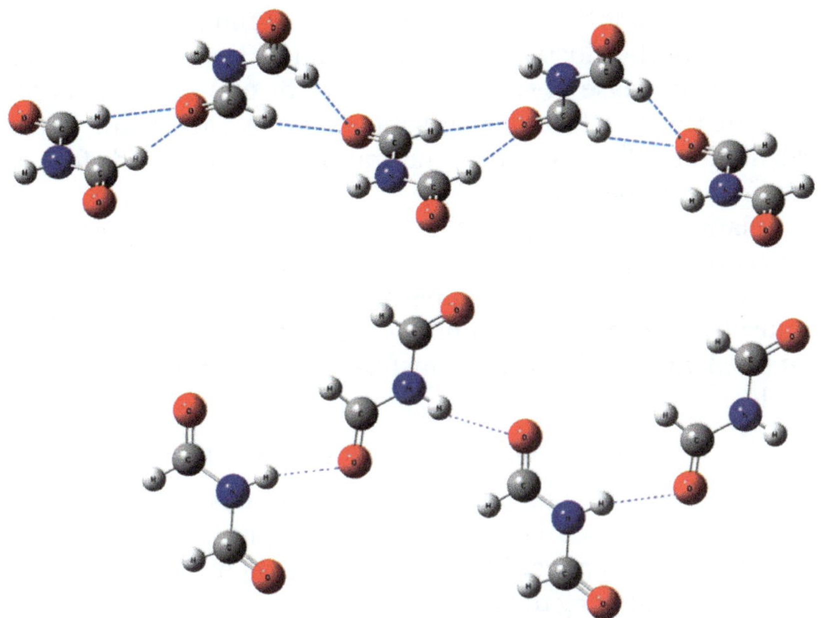

Figure 19.8 Two patterns observed in the crystal structure of *N*-formylformamide,[66] H-bonds are designated by blue broken lines; CSD refcode: LUHSIR.

the cluster of size n and ΔQ_2 is that of the dimer. For the Z,Z conformer of *N*-formylformamide the cooperativity effect increases with the increase of the cluster size.[65] For the dissociation energy calculated with the zero-point vibrational energy correction, the co-operativity (eqn (19.7)) is equal to 3.03 and 5.01 kcal mol^{-1} for clusters containing 3 and 10 molecules, respectively. It is worth mentioning that the cooperativity increase is smaller if the number of units in the cluster increases. For example, the difference in the cooperativity between clusters containing 3 and 2 molecules amounts 0.67 kcal mol^{-1} while between clusters containing 10 and 9 units it is equal to 0.11 kcal mol^{-1}. This means that for a greater number of units in the cluster one may expect "a saturation", beyond which the adding of the next unit does not influence the parameters, among them energetic ones.

Parra has analyzed other parameters, stretching frequencies of N–H and C–H bonds as well as geometrical parameters (calculated at B3LYP/6-31 + G(d) level).[65] For example, bond lengths associated with the strongest H-bond interactions in the clusters were summarized and discussed. If the number of molecules in the chain (Figure 19.6) increases an elongation of the N–H proton donating bond is observed;

the N–H bond length is equal to 1.015 Å for monomer, 1.019 Å for dimer while for the cluster containing 9 molecules it amounts 1.026 Å and the next increase of size of the cluster by one unit results in the increase of the N–H bond length by 0.0001 Å (below the accuracy of calculations).[68] A similar "saturation" is observed for other geometrical parameters.

The relation (eqn (19.7)) applied as a measure of cooperativity for clusters containing the same kind units is slightly modified for cyclic systems, where the number of intermolecular (inter-unit) connections is greater by one than for chains. The modified relation (eqn (19.8)) was applied to describe cyclic clusters built up from carbonic acid molecules.[69]

$$\text{cooperativity} = [\Delta Q_n - n\Delta Q_2]/(n-1) \tag{19.8}$$

The measures of cooperativity (nonadditivity) described earlier concern the whole cluster analyzed. A deeper insight into all interactions in a many-body system could be obtained by their comparison in the cluster considered or by the comparison of interactions in a sample of clusters differing in the number of units. Hence other approaches are sometimes applied in order to measure cooperative effects for single inter-unit links. For example, *cis-N*-methylformamide oligomers were analyzed theoretically (B3LYP/cc-pVTZ level) up to pentamers – the chains connected by the N–H···O hydrogen bonds were considered.[70] The following relation was applied to evaluate the single hydrogen bond energy within the cluster (eqn (19.9)).

$$E_{\text{HB}m} = E_{\text{total}} - E_{1,2,3\ldots m} - E_{m+1,m+2\ldots n} \quad m = 1,2,\ldots,n-1 \tag{19.9}$$

$E_{\text{HB}m}$ is defined as the difference between the total energy of the chain cluster containing n monomers and the sum of the energies of two components of the cluster, the first component contains m units, from the first one to the m-th unit in the chain, while the second component consists of the remaining units, from the $(m+1)$-th to the n-th one. In other words, the contact between units corresponding to the interaction of interest divides the cluster into two components. The cooperative energy of the m-th hydrogen bond, $E_{\text{coop}m}$, was also defined (eqn (19.10)).[70]

$$E_{\text{coop}m} = E_{\text{HB}m} - \sum_{i=1}^{m} \sum_{j=m+1}^{n} IE_{ij} \quad m = 1, 2, \ldots, n-1 \tag{19.10}$$

$E_{\text{coop}m}$ is the part of the HB energy which excludes some of two-body terms, *i.e. IE_{ij}*'s which are defined as the interaction energies between monomers where one of monomers comes from the first component and the other one from the second component. Two latter eqn (19.9 and 19.10) were also applied to analyze the cooperativity effects in triads consisting of an N-methylformamide dimer and an extra amino acid residue.[71]

The eqn (19.9) is often applied to analyze clusters containing two types of species that are linked by hydrogen bonds. It means that different molecules or ions play roles of Lewis acid and Lewis base. For example, the $C_6H_6\cdots HF$ complex was analyzed. Benzene acts as the proton acceptor by its π-electrons and the HF molecule plays the role of the proton donor (Figure 19.9).[72] The MP2/6-311++G(d,p) interaction energy (with the BSSE correction included) for this complex is equal to -3.2 kcal mol^{-1}. The next HF molecules were added to the system and the clusters containing enlarging chain of HF molecules were fully optimized; *i.e.* the $C_6H_6\cdots HF\cdots HF\cdots$ arrangements were formed. The interaction energies for such clusters were calculated for the contact between the benzene molecule and the neighbouring hydrogen fluoride according to eqn (19.9). Figure 19.9 shows the $C_6H_6\cdots(HF)_4$ cluster as an example. The interaction energies for clusters containing 2, 3 and 4 HF molecules are equal to -5.3, -7.3 and -8.2 kcal mol^{-1}, respectively. Thus an enhancement of the F–H$\cdots$ π(benzene) hydrogen bond is observed if the number of HF molecules increases. The (F)H$\cdots$C intermolecular distance for the $C_6H_6\cdots HF$ dyad is equal to 2.50 Å (the shortest contact among the H$\cdots$C ones is considered). The latter (F)H$\cdots$C distance decreases with the increase of HF molecules in the cluster to 2.30, 2.24 and finally to 2.18 Å for the $C_6H_6\cdots(HF)_4$ cluster. There are similar findings concerning energetic and geometrical parameters for the analogues $C_2H_2\cdots(HF)_n$ and $C_2H_4\cdots(HF)_n$ clusters of acetylene and ethylene.[73,74]

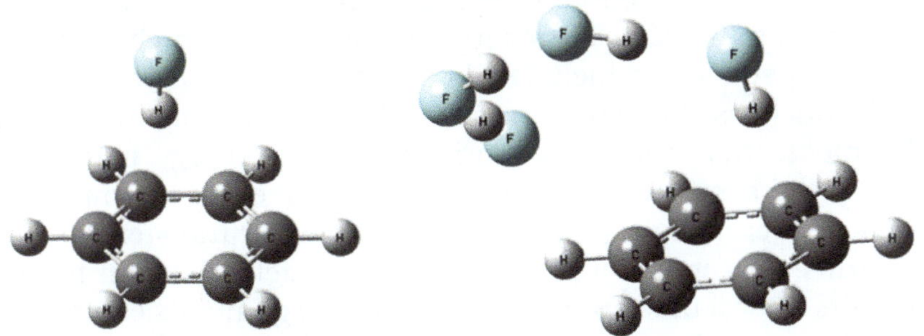

Figure 19.9 The $C_6H_6\cdots HF$ complex (left) and the $C_6H_6\cdots(HF)_4$ cluster (right) (based on the results from ref. 72).

The $H_2CO\cdots(HF)_n$ $(n=1,2,3)$ clusters are other interesting examples.[75] The O-centre of formaldehyde plays the role of proton acceptor and subsequent HF molecules play the dual role of Lewis acid (by H-atom) and Lewis base (by F-centre), excluding the last HF molecule in the sequence which sometimes acts only as a Lewis acid. HF molecules in $H_2CO\cdots(HF)_n$ systems play a similar role as in the clusters of benzene, acetylene and ethylene described earlier. For example, for the $C_6H_6\cdots(HF)_4$ cluster presented in Figure 19.9, the HF molecule in contact with benzene acts as the Lewis acid through its H-atom and the F-atom of the same molecule is the Lewis base being in contact with the next HF molecule.

19.3.2 Decomposition of the Energy of Interaction for Many-body Systems

For the $H_2CO\cdots(HF)_n$ clusters described in the previous section the interaction energy (eqn (19.9)) is equal to -7.7, -13.3 and -16.2 kcal ml^{-1}, for $n=1$, 2 and 3, respectively (MP2/aug-cc-pVDZ level). The decomposition of the energy of interaction was also performed to deepen the understanding of the nature of interactions between the H_2CO unit and the second part consisting of the HF molecules.[75] There are the following interaction energy components in the partitioning scheme applied (eqn (19.11)).[44,45]

$$\Delta E = E_{EL}^{(1)} + E_{EX}^{(1)} + E_{DEL}^{(R)} + E_{CORR} \tag{19.11}$$

$E_{EL}^{(1)}$ is the first-order electrostatic term describing the Coulomb interaction of static charge distributions of both units, $E_{EX}^{(1)}$ is the repulsive first-order exchange component resulting from the Pauli exclusion principle, while $E_{DEL}^{(R)}$ and E_{CORR} correspond to higher order delocalization and correlation terms. The delocalization term contains all classical induction, exchange-induction, *etc.*, from the second order up to infinity. The charge transfer term, which is strongly basis set dependent, is included in the delocalization contribution, which is much less basis set sensitive.[44,45] It is worth mentioning that in the most often applied Kitaura–Morokuma energy partitioning scheme[76] there are the following interaction energy components: electrostatic, exchange, polarization and charge transfer. The two latter terms correspond approximately to the delocalization term. In the Kitaura–Morokuma approach the total energy and its components are not free of basis set superposition error (BSSE), while in the partitioning applied for the $H_2CO\cdots(HF)_n$ clusters the BSSE is mostly cancelled.[44,45] In the Kitaura–Morokuma scheme the polarization interaction energy may be approximately described as connected with the internal redistribution of electron charge in the interacting units as a result of complexation, whereas the charge transfer term is connected with the density shifts from one unit to the other one. The delocalization term of eqn (19.11) is approximately connected with the energy of interaction related to the electron charge density shifts (for the whole cluster) as a result of the complex formation. The delocalization interaction energy term (eqn (19.11)) or the other related terms of other decomposition schemes (charge transfer, polarization, induction *etc.*) are often attributed in the literature to the covalency of interaction;[32] this means that the increase of the covalent character of interaction is manifested in the increase of such energy components.

The partitioning of energies of interaction (eqn (19.11)) for the $H_2CO\cdots(HF)_n$ clusters shows that effects connected with the electron charge density shifts increase, *i.e.* the covalency of interaction increases, with the increase of the number of HF molecules in the cluster.[75] The correlation term, E_{CORR}, is not important (for all systems analysed, correlation energies do not exceed 0.2 kcal mol^{-1}) and the electrostatic and delocalization attractive interactions govern the stability of the cluster. However the $E_{DEL}^{(R)}/E_{EL}^{(1)}$ ratio is equal to 0.43, 0.53 and 0.61 for 1, 2 and 3 HF molecules in the cluster, respectively.[75] Similarly, referring to the Kitaura–Morokuma decomposition,[76] the ratio between the sum of charge transfer and polarization terms and the electrostatic one for the above clusters is equal to 0.63, 0.79 and 0.97, respectively.[75] This clearly shows the increase in the importance of the effects of the electron charge density shifts as the number of HF molecules in the cluster increases. This is also connected with the increase of the strength of F–H$\cdots$O hydrogen bond in the cluster. It was justified that in the case of hydrogen bond, the increase of its strength is strongly connected with the increase of the importance of the delocalization energy term.[77] This is in line with the DFT/NEDA results for clusters of water described earlier here that the cooperativity in water clusters is governed by the charge transfer and polarization interactions.[61] It is worth mentioning that the Kitaura–Morokuma partitioning of the energy of interaction concerns the Hartree–Fock energy of interaction, and correlation is not included.[76] The physical meaning of the energy terms for the latter partitioning is slightly different than the meaning of terms in eqn (19.11) (for the review of different decomposition schemes see ref. 78). This is why the ratios presented for both decompositions differ.

19.3.3 QTAIM Approach to Analyze Cooperativity Effects

The Quantum Theory of Atoms in Molecules (QTAIM) approach[46,47] is a useful tool to analyze inter- and intramolecular interactions, among them hydrogen bonds. One of the basic concepts of QTAIM concerns the partitioning of 3D electron density space into basins attributed to atoms. It was pointed out that atomic basins are characterized by specific properties and that these properties are often transferable from one chemical moiety to another one. However, the QTAIM approach also allows analysis of the electron density of the system considered (atom, ion, molecule, cluster, even crystal). There are special points of the considered electron density, these are critical points (CPs), for which the gradient of electron density ($\nabla\rho$) vanishes. A detailed description of CPs is presented in numerous monographs and review articles.[46,47,79–81] It is possible to distinguish between various CPs if the second derivatives of the electron density are considered. Maxima of the electron density are attributed to attractors which correspond to positions of atoms. Minima correspond to so-called cage critical points (CCPs) while the bond critical points (BCPs) and the ring critical points (RCPs) correspond to the saddle points. The BCPs are situated on the bond paths. The bond path is attributed to the stabilized interaction (chemical bond or other inter-atomic interaction).[82–84] It is the line linking attractors that is characterized by the maximum electron density if compared with other links.

The BCP situated on the bond path (BP) is characterized by the lowest electron density in comparison with the other points of the BP. The BCP characteristics often correlate with numerous energetic and geometrical parameters. For example, the electron density at BCP, ρ_{BCP}, corresponding to the intermolecular link, often correlates with the binding energy, with the intermolecular atom-atom distance *etc.* These relations are especially in force for the same pairs of atoms in related complexes or species.[14] There are other characteristics of BCPs: the Laplacian of the electron density at BCP, $\nabla^2\rho_{\,BCP}$, the total electron energy density at BCP, H_{BCP}, and the components of the latter, the kinetic electron energy density, G_{BCP}, and the potential electron energy density, V_{BCP}. The $\nabla^2\rho_{BCP}$ is usually negative for covalent bonds while it is positive for closed-shell interactions (ionic bonds, hydrogen bonds and other noncovalent interactions).[47] The following relation (eqn (19.12)) between the parameters mentioned above results from the virial theorem (in atomic units).[47]

$$\tfrac{1}{4}\,\nabla^2\rho_{CP} = 2G_{CP} + V_{CP} \quad \text{where } H_{CP} = V_{CP} + G_{CP} \tag{19.12}$$

The subscript CP refers to the critical point. It is worth mentioning that G_{CP} is always positive while V_{CP} is negative. If the absolute value of V_{CP} is greater than twice the value of G_{CP}, the corresponding Laplacian is negative; the latter is attributed to the covalency of interaction. However, there are such interactions where the modulus of the potential energy outweighs the kinetic energy only one time, in such a case the Laplacian is positive but H_{CP} is negative; and the interaction is often considered as partially covalent in nature.[85–87]

The QTAIM approach was also applied to analyze cooperative effects in the $H_2CO\cdots(HF)_n$ clusters described earlier.[75] Figure 19.10 presents the molecular graphs for the fully optimized clusters containing HF molecules up to 3. The molecular graph is usually understood in the literature[47] as one where the positions of attractors (atoms) are presented with bond paths linking attractors and with critical points. One can see that for clusters containing 2 and 3 HF molecules the cyclic structure is formed since the "terminal" HF molecule plays the role of the Lewis base through the F-centre – a $C–H\cdots F$ hydrogen bond (or $F\cdots C$ contact) is formed closing the ring.

The BCPs of the $H\cdots O$ bond paths correspond to the interactions described in the previous section. As was pointed out earlier, the strength of this interaction increases if the number of HF molecules increases in the cluster. The electron density at $H\cdots O$ BCP follows this trend since it is equal to 0.036, 0.049 and 0.059 au for n equal to 1, 2 and 3, respectively (MP2/aug-cc-pVDZ level).[75] There are other interesting QTAIM results on these clusters. The H_{BCP} value for the $H\cdots O$ BCP of the $H_2CO\cdots HF$ complex is positive and equal to $+0.004$ au. However this value for the corresponding BCP in the $H_2CO\cdots(HF)_2$ cluster is equal to -0.003 au, which may indicate a partially covalent character for this interaction. For the $H_2CO\cdots(HF)_3$ cluster this value is even "more negative" and amounts to -0.011 au. This is in line with the partitioning scheme results (eqn (19.11)) presented in the previous section.

The QTAIM approach was applied to analyze other clusters, for example, the $C_2H_2\cdots(HF)_n$ and $C_2H_4\cdots(HF)_n$ ones [73,74] described in the previous section, $H\cdots\pi$ interactions and dihydrogen bonds.[32,77,88,89]

19.3.4 π-Electron Delocalization in Intramolecular Hydrogen Bonds

It was described in the previous sections that the term π-bond cooperativity was introduced by Jeffrey for systems involving hydrogen

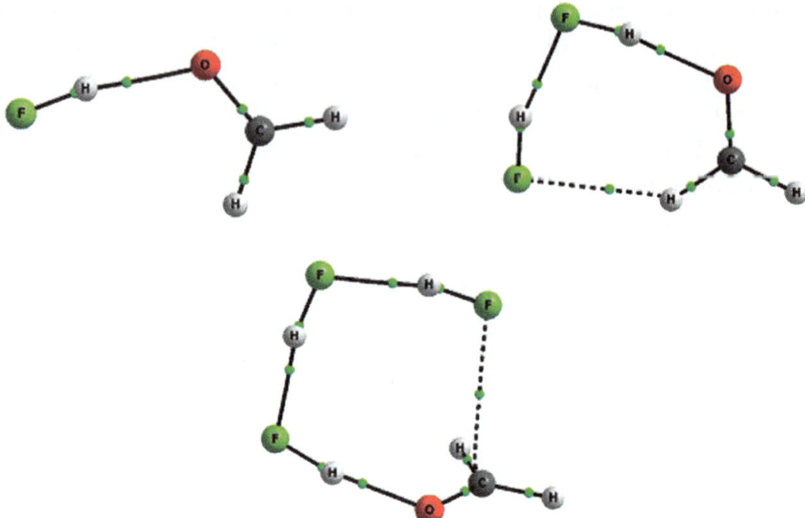

Figure 19.10 Molecular graphs of the $H_2CO\cdots(HF)_n$ clusters; attractors are designated by big circles, bond critical points by small green circles while the bond paths by solid or broken lines (based on the results from ref. 75).

bonds between units with conjugated multiple π-bonds.[10] Some of these systems are sometimes classified as resonance assisted hydrogen bonds (RAHBs),[38,39] where the term RAHB was first applied for the $HOCR_1=CR_2-CR_3=O$ fragment (R_1, R_2 and R_3 designate substituents) contained in the enol form of acetylacetone and other related β-diketones and β-keto esters. The RAHB model was tested firstly for the enolones linked through the intramolecular $O-H\cdots O$ hydrogen bonds.[38,39] However it was also analyzed for enaminones, enaminoimines, enolimines and other similar systems linked through intramolecular HBs. It was pointed out that this model may be applied for systems linked through intermolecular hydrogen bonds and a few classes of compounds were mentioned,[38] among them carboxylic acid and amide dimers, both linked through double HBs.

The basic assumptions of the RAHB model if applied for malonaldehyde (malondialdehyde) and its derivatives with intramolecular HBs (see Scheme 19.2) may be summarized in the following statements;[38,39] (i) there is a synergistic reinforcement of the hydrogen bond and π-electron delocalization, (ii) the latter leads to the $O\cdots O$ distance shortening accompanied by the $O-H$ bond lengthening and thus to the H-atom movement to the centre of $O\cdots O$ distance, (iii) the real structure observed is the mixture of resonance forms (see Scheme 19.2) which leads to the equalization of C–C and C–C bonds as well as to the equalization of C–O and C–O bonds.

This model was contested; on one hand the conjugated π-electron system does not guarantee enhancement of H-bond strength,[40] while on the other hand recent calculations show that the systems usually attributed to the RAHB mechanism are not a mixture of resonance forms.[41] The latter conclusion results from Complete Active Space Valence Bond (CASVB)[42,43] calculations performed on malonaldehyde and its derivatives, as well as on the carboxylic acid and amide dimers. Thus it seems that the systems usually referred to as having resonance assistance are, in fact, π-electron delocalization assisted. It seems that the enhancement of the HB strength corresponds to the "degree" of electron delocalization for systems characterized by conjugated multiple π-bonds. The greater delocalization (expressed by different indices; see eqn (19.13)–(19.15) below) correlates with greater enhancement of the HB strength. However, such enhancement is not connected with the existence of the conjugated π-bonds since a comparison of HBs in the latter systems with HBs in corresponding species characterized by a single bond framework show that in the latter case the HBs are often stronger.[40]

A few indices describing the π-electron delocalization were introduced, especially for species with the ring closed by an intramolecular HB, for example the Q parameter (eqn 19.13); for the designations, $d_1 \cdots d_4$, see Scheme 19.3).[38]

$$Q = q_1 + q_2 \quad \text{where} \quad q_1 = d_4 - d_1 \text{ and } q_2 = d_2 - d_3 \tag{19.13}$$

In other words, the Q index informs of the equalization of single and double bonds in the system considered, with greater equalization giving a Q-parameter closer to zero; for the totally delocalized system the Q parameter is equal to zero.

Scheme 19.3 (based on ref. 38) shows that the system with conjugated π-bonds may be treated as the mixture of two tautomeric forms (enol-keto and keto-enol);[38] as it was mentioned earlier, in another

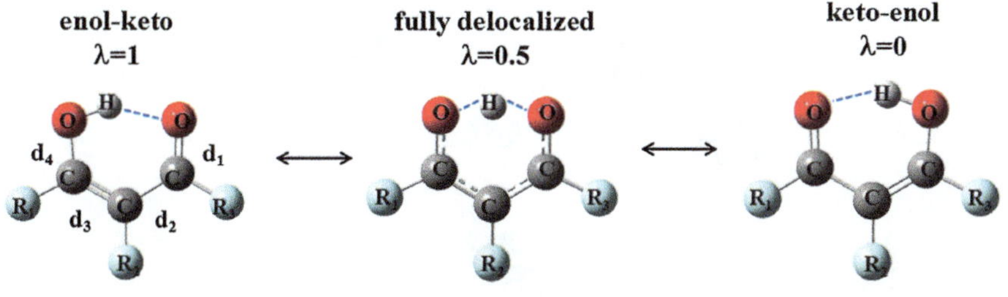

Scheme 19.3

study Gilli *et al.* claim that it is a mixture of two resonance forms.[39] A λ coupling parameter referring to the π-electron delocalization was introduced (eqn (19.14)).[38]

$$\lambda = (1 - Q/Q_0)/2 \tag{19.14}$$

The Q parameter is defined by eqn (19.13) while Q_0 corresponds to the standard single and double CO and CC bonds, $d_1 = 1.20$ Å, $d_2 = 1.48$ Å, $d_3 = 1.33$ Å and $d_4 = 1.37$ Å. Hence Q_0 is equal to 0.320 Å, *i.e.* corresponds to the model structure containing the conjugated system where there is no electron delocalization.

Another delocalization parameter, Δ_{dp}, describing the π-electron delocalization as an effect of the formation of the intramolecular HB was proposed (eqn (19.15)).[90,91]

$$\Delta_{\mathrm{dp}} = \frac{1}{2}[(q_1' - q_1)/q_1' + (q_2' - q_2)/q_2'] \tag{19.15}$$

The q_1 and q_2 parameters were defined by eqn (19.13) and correspond to the closed system with the intramolecular HB (enol-keto or keto-enol form); the q_1' and q_2' parameters correspond to the conformer where such an HB does not exist – the so-called open-form (see Scheme 19.4). In other words, the Δ_{dp} parameter refers to the changes between two conformations, closed and open, and it describes the enhancement of electron delocalization as an effect of the closing of the system (formation of the HB). In the case of the λ parameter, the differences between the reference system not perturbed by delocalization and the analyzed system with intramolecular HBs are considered.

One can see that the parameters described above, Q, λ and Δ_{dp}, may be treated as those which describe the cooperativity effects for cases

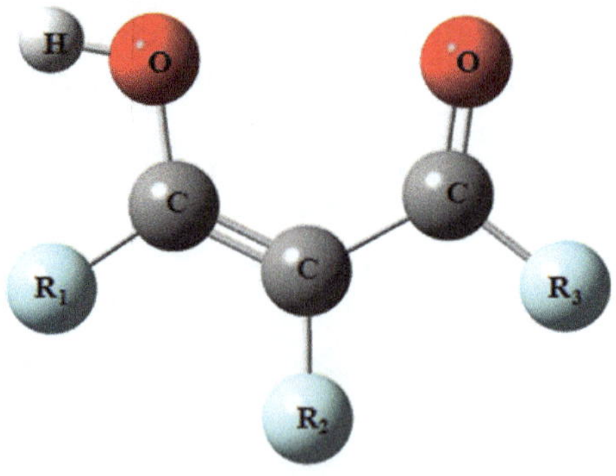

Scheme 19.4

classified as π-bond cooperativity. It is worth mentioning that they have been mostly applied to intramolecular hydrogen bonds, and mainly for O–H$\cdots$O hydrogen bonds. Often modifications of the above-presented parameters are required for application to the other types of hydrogen bonds.[92,93]

One may ask if there is any experimental evidence of the relations mentioned above, especially between the π-electron delocalization and the strength of hydrogen bonding. For example, the linear correlation between q_1 and q_2 parameters (eqn (19.13)) has been presented for the crystal structures containing the β-diketone fragment;[38] q_1 concerns the equalization of the CO bonds while q_2 relates to the CC bonds, thus this correlation shows interrelations of the delocalization effects. For the same sample of species, the dependence between the λ parameter and the O$\cdots$O distance has been presented.[38] Since the O$\cdots$O distance is sometimes treated as a rough measure of the strength of O–H$\cdots$O HB,[10] thus it could provide evidence of the interrelation between π-electron delocalization and the strength of the HB. However, the correlation is rather poor, and the only tendency observed is that for species with the λ-value close to 0.5 short O$\cdots$O distances are observed, while if the λ parameter decreases to 0 or increases to 1 the O$\cdots$O distance increases.[38] The other example (8 species in the sample) concerns the intramolecular HBs corresponding to the tautomeric forms presented in Scheme 19.3; the linear correlation coefficient for the relation between the O$\cdots$O distance and the Q parameter is equal to 0.82.[39] Much better correlations are observed for theoretical results where, in contrast to the crystal structures, the environmental effects are not so complex. For example, for malonaldehyde and its simple derivatives (R-substituents are F or Cl atoms – see Scheme 19.3), calculated at the MP2/6-311++G(d,p) level, the second order polynomial correlation between the Q-parameter and the hydrogen bond energy is observed ($r^2 = 0.94$).[94] For a similar sample of fluoro and chloro derivatives of malonaldehyde, the linear correlation between Δ_{dp} and the electron density of the H$\cdots$O BCP (corresponding to the HB interaction and expressing its strength) is observed and $r^2 = 1.00$![90]

The indices described above (eqn (19.13)–(19.15)) were applied in numerous experimental and theoretical studies to describe systems with conjugated π-bonds. However, single experimental crystal structure determinations are also sometimes presented to show the influence of the π-electron delocalization on the HB strength. One can mention the crystal structure of benzoylacetone (Figure 19.11) where the analysis of the neutron diffraction data shows the H-atom placed

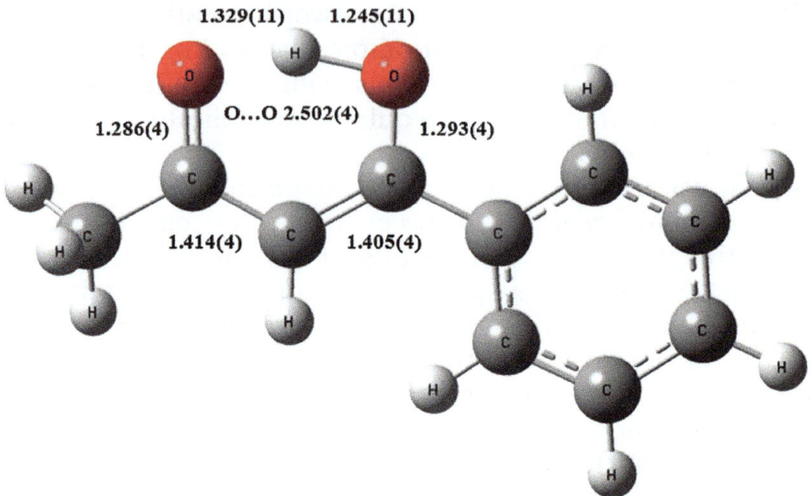

Figure 19.11   The molecule of benzoylacetone in the crystal structure;[95] CSD refcode: BZOYAC03, the bond lengths and O···O distance in Å, e.s.d values in parentheses.

asymmetrically in a large flat potential well.[95] Figure 19.11 shows the equalization of CO and CC bonds in the ring closed by the intramolecular HB. The position of the H-atom is not as precise as those of heavy atoms (Figure 19.11), partly because of its large thermal motions (that are confirmed by the difference map from the neutron refinement and the other crystal structure data).[95]

It is interesting that the low-temperature neutron diffraction was also performed on the deuterated benzoylacetone[96] and it was found that the enol deuteron has a bimodal probability density distribution. The latter studies supported by *ab initio* calculations show that the deuteron in benzoylacetone is an atom tunnelling between two possible localized positions, while in the previous studies on non-deuterated species the proton has sufficient energy to shuttle over the low potential energy barrier, even at the extremely low temperature of 15 K.[95,96]

19.3.5   π-Electron Delocalization in Intermolecular Hydrogen Bonds

There are not any special indices describing π-bond cooperativity for intermolecular hydrogen bonds. Sometimes the Q and λ parameters are used for these systems, as they were for structures containing a β-diketone fragment with intermolecular HB links.[38,39] It is worth mentioning that the cooperativity effects in crystal structures are often very complex ones; the crystal structure of diformylamide[66] presented earlier here is an example (Figure 19.8).

Similarly, complex interactions are observed in the crystal structures of carboxylic acids.[7] The carboxylic groups are often linked by two equivalent HBs in centrosymmetric dimers that lead to the enhancement of the π-electron delocalization within the carboxylic groups, expressed by the equalization of C–O and C–O(H) bonds. It seems that the latter effect of delocalization is connected with the strength of the HBs. For example, dimers of formic acid, acetic acid, formamide and pyrrole-2-carboxylic acid (PCA) were analyzed at MP2/aug-cc-pVDZ level; the *ab initio* calculations were supported by QTAIM and partitioning of the energy of interaction analyses (eqn (19.11)).[97] For each dimer there are two equivalent hydrogen bonds and the dimers are centrosymmetric. For two dimers, the intermolecular connections are through the N–H$\cdots$O bonds (the C dimer of PCA and the formamide dimer), while in the remaining dimers, double O–H$\cdots$O bonds between the carboxylic groups are formed. There is a difference between the PCA dimers; dimer A consists of *s-trans* conformers of the PCA molecules, the B and C complexes consist of *s-cis* conformers (see Figure 19.12).

There is the rough correlation between the π-electron delocalization effects and the strength of the interaction for the above mentioned dimers.[97] For links between carboxylic groups (acetic and formic acids and forms A and B of PCA), an elongation of the C=O ($\sim$0.02 Å) and a

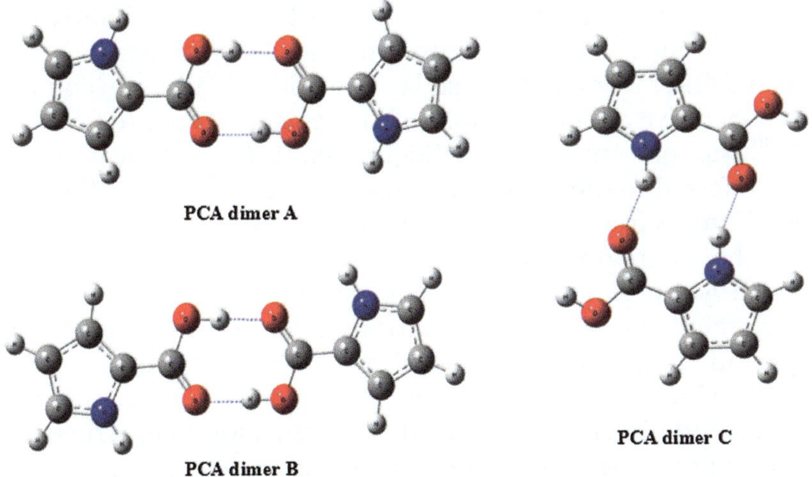

Figure 19.12   Three types of centrosymmetric dimers linked through HBs for pyrrole-2-carboxylic acid.[97,98]

shortening of C–O (~0.03–0.04 Å) bonds are observed in dimers if they are compared with isolated monomers. The greatest changes are observed for the strongest HB interaction of −8.2 kcal mol^{-1} (energy for a single HB) for the PCA-A dimer. However, similar but slightly smaller changes are observed for the PCA-B dimer where the HB energy is equal to −7.5 kcal mol^{-1}; the latter arrangement is observed in the crystal structure of PCA.[98] In the case of the formamide and PCA-C dimers, double N–H$\cdots$O HB links are observed; the HBs are weaker here than for the carboxylic groups' links which is connected to smaller changes in the π-conjugated fragments, shortening of C–N and lengthening of C=O bonds by up to ~0.02 Å.

The latter observations are supported by other results;[97] interaction difference density maps show greater electron density shifts for the O–H$\cdots$O HBs than for the N–H$\cdots$O ones and greater ρ_{BCP} values for the (O)H$\cdots$O BCPs than for (N)H$\cdots$O ones. Additionally, the H_{BCP} values for the H$\cdots$O contacts corresponding to the O–H$\cdots$O hydrogen bonds are negative while for the N–H$\cdots$O HBs they are positive. The negative H_{BCP} values indicate that the interactions are partly covalent in nature[85–87] and this is connected with the greater electron density shifts as the result of complexation. It is in agreement with the more important contribution of the delocalization interaction energy (eqn (19.11)) for O–H$\cdots$O HBs than for N–H$\cdots$O ones.[97]

One can see that the latter results show the interrelation between π-electron delocalization and the strength of the hydrogen bonds. In carboxylic acid dimers in the solid phase, the situation is often more complex, since effects of static or dynamic disorder may often exist which also lead to the equalization of carbon–oxygen bonds in the carboxylic groups.[7] Besides, centrosymmetric dimers are involved in numerous interactions in crystals that additionally are connected with cooperativity. The example of the crystal structure of β-oxalic acid[49,50] was presented earlier (Figure 19.3). Here the cooperativity may be referred to the existence of $R_2^2(8)$ motifs,[51,52] as well as to the chain of the β-oxalic acid molecules (the O=C–C–O–H$\cdots$O=C–C–O–H$\cdots$ chains are observed).

Even early experimental studies on gas phase carboxylic acids show the interrelations between π-electron delocalization in carboxylic groups and the hydrogen bonds. For example, the gas electron diffraction technique was applied to investigate the structure of the monomer and dimer of formic acid.[99] For the monomer structure the C=O, C–O and O–H bond lengths are equal to 1.217(1), 1.361(1) and 0.984(9) Å, respectively; corresponding values for dimer are equal to 1.220(1), 1.323(1) and 1.036(1) Å. One can see an important

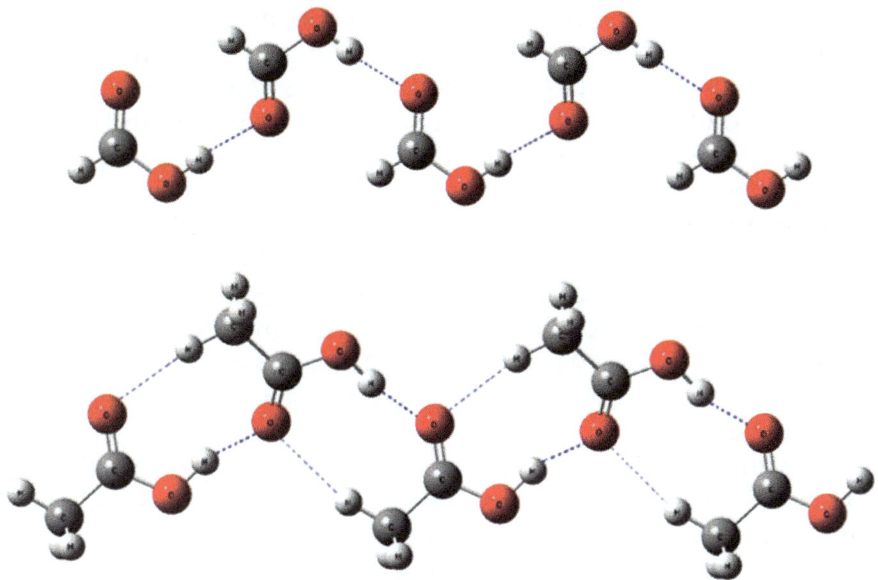

Figure 19.13   The catemer motifs of the crystal structures of formic acid (top)[100] and of the acetic acid (bottom);[101] the (O)H···O and (C)H···O hydrogen bonds designated by blue broken lines; CSD refcodes: FORMAC01 and ACETAC03, respectively.

shortening of the C–O bond length as an effect of the formation of HBs but the elongation of C=O bond in negligible; this may be connected with the weak HB since the O···O intermolecular distance in the ring dimer structure amounts to 2.703(8) Å. In the formic acid crystal structure, the C=O and C–O bond lengths are equal to 1.222(2) and 1.308(2) Å while the O···O distance for the H-bond bridge is equal to 2.625(4) Å.[100] If one assumes that the O···O distance corresponds roughly to the strength of HB thus one may conclude that the further equalization of CO bonds in the crystal structure is related to the enhancement of HB compared with the gas phase structure. However, the HCOOH molecules in the crystal structure are arranged in chains forming a so-called catemer motif (Figure 19.13) which is not directly comparable with the gas phase ring structure of the dimer.

Theoretical studies on the structure of acetic acid were performed by Rovira and Novoa.[102] The crystal structure of acetic acid contains chains of *syn* molecules which form a catemer motif, similarly as in the crystal structure of formic acid (Figure 19.13). There are additional very weak C–H···O HBs in the crystal structure of acetic acid[101,102] since the (C)H···O distance is equal to 2.41 Å. A similar arrangement may be found for the formic acid crystal structure

(Figure 19.13), however here the (C)H$\cdots$O distance is equal to 2.66 Å which is slightly longer than the corresponding sum of van der Waals radii of H and O atoms (2.6 Å, according to the Pauling scale, see ref. 103). The variation of the bond energy with the change of the number of acetic acid molecules in the catemer chain was considered.[102] The chain was fixed to the experimental structure.[101] For two acetic acid molecules this energy is equal to -7.8 kcal mol^{-1} while when there are 9 molecules in the chain it amounts ~ -9 kcal mol^{-1} (these energies correspond to a single O–H$\cdots$O hydrogen bond). The limit value for an infinite chain, calculated with the use of boundary conditions is equal to -9 kcal mol^{-1}.[102] Thus an extra stabilization of 1.2 kcal mol^{-1} is gained on going from an isolated dimer to an infinite chain.

This means that the theoretical calculations may reproduce the real crystal structure properties if a sufficient number of molecules are taken into account and that such a number is not so great; for the calculations presented here on acetic acid, nine molecules in the chain is sufficient to reproduce the cooperativity in the crystal structure.[102] It was also mentioned earlier that for the sufficiently large clusters "saturation" is observed, and further enlargement of the structure does not influence its parameters.

19.4 Cooperativiy Effects for Halogen Bonded Systems

19.4.1 Theoretical Analysis of Cooperativity for Halogen Bonded Systems

The hydrogen bond is one of the most often analyzed interactions. However in recent years the other so-called noncovalent interactions have been investigated intensively due to their significance in numerous chemical, physical and biological processes.[104,105] Among them there is the halogen bond,[55–57] designated hereafter as XB, which may be defined as the interaction of halogen atom, X, acting as a Lewis acid centre with an electron rich region, *i.e.* with a Lewis base centre. The halogen atom covalently bonded with another atom, A, often has a strong directional preference to act as a Lewis acid since A–X$\cdots$B linear, or nearly so, arrangements are observed (B is the Lewis base centre).[57,106] These A–X$\cdots$B links result from the depletion of the electron charge density in the extension of A–X bond.[106] The X-centre often acts as the Lewis acid in the A–X direction while in the direction perpendicular to the bond, or nearly so, it acts as the Lewis base.[106] Such dual properties of the X-centre are usually detected for C–X bonds but also for other ones, like for example Sb–X[107] or Si–X.[108] The nature of the halogen bond interaction is described in other chapters of this volume. It may be mentioned here that the Lewis acid properties in the elongation of the A–X bonds may be explained in terms of the σ-hole concept.[104,105] The halogen bond is often competitive with the hydrogen bond; for example, Metrangolo *et al.* have pointed out that in crystal structures the former interactions are often preferred to the latter ones.[109]

The cooperativity effects described so far for hydrogen bonds are observed for other interactions, like halogen bonds. For example, these effects for the halogen bonded $H_2CO\cdots(ClF)_n$ clusters were analyzed (MP2/6-311++G(d,p) level).[110] The linear $H_2CO\cdots(ClF)_n$ systems were analyzed ($n=1$, 2,...6), *i.e.* clusters with symmetry constraints where the $C{=}O\cdots Cl{-}F\cdots Cl{-}F...$ atoms are linear which implies C_{2v} symmetry (Figure 19.14 shows the $H_2CO\cdots(ClF)_3$ cluster as an example). Additionally, the full optimization was performed at the same level for the $H_2CO\cdots ClF$ and $H_2CO\cdots(ClF)_2$ moieties (Figure 19.15 shows the $H_2CO...(ClF)_2$ fully optimized cluster).

The interaction energy for the systems mentioned above was calculated as the difference between the energy of the cluster and the sum of energies of the monomers. In this study 'monomers' were understood in the following way. Since the strength of the $Cl\cdots O$ interaction was analyzed, thus H_2CO was treated as one component and the remaining ClF molecules (one molecule if $n=1$) as the other component – *i.e.* the chain of ClF molecules in the cluster. The deformation energy (relaxation) as an effect of complexation was not included here since the geometries of chains of ClF molecules with

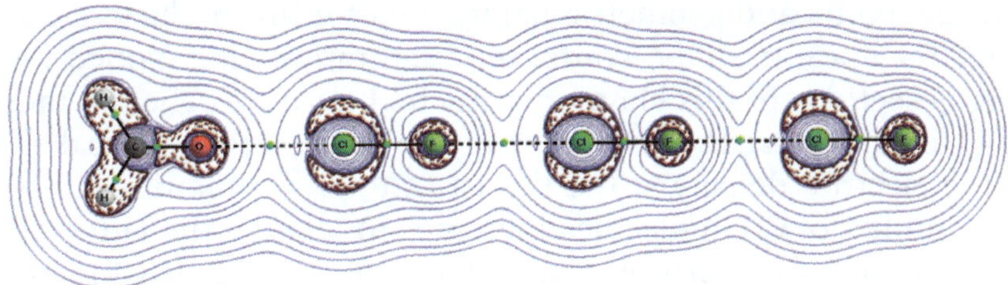

Figure 19.14 The molecular graph of the $H_2CO\cdots(ClF)_3$ linear cluster, solid and broken lines correspond to bond paths, big circles to attractors and small green circles to bond critical points, the isolines of the Laplacian of electron density are presented; positive values are depicted in solid lines and negative values in broken lines; the Laplacian isodensity lines in the plane of the cluster (based on the results of ref. 110).

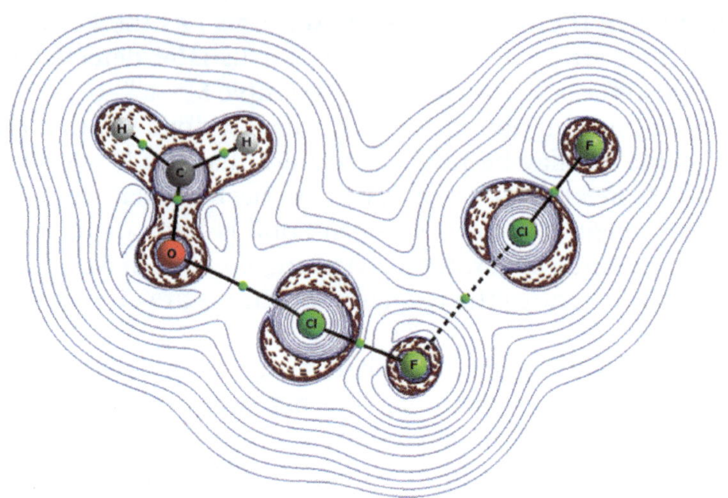

Figure 19.15 The molecular graph of the $H_2CO\cdots(ClF)_2$ fully optimized cluster, solid and broken lines correspond to bond paths, big circles to attractors and small green circles to bond critical points, the isolines of the Laplacian of electron density are presented; positive values are depicted in solid lines and negative values in broken lines; the Laplacian isodensity lines in the plane of the cluster (based on the results of ref. 110).

corresponding energies were taken from the clusters. Hence the interaction energy corresponding to the $O\cdots Cl$ contact was calculated according to eqn (19.9). It is worth mentioning that for the halogen bonded clusters the same measures of cooperativity as those applied for the hydrogen bonded systems may be applied (eqn (19.1)–(19.10)).

The results clearly show the increase of the strength of the $O\cdots Cl$ interaction with the increase of the number of the ClF molecules in the cluster. For linear systems the $O\cdots Cl$ distance is equal to 2.737 Å for the $H_2CO\cdots ClF$ complex and it decreases to 2.685 Å for $n = 5$ and 6. The energetic and geometric parameters for linear clusters do not change for $n > 5$; in other words, the "saturation" mentioned in the previous sections of this chapter is reached for $n = 5$. The interaction energy for the linear $H_2CO\cdots ClF$ complex is equal to -3.2 kcal mol^{-1} and the electron density at the $O\cdots Cl$ BCP, ρ_{BCP}, amounts 0.013 au. If the number of ClF molecules increases to 5 and 6 the latter values are equal to -4.1 kcal mol^{-1} and 0.015 au, respectively. Thus the "strength increase" amounts to only ~0.9 kcal mol^{-1}, and the electron density counterpart is equal to 0.002 au. A more significant strength increase is observed for the fully optimized systems. For the $H_2CO\cdots ClF$ complex the $O\cdots Cl$ distance is equal to 2.558 Å, the interaction energy amounts to -4.1 kcal mol^{-1} and the $O\cdots Cl$ ρ_{BCP} is equal to 0.024 au; the corresponding values for the $H_2CO \cdots (ClF)_2$ cluster are equal to 2.481 Å, -5.20 kcal mol^{-1} and 0.029 au, respectively. This means that the cooperative strengthening of the $O \cdots Cl$ interaction is equal to 1.1 kcal mol^{-1} and 0.005 au, in spite of the fact that the fully optimized $H_2CO \cdots (ClF)_2$ cluster is far from the "saturated" case.

Figures 19.14 and 19.15 show the isolines of the Laplacian of electron density; the negative values of the Laplacian ($\nabla^2 \rho$) show the regions of the concentration of the electron density. There is no such concentration at Cl-centres in the elongation of F–Cl bonds. This is why the electron density depleted regions of Cl-centres are in contact with those which play the role of Lewis bases. The above mentioned 'electron poor' region of the Cl-centre corresponds to the positive electrostatic potential (EP) area which is observed in the extension of F–Cl bond (Figure 19.16).

The interrelations between different types of intermolecular interactions have recently been the subject of numerous studies; *i.e.* cooperativity effects for clusters linked by distinct noncovalent bonds have been analyzed. For example, triads linked by hydrogen and halogen bonds were analyzed theoretically and the electron charge density shifts were described.[111,112] Figure 19.17 shows the molecular graph of such a triad, as an example, *i.e.* $FH \cdots HCCCl \cdots Cl^-$ while Figure 19.18 presents the molecular graphs of its components; dyads and monomers.

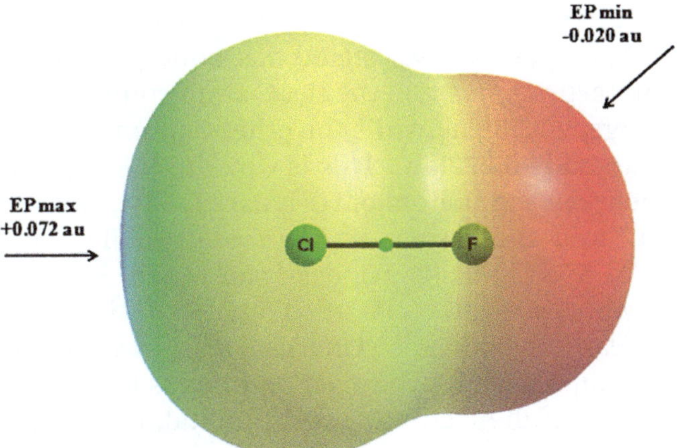

Figure 19.16 The map of the electrostatic potential calculated at the 0.001 au molecular electron density surface for the ClF molecule; red and blue colors correspond to negative and positive EP, respectively. The maximum and minimum EP are presented; MP2/6-311++G(d,p) level of calculations (based on results of ref. 110).

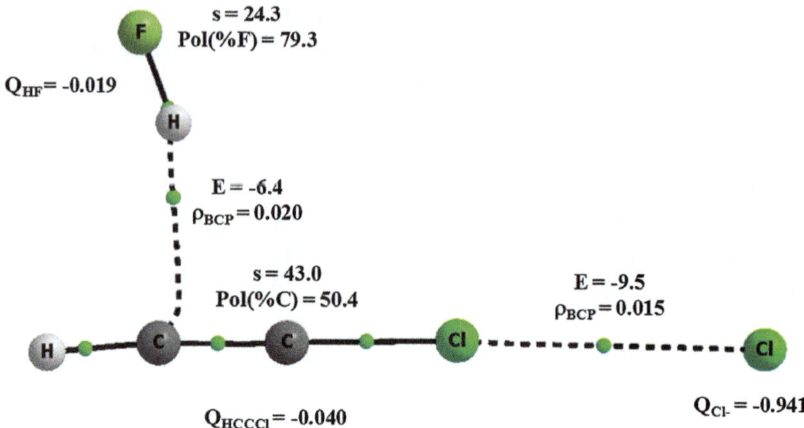

Figure 19.17 The molecular graph of the HF$\cdots$HCCCl$\cdots$Cl$^-$ cluster, s-characters, polarizations of bonds, Pols, participating in interactions, and NBO charges of components of the cluster (in au) are given. The electron density at BCP, ρ_{BCP} (au), and the interaction energy, E (kcal mol^{-1}), are included (based on results of ref. 111).

A positive cooperative effect is observed for this triad (Figures 19.17 and 19.18). For example one can observe the electron charge density shifts mentioned earlier as the result of HB formation. The HF neutral molecule ($Q_{HF} = 0$), if involved in HB interaction with HCCCl, withdraws electron charge and is negatively charged ($Q_{HF} = -0.011$ au) in the FH$\cdots$HCCCl complex linked through FH$\cdots\pi$ hydrogen bonds. To be more precise, it is rather a FH$\cdots$C hydrogen bond since the bond path links H and C attractors (Figure 19.18). This is in line with the properties of bond paths. It was stated that bond paths correspond to preferable interactions,[82,83] and the the C-atom connected with Cl has a positive charge, +0.139 au while the other C-atom in the complex is negatively charged (-0.272 au). This is why the H$\cdots$C preferable link is observed between the H-atom of HF (charge of the H-centre is equal to +0.554 au) and the carbon negative centre (NBO charges were considered[111]). Such AH$\cdots$C (C designates here a carbon atom) interactions, which are often classified as AH$\cdots\pi$ interactions, were discussed in earlier studies.[113] One can also observe that for the FH$\cdots$HCCCl complex the rehybridization process leads to an increase of the s-character of the F atomic hybrid orbital, from 20.7 in the HF monomer to 23.1 in the complex. Also the polarization of the HF bond, calculated as the percentage of electron density at the F-atom, increases from 77.6% in the HF monomer to 78.4% in the complex.

We have a similar situation for the HCCCl$\cdots$Cl$^-$ complex linked through an XB interaction. The HCCCl molecule, acting here as a

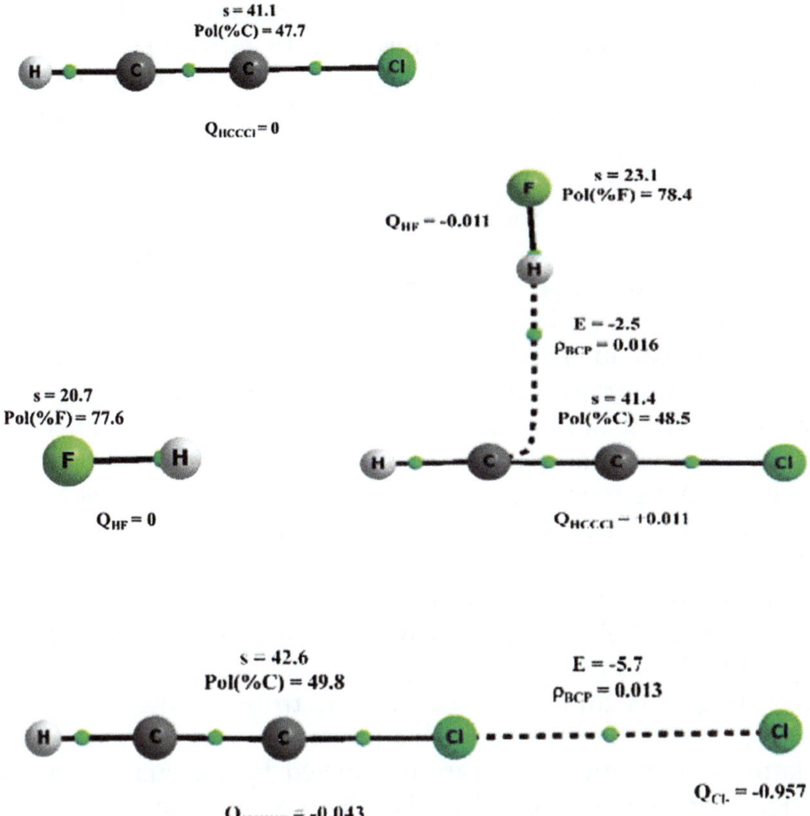

Figure 19.18 The molecular graphs of the HF and HCCCl monomers as well as for the FH···HCCCl and HCCCl···Cl⁻ dyads, s-characters, polarizations of bonds, Pols, participating in interactions, and NBO charges of monomers in species considered (in au) are given. The electron density at BCPs, ρ_{BCP} (au), and the interaction energy, E (kcal mol⁻¹), are included (based on results of ref. 111).

Lewis acid, withdraws electron charge density from the chlorine anion and it is negatively charged in the complex ($Q_{HCCCl} = -0.043$ au). Rehybridization leads to the increase of s-character of the C-atom connected with chlorine and an increase of the polarization of the C–Cl bond (% at C atom); from 41.1 to 42.6 and from 47.7% to 49.8%, respectively.

In the triad (Figure 19.17), the central HCCCl molecule plays simultaneously the roles of Lewis base interacting with HF and Lewis acid, interacting with the Cl⁻ anion. The values of parameters for the F and C atoms (s-character) and for the H–F and C–Cl bonds (polarization) are higher here than the corresponding ones in the dyads and monomers (see Figure 19.18). This shows the positive cooperative effect. It means that the halogen bond is enhanced if the HCCCl$\cdots$Cl$^-$ complex interacts additionally through an HB with an HF species; the interaction energy amounts -5.7 kcal mol^{-1} for the dyad while it is equal to -9.5 kcal mol^{-1} for the triad; also ρ_{BCP} for the Cl$\cdots$Cl$^-$ contact increases from 0.013 au in the dyad to 0.015 au in the triad. On the other hand, the HB interaction is enhanced in the FH$\cdots$HCCCl complex if the latter one interacts additionally with the Cl$^-$ anion. The interaction energy for the FH$\cdots$HCCCl complex is equal to -2.5 kcal mol^{-1} while the corresponding interaction in the triad amounts -6.4 kcal mol^{-1}; the corresponding ρ_{BCP} increases from 0.016 au to 0.020 au. The interaction energies mentioned above were calculated for the triad according to the eqn (19.9) described earlier.

The number of studies on cooperativity effects for noncovalent interactions other than HBs has increased in recent years. Numerous examples may be mentioned here. For example, the triangular halogen trimers (RX)$_3$, where X $=$ Br and I while R $=$ H, H$_3$C, H$_2$FC, HF$_2$C, F$_3$C, CH$_2$–CH, CH$\equiv$C and phenyl have been analyzed theoretically and it was found that weak cooperativity effects are observed for some of iodine trimers.[114] The cooperativity effects of hydrogen, lithium and halogen bonds in the F–H/Li$\cdots$OH$_2$ complexes were described.[115] The F$_3$CX$\cdots$HMgH$\cdots$Y and F$_3$CX$\cdots$Y$\cdots$HMgH trimers, where X $=$ Cl and Br, while Y $=$ HCN and HNC were analyzed[116] – one can see that for the latter systems the triads are connected by various interactions, hydrogen or halogen bonds; also by so-called halogen-hydride bonds.[117] Another interesting study refers to the halogen analogues of the HB systems assisted by π-electron delocalization,[118] which were classified by Gilli *et al.* as resonance assisted hydrogen bonds (RAHBs)[38,39] – a resemblance between hydrogen bonded clusters and their halogen analogues was detected.[118] Numerous other studies may also be mentioned here.

19.4.2 Cooperativity in Halogen Bonded Systems in Crystals

It is obvious that, similarly to the case of hydrogen bonds, in crystal structures the chains of molecules related by translation symmetry should be observed where the units of chains are linked through other interactions, for example, through halogen bonds. Figure 19.19 presents the example of the 1-methylpyrrol-2-yl trichloromethyl ketone structure[119] where one can observe the Cl–C–C=O$\cdots$Cl–C–C=O$\cdots$ chains linked through O$\cdots$Cl halogen bonds. The O$\cdots$Cl distances amount to 3.047(2) Å, slightly less than

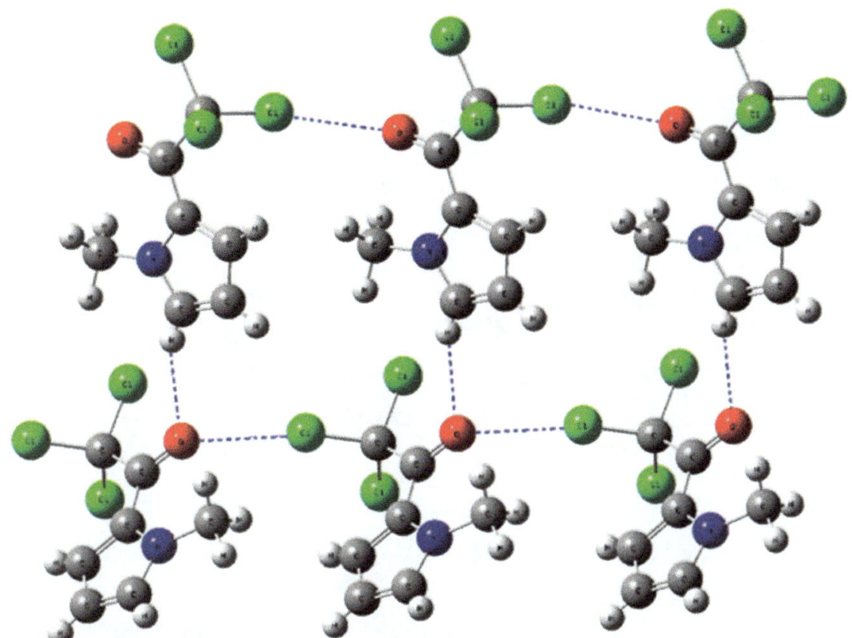

Figure 19.19 A fragment of the crystal structure of 1-methylpyrrol-2-yl trichloromethyl ketone;[119] the O···Cl and H···O intermolecular links are designated by broken blue lines; CSD refcode: WEYYUV.

the corresponding sum of the van der Waals radii – 3.2 Å.[103] The C–Cl···O angle is equal to 168.6°, close to linearity, which is in agreement with the σ-hole concept [104,105] which states that XB interactions are usually linear due to the small area of positive EP situated in the extension of A–X bond (C–Cl bond for the structure presented here). Weak C–H···O interactions are also observed in this structure; the H···O distance of 2.64 Å is close to the corresponding sum of the van der Waals radii (2.6 Å).[103]

The Cambridge Structural Database [16] was searched for the other structures containing the Cl–C–C=O motif repeated by the translation symmetry to form halogen bonded chains.[119] Few structures containing such chains were found (besides the 1-methylpyrrol-2-yl trichloromethyl crystal, only five others) with the O···Cl contacts between 2.89 Å and 3.17 Å and the C–Cl···O angle in the range 145.3–165.9°. Figure 19.20 shows a fragment of a crystal structure [120] where trichloroacetic acid molecules are linked in Cl–C–C=O···Cl–C–C=O··· chains. However, there are also motifs of two carboxylic groups linked by double HBs and forming the eight-member $R_2^2(8)$ ring.[51,52] The O···O distances are equal to 2.662 Å for those rings.

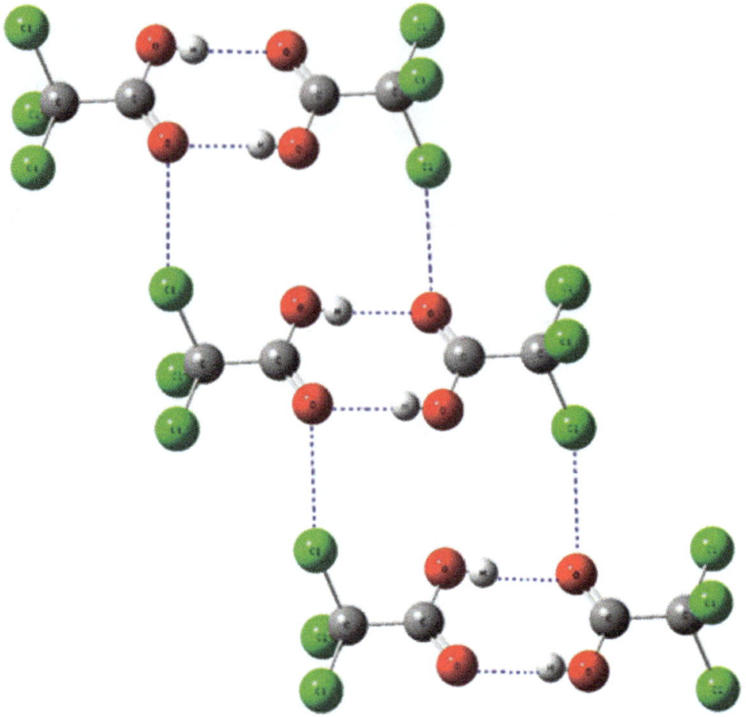

Figure 19.20 A fragment of the crystal structure of trichloroacetic acid;[120] the O···Cl and H···O intermolecular links are designated by broken blue lines; CSD refcode: TCACAD01.

One may also mention studies on so-called dihalogen bonds, *i.e.* on intermolecular C–Br···Br–C and C–Cl···Cl–C interactions where properties in crystal structures were analyzed.[121,122] The X···X intermolecular contacts were presented in these studies and detailed theoretical analysis on the nature of the interactions in the corresponding dimers was performed.[121,122] The crystal structures analyzed are characterized by numerous properties which show the significance of cooperative effects. For example, Figure 19.21 presents a fragment of the crystal structure of methyl (2*R*,3*S*)-3-bromo-2-chloro-3-phenylpropanoate,[123] analyzed by Novoa *et al.*,[121] where the motifs linked by Br···Br and Cl···Cl dihalogen bonds are observed. The Br···Br distances of 3.390 Å are smaller than the corresponding sum of the van der Waals radii (3.9 Å[103]) while the Cl···Cl distances of 3.397 Å differ less from the corresponding sum (3.6 Å[103]). This means that the interactions, and probably also the cooperative effects, for the bromine contacts are more distinct than for chlorine ones. This is partly connected with the greater σ-hole effect for the Br-centre than for the Cl one.[104,105,108]

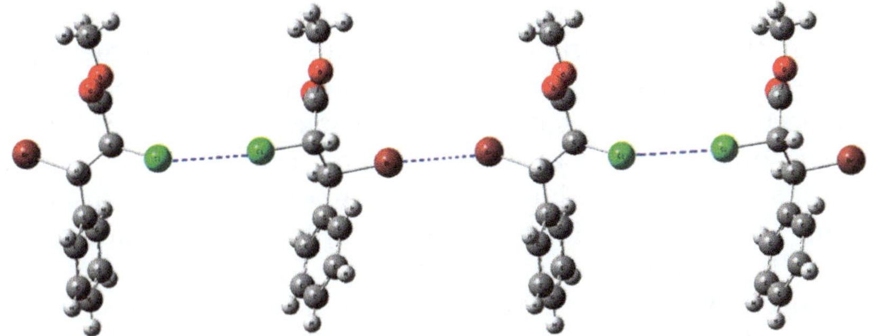

Figure 19.21 A fragment of the crystal structure of methyl (2R,3S)-3-bromo-2-chloro-3-phenylpropanoate;[123] the Cl···Cl (3.397 Å) and Br···Br (3.390 Å) intermolecular links are designated by broken blue lines; CSD refcode: LIKFIU.

19.5 Summary

The cooperativity effect is mainly understood as the enhancement of the first hydrogen bond (HB) between the proton donor and the proton acceptor when an additional HB is formed between one of these units and the next species.[1] However the term cooperativity also describes other effects, giving rise to its other numerous meanings. This is why measures of cooperativity, based on the equations expressing or related to the interaction energy parameters, allow one to define it more precisely.

For example, the cooperativity may be defined as the nonadditive energy of the cluster; *i.e.* it is the many-body interaction energy term's contribution to the total interaction energy of the system. In other words, it is defined as the difference between the total interaction energy and the sum of two-body terms. It is important that the latter definition of cooperativity (nonadditivity) concerns not only hydrogen bond interactions but may also be applied to other intermolecular interactions.

The other precisely described index concerns the specific interaction within the cluster considered (eqn (19.9)); this is why one can understand this effect as the enhancement of the interaction in the complex due to interactions with additional moieties. One may decompose the cluster into two parts exactly at the contact between them, *i.e.* for the interaction of interest. Further, this interaction may be analyzed when the sizes of components change. This approach may be also applied not only to hydrogen bonded systems but to other types of interactions.

Another approach to expressing the cooperativity effect may be applied if the system analyzed is built up from the same type of molecules; in such a way the mean energy of interaction between the two molecules in contact in the system may be analyzed. This approach is often applied to water clusters. Table 19.1 presents the mean theoretical interaction energies (for convenience of discussion the sign ' − ' is omitted for these energies) and corresponding mean distances between neighbouring molecules for water clusters linked by hydrogen bonds (see Figure 19.5); some experimental data on water, as well as theoretical results on FBr clusters linked by halogen bonds are also presented for comparison.

The bold values presented in Table 19.1 clearly indicate the trend described in this chapter that the cooperativity effect is connected with the increase of the strength of interaction as the number of moieties in the system analyzed increases. The MP2/aug-cc-pV5Z results[127] show that the HB energy for the dimer (5 kcal mol^{-1}) increases for the cyclic trimer to 5.3 kcal mol^{-1} and to 6.9 and 7.3 kcal mol^{-1} for tetramer and pentamer, respectively. This trend is confirmed by other theoretical results (MP2/6-311++G(2d,2p))[128] where the hydrogen bond strength reaches 7.6 kcal mol^{-1} for the

Table 19.1 The cooperativity effects for water clusters and for the clusters of FBr; energies in kcal mol^{-1}, distances in Å (O$\cdots$O distances for water clusters and F$\cdots$Br distances for halogen bonded FBr clusters); the negative energies correspond to the interactions presented – for the convenience of discussion the sign '–' is omitted in this table and the corresponding text in the chapter.

HB cooperativity				XB cooperativity[h]		
System	O$\cdots$O	HB$_{energy}$	HB$_{energy}$	System	F$\cdots$Br	XB$_{energy}$
$(H_2O)_2$ exp. gas	~2.98^a	3.16 ± 0.03^b		no experimental results		
$(H_2O)_2$ theo.		3.15 ± 0.01^c		**(FBr)$_2$ chain**	**2.686**	**3.7**
$(H_2O)_2$	**2.907^d**	**5.0^d**	5.4^e	**(FBr)$_3$ chain**	**2.637**	**4.1**
$(H_2O)_3$ cycle	**2.787^d**	**5.3^d**	5.5^e	**(FBr)$_4$ chain**	**2.608**	**4.3**
$(H_2O)_4$ cycle	**2.732^d**	**6.9^d**	7.2^e	**(FBr)$_3$ cycle**	**3.095**	**2.3**
$(H_2O)_5$ cycle	**2.716^d**	**7.3^d**	7.6^e	**(FBr)$_4$ cycle**	**2.665**	**4.1**
crystal	~2.75^f	7.8^g		no crystal structure results		
liquid	~2.81^f			no results for liquid		

^{a}Ref. 124.
^{b}Ref. 125.
^{c}Ref. 126.
^{d}Ref. 127.
^{e}Ref. 128.
^{f}Ice *Ih*, ref. 129.
^{g}Ref. 130.
^{h}Ref. 131.

pentamer. One may observe here the slight increase of the strength of interaction if the trimer is compared with the dimer; this is connected with the strain effects occurring for the former system, such strain effects are not so important for the larger cycles of water molecules (tetramer and pentamer). The H-bond strength between neighbouring water molecules in the ice structure was calculated (B3LYP/ 6-31++(df))[130] to be equal to 7.8 kcal mol^{-1}. The H-bond experimental energy of 3.16 kcal mol^{-1} for the water dimer (Table 19.1) is much lower than the theoretical ones for the most stable (*trans*-linear) water dimer configuration (5 and 5.4 kcal mol^{-1} Table 19.1). However, the experimental evaluation concerns all water dimer configurations in the gas phase[125] and the corresponding benchmark calculations where *ab initio* full-dimensional potential energy surfaces were considered are in excellent agreement (3.15 kcal mol^{-1})[126] with the experiment. The O$\cdots$O distances follow the strength of the hydrogen bond, *i.e.* shorter distances are observed for stronger interactions. The experimental O$\cdots$O distance between neighbouring water molecules in the gas phase is equal to $\sim$2.98 Å;[124] theoretical calculations show that this distance is shorter for the *trans*-linear water dimer (2.907 Å) and it decreases for the trimer, tetramer and pentamer; to 2.787, 2.732 and 2.716 Å, respectively.[127] The experimental values in the crystal (*Ih* ice) and liquid (2.75 and 2.81 Å)[129] are shorter than in the gas phase (2.98 Å).[124]

A cooperativity enhancement of the interaction is also observed for the halogen bond as described in this chapter. Table 19.1 shows the results for (FBr)$_n$ clusters; for the chain structures the XB strength increases from 3.7 for the dimer to 4.1 and 4.3 kcal mol^{-1} for trimer and tetramer, respectively, accompanied by a decrease of F$\cdots$Br distance.[131] For the cyclic (FBr)$_3$ trimer the XB interaction energy is equal to 2.3; such a decrease of the strength of interaction in comparison with the dimer is the result of strain effects which are much more important here than in the case of the water cycle trimer; for the cyclic (FBr)$_4$ cluster the XB interaction energy is equal to 4.1 kcal mol^{-1}.[131]

Mainly hydrogen bonded systems were described in this chapter, followed by halogen bonded ones, since it seems that theoretical and experimental analyses on cooperativity are most often connected with these interactions. However, in recent years, numerous studies on cooperativity for other noncovalent bonds have been performed; on cooperativity of chalcogen,[132,133] pnicogen [133,134] or triel bonds (triel designates the atom of the 13th group of elements);[135] and very recently the analysis of cooperativity in chains linked through tetrel bonds was carried out.[136,137]

Finally, it seems that cooperativity effects are very important when describing real systems which are usually constituted of numerous sub-units. For example, crystal structures contain large chains of molecules or ions and even more, the 3D network of units observed in crystal structures is connected by distinct interactions.

This chapter presents a selective choice of results to describe the cooperativity phenomenon and the author is aware that due to the huge number of studies some of them may have been overlooked.

Acknowledgements

Financial support comes from Eusko Jaurlaritza (GIC IT-588-13) and the Spanish Office for Scientific Research (CTQ2012-38496-C05-04). Technical and human support provided by Informatikako Zerbitzu Orokora – Servicio General de Informática de la Universidad del País Vasco (SGI/IZO-SGIker UPV/EHU), Ministerio de Ciencia e Innovación (MICINN), Gobierno Vasco Eusko Jaurlanitza (GV/EJ), European Social Fund (ESF) is gratefully acknowledged.

References and Notes

1. O. Mó, M. Yañez, J. E. Del Bene, I. Alkorta and J. Elguero, *ChemPhysChem*, 2005, **6**, 1411.
2. H. S. Frank and W. Y. Wen, *Discuss. Faraday Soc.*, 1957, **24**, 133.
3. J. Del Bene and J. A. Pople, *Chem. Phys. Lett.*, 1969, **4**, 426.
4. E. Clementi, W. Kołos, G. C. Lie and G. Ranghino, *Int. J. Quant. Chem.*, 1980, **17**, 377.
5. J. D. Dunitz, *X-Ray Analysis and the Structure of Organic Molecules*, Cornell University Press, Ithaca, 1979.
6. *Engineering of Crystalline Materials Properties*, ed. J. J. Novoa, D. Braga and L. Addadi, Springer, Dordrecht, The Netherlands, 2007.
7. L. Leiserowitz, *Acta Crystallogr. Sect. B*, 1976, **32**, 775.
8. C. Ceccarelli, G. A. Jeffrey and R. Taylor, *J. Mol. Struct.*, 1981, **70**, 255.
9. In the case of bond lengths, intermolecular distances, valence angles, *etc.* taken from crystal structures, the estimated standard deviations (e.s.d.'s), if available, are given in parentheses thereafter, for example, 1.805(9)Å and 1.869(23)Å mean that there are distances of 1.805 and 1.869 Å, with the corresponding e.s.d.'s of 0.009 and 0.023 Å, respectively.
10. G. J. Jeffrey, *An Introduction to Hydrogen Bonding*, Oxford University Press, New York, 1997.
11. S. J. Grabowski, *J. Phys. Chem. A*, 2001, **105**, 10739.
12. G. A. Jeffrey and W. Saenger, *Hydrogen Bonding in Biological Structures*, Springer, Berlin, 1991, p. 51.
13. P. Gilli, V. Bertolasi, V. Ferretti and G. Gilli, *J. Am. Chem. Soc.*, 1994, **116**, 909.
14. S. J. Grabowski, *J. Phys. Org. Chem.*, 2004, **17**, 18.

15. Y. B. R. D. Rajesh, S. Ranganathan, R. D. Gilardi and I. L. Karle, *J. Chem. Crystallogr.*, 2008, **38**, 39.
16. R. Wong, F. H. Allen and P. Willett, *J. Appl. Crystallogr.*, 2010, **43**, 811.
17. GaussView, Version 5, R. Dennington, T. Keith and J. Millam, Semichem Inc., Shawnee Mission, KS, 2009.
18. W. D. Kumler, *J. Am. Chem. Soc.*, 1935, **57**, 600.
19. W. F. Giauque and R. A. Ruehrwein, *J. Am. Chem. Soc.*, 1939, **61**, 2626.
20. L. Pauling, *The Nature of the Chemical Bond*, Cornell University Press, New York, 3rd edn, 1960, p. 459.
21. W. J. Dulmage and W. N. Lipscomb, *Acta Crystallogr.*, 1951, **4**, 330.
22. G. A. Hopkins, M. Maroncelli and J. W. Nibler, *Chem. Phys. Lett.*, 1985, **114**, 97.
23. M. Kofranek, A. Karpfen and H. Lischka, *Chem. Phys.*, 1987, **113**, 53.
24. S. Scheiner, *Hydrogen Bonding: A Theoretical Perspective*, Oxford University Press, New York, 1997, pp. 230–290.
25. Z. Latajka and S. Scheiner, *Chem. Phys.*, 1988, **122**, 413.
26. C. Rovira, P. Constans, M.-H. Whangbo and J. J. Novoa, *Int. J. Quantum Chem.*, 1994, **52**, 177.
27. G. R. Desiraju and T. Steiner, *The Weak Hydrogen Bond in Structural Chemistry and Biology*, Oxford University Press, New York, 1999, pp. 66–68.
28. L. Sobczyk, S. J. Grabowski and T. M. Krygowski, *Chem. Rev.*, 2005, **105**, 3513.
29. L. Chęcińska and S. J. Grabowski, *J. Phys. Chem. A*, 2005, **109**, 2942.
30. C. S. Hawes and P. E. Kruger, *Dalton Trans.*, 2014, **43**, 16450.
31. A. J. Bortoluzzi, D. Sebrao, M. M. Sa and M. G. Nascimento, *Acta Crystallogr., Sect. E*, 2011, **67**, o2778.
32. S. J. Grabowski, *Chem. Rev.*, 2011, **111**, 2597.
33. E. Reed, L. A. Curtiss and F. Weinhold, *Chem. Rev.*, 1988, **88**, 899.
34. F. Weinhold and C. Landis, *Valency and Bonding, A Natural Bond Orbital Donor – Acceptor Perspective*, Cambridge University Press, Cambridge, 2005.
35. I. V. Alabugin, M. Manoharan, S. Peabody and F. Weinhold, *J. Am. Chem. Soc.*, 2003, **125**, 5973.
36. G. A. Jeffrey, M. E. Gress and S. Takagi, *J. Am. Chem. Soc.*, 1977, **99**, 609.
37. W. Saenger, *Nature*, 1979, **279**, 343.
38. G. Gilli, F. Bellucci, V. Ferretti and V. Bertolasi, *J. Am. Chem. Soc.*, 1989, **111**, 1023.
39. V. Bertolasi, P. Gilli, V. Ferretti and G. Gilli, *J. Am. Chem. Soc.*, 1991, **113**, 4917.
40. I. Alkorta, J. Elguero, O. Mó, M. Yañez and J.E. Del Bene, *Mol. Phys.*, 2004, **102**, 2563.
41. R. W. Góra, M. Maj and S. J. Grabowski, *Phys. Chem. Chem. Phys.*, 2013, **15**, 2514.
42. T. Thorsteinsson, D. L. Cooper, J. Gerratt, P. B. Karadakov and M. Raimondi, *Theor. Chim. Acta*, 1996, **93**, 343.
43. T. Thorsteinsson and D. L. Cooper, *J. Math. Chem.*, 1998, **23**, 105.
44. W. A. Sokalski, S. Roszak and K. Pecul, *Chem. Phys. Lett.*, 1988, **153**, 153.
45. W. A. Sokalski and S. Roszak, *J. Mol. Struct.: THEOCHEM*, 1991, **234**, 387.
46. R. F. W. Bader, *Acc. Chem. Res.*, 1985, **18**, 9.
47. R. F. W. Bader, *Atoms in Molecules, A Quantum Theory*, Oxford University Press, Oxford, 1990.
48. M. L. Huggins, *J. Org. Chem.*, 1936, **1**, 405.
49. C. A. Coulson, *Valence*, Oxford University Press, Oxford, 1952.
50. S. Bhattacharya, V. G. Saraswatula and B. K. Saha, *Cryst. Growth Des.*, 2013, **13**, 3651.
51. M. C. Etter, J. C. MacDonald and J. Bernstein, *Acta Crystallogr. Sec. B*, 1990, **46**, 256.
52. M. C. Etter, *Acc. Chem. Res.*, 1990, **23**, 120.
53. S. J. Grabowski and T. M. Krygowski, *Acta Crystallogr. Sec. C*, 1985, **41**, 1224.

54. I. Alkorta, F. Blanco, P. M. Deyà, J. Elguero, C. Estarellas, A. Frontera and D. Quiñonero, *Theor. Chem. Acc.*, 2010, **126**, 1.
55. A. C. Legon, *Angew.Chem., Int. Ed. Engl.*, 1999, **38**, 2686.
56. P. Metrangolo and G. Resnati, *Chem. – Eur. J.*, 2001, **7**, 2511.
57. M. Formigué and P. Batail, *Chem. Rev.*, 2004, **104**, 5379.
58. D. Hankins, J. W. Moskowitz and F. H. Stillinger, *J. Chem. Phys.*, 1970, **53**, 4544.
59. S. S. Xantheas, *Chem. Phys.*, 2000, **258**, 225–231.
60. P. Salvador and M. M. Szczęśniak, *J. Chem. Phys.*, 2003, **118**, 537.
61. E. D. Glendening, *J. Phys. Chem A*, 2005, **109**, 11936.
62. E. D. Glendening and A. Streitwieser Jr., *J. Chem. Phys.*, 1994, **100**, 2900.
63. G. K. Schenter and E. D. Glendening, *J. Phys. Chem.*, 1996, **100**, 17152.
64. E. D. Glendening, *J. Am. Chem. Soc.*, 1996, **118**, 2473.
65. R. D. Parra, *J. Chem. Phys.*, 2003, **118**, 3499.
66. H. Brand, J. Martens, P. Mayer, A. Schulz, M. Seibald and T. Soller, *Chem. – Asian J.*, 2009, **4**, 1588.
67. W. E. Steinmetz, *J. Am. Chem. Soc.*, 1973, **95**, 2777.
68. R. Hoffmann, P. V. R. Schleyer and H. F. Schaefer, *Angew. Chem., Int. Ed.*, 2008, **47**, 7164.
69. R. D. Parra, S. Bulusu and X. C. Zeng, *J. Chem. Phys.*, 2005, **122**, 184325.
70. H. Tan, W. Qu, G. Chen and R. Liu, *J. Phys. Chem. A*, 2005, **109**, 6303.
71. X. Li, W. Liu, K. Sun, Y. Wang, H. Tan and G. Chen, *Phys. Chem. Chem. Phys.*, 2008, **10**, 5607.
72. S. J. Grabowski and J. M. Ugalde, *Can. J. Chem.*, 2010, **88**, 769.
73. B. G. Oliveira, R. C. M. U. Araujo, A. B. Carvalho, E. F. Lima, W. L. V. Silva, M. N. Ramos and A. M. Tavares, *J. Mol. Struct.: THEOCHEM*, 2006, **775**, 39.
74. S. J. Grabowski and J. Leszczynski, *Chem. Phys.*, 2009, **355**, 169.
75. M. Ziółkowski, S. J. Grabowski and J. Leszczynski, *J. Phys. Chem. A*, 2006, **110**, 6514.
76. K. Kitaura and K. Morokuma, *Int. J. Quantum Chem.*, 1976, **10**, 325.
77. S. J. Grabowski, W. A. Sokalski, E. Dyguda and J. Leszczynski, *J. Phys. Chem. B*, 2006, **110**, 6444.
78. A. M. Pendas, M. A. Blanco and E. Francisco, *J.Chem.Phys.*, 2006, **125**, 184112.
79. C. F. Matta and R. J. Boyd, in *Quantum Theory of Atoms in Molecules: Recent Progress in Theory and Application*, ed. C. J. Matta and R. J. Boyd, Wiley-VCH: New York, 2007, ch. I.
80. C. F. Matta, in *Hydrogen Bonding-New Insights*, ed. S. J. Grabowski, Springer, New York, 2006, ch. 9.
81. R. F. W. Bader, *Monatsh. Chem.*, 2005, **126**, 819.
82. R. F. W. Bader, *J. Phys. Chem. A*, 1998, **102**, 7314.
83. R. F. W. Bader, *J. Phys. Chem. A*, 2009, **113**, 10391.
84. G. R. Runtz, R. F. W. Bader and R. R. Messer, *Can. J. Chem.*, 1977, **55**, 3040.
85. S. Jenkins and I. Morrison, *Chem. Phys. Lett.*, 2000, **317**, 97.
86. I. Rozas, I. Alkorta and J. Elguero, *J. Am. Chem. Soc.*, 2000, **122**, 11154.
87. W. D. Arnold and E. Oldfield, *J. Am. Chem. Soc.*, 2000, **122**, 12835.
88. S. J. Grabowski and P. Lipkowski, *J. Phys. Chem. A*, 2011, **115**, 4765.
89. B. G. de Oliveira, *Phys. Chem. Chem. Phys.*, 2013, **15**, 37.
90. S. J. Grabowski, *J. Phys. Org. Chem.*, 2003, **16**, 797.
91. Originally this parameter was named as the "resonance parameter"; however, regarding other studies (refs. 40, 41) it seems that the term "delocalization parameter" better fits to the processes which take place for the systems with conjugated π-bonds.
92. T. Steiner, *Chem. Commun.*, 1998, 411.
93. P. Gilli, V. Bertolasi, V. Ferretti and G. Gilli, *J. Am. Chem. Soc.*, 2000, **122**, 10405.
94. S. J. Grabowski, *J. Mol. Struct.*, 2001, **562**, 137.

95. G. K. H. Madsen, B. B. Iversen, F. K. Larsen, M. Kapon, G. M. Reisner and F. H. Herbstein, *J. Am. Chem. Soc.*, 1998, **120**, 10040.

96. G. K. H. Madsen, F. K. Larsen, B. Schiøt and F. K. Larsen, *Chem. – Eur. J.*, 2007, **13**, 5539.

97. R. Gora, S. J. Grabowski and J. Leszczynski, *J. Phys. Chem. A*, 2005, **109**, 6397.

98. S. J. Grabowski, A. T. Dubis, D. Martynowski, M. Główka, M. Palusiak and J. Leszczynski, *J. Phys. Chem. A*, 2004, **108**, 5815.

99. A. Almenningen, O. Bastiansen and T. Motzfeldt, *Acta Chem. Scand.*, 1969, **23**, 2848.

100. I. Nahringbauer, *Acta Crystallogr. Sec. B*, 1978, **34**, 315.

101. P.-G. Jönsson, *Acta Crystallogr. Sec. B*, 1971, **27**, 893.

102. C. Rovira and J. J. Novoa, *J. Chem. Phys.*, 2000, **113**, 9208.

103. L. Pauling, *The Nature of the Chemical Bond*, Cornell University Press, New York, 3rd edn, 1960, p. 260.

104. P. Politzer, J. S. Murray and T. Clark, *Phys. Chem. Chem. Phys.*, 2010, **12**, 7748.

105. P. Politzer, J. S. Murray and T. Clark, *Phys. Chem. Chem. Phys.*, 2013, **15**, 11178.

106. F. Zordan, L. Brammer and P. Sherwood, *J.Am.Chem.Soc.*, 2005, **127**, 5979.

107. C. Rømming, *Acta Chem. Scand.*, 1960, **14**, 2145.

108. P. Politzer and J. S. Murray, *ChemPhysChem*, 2013, **14**, 278.

109. P. Metrangolo, H. Neukirch, T. Pilati and G. Resnati, *Acc. Chem. Res.*, 2005, **38**, 386.

110. S. J. Grabowski and E. Bilewicz, *Chem. Phys. Lett.*, 2006, **427**, 51.

111. S. J. Grabowski, *Theor. Chem. Acc.*, 2013, **132**, 1347.

112. S. J. Grabowski, *Cooperativity of hydrogen and halogen bond interactions*, in 8th Congress on Electronic Structure: Principles and Applications (ESPA 2012), eds. J. J. Novoa and M. F. Ruiz-López, Springer, Berlin, Heidelberg, 2014, vol. 5 of Series; Highlights in Theoretical Chemistry, series eds. Ch. J. Cramer and D. G. Truhlar.

113. S. J. Grabowski and J. M. Ugalde, *J. Phys. Chem. A*, 2010, **114**, 7223.

114. Y. Lu, J. Zou, H. Wang, Q. Yu, H. Zhang and Y. Jiang, *J. Phys. Chem. A*, 2005, **109**, 11956.

115. S. A. C. McDowell and H. K. Yarde, *Phys. Chem. Chem. Phys.*, 2012, **14**, 6883.

116. M. Solimannejad, M. Malekani and I. Alkorta, *J. Phys. Chem. A*, 2010, **114**, 12106.

117. P. Lipkowski, S. J. Grabowski and J. Leszczynski, *J. Phys. Chem. A*, 2006, **110**, 10296.

118. L. P. Wolters, N. W. C. Smits and C. F. Guerra, *Phys. Chem. Chem. Phys.*, 2015, **17**, 1585.

119. E. Bilewicz, A. J. Rybarczyk-Pirek, A. T. Dubis and S. J. Grabowski, *J. Mol. Struct.*, 2007, **829**, 208.

120. K. Rajagopal, A. Mostad, R. V. Krishnakumar, M. S. Nandhini and S. Natarajan, *Acta Crystallogr. Sec. E*, 2003, **59**, o316.

121. M. Capdevila-Cortada and J. J. Novoa, *CrystEngComm*, 2015, **17**, 3354.

122. M. Capdevila-Cortada, J. Castelló and J. J. Novoa, *CrystEngComm*, 2014, **16**, 8232.

123. J. P. Shaw, E. W. Tan and A. G. Blackman, *Acta Crystallogr. Sec. C*, 1995, **51**, 134.

124. A. Mukhopadhyay, W. T. S. Cole and R. J. Saykally, *Chem. Phys. Lett.*, 2015, **633**, 13.

125. B. E. Rocher-Casterline, L. C. Ch'ng, A. K. Mollner and H. Reisler, *J. Chem. Phys.*, 2011, **134**, 211101.

126. A. Shank, Y. Wang, A. Kaledin, B. J. Braams and J. M. Bowman, *J. Chem. Phys.*, 2009, **130**, 144314.

127. B. Santra, A. Michaelides and M. Scheffler, *J. Chem. Phys.*, 2007, **127**, 184104.

128. S. Maheshwary, N. Patel, N. Sathyamurthy, A. D. Kulkarni and S. R. Gadre, *J. Phys. Chem. A*, 2001, **105**, 10525.

129. U. Bergmann, A. Di Cicco, P. Wernet, E. Principi, P. Glatzel and A. Nilsson, *J. Chem. Phys.*, 2007, **127**, 174504.
130. M. Huša and T. Urbic, *J. Chem. Phys.*, 2012, **136**, 144305.
131. I. Alkorta, F. Blanco and J. Elguero, *Struct. Chem.*, 2009, **20**, 63.
132. L. M. Azofra and S. Scheiner, *J. Chem. Phys.*, 2014, **140**, 034302.
133. J. George, V. L. Deringer and R. Dronskowski, *J. Phys. Chem. A*, 2014, **118**, 3193.
134. I. Alkorta, G. Sánchez-Sanz, J. Elguero and J. E. Del Bene, *J. Chem. Theory Comput.*, 2012, **8**, 2320.
135. S. J. Grabowski, *ChemPhysChem*, 2014, **15**, 2985.
136. M. D. Esrafili, N. Mohammadirad and M. Solimannejad, *Chem. Phys. Lett.*, 2015, **628**, 16.
137. M. Marín-Luna, I. Alkorta and J. Elguero, *J. Phys. Chem. A*, 2016, **120**, 648.

20 Crystal Engineering: State of the Art and Open Challenges

D. Braga* and F. Grepioni

Department of Chemistry G. Ciamician, Via F. Selmi 2, 40126 Bologna, Italy
*Email: dario.braga@unibo.it

20.1 Introduction

The design, construction and exploitation of functional, molecular crystalline materials – namely crystal engineering – has become the new frontier of solid state chemistry.

Crystal engineering makes use of self-recognition and self-assembly processes in order to generate crystalline materials, with properties resulting from the convolution of the molecular building block properties with intermolecular bonding and crystal periodicity.[1]

Born in the field of organic solid state chemistry, crystal engineering arose to the level of world-wide-spread field of research in the late eighties, following the success of supramolecular chemistry.[2]

At about the same time, coordination chemistry experienced a similar evolution. The focus of research shifted from the synthesis and characterization of "zero dimensional" molecular complexes to that of three-dimensional networks. Substituting ligand polydentation on the same metal with polydentation on different metal atoms allowed movement from intra- to intermetallic coordination, *i.e.* from 0D complexes to 1–3D networks, and opened the way to the construction of metallorganic and organometallic framework structures,

Intermolecular Interactions in Crystals: Fundamentals of Crystal Engineering
Edited by Juan J. Novoa
© The Royal Society of Chemistry 2018
Published by the Royal Society of Chemistry, www.rsc.org

with properties depending on the nature of the ligands and on the coordination geometry, electronic, spin and charge state of the metal centres.[3] The confluence of these two streams of research, namely periodical coordination chemistry *via* coordination bonds and supramolecular self-assembly *via* directional intermolecular interactions, such as hydrogen bonds (see Chapter 16 and ref. 4) and, more recently, halogen bonds (see Chapter 19 and ref. 5), led rapidly to the explosive growth of a new interdisciplinary area of chemistry where traditional disciplinary barriers between organic, organometallic, metallorganic and inorganic chemistry were no longer meaningful. As pointed out by Desiraju, crystal engineering brought about a novel multidisciplinary – holistic – view of crystal chemistry.[6]

Hydrogen bonding, electrostatic interactions and ligand–metal coordination bonds *etc.* – exploited in crystal engineering exercises and amply discussed in this book – were investigated long before the idea of using them to obtain crystalline materials by design became popular.

Why did this not happen before? In other words, why did it take so long to pass from a structural chemistry essentially focused on the single molecule to a structural chemistry focused on aggregates of molecules?

There are a number of reasons. Certainly the "supramolecular way" of looking at interactions between molecules had a most significant role, bringing about the awareness of a molecular crystal as an "organized entity of higher complexity held together by intermolecular forces", rather than a mere container for molecules. But it cannot be denied that other factors concurred in a significant manner: (i) the increase in computational facilities for theoretical assessments, database analysis and crystal structure prediction, (ii) the decrease in price of, hence the increase in accessibility to, instrumentations to investigate the solid state (area detectors for diffractometers, variable temperature devices, solid state NMR, *etc.*) and, last but not least, the increasing need for the academic community of utilitarian objectives to support funding applications and to direct basic research efforts and investments.[7] All these aspects concurred to boost crystal engineering as the most promising approach to novel multifunctional molecular materials. The contributions collected in this book all deal with supramolecular interactions, whether weak or strong, directional or diffuse, repulsive or attractive. Making crystals by "using" these interactions by design to construct crystal structures with desired structure–function relationships is the ultimate goal of crystal engineering.

We have made considerable progress in the understanding of supramolecular interactions and this is well demonstrated by this book.

Weak and strong hydrogen bonding, aromatic interactions, dipole–dipole interactions, halogen bonding, Coulombic interactions between ions and "charge assistance" to weaker bonds in polarized molecules (*e.g.* zwitterions), interactions with delocalized electron systems, *etc.* have been analyzed, evaluated and extracted from the plethora of data accumulated in the CSD and ranked in terms of strength, directionality and competition.

Crystals have been and still are being dissected in all possible ways down to the minute contact points between molecules in the crystal lattice, often forgetting that a crystal is the global result of a very large number of such interactions and also that most molecules are flexible and/or possess degrees of internal structural freedom (rotation about bonds, angular torsion and bond stretching *etc.*) which also come into play in defining the free energy landscape of the "molecule in a crystal" system.[8] This is probably why, in spite of the depth reached in the knowledge of the geometrical features and energetic requirements of supramolecular bonding in the crystals, we are still quite far from being able to actually *apply* the crystal engineering paradigm, *i.e.* making crystals by design, with success in a large number of cases. This aspect is less problematic in the crystal engineering of 1-D, 2-D and 3-D metallorganic frameworks, which are based on coordination bonds. An adequate choice of linkers and nodes in MOFs, made on the basis of the coordination preference of the metal centers and of ligand–metal dative bonds, does usually permit a higher level of engineering, that is to say a higher level of predictability, than in the case of molecular systems. In molecular systems it is usually possible to single out specific supramolecular bonding contributions (synthons) that are sufficiently robust as to be transferable from crystal to crystal. What becomes "fuzzy", and increases the uncertainty level of design strategy, is what the rest of the molecule does. Once the important interaction is satisfied, the dispersive (attractive or repulsive) interactions between neighboring peripheral atoms, the more or less rigid and adaptable molecular shapes, the charge distribution over the molecular volumes and the longer range interactions between dipoles, all come into play in finding ways – often many ways as demonstrated by the phenomenon of crystal polymorphism (*vide infra*) – to minimize the global energy of the molecule–crystal system.[8] Hence, our level of confidence in predicting what molecules will do, when they recognize each other and generate the first crystallization nuclei, is very high as far as popular (*i.e.* energetically favored) synthons are concerned, but is from low to very low when the design strategy has to work out the most stable arrangements of the mobile atoms distributed over the surface of organic molecules.

Furthermore, the variability increases, hence the level of predictability decreases further, if the crystallization process takes place in solution where solvent molecules are involved in the nucleation and growth steps. If stable, or unstable but longer living, crystal nuclei are formed with the participation of solvent molecules, the unpredictable outcome of the crystallization process will be a solvate.

This chapter deals with three situations that challenge the crystal engineering paradigm: crystal hydrates, co-crystals and polymorphism. The unpredictable formation of a solvate upon crystallization from a solvent, the unpredictable occurrence of polymorphs, the failure or success in the construction of a co-crystal and the inherent uncertainty about solvation and polymorphism challenge both the experimentalist and the theoretician. Examples coming from various sources will be discussed in the following. It is important to point out that for each of these situations a number of extremely valuable review articles have been recently published. The interested reader is directed to these for details and for literature coverage. Some recent ones will be cited throughout in the following.

20.2 Hydrates

Crystallization from a solution is probably the most common way to obtain crystals. Precipitation from a solvent either by solvent evaporation or by temperature gradient are often the methods of choice. In these conditions a solvate *can* be formed, often as an unpredictable, though unsurprising event. In the case of hydrates, the situation is further complicated by the ubiquity of water.[9] Unless one adopts specific precautions, all materials, glassware, reactants, solvents are invariably "wet" to some extent. This is to say that the formation of hydrates, though not certain, is common.[10] Sometimes, if water is undesired and the hydration process is difficult to avoid, it can be convenient to *first* crystallize the hydrate(s) and *subsequently* remove water by thermal/vacuum treatment. The dehydration process, however, is often accompanied by a change in crystal structure or by a collapse of the whole crystal structure, followed by formation of amorphous material; depending on the atmospheric humidity, the anhydrous phase can show instability towards water re-uptake, and the original, or a different hydrated form can be obtained in these cases (see Figure 20.1).

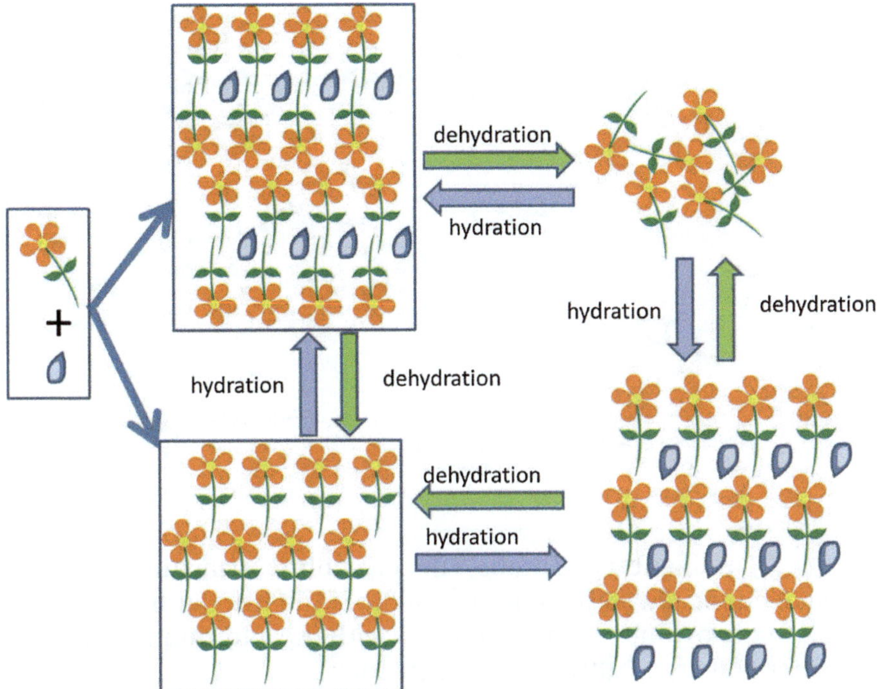

Figure 20.1 A cartoon representing the relationship between an anhydrous crystal (bottom left), stoichiometric hydrates (top left and bottom right) and amorphous material (top right).

Moreover, when crystalline materials are being used by living beings, the permitted solvents are often restricted to water and very few bio-compatible (GRAS: Generally Regarded As Safe)[11] solvents. Water is also the biological solvent and hence, everything affecting solubility or intrinsic dissolution rate in water of compounds entering the life cycle of living organisms (agrochemicals food industry, nutraceuticals, pharmaceuticals but also cosmetics and possibly others) is important. Other solvents are usually treated in terms of impurity profile or limited dosage.

The hydrate–anhydrate equilibria in organic solvents have been extensively studied by Grant.[12] For instance, theophylline forms an anhydrate and a monohydrate which interconvert in organic solvents, depending on the water activity. At low water activity (<0.25), the anhydrate is the only species present, whereas at higher water activity the monohydrate is the most stable form[13] (see Figure 20.2).

In the case of ampicillin, on the other hand, it has been shown that anhydrous crystalline ampicillin is kinetically stable for several days in $MeOH/H_2O$ mixtures over the whole range of water activity, even

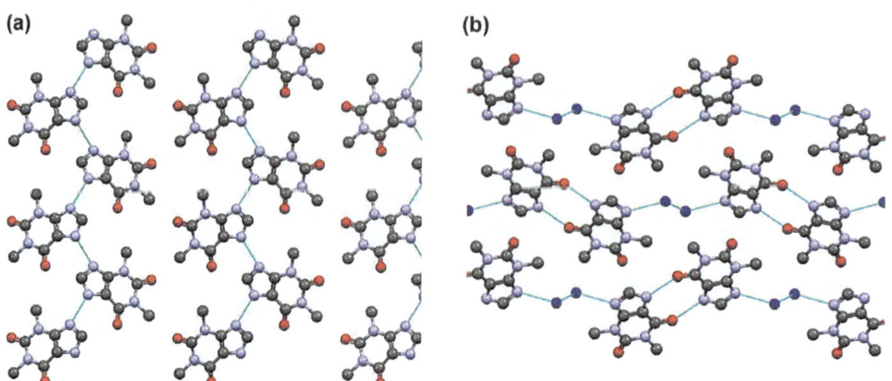

Figure 20.2   The structures of the anhydrate (a, refcode BAPLOT01) and of the monohydrate (b, refcode THEOPH01) crystals of theophylline. The anhydrate is obtained from a mixture of an organic solvent and water at low water activity.

though at ambient conditions the crystalline trihydrate is more thermodynamically stable and less soluble than the anhydrate: conversion to the hydrate form occurs only with the addition of seeds and at a water activity >0.381[14] (see Figure 20.3).

There is, obviously, a large number of examples of hydrate–anhydrate systems. We can cite here only few representative cases.

The compound 3-(4-dibenzo[*b*,*f*][1,4]oxepin-11-ylpiperazin-1-yl)-2,2-dimethylpropanoic acid (DB7z) possesses two zwitterionic hydrate phases with disordered water molecules, which differ remarkably in stability and hydration/dehydration mechanisms.[15] The dihydrate Hy2 is the precursor to Form III, a high energy disordered anhydrate, with the level of disorder depending on the drying conditions. Gravimetric moisture sorption and desorption curves of DB7z solid forms at 25 °C are shown in Figure 20.4.

Hydrates are generally expected to be thermodynamically more stable, hence less soluble and slower to dissolve, than anhydrate forms above the critical water activity for hydrate formation. Hence dehydrated hydrates tend to be metastable and easily uptake water (or other solvents) to fill volume in. Non stoichiometric hydrates, on the other hand, often become amorphous when dried.[16]

In yet another example, two hydrates of diatrizoic acid have been characterized together with three anhydrous and nine solvated solid forms.[17] The water molecules in the dihydrate are encapsulated between pairs of host molecules (see Figure 20.5). Calculations have shown that these molecules are strongly linked because of the hydrogen bonds between the individual enclosed water squares.

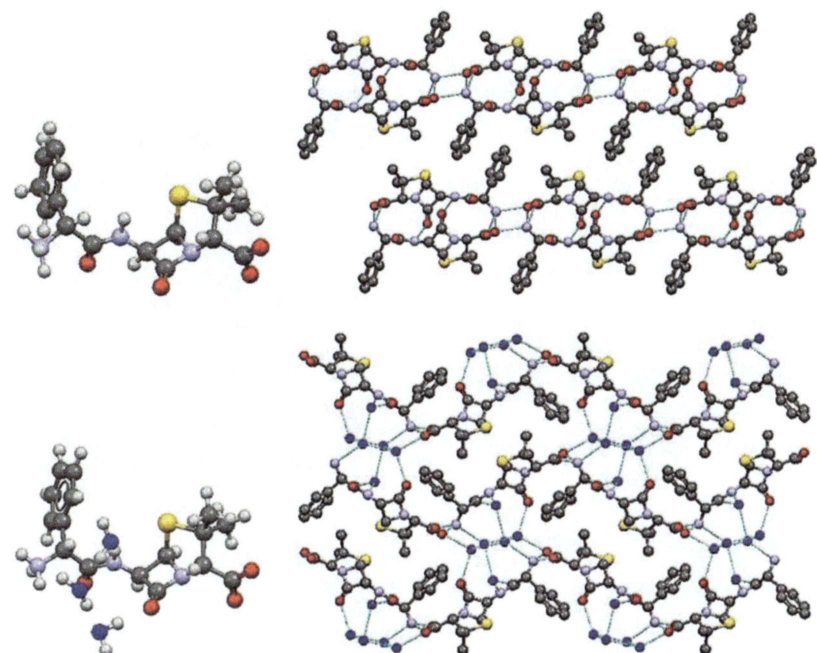

Figure 20.3 Anhydrous crystalline ampicillin (top, refcode AMCILL) has been shown to be kinetically stable for several days in MeOH/H$_2$O mixtures. The trihydrate (bottom, refcode AMPCIH01) can be obtained only by seeding.

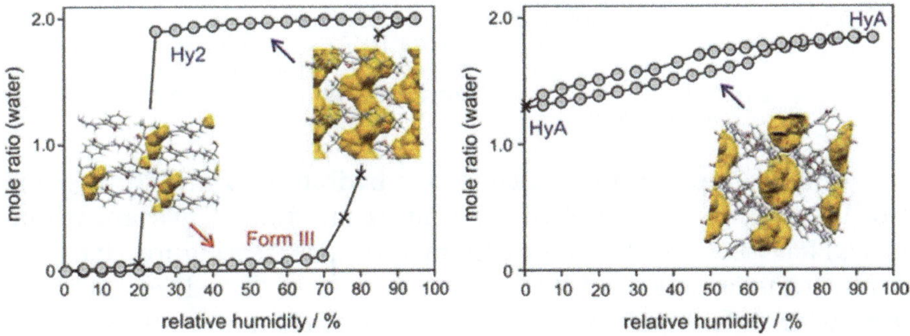

Figure 20.4 Gravimetric moisture sorption and desorption curves of DB7z solid forms at 25 °C: Form III/Hy2, and HyA. The gray circles represent data points that fulfill the preset equilibrium conditions, whereas the crosses mark measurement values that did not reach equilibrium within the allowed time limit (48 h).
(Reproduced with permission from D. E. Braun, L. H. Koztecki, J. A. McMahon, S. L. Price and S. M. Reutzel-Edens, *Mol. Pharmaceutics*, 2015, **12**, 3069–3088. Copyright 2015 American Chemical Society. DOI: 10.1021/acs.molpharmaceut.5b00357).

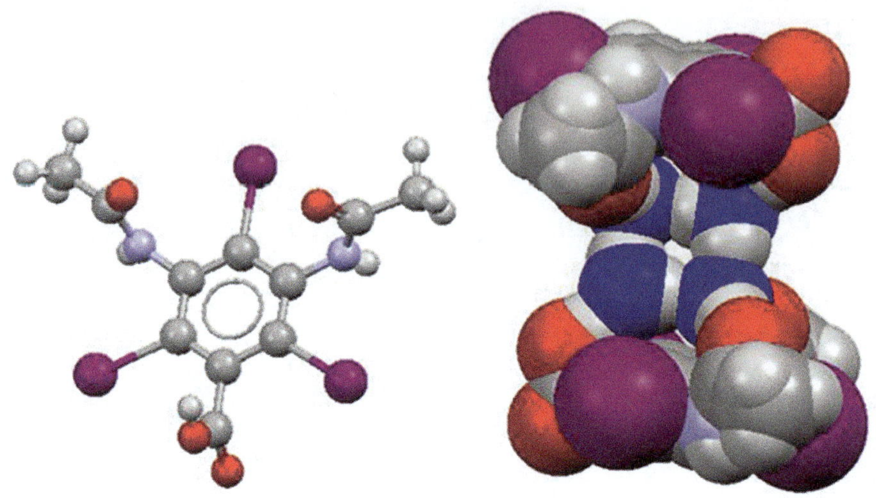

Figure 20.5   (a) Dihydrated diatrizoic acid; (b) the water molecules (oxygens in blue) in dihydrated diatrizoic acid (refcode PUFGUU) are encapsulated between a pair of host molecules.

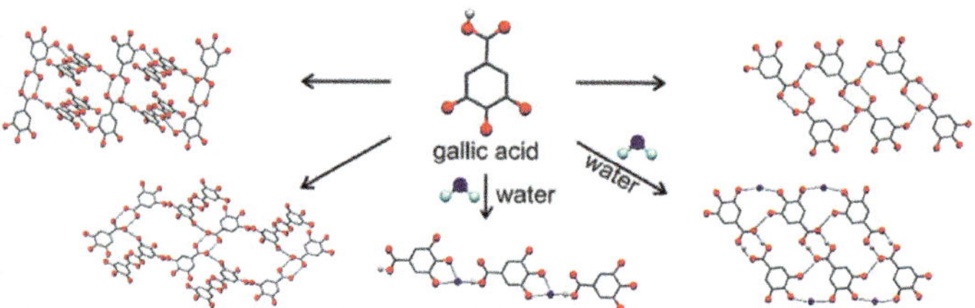

Figure 20.6   Polymorphs of anhydrous and hydrated gallic acid.
(Reproduced with permission from D. E. Braun; R. M. Bhardwaj; A. J. Florence, D. A. Tocher and S. L. Price, *Cryst. Growth & Des.*, 2013, 13, 19–23. Copyright 2012 American Chemical Society. DOI: 10.1021/cg301506x).

A recent example of polymorphs of a hydrate is provided by gallic acid, for which five monohydrate phases (as well as two anhydrate phases) have been isolated and characterized[18] (see Figure 20.6).

Anhydrous crystals and hydrates may possess very different properties, with differences that are in general larger than those between single molecule polymorphs, especially in terms of solubility in water and intrinsic dissolution rates, but also in terms of hygroscopicity, vapor uptake/release *etc.* This is one of the reasons for the widespread interest in the pharmaceutical field.

An egregious example is provided by the family of hydrates of the antibiotic rifaximin (4-deoxy-40-methylpyrido[10,20-1,2] imidazo-[5,4-*c*]rifamycin SV), manufactured by Alfa Wassermann S.p.A.

(now Alfa-Sigma). Four distinct crystal forms of rifaximin are known (α, β, δ, ε) and a poorly crystalline, mostly amorphous form γ, have been characterized.[19] The distinct forms are obtained by varying the drying conditions or omitting washing. Diffraction patterns are shown in Figure 20.7.

When exposed to humidity, the α, δ, γ and ε forms completely transform into the β-form; the α-form is the fastest, while the γ-form is the slowest. The transition from the α- to the δ-form is obtained by keeping the α-form at 11.3% RH for about three months. Transformation of the various crystal forms is schematically summarized in Figure 20.8.

Structure determination by single crystal and powder diffraction has allowed identification of the main differences between these crystalline forms, with the exception of form γ, which is poorly crystalline. The four crystalline hydrates share a common feature. Rifaximin molecules are organized in pairs of molecules hydrogen bonded *via* an amido group. A first water molecule is completely embedded within the molecular *ansa* bridge (see Figure 20.9). The remaining water molecules appear to play a crucial role in stabilizing the different structures. All four hydrates are actually non stoichiometric, with intervals of stability depending on the hydration state of each form.

The stoichiometric *versus* non stoichiometric nature of the hydrates has been amply discussed by Griesser and others.[20] In simple words, stoichiometric hydrates generally show structural changes upon uptake or removal of water. Dehydration processes are usually shown on a DSC graph as endothermic peaks, and correspond to mass loss in TGA measurements. In the hydration process, water uptake in vapour sorption experiments generally takes place when the threshold critical humidity is reached and is revealed by an abrupt change in the sample mass. An example is provided by Figure 20.10, which shows the hydration/dehydration process involving two concomitant, anhydrous polymorphs of β-resorcylic acid (2,4-dihydroxybenzoic acid) and the stable hemihydrate form.[21]

On the other hand, nonstoichiometric hydrates are generally characterized by a robust crystal structure that can sustain water uptake and release, usually because of the presence of channels and accessible cavities, without severe structural changes, in a continuum of hydration states. In DVS these hydrates do show an increasing uptake capacity as a function of humidity.

An interesting example is provided by erythromycin A, which exists as a stable dihydrate crystalline phase.[22a] Two anhydrous phases are

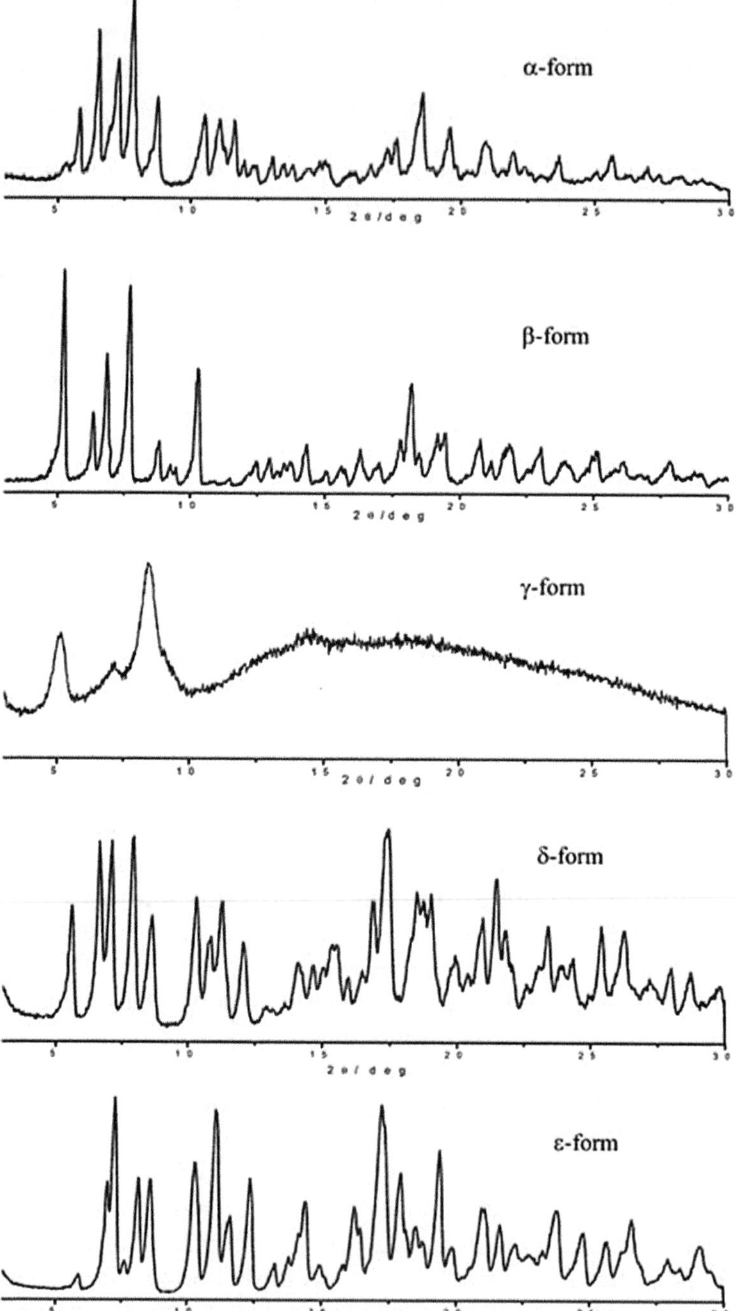

Figure 20.7 X-ray powder diffraction patterns of the rifaximin crystal forms (Reproduced from ref. 19 with permission from the Royal Society of Chemistry).

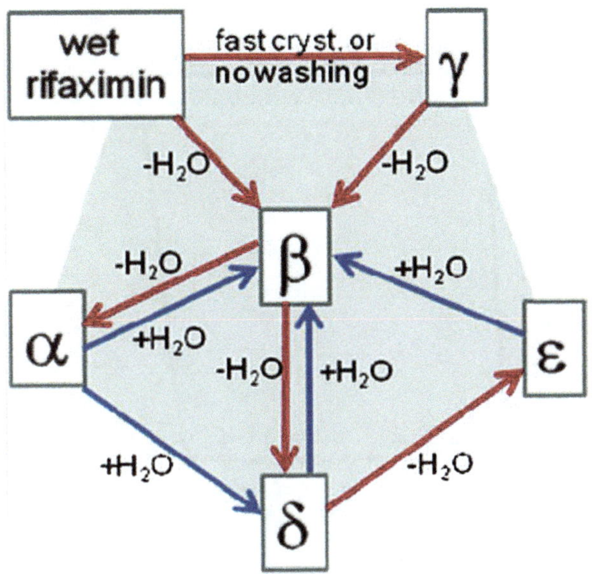

Figure 20.8 The relationship between the various hydrated crystal forms of rifaximin (Reproduced from ref. 19 with permission from the Royal Society of Chemistry).

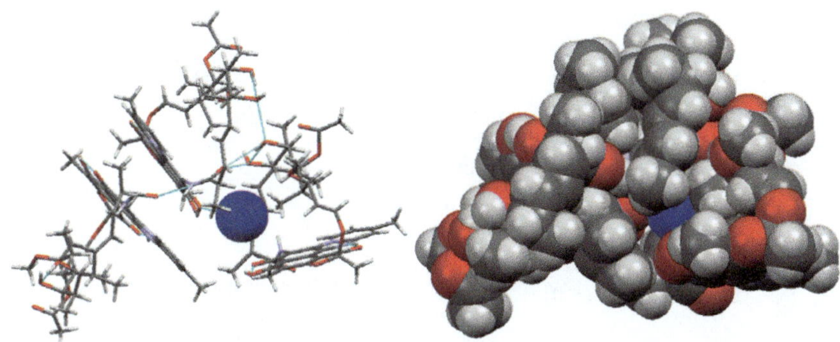

Figure 20.9 The water molecule embedded in the "niche" of the *ansa* bridge in crystalline rifaximin (Reproduced from ref. 19 with permission from the Royal Society of Chemistry).

known for this system: form I, obtained by dehydration of the dihydrate, converts back to the dihydrate upon hydration, while form II, obtained *via* melting, shows nonstoichiometric hydration to give a humidity dependent nonstoichiometric hydrate phase (see Figure 20.11).[22b]

The presence of accessible voids in both anhydrous forms is the basis of their hydration behavior, as the voids are sufficiently large to make water absorption possible without major structural changes.

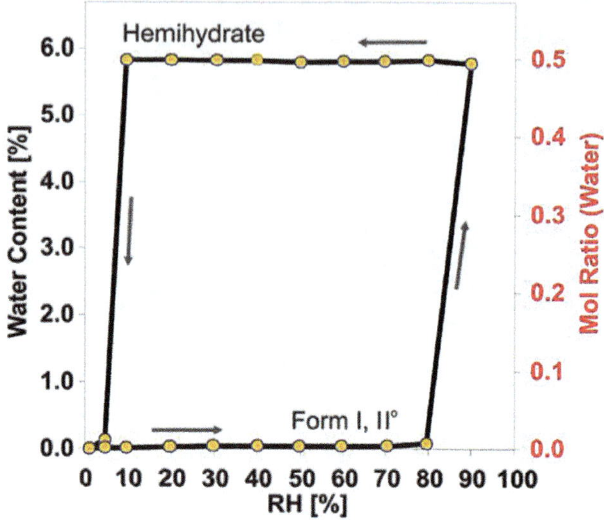

Figure 20.10 Moisture sorption isotherm of β-resorcylic acid. (Reproduced with permission from D. E. Braun, P. G. Karamertzanis, J-B. Arlin, A. J. Florence, V. Kahlenberg, D. A. Tocher, U. J. Griesser and S. L. Price, *Cryst. Growth & Des.*, 2011, 11, 210–220. Copyright 2010 American Chemical Society. DOI: 10.1021/cg101162a).

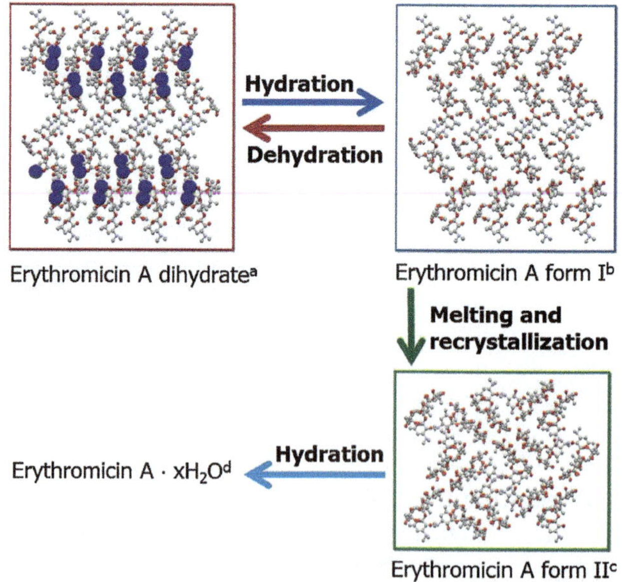

Figure 20.11 Hydration and dehydration processes in erythromycin A. (Refcodes: [a] NAVTAF, [b] QIFKEX01, [c] QIFKEX; [d] non-stoichiometric hydrate). (Figure adapted from Kotaro Fujii, Masahide Aoki, and Hidehiro Uekusa Cryst. Growth Des., 2013, 13, 2060–2066. Copyright 2013 American Chemical Society. DOI: 10.1021/cg400121u).

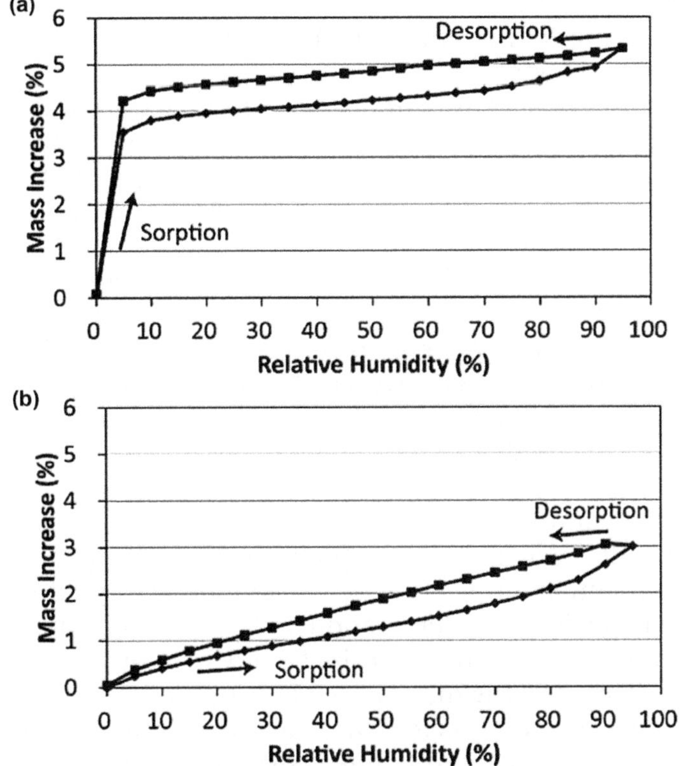

Figure 20.12　DVS plot of erythromycin A anhydrous phases (a) I and (b) II. (Reproduced with permission from Kotaro Fujii, Masahide Aoki, and Hidehiro Uekusa, *Cryst. Growth Des.*, 2013, 13, 2060–2066. Copyright 2013 American Chemical Society. DOI: 10.1021/cg400121u).

The different behaviour of forms I and II upon hydration is evident from the DVS measurements shown in Figure 20.12.[22b]

Hydrates are often patentable. Several interesting cases of litigation concern hydrates and the innovative step leading from anhydrous to hydrate crystals. Indeed, the relevant criteria of patentability, namely non-obviousness, novelty and utility are easily justified. First, they meet the requirement of non-obviousness, because it is as yet almost impossible to predict with confidence the formation of a hydrate from a given API, even when crystallization takes place from organic solvents, because of residual water content (see above), and whether they will or will not form stable mono-hydrates or polyhydrates. The chemical composition of a hydrate will be different from that of the single molecule API crystal. Also hydrates are novel because their different composition and structure will generate properties that are distinctive to those of polymorphs and other solvates of the same API. As mentioned above, solubility is the most obvious one.

It is useful to note that hydrates tend to be less soluble than the corresponding anhydrous crystals. For instance, the solubilities of anhydrous and hydrated crystals of caffeine are 49.7 and 21.8, those of carbamazepine 0.424 and 0.139, and those of sulfaguanidine 1.38 and 1.07 mg mL^{-1}, respectively.[23]

Water plays a critical role in pharmaceutical formulations. It may enter the system *via* the active ingredient, the excipients, or absorbed from the atmosphere. Water absorption and release *etc.* impact also on the physical aspects and workability of samples. It can induce phase transitions, dissolve soluble components, and increase interactions between the drug and excipients, all of which can adversely affect the physical and chemical stability of the drug substance. Water needs to be accounted for at all stages of drug substance and product manufacturing. Even the drying processes commonly used during large scale manufacturing have to be carefully designed and monitored to avoid possible agglomeration and dehydration, and allow the production of an API with the desired solid state properties.[24]

Not surprisingly, of the many possible critical quality attributes of a drug substance, hygroscopicity, a measure of the water vapor taken up by a solid with the potential to effect surface and bulk properties, is routinely evaluated, typically by moisture sorption analysis, at the earliest stages of drug product development. However, it is useful to recall that crystallization from water does not necessarily lead to formation of hydrate species, therefore it is not inevitable that materials crystallized from water are hydrates or would form hydrates under different conditions. Solvate formation upon crystallization from solution of a novel system is still an unpredictable event.

20.3 Co-crystals

Co-crystals represent a step ahead on the route towards design of materials. It is by a judicious choice of the "extramolecular" interactions (reliable synthons, functional groups, polarity, *etc.*) that the crystal explorer attempts construction of novel aggregates.

Co-crystallization is a means to attain new and different solid state properties or to alter the properties of a given single molecule crystal or solvate by crystallizing it together with one or more co-formers, which interact with the molecule of interest *via* non covalent bonding.

The definition of co-crystals is not straightforward.[25] However, it is generally assumed that co-crystals are formed by two or more components that possess a chemical identity on their own and that form stable aggregates at room temperature; the molecule of interest and the selected co-former interact *via* non covalent bonds.

The topic of co-crystals has been recently addressed by a number of authors. A book by Wouters and Quéré is dedicated exclusively to pharmaceutical co-crystals,[26] while comprehensive updates of the subject matter, also with a focus on pharmaceutical co-crystals, have recently been published.[27] Furthermore, the occurrence of polymorphism for multicomponent systems, including co-crystals, has been reviewed recently.[28]

The interest in co-crystals stems from their potential uses in the food and pharma areas. Indeed, co-crystals may offer new ways to design or to alter the properties of solid active pharmaceutical ingredients (APIs) including thermal stability, shelf life, solubility, dissolution rate, compressibility, *etc.*, by linking an ancillary GRAS molecule.[29] If the co-former happens to be another API, one can speak of a co-drug.

The number of possible combinations between API and co-formers is, in principle, without limit. In general terms, co-crystals show differences in physico-chemical properties with respect to the parent single molecule crystal that are usually larger than those between polymorphs and often also than those between the crystalline API and its solvates.

Ternary co-crystals have also been reported,[30] see Figure 20.13.

A thorough exploration of the co-crystal domain of carbamazepine has been reported recently.[31] The authors not only explored a large number of co-formers, but also tested and compared different preparation techniques (Figure 20.14).

20.3.1 Ionic Co-crystals

A new class of co-crystals, called ionic co-crystals, containing ionized components together with neutral organic molecules has been reported.[32] The ionic co-formers may be either salts of ionizable molecules (carboxylic acids, amines, *etc.*), or inorganic salts. Figure 20.15 gives a new scheme showing the two types of ionic co-crystals.

When hydrogen bonding is involved it is often difficult, and sometimes meaningless, to distinguish between a co-crystal and a salt because in the cases of hydrogen bonded systems the proton may

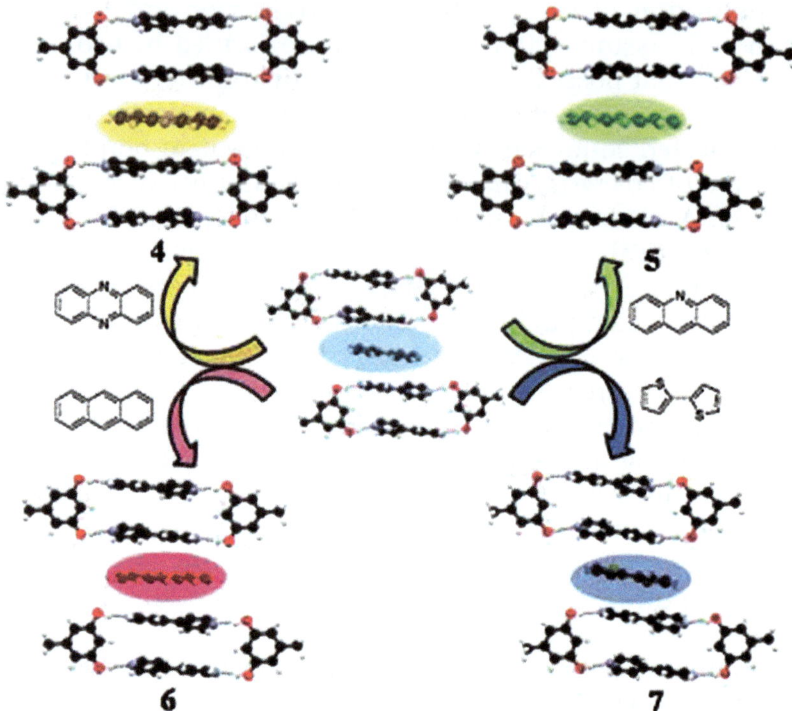

Figure 20.13   Strategy of replacing the "free" 4,4′-bipyridine molecule in orcinol: 4,4′-bipyridine crystal structure by suitable guests, to give co-crystals 4–7. (Reproduced from ref. 30 with permission from the Royal Society of Chemistry).

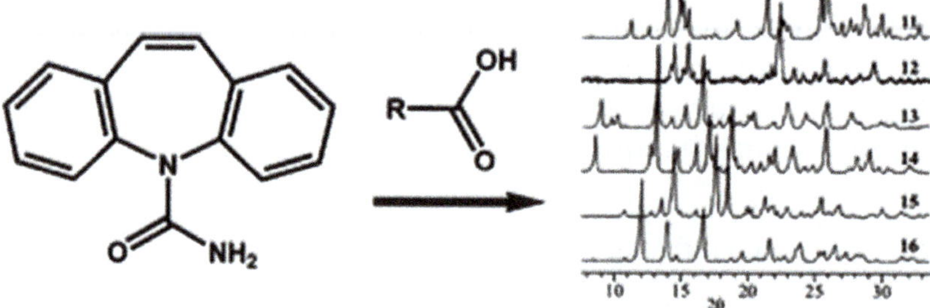

Figure 20.14   Co-crystal formation between carbamazepine and organic carboxylic acids. (Reproduced from ref. 31 with permission from the Royal Society of Chemistry).

move along a hydrogen bond and occupy different sites depending on the temperature, see Figure 20.16.[33]

It is well known that in many instances the salt-neutral nature depends only on the position of the proton along the D–H–A vector,

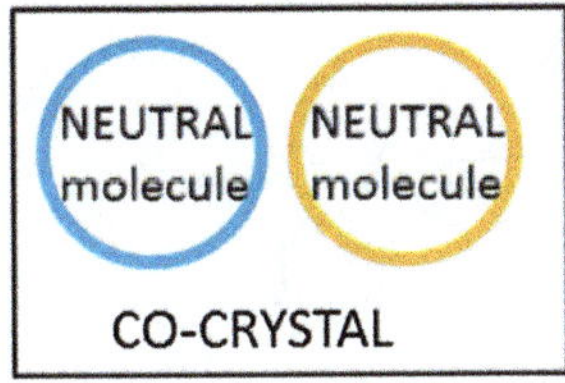
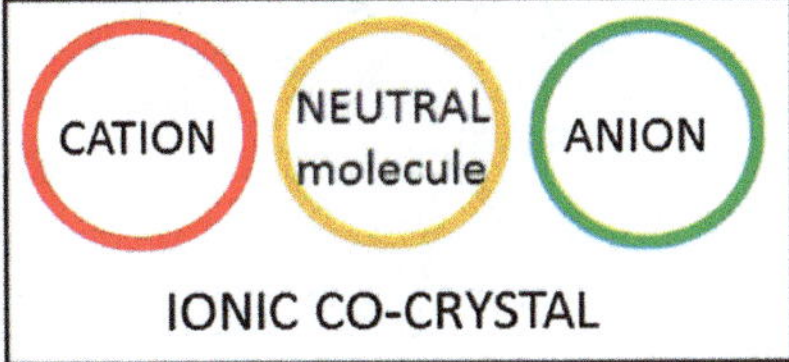

Figure 20.15 Representation of co-crystals and ionic co-crystals (ICCs). Neutral molecules in their pure form are solid at ambient conditions.

Figure 20.16 Temperature dependence of the proton position along a O–H···N bond. At approximately 90 K, the H atom is exactly centred between the O and the N atoms.

which, in turn, depends on the relative acid/base strength and on the temperature.[34] This has been demonstrated in a series of studies on the dependence of melting point on the carbon atom chain length and on the proton position along the O–H–N hydrogen bond in acid–base co-crystals. In mechanochemically prepared 1:1 co-crystals of the dicarboxylic acids $HOOC(CH_2)nCOOH$ ($n = 1–7$) with 1,4-diazabicyclo[2.2.2]octane (DABCO), the melting points of the co-crystals were found to alternate as in the corresponding diacids, irrespective of the salt/molecular nature of the co-crystals, as investigated by X-ray diffraction and solid state NMR spectroscopy.[35] This behaviour is also exhibited by co-crystals of the diacids with dipyridyl molecules 4,4′-bipyridine (bipy), 1,2-bis(4-pyridyl)ethane (bpa), and 1,2-(di-4-pyridyl)ethylene (bpe) which contain an even number of C atoms between the two N atoms, while it is reversed in co-crystals with 1,2-bis(4-pyridyl)propane (bpp), which contains an odd number of (CH_2) groups,[36] see Figure 20.17.

Pharmaceutical ionic co-crystals are of great interest because of the often dramatic change in physico-chemical properties (especially thermal stability, dissolution rate, hygroscopicity), but also because they could provide the route to inorganic–organic co-drugs. The stability of the ionic co-crystals depends on the interactions established

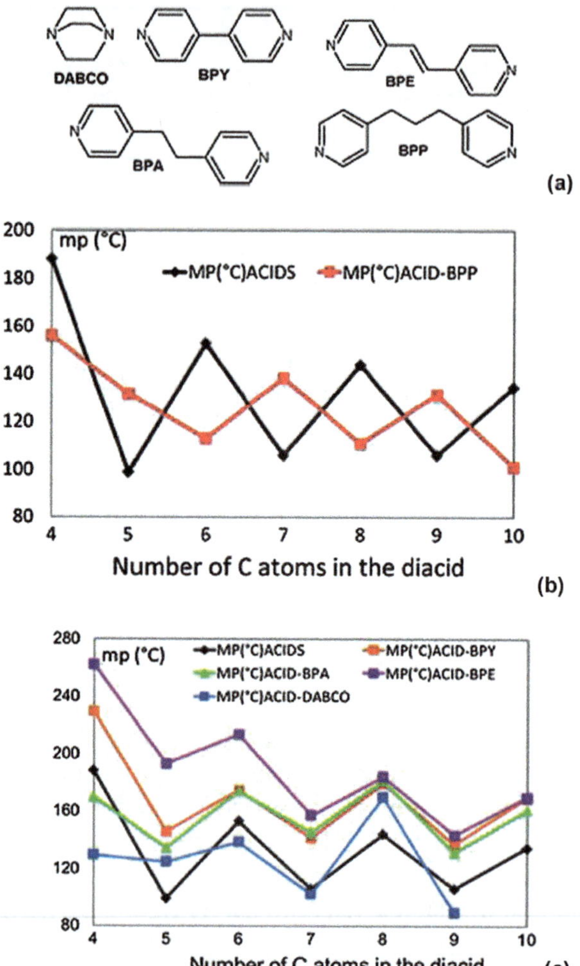

Figure 20.17   Comparison of alternating melting points for pure *dicarboxylic acids* and for co-crystals of *dicarboxylic acids* with dinitrogen bases (a) containing an odd (b) or an even (c) number of carbon atoms. (Reproduced from ref. 37 with permission from the Royal Society of Chemistry).

by the organic moiety with cations and anions; usually oxygen or nitrogen atoms donate electrons towards the cation, while hydrogen bonds are formed between hydrogen donor groups on the organic moiety and the anions. Therefore, they resemble the interactions that a solvent molecule might establish with ions in solution. We have investigated together with researchers at the UCB[38] the preparation of lithium chloride (a drug used a mood stabilizer in patients affected by bipolar disorder) together with piracetam, a cognition enhancing

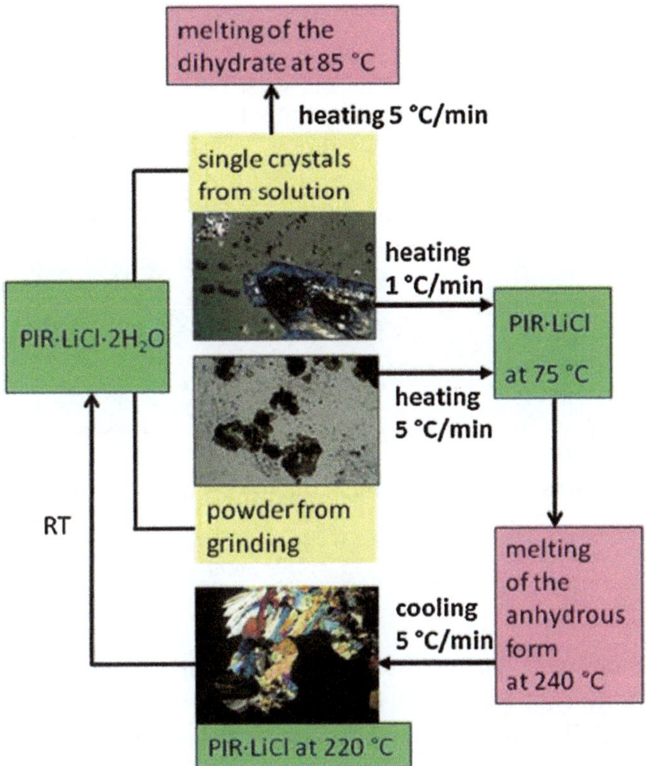

Figure 20.18 The complex phase relationship between hydrate and anhydrous crystals of the ionic co-crystals formed between LiCl and piracetam, and dependence of thermal behaviour on sample crystallinity and heating conditions.
(Reproduced from ref. 38 with permission from the Royal Society of Chemistry).

medicine (Nootropil®). Ionic co-crystals can easily be formed by mechanical treatment of piracetam with the lithium salts LiCl and LiBr as well as with other bio-compatible inorganic salts.[39]

In the case of the ionic co-crystal of piracetam with lithium chloride the solid was prepared in its dihydrate form, and water removal by thermal treatment allowed formation and characterization of the anhydrous form (see Figure 20.18).

More examples of lithium salts co-crystallized with amino acids have also been reported.[40]

20.3.2 Co-crystals *Versus* Solid Solutions

When molecular size and shape are very similar, it is possible that the components become miscible in a range of compositions, as in the

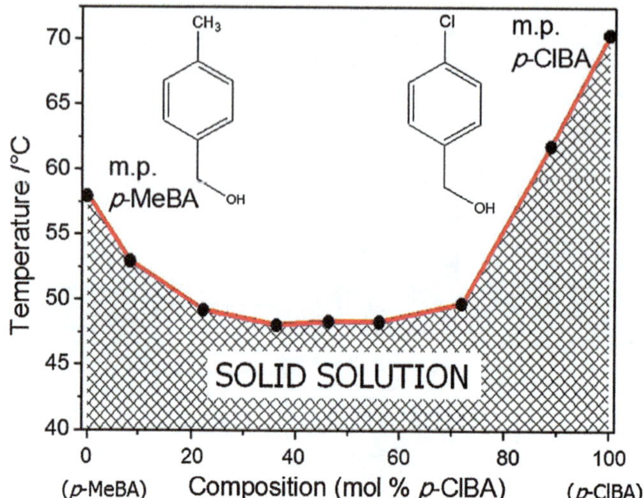

Figure 20.19 *Solidus* line in the phase diagram of the binary isomorphous system p-ClBA–p-MeBA, showing the melting points (indicated by the black dots) of the solid solution at various compositions.

case of alloys. An interesting example is provided by the pair of molecules *p*-methyl benzyl alcohol (*p*-MeBa) and *p*-chloro benzyl alcohol (*p*-ClBa).[41] The two compounds differ in the substitution of the methyl group for a chlorine atom in the molecular structure. In terms of crystal structure, they form different crystals and no polymorphic modifications are known. However, a new polymorph of *p*-MeBa has been obtained by seeding a hexane solution of *p*-MeBa with crystals of the less soluble *p*-ClBa. The new form of *p*-MeBa is thus isomorphous with crystals of *p*-ClBa. Once in possession of the isomorphous crystals of *p*-MeBa and *p*-ClBa, it was possible to mix the two compounds by co-melting and to measure the variation in melting points as a function of the composition (see Figure 20.19), obtaining a trend characteristic of molecular alloys.

In an analogous case of quasi-isostructurality it has been otherwise possible to construct co-crystals of *ortho*-toluic and *ortho*-chlorobenzoic acids in a 50:50 ratio, while no alloying was observed.[42]

The importance of non-stoichiometric molecular mixed crystals as novel functional materials has been recently emphasized in a study of mixed crystals of acridine and phenazine prepared mechanochemically.[43] This approach has been expanded to the preparation of ternary mixed crystals of anthracene, phenazine and acridine.[44] Mixed crystals will certainly provide access to a broad variety of new properties, as it has been amply demonstrated in other cases of molecular alloying with organometallic molecules and salts.[45]

20.4   Crystal Polymorphism

Crystal polymorphism is the existence of more than one crystal structure for a given compound. The phenomenon is known and has been amply investigated. Recent excellent reviews and books are available.[46] In the contest of this book, however, it is important to ask oneself if the crystal engineering paradigm might be undermined by the fact that a design strategy, based on synthon selection and intermolecular interactions, often yields diverse and unexpected results. Does polymorphism, together with solvate formation, contradict the deterministic approach behind crystal engineering? If one considers (see previous section) that co-crystals can also be polymorphic, the implications become significant.

It is by now well accepted that the variety of phenomena related to polymorphism requires a thorough mapping of the crystal/energy landscape of a substance that is ultimately intended for some specific applications (*e.g.* drugs, pigments, agrochemicals and food additives, explosives, *etc.*). The industrial production and marketing lines need to know not only the exact nature of the material in the process, but also its stability with time, the variability of its chemical and physical properties as a function of the crystal form, *etc.* The search for and characterization of crystal forms is therefore a crucial step in the development of new chemicals, with relevant consequences for intellectual property issues.

Due to difficult predictability, polymorphism is viewed often as a threat rather than a possibility for organic materials industries, because of the possible problems that can occur if a new polymorph appears during scale-up, production or storage. Examples of such cases will be discussed below. To avoid such risks, it is best practice to perform a polymorph screening on industrially relevant compounds. There are many approaches and strategies in screening for polymorphs (conventionally various solvent crystallizations) but there is clearly no real conception, recipe or strategy that guarantees the finding of all possible forms. Examples in the past show that even expensive high throughput strategies, performing thousands of experiments, may fail in this task and, unfortunately, no guarantee can be given that all relevant crystalline forms have been found.

What is truly unpredictable is the interplay of kinetic and thermo-dynamic factors in a crystallization process, with consequences that are well known to the practitioners in the area, as efficaciously pointed out by Bernstein[47] "it is sometimes difficult to comprehend why and how new polymorphs still emerge (while others disappear) long after crystal-form screens presumably have been completed. [...] The point is that it can never be stated with certainty that the most stable form has been found; at best it can be determined which of the known forms is the most stable. [...] a new (and most often more-stable) form can appear at any stage in the history of a compound (or life-cycle of a drug)".

Investigation of the crystal stability landscape is required in order to reduce (though not eliminate) the risk of the appearance of a new, more stable, crystal form of a given compound at a later stage of marketing, authorization and distribution.

An emblematic case, and one of the best known, is that of ritonavir (Norvir®), used for the treatment of HIV patients, which has been thoroughly described in ref. 47. Subsequent investigations led to the discovery of four additional crystalline forms of ritonavir.[48]

A more recent case is that of rotigotine (Neupro®), a Parkinson's disease drug by UCB, administered as skin patches. In 2008 a new form suddenly appeared, which crystallised in these patches and re-duced their efficacy. "Surprisingly, a further crystalline form of roti-gotine (polymorphic form (II)) has now been identified and found to show a greatly enhanced thermodynamic stability and an improved shelf-life as well as a cubic crystal shape that represents an advantage over the needle-like particles of polymorphic form (I) regarding its handling properties such as filtering properties, flowability, electrostatic behavior, *etc.*"[49] The product had to be withdrawn, with considerable impact on the patients – and on the company (see Figure 20.20). The new crystal form was described in a patent filed in November 2008 and granted in July 2012.[50]

An additional example is that of the blood pressure medica-tion Avalide, which combines the two anti-hypertensives hydro-chlorothiazide and irbesartan. In 2010 there was a drug product recall of over 60 million tablets produced by BMS, because of concerns over possible variability in the amounts of the less soluble polymorph of irbesartan, which might result in slower dissolution.[51]

In spite of the efforts of a great number of research groups world-wide, and of a familiarity with the experimental factors that can lead to multiple crystal forms, our ability to predict the occurrence of polymorphism is still embryonic. In many cases the crystallization of

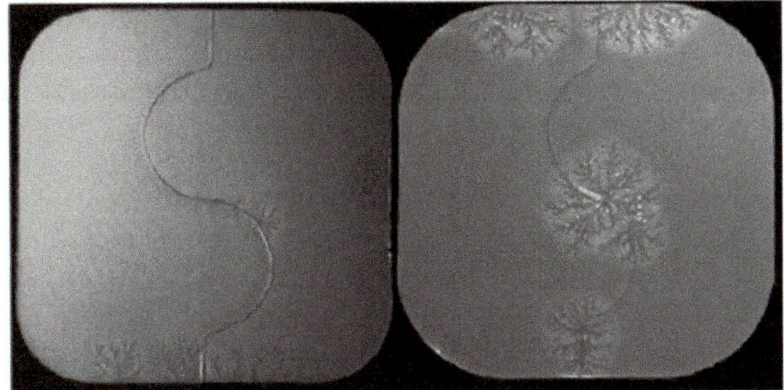

Figure 20.20 Photographs of Neupro™ patches demonstrating the growth of snowflake-like crystals of the less soluble polymorph of rotigotine during storage.
(Reproduced from *J. Pharm. Sci.*, **104**, I. B. Rietveld, R. Céolin, Rotigotine: Unexpected Polymorphism with Predictable Overall Monotropic Behavior, 4117–22. Copyright 2015 Wiley Periodicals, Inc. and the American Pharmacists Association, with permission from Elsevier).

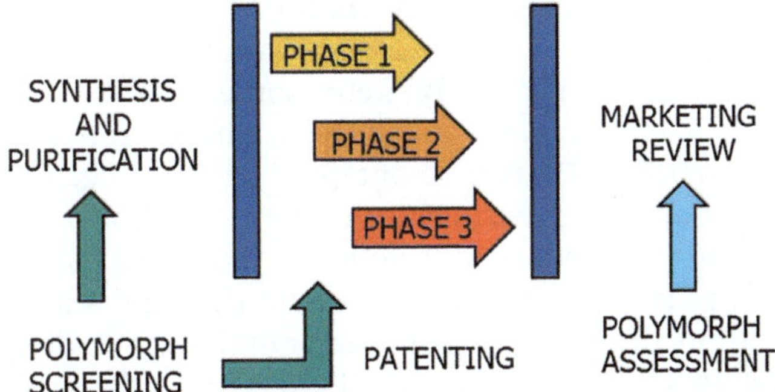

Figure 20.21 A flow-chart showing how polymorph screening ought to be associated with the API selection before entering trial stage, as well as with the quality control process once the API is on the market.

a new crystal form or of an amorphous phase of a given substance turns out to be the result of *serendipity* rather than a process under complete human control.[52]

Polymorph assessment, on the other hand, is part of the system of quality control. It is necessary to make sure that the scale-up from laboratory preparation to industrial production does not introduce variations in crystal forms. Polymorph assessment also guarantees that the product conforms to the guidelines of the appropriate regulatory agencies and does not infringe the intellectual property protection that may cover other crystal forms, see Figure 20.21.

Although the motivation of such studies arises essentially from safety and commercial necessities, the consequences are important also for fundamental science. Litigations over polymorphs and hydrates, in addition to famous polymorphism "incidents" have fueled research in solid state chemistry, and oriented the experience and competence of many research groups worldwide. The birth of many spinoff companies, meeting the need of industrial outsourcing of accurate solid state investigation, ought also to be mentioned.

Even though "certainty" is out of reach, one can reduce the probability of having missed THE crystal form by carrying out a series of well thought over experiments that will guarantee that most if not all of the possible variables have been taken into account.

20.4.1   Tautomeric Polymorphism

A special, often intriguing, case is represented by structurally non rigid molecules such as tautomers, which may adopt different structures in the solid state depending upon the balance between crystal energy and molecular energy optimization.[53,54]

A text-book example of such interplay is afforded by barbituric acid, which is always presented as the keto isomer, which is indeed preferred in solution and in most polymorphic phases. It has been shown that a new form (form IV) can be obtained by grinding. The resulting crystal consists of molecules in the enol form, as shown by solid state NMR and neutron powder diffraction.[55] Form IV is the thermodynamic minimum at room temperature in the solid state because of the stabilization provided by a higher number of hydrogen bonds with respect to the keto form. See Figure 20.22 for a view of the packing of the new form.

Another interesting example of interplay between molecular and crystal structure is provided by 2-thiobarbituric acid, for which five polymorphs and one hydrated form have been isolated and characterized by solid-state methods.[56] In both the crystalline form **II** and in the hydrate form, the 2-thiobarbituric acid molecules are present in the enol form, whereas only the keto isomer is present in crystalline forms **I** (reported in 1967 by Calas and Martinez)[57], **III**, **V** and **VI**. In form **IV**, prepared mechanochemically as in the case of barbituric acid on the other hand, a 50:50 ordered mixture of enol/keto molecules is present. All new forms have been characterised by single-crystal X-ray diffraction, 1D and 2D (^{1}H, ^{13}C, and ^{15}N) solid-state NMR spectroscopy, Raman spectroscopy and X-ray powder diffraction at variable temperature. Both isomers in form **IV** maximise the number

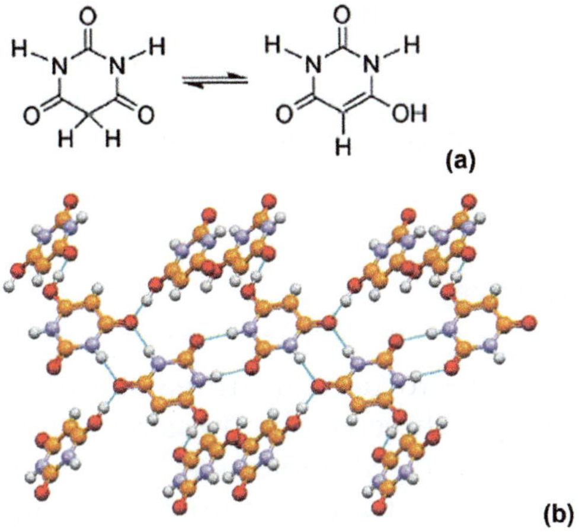

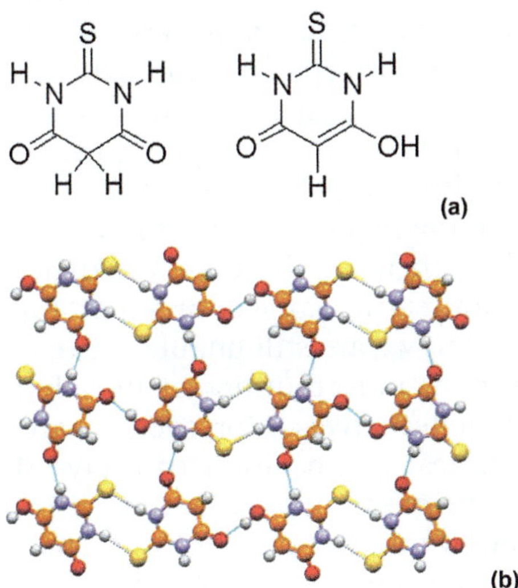

Figure 20.22 (a) Keto/enol tautomerism of barbituric acid (b) and crystal structure at room temperature of the enol form.
(Figure (a) reproduced from M. R. Chierotti, R. Gobetto, L. Pellegrino, L. Milone, P. Venturello, *Cryst. Growth Des.*, 2008, 8, 1454–1457. Copyright 2008 American Chemical Society), (Figure (b) reproduced from M. U. Schmidt, J. Brüning, J. Glinnemann, M. W. Hützler, P. Mörschel, S. N. Ivashevskaya, J. van de Streek, D. Braga, L. Maini, M. R. Chierotti, R. Gobetto, *Angew. Chem. Int Ed.*, 2011, 50, 7924–7926. Copyright 2011 WILEY-VCH Verlag GmbH & Co. KGaA, Weinheim, with permission from Wiley).

Figure 20.23 (a) Keto/enol tautomerism of thiobarbituric acid (b) and the crystal structure of the most stable form of thiobarbituric acid at room temperature, showing the presence of both the keto and enol forms. (Reproduced from M. R. Chierotti, L. Ferrero, N. Garino, R. Gobetto, L. Pellegrino, D. Braga, F. Grepioni and L. Maini, *Chem. Eur. J.*, 2010, **16**, 4347–4358. Copyright 2010 WILEY-VCH Verlag GmbH & Co. KGaA, Weinheim, with permission from Wiley.)

of hydrogen bonds, which is higher than in all of the other crystalline structures. As a matter of fact, one can be tempted to describe the most stable crystal form of thiobarbituric acid as a sort of co-crystal formed by the two extremes of the tautomeric equilibrium (see Figure 20.23).

20.5 Conclusions

In this chapter we have focused on three aspects of crystal engineering that challenge directly the possibility of predicting the outcome of a crystal-directed molecular self-assembly process on the basis of a knowledge of the supramolecular interactions acting between molecules at crystal nucleation stage. Namely, hydrates, co-crystals and polymorphs. The formation of a hydrate (or any solvate), its composition, its stoichiometric or non stoichiometric nature, as well as the outcome of a co-crystallization experiment, whether in solution or in the solid state, and the possibility of polymorphism of hydrates and co-crystals, the formation of co-crystal solvates, and their possible polymorphs, and polymorphism itself still escape design strategies. Even automated high throughput methods cannot guarantee the complete (complete = 100%) exploration of the crystal energy landscape. On the other hand, crystal structure prediction, CSP, in spite of the gigantic steps forward made in recent years, is still unable to reassure chemists, especially materials scientists in industry, on the chances of falling into "polymorphism traps".[58]

One may predict with a high level of confidence, also based on the statistical analysis of the behavior of the same group in thousands of other crystals by interrogating the CSD, whether a given functional group will link to another (the same or different) functional group in a predetermined way, but we are still unable to predict whether we will obtain from solution the most thermodynamically stable form or only one of many kinetic alternatives. Moreover, we are unable to predict even the exact chemical composition of a crystalline product precipitated for the first time from a solvent or mixture of solvents well enough that no one would bet on the formation of a solvate or an unsolvate and/or on the number of solvent molecules that might be brought into the crystal.

Therefore, the target of being able to obtain a pre-determined solid state property (colour, magnetism, dissolution rate, stability to water uptake/loss, *etc.*) by choosing at will molecules possessing the appropriate bonding attributes is still far over the horizon and there it will reside until we are able to design the structure of the most thermodynamically stable aggregate whether single molecule or solvate crystal and … make it. If this level of confidence is ever reached it will eliminate *tout-court* all those problems that might arise from the sudden appearance of undesired more stable forms and would reduce, if not cancel altogether, the chances of discovering alternative crystal forms with improved properties. This objective is not at hand, which is good. It means that we need to work harder and study and experiment more.

References

1. G. R. Desiraju, *Crystal Engineering: The Design of Organic Solids*, Elsevier, Amsterdam, 1989. This book remains, to date, the only single author book on the subject; See also D. Braga, F. Grepioni and A. G. Orpen, *Crystal Engineering: from Molecules and Crystals to Materials*, Kluwer Academic Publishers, Dordrecht, 1999.
2. (a) J. M. Lehn, *Angew. Chem., Int. Ed. Engl.*, 1990, **29**, 1304; (b) *Supramolecular Chemistry: Concepts and Perspectives*, ed. J. M. Lehn, VCH, Weinheim, 1995.
3. (a) P. J. Stang and B. Olenyuk, *Acc. Chem. Res.*, 1997, **30**, 502–518; (b) B. F. Hoskins and R. Robson, *J. Am. Chem. Soc.*, 1990, **112**, 1546–1554; (c) R. B. Moulton and M. J. Zaworotko, *Chem. Rev.*, 2001, **101**, 1629; (d) M. Fujita, *Chem. Soc. Rev.*, 1998, **27**, 417; (e) O. R. Evans and W. Lin, *Acc. Chem. Res.*, 2002, **35**, 511; (f) M. Oh, G. B. Carpenter and D. A. Sweigart, *Acc. Chem. Res.*, 2004, **37**, 1; (g) L. Carlucci, G. Ciani and D. M. Proserpio, *CrystEngComm*, 2003, **5**, 269; (h) C. Janiak, *Dalton Trans.*, 2003, 2781; (i) R. Robson, *J. Chem. Soc., Dalton Trans.*, 2000, 3735; (l) S. R. Batten and R. Robson, *Angew. Chem., Int. Ed.*, 1998, **37**, 1461; (k) G. S. Papaefstathiou and L. R. MacGillivray, *Coord. Chem. Rev.*, 2003, **246**, 169.
4. (a) M. C. Etter, *Acc. Chem. Res.*, 1990, **23**, 120–126; (b) M. D. Ward, in *Molecular Networks, Book Series: Structure and Bonding*, ed. M. W. Hosseini, 2009, vol. 132, pp. 1–23; K. Biradha, *CrystEngComm*, 2003, **5**, 374–384.
5. (a) P. Metrangolo and G. Resnati, *Struct. Bonding*, 2008, **126**, 1–221; (b) P. Metrangolo, F. Meyer, T. Pilati, G. Resnati and G. Terraneo, *Angew. Chem., Int. Ed.*, 2008, **47**, 6114–6127; (c) P. Metrangolo and G. Resnati, *Science*, 2008, **321**, 918–919; (d) P. Metrangolo, T. Pilati, G. Terraneo, S. Biella and G. Resnati, *CrystEngComm*, 2009, **11**, 1187–1196; (e) A. C. Legon, *Phys. Chem. Chem. Phys.*, 2010, **12**, 7736–7747; (f) L. Brammer, G. M. Espallargas and S. Libri, *CrystEngComm*, 2008, **10**, 1712–1727.
6. G. R. Desiraju, *Angew. Chem., Int. Ed.*, 2007, **46**, 8342–8356.
7. D. Braga, *Chem. Commun.*, 2003, 2751.
8. (a) A. Gavezzotti, *Cryst. Res. Technol.*, 2013, **48**, 793–810; (b) A. Gavezzotti, *CrystEngComm*, 2013, **15**, 4027–4035.
9. D. E. Braun, L. H. Koztecki, J. A. McMahon, S. L. Price and S. M. Reutzel-Edens, *Mol. Pharmaceutics*, 2015, **12**, 3069–3088.
10. U. J. Griesser, in *Polymorphism: In the Pharmaceutical Industry*, ed. R. Hilfiker, Wiley-VCH, Germany, 2006, pp. 211–233.
11. http://www.fda.gov/Food/IngredientsPackagingLabeling/GRAS/.
12. R. K. Khankari and D. J. W. Grant, *Thermochim. Acta*, 1995, **48**, 61–79.
13. H. Zhu, C. Yuen and D. J. W. Grant, *Int. J. Pharm.*, 1996, **135**, 151.
14. H. Zhu and D. J. W. Grant, *Int. J. Pharm.*, 1996, **139**, 33.

15. D. E. Braun, L. H. Koztecki, J. A. McMahon, S. L. Price and S. M. Reutzel-Edens, *Mol. Pharmaceutics*, 2015, **12**, 3069–3088.
16. T. Fang, Q. Haiyan, A. Zimmermann, T. Mtunk, A. C. Joergensen and J. Rantanen, *J. Pharm. Pharmacol.*, 2010, **62**, 1534–1546.
17. K. Fucke, G. J. McIntyre, M. H. Lemee-Cailleau, C. Wilkinson, A. J. Edwards, J. A. Howard and J. W. Steed, *Chem. – Eur. J.*, 2015, **21**, 1036–1047.
18. D. E. Braun, R. M. Bhardwaj, A. J. Florence, D. A. Tocher and S. L. Price, *Cryst. Growth Des.*, 2013, **13**, 19–23.
19. D. Braga, F. Grepioni, L. Chelazzi, M. Campana, D. Confortini and G. C. Viscomi, *CrystEngComm*, 2012, **14**, 6404–6411.
20. U. J. Griesser, in *Polymorphism in the Pharmaceutical Industry*, ed. R. Hilfiker, WILEY-VCH Verlag GmbH & Co. KGaA, Weinh, 2006.
21. D. E. Braun, P. G. Karamertzanis, J.-B. Arlin, A. J. Florence, V. Kahlenberg, D. A. Tocher, U. J. Griesser and S. L. Price, *Cryst. Growth Des.*, 2011, **11**, 210–220.
22. (a) G. A. Stephenson, J. G. Stowell, P. H. Toma, R. R. Pfeiffer and S. R. Byrn, *J. Pharm. Sci.*, 1997, **86**, 1239; (b) K. Fujii, M. Aoki and H. Uekusa, *Cryst. Growth Des.*, 2013, **13**, 2060–2066.
23. A. D. Gift, P. E. Luner, L. Luedeman and L. S. Taylor, *J. Pharm. Sci.*, 2008, **97**, 5198–5211.
24. J. Adamson, N. Faiber, A. Gottlieb, L. Hamsmith, F. Hicks, C. Mitchell, B. Mittal, K. Mukai and C. D. Papageorgiou, *Org. Process Res. Dev.*, 2016, **20**, 51–58.
25. (a) O. Almarsson and M. J. Zaworotko, *Chem. Commun.*, 2004, 1889; (b) N. K. Duggirala, M. L. Perry, O. Almarsson and M. J. Zaworotko, *Chem. Commun.*, 2016, **52**, 640–655; (c) J. W. Steed, *Trends Pharmacol. Sci.*, 2013, **34**, 3; (d) H. G. Brittain, *Cryst. Growth Des.*, 2012, **12**, 5823–5832.
26. *Pharmaceutical Salts and Co-Crystals*, ed. J. Wouters and L. Quéré, RSC Drug Discovery Series, 2012.
27. (a) D. Braga, L. Maini and F. Grepioni, *Chem. Soc. Rev.*, 2013, **42**, 7638–7648; (b) S. A. Ross, D. A. Lamprou and D. Douroumis, *Chem. Commun.*, 2016, **52**, 8772–8786; (c) G. Bolla and A. Nangia, *Chem. Commun.*, 2016, **52**, 8342–8360.
28. A. Cruz-Cabeza, S. Reutzel-Edens and J. Bernstein, *Chem. Soc. Rev.*, 2015, **44**, 8619–8635.
29. J. Lu and S. Rohani, *Org. Process Res. Dev.*, 2009, **13**, 1269–1275.
30. S. Tothadi, A. Mukherjee and G. R. Desiraju, *Chem. Commun.*, 2011, **47**, 12080–12082.
31. S. L. Childs, N. Rodríguez-Hornedo, L. S. Reddy, A. Jayasankar, C. Maheshwari, L. McCausland, R. Shipplett and B. C. Stahly, *CrystEngComm*, 2008, **10**, 856–864.
32. D. Braga, F. Grepioni, L. Maini, S. Prosperi, R. Gobetto and M. R. Chierotti, *Chem. Commun.*, 2010, **46**, 7715–7717.
33. T. Steiner, I. Majerz and C. C. Wilson, *Angew. Chem., Int. Ed.*, 2001, **40**, 2651–2654.
34. (a) S. Mohamed, D. A. Tocher, M. Vickers, P. G. Karamertzanis and S. L. Price, *Cryst. Growth Des.*, 2009, **9**, 2881–2889; (b) S. L. Childs, G. P. Stahly and A. Park, *Mol. Pharmaceutics*, 2007, **4**, 323–338; (c) G. P. Stahly, *Cryst. Growth Des.*, 2007, **7**, 1007–1026; (d) C. B. Aakeroy and D. J. Salmon, *CrystEngComm*, 2005, **7**, 439–448; (e) P. Vishweshwar, J. A. McMahon, J. A. Bis and M. J. Zaworotko, *J. Pharm. Sci.*, 2006, **95**, 499–516; (f) A. D. Bond, *CrystEngComm*, 2007, **9**, 833–834.
35. D. Braga, L. Maini, G. de Sanctis, K. Rubini, F. Grepioni, M. R. Chierotti and R. Gobetto, *Chem. – Eur. J.*, 2003, **9**, 5538–5548.
36. D. Braga, E. Dichiarante, G. Palladino, F. Grepioni, M. R. Chierotti, R. Gobetto and L. Pellegrino, *CrystEngComm*, 2010, **12**, 3534–3536.
37. D. Braga, L. Maini and F. Grepioni, *Chem. Soc. Rev.*, 2013, **42**, 7638.
38. D. Braga, F. Grepioni, L. Maini, D. Capucci, S. Nanna, J. Wouters, L. Aerts and L. Quéré, *Chem. Commun.*, 2012, **48**, 8219–8221.

39. F. Grepioni, J. Wouters, D. Braga, S. Nanna, B. Fours, G. Coquerel, G. Longfils, S. Rome, L. Aerts and L. Quéré, *CrystEngComm*, 2014, **16**, 5887–5896.
40. T. T. Ong, P. Kavuru, T. Nguyen, R. Cantwell, L. Wojtas and M. J. Zaworotko, *J. Am. Chem. Soc.*, 2011, **133**, 9224–9227.
41. D. Braga, F. Grepioni, L. Maini, M. Polito, K. Rubini, M. Chierotti and R. Gobetto, *Chem. – Eur J.*, 2009, **15**, 1508–1515.
42. M. Polito, E. D'Oria, L. Maini, P. G. Karamertzanis, F. Grepioni, D. Braga and S. L. Price, *CrystEngComm*, 2008, **10**, 1848–1854.
43. E. Schur, E. Nauha, M. Lusi and J. Bernstein, *Chem. – Eur J.*, 2015, **21**, 1735.
44. M. Lusi, I. J. Votorica-Yrezabal and M. J. Zaworotko, *Cryst. Growth Des.*, 2015, **15**, 4098–4103.
45. (a) D. Braga, G. Cojazzi, D. Paolucci and F. Grepioni, *Chem. Commun.*, 2001, 803–804; (b) J. W. Steed, A. E. Goeta, J. Lipkowski, D. Swierczynski, V. Panteleonc and S. Handad, *Chem. Commun.*, 2007, 813–815.
46. (a) J. Bernstein, *Polymorphism in Molecular Crystals*, Oxford UniversityPress, United States, 2002; (b) H. G. Brittain, *Polymorphism in Pharmaceutical Solids*, Marcel Dekker, Inc., New York, USA, 1999; (c) A. Cruz-Cabeza, S. Reutzel-Edens and J. Bernstein, *Chem. Soc. Rev.*, 2015, **44**, 8619–8635.
47. L. Bucar and J. Bernstein, *Angew. Chem.*, 2015, **54**, 6972–6993.
48. J. Bauer, S. Spanton, R. Henry, J. Quick, W. Dziki, W. Porter and J. Morris, *J. Pharm. Res.*, 2001, **6**, 59–66.
49. H.-M. Wolff, L. Queré and J. Riedner, 2008, EP2215072B1.
50. (a) H.-M. Wolff, L. Queré and J. Riedner, 2012, US 8232414 B2; (b) I. B. Rietveld and R. Céolin, *J. Pharm. Sci.*, 2015, **104**, 4117–4222.
51. A. Y. Lee, D. Erdemir and A. S. Myerson, *Annu. Rev. Chem. Biomol. Eng.*, 2011, 2.
52. D. Braga and F. Grepioni, *Chem. Commun.*, 2005, 3635–3645.
53. P. M. Bhatt and G. R. Desiraju, *Chem. Commun.*, 2007, 2057–2059.
54. A. J. Cruz-Cabeza and C. R. Groom, *CrystEngComm*, 2011, **13**, 93–98.
55. (a) M. R. Chierotti, R. Gobetto, L. Pellegrino, L. Milone and P. Venturello, *Cryst. Growth Des.*, 2008, **8**, 1454–1457; (b) M. U. Schmidt, J. Brüning, J. Glinnemann, M. W. Hützler, P. Mörschel, S. N. Ivashevskaya, J. van de Streek, D. Braga, L. Maini, M. R. Chierotti and R. Gobetto, *Angew. Chem., Int. Ed.*, 2011, **50**, 7924–7926.
56. M. R. Chierotti, L. Ferrero, N. Garino, R. Gobetto, L. Pellegrino, D. Braga, F. Grepioni and L. Maini, *Chem. – Eur. J.*, 2010, **16**, 4347–4358.
57. M. R. Calas and J. Martinez, *C. R. Acad. Sci. Ser. C*, 1967, **265**, 631.
58. S. L. Price, *Chem. Soc. Rev.*, 2014, **43**, 2098–2111.

Subject Index

maxima 165
    non-bonded 653–4, 657
Mehring notation 312–13
metal–organic frameworks
  (MOFs) 366–7
methane dimer 61–2
    in crystal packing 568
    methane–benzene 462–3
    water–methane 422–3
methanol–PHA dimer 278–9
methyl groups, charge effects 433–6
methylamine–PHA dimer 279–80
microwave spectroscopy 265–7, 282–3
    carbon bonds 300–3
    CH–π bonds 460–1
    halogen bonds 296–7
    molecular beam 265, 271–5, 296
minimal molecular unit (MMU) 465
models
    flat and network database 350
    mathematical 388–93
    multipolar 622–3, 662–3, 666–7
    QC-CPA 104–5
    see also computational tools;
      simulation methods
moderate hydrogen bonds 88, 92–3
MOFs (metal–organic frameworks)
  366–7
Mogul 370
molecular aggregates 251–3
molecular beam microwave
  spectroscopy 265, 271–5, 296
molecular crystals
    acetic acid 105–9
    cation–π interactions in 522–7
    intermolecular bonds 71–6,
      617–67
        properties of 81–9
    intermolecular interactions 71–6
      electron density
        applications 635–67
      electron density methods
        624–35
      properties of 81–9
    rationalizing structure of 69–71
    see also crystal packing; crystals
molecular densities 248
molecular dipole moments 160, 652–3,
  657–8
molecular geometry 649–58
molecular graphs 157–8, 625
    beryllium bonds 535, 536–7
molecular orbital theory (MOT) 6, 15
    bonding in 25–9

electron pair bonding motifs
  35–40
    FMO method 474
    formulation 16, 19–21
molecular orbitals
    fragment molecular orbital 474
    hydrogen–hydrogen
      bonding 574–5
    single occupied 87, 599–600,
      606–9
molecular trapping 251
molecules, and electricity 52–3
monodeterminantal methods 83–4
monoelectronic basis set 82–5
monovariate statistics 360
Monte Carlo simulation 122, 125, 153
Moore's law 71
MOT *see* molecular orbital theory (MOT)
MP2 method 83–4, 88–9, 96, 198
MRMP2 calculations 600, 602–3, 605,
  608, 610
Mulliken charge distribution 104
Mulliken charge-transfer (CT) theory 59
Mulliken–Hund MO theory 16, 19
Mulliken population analysis 152
multi-centered bonds 41–3
    see also long, multicenter bonds
      (LMBs)
multideterminantal methods 83–4
multiple-quantum magic angle spinning
  (MQMAS) 316
multipolar models 622–3, 662–3, 666–7
multipole expansion 163, 182, 204–6

N-methylacetamide (NMA) 433, 436–7
n–π^* interactions 303–4
^{15}N SSNMR 323, 326, 328–30, 335,
  339–41
nanoscale structures 527
natural atomic orbitals (NAOs) 189
natural bond orbital (NBO) method 189,
  416–17, 492
    beryllium bonds 535–6, 540
natural hybrid orbitals (NHOs) 492–4
natural resonance theory (NRT)
  540–1, 543
Nature of the Chemical Bond
  (Pauling) 262
NCI *see* non-covalent index (NCI)
NEDA method 189–90, 197, 206,
  215–16, 415
    charge transfer 211
    real space partitionings 217
    results 200